FLORA & FAUNA HANDBOOK No. 9

A Systematic Catalogue of the SOFT SCALE INSECTS of the World

(Homoptera: Coccoidea: Coccidae)

with data on geographical distribution,
host plants, biology and economic importance

Flora & Fauna Handbooks

This series of handbooks provides for the publication of book length working tools useful to systematists for the identification of specimens, as a source of ecological and life history information, and for information about the classification of plant and animal taxa. Each book is sequentially numbered, starting with Handbook No. 1, as a continuing series. The books are available on a standing order basis, or singly.

Each book treats a single biological group of organisms (*e.g.*, family, subfamily, single genus, etc.) or the ecology of certain organisms or certain regions. Catalogs and checklists of groups not covered in other series are included in this series.

The books are complete by themselves, not a continuation or supplement to an existing work, or requiring another work in order to use this one. The books are comprehensive, and therefore, of general interest.

The books in this series to date are:

Handbook No. 1, 1985. **THE SEDGE MOTHS,** by John B. Heppner

Handbook No. 2, 1992, 2nd printing, slightly revised. **INSECTS AND PLANTS: Parallel Evolution and Adaptation,** by Pierre Jolivet

Handbook No. 3, 1988. **THE POTATO BEETLES,** by Richard L. Jacques, Jr.

Handbook No. 4, 1988. **THE PREDACEOUS MIDGES OF THE WORLD,** by Willis W. Wirth and William L. Grogan, Jr.

Handbook No. 5, 1990. **A CATALOG OF THE NEOTROPICAL COLLEMBOLA,** by José A. Mari Mutt and Peter F. Bellinger

Handbook No. 6, 1990. **THE ENDANGERED ANIMALS OF THAILAND,** by Stephen R. Humphries and James R. Bain

Handbook No. 7, 1990. **A REVIEW OF THE GENERA OF NEW WORLD MYMARIDAE,** by Carl M. Yoshimoto

Handbook No. 8, 1990. **MAYFLIES OF THE WORLD: A Catalog of the Family and Genus Group Taxa,** by Michael D. Hubbard

Handbook No. 9, 1993. **CATALOG OF THE SOFT SCALE INSECTS OF THE WORLD,** by Yair Ben-Dov

Handbook No. 10, 1993. **NORTH AMERICAN PSOCOPTERA,** by Edward L. Mockford

Handbook No. 11, 2nd Edition, 1993. **INSECT AND SPIDER COLLECTIONS OF THE WORLD,** by Ross H. Arnett, Jr. and G. Allan Samuelson

Handbook No. 12, 1993. **SPIDERS OF PANAMA** by W. Nentwig

FLORA & FAUNA HANDBOOK No. 9

A Systematic Catalogue of the SOFT SCALE INSECTS of the World

(Homoptera: Coccoidea: Coccidae)
with data on geographical distribution,
host plants, biology and economic importance

by
Yair Ben-Dov

*Agricultural Research Organization
The Volcani Center, Bet Dagan, Israel*

CRC Press
Taylor & Francis Group
Boca Raton London New York

CRC Press is an imprint of the
Taylor & Francis Group, an **informa** business

TABLE OF CONTENTS

SYNOPSIS

In this catalogue of the soft scale insects family (Insecta: Homoptera: Coccoidea: Coccidae) of the world, are listed 162 genera comprising 1090 species and subspecies which have been described since Linnaeus (1758) until the cutoff date of December, 1991. Extensive data is presented on taxonomy, nomenclature, synonyms, geographical distribution, host plants, biology and economic importance of the species. New Combinations are established for 40 species. One species, namely *Filippia subterranea* Gomez-Menor Ortega, is newly synonymized with *Lecanopsis formicarum* Newstead. New Names are given to 4 species. The catalogue also contains data on depositories of type-series, discussion on suprageneric-groups in the family and biographical data on deceased coccidologists. Indices to genera and species are also included.

Key Words: Coccoidea, Coccidae, Soft Scale Insects, Systematics, Catalogue

INTRODUCTION

Scale insects are terrestrial, plant-feeding animals that generally are classified in the superfamily Coccoidea of the Homoptera, however, several students regard the group as the suborder Coccinea.

The soft scale insects, also named coccids (as well as Lecaniine scales during earlier decades), comprise the family Coccidae, which is one of 16-20 families recognized among scale insects. Members of the family are widely distributed in all zoogeographical regions, predominantly in the tropics and subtropics. Many species are notorious plant pests of great economic importance in crops, such as edible fruit trees, ornamentals and fruit. However, a few species have some beneficial value.

The genera and species included in the homopterous family Coccidae, were catalogued for the first time, about a century after Linnaeus (1758), by Targioni Tozzetti (1868) and by Signoret (1869a, 1877a). These authors analyzed 13 genera of Coccidae comprising about 90 species. Four decades later, Fernald (1903) registered about 400 species, belonging to 48 genera of the family, in the Catalogue of the superfamily Coccoidea of the World. During the twentieth century the number of species and genera in the family has increased considerably, and now totals 162 genera and 1090 species, which are discussed in this treatise.

This catalogue has been compiled to provide a data base for all species and genera described in the family since Linnaeus (1758) till December (1991), the cutoff date of literature survey for the present compilation.

ACKNOWLEDGEMENTS

This study would have not been finalized without the kind cooperation of many colleagues to all of whom I am very grateful.

First to Mr. David French, Computing Section, The Natural History Museum, London, England, who made available the computer program of the Museum Cataloguing System, and advised me about its use.

Secondly to all colleagues who so kindly responded to my requests for publications and information at their disposal. Special thanks are due to my colleagues for their untiring efforts to trace and obtain reprints of those old and "forgotten" publications in periodicals that are not available in ordinary science libraries.

Dr. John W. Beardsley, Honolulu, HAWAII;
Ms. Helen M. Brookes, Glen Osmond, AUSTRALIA;
Dr. A. Blay, Madrid, SPAIN;
Dr. Michel Canard, Toulouse, FRANCE;
Dr. Jennifer M. Cox (formerly of the BMNH), London, ENGLAND;
Dr. Trevor K. Crosby, DSIR, Auckland, NEW ZEALAND;
Dr. Evelyna M. Danzig, St. Petersburg, RUSSIA;
Dr. Lewis L. Deitz, Raleigh, North Carolina, U.S.A.;
Dr. Francisca C. do Val, Sao Paulo, BRAZIL;
Dr. Idinha M. Fernandes, Lisbon, PORTUGAL;
Dr. Imre Foldi, Paris, FRANCE;
Ms. Pamela Gilbert, Library, BMNH, London, ENGLAND;
Mr. Raymond J. Gill, Sacramento, California, U.S.A.;
Dr. Roberto H. Gonzalez, Santiago, CHILE;
Dr. Maria Cristina Granara de Willink, San Miguel de Tucumán, ARGENTINA;
Dr. Penny J. Gullan, Canberra ACT, AUSTRALIA;
Dr. Zoya Hadzibejli, Tbilisi, Georgia, USSR;
Dr. Avas B. Hamon, Gainesville, Florida, U.S.A.;
Ms. Julie Harvey, Library, BMNH, London, ENGLAND;
Dr. Christopher J. Hodgson, Wye, Kent, ENGLAND;
Mr. David Hollis, BMNH, London, ENGLAND;
Dr. I. Izquierdo, Madrid, SPAIN;
Dr. Gunther Köhler, Jena, GERMANY;
Dr. Michael Kosztarab, Blacksburg, Virginia, U.S.A.;
Dr. Jan Koteja, Krakow, POLAND;
Dr. Ferenc Kozár, Budapest, HUNGARY;
Dr. Paris L. Lambdin, Knoxville, Tennessee, U.S.A.;
Mr. Ireno L. Lit, Los Banos, PHILIPPINES;
Dr. Raymond J. Mamet, Paris, FRANCE;
Dr. Salvatore Marotta, Portici, ITALY;
Mr. Jon Martin, BMNH, London, ENGLAND;
Mme. Daniele Matile-Ferrero, Paris, FRANCE;
Mr. Ian M. Millar, Pretoria, SOUTH AFRICA;
Dr. Douglass R. Miller, Beltsville, Maryland, U.S.A.;
Dr. Gary L. Miller, Beltsville, Maryland, U.S.A.;
The Late, Dr. Clare Morales, Auckland, NEW ZEALAND;
Dr. Laurence A. Mound, BMNH, London, ENGLAND;
Mr. Steve Nakahara, Beltsville, Maryland, U.S.A.;
Dr. Uzi Nur, Rochester, New York, U.S.A.;

Mr. Marcelo Guena de Oliveira, Sao Paulo, BRAZIL;
Dr. Frej Ossiannilsson, Uppsala, SWEDEN;
Dr. Giuseppina Pellizzari Scaltriti, Padova, ITALY;
Dr. Gerhard L. Prinsloo, Pretoria, SOUTH AFRICA;
Mr. Ting Kui Qin, Canberra ACT, AUSTRALIA
Dr. Salvatore Ragusa di Chiara, Palermo, ITALY;
Dr. Agatino Russo, Catania, ITALY;
The late Dr. S. Adam Shafee, Aligarh, INDIA;
Dr. J.R. Steffan, Paris, FRANCE;
Dr. Sadao Takagi, Sapporo, JAPAN;
Dr. Tang Fang-teh, Taiku, Shansi, CHINA;
Dr. Charles Chia-chu Tao, Taipei, TAIWAN;
Dr. Arturo Teran, Tucumán, ARGENTINA;
Dr. Antonio Tranfaglia, Potenza, ITALY;
Dr. Ermengildo Tremblay, Portici, ITALY;
Dr. R.K. Varshney, Calcutta, INDIA;
Dr. Wang Tze-ching, Beijing, CHINA;
Dr. Gilian M. Watson, London, ENGLAND;
Dr. Douglas J. Williams, London, ENGLAND;
Dr. Michael L. Williams, Auburn, Alabama, U.S.A.;
Dr. Yang Bain-ley, Shanghai, CHINA;
Dr. Jiri Zahradnik, Prague, CZECHOSLOVAKIA.

Thanks are due to Prof. Y. Folman, Chief Scientist, Ministry of Agriculture, Israel, for providing partial financial support towards the publication of this catalogue.

ARRANGEMENT OF THE CATALOGUE

Genera, and the species included in each, are listed alphabetically.

GENUS

Generic headings are given in this catalogue to senior synonyms. The genera, which are here interpreted as subjective synonyms, are treated as senior synonyms and therefore each is also recorded under a separate heading. An objective synonym is not registered in a separate entry, but under its respective senior synonym. The Index to Genera will assist the user to locate an objective synonym.

Each generic entry includes data in the following sequence:

a. The valid generic name and its author name.

b. List of synonyms arranged chronologically. It includes the first proposed name of the senior synonym, objective synonyms, homonyms, nomina nuda, replacement names, unjustified or unnecessary replacement names, changes in hierarchal positions (genus, subgenus), mis-spellings and erroneous authorships.The data for each objective synonym include the genus name and author, author and publication year (as cited in the References) and the page number on which it was first proposed. The first proposal of a new and available name is indicated by presenting it without a semicolon between the taxon and the author's name. Citations of other, subsequent, uses are given with a semicolon between the author/s' name of the taxon and the author/s' name of the relevant publication.

A homonym, nomen nudum, replacement name, unjustified or unnecessary replacement name, mis-spelling and erroneous authorship are marked with square brackets that enclose the respective category after the page number in the relevant publication, e.g. [HOMONYM]. A mis-spelling is also indicated by presenting it without a semicolon between the taxon and the name of author/s of the mis-spelling.

Changes in the hierarchal position are documented and cited only by the first and earliest reference in which it was introduced.

c. The type-species and how it was designated, are given for senior synonyms and objective synonyms, e.g. original designation, monotypy or designated by Fernald (1903).

d. Remarks. This sub-section presents: 1) Reference to publications in which the generic characters were discussed. 2) Reference to publications that include a key to species of the genus. 3) In the case of a genus here regarded as a subjective synonym, a short discussion on the synonymy is provided. 4) Clarifying comments are given for genera with a confused nomenclatural history.

SPECIES AND SUBSPECIES

Each entry of a species or subspecies includes data in the following sequence:

a. Species or subspecies name and author
b. Synonyms

c. Type data
d. Geographical distribution
e. Host plants
f. Remarks
g. Biology
h. Economic importance

a. The name of the senior synonym and author, in its currently accepted generic combination and species name ending, is listed at the beginning of the entry. The generic combination of several species is amended for the first time. These are denoted **n.comb.** in the species heading. Several species names that proved to be junior homonyms are replaced by a New Name, and indicated **n. name.** in the species heading.

b. The list of synonyms, arranged chronologically, includes the senior synonym, junior synonyms, homonyms, nomina nuda, replacement names, different endings of the specific name, unjustified or un-necessary replacement names, different generic combinations, mis-spellings and erroneous authorships.

The first proposal of a new and available name of a Senior Synonym is given in the following sequence: the species name (as published in the original description); author name separated by a comma from the publication year (as cited in the References); the page number on which it was first described. In other words, a Senior Synonym is indicated by presenting it without a semicolon between the taxon and the author/s' name of the relevant publication. The type data of a senior synonym are presented under the type data heading; see below.

The data for a Junior Synonym are presented in the following sequence: the species name (as published in the original description); author name separated with a comma from the publication year; page number on which it was first described; status of type-series, i.e. holotype, lectotype, neotype or syntype; if a lectotype or neotype have been designated, reference is given to the publication where it was selected; type-locality and host plant of the type-series; for species described from syntypes originating from various localities, data is given for all locations; the abbreviation (in parenthesis) for depository of the type-series; reference to the publication in which the synonymy was first introduced. Citations for subsequent uses are given with a semicolon between the author's name of the species and the author's name of the publication.

A homonym, nomen nudum, replacement name, unjustified or unnecessary replacement name, mis-spelling and erroneous authorship are marked with square brackets enclosing the respective category after the page number of the reference, e.g. [RE-PLACEMENT NAME]. Changes in the generic combination are documented and cited only by the first and earliest reference in which it was introduced.

c. The type-data for a Senior Synonym include: status of type-series, i.e. holotype, lectotype, neotype or syntypes; if a lectotype or neotype have been designated reference is given to the publication in which it was selected; type-locality and host plant of the type-series; for species described from syntypes originating from various localities, data are given for all locations; and the abbreviation (in parenthesis) for depository (see below) of the type-series. The information on availability of type-series in the various depositories is based on the original publications, on later publications and on information which I received from the curators. Species, for which it was clarified that the type-series is lost, are indicated lost (in parenthesis) followed by the name of the colleague who provided that information.

d. The geographical distribution of each species is recorded by countries. These are arranged in nine zoogeographical regions (see Map 1) - Palaearctic, Nearctic, Neotropical, New Zealand & South Pacific, Australian, Austro-Oriental, Oriental, Madagasian and Ethiopian. The names of countries have been converted to modern usage, e.g. Ghana instead of Gold Coast. Records in Australia and the U.S.A. are presented by the State name only, e.g. Florida, Queensland. Records in Argentina, Brazil and India are given as Country followed, without a comma, by the State, Province or Region, e.g. Brazil Sao Paulo; India Kashmir. Records in the former Soviet Union are given as USSR followed, without a comma, by the Republic or Region, e.g. USSR Georgia. Most of the published records are here accepted as verified and accordingly documented. However, with regard to several cosmopolitan species (e.g. *Ceroplastes* spp., *Coccus hesperidum*, *C. longulus*, *Saissetia oleae*), where taxonomic identity has been clarified and established only in recent decades, some selectivity has been applied by me.

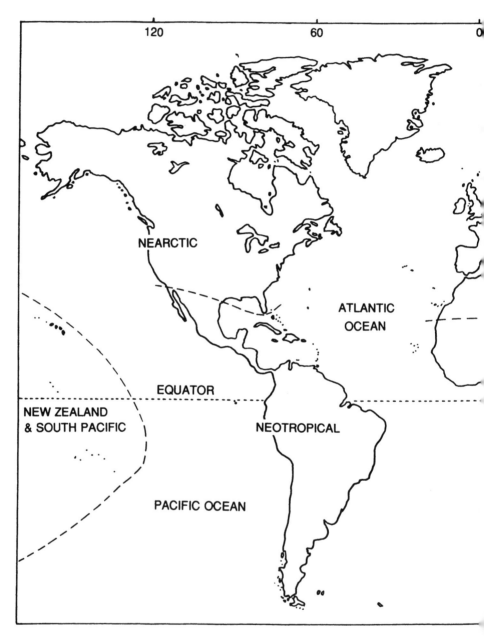

Map 1. Zoogeographical regions of the world as used in this catalogue.

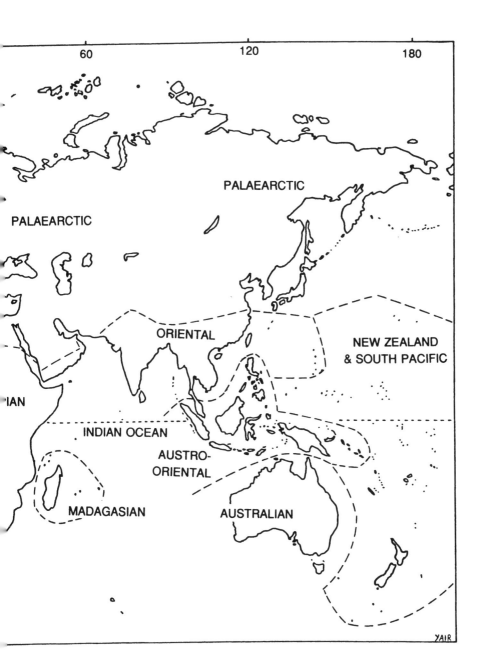

PALAEARCTIC

PALAEARCTIC

ORIENTAL

NEW ZEALAND
& SOUTH PACIFIC

IAN

INDIAN OCEAN

AUSTRO-
ORIENTAL

MADAGASIAN

AUSTRALIAN

YAIR

Consequently, some early, probably doubtful, records have been omitted.

e. Recorded host plants are listed alphabetically by family. The name of host-plants are given here as printed in the original publication, except for obvious typographical errors which are here amended. The plant classification and names of families and genera follow Willis (1988). However, regarding the families - Compositae [=Asteraceae], Cruciferae [=Brassicaceae], Graminae [=Poaceae] and Leguminosae [=Fabaceae, Mimosaceae and Caesalpiniaceae], I am using these permitted alternative names (Cronquist, 1988), rather than their formally accepted names, given in square brackets above.

f. The Remarks section lists references to publications on the following topics: 1) Common Name. This is given only for the cosmopolitan and noxious species. 2) Taxonomic redescriptions and illustrations of the adult female, adult male and larval instars. 3) Colour photographs of the various instars. 4) Other aspects of morphology, including SEM micrographs. 5) Significant data on nomenclature and synonymy. 6) Anatomy and histology. 7) Cytology. 8) Chemistry of pigments and the cover. 9) Records of distribution and host plants, published later than the original description of a species.

g. The biology section presents brief discussion, together with references, on various aspects of the biology, life history and ecology of the species.

h. The economic importance of various species is outlined in this section. Special emphasis is given to provide the user with data on the pest species.

INDICES

The catalogue is provided with an index to species names and a second for generic names. Each index is arranged in two columns, the left including all specific or generic names, either valid or not, and the right one shows the current, valid status of the taxon. The taxa listed in these indices are not indicated with the relevant page number. However, since the catalogue is arranged in alphabetical order, the user should easily locate any taxon.

DEPOSITORIES

The type-data of species listed in this catalogue also includes the depository in which the type series is deposited. The name of each depository is abbreviated, as upper case characters in parenthesis, at the end of the type-data. The abbreviations together with full data on the depositories are listed below.

AEBI-Department of Agricultural Entomology, Bihan Chandra Krishi Viswavidyalaya, Kalyani, West Bengal, INDIA

AESB-Agricultural Experiment Station, Bogar, INDIA

AMNY-American Museum of Natural History, New York, U.S.A.

AMUI-Zoological Museum, Aligarh Muslim University, INDIA

ANIC-Australian National Insect Collection, C.S.I.R.O, Canberra, AUSTRALIA

ASAR-Academie des Sciences Agricoles et Forestieres, Bucarest, ROMANIA

BCR-Bodenheimer Collection of Coccoidea, Department of Entomology, Faculty of Agriculture, Rehovot, ISRAEL

BCRI-Biological and Chemical Research Institute, Rydalmere, AUSTRALIA

BMNH-The Natural History Museum (formerly British Museum, Natural History) London, UK

BP -Bishop Museum, Honolulu, HAWAII

CMNZ-Canterbury Museum, Christchurch, NEW ZEALAND

CNOC-Canadian National Collection, Ottawa, CANADA

CSDA-California State Department of Agriculture, Sacramento, California, U.S.A.

CTBC-Coleccion Taxonomica Nacional del Programa de Entomologia del ICA en el CNIA, Bogota, COLOMBIA

CZLP-Coleccoes do Centro de Zoologia do Instituto de Investigacao Cientifica Tropical, Lisboa, PORTUGAL

DAMU-Defensa Agricola, Montevideo, URUGUAY

DPBA-Departamento de Patologia Vegetal, INTA, C.C. no. 25, Castelar, Provincia de Buenos Aircs, ARGENTINA

DPNI-Directorate of Plant Protection, Quarantine and Storage, New Delhi, INDIA

ECMP-Entomological Collection, Bureau of Science, Manila, PHILIPPINES

EISC-Entomological Institute, Shanxi Agricultural University, Taigu, Shanxi, CHINA

FDA-Florida Department of Agriculture, Gainesville, U.S.A.

FML-Fundacion Miguel Lillo, Tucuman, ARGENTINA

FOCB-Collection of Fundacao Oswaldo Cruz, Manguinhos, Rio de Janeiro, BRAZIL

GOUX-L. Goux Collection, Marseille, FRANCE
GPPT-Plant Protection Institute, Tbilisi, REPUBLIC OF GEORGIA
HMNH-Museum of Natural History, Hamburg, GERMANY
HNHB-Hungarian Natural History Museum, Budapest, HUNGARY
HUSJ-Entomological Institute, Hokkaido University, Sapporo, JAPAN
IBEM-Coleccion Entomologica, Instituto de Biologia, Universidad Nacional
   Autonoma de Mexico, Mexico, D.F. MEXICO
IBPI-Institute of Plant Diseases and Pests, Bogor, Java, INDONESIA
IBSP-Instituto Biologico de Sao Paulo, Sao Paulo, BRAZIL
ICV-Department of Entomology, The Volcani Centre, Bet Dagan, ISRAEL
IEAB-Istituto di Entomologia Agraria, Bari, ITALY
IEAP-Istituto di Entomologia Agraria, Portici, ITALY
IEBC-Institute of Entomology, Academy of Sciences, Beijing, CHINA
IECI-Istituto di Entomologia Agraria, Catania, ITALY
IMZT-Insect Museum, Department of Applied Zoology, Taiwan Agricul-
   tural Research Institute, Wufeng, Taichung. (Several type-series of
   the species described by R. Takahashi are deposited in this collec-
   tion), TAIWAN
INPA-Instituto Nacional de Pesquisas da Amazonias, BRAZIL
ISZF-Istituto Sperimentale per la Zoologia, Firenze, ITALY
ITLJ-Insect Taxonomy Laboratory, National Institute of Agricultural
   Environmental Sciences, Konnon-dai, Yatabe, Tsukuba-shi, Ibaraki-
   ken, (I. Kuwana Collection), JAPAN
KAYJ-S. Kanda Collection, Asano Senior High School, Kanagawa-ku,
   Yokohama, JAPAN
MLPA-Museo Argentino de Ciencias Naturales y Museo, La Plata,
   ARGENTINA
MMBC-Moravian Museum (K. Sulc Collection), Brno, CZECHOSLOVAKIA
MNCN-Museo Nacional de Ciencias Naturales, Madrid, SPAIN
MNHN-Museum National d'Histoire Naturelle, Paris, FRANCE
MZSP-Museu de Zoologia, Universidade de Sao Paulo, BRAZIL
NZAC-New Zealand Arthropod Collection, Entomology Division, DSIR,
   Mt. Albert Research Centre, Auckland, NEW ZEALAND
PASK-Institute of Systematic and Experimental Zoology, Polish Acad-
   emy of Sciences, Krakow, POLAND
PUSA-Pusa Collection, Indian Agricultural Research Institute, New
   Delhi, INDIA
SANC-South African National Collection of Insects, Plant Protection
   Research Institute, Pretoria, SOUTH AFRICA
SENC-H. Schmutterer Collection of scale insects, Wettenberg 1, Wiesenstr.
   55, GERMANY. This collection will be deposited in due course in the
   Senckenberg Museum, Frankfurt/M, GERMANY
SIEC-Shanghai Institute of Entomology, Shanghai, CHINA
SMKM-Selangor Museum, Kuala Lumpur, MALAYSIA

TAEJ-Tokyo Agricultural Experiment Station, Tachikawa, Tokyo, JAPAN

UCD-Department of Entomology, University of California, Davis, California, (The Ferris & McKenzie Collection), U.S.A.

USNM-US National Museum of Natural History, Washington, D.C., U.S.A.

VMNH-Vienna Museum of Natural History, Vienna, AUSTRIA

VCCB-Departamento de Zoologia, Centro Politecnico, Universidade Federal do Parana, Curitiba, BRAZIL

ZAWP-Department of Zoology, Agricultural University, Warsaw, POLAND

ZIAE-Zoological Institute, Academy of Sciences of Armenia, Erevan, Armenia, USSR

ZIAS-Zoological Museum, Academy of Sciences, St. Petersburg, RUSSIA

ZIKU-Zoological Institute, Academy of Sciences of Ukraine, Kiev, UKRAINE

ZMAN-Zoological Museum, Amsterdam, NETHERLANDS

ZSCI-Zoological Survey of India, Calcutta, INDIA

SUPRAGENERIC GROUP CLASSIFICATION IN THE COCCIDAE

Fallen (1814) was the first to use the group name Coccides, which was based on the nominal genus *Coccus*. However, the first attempt to cluster taxa, now placed in the family Coccidae, in generic groups appears to be associated with Targioni Tozzetti (1868), who divided his family Coccides, comparable to the Coccoidea of current concept, into four tribes. Among these he distinguished the Coccites, comprising the genus *Coccus*, and the Lecanites. Both tribes roughly compare to the family Coccidae of modern interpretation. In the Lecanites, Targioni Tozzetti recognized the following four sections: A. Eriophori demum folliculares, comprising *Filippia* and *Luzulaspis*. B. Pulvinati, comprising *Pulvinaria*. C. Ceriferi, comprising *Ceroplastes, Columnea* and *Ericerus*. D. Nudi, comprising *Lecanopsis* and *Lecanium*.

Atkinson (1886), placed the soft scales in the subfamily Lecanina, and recognized 5 subdivisions: Lecaniodiaspiaria, Signoretiaria, Ceroplastaria, Pulvinariaria and the Lecaniaria.

Steinweden (1929) studied the generic classification of the family and proposed three groups of genera, namely the *Coccus, Toumeyella* and *Exaeretopus* groups. However, it should be pointed out that a great number of genera where regarded by him as "ungroupable".

Bodenheimer (1953) distinguished four subfamilies, namely Ceroplastiinae, Coccinae, Eriopeltiinae and the Filippiinae. Borchsenius (1957) recognized three subfamilies as follows:

Filippiinae
Coccinae
  Pulvinariini
  Coccini
Ceroplastinae

Gilliomee (1967) divided the family into four groups of genera, based on male morphology - *Eulecanium*, *Eriopeltis*, *Inglisia* and the *Coccus* groups. Ali (1971) and Varshney (1985) adopted the classification introduced by Borchsenius (1957). Koteja (in: Kosztarab & Kozar, 1988) elevated three of the genera-groups, which had been proposed by Gilliomee (1967), to subfamily level, namely Eriopeltinae, Coccinae and Eulecaniinae.

Tang *et al.* (1990) and Tang (1991) proposed a detailed division into 4 subfamilies with several tribes and subtribes:

Pseudopulvinariinae
Filippiinae
  Lecanopsiini
  Filippiini
    Ceronemina
    Ceroplastodina
    Eriopeltina
    Filippiina
Coccinae
  Coccini
    Coccina
    Eulecaniina
  Pulvinariini
    Pulvinariina
    Takahashiina
Ceroplastinae
  Ceroplastini
  Ctenochitonini
    Cardiococcina
    Ctenochitonina

SUPRAGENERIC GROUP NAMES IN THE COCCIDAE

The suprageneric group names, which have been proposed in the Coccidae, are listed below in alphabetical order.

CARDIOCOCCINA Tang *et al.*, 1990: 76; Tang, 1991: 313.

CERONEMINA Tang *et al.*, 1990: 75; Tang, 1991: 29.

CEROPLASTARIA Atkinson, 1886: 276.

CEROPLASTINAE Atkinson; Bodenheimer, 1952: 317.

CEROPLASTIINAE Atkinson; Bodenheimer, 1953: 93.

CEROPLASTINAE Atkinson; Borchsenius, 1957: 447; Williams, 1969: 321; Ali, 1971: 15; Varshney, 1985: 28; Tang *et al.*, 1990:76; Tang, 1991: 294.

CEROPLASTINI Atkinson; Tang *et al.*, 1990: 76; Tang, 1991: 295.

CEROPLASTODINA Tang *et al.*, 1990: 75 (originally mis-spelled as Ceroplstina); Tang, 1991: 32.

CISSOCOCCINI Brain, 1918: 107; Williams, 1969: 322.

COCCARIA Fallen; Atkinson, 1886: 285.

COCCIDES Fallen, 1814: 23.

COCCINA Fallen; Tang *et al.*, 1990: 75; Tang, 1991: 69.

COCCINAE Fallen; Borchsenius, 1957: 199; Tang *et al.*, 1990: 75; Tang, 1991: 67.

COCCINI Fallen; Borchsenius, 1957: 290; Ali, 1971: 19; Varshney, 1985: 26; Tang *et al.*, 1990: 75; Danzig, 1980c: 268; Tang, 1991:68.

Borchsenius (1957), Ali (1971), Danzig (1980c), Varshney (1985) and Tang (1990, 1991) placed this tribe and the Pulvinariini in the subfamily Coccinae.

CTENOCHITONINA Cockerell; Tang *et al.*, 1990: 76; Tang, 1991: 318.

CTENOCHITONINI Cockerell, 1889n: 16; Williams, 1969: 323; Tang *et al.*, 1990: 76; Tang, 1991: 312.

ERIOPELTIINA Sulc; Tang *et al.*, 1990: 75; Tang, 1991: 40.

ERIOPELTINAE Sulc; Bodenheimer, 1952: 317; Danzig, 1980c: 245.

ERIOPELTINI Sulc, 1941: 3; Koteja, 1978: 311.

The group names SIGNORETIARIA Atkinson, 1886: 276, and SIGNORETIINI Ashmead, 1891: 98, were based on the name *Signoretia* Targioni Tozzetti, a junior homonym which was replaced by *Luzulaspis* Cockerell. However, there is no need for a new group name, because the latter belongs to the group Eriopeltini Sulc.

EULECANIINA Koteja; Tang *et al.*, 1990: 75; Tang, 1991: 145.

EULECANIINAE Koteja in Kosztarab & Kozar, 1988: 173.

FILIPPIINA Bodenheimer; Tang *et al.*, 1990: 75; Tang, 1991: 46.

FILIPPIINAE Bodenheimer; Bodenheimer, 1952: 317; Bodenheimer, 1953: 93; Borchsenius, 1957: 88; Tang *et al.*, 1990: 75; Tang, 1991: 19.
FILIPPIINI Bodenheimer; Tang *et al.*, 1990: 75; Tang, 1991: 28.

LECANIINI - The group name Lecaniini and or Lecaniini has been frequently used in the Coccidae. However, the genus *Lecanium* is an objective synonym of *Coccus*, and in addition has been placed (Opinion 1303, 1985) on the Official Index of Rejected and Invalid Generic Names. Therefore this group name should not be used. Generally, the interpretation of Lecaniini by earlier authors was comparable to the Coccinae of current use.
LECANOPSIINI Tang *et al.*, 1990: 75; 1991: 20.

PARALECANIINI Williams, 1969: 333.
PHYSOKERMINI Sulc, 1912: 35.
PSEUDOPULVINARIINAE Tang *et al.*, 1990: 75; 1991: 13.
PULVINARIARIA Targioni Tozzetti; Atkinson, 1886: 277.
PULVINARIEAE Targioni Tozzetti; Maskell, 1879: 205.
PULVINARIINA Targioni Tozzetti; Tang *et al.*, 1990: 75; Tang, 1991: 223.
PULVINARIINI Targioni Tozzetti; Ashmead, 1891: 98; Borchsenius, 1957: 202; Varshney, 1985: 25; Tang *et al.*, 1990: 75; Tang, 1991: 223.
Borchsenius (1957), Danzig (1980c), Varshney (1985), Tang *et al.* (1990) and Tang (1991) placed this tribe and the Coccini in the subfamily Coccinae.
PULVINATI Targioni Tozzetti, 1868: 727.

TAKAHASHIINA Tang *et al.*, 1990: 75; Tang, 1991: 269.

OPINIONS OF THE INTERNATIONAL COMMISSION ON ZOOLOGICAL NOMENCLATURE RELATING TO THE FAMILY COCCIDAE

Opinion 228. 1954. Rejection for nomenclatorial purposes of Geoffroy, 1762, Histoire abregee des Insectes qui se trouvent aux environs de Paris. *Opinions and Declarations Rendered by the International Commission on Zoological Nomenclature* 4:209-220.

Opinion 1192. 1981. *Lecanium acuminatum* Signoret, 1873 (Insecta, Homoptera, Coccidae): Neotype designated. *Bulletin of Zoological Nomenclature* 38: 252-253.

Opinion 1303. 1985. *Coccus* Linnaeus, 1758 and *Parthenolecanium* Sulc, 1908 (Insecta, Hemiptera, Homoptera): Type species designated. *Bulletin of Zoological Nomenclature* 27: 139- 141.

Opinion 1627. 1991. *Saissetia* Deplanche, 1859 (Insecta, Homoptera): *Lecanium coffeae* Walker, 1852 designated as the type species. *Bulletin of Zoological Nomenclature* 48: 72-73.

NOMINA NUDA

A *nomen nudum* is not an available name, therefore should not be regarded a synonym. However, several *nomina nuda*, which have been associated by authors with valid species, are listed in this catalogue as "synonyms" of the latter. In addition, the user may find in the index these *nomina nuda* respectively placed to their "senior synonyms". Other *nomina nuda*, which cannot be linked to available names, are listed in this chapter.

Ceroplastes bussei Newstead; Newstead, 1906: 74 [NOMEN NUDUM]. See De Lotto, 1965: 181.

Ceroplastodes guilliermondi Mahdihassan; Mahdihassan, 1933: 561 [NOMEN NUDUM].

Coccus poterii Walker; Walker, 1852: 1082 [NOMEN NUDUM].

Eriopeltis brachypodii Giard; Giard, 1894: cxcix [NOMEN NUDUM]. See Lindinger, 1912: 85.

Eulecanium tivoni Sternlicht; Sternlicht in Bytinski-Salz & Sternlicht, 1967: 126 [NOMEN NUDUM].

Exaeretopus hellenicus Green; Bodenheimer, 1928: 192 [NOMEN NUDUM].

Lacca Signoret; Signoret, 1869b: 848 [NOMEN NUDUM].

Lacca alba Signoret; Signoret, 1869b: 848 [NOMEN NUDUM].

Lecanium limnanthemi Goury; Goury, 1905: 62 [NOMEN NUDUM].

Lecanium (Eulecanium) lustneri King; Reh, 1903:409 [NOMEN NUDUM].

Lecanium mercarae Ramakrishna Ayyar; Ramakrishna Ayyar, 1919: 36 [NOMEN NUDUM].

Lecanium monotonae Newstead; Newstead, 1906: 69 [NOMEN NUDUM].

Lichtensia dubia Cockerell; Cockerell, 1895l: 255 [NOMEN NUDUM].

Luzulaspis erianthi Green; Rehacek, 1954: 144 [NOMEN NUDUM]. See Koteja, 1978: 325.

Stozia fuscata Wang; Yang, 1982: 155 [NOMEN NUDUM].

Pulvinaria goethei King; Reh, 1903: 460 [NOMEN NUDUM].

Pulvinaria rehi King; Reh, 1903: 460 [NOMEN NUDUM].

Pulvinaria subterranea Bodenheimer; Bodenheimer, 1935: 249 [NOMEN NUDUM]. See Furth *et al.*, 1983: 108.

Pulvinaria vitis opacus King; King, 1903: 461 [NOMEN NUDUM].
Pulvinaria vitis sorbusae King; King, 1903: 461 [NOMEN NUDUM].

Pulvinaria vitis verrucosae King; King, 1903: 461 [NOMEN NUDUM].

Pulvinaria vitis opacus King; Reh, 1903: 463 [NOMEN NUDUM].

Pulvinaria vitis sorbusae King; Reh, 1903: 463 [NOMEN NUDUM].

Pulvinaria vitis verrucosae King; Reh, 1903: 463 [NOMEN NUDUM].

BIOGRAPHICAL DATA
ON DECEASED COCCIDOLOGISTS

The names of deceased entomologists, who during their life time have described new taxa in the family Coccidae, are listed here. The list presents data in the following order: Surname, Private Name/s, Year of birth and of death (in parentheses); the references from which these data were taken, are given after the column. The information available in each reference, listed here, is denoted by any of the following abbreviations: B = Biography, L = List of Publications, O = Obituary. For more data the user may consult Hagen (1862), Carpenter (1945), Ferris (1957) and Gilbert (1977).

The names of deceased coccidologists, for whom no biographical data became available, are followed by blank space between the parentheses.

ARCHANGELSKAYA, A.D. ( ).
ASHMEAD, William Harris (1855-1908): Carpenter (1945).
ATKINSON, Edwin Felix Thomas (1840-1890): Carpenter (1945).

BALACHOWSKY, Alfred Serge (1901-1983): Matile-Ferrero (1983) (B); Pesson (1984) (B,L,O); Remaudiere (1984) (O).
BANKS, C.S. ( ).
BECCARI, Odorado (1843-1920): Carpenter (1945).
BELLIO, Giuseppe (1901-1949); Gilbert (1977).
BERLESE, Antonio (1863-1927); Melis (1928) (B,L).
BERNARD, Pons-Joseph (1748-1816).
BODENHEIMER, Fritz Simon (1897-1959): Harpaz (1984) (B).
BOISDUVAL, Jean Baptiste Alphonse Dechauffour de (1799-1879): Carpenter (1945).
BONDAR, Gregorio (1881-1959); Gilbert (1977).
BORATYNSKI, Kajetan Ludwik (1907-1980): Davies & Williams (1982) (B,O).
BORCHSENIUS, Nikolay Sergeyevitch (1906-1965): Kryzhanovskiy (1965) (B,L,O).
BOUCHE, Peter Friedrich (1784-1856): Sachtleben (1944) (B).
BRAIN, Charles Kimberlin (1886-1954).
BRETHES, Juan (1871-1928): Carpenter (1945).
BRITTIN, G.F.J.M. (1878-1940): Miller (1940) (B).

CAMPBELL, R.E. ( ).
CASTEL-BRANCO, Armando Jacques Favre (1909-1977): Nunes (1977) (B).
CHARMOY d'EMMEREZ, Paul Donald (1873-1930). Halais & d'Emmerez de Charmoy (1941) (B).
CHAVANNES, A. (1810-1879): Carpenter (1945).
CHEN, Fang-G. ( ).
COCKERELL, Theodore Dru Alison (1866-1948): Weber (1965) (B,L).
COLEMAN, George Albert (1866-1932): Carpenter (1945).
COMSTOCK, John Henry (1849-1931): Carpenter (1945).
COOK, Albert John (1842-1916): Carpenter (1945).
COQUILLETT, Daniel William (1856-1911): Carpenter (1945).

COSTA, Achille (1828-1898): Carpenter (1945).
COSTA, Oronzio Gabriele (1787-1867): Del Giudice (1868) (B,L,O).
COSTA-LIMA, Angelo Moreira da (1887-1964): Gilbert (1977).
CRAW, Alexander (1850-1908): Carpenter (1945).
CURTIS, John (1791-1862): Carpenter (1945).

DALMAN, Johann Wilhelm (1787-1828): Carpenter (1945).
DE LOTTO, Giovanni (1912-1990).
DEL GUERCIO, Giacomo (1863-1954): Gilbert (1977).
DOUGLAS, John William (1814-1905): Carpenter (1945).
DOZIER, H.L. ( ).
DUFOUR, Leon (1780-1865): Carpenter (1945).

EHRHORN, Edward Macfarlaine (1862-1941): Carpenter (1945).
EZZAT, Yehia M. (-1991).
FABRICIUS, Johann Christian (1745-1808): Carpenter (1945).
FALLEN, C.F. (1764-1830).
FERNALD, Maria E. (1839-1919): Carpenter (1945).
FERRIS, Gordon Floyd (1893-1958): McKenzie (1959) (B,O); Wiggins
    (1958) (B,O).
FITCH, Asa (1809-1879): Carpenter (1945); Russell (1960) (B,L); Barnes
    (1988) (B,L).
FONSCOLOMBE, Etienne Laurent Joseph Hippolyte Boyer de (1772-
    1853): Carpenter (1945).
FONSECA, Jose Pinto da (-1982).
FOURCROY, Antoine Francois (1755-1809): Carpenter (1945).
FROGGATT, Walter Wilson (1858-1937): Carpenter (1945).
FULLER, Claude (1872-1928): Carpenter (1945).

GENNADIUS, P. ( ).
GIANNOTTI, O. ( ).
GIARD, Alfred (1846-1908): Carpenter (1945).
GMELIN, Johann Friedrich (1748-1804): Carpenter (1945).
GOMEZ MENOR ORTEGA, Juan (1903-1983).
GOUREAU, Claude Charles (1790-1879): Carpenter (1945).
GRANDPRES Daruty, Jean Marie Rose Albert (1853-1928). d'Emmerez
    de Charmoy (1941) (B).
GRAY, John Edward (1800-1875): Carpenter (1945).
GREEN, Edward Ernest (1861-1949): Gilbert (1977).
GUERIN-MENEVILLE, Felix Edouard (1799-1874): Carpenter (1945).

HALL, Wilfrid John (1893-1965): Pearson (1965) (B,O); Gilbert (1977).
HAWORTH, Adrian Hardy (1767-1833): Carpenter (1945).
HEMPEL, Adolfo (-1949): Anonymous (1949).
HOLLINGER, A.H. ( ).

HUNTER, S.J. ( ).

IHERING, Hermann von (1850-1930): Carpenter (1945).
ISHII, Tei (1894-1959).

JOUBERT, C.J. ( ).

KALTENBACH, Johann Heinrich (1807-1876): Gilbert (1977).
KANDA, Shigeo (1898-1961): Gilbert (1977).
KAWECKI, Zbigniew (1908-1981): Dziedzicka (1984) (B,O).
KING, George B. (1848-1916): Carpenter (1945).
KIRITSHENKO, Alexis Nikolayevitch (1884-1971): Gilbert (1977).
KIRKALDY, George Willis (1873-1910): Carpenter (1945).
KOTINSKY, Jacob (1873-1928): Carpenter (1945).
KUWANA, Shinkai Inokichi (1872-1933): Carpenter (1945).

LAHILLE, Fernando (1861-1940): Carpenter (1945).
LAING, Frederick (1890-1965).
LAWSON, Paul Brown (1888-1954): Gilbert (1977).
LEONARDI, Gustavo (1869-1918): Silvestri (1918) (B,L,O).
LEPAGE, H.S. ( ).
LICHTENSTEIN, Wilhelm Auguste Jules (1818-1886): Carpenter (1945).
LIDGETT, James (-1941); Gilbert (1977).
LINDINGER, Leonhard (1879-1965): Weidner & Wagner (1968) (B,L).
LINNAEUS, Carl von (1707-1778): Carpenter (1945).
LIZER Y TRELLES, Carlos A. (1887-1959): Biraben (1959) (O).
LOEW, Franz (1829-1889): Carpenter (1945).
LULL, R.S. ( ).

MACFIE, J.W.S. ( ).
MACGREGOR LOAEZA, Raul (1926-1983): Sampedra (1984) (B,O).
MARCHAL, Paul (1862-1942): Gilbert (1977).
MASKELL, William Miles (1840-1898): Dietz & Tocker (1980) (B,L).
McCONNELL, Harold Sloan (1893-1958): Gilbert (1977).
McKENZIE, Howard Lester (1910-1968): Miller *et al.* (1969) (B,L,O).
MORRISON, Harold (1890-1963): Reyne (1963) (O); Russell (1963) (B,O).
MOSQUERA, Luis Felipe (1945-1987).

NASSONOV, Nikolai Viktorovitch (1855-1939): Gilbert (1977).
NEWSTEAD, Robert (1859-1947): Gilbert (1977).
NIETNER, J. ( ).

OLIVIER, Guillaume Antoine (1756-1814): Carpenter (1945).
OLLIFF, Arthur Sydney (1865-1895): Gilbert (1977).
PAOLI, Guido (1881-1947): Gilbert (1977).

xxvii

PARROTT, P.J. ( ).
PERGANDE, Theodore (1840-1916): Carpenter (1945).
PERINGUEY, Louis Albert (-1924).
PLANCHON, Gilbert (1823-1888).

RAO, V.P. (1910-1983).
RATHVON, Simon Snyder (1812-1891): Carpenter (1945).
RATZEBURG, Julius Theodor Christian (1801-1871): Carpenter (1945).
REH, Ludwig (1867-1940): Carpenter (1945).
REYNE, Adriaan (1890-1967): Tammes & van Eyndhoven (1967, 1985)
    (B,O).
RILEY, Charles Valentine (1843-1895): Carpenter (1945).
RUTHERFORD, Andrew (-1915): Petch (1915) (O).

SCHRANK, Franz von Paula (1747-1835): Carpenter (1945).
SHAFEE, Sheik Adam (1941-1992).
SHINJI, Orihei (1885-1951): Gilbert (1977).
SIGNORET, Victor Antoine (1816-1889): Carpenter (1945).
SILVESTRI, Filippo (1873-1949): Russo (1949) (B,L,O); Viggiani (1973).
SIRAIWA, Hideo (1897-1946).
SULC, Karl (1872-1932): Obenberger (1932) (B,L,O).
SULZER, Johann Heinrich (1735-1813): Carpenter (1945).

TAKAHASHI, Ryoichi (1898-1963); Ito & Sorin (1963) (B,L).
TARGIONI TOZZETTI, Adolfo (1823-1902): Bargagli (1902) (B,L,O).
THRO, W.C. ( ).
TOWNSEND, Charles Henry Tyler (1863-1944): Gilbert (1977).
TYRRELL, W.M. ( ).

VALLOT, Jacques Nicolas (1771-1860).

WALKER, Francis (1809-1874): Carpenter (1945).
WALSH, Benjamin Dann (1808-1869).
WESTWOOD, John Obadiah (1805-1893): Carpenter (1945).
WHITE, Adam (1817- 1879): Carpenter (1945).
WUNN, H. ( ).

ZEHNTNER, Leo (1864-1961): Hauser (1972) (B).

A Systematic Catalogue of the Soft Scale Insects of the World
(Homoptera: Coccoidea: Coccidae)

Family **COCCIDAE** Fallen, 1814

Type-genus: *Coccus* Linnaeus, 1758.

The group of soft scale insects was first ranked a family status by Steinweden (1929). However, his formal act was preceded by similar interpretations of earlier taxonomists. The following list of suprageneric names elucidates the evolution of distinguishing the soft scales from other groups until their establishment as a distinct family.

Coccides Fallen, 1814: 23
Coccites Targioni Tozzetti, 1868: 722
Lecanites Targioni Tozzetti, 1868: 722
Lecanides Signoret, 1869c: 100
Lecanidae Maskell, 1879: 203
Lecaninae Comstock, 1881: 330
Lecanina Atkinson, 1886: 268
Lecaniini Ashmead, 1891: 98
Lecaniinae Cockerell, 1896b: 329
Coccinae Fernald, 1903: 127
Coccinae MacGillivray, 1921: 99
Lecaniidae Cockerell, 1929: 150
Coccidae Steinweden, 1929: 197
Lecanoidae Balachowsky, 1942: 37
Lecanini Balachowsky, 1948: 255

COMPREHENSIVE PUBLICATIONS ON THE FAMILY

GENERAL MORPHOLOGY OF THE ADULT FEMALE
Newstead (1903); Green (1904d); Kuwana (1917a); Steinweden (1929); Borchsenius (1937, 1957); Takahashi (1955d); Gómez-Menor Ortega (1958c); Ezzat & Hussein (1969); Williams & Kosztarab (1972); Schmutterer (1972); Gimpel et al., (1974); Gill et al. (1977); Kosztarab & Kozár (1978, 1988); Kawai (1980); Danzig (1980c, 1988); Wang (1980); Tereznikova (1981); Tao et al., (1983); Hamon & Williams (1984); Gill (1988); Williams & Watson (1990); Tang (1991); Qin & Gullan (1992).

GENERAL MORPHOLOGY OF THE ADULT MALE
Newstead (1903); Green (1904d); Borchsenius (1957); Giliomee (1967).

General morphology of the male test with key to species of North America based on this stage given by Miller & Williams (1990).

GENERAL MORPHOLOGY OF THE FIRST INSTAR
Borchsenius (1957); Miller (1991).

STRUCTURE OF WAX GLANDS
MALE - Sulc (1931). FEMALE - Foldi (1978, 1991); Foldi & Cassier (1985); Foldi & Pearce (1985).

MORPHOLOGY OF MOUTHPARTS
Koteja (1974a, 1974c, 1976); Koteja & Liniowska (1976).

SENSORY ORGANS
TARSUS - Koteja (1974b). ANTENNA - Koteja (1980a); Rosciszewska (1989).

MORPHOLOGY OF HINDLEG
Dziedzicka (1977).

KEY TO GENERA
BRAZIL - Hempel (1900b, 1920a).
CALIFORNIA - Steinweden (1929); Gill (1988).
CHINA - Wang (1980); Yang (1982); Tang (1991).
CONNECTICUT - Britton (1923).
CZECHOSLOVAKIA - Zahradnik (1959a).
EGYPT - Ezzat & Hussein (1969).
ENGLAND - Newstead (1903).
ERIOPELTINI - Koteja (1970, 1978).
EUROPE - Kosztarab & Kozár (1978, 1988).
FLORIDA - Hamon & Williams (1984).
HAWAII - Zimmerman (1948).
INDIA - Avasthi & Shafee (1991b).
INDIANA, U.S.A - Dietz & Morrison (1916).
ITALY - Leonardi (1920).
JAPAN - Takahashi (1955d).
KOREA - Paik (1978).
MICRONESIA - Beardsley (1966a).
PHILIPPINES - Morrison (1920).
PULVINARIINI OF NORTH AMERICA - Nakahara & Gill (1985).
SPAIN - Gómez-Menor Ortega (1937, 1948, 1958c).
SRI LANKA - Green (1904d).
TAIWAN - Tao et al., (1983).
TROPICAL SOUTH PACIFIC - Williams & Watson (1990).
UKRAINE - Tereznikova (1981).
USSR (FAR EAST) - Danzig (1980c, 1988).
USSR - Borchsenius (1937, 1957), Danzig (1964).
VIRGINIA - Williams & Kosztarab (1972).
ZIMBABWE - Hodgson (1969b).

CHECKLISTS OF SPECIES
ANGOLA - Almeida (1969, 1973).
ARGENTINA - Lizer y Trelles (1939).
ARMENIA - Ter-Grigorian (1962).

AUSTRALIA - Maskell (1895a).
BERMUDA - Simmonds (1957); Hodgson & Hilburn (1991a, 1991b).
BRAZIL - Gomes Costa (1949); Vernalha (1953); Silva et al. (1968);
    Corseuil & Barbosa (1971).
CALIFORNIA - Carnes (1907); Gill (1988).
CANARY ISLANDS - Lindinger (1918); Perez Guerra & Carnero
    Hernandez (1985); Carnero Hernandez & Perez Guerra (1986).
CARIBBEAN ISLANDS - Pollard & Alleyne (1986).
CHINA - Chang (1929), Wang (1980); Yang (1982); Tang (1991).
COMORES - Matile-Ferrero (1978).
CUBA - Bruner et al. (1975).
CYPRUS - Georghiou (1977).
CZECHOSLOVAKIA - Zahradnik (1959b, 1977).
DENMARK - Kozárzhevskaya & Reitzel (1975).
EUROPE - Kosztarab & Kozár (1988).
FLORIDA - Riddick (1955); Hamon & Williams (1984).
GERMANY - Wunn (1937); Schmutterer (1980).
GUYANA - Bodkin (1922).
HAWAII - Zimmerman (1948); Nakahara (1981b).
HUNGARY - Kosztarab & Kozár (1978); Kozár & Drozdjak (1990).
INDIA - Green (1908); Ramakrishna Ayyar (1919, 1930); Ghose
    (1961); Varshney (1985); Shafee et al. (1989); Avasthi & Shafee
    (1991b).
INDIA, ORISSA STATE - Varshney & Moharana (1987).
INNER MONGOLIA, CHINA - Tang & Li (1988).
ISRAEL - Bodenheimer (1935); Ben-Dov (1971).
ITALY - Leonardi (1920); Marotta (1987).
IVORY COAST - Couturier et al. (1985).
JAPAN - Takahashi & Tachikawa (1956); Kawai (1972, 1980).
KOREA - Paik (1978).
MADEIRA ISLANDS - Vieira et al. (1983).
MASCARENE ISLANDS - Mamet (1943); Williams & Williams (1988).
MAURITIUS - Mamet (1948, 1949).
NEW ZEALAND - Wise (1977).
OGASAWARA ISLANDS - Kawai et al. (1971).
ORIENTAL REGION - Ali (1971).
PALAEARCTIC REGION - Borchsenius (1957); Kozár & Walter (1985);
    Kosztarab & Kozár (1988).
POLAND - Koteja & Zak-Ogaza (1983).
PUERTO RICO - Nakahara & Miller (1981).
SICILY (Italy) - Russo (1989).
SOUTH AFRICA - Brain (1920b); Munro & Fouche (1936); De Lotto
    (1967b, 1970a, 1978, 1979).
SOUTH AMERICA - Cockerell (1902i).
SPAIN - Gómez-Menor Ortega (1937); Martin Mateo (1984).
SRI LANKA - Green (1937).
TAIWAN - Tao (1978, 1989); Tao et al. (1983).
TIBET - Wang (1981).
UKRAINE - Tereznikova (1981).
USSR - Borchsenius (1957); Danzig (1964, 1972a).
VIRGIN ISLANDS - Nakahara (1983).
YUGOSLAVIA - Atanasov (1959).

ANNOTATED LIST OF GENERIC NAMES
Morrison & Morrison (1966); Russell (1970) (First Supplement); Kosztarab & Russell (1974) (Second Supplement); Kosztarab et al., (1986) (Third Supplement).

BIBLIOGRAPHY
Morrison & Renk (1957); Morrison & Morrison (1965) (First Supplement); Russell et al. (1974) (Second Supplement); Kosztarab & Kosztarab (1988) (Third Supplement).

Acantholecanium Borchsenius

Acantholecanium Borchsenius, 1949b: 339.
TYPE-SPECIES: *Ctenochiton haloxyloni* Hall, by original designation and monotypy.
REMARKS. Generic characters discussed by Borchsenius (1950, 1957) and by Tang (1991).

Acantholecanium haloxyloni (Hall)

Ctenochiton haloxyloni Hall, 1926: 17.
Ctenochiton haloxyli Hall; Lindinger, 1932f: 197 [UNJUSTIFIED EMENDATION].
Acantholecanium haloxyloni (Hall); Borchsenius, 1949b: 340.
TYPE DATA. Syntypes female, EGYPT: Suez Road (between 5th and 6th Towers), on roots of *Haloxylon schweinfurthii* (BMNH).
DISTRIBUTION. PALAEARCTIC REGION: Egypt, Jordan, USSR Tadzhikistan, USSR Turkmenia.
HOST PLANTS. Chenopodiaceae: *Haloxylon schweinfurthii*, *Hammada salicornium*, *Salsola richteri*, *Suaeda vermiculata*.
REMARKS. Adult female redescribed and illustrated by Ezzat & Hussein (1969) and by Tang (1991). Adult female redescribed by Hosny (1939).
BIOLOGY. Living on roots of its host plant.

Acanthopulvinaria Borchsenius

Acanthopulvinaria Borchsenius, 1952: 301.
TYPE-SPECIES: *Pulvinaria orientalis* Nassonov, by original designation.
REMARKS. Generic characters discussed by Borchsenius (1957), Hadzibejli (1983) and by Tang (1991). Key to species given by Borchsenius (1957).

Acanthopulvinaria discoidalis (Hall)

Pulvinaria discoidalis Hall, 1926: 16.
Acanthopulvinaria discoidalis (Hall); Borchsenius, 1952: 301.
TYPE DATA. Syntypes female, EGYPT: Suez Road (at the 7th Tower), on undetermined plant (BMNH).
DISTRIBUTION. PALAEARCTIC REGION: Egypt.
HOST PLANTS. Chenopodiaceae: *Haloxylon*.
REMARKS. Adult female redescribed and illustrated by Borchsenius

(1957) and by Ezzat & Hussein (1969). Adult female redescribed by Tang (1991).

Acanthopulvinaria orientalis (Nassonov)

Pulvinaria orientalis Nassonov, 1909: 493.
Acanthopulvinaria orientalis (Nassonov); Borchsenius, 1952: 301.
Rhizopulvinaria iljiniae Danzig; Danzig, 1972c: 341. Holotype female, MONGOLIA: Bayan-Khongorsk Aymak, Yekhin-Gol Oasis, on *Iljinia regelii* (ZIAS). Syn. by Danzig, 1974: 70.
TYPE DATA. Syntypes adult female and larvae, USSR: Syr-Darjensi Province, on *Haloxylon ammodendri* (ZIAS).
DISTRIBUTION. PALAEARCTIC REGION: China, Iran, Mongolia, USSR Georgia, USSR Kazakhstan, USSR Tadzhikistan, USSR Turkmenia, USSR Uzbekistan.
HOST PLANTS. Chenopodiaceae: *Anabasis aphilla, Haloxylon, Iljinia regelii, Kochia, Salsola ericoides, S. glauca, S. richteri.*
Compositae: *Artemisia* . **Tamaricaceae:** *Reaumuria.*
REMARKS. Adult female redescribed and illustrated by Borchsenius (1937, 1957), Hadzibejli (1983) and by Tang (1991). Adult female redescribed by Hadzibejli (1977). Distribution and host plant records given by Borchsenius (1937, 1957), Hadzibejli (1977) and by Tang (1991).
BIOLOGY. Lives on roots of the host plant.

Akermes Cockerell

Akermes Cockerell, 1902e: 89.
TYPE-SPECIES: *Akermes bruneri* Cockerell, by original designation and monotypy.
REMARKS. Generic characters discussed by De Lotto (1965).

Akermes bruneri Cockerell

Akermes bruneri Cockerell, 1902e: 89.
TYPE DATA. Syntypes female, PARAGUAY: San Bernardino, on spring plant (BMNH, USNM).
DISTRIBUTION. NEOTROPICAL REGION: Argentina, Paraguay, Uruguay.
HOST PLANTS. Ulmaceae: *Celtis tala.*
REMARKS. Adult female redescribed and illustrated by De Lotto (1968b).

Akermes colae Green & Laing

Lecanium (Akermes) colae Green & Laing, 1924: 419.
TYPE DATA. Syntypes female, GHANA [=GOLD COAST]: Aburi, on *Cola acuminata* (BMNH).
DISTRIBUTION. ETHIOPIAN REGION: Ghana.
HOST PLANTS. Sterculiaceae: *Cola acuminata.*

Akermes colimae Cockerell

Akermes colimae Cockerell, 1903c: 47.
TYPE DATA. Syntypes female, MEXICO: Cualatan, Colima, on undetermined tree (USNM).

DISTRIBUTION. NEOTROPICAL REGION: Mexico.
BIOLOGY. Cockerell (1903c) found it in large, hollow, pyriform twig-galls (about 18 mm diameter), in association with the ant *Azteca longiceps*.

Akermes cordiae Morrison

Akermes cordiae Morrison, 1929: 45.
TYPE DATA. Holotype female, PANAMA: Canal Zone, Ancon, on *Cordia alliodora* (USNM).
DISTRIBUTION. NEOTROPICAL REGION: Panama.
HOST PLANTS. Ehretiaceae: *Cordia alliodora*.

BIOLOGY. Develops in hollow swellings of the host and attended by the ant *Cryptocerus* sp. (Morrison, 1929).

Akermes levis (Maskell)

Lecanium scrobiculatum leve Maskell, 1896b: 392.
Akermes levis (Maskell); Cockerell, 1902f: 453.
Lecanium levis Maskell; Froggatt, 1915: 608.
TYPE DATA. Syntypes female, AUSTRALIA: New South Wales, Sydney, on *Acacia longifolia* (NZAC).
DISTRIBUTION. AUSTRALIAN REGION: New South Wales.
HOST PLANTS. Leguminosae: *Acacia longifolia*.
REMARKS. Adult female redescribed by Froggatt (1915).

Akermes monilis (Cockerell)

Lecanium monile Cockerell, 1895p: 203.
Akermes monilis (Cockerell); Cockerell, 1902f: 453.
TYPE DATA. Syntypes female, BRAZIL: Sao Paulo, on undetermined tree (USNM).
DISTRIBUTION. NEOTROPICAL REGION: Brazil, Sao Paulo.
REMARKS. Adult female redescribed and illustrated by Hempel (1900b).

Akermes montanus (Green)

Lecanium montanum Green, 1908: 30.
Akermes montanus (Green); Sanders, 1909b: 46.
TYPE DATA. Syntypes female, INDIA: Janusai, Himalayas, on undetermined shrub (USNM).
DISTRIBUTION. ORIENTAL REGION: India.

Akermes punctatus (Cockerell)

Lecanium punctatum Cockerell, 1895k: 194.
Akermes punctatus (Cockerell); Cockerell, 1902f: 453.
TYPE DATA. Syntypes female, GRENADA: Botanic Gardens, on *Citrus medica* var. *acida* (USNM).
DISTRIBUTION. NEOTROPICAL REGION: Grenada.
HOST PLANTS. Rutaceae: *Citrus medica acida*.

Akermes riograndensis Hempel

Akermes riograndensis Hempel, 1932: 331.

Mesolecanium riograndense (Hempel); Lindinger, 1957: 544.
TYPE DATA. Syntypes female, BRAZIL: Rio Grande do Sul, Lavras, on *Schinus dependens* (IBSP).
DISTRIBUTION. NEOTROPICAL REGION: Brazil Rio Grande do Sul.
HOST PLANTS. Anacardiaceae: *Schinus dependens.*

Akermes scrobiculatus (Maskell)

Lecanium scrobiculatum Maskell, 1893b: 221.
Lecanium pingue Maskell; Maskell, 1895b: 58. Syntypes female and larva, AUSTRALIA: New South Wales, Bankstown near Sydney, on *Dillwynia juniperina* (NZAC, USNM). Syn. by Froggatt, 1915: 612.
Lecanium scrobiculatum pingue Maskell; Maskell, 1896b: 392.
Akermes pinguis (Maskell); Cockerell, 1902f: 453.
Akermes scrobiculatus (Maskell); Fernald, 1903: 178.
TYPE DATA. Syntypes female and male, AUSTRALIA: New South Wales, Whitton, on *Acacia* sp. (BMNH, NZAC, USNM).
DISTRIBUTION. AUSTRALIAN REGION: New South Wales.
HOST PLANTS. Leguminosae: *Acacia decurrens, Dillwynia juniperina.*
REMARKS. Adult female redescribed and illustrated by Froggatt (1915).

Akermes townsendi (Cockerell)

Lecanium townsendi Cockerell, 1898f: 433.
Akermes townsendi (Cockerell); Cockerell, 1902f: 453.
TYPE DATA. Syntypes female, MEXICO: Frontera, Tabasco, on orange tree (USNM).
DISTRIBUTION. NEOTROPICAL REGION: Mexico.
HOST PLANTS. Rutaceae: *Citrus.*

Akermes verrucosus (Signoret)

Lecanium verrucosum Signoret, 1874a: 442.
Saissetia (Megasaissetia) verrucosa (Signoret); Cockerell, 1901d: 33.
Akermes verrucosus (Signoret); Cockerell, 1902e: 90.
Mesolecanium verrucosum (Signoret); Lindinger, 1957: 544.
TYPE DATA. Syntypes female, URUGUAY: Montevideo, on undetermined plant (VMNH).
DISTRIBUTION. NEOTROPICAL REGION: Uruguay.

Alecanium Morrison

Alecanium Morrison, 1921: 648.
TYPE-SPECIES: *Alecanium hirsutum* Morrison, by original designation and monotypy.
REMARKS. Generic characters discussed by Hodgson (1990) and by Tang (1991).

Alecanium hirsutum Morrison

Alecanium hirsutum Morrison, 1921: 648.

TYPE DATA. Syntypes female, larva and male, SINGAPORE: Botanic Gardens, on *Alsodeia echinocarpa* (USNM).
DISTRIBUTION. AUSTRO ORIENTAL REGION: Malaysia, Singapore.
HOST PLANTS. Annonaceae: *Annona squamosa.* **Guttiferae:** *Mesua.* **Leguminosae:** *Dialium laurinum.* **Myrtaceae:** *Eugenia aquea.* **Violaceae:** *Alsodeia echinocarpa.*
REMARKS. Adult female redescribed and illustrated by Tang (1991).

Alecanochiton Hempel

Alecanochiton Hempel, 1921: 144.
TYPE-SPECIES: *Alecanochiton marquesi* Hempel, by original designation and monotypy.
Alecaniochiton Lindinger; Lindinger, 1937: 178 [UNJUSTIFIED EMENDATION].

Alecanochiton marquesi Hempel

Alecanochiton marquesi Hempel, 1921: 144.
TYPE DATA. Syntypes female, BRAZIL: Sao Paulo, Angatuba, on coffee (MZSP).
DISTRIBUTION. NEOTROPICAL REGION: Brazil Sao Paulo.
HOST PLANTS. Rubiaceae: *Coffea.*

Alecanopsis Cockerell

Alecanopsis Cockerell in Cockerell & Parrott, 1901: 58.
Alecaniopsis Lindinger; Lindinger, 1932f: 178 [UNJUSTIFIED EMENDATION].
TYPE-SPECIES: *Lecanopsis filicum* Maskell, by original designation and monotypy.
REMARKS. Generic characters discussed by Morrison & Morrison (1922) and by Green (1924). Key to species given by Green (1924).

Alecanopsis callitris (Froggatt)

Lecanium callitris Froggatt, 1925: 379.
Alecanopsis callitris (Froggatt); Froggatt, 1933: 365.
TYPE DATA. Syntypes female, AUSTRALIA: New South Wales, Forbes, on *Callitris calcarata* (BCRI).
DISTRIBUTION. AUSTRALIAN REGION: New South Wales.
HOST PLANTS. Cupressaceae: *Callitris calcarata.*
BIOLOGY. Occurs in cavities under the bark, in nests of the ant *Podomyrma bimaculata* (Froggatt, 1925).

Alecanopsis casuarinae (Maskell)

Lecanium casuarinae Maskell, 1898: 240.
Alecanopsis (Lecanopsis) casuarinae (Maskell); Froggatt, 1933: 364.
TYPE DATA. Syntypes female, AUSTRALIA: Victoria, Myrniong, on *Casuarina* sp. (NZAC).
DISTRIBUTION. AUSTRALIAN REGION: Victoria.
HOST PLANTS. Casuarinaceae: *Casuarina.* **Cupressaceae:** *Frenela robusta.*
REMARKS. Adult female redescribed by Froggatt (1915).

BIOLOGY. Found in deserted burrow of a wood moth in the centre of a stem of *Casuarina* (Maskell, 1898).

Alecanopsis dixoni (Froggatt)
Lecanium dixoni Froggatt, 1925: 379.
Alecanopsis dixoni (Froggatt); Froggatt, 1933: 365.
TYPE DATA. Syntypes female, AUSTRALIA: Victoria, Eltham, in stem of *Loranthus* sp. parasitic on *Eucalyptus* sp. (BCRI).
DISTRIBUTION. AUSTRALIAN REGION: Victoria.
HOST PLANTS. Loranthaceae: *Loranthus*.

Alecanopsis eucalypti (Froggatt)
Lecanium eucalypti Froggatt 1925: 378.
Alecanopsis eucalypti (Froggatt); Froggatt, 1933: 365.
TYPE DATA. Syntypes female, AUSTRALIA: New South Wales, Dorrigo, Brooklana, on stem of *Eucalyptus* sp. (BCRI).
DISTRIBUTION. AUSTRALIAN REGION: New South Wales.
HOST PLANTS. Myrtaceae: *Eucalyptus*.

Alecanopsis filicum (Maskell)
Lecanopsis filicum Maskell, 1894b: 225.
Alecanopsis filicum (Maskell); Cockerell & Parrott, 1901: 58.
TYPE DATA. Syntypes female, AUSTRALIA: New South Wales, Kurrajong Heights, near Richmond, on *Doodia aspera* (NZAC).
DISTRIBUTION. AUSTRALIAN REGION: New South Wales.
HOST PLANTS. Blechnaceae: *Doodia aspera*.
REMARKS. Adult female redescribed by Froggatt (1915). Adult female redescribed and illustrated by Morrison & Morrison (1922).
BIOLOGY. Lives on roots of the host plant (Maskell, 1894b).

Alecanopsis grandis Green
Alecanopsis grandis Green, 1924: 44.
TYPE DATA. Syntypes female, AUSTRALIA: New South Wales, Bundarra, on a fern rhizome (BMNH).
DISTRIBUTION. AUSTRALIAN REGION: New South Wales.
HOST PLANTS. Pteridophyta.
BIOLOGY. Found in a burrow in the rhizome made by the ant *Camponotus intrepidus* (Green, 1924).

Alecanopsis mirus Green
Alecanopsis mirus Green, 1924: 42.
TYPE DATA. Syntypes female, AUSTRALIA: Queensland, Townsville, from nests of *Crematogaster australis* (BMNH).
DISTRIBUTION. AUSTRALIAN REGION: Queensland.
BIOLOGY. Collected from nests of the ant *Crematogaster australis* (Green, 1924).

Alecanopsis tenuis Green
Alecanopsis tenuis Green, 1924: 41.
TYPE DATA. Syntypes female, AUSTRALIA: Victoria, Beaumaris, on *Banksia integrifolia* (BMNH).

DISTRIBUTION. AUSTRALIAN REGION: Victoria.
HOST PLANTS. Proteaceae: *Banksia integrifolia.*
BIOLOGY. Found in hollows in the stem of the host plant (Green, 1924).

Alichtensia Cockerell

Alichtensia Cockerell, 1902f: 451.
TYPE-SPECIES: *Lichtensia attenuata* Hempel, by original designation.

Alichtensia attenuata (Hempel)

Lichtensia attenuata Hempel, 1900b: 494.
Alichtensia attenuata (Hempel); Cockerell, 1902f: 451.
TYPE DATA. Syntypes female, BRAZIL: Sao Paulo, Ypiranga, on *Baccharis genistelloides* var. *trimera* (MZSP).
DISTRIBUTION. NEOTROPICAL REGION: Brazil Sao Paulo.
HOST PLANTS. Compositae: *Baccharis genistelloides trimera.*
REMARKS. Adult female redescribed by Hempel (1901b).

Alichtensia orientalis Lahille

Alichtensia orientalis Lahille, 1924: 105.
TYPE DATA. Syntypes adult female, adult male and larva, URUGUAY: Colonia, on *Baccharis refracta* (DAMU).
DISTRIBUTION. NEOTROPICAL REGION: Uruguay.
HOST PLANTS. Compositae: *Baccharis refracta.*

Allopulvinaria Brain

Allopulvinaria Brain, 1920b: 16.
TYPE-SPECIES: *Allopulvinaria subterranea* Brain, by original designation and monotypy.

Allopulvinaria subterranea Brain

Allopulvinaria subterranea Brain, 1920b: 16.
TYPE DATA. Syntypes female, SOUTH AFRICA: Cape Province, Stellenbosch, Jonkershoek, on stems of "quick" grass (SANC).
DISTRIBUTION. ETHIOPIAN REGION: South Africa.
HOST PLANTS. Gramineae.

Anapulvinaria Borchsenius

Anapulvinaria Borchsenius, 1952: 300.
TYPE-SPECIES: *Pulvinaria pistaciae* Bodenheimer, by original designation and monotypy.
REMARKS. This genus is here regarded as a subjective synonym of *Pulvinaria*. Generic characters discussed by Borchsenius (1957), Hadzibejli (1983) and by Tang (1991).

Anopulvinaria Fonseca

Anopulvinaria Fonseca, 1972b: 195.
TYPE-SPECIES: *Anopulvinaria cephalocarinata* Fonseca, by original designation and monotypy.

Anopulvinaria cephalocarinata Fonseca

Anopulvinaria cephalocarinata Fonseca, 1972b: 195.
Anapulvinaria cephalocarinata Fonseca; Foldi, 1991: 174 [MIS-SPELLING of *Anopulvinaria*]
TYPE DATA. Syntypes female, BRAZIL: Rio Grande do Sul, Caxias do Sul, on *Annona* sp. (MZSP).
DISTRIBUTION. NEOTROPICAL REGION: Brazil Rio Grande do Sul.
HOST PLANTS. Annonaceae: *Annona*.
REMARKS. SEM micrograph of tubular duct given by Foldi (1991).

Antandroya Mamet

Antandroya Mamet, 1959b: 410.
TYPE-SPECIES: *Antandroya euphorbiae* Mamet, by original designation.

Antandroya euphorbiae Mamet

Antandroya euphorbiae Mamet, 1959b: 412.
TYPE DATA. Holotype female, MADAGASCAR: Faux Cap, on *Euphorbia stenoclada* (MNHN).
DISTRIBUTION. MADAGASIAN REGION: Madagascar.
HOST PLANTS. Euphorbiaceae: *Euphorbia oncoclada, E. stenoclada.*

Antandroya tulearensis Mamet

Antandroya tulearensis Mamet, 1959b: 416.
TYPE DATA. Holotype female, MADAGASCAR: Tulear, on *Euphorbia* sp. (MNHN).
DISTRIBUTION. MADAGASIAN REGION: Madagascar.
HOST PLANTS. Euphorbiaceae: *Euphorbia.*

Anthococcus Williams & Watson

Anthococcus Williams & Watson, 1990: 61.
TYPE-SPECIES: *Anthococcus keravatae* Williams & Watson, by original designation and monotypy.

Anthococcus keravatae Williams & Watson

Anthococcus keravatae Williams & Watson, 1990: 63.
TYPE DATA. Holotype female, PAPUA NEW GUINEA: Keravat, on *Eugenia malaccensis* (BMNH).
DISTRIBUTION. AUSTRO ORIENTAL REGION: Papua New Guinea.
HOST PLANTS. Annonaceae: *Annona muricata.* **Leguminosae:** *Gliricidia.* **Moraceae:** *Artocarpus integrifolia.* **Myrtaceae:** *Eugenia malaccensis.* **Rutaceae:** *Citrus.* **Sterculiaceae:** *Theobroma cacao.*

Austrolichtensia Cockerell

Austrolichtensia Cockerell; Cockerell, 1902f: 451.
TYPE-SPECIES: *Lecaniodiaspis hakearum* Fuller, by original designation and monotypy.

Austrolichtensia hakearum (Fuller)

Lecaniodiaspis hakearum Fuller, 1897b: 1345.
Lichtensia hakearum Fuller; Fuller, 1899: 457. Syntypes female,
   AUSTRALIA: Western Australia, Pinjarrah, on *Hakea media*
   (probably lost; P. Gullan, 1990, personal communication). Syn.
   by Fuller, 1899: 458.
Austrolichtensia hakearum (Fuller); Cockerell, 1902f: 451.
TYPE DATA. Syntypes female, AUSTRALIA: Western Australia,
   Pinjarrah, on *Hakea media* (probably lost; P. Gullan, 1990,
   personal communication).
DISTRIBUTION. AUSTRALIAN REGION: Western Australia.
HOST PLANTS. Proteaceae: *Hakea media.*
REMARKS. Fuller (1899) described the junior synonym as a species
in the genus *Lichtensia,* but clearly indicated that its type-data are
the same as those of the senior synonym. Adult female redescribed
and illustrated by Froggatt (1915).

Avricus De Lotto

Avricus De Lotto, 1975: 61.
TYPE-SPECIES: *Ceroplastodes psychotriae* De Lotto, by original
   designation.
REMARKS. Key to species given by De Lotto (1975).

Avricus adspersus (De Lotto)

Ctenochiton adspersus De Lotto, 1958b: 165.
Avricus adspersus (De Lotto); De Lotto, 1975: 62.
TYPE DATA. Holotype female, SOUTH AFRICA: Transvaal, Nelspruit,
   on *Trichilia roka* (BMNH).
DISTRIBUTION. ETHIOPIAN REGION: South Africa, Zimbabwe.
HOST PLANTS. Euphorbiaceae: *Antidesma venosum.* **Meliaceae:**
   Trichilia roka.

Avricus amoenus (De Lotto)

Ctenochiton amoenus De Lotto, 1958b: 168.
Avricus amoenus (De Lotto); De Lotto, 1975: 62.
TYPE DATA. Holotype female, ZIMBABWE: Melsetter, on *Coffea
   arabica* (BMNH).
DISTRIBUTION. ETHIOPIAN REGION: Zimbabwe.
HOST PLANTS. Rubiaceae: *Coffea arabica.*

Avricus arborescens (Laing)

Ctenochiton arborescens Laing, 1928: 215 [NOMEN NUDUM].
Ctenochiton arborescens Laing; Laing, 1929: 478.
Avricus arborescens (Laing); De Lotto, 1975: 62.
TYPE DATA. Syntypes female, GHANA: Aburi, on *Stenophylla coffee*
   (BMNH).
DISTRIBUTION. ETHIOPIAN REGION: Angola, Ghana, Kenya, Sao
   Tome, Sudan, Tanzania, Uganda, Zimbabwe.
HOST PLANTS. Apocynaceae: *Tabernaemontana.* **Loganiaceae:**
   Strychnos angolensis. **Rubiaceae:** *Coffea arabica, C. canephora,
   C. robusta.*

REMARKS. Adult female redescribed and illustrated by De Lotto (1958b). Distribution and host plant records given by De Lotto (1958b, 1968b, 1975), Hodgson (1969b) and by Almeida (1973).

Avricus castaneus (De Lotto)

Ctenochiton castaneus De Lotto, 1958b: 171.
Avricus castaneus (De Lotto); De Lotto, 1975: 62.
TYPE DATA. Holotype female, KENYA: Thika, on *Pistacia aethiopica* (BMNH).
DISTRIBUTION. ETHIOPIAN REGION: Kenya.
HOST PLANTS. Anacardiaceae: *Pistacia aethiopica.* **Apocynaceae:** *Acokanthera schimperi.*

Avricus pluvialis (Hodgson)

Ctenochiton pluvialis Hodgson, 1969b: 11.
Avricus pluvialis (Hodgson); De Lotto, 1975: 62.
TYPE DATA. Holotype female, ZIMBABWE: Victoria Falls, on *Ficus capensis* (BMNH).
DISTRIBUTION. ETHIOPIAN REGION: Zimbabwe.
HOST PLANTS. Moraceae: *Ficus capensis.*

Avricus psychotria (De Lotto)

Ceroplastodes psychotriae De Lotto, 1956b: 310.
Avricus psychotriae (De Lotto); De Lotto, 1975: 61.
TYPE DATA. Holotype female, KENYA: Nairobi, on *Psychotria nairobensis* (BMNH).
DISTRIBUTION. ETHIOPIAN REGION: Kenya.
HOST PLANTS. Rubiaceae: *Psychotria nairobensis.*
REMARKS. Adult female redescribed and illustrated by Hodgson (1971).

Baccacoccus Brain

Baccacoccus Brain, 1920a: 127.
Bacococcus Lindinger; Lindinger, 1937: 180 [UNJUSTIFIED EMEN-DATION].
TYPE-SPECIES: *Baccacoccus elytropappi* Brain, by original designation and monotypy.
REMARKS. De Lotto (1971c) studied material of the type-species and confirmed that it is a wax scale. The genus is here regarded a subjective synonym of *Ceroplastes* as suggested by De Lotto (1971c).

Bernardia Ashmead

Bernardia Ashmead, 1891: 100.
Neobernardia Cockerell, 1892b: 333 [UNJUSTIFIED REPLACEMENT NAME].
TYPE-SPECIES: *Coccus oleae* Olivier, by subsequent designation of Marlatt (1892).
REMARKS. This genus is here regarded as a subjective synonym of *Saissetia.* See Ben-Dov (1989a).

Bodenheimera Bodenheimer

Bodenheimera Bodenheimer, 1935: 251.
TYPE-SPECIES: *Lecanium (Eulecanium) racheli* Bodenheimer, by monotypy.
REMARKS. Generic characters and nomenclature discussed by Ben-Dov (1969). Generic characters discussed by Tang (1991).

Bodenheimera rachelae (Bodenheimer)

Lecanium (Eulecanium) racheli Bodenheimer, 1924: 68.
Bodenheimera racheli (Bodenheimer); Bodenheimer, 1935: 251.
Bodenheimera rachelis (Bodenheimer); Bodenheimer, 1943: 13.
Bodenheimera rachelae (Bodenheimer); Ben-Dov, 1969: 70.
TYPE DATA. Lectotype female designated by Ben-Dov (1980), ISRAEL: Wadi Auje near Jericho, on *Vitex agnus-castus* (BCR).
DISTRIBUTION. PALAEARCTIC REGION: Cyprus, Iran, Iraq, Israel, Turkey.
HOST PLANTS. Verbenaceae: *Vitex agnus-castus, V. negundo, V. pseudonegundo.*
REMARKS. Adult female redescribed and illustrated by Ben-Dov (1969) and by Tang (1991). First instar larva described and illustrated by Ben-Dov (1969).
BIOLOGY. A biparental species, which develops in Israel two annual generations. Reproducing females appear in May-June and in September-October.

Cajalecanium Gómez-Menor Ortega

Cajalecanium Gómez-Menor Ortega, 1965: 108.
TYPE-SPECIES: *Cajalecanium salicorniae* Gómez-Menor Ortega, by original designation and monotypy.

Cajalecanium salicorniae Gómez-Menor Ortega

Cajalecanium salicorniae Gómez-Menor Ortega, 1965: 108.
TYPE DATA. Syntypes female, SPAIN: Cabo Huerta de Alicante, on *Salicornia* sp. (MNCN).
DISTRIBUTION. PALAEARCTIC REGION: Spain.
HOST PLANTS. Chenopodiaceae: *Salicornia.*

Calypticus Costa

Calypticus Costa, 1829: 8.
Calypticus Costa; 1835: 1.
Calymmatus Costa, 1835: 50 [LAPSUS of *Calypticus*].
Calypticus Costa, 1840: 28.
Calymmatus Costa, 1840: 50 [LAPSUS of *Calypticus*].
Calymnatus Signoret; Signoret, 1868: 511 [MIS-SPELLING of *Calymmatus*].
Calymnatus Signoret; Signoret, 1869a: 856 [MIS-SPELLING of *Calymmatus*].
TYPE-SPECIES: *Coccus hesperidum* Linnaeus [as a misidentification of *Saissetia oleae* (Olivier)], by subsequent designation of Costa (1835) and by Fernald (1902b).
REMARKS. After Fernald (1903) designated *Coccus hesperidum* as

the type species of this genus, it was generally accepted as an objective synonym of *Coccus*. However, De Lotto (1970b) showed that the nominal type species of *Calypticus* is a misidentification by Costa (1829) of *Saissetia oleae* (Olivier). *Saissetia oleae* is also the type species of *Bernardia*. Therefore the case should be referred to ICZN. Morrison & Morrison (1966) regarded the generic name *Calymmatus* Costa, 1835, a lapsus of *Calypticus*, an interpretation accepted in this catalogue.

Cardiococcus Cockerell

Cardiococcus Cockerell, 1903b: 155.
TYPE-SPECIES: *Cardiococcus umbonatus* Cockerell, by original designation.
REMARKS. Generic characters discussed by Steinweden (1929), Green (1922) and by Tang (1991).

Cardiococcus bivalvata (Green)

Inglisia bivalvata Green, 1903: 95.
Cardiococcus bivalvata (Green); Green, 1922b: 1034.
TYPE DATA. Syntypes female, INDIA: South India, Rameswaram Island, on *Thespesia populnea* (BMNH).
DISTRIBUTION. ORIENTAL REGION: India.
HOST PLANTS. Leguminosae: *Pongamia*. **Malvaceae:** *Thespesia populnea*.

Cardiococcus foraminifer (Maskell)

Inglisia foraminifer Maskell, 1893b: 213.
Cardiococcus foraminifer (Maskell); Cockerell, 1903b: 156.
TYPE DATA. Syntypes female, AUSTRALIA: South Australia, Semaphore near Adelaide, on *Santalum acuminatum* (NZAC).
DISTRIBUTION. AUSTRALIAN REGION: South Australia, Western Australia.
HOST PLANTS. Santalaceae: *Santalum acuminatum*.
REMARKS. Adult female redescribed and illustrated by Froggatt (1915).

Cardiococcus foraminifer loranthi (Fuller)

Inglisia foraminifer loranthi Fuller, 1897b: 1345.
Cardiococcus foraminifer loranthi (Fuller); Fernald, 1903: 161.
TYPE DATA. Syntypes female, AUSTRALIA: Western Australia, Geraldton, on *Loranthus* sp. growing on "Quandong" (Probably lost; P. Gullan, 1990, personal communication).
DISTRIBUTION. AUSTRALIAN REGION: Western Australia.
HOST PLANTS. Loranthaceae: *Loranthus*.
REMARKS. Adult female redescribed and illustrated by Fuller (1899). Froggatt (1915) regarded this variety as a synonym of *C. foraminifer*.

Cardiococcus formosanus (Takahashi)

Inglisia formosana Takahashi, 1930: 34.
Cardiococcus formosanus (Takahashi); Tang, 1991: 315.

TYPE DATA. Syntypes female, TAIWAN: Shirin, on *Pyrus serotina*
(IMZT).
DISTRIBUTION. ORIENTAL REGION: Taiwan.
HOST PLANTS. Rosaceae: *Pyrus serotina.*
REMARKS. Adult female redescribed and illustrated by Tao *et al.*
(1983) and by Tang (1991).

Cardiococcus fossilis (Maskell)

Inglisia fossilis Maskell, 1897a: 308.
Cardiococcus fossilis (Maskell); Cockerell, 1903b: 156.
TYPE DATA. Syntypes female, AUSTRALIA: Western Australia,
Darling Range, on *Acacia* sp. (NZAC, USNM).
DISTRIBUTION. AUSTRALIAN REGION: Victoria, Western Austra-
lia.
HOST PLANTS. Leguminosae: *Acacia, Templetonia.* **Polygonaceae:**
Muehlenbeckia adpressa.
REMARKS. Adult female redescribed and illustrated by Fuller
(1899). Adult female redescribed by Froggatt (1915).

Cardiococcus umbonatus Cockerell

Cardiococcus umbonatus Cockerell, 1903b: 155.
TYPE DATA. Syntypes female, MEXICO: Zapotlan, Jalisco, on wild
guava (BMNH, USNM).
DISTRIBUTION. NEOTROPICAL REGION: Mexico.
HOST PLANTS. Myrtaceae: *Psidium.*

Ceronema Maskell

Ceronema Maskell, 1895b: 55.
Ceronesera Watt & Mann, 1903: 310 [MIS-SPELLING].
Cerontna Kuwana, 1917b: 172 [MIS-SPELLING].
TYPE-SPECIES: *Ceronema banksiae* Maskell, by monotypy.
REMARKS. Generic characters discussed by Green (1909), Froggatt
(1915), Morrison & Morrison (1922), Steinweden (1929), Hodgson
(1967c), De Lotto (1978). Key to species: ETHIOPIAN REGION -
Hodgson (1967c); ZIMBABWE - Hodgson (1969b).

Ceronema africana Macfie

Ceronema africana Macfie, 1913: 31.
Ceronema acaciae Hall, 1923: 13. Syntypes female, EGYPT: Luxor,
on *Acacia arabica nilotica* (BMNH) Syn. by Hall, 1932: 185.
TYPE DATA. Syntypes female, NIGERIA: Northern Nigeria, Shonga,
on *Caesalpinia pulcherrima* (BMNH).
DISTRIBUTION. ETHIOPIAN REGION: Mauritania, Nigeria, Zimbabwe. **PALAEARCTIC REGION:** Egypt.
HOST PLANTS. Leguminosae: *Acacia nilotica, Caesalpinia pulcherrima.* **Rhamnaceae:** *Ziziphus.*
REMARKS. Adult female redescribed and illustrated by Hodgson
(1967c) and by Ezzat & Hussein (1969). Distribution and host plant
records given by Hall (1923, 1932), Hodgson (1967c), Ezzat & Hussein (1969) and by Balachowsky & Matile-Ferrero (1970).

Ceronema asparagi (Brain)

Lichtensia asparagi Brain, 1920b: 23.
Filippia asparagi (Brain); Lindinger, 1928: 107 [HOMONYM of *Filippia asparagi* Giard].
Filippia braini Lindinger, 1928: 107 [UNNECESSARY REPLACEMENT NAME].
Ceronema asparagi (Brain); Matile-Ferrero, 1978: 44.
TYPE DATA. Syntypes female, SOUTH AFRICA: Eastern Cape Province, on *Asparagus capensis* (SANC).
DISTRIBUTION. ETHIOPIAN REGION: South Africa.
HOST PLANTS. Liliaceae: *Asparagus capensis*.

Ceronema banksiae Maskell

Ceronema banksiae Maskell, 1895b: 56.
TYPE DATA. Syntypes female and larva, AUSTRALIA: New South Wales, near Sydney, on *Banksia serrata* (NZAC).
DISTRIBUTION. AUSTRALIAN REGION: New South Wales, Western Australia.
HOST PLANTS. Proteaceae: *Banksia attenuata*, *B. ilicifolia*, *B. menziesii*, *B. serrata*.
REMARKS. Adult female redescribed and illustrated by Fuller (1899), Froggatt (1915) and by Morrison & Morrison (1922). First instar larva described and illustrated by Morrison & Morrison (1922).

Ceronema brachystegiae Hall

Ceronema brachystegiae Hall, 1941: 232.
TYPE DATA. Syntypes female, ZIMBABWE: Inyanga, on *Brachystegia* sp. (BMNH).
DISTRIBUTION. ETHIOPIAN REGION: Zambia, Zimbabwe.
HOST PLANTS. Leguminosae: *Brachystegia*, *Cassia florida*.
REMARKS. Adult female redescribed and illustrated by Hodgson (1967c).

Ceronema caudata Froggatt

Ceronema caudata Froggatt, 1915: 412.
TYPE DATA. Syntypes female and male, AUSTRALIA: New South Wales, Thirroul and Lake Toronto near Newcastle, on *Eucalyptus robusta* (BCRI).
DISTRIBUTION. AUSTRALIAN REGION: New South Wales.
HOST PLANTS. Myrtaceae: *Eucalyptus robusta*.

Ceronema dryandrae Fuller

Ceronema dryandrae Fuller, 1897b: 1345.
TYPE DATA. Syntypes female, AUSTRALIA: Western Australia, on *Dryandra floribunda* and *D. nivea* (Probably lost; P. Gullan, 1990, personal communication).
DISTRIBUTION. AUSTRALIAN REGION: Western Australia.
HOST PLANTS. Proteaceae: *Dryandra floribunda*, *D. nivea*.
REMARKS. Adult female redescribed and illustrated by Fuller (1899).

Ceronema fryeri Green

Ceronema fryeri Green, 1922b: 1028.
TYPE DATA. Syntypes female, SRI LANKA: Maha Illuppalama, on
undetermined shrub (BMNH).
DISTRIBUTION. ORIENTAL REGION: Sri Lanka.

Ceronema gowdeyi (Newstead)

Ceroplastodes gowdeyi Newstead, 1911a: 98.
Ceroplastes gowdeyi Newstead; Sasscer, 1912: 88 [MIS-SPELLING of
Ceroplastodes].
Ceronema gowdeyi (Newstead); Matile-Ferrero, 1978: 45.
TYPE DATA. Syntypes female, UGANDA: Entebbe, on *Ficus syco-
morus* (BMNH).
DISTRIBUTION. ETHIOPIAN REGION: Uganda.
HOST PLANTS. Moraceae: *Ficus sycomorus*.
REMARKS. Adult female redescribed and illustrated by Hodgson
(1971).

Ceronema iceryoides Green

Ceronema iceryoides Green, 1922b: 1029.
Marsipococcus iceryoides (Green); Tang, 1991: 116.
TYPE DATA. Syntypes female, SRI LANKA: Putalam, on undeter-
mined plant (BMNH).
DISTRIBUTION. ORIENTAL REGION: Sri Lanka.
REMARKS. Adult female redescribed by Tang (1991).

Ceronema koebeli Green

Ceronema koebeli Green, 1909: 256.
Mametia koebeli (Green); Tang, 1991: 62.
TYPE DATA. Syntypes female and male, SRI LANKA: Kandy, on
Sapium sebiferum (BMNH).
DISTRIBUTION. ORIENTAL REGION: India, Sri Lanka.
HOST PLANTS. Euphorbiaceae: *Sapium sebiferum*. **Leguminosae:**
Caesalpinia coriaria, Pithecellobium saman.
REMARKS. Adult female redescribed by Rutherford (1914) and by
Tang (1991). Adult female redescribed and illustrated by Venkatra-
man (1941).
BIOLOGY. Rutherford (1914) presented observations on copulatory
behaviour of the male.

Ceronema mobile Brain

Ceronema mobilis Brain, 1920b: 22.
Ceronema mobile Brain; De Lotto, 1978: 135.
TYPE DATA. Syntypes female, SOUTH AFRICA: Natal, Illovo River,
on *Celastrus cordata* (SANC).
DISTRIBUTION. ETHIOPIAN REGION: South Africa.
HOST PLANTS. Celastraceae: *Celastrus cordata*.
REMARKS. Adult female redescribed and illustrated by De Lotto
(1978).

Ceroplastes Gray

Coccus (Ceroplastes) Gray, 1828: 7.
Ceroplastus Westwood; Westwood, 1840: 449 [MIS-SPELLING].
Ceroplastes Gray; Signoret, 1872b: 35.
Ceroplasses Sankaran; Sankaran, 1954: 100 [MIS-SPELLING].
TYPE SPECIES: *Coccus (Ceroplastes) janetrensis* Gray, by subsequent designation of Fernald, 1903: 147.
REMARKS. Generic characters discussed by Hempel (1900b), Newstead (1903), Green (1909), Froggatt (1915), Steinweden (1929), Gómez-Menor Ortega (1937, 1958c), Zimmerman (1948), Borchsenius (1957), De Lotto (1965, 1969b, 1971c, 1978), Hodgson (1969a), Williams & Kosztarab (1972), Gimpel *et al.* (1974), Paik (1978), Mosquera (1979), Kawai (1980), Wang (1980), Yang (1982), Tao *et al.* (1983), Hadzibejli (1983), Hamon & Williams (1984), Avasthi & Shafee (1986), Gill (1988), Williams & Watson (1990) and by Tang (1991). The genera *Baccacoccus* Brain, *Ceroplastidia* Cockerell, *Ceroplastina* Cockerell, *Cerostegia* De Lotto, *Columnea* Targioni Tozzetti, *Gascardia* Targioni Tozzetti and *Paracerostegia* Tang are here regarded as subjective synonyms of *Ceroplastes*. Key to species: ASIA - Tang (1991). CALIFORNIA - Gill (1988). CHINA - Wang (1980), Tang (1991). COLOMBIA - Mosquera (1979, 1984). FLORIDA - Hamon & Williams (1984). INDIA - Avasthi & Shafee (1986). JAPAN - Kawai (1980). KOREA - Paik (1978). MICRONESIA - Beardsley (1966a). SOUTH AFRICA - De Lotto (1978). SOUTHERN AFRICA - De Lotto (1965). SRI LANKA - Green (1909). TAIWAN - Tao *et al.* (1983). TROPICAL SOUTH PACIFIC - Williams & Watson (1990). U. S. A. - Gimpel *et al.* (1974). USSR - Borchsenius (1957). VIRGINIA - Williams & Kosztarab (1972). ZIMBABWE - Hall (1931), Hodgson (1969a, 1969b).

Ceroplastes actiniformis Green

Ceroplastes actiniformis Green, 1896: 8.
Ceroplastes actiniformes Green; Moharana, 1990: 48 [MIS-SPELLING].
TYPE DATA. Syntypes female, SRI LANKA: Punduloya and Kandy, on coconut palm (BMNH).
DISTRIBUTION. AUSTRO ORIENTAL REGION: Java, Sumatra. **ORIENTAL REGION:** India, Sri Lanka.
HOST PLANTS. Anacardiaceae: *Mangifera indica.* **Apocynaceae:** *Alstonia scholaris.* **Cannaceae:** *Canna.* **Euphorbiaceae:** *Sapium sebiferum.* **Guttiferae:** *Calophyllum inophyllum.* **Loranthaceae:** *Loranthus.* **Moraceae:** *Ficus carica.* **Myrtaceae:** *Psidium guajava.* **Palmae:** *Areca catechu, Cocos nucifera.* **Santalaceae:** *Santalum album.*
REMARKS. Adult female redescribed and illustrated by Green (1909). Distribution and host plant records given by Ramakrishna Ayyar (1919, 1930), Green (1930c, 1937), Ali (1971, 1973), Varshney & Moharana (1987) and by Shafee *et al.* (1989). Parida & Moharana (1982) and Moharana (1990) reported on chromosome number 2n=32 in India.

Ceroplastes agrestis Hempel

Ceroplastes agrestis Hempel, 1932: 322.
TYPE DATA. Syntypes female, BRAZIL: Sao Paulo, Pirapitinguy, on undetermined tree (IBSP).
DISTRIBUTION. NEOTROPICAL REGION: Brazil Sao Paulo.

Ceroplastes ajmerensis (Avasthi & Shafee) n. comb.

Cerostegia ajmerensis Avasthi & Shafee, 1979: 37.
Ceroplastes neoceriferus Yousuf & Shafee, 1988: 61. Holotype female, INDIA: Himachal Pradesh, Solan, on *Citrus* sp. (AMUI). Syn. by Avasthi & Shafee, 1991b: 22.
Paracerostegia ajmerensis (Avasthi & Shafee); Tang, 1991: 304.
TYPE DATA. Holotype female, INDIA: Rajasthan, Ajmer, Hathi Bhata, on *Cassia fistula* (AMUI).
DISTRIBUTION. ORIENTAL REGION: India.
HOST PLANTS. Leguminosae: *Cassia fistulata*. **Myrtaceae:** *Psidium guajava*. **Rutaceae:** *Citrus*.
REMARKS. Adult female redescribed by Tang (1991).

Ceroplastes alamensis Avasthi & Shafee

Ceroplastes alamensis Avasthi & Shafee, 1986: 328.
Ceroplastes alami Avasthi & Shafee; Shafee *et al.*, 1989: 47 [MIS-SPELLING].
TYPE DATA. Holotype female, INDIA: Tamil Nadu, Coimbatore, Mettupalaiyam, on undetermined wild plant (AMUI).
DISTRIBUTION. ORIENTAL REGION: India.
HOST PLANTS. Leguminosae: *Dalbergia sissoo*.

Ceroplastes albolineatus Cockerell

Ceroplastes albolineatus Cockerell, 1894d: 157.
TYPE DATA. Syntypes female, JAMAICA: Kingston, on ornamental shrub (USNM).
DISTRIBUTION. NEOTROPICAL REGION: Brazil Sao Paulo, Jamaica.
HOST PLANTS. Anacardiaceae: *Schinus*. **Celastraceae:** *Maytenus*. **Compositae:** *Baccharis*. **Onagraceae:** *Fuchsia*. **Platanaceae:** *Platanus*.
REMARKS. Adult female redescribed by Hempel (1900b). Distribution and host plant records given by Hempel (1900b, 1912) and by Silva *et al.* (1968). Biochemistry of the cover wax studied by Rios (1966), Rios & Colunga (1965), Rios & Gomez (1969), Rios & Perez (1969) and by Rios & Quijano (1969). Chemistry of the wax cover discussed by Brown (1975).

Ceroplastes albolineatus vulcanicus Cockerell

Ceroplastes albolineatus vulcanicus Cockerell, 1903b: 160.
TYPE DATA. Syntypes female, MEXICO: Volcan de Colima, on low bush below pines (BMNH, USNM).
DISTRIBUTION. NEOTROPICAL REGION: Mexico.

Ceroplastes amazonicus Hempel

Ceroplastes amazonicus Hempel, 1900b: 454.
TYPE DATA. Syntypes female, BRAZIL: Amazonas, Manaos, on undetermined tree (MZSP).
DISTRIBUTION. NEOTROPICAL REGION: Brazil Amazonas.
REMARKS. Adult female redescribed by Hempel (1901a).

Ceroplastes angulatus Cockerell

Ceroplastes angulatus Cockerell, 1898f: 434.
TYPE DATA. Syntypes female, MEXICO: Frontera, on twig of native tree (USNM).
DISTRIBUTION. NEOTROPICAL REGION: Mexico.

Ceroplastes argentinus Brethes

Ceroplastes argentinus Brethes, 1921: 79.
TYPE DATA. Syntypes female, ARGENTINA: near Parana, on undetermined plant (MLPA).
DISTRIBUTION. NEOTROPICAL REGION: Argentina.

Ceroplastes avicenniae Newstead

Ceroplastes avicenniae Newstead, 1917b: 24.
TYPE DATA. Syntypes female, GUYANA: Mahaica Creek, on *Avicennia nitida* (BMNH).
DISTRIBUTION. NEOTROPICAL REGION: Guyana.
HOST PLANTS. Verbenaceae: *Avicennia nitida*.

Ceroplastes bergi Cockerell

Ceroplastes bergi Cockerell, 1901a: 288.
TYPE DATA. Syntypes female, ARGENTINA: Buenos Aires, on *Schinus molle, Ligustrum japonicum* and *Citrus aurantium* (USNM).
DISTRIBUTION. NEOTROPICAL-REGION: Argentina.
HOST PLANTS. Anacardiaceae: *Schinus molle.* **Oleaceae:** *Ligustrum japonicum.* **Rutaceae:** *Citrus aurantium.*

Ceroplastes bernardensis Cockerell

Ceroplastes bernardensis Cockerell, 1902e: 93.
TYPE DATA. Syntypes female, PARAGUAY: San Bernardino, on twigs of undetermined plant (USNM).
DISTRIBUTION. NEOTROPICAL REGION: Paraguay.

Ceroplastes bicolor Hempel

Ceroplastes bicolor Hempel, 1901c: 390.
TYPE DATA. Syntypes female and male, BRAZIL: Sao Paulo, Campinas, on undetermined tree (MZSP).
DISTRIBUTION. NEOTROPICAL REGION: Brazil Sao Paulo.

Ceroplastes bipartitus Newstead

Ceroplastes bipartitus Newstead, 1917b: 25.
Gascardia bipartita (Newstead); De Lotto, 1965: 195.

TYPE DATA. Syntypes female, SOUTH AFRICA: Locality and host plant not indicated (BMNH).
DISTRIBUTION. ETHIOPIAN REGION: South Africa, Zambia, Zimbabwe.
HOST PLANTS. Acanthaceae: *Barleria*. **Bignoniaceae:** *Markhamia acuminata*. **Euphorbiaceae:** *Croton sylvaticus*. **Rubiaceae:** *Hymenodictyon floribunda*.
REMARKS. Adult female redescribed and illustrated by De Lotto (1965). Distribution and host plant records given by Brain (1920b), Hall (1931), De Lotto (1965) and by Hodgson (1969a).

Ceroplastes boyacensis Mosquera

Ceroplastes boyacensis Mosquera, 1979: 599.
TYPE DATA. Holotype female, COLOMBIA: Umbita (Boyaca), on *Baccharis tricuneata* (CTBC).
DISTRIBUTION. NEOTROPICAL REGION: Colombia.
HOST PLANTS. Compositae: *Baccharis tricuneata*.

Ceroplastes brachystegiae Hodgson

Ceroplastes brachystegiae Hodgson, 1969a: 3.
TYPE DATA. Holotype female, ZIMBABWE: Umtali, on *Brachystegia* sp. (BMNH).
DISTRIBUTION. ETHIOPIAN REGION: Zimbabwe.
HOST PLANTS. Leguminosae: *Brachystegia*.

Ceroplastes brachyurus Cockerell

Ceroplastes brachyurus Cockerell, 1903b: 157.
TYPE DATA. Lectotype female designated by Gimpel *et al.* (1974), MEXICO: Zapotlan, Jalisco, on *Rhus*-like shrub (USNM).
DISTRIBUTION. NEOTROPICAL REGION: Mexico. **NEARCTIC REGION:** Alabama, Arizona.
HOST PLANTS. Aquifoliaceae: *Ilex*. **Rubiaceae:** *Bouvardia*. **Rutaceae:** *Citrus*.
REMARKS. Adult female redescribed and illustrated by Gimpel *et al.* (1974) and by Hamon & Williams (1984).

Ceroplastes brevicauda Hall

Ceroplastes destructor brevicauda Hall, 1931: 293.
Ceroplastes brevicauda Hall; De Lotto, 1955: 267.
Ceroplastes luteolus De Lotto, 1955: 268. Holotype female, KENYA: Nairobi, on *Coffea arabica* (BMNH). Syn. by De Lotto, 1965: 196.
Gascardia brevicauda (Hall); De Lotto, 1965: 196.
TYPE DATA. Syntypes female, ZIMBABWE: Mazoe, Sinoia and Umtali on *Citrus aurantium, Toddalia asiatica* and on *Cedrela toona* (BMNH).
DISTRIBUTION. ETHIOPIAN REGION: Angola, Eritrea, Ivory Coast, Kenya, South Africa, Uganda, Zimbabwe.
HOST PLANTS. Anacardiaceae: *Rhus anchietae, R. dentata, Schinus molle, Sclerocarya caffra*. **Apocynaceae:** *Acokanthera longiflora, Allamanda, Nerium oleander, Rauwolfia caffra*. **Aquifoliaceae:** *Ilex*. **Bignoniaceae:** *Markhamia platycalyx*. **Compositae:** *Bidens pilosa*. **Euphorbiaceae:** *Uapaca kirkiana*. **Meliaceae:** *Cedrela*

toona, Khaya nyasica, Melia azedarach. **Moraceae:** *Ficus lutea.*
Myricaceae: *Myrica serrata.* **Myrtaceae:** *Syzygium cordatum.*
Pittosporaceae: *Pittosporum viridiflorum.* **Plumbaginaceae:**
Plumbago. **Rubiaceae:** *Coffea arabica, C. canephora, C. robusta,
C. stenophylla, Gardenia, Pentas schimperana.* **Rutaceae:** *Citrus
aurantium, C. maxima, C. sinensis, Toddalia asiatica.*
REMARKS. Adult female redescribed and illustrated by De Lotto
(1965) and by De Lotto (1955) (as *C. luteolus*). Cilliers (1967) de-
scribed the wax cover of the adult female and larval instars. Distri-
bution and host plant records given by De Lotto (1955, 1965,
1967a), Hodgson (1969a), Cilliers (1967), Almeida (1973) and by
Couturier *et al.* (1985).
BIOLOGY. Cilliers (1967) studied the biology and natural enemies in
South Africa.
ECONOMIC IMPORTANCE. Biology and pest status on coffee presented by Le Pelley (1968).

Ceroplastes breviseta Leonardi

Ceroplastes breviseta Leonardi, 1911: 264.
TYPE DATA. Syntypes female, ARGENTINA: Cacheuta, on *Atriplex
lampa* (IEAP).
DISTRIBUTION. NEOTROPICAL REGION: Argentina.
HOST PLANTS. Chenopodiaceae: *Atriplex lampa.*

Ceroplastes bruneri Cockerell & Cockerell

Ceroplastes bruneri Cockerell & Cockerell in Cockerell, 1902e: 91.
Ceroplastes bruneri Cockerell; Fernald, 1903: 149 [ERRONEOUS
AUTHORSHIP].
Ceroplastes (Ceroplastidia) bruneri Cockerell; Lizer y Trelles, 1939:
195 [ERRONEOUS AUTHORSHIP].
TYPE DATA. Syntypes female, PARAGUAY: San Bernardino, on
unspecified host plant (BMNH, USNM).
DISTRIBUTION. NEOTROPICAL REGION: Argentina, Bolivia,
Paraguay.
HOST PLANTS. Leguminosae: *Acacia bonariensis, A. retinodes, A.
riparia, Ceratonia siliqua, Manganaroa furcata, M. platensis,
Parkinsonia aculeata.*
REMARKS. Distribution and host plant records given by Lizer y
Trelles (1939).

Ceroplastes caesalpiniae Reyne

Ceroplastes caesalpiniae Reyne, 1964: 114.
TYPE DATA. Syntypes female and larva, CURACAO: Noordkant, on
Caesalpinia coriaria (ZMAN).
DISTRIBUTION. NEOTROPICAL REGION: Curacao.
HOST PLANTS. Leguminosae: *Caesalpinia coriaria.*
REMARKS. Reyne (1964) indicated that this species is allied to *C.
bruneri* and to *C. utilis.*

Ceroplastes campinensis Hempel

Ceroplastes campinensis Hempel, 1901c: 389.

TYPE DATA. Syntypes female and male, BRAZIL: Sao Paulo, Botuca-
tu and Campinas, on Myrtaceae and *Psidium guajava* (MZSP).
DISTRIBUTION. NEOTROPICAL REGION: Brazil Sao Paulo.
HOST PLANTS. Myrtaceae: *Psidium guajava.*
REMARKS. Adult female redescribed by Hempel (1920a).

Ceroplastes candella Cockerell & King

Ceroplastes candella Cockerell & King in Cockerell, 1902c: 113.
TYPE DATA. Syntypes female, SOUTH AFRICA: Natal, Richmond,
host plant not specified (USNM).
DISTRIBUTION. ETHIOPIAN REGION: South Africa.

Ceroplastes cassiae (Chavannes)

Coccus cassiae Chavannes, 1848: 141.
Columnea gray Targioni Tozzetti, 1866: 145 [UNJUSTIFIED RE-
PLACEMENT NAME].
Ceroplastes gray (Targioni Tozzetti); Targioni Tozzetti, 1868: 728.
Ceroplastes cassiae (Chavannes); Signoret, 1869a: 848.
TYPE DATA. Syntypes female, BRAZIL: Rio de Janeiro, Bosafogo
Bay, on *Cassia* sp. (depository unknow).
DISTRIBUTION. NEOTROPICAL REGION: Brazil.
HOST PLANTS. Leguminosae: *Cassia.*
REMARKS. Adult female redescribed and illustrated by Westwood
(1853b) and by Signoret (1872b). Adult female redescribed by
Hempel (1900b).

Ceroplastes castelbrancoi Almeida

Ceroplastes aff. longicauda Brain; Almeida, 1969: 17 [MISIDENTIFI-
CATION]
Ceroplastes castelbrancoi Almeida, 1973: 2.
TYPE DATA. Syntypes female, ANGOLA: Sa da Bandeira, on twigs of
quince [-*Cydonia oblonga*] (CZLP).
DISTRIBUTION. ETHIOPIAN REGION: Angola.
HOST PLANTS. Rosaceae: *Cydonia oblonga.*

Ceroplastes centroroseus Chen

Ceroplastes centroroseus Chen, 1974: 325.
Paracerostegia centroroseus (Chen); Tang, 1991: 305.
TYPE DATA. Syntypes female, CHINA: Sichuan Province, on *Citrus*
sp. (lost; F. T. Tang, 1989, personal communication).
DISTRIBUTION. PALAEARCTIC REGION: China.
HOST PLANTS. Agavaceae: *Yucca filamentosa.* **Rutaceae:** *Citrus.*
Theaceae: *Thea sinensis.*
REMARKS. Adult female redescribed and illustrated by Tang (1991).
Distribution and host plant records given by Tang (1991).

Ceroplastes ceriferus (Fabricius)

Coccus ceriferus Fabricius, 1798: 546.
Coccus (Ceroplastes) chilensis Gray, 1828: 7. Syntypes preadult
female, CHILE: on branches and peduncles of unidentified tree
(BMNH). Syn. by Green, 1899: 191.

Ceroplastes ceriferus (Fabricius); Walker, 1852: 1087.
Ceroplastes australiae Walker, 1852: 1087. Syntypes female, AUSTRALIA: (BMNH). Syn. by Green, 1899: 191.
Columnea cerifera (Fabricius); Targioni Tozzetti, 1866: 144.
Columnea chilensis (Gray); Targioni Tozzetti, 1866: 145.
Ceroplastes ceriferus (Anderson); Signoret, 1869a: 848 [ERRONEOUS AUTHORSHIP].
Lacca alba Signoret; Signoret, 1869a: 848 [NOMEN NUDUM].
Ceroplastes vayssierei Mahdihassan, 1933: 561 [NOMEN NUDUM and UNJUSTIFIED REPLACEMENT NAME].
Gascardia cerifera (Anderson); De Lotto, 1965: 198 [ERRONEOUS AUTHORSHIP].
Ceroplastes ceriferens (Anderson); Tao, 1978: 79 [MIS-SPELLING and ERRONEOUS AUTHORSHIP].
Ceroplastes ceriferens (Anderson); Su, 1982: 61 [MIS-SPELLING and ERRONEOUS AUTHORSHIP].
Ceroplastes cerifera (Fabricius); Gill, 1988: 18.
TYPE DATA. Syntypes female, INDIA: Coromandel Coast, probably on *Maytenus emarginatus* (lost; see De Lotto, 1971c).
DISTRIBUTION. AUSTRO ORIENTAL REGION: Indonesia, Malaysia, Papua New Guinea, Philippines. **AUSTRALIAN REGION:** New South Wales, Queensland. **NEOTROPICAL REGION:** Chile, Panama, Virgin Islands. **NEW ZEALAND and PACIFIC REGION:** Cook Islands, Fiji, Guam, New Caledonia, Tonga, Vanuatu. **NEARCTIC REGION:** Alabama, Arkansas, Florida, Georgia, Illinois, Louisiana, Maryland, Mississippi, New Jersey, New York, North Carolina, Oklahoma, South Carolina, Tennessee, Texas, Virginia, Washington, D. C. **ORIENTAL REGION:** India, Sri Lanka, Thailand, Vietnam. **PALAEARCTIC REGION:** China, England, Japan.
HOST PLANTS. Amaranthaceae: *Amaranthus.* **Anacardiaceae:** *Rhus succedanea.* **Apocynaceae:** *Thevetia peruviana.* **Aquifoliaceae:** *Ilex aquifolium, I. cornuta, I. crenata, I. japonicus, I. latifolia, I. opaca, I. serrata, I. vomitoria.* **Araliaceae:** *Fatsia japonica.* **Berberidaceae:** *Berberis julianae, B. tricanthophora, Mahonia.* **Betulaceae:** *Betula pendula.* **Bignoniaceae:** *Pandora pandorana.* **Buxaceae:** *Buxus sempervirens.* **Caprifoliaceae:** *Viburnum.* **Celastraceae:** *Euonymus europaeus, E. japonicus.* **Compositae:** *Artemisia abrotanum.* **Cucurbitaceae:** *Cucurbita moschata.* **Ebenaceae:** *Diospyros kaki.* **Epacridaceae:** *Monotoca elliptica.* **Ericaceae:** *Azalea, Vaccinium arboreum.* **Icacinaceae:** *Phytocrene.* **Lauraceae:** *Persea gratissima.* **Lythraceae:** *Lagerstroemia indica.* **Magnoliaceae:** *Magnolia grandiflora, M. virginiana.* **Malvaceae:** *Abutilon indicum.* **Moraceae:** *Ficus hawii, Morus alba.* **Myrtaceae:** *Eucalyptus deglupta, Eugenia malaccensis.* **Philadelphaceae:** *Deutzia.* **Pinaceae:** *Tsuga canadensis.* **Piperaceae:** *Piper.* **Pittosporaceae:** *Pittosporum.* **Platanaceae:** *Platanus.* **Podocarpaceae:** *Podocarpus macrophyllus, P. nagi.* **Polypodiaceae:** *Pteridium esculentum, Pteris aquilina, Pyrrosia lanceolata.* **Rosaceae:** *Armeniaca vulgaris, Chaenomeles japonica, Cormus tschonoskii, Crataegus Malus sylvestris, Persica vulgaris, Prunus domestica, P. yedoensis, Pyracantha coccinea, Spiraea.* **Rubiaceae:** *Gardenia florida, Ixora triflora.* **Rutaceae:** *Citrus.* **Salicaceae:** *Salix.* **Sapindaceae:** *Euphoria longana, Nephelium*

lappaceum. **Tamaricaceae:** *Tamarix gallica*. **Theaceae:** *Camellia japonica, C. sasanqua, C. sinensis.* **Ulmaceae:** *Celtis occidentalis, Ulmus.* **Verbenaceae:** *Callicarpa.*

REMARKS. Common name - Indian wax scale, or Indian white wax scale; however, various common names were used (see Gimpel *et al.,* 1974). De Lotto (1971c) presented a critical account on the nomenclature and proved that the authorship should be credited to Fabricius (1798) and not to Anderson, as erroneously ascribed by Signoret (1869). Green (1899) discussed the synonymy of *C. chilensis* with this species and the type locality of the former. Signoret (1869a: 848) published for the first time the binomen *Lacca alba* [NOMEN NUDUM] which he attributed to Pearson (1794). However, Pearson (1794) did not use such a binomen in his article. The latter publication presents an analysis of the chemical and physical properties of a wax, produced by a scale insect in Madras, which Pearson consistently named White Lac. The binomen *Lacca alba*, introduced by Signoret (1869a), is a Latin form of the term 'white lac'. The above binomen is here regarded as a NOMEN NUDUM of *Ceroplastes ceriferus*. Adult female redescribed and illustrated by Westwood (1853a), Green (1899), Green (1909), Froggatt (1915), Kuwana (1917a, 1923c), Morrison (1920), Borchsenius (1957), De Lotto (1971c), Williams & Kosztarab (1972), Gimpel *et al.* (1974), Kawai (1980), Hamon & Williams (1984), Avasthi & Shafee (1986), Williams & Watson (1990) and by Tang (1991). Adult male described and illustrated by Kuwana (1923c), Gimpel *et al.* (1974). Colour photograph given by Kawai (1980) and by Hamon & Williams (1984). Distribution and host plant records given by Green (1909, 1921b), Morrison (1920), Kuwana (1923c), Takahashi (1942a, 1952), Ferris (1950), De Lotto (1971c), Williams & Kosztarab (1972), Gimpel *et al.* (1974), Lambdin & Watson (1980), Hamon & Williams (1984), Avasthi & Shaffee (1986), Beardsley (1986), Shafee *et al.* (1989), Williams & Watson (1990), Danzig & Konstantinova (1990) and by Tang (1991). Tamaki (1964) reported (as *C. pseudoceriferus*) on the carbohydrates in the honeydew, and Tamaki (1966) analysed (as *C. pseudoceriferus*) the chemical composition of the wax cover. Komura *et al.* (1982) identified ceroalbolinic acid as a body pigment of the female. Pawlak *et al.* (1983) isolated and determined macrocyclic sesterpenoids from secretion of the scale.

BIOLOGY. Ohgushi & Nishino (1975) studied and formed the life tables in Japan. Males have been reported in Japan (Kuwana, 1923c), however, Gimpel *et al.* (1974) found small numbers (less than 2%) of males among greenhouse populations in Maryland, U.S.A.

ECONOMIC IMPORTANCE. A pest of economic importance to many ornamentals in U. S. A. (Gimpel *et al.,* 1974).

Ceroplastes circumdatus Green

Ceroplastes circumdatus Green, 1923a: 95.

TYPE DATA. Syntypes female, GUYANA: Demerara, on *Triphasia* sp. (BMNH).

DISTRIBUTION. NEOTROPICAL REGION: Guyana.

HOST PLANTS. Rutaceae: *Triphasia.*

Ceroplastes cirripediformis Comstock

Ceroplastes cirripediformis Comstock, 1881: 333.
Ceroplastes plumbaginis Cockerell, 1893c: 82. Syntypes female and larva, ANTIGUA: on *Plumbago capensis* (USNM). Syn. by Gimpel *et al.*, 1974: 29.
Ceroplastes euphorbiae Cockerell, 1896i: 17. Syntypes female, JAMAICA: Red Hill District, on *Euphorbia hypericifolia* (USNM). Syn. by Gimpel *et al.*, 1974: 29.
Ceroplastes mexicanus Cockerell, 1896e: 20. Syntypes female, MEXICO: San Luis Potosi, on *Catalpa* sp. (USNM). Syn. by Gimpel *et al.*, 1974: 29.
Ceroplastes cerripidiformis Comstock; Houser, 1918: 159 [MISSPELLING].
TYPE DATA. Lectotype female designated by Gimpel *et al.* (1974). U.S.A.: Florida, Sanford, on Eupatorium sp. (USNM).
DISTRIBUTION. AUSTRO ORIENTAL REGION: Philippines. **NEOTROPICAL REGION:** Antigua, Bermuda, Brazil, Chile, Colombia, Cuba, Galapagos Islands, Guyana, Jamaica, Lower California, Mexico, Puerto Rico, Trinidad, Virgin Islands. **NEW ZEALAND and PACIFIC REGION:** Hawaii, Marshall Islands, Wake Island. **NEARCTIC REGION:** Alabama, Arizona, Arkansas, California, Florida, Georgia, Louisiana, Maryland, Mississippi, Missouri, North Carolina, Ohio, Pennsylvania, South Carolina, Texas, Washington, D. C. **PALAEARCTIC REGION:** Italy.
HOST PLANTS. Acanthaceae: *Asystasia gangetica, Graptophyllum pictum, Pseuderanthemum atropurpureum, Strobilanthes anisophyllus.* **Agavaceae:** *Agave americana.* **Amaranthaceae:** *Alternathera amoena, Amaranthus.* **Anacardiaceae:** *Mangifera indica, Schinus terebinthifolia.* **Apocynaceae:** *Carissa carandas, Plumeria.* **Aquifoliaceae:** *Ilex aquifolium.* **Araceae:** *Philodendron.* **Balanitaceae:** *Balanites.* **Bignoniaceae:** *Catalpa, Tabebuia donellsmithi.* **Caprifoliaceae:** *Viburnum suspensum.* **Celastraceae:** *Euonymus, Maytenus.* **Compositae:** *Ageratum conyzoides, Artemisia, Baccharis halimifolia, Borrichia frutescens, Chrysanthemum, Eupatorium capillifolium, Iva frutescens, Palafoxia feayi, Pluchea indica, Pluchea odorata, Wedelia trilobata.* **Convolvulaceae:** *Argyreia nervosa, Ipomoea batatas.* **Ebenaceae:** *Diospyros kaki.* **Ehretiaceae:** *Ehretia anacua.* **Euphorbiaceae:** *Codiaeum.* **Euphorbiaceae:** *Euphorbia hypericifolia, E. pulcherrima, Hura crepitans.* **Guttiferae:** *Mammea.* **Labiatae:** *Rosmarinus.* **Leguminosae:** *Cassia corymbosa, Parkinsonia aculeata.* **Malpighiaceae:** *Malpighia coccigera.* **Meliaceae:** *Melia azedarach.* **Musaceae:** *Musa paradisiaca.* **Myrsinaceae:** *Ardisia pyramidalis.* **Myrtaceae:** *Eugenia, Feijoa sellowiana, Myrtus, Pimenta officinalis, Psidium guajava.* **Nyctaginaceae:** *Pisonia.* **Onagraceae:** *Gaura.* **Orchidaceae:** *Vanda aurora.* **Passifloraceae:** *Passiflora edulis, P. ligularis, P. quadrangularis.* **Plumbaginaceae:** *Plumbago capensis.* **Punicaceae:** *Punica granatum.* **Rhamnaceae:** *Karwinskia humboltiana.* **Rhizophoraceae:** *Rhizophora mangle.* **Rosaceae:** *Cotoneaster.* **Rubiaceae:** *Gardenia jasminoides, G. taitensis.* **Rutaceae:** *Citrus limon, C. sinensis.* **Salicaceae:** *Salix.* **Sapindaceae:** *Dodonaea viscosa.* **Sapotaceae:** *Chrysophyllum cainito.* **Tamaricaceae:** *Tamarix.* **Theaceae:** *Cleyera japonica.*

Ulmaceae: *Celtis laevigata, Ulmus parvifolia, U. pumila.* **Verbenaceae:** *Citharexylum spinosum, Clerodendrum fragrans, Duranta repens.*
REMARKS. Common Name - Barnacle wax scale. Adult female redescribed and illustrated by Gimpel *et al.* (1974), Hamon & Williams (1984), Gill (1988). Male described and illustrated by Gimpel *et al.* (1974). Colour photograph given by Gill (1988) and by Johnson & Lyon (1988). Distribution and host plant records given by Newstead (1917c), Houser (1918), Ferris (1921a), Bibby (1931), Silva *et al.* (1968), Gimpel *et al.* (1974), Nakahara & Miller (1981), Hamon & Williams (1984), Gonzalez (1989) and by Hodgson & Hilburn (1991a, 1991b).
ECONOMIC IMPORTANCE. This soft scale is a pest of Citrus and many ornamentals (Gimpel *et al.*, 1974).

Ceroplastes cistudiformis Cockerell
Ceroplastes psidii cistudiformis Cockerell, 1893e: 104.
Ceroplastes cistudiformis Townsend & Cockerell; Cockerell, 1896b: 331 [ERRONEOUS AUTHORSHIP].
Ceroplastes cistudiformis Cockerell; Cockerell, 1898n: 141.
Ceroplastes cistudiformis Townsend & Cockerell; Fernald, 1903: 150 [ERRONEOUS AUTHORSHIP].
TYPE DATA. Lectotype female designated by Gimpel *et al.*, (1974), MEXICO: Guanajuato, on *Bignonia* sp. and *Chrysanthemum* sp. (USNM).
DISTRIBUTION. NEOTROPICAL REGION: Cuba, Mexico. **NEARCTIC REGION:** California.
HOST PLANTS. Acanthaceae: *Beloperone gutata.* **Anacardiaceae:** *Schinus.* **Apocynaceae:** *Nerium oleander.* **Bignoniaceae:** *Bignonia.* **Compositae:** *Chrysanthemum, Parthenium.* **Convolvulaceae:** *Porana paniculata.* **Ehretiaceae:** *Cordia boissieri.* **Orchidaceae:** *Chysis aurea.* **Pandaceae:** *Panda.* **Passifloraceae:** *Passiflora.* **Rubiaceae:** *Bouvardia.* **Rutaceae:**
REMARKS. Adult female redescribed and illustrated by Gimpel *et al.* (1974), and by Gill (1988). Colour photograph given by Gill (1988).

Ceroplastes coloratus Cockerell
Ceroplastes coloratus Cockerell, 1898f: 435.
TYPE DATA. Syntypes female, MEXICO: Las Minas, Tabasco, on "crucetilla" (BMNH, USNM).
DISTRIBUTION. NEOTROPICAL REGION: Mexico.

Ceroplastes combreti Brain
Ceroplastes combreti Brain, 1920b: 27.
Gascardia combreti (Brain); De Lotto, 1970a: 145.
TYPE DATA. Syntypes female, SOUTH AFRICA: De Wildt, Pretoria District, on *Combretum* sp. (SANC).
DISTRIBUTION. ETHIOPIAN REGION: South Africa.
HOST PLANTS. Combretaceae: *Combretum.* **Sterculiaceae:** *Dombeya rotundifolia.*
REMARKS. Adult female redescribed and illustrated by De Lotto (1970a).

Ceroplastes communis Hempel

Ceroplastes communis Hempel, 1900b: 459.
TYPE DATA. Syntypes female, BRAZIL: Sao Paulo, Ypiranga, on *Maytenus* sp. (MZSP).
DISTRIBUTION. NEOTROPICAL REGION: Brazil Sao Paulo.
HOST PLANTS. Casuarinaceae: *Casuarina.* **Celastraceae:** *Maytenus.*
REMARKS. Adult female redescribed by Hempel (1901a).
BIOLOGY. Attended by a *Crematogaster* ant (Hempel, 1900b).

Ceroplastes confluens Cockerell & Tinsley

Ceroplastes confluens Cockerell & Tinsley, 1898: 468.
TYPE DATA. Syntypes female, JAMAICA: on unknown host plant (USNM).
DISTRIBUTION. NEOTROPICAL REGION: Argentina, Brazil Rio Grande do Sul, Jamaica, Uruguay.
HOST PLANTS. Compositae: *Vernonia polyanthes.* **Leguminosae:** *Acacia bonariensis, A. dealbata, A. decurrens mollis, A. melanoxylon, Calliandra tweediei, Inga, Mimosa saepiaria, M. serrana.* **Myrsinaceae:** *Myrsine umbellata.* **Myrtaceae:** *Eugenia.* **Ulmaceae:** *Celtis tala.*
REMARKS. Adult female redescribed by Hempel (1900b) and by Gomes Costa (1949). Distribution and host plant records given by Lizer y Trelles (1939), Gomes Costa (1949), Silva *et al.* (1968) and by Corseuil & Barbosa (1971).

Ceroplastes coniformis Newstead

Ceroplastes coniformis Newstead, 1913: 72.
TYPE DATA. Syntypes female, UGANDA: Entebbe, Botanic Gardens, on *Ficus* sp. (BMNH).
DISTRIBUTION. ETHIOPIAN REGION: Uganda.
HOST PLANTS. Moraceae: *Ficus.*

Ceroplastes constricta (De Lotto) n. comb.

Gascardia constricta De Lotto, 1969a: 417.
TYPE DATA. Holotype female, ANGOLA: Novo Redondo, on *Elaeis guineensis* (SANC).
DISTRIBUTION. ETHIOPIAN REGION: Angola.
HOST PLANTS. Palmae: *Elaeis guineensis.*

Ceroplastes cultus Hempel

Ceroplastes cultus Hempel, 1900b: 470.
TYPE DATA. Syntypes female, BRAZIL: Sao Paulo, Ypiranga, on twigs of *Erigeron canadensis* (MZSP).
DISTRIBUTION. NEOTROPICAL REGION: Brazil Sao Paulo.
HOST PLANTS. Compositae: *Erigeron canadensis.*
REMARKS. Adult female redescribed by Hempel (1901b).

Ceroplastes cundinamarcensis Mosquera

Ceroplastes cundinamarcensis Mosquera, 1979: 605.

TYPE DATA. Holotype female, COLOMBIA: Bogota (Cundinamarca), on *Schinus molle* (CTBC).
DISTRIBUTION. NEOTROPICAL REGION: Colombia.
HOST PLANTS. Anacardiaceae: *Schinus molle.*

Ceroplastes cuneatus Hempel

Ceroplastes cuneatus Hempel, 1900b: 471.
TYPE DATA. Syntypes female, BRAZIL: Sao Paulu, Ypiranga, on twigs of *Erigeron canadensis* (MZSP).
DISTRIBUTION. NEOTROPICAL REGION: Brazil Sao Paulo.
HOST PLANTS. Compositae: *Erigeron canadensis.*
REMARKS. Adult female redescribed by Hempel (1901b).

Ceroplastes deceptrix (De Lotto) n. comb.

Gascardia deceptrix De Lotto, 1965: 200.
TYPE DATA. Holotype female, SOUTH AFRICA: Cape Province, Clanwilliam District, on *Rhus undulata* (BMNH).
DISTRIBUTION. ETHIOPIAN REGION: South Africa.
HOST PLANTS. Anacardiaceae: *Rhus undulata.*

Ceroplastes deciduosus Morrison

Ceroplastes deciduosus Morrison, 1919: 79.
TYPE DATA. Syntypes female, ARGENTINA: Buenos Aires, on *Sapium biglandulosum* (USNM).
DISTRIBUTION. NEOTROPICAL REGION: Argentina, Brazil Rio Grande do Sul.
HOST PLANTS. Euphorbiaceae: *Sapium aucuparium, S. aucuparium lanceolatum, S. aucuparium salicifolia, S. biglandulosum, S. haematospermum.* **Punicaceae:** *Punica granatum.*
REMARKS. Distribution and host plant records given by Lizer y Trelles (1939) and by Corseuil & Barbosa (1971).

Ceroplastes deodorensis Hempel

Ceroplastes deodorensis Hempel, 1937: 9.
TYPE DATA. Syntypes female, BRAZIL: Rio de Janeiro State, Deodoro, on Annonaceae (IBSP).
DISTRIBUTION. NEOTROPICAL REGION: Brazil Rio de Janeiro.
HOST PLANTS. Annonaceae.

Ceroplastes depressus Cockerell

Ceroplastes depressus Cockerell, 1893c: 81.
TYPE DATA. Syntypes female, JAMAICA: Kingston, under bark of a lignum-vitae tree (USNM).
DISTRIBUTION. NEOTROPICAL REGION: Cuba, Jamaica.
HOST PLANTS. Burseraceae: *Bursera gummifera.*

Ceroplastes destructor Newstead

Ceroplastes ceriferus (Anderson); Newstead, 1910a: 66 [MISIDENTIFICATION].
Ceroplastes ceriferus (Anderson); Newstead, 1910c: 195 [MISIDENTIFICATION].

Ceroplastes ceriferus (Anderson); Newstead, 1911b: 167 [MISIDEN-TIFICATION].
Ceroplastes ceriferus (Anderson); Lindinger, 1913a: 80 [MISIDEN-TIFICATION].
Ceroplastes destructor Newstead, 1917b: 26.
Gascardia destructor (Newstead); De Lotto, 1965: 200.
TYPE DATA. Lectotype female designated by Williams & Watson (1990), UGANDA: Entebbe, Botanical Gardens, on *Antigonon* sp. (BMNH).
DISTRIBUTION. AUSTRO ORIENTAL REGION: Papua New Guinea. **AUSTRALIAN REGION:** New South Wales, Queensland. **ETHIOPIAN REGION:** Angola, Cameroon, Congo, Ivory Coast, Kenya, Mozambique, South Africa, Uganda, Zambia, Zimbabwe. **MADAGASIAN REGION:** Madagascar. **NEW ZEALAND and PACIFIC REGION:** New Zealand, Norfolk Island, Solomon Islands. **ORIENTAL REGION:** India.
HOST PLANTS. Acanthaceae: *Dicliptera*. **Anacardiaceae:** *Rhus simarubaefolia*. **Apocynaceae:** *Plumeria*. **Araliaceae:** *Cussonia spicata, Schefflera*. **Celastraceae:** *Elaeodendron capense, Gymnosporia buxifolia, Maytenus senegalensis*. **Compositae:** *Conyza*. **Ebenaceae:** *Euclea crispa*. **Ericaceae:** *Philippia lecana*. **Euphorbiaceae:** *Uapaca*. **Hippocrateaceae:** *Hippocratea parvifolia*. **Lauraceae:** *Persea americana*. **Loganiaceae:** *Nuxia oppositifolia, N. viscosa*. **Magnoliaceae:** *Magnolia*. **Meliaceae:** *Dysoxylum patersoni, Melia azedarach*. **Myrsinaceae:** *Maesa*. **Myrtaceae:** *Eugenia malaccensis, Psidium guajava, Syzygium cordatum*. **Pittosporaceae:** *Pittosporum crassifolium*. **Rubiaceae:** *Aida micrantha, Coffea arabica, C. canephora, C. robusta, Gardenia, Otiophora tryangana parvifolia, Platanocephalus morindaefolius*. **Rutaceae:** *Citrus maxima, C. sinensis, Poncirus trifoliata*. **Sapindaceae:** *Dodonaea viscosa*. **Sterculiaceae:** *Theobroma cacao*. **Theaceae:** *Camellia sinensis*.
REMARKS. Adult female redescribed and illustrated by De Lotto (1965), Almeida (1969) and by Williams & Watson (1990). Hackman (1951) analyzed the chemical composition of the wax cover. Hackman & Trikojus (1952) analyzed the chemical composition of the honeydew. Cilliers (1967) described the wax cover of the adult female and larval instars. Distribution and host plant records given by Brain (1920b), Hall (1931), Strickland (1947), Mamet (1959b), De Lotto (1965, 1967a), Snowball (1969), Hodgson (1969a), Almeida (1969, 1973), Wise (1977), Couturier *et al.* (1985), Matile-Ferrero & Nonveiller (1986), Avasthi & Shafee (1986), Fernandes (1989) and by Williams & Watson (1990).
BIOLOGY. Cilliers (1967) studied the biology and natural enemies in South Africa. Two generations were found to develop annually on citrus in Queensland, Australia (Smith, 1970; Smith & Ironside, 1974). De Lotto (1971c) noted that the development of the caudal process is gradual and the organ attains its full size some time after the last moult has taken place.
ECONOMIC IMPORTANCE. A major citrus pest in Queensland, Australia (Sabine, 1969; Smith & Ironside, 1974). Biology and pest status on coffee presented by Le Pelley (1968). Snowball (1969) surveyed the natural enemies in South Africa.

Ceroplastes diospyros Hempel

Ceroplastes diospyros Hempel, 1928: 236.
TYPE DATA. Syntypes female, BRAZIL: Sao Paulo State, Itupararan-
ga, near Sorocaba, on *Diospyros kaki* (IBSP).
DISTRIBUTION. NEOTROPICAL REGION: Brazil Sao Paulo.
HOST PLANTS. Ebenaceae: *Diospyros kaki.*

Ceroplastes dugesii Lichtenstein

Ceroplastes dugesii Lichtenstein, 1885: cxli.
Ceroplastes ceriferus (Anderson); Cockerell, 1893g: 373 [MISIDEN-
TIFICATION].
Ceroplastes dugesii Townsend; Cockerell, 1893t: 100 [ERRONEOUS
AUTHORSHIP].
Ceroplastes roseatus Townsend & Cockerell, 1898: 176. Syntypes
female, MEXICO: El Cuyu del Chicosapote, near Frontera,
Tabasco, on branches of wild fruit tree "cojon de venado" (BMNH,
USNM). Syn. by Gimpel *et al.*, 1974: 39.
Ceroplastes townsendi Cockerell, 1899g: 18. Syntypes female,
MEXICO: Arroyo San Isidro, near Frontesa, Tabasco, on bark of
a small shrub (USNM). Syn. by Gimpel *et al.*, 1974: 39.
Ceroplastes roseatus var. *B* Cockerell, 1903b: 157. Syntypes female,
MEXICO: Base of Volcan de Colima, on stems of herbaceous
Compositae (USNM). Syn. by Gimpel *et al.*, 1974: 39.
Ceroplastes townsendi percrassus Cockerell, 1903b: 159. Syntypes
female, MEXICO: Zapotlan, on *Ficus* sp. (USNM). Syn. by Gimpel
et al., 1974: 39.
Ceroplastes dugesii Townsend; De Lotto, 1971c: 140 [ERRONEOUS
AUTHORSHIP].
TYPE DATA. Syntypes female, MEXICO: Guanajuato, on *Hybiscus*
[sic], *Ficus sphaerocarpa*, laurie and rose (probably lost).
DISTRIBUTION. NEOTROPICAL REGION: Barbados, Cuba, Mexico,
Guyana, Panama, Puerto Rico, Virgin Islands. **NEARCTIC
REGION:** Florida.
HOST PLANTS. Anacardiaceae: *Schinus molle.* **Annonaceae:**
Annona. **Apocynaceae:** *Nerium oleander.* **Burseraceae:** *Bursera
gummifera*, *B. simaruba.* **Compositae.** **Ebenaceae:** *Diospyros
silvestris.* **Malvaceae:** *Malva, Malvaviscus acerifolius, M. arboreus.* **Moraceae:** *Ficus.* **Nyctaginaceae:** *Torrubia bracei.* **Piperaceae:** *Piper medium.* **Ulmaceae:** *Trema mollis.*
REMARKS. Adult female redescribed and illustrated by Gimpel *et al.*
(1974) and by Hamon & Williams (1984). Colour photograph given by
Hamon & Williams (1984). Distribution and host plant records given
by Townsend (1892b), Gimpel *et al.* (1974), Nakahara & Miller (1981)
and by Hamon & Williams (1984).

Ceroplastes elytropappi (Brain)

Baccacoccus elytropappi Brain, 1920a: 127.
Ceroplastes adustus De Lotto, 1967b: 781. Holotype female, SOUTH
AFRICA: Cape Province, Ceres, on *Passerina* sp. (SANC).
Ceroplastes elytropappi (Brain); De Lotto, 1971c: 141.
TYPE DATA. Syntypes female, SOUTH AFRICA: Cape Province,
French Hoek, on *Elytropappus rhinocerotis* (SANC).

DISTRIBUTION. ETHIOPIAN REGION: South Africa.
HOST PLANTS. Compositae: *Elytropappus rhinocerotis*. **Thyme-laeaceae:** *Passerina*.
REMARKS. Adult female redescribed and illustrated (as *C. adustus*) by De Lotto (1967b).

Ceroplastes eucleae Brain

Ceroplastes eucleae Brain, 1920b: 30.
TYPE DATA. Syntypes female, SOUTH AFRICA: Transvaal, Pretoria, on *Euclea* sp., *Ochna* sp. and *Pavetta* sp. (SANC).
DISTRIBUTION. ETHIOPIAN REGION: South Africa.
HOST PLANTS. Ebenaceae: *Euclea*. **Ochnaceae:** *Ochna*. **Rubiaceae:** *Pavetta*.

Ceroplastes eugeniae Hall

Ceroplastes rusci eugeniae Hall, 1931: 298.
Ceroplastes eugeniae Hodgson, 1969a: 4.
TYPE DATA. Syntypes female, ZIMBABWE: Mazoe on *Eugenia owariensis* and Banket on *Diplorhynchus mosambicensis* (BMNH).
DISTRIBUTION. ETHIOPIAN REGION: Mozambique, Zimbabwe. **MADAGASIAN REGION:** Comores.
HOST PLANTS. Annonaceae: *Cananga odorata*. **Apocynaceae:** *Diplorhynchus mosambicensis*. **Guttiferae:** *Garcinia huillensis*. **Loranthaceae:** *Desrousseauxia, Loranthus quequensis*. **Myrtaceae:** *Eugenia owariensis, Psidium guajava, Syzygium guiniense*. **Rosaceae:** *Cliffortia nitidula, Cydonia oblonga, Malus sylvestris*. **Thymelaeaceae:** *Synaptolepis alternifolia*.
REMARKS. Adult female redescribed and illustrated by Hodgson (1969a). Adult female redescribed by Matile-Ferrero (1978).

Ceroplastes excaecariae Hempel

Ceroplastes excaecariae Hempel, 1912: 66.
TYPE DATA. Syntypes female, BRAZIL: Sao Paulo, Ypiranga, on *Excaecaria biglandulosa* (MZSP).
DISTRIBUTION. NEOTROPICAL REGION: Brazil Sao Paulo.
HOST PLANTS. Euphorbiaceae: *Excaecaria biglandulosa*.

Ceroplastes fairmairii Signoret

Columnea fairmairei Targioni Tozzetti, 1866: 146 [NOMEN NUDUM].
Ceroplastes fairmairei (Targioni Tozzetti); Targioni Tozzetti, 1868: 728 [NOMEN NUDUM].
Ceroplastes fairmairii Signoret, 1872b: 43.
Ceroplastes fairmairii Targioni Tozzetti; Fernald, 1903: 152 [ERRONEOUS AUTHORSHIP].
TYPE DATA. Syntypes female, URUGUAY: Montevideo, on Myrtaceae (VMNH).
DISTRIBUTION. NEOTROPICAL REGION: Brazil, Uruguay.
HOST PLANTS. Myrtaceae.
REMARKS. Maskell (1893b) synonymized this species with *C. certferus*, but it was not endorsed by subsequent authors.

Ceroplastes ficus Newstead

Ceroplastes ficus Newstead, 1910c: 190.
Ceroplastes pallidus Brain, 1920b: 33. Syntypes female, SOUTH
    AFRICA: Transvaal, Pretoria, on fig (SANC). Syn. by Hall, 1931:
    294.
TYPE DATA. Syntypes female, TANZANIA: Bukoba, on *Ficus* sp.
    (BMNH).
DISTRIBUTION. ETHIOPIAN REGION: Angola, Ghana, Malawi,
    South Africa, Tanzania, Uganda, Zimbabwe.
HOST PLANTS. Annonaceae: *Annona, Artobotrys brachypetalus*.
    Celastraceae: *Maytenus senegalensis*. **Euphorbiaceae:** *Euphor-
    bia pulcherrima*, **Moraceae:** *Ficus*. **Ochnaceae:** *Ochna pulchella*.
    Proteaceae: *Grevillea robusta*. **Rosaceae:** *Parinari curatellifolia*.
REMARKS. Adult female redescribed and illustrated by De Lotto
    (1965). Distribution and host plant records given by Newstead
    (1917c), Brain (1920b), Hall (1931), De Lotto (1965, 1967a), Hodgson
    (1969a) and by Almeida (1973).

Ceroplastes floridensis Comstock

Ceroplastes floridensis Comstock, 1881: 331.
Cerostegia floridensis (Comstock); De Lotto, 1969b: 211.
Paracerostegia floridensis (Comstock); Tang, 1991: 306.
TYPE DATA. Lectotype female designated by Gimpel *et al.* (1974),
    U.S.A.: Florida, Jacksonville, on Tangerine orange (USNM).
DISTRIBUTION. AUSTRO ORIENTAL REGION: Hong Kong, Irian
    Jaya. **ETHIOPIAN REGION:** Zanzibar. **MADAGASIAN REGION:**
    Comores, Madagascar, Mauritius, Reunion, Seychelles. **NEO-
    TROPICAL REGION:** Bermuda, Brazil Rio Grande do Sul,
    Colombia, Cuba, Curacao, Ecuador, Guyana, Honduras, Mexico,
    Montserrat, Nicaragua, Puerto Rico, Trinidad, Virgin Islands.
    NEW ZEALAND and PACIFIC REGION: Mariana Islands,
    Ogasawara Islands, Palaus. **NEARCTIC REGION:** Florida, Geor-
    gia, Louisiana, Maryland, Mississippi, New Mexico, New York,
    North Carolina, South Carolina, Tennessee, Texas, Virginia,
    Washington, D. C. **ORIENTAL REGION:** India, Taiwan, Vietnam.
    PALAEARCTIC REGION: China, Cyprus, Egypt, France, Greece,
    Israel, Italy, Lebanon, Madeira, Turkey.
HOST PLANTS. Anacardiaceae: *Anacardium occidentale, Mangifera
    indica, Pistacia lentiscus, P. palestina, Schinus molle, S. tere-
    binthifolius*. **Annonaceae:** *Annona cherimolia, A. muricata, A.
    squamosa*. **Apocynaceae:** *Carissa carandas, C. grandiflora,
    Nerium oleander, Plumeria rubra, Thevetia peruviana, Trachelos-
    permum jasminoides*. **Aquifoliaceae:** *Ilex canariensis, I. cornuta,
    I. crenata, I. perado, I. pulviflora, I. vomitoria*. **Araceae:** *Philoden-
    dron*. **Araliaceae:** *Aralia, Dizygotheca veitchii, Hedera canarien-
    sis, H. helix*. **Asclepiadaceae:** *Periploca gracea*. **Bignoniaceae:**
    Kiggelia pinnata, Stenolobium stans, Tecomaria capensis. **Bur-
    seraceae:** *Bursera simaruba*. **Celastraceae:** *Elaeodendron,
    Euonymus japonicus*. **Combretaceae:** *Bucida buceras, Terminalia
    arjuna*. **Compositae:** *Chrysanthemum indicum, Erigeron crispus,
    Psiadia altissima*. **Convolvulaceae:** *Convolvulus*. **Cycadaceae:**
    Cycas revoluta. **Ebenaceae:** *Diospyros discolor, D. kaki*. **Elaeag-
    naceae:** *Elaeagnus angustifolia*. **Ephedraceae:** *Ephedra alte*.

Ericaceae: *Agauria salicifolia*, **Ericaceae:** *Arbutus andrachne*, *A. unedo*, *Pernettya*, *Vaccinium*. **Euphorbiaceae:** *Antidesma bunius*, *Phyllanthus*. **Guttiferae:** *Calophyllum inophyllum*. **Heliconiaceae:** *Heliconia*. **Lauraceae:** *Cinnamomum pseudopedunculatum*, *Laurus azorica*, *L. nobilis*, *Lindera benzoin*, *Machilus*, *Persea americana*, *P. borbonia*. **Leguminosae:** *Acacia farnesiana*, *Ceratonia siliqua*, *Retama roetam*. **Loganiaceae:** *Strychnos spinosa*. **Lythraceae:** *Lagerstroemia indica*. **Magnoliaceae:** *Michelia champaca*. **Meliaceae:** *Melia azedarach*. **Moraceae:** *Artocarpus integrifolia*, *Ficus benghalensis*, *F. carica*, *F. obliqua*, *F. retusa*, *F. sycomorus*, *Morus alba*, *Treculia perrieri*. **Musaceae:** *Musa*. **Myoporaceae:** *Myoporum acuminatum*, *M. laetum*. **Myrsinaceae:** *Ardisia sieboldii*, *Maesa japonica*. **Myrtaceae:** *Callistemon phoeniceus*, *Eugenia compacta*, *E. jambolana*, *Feijoa sellowiana*, *Melaleuca armillaris*, *Metrosideros boninensis*, *Myrtus communis*, *Psidium cattleyanum*, *P. guajava*, *P. littorale*, *P. pomiferum*, *Rhodomyrtus tomentosa*. **Oleaceae:** *Ligustrum micranthum*. **Oleandraceae:** *Nephrolepis cordifolia*. **Palmae:** *Washingtonia filifera*, **Pinaceae:** *Pinus elliotti*, **Pinaceae:** *Pinus halepensis*, *Tsuga canadensis*, **Pittosporaceae:** *Pittosporum chichijimense*, *P. undulatum*. **Platanaceae:** *Platanus*. **Polygonaceae:** *Coccoloba diversifolia*, *Polygonum baldschuanicum*. **Polypodiaceae:** *Nephrolepis exaltata*. **Punicaceae:** *Punica granatum*. **Rhamnaceae:** *Rhamnus alaternus*. **Rosaceae:** *Amygdalus communis*, *Cotoneaster pannosa*, *Crataegus azarolus*, *Cydonia oblonga*, *Eriobotrya japonica*, *Malus sylvestris*, *Persica vulgaris*, *Prunus armeniaca*, *P. salicina*, *Pyrus communis*, *Raphiolepis integerrima*, *R. ovata*, *R. umbellata*. **Rubiaceae:** *Casimiroa edulis*, *Coffea arabica*, *Gardenia jasminoides*, *Ixora*, *Psychotria boninensis*, *P. rubra*. **Rutaceae:** *Citrus aurantium*, *C. histrix*, *C. mitis*, *C. paradisi*, *Poncirus trifoliata*. **Salicaceae:** *Populus deltoides*. **Santalaceae:** *Santalum boninense*. **Sapindaceae:** *Dodonaea viscosa*, *Melicoccus bijugatus*, *Nephelium lappaceum*. **Sapotaceae:** *Achras sapota*, *Chrysophyllum cainito*, *Manilkara zapota*. **Schizaeaceae:** *Actinostachys boninensis*. **Scrophulariaceae:** *Russelia equisetiformis*. **Sebestenaceae:** *Cordia myxa*. **Serraceniaceae:** *Serracenia minor*. **Solanaceae:** *Solanum villosum*. **Theaceae:** *Camellia sinensis*, *Schima mertensiana*, *S. superba*, *Ternstroemia japonica*. **Urticaceae:** *Boehmeria boninensis*. **Verbenaceae:** *Duranta repens*. **Zamiaceae:** *Encephalartos*.

REMARKS. Common Name - Florida wax scale. Adult female redescribed and illustrated by Kuwana (1917a, 1923c), Ferris (1950), Borchsenius (1957), De Lotto (1969b), Ezzat & Hussein (1969), Ben-Dov (1970a), Williams & Kosztarab (1972), Gimpel *et al.* (1974), Wang (1980), Kawai (1980), Yang (1982), Mosquera (1984), Hamon & Williams (1984), Williams & Watson (1990) and by Tang (1991). Colour photograph given by Delucchi (1975), Kawai (1980) and by Johnson & Lyon (1988). Intraspecific variation of taxonomic characters of the adult female given by Ben-Dov (1970a). Amitai (1969) and Ben-Dov (1970b) observed and described three larval instars in the development of the female, which is typical to species of *Ceroplastes*, as well as in the Coccidae. However, Ezzat & Fayez (1980), erroneously noted that the female develops through four nymphal instars. Green (1937) and Williams & Williams (1988) indicated that *C.*

vinsonii Signoret is likely identical with this species. Parida & Moharana (1982) and Moharana (1990) reported chromosome number 2n=36 in India. Distribution and host plant records given by Green (1907, 1916b), Hall (1922), Bodenheimer (1924, 1926a, 1951b, 1953), Ballou (1926), Balachowsky (1933a, 1939a), Gomes Costa (1949), Ferris (1950), Mamet (1951), Beardsley (1966a), Ben-Dov (1970d), Corseuil & Barbosa (1971), Kawai *et al.* (1971), Gimpel *et al.* (1974), Matile-Ferrero (1978), Argyriou & Kourmadas (1980), Lambdin & Watson (1980), Nakahara (1983), Tao *et al.* (1983), Vieira *et al.* (1983), Hamon & Williams (1984), Mosquera (1984), Sinha & Denish (1984), Avasthi & Shaffee (1986), Marotta (1987), Varshney & Moharana (1987), Williams & Williams (1988), Shafee *et al.* (1989), Williams & Watson (1990), Danzig & Konstantinova (1990) and by Tang (1991).

BIOLOGY. Bodenheimer (1951b) discussed the biology and economic importance. Develops two generations in a year on Citrus, in Israel (Ben-Dov, 1976c; Podoler *et al.*, 1981), and in Greece (Argyriou & Kourmadas, 1980). Methods for laboratory rearing given by Ben-Dov (1970c) and mass rearing by Argov *et al.* (1987). Population dynamics on Citrus in Israel studied by Podoler *et al.* (1981) and by Schneider *et al.* (1987a, 1987b). Yardeni (1987) and Yardeni & Rosen (1990) studied the wind dispersal of crawlers.

ECONOMIC IMPORTANCE. A major citrus pest in Israel (Bodenheimer, 1951b; Ben-Dov, 1976c). A pest of many ornamentals in U. S. A. (Gimpel *et al.*, 1974). The adverse effects of IGR on this pest studied by Peleg & Gothilf (1981) and by Eisa *et al.* (1990). Peleg (1987) reported on resistance to commercial formulations of Carabaryl in Israel.

Ceroplastes formicarius Hempel

Ceroplastes formicarius Hempel, 1900b: 472.

TYPE DATA. Syntypes female, BRAZIL: Sao Paulo, Ypiranga, on *Maytenus* sp. (MZSP).

DISTRIBUTION. NEOTROPICAL REGION: Brazil Sao Paulo.

HOST PLANTS. Celastraceae: *Maytenus*.

REMARKS. Adult female redescribed by Hempel (1901b).

BIOLOGY. Attended by a species of *Camponotus* (Hempel, 1900b).

Ceroplastes formosus Hempel

Ceroplastes formosus Hempel, 1900b: 468.

TYPE DATA. Syntypes female, BRAZIL: Minas Gerais, Pocos de Caldas, on *Eugenia* sp. (MZSP).

DISTRIBUTION. NEOTROPICAL REGION: Brazil Minas Gerais.

HOST PLANTS. Myrtaceae: *Eugenia*.

REMARKS. Adult female redescribed by Hempel (1901b).

Ceroplastes fumidus De Lotto

Ceroplastes quadrilineatus simplex Brain, 1920b: 33.

Ceroplastes simplex Brain; De Lotto, 1965: 187 [HOMONYM of *Ceroplastes simplex* Hempel, 1900].

Ceroplastes fumidus De Lotto, 1978: 138 [REPLACEMENT NAME].

TYPE DATA. Syntypes female, SOUTH AFRICA: Cape Province, Victoria West, on *Rhus* sp. (SANC).
DISTRIBUTION. ETHIOPIAN REGION: South Africa.
HOST PLANTS. Anacardiaceae: *Rhus.*
REMARKS. Adult female redescribed by De Lotto (1978).

Ceroplastes galeatus Newstead

Ceroplastes galeatus Newstead, 1911a: 95.
TYPE DATA. Syntypes female, UGANDA: Entebbe, on undetermined plant (BMNH).
DISTRIBUTION. ETHIOPIAN REGION: Uganda.
BIOLOGY. Biology and pest status on coffee presented by Le Pelley (1968).

Ceroplastes giganteus Dozier

Ceroplastes giganteus Dozier, 1931: 2.
TYPE DATA. Syntypes female, HAITI: Source Cazeau, on *Ficus rubricosta* (AMNY, USNM).
DISTRIBUTION. NEOTROPICAL REGION: Haiti.
HOST PLANTS. Bombacaceae: *Neobuchia paulinae.* **Moraceae:** *Ficus rubricosta.*

Ceroplastes gigas Cockerell

Ceroplastes gigas Cockerell, 1914: 331.
TYPE DATA. Syntypes female, PHILIPPINES: near Los Banos, host plant not recorded (USNM).
DISTRIBUTION. AUSTRO ORIENTAL REGION: Philippines.

Ceroplastes grandis Hempel

Ceroplastes grandis Hempel, 1900b: 455.
TYPE DATA. Syntypes female, BRAZIL: Ypiranga and Sao Paulo, on *Zanthoxylum* sp., *Ilex* sp., *Psidium* sp., *Mechilia flava, Baccharis* sp. and various Myrtaceae (MZSP).
DISTRIBUTION. NEOTROPICAL REGION: Argentina, Brazil Rio Grande do Sul, Brazil Sao Paulo.
HOST PLANTS. Apocynaceae: *Nerium oleander.* **Aquifoliaceae:** *Ilex paraguariensis.* **Bignoniaceae:** *Jacaranda.* **Combretaceae:** *Laguncularia recemosa.* **Compositae:** *Baccharis.* **Ebenaceae:** *Diospyros kaki.* **Leguminosae:** *Acacia decurrens mollis.* **Lythraceae:** *Lagerstroemia indica.* **Myrtaceae:** *Eucalyptus, Eugenia guabiju, Psidium guajava.* **Platanaceae:** *Platanus orientalis.* **Punicaceae:** *Punica granatum.* **Rhizophoraceae:** *Rhizophora.* **Rosaceae:** *Cydonia vulgaris, Mespilus germanica, Pyrus communis, P. germanica.* **Rutaceae:** *Citrus deliciosa, C. nobilis, Zanthoxylum.* **Tiliaceae:** *Luehea divaricata.*
REMARKS. Adult female redescribed by Hempel (1901a, 1920a). Adult female redescribed and illustrated by Gomes Costa (1949) and by Vernalha *et al.* (1974). Vernalha *et al.* (1974) assigned this species to the subgenus *Ceroplastes (Octoceroplastes),* however, this post-1930 name is not available according to the International Code (1985). Distribution and host plant records given by Lizer y Trelles

(1939), Gomes Costa (1949), Corseuil & Barbosa (1971) and by Silva et al. (1968).

Ceroplastes gregarius Hempel

Ceroplastes gregarius Hempel, 1932: 323.
TYPE DATA. Syntypes female, BRAZIL: Sao Paulo State, Sao Paulo, on Gomphrena nagansellarti (IBSP).
DISTRIBUTION. NEOTROPICAL REGION: Brazil Sao Paulo.
HOST PLANTS. Amaranthaceae: Gomphrena nagansellarti.

Ceroplastes hawanus Williams & Watson

Ceroplastes hawanus Williams & Watson, 1990: 73.
TYPE DATA. Holotype female, SOLOMON ISLANDS: San Cristobal, Hawa, on Barringtonia asiatica (BMNH).
DISTRIBUTION. AUSTRO ORIENTAL REGION: Solomon Islands.
HOST PLANTS. Leycithidaceae: Barringtonia asiatica.

Ceroplastes helichrysi Hall

Ceroplastes helichrysi Hall, 1931: 295.
Gascardia helichrysi (Hall); De Lotto, 1965: 181.
TYPE DATA. Syntypes female, ZIMBABWE: Bromley, on Helichrysum sp. (BMNH).
DISTRIBUTION. ETHIOPIAN REGION: Zimbabwe.
HOST PLANTS. Compositae: Helichrysum.
REMARKS. Adult female redescribed and illustrated by Hodgson (1969a).

Ceroplastes hempeli Lizer y Trelles

Ceroplastes grandis hempeli Lizer y Trelles, 1919: 381.
Ceroplastes hempeli Lizer y Trelles; Vernalha et al., 1974: 127.
TYPE DATA. Syntypes female, ARGENTINA: Misiones, on Ilex paraguariensis (DPBA).
DISTRIBUTION. NEOTROPICAL REGION: Argentina.
HOST PLANTS. Aquifoliaceae: Ilex paraguariensis.
REMARKS. Adult female redescribed and illustrated by Vernalha et al. (1974). Vernalha et al. (1974) actually placed this species in the subgenus Ceroplastes (Octoceroplastes). However, according to the International Code (1985) this post-1930 generic name is not available.

Ceroplastes hodgsoni (Matile-Ferrero & Le Ruyet) n. comb.

Gascardia hodgsoni Matile-Ferrero & Le Ruyet, 1985: 262.
TYPE DATA. Holotype female, IVORY COAST: Tai, on Cleistanthus polystachyus (MNHN).
DISTRIBUTION. ETHIOPIAN REGION: Ivory Coast.
HOST PLANTS. Euphorbiaceae: Cleistanthus polystachyus.
BIOLOGY. Protected under soil shelter constructed by ants of a Crematogaster sp. (Matile-Ferrero & Le Ruyet, 1985).

Ceroplastes hololeucus De Lotto

Ceroplastes hololeucus De Lotto, 1969a: 413.
TYPE DATA. Holotype female, ANGOLA: Novo Redondo, on *Elaeis guineensis* (SANC).
DISTRIBUTION. ETHIOPIAN REGION: Angola.
HOST PLANTS. Palmae: *Elaeis guineensis.*

Ceroplastes iheringi Cockerell

Ceroplastes iheringi Cockerell, 1895f: 100.
TYPE DATA. Syntypes female, BRAZIL: Rio Grande do Sul, on *Baccharis platensis* (USNM).
DISTRIBUTION. NEOTROPICAL REGION: Brazil Rio Grande do Sul, Brazil Sao Paulo.
HOST PLANTS. Compositae: *Baccharis dracunculifolia, B. genisteloides, B. platensis, Hetherothalamus brunioides.* **Leguminosae:** *Mimosa bracatinga.*
REMARKS. Adult female redescribed and illustrated by Hempel (1900b) and by Gomes Costa (1949).

Ceroplastes immanis Green

Ceroplastes immanis Green, 1935b: 274.
TYPE DATA. Syntypes female, BRAZIL: Goyaz, on branches of undetermined plant (BMNH).
DISTRIBUTION. NEOTROPICAL REGION: Brazil.

Ceroplastes insulanus De Lotto

Ceroplastes insulanus De Lotto, 1971c: 141.
TYPE DATA. Holotype female, AUSTRALIA: New South Wales, Lord Howe Island, on areal roots of an unidentified plant (ANIC).
DISTRIBUTION. AUSTRALIAN REGION: Lord Howe Island.
HOST PLANTS. Myrsinaceae: *Rapanea.*
REMARKS. Adult female redescribed by Williams & Watson (1990).

Ceroplastes irregularis Cockerell

Ceroplastes irregularis Cockerell, 1893r: 351.
Ceroplastes artemisiarum Cockerell; Cockerell, 1893n: 160 [NOMEN NUDUM].
Ceroplastes irregularis rubidus Cockerell, 1896h: 203. Syntypes female, U.S.A.: New Mexico, Whitewater, on *Atriplex canescens* (USNM). Syn. by Gimpel *et al.*, 1974: 49.
TYPE DATA. Lectotype female designated by Gimpel *et al.* (1974), MEXICO: State of Chihuahua, 6 miles north of Montezuma Railroad Station, on *Artemisia* sp. (USNM).
DISTRIBUTION. NEOTROPICAL REGION: Mexico, Mexico Baja California. **NEARCTIC REGION:** Arizona, California, Idaho, Nevada, New Mexico, Texas.
HOST PLANTS. Chenopodiaceae: *Atriplex canescens, Eurotia lanata, Suaeda.* **Compositae:** *Artemisia, Chrysothamnus.*
REMARKS. Adult female redescribed and illustrated by Gimpel *et al.* (1974) and by Gill (1988). Male test described and illustrated by Miller & Williams (1990). Colour photograph given by Gill (1988).

Distribution and host plant records given by Ferris (1921a), Gimpel
et al. (1974) and by Gill (1988).

Ceroplastes itatiayensis Hempel

Ceroplastes itatiayensis Hempel, 1938: 263.
TYPE DATA. Syntypes female, BRAZIL: Rio de Janeiro State, Ita-
tiaya, on undetermined tree (MZSP).
DISTRIBUTION. NEOTROPICAL REGION: Brazil Rio de Janeiro.

Ceroplastes jamaicensis White

Ceroplastes jamaicensis White, 1846: 333.
Columnea iamaicensis (White); Targioni Tozzetti, 1866: 145 [MIS-
SPELLING].
TYPE DATA. Syntypes female, JAMAICA: on trunk of Lance-wood
tree [-*Calophyllum* sp.] (USNM).
DISTRIBUTION. NEOTROPICAL REGION: Jamaica.
HOST PLANTS. Guttiferae: *Calophyllum.*

Ceroplastes janeirensis Gray

Coccus (Ceroplastes) janeirensis Gray, 1828: 7.
Columnea ianeirensis Targioni Tozzetti; Targioni Tozzetti, 1866: 145
[MIS-SPELLING].
Ceroplastes janeirensis Gray; Signoret, 1869a: 858.
TYPE DATA. Syntypes female, BRAZIL: on *Solanum* sp. (probably
lost).
DISTRIBUTION. NEOTROPICAL REGION: Brazil Rio Grande do Sul.
HOST PLANTS. Leguminosae: *Acacia bonariensis.* **Moraceae:** *Ficus
retusa.* **Myrtaceae:** *Eugenia uniflora, Phyllocalyx laevigatus,
Psidium guajava, Stenocalyx pitanga.* **Solanaceae:** *Solanum.*
REMARKS. Adult female redescribed and illustrated by Westwood
(1853b), Gomes Costa (1949) and by De Lotto (1965). Adult female
redescribed by Hempel (1900b, 1920a). Distribution and host plant
records given by Westwood (1853a), Gomes Costa (1949), Corseuil &
Barbosa (1971) and by Silva *et al.* (1968).

Ceroplastes japonicus Green

Ceroplastes floridensis japonicus Green, 1921b: 258.
Ceroplastes japonicus Green; Borchsenius, 1949c: 181.
Cerostegia japonica (Green); De Lotto, 1969b: 213.
Paracerostegia japonica (Green); Tang, 1991: 308.
TYPE DATA. Syntypes female, ENGLAND: Herts, St. Albans, on
smaller branches of Japanese Maple (imported from Japan)
(BMNH).
DISTRIBUTION. PALAEARCTIC REGION: China, England, Italy,
Japan, Korea, USSR Georgia.
HOST PLANTS. Aceraceae: *Acer japonicus.* **Apocynaceae:** *Nerium
oleander, Trachelospermum asiaticum.* **Aquifoliaceae:** *Ilex inte-
gra.* **Araliaceae:** *Hedera helix.* **Berberidaceae:** *Berberis, Epimed-
ium colchicum.* **Buxaceae:** *Buxus.* **Celastraceae:** *Euonymus.*
Cornaceae: *Cornus mas, Svida.* **Cycadaceae:** *Cycas revoluta.*
Ebenaceae: *Diospyros kaki.* **Ehretiaceae:** *Ehretia acuminata.*
Elaeagnaceae: *Elaeagnus pungens.* **Elaeocarpaceae:** *Elaeocar-*

pus decipiens. **Lauraceae:** *Laurus nobilis, Machilus thunbergii.*
Magnoliaceae: *Magnolia grandiflora.* **Moraceae:** *Morus, Myrtus communis, Feijoa sellowiana.* **Pittosporaceae:** *Pittosporum tobira.* **Podocarpaceae:** *Podocarpus nagi.* **Rhamnaceae:** *Ziziphus.* **Rosaceae:** *Cerasus avium, Cerasus vulgaris,* **Rosaceae:** *Crataegus, Cydonia vulgaris, Eriobotrya japonica, Malus, Persica vulgaris, Prunus laurocerasus, P. mume, P. yedoensis, Pyrus sinensis.* **Rutaceae:** *Citrus, Poncirus trifoliata.* **Salicaceae:** *Salix glandulosa, S. saidaeana.* **Theaceae:** *Camellia japonica, Eurya japonica, Thea sinensis.*
REMARKS. Adult female redescribed and illustrated by Borchsenius (1957), De Lotto (1969b), Paik (1978), Kawai (1980), Wang (1980), Longo (1985) and by Tang (1991). Intraspecific variation of taxonomic characters of the adult female given by Longo (1985). Adult male described and illustrated by Borchsenius (1957) and by Paik (1978). Colour photograph given by Kawai (1980) and by Tranfaglia & Viggiani (1988). Distribution and host plant records given by Borchsenius (1949a, 1957, 1960), Takahashi & Tachikawa (1956), Longo (1985) and by Tang (1991). Tamaki & Kawai (1967) analyzed fatty acids, alcohols and hydrocarbons in the body lipids.
BIOLOGY. Ohgushi & Nishino (1975) studied the biology in Japan and constructed life tables. Develops one annual generation in Northern Italy; overwintering as mated female (Longo, 1985).

Ceroplastes kunmingensis (Tang & Xie) n. comb.

Paracerostegia kunmingensis Tang & Xie in Tang, 1991: 310.
TYPE DATA. Holotype female, CHINA: Yunnan Province, Kunming City, on *Pittosporum glabratum* (EISC).
DISTRIBUTION. PALAEARCTIC REGION: China.
HOST PLANTS. Pittosporaceae: *Pittosporum glabratum, P. tobira.*

Ceroplastes lahillei Cockerell

Ceroplastes lahillei Cockerell, 1910a: 74.
Ceroplastes (Ceroplastina) lahillei Cockerell; Cockerell, 1910a: 76.
TYPE DATA. Syntypes female, ARGENTINA: Tucuman, Santa Ana, on undetermined plant (USNM).
DISTRIBUTION. NEOTROPICAL REGION: Argentina Tucuman.

Ceroplastes lamborni Newstead

Ceroplastes lamborni Newstead, 1917b: 29.
TYPE DATA. Syntypes female and larva, NIGERIA: Ibadan, on cacao and on climber on bush tree (BMNH).
DISTRIBUTION. NEOTROPICAL REGION: Nigeria.
HOST PLANTS. Sterculiaceae: *Theobroma cacao.*

Ceroplastes leonardianus Lizer y Trelles

Ceroplastes irregularis Leonardi, 1911: 271 [HOMONYM of *Ceroplastes irregularis* Cockerell].
Ceroplastes leonardianus Lizer y Trelles, 1939: 194 [REPLACEMENT NAME].
Ceroplastes leonardianus (Leonardi); Teran, 1973: 190 [ERRONEOUS AUTHORSHIP].

TYPE DATA. Syntypes female, ARGENTINA: Cacheuta, on *Larrea cuneata* (IEAP).
DISTRIBUTION. NEOTROPICAL REGION: Argentina.
HOST PLANTS. Compositae: *Eupatorium buniifolium, Tessaria absinthioides.* **Tiliaceae:** *Heliocarpus.* **Zygophyllaceae:** *Larrea cuneata, L. divaricata.*
REMARKS. Adult female redescribed and illustrated by Teran (1973). Distribution and host plant records given by Lizer y Trelles (1939) and by Teran (1973).

Ceroplastes lepagei Costa Lima

Ceroplastes lepagei Costa Lima, 1940: 9.
TYPE DATA. Syntypes female, BRAZIL: Rio de Janeiro, Manguinhos, on *Moquilea tomentosa* (FOCB).
DISTRIBUTION. NEOTROPICAL REGION: Brazil Rio de Janeiro.
HOST PLANTS. Chrysobalanaceae: *Moquilea tomentosa.*

Ceroplastes longicauda Brain

Ceroplastes longicauda Brain; Brain, 1920b: 31.
Gascardia longicauda (Brain); De Lotto, 1965: 202.
TYPE DATA. Syntypes female, SOUTH AFRICA: Natal Coast, on stems of native shrub (BMNH).
DISTRIBUTION. ETHIOPIAN REGION: Angola, Kenya, South Africa, Zambia, Zimbabwe.
HOST PLANTS. Bignoniaceae: *Jacaranda mimosaefolia, J. ovalifolia.* **Euphorbiaceae:** *Euphorbia pulcherrima.* **Meliaceae:** *Melia.* **Tiliaceae:** *Grewia flavescens, G. monticola.* **Verbenaceae:** *Vitex petersiana.*
REMARKS. Adult female redescribed and illustrated by De Lotto (1965). Distribution and host plant records given by De Lotto (1965, 1967a), Hodgson (1969a) and by Almeida (1973).

Ceroplastes longicauda sapii Hall

Ceroplastes longicauda sapii Hall, 1931: 296.
TYPE DATA. Syntypes female, ZIMBABWE: Embeza, on *Sapium* sp. (BMNH)
DISTRIBUTION. ETHIOPIAN REGION: Zimbabwe.
HOST PLANTS. Euphorbiaceae: *Sapium.*
REMARKS. De Lotto (1965) and Hodgson (1969a) supposed that this species is a synonym of *C. longicauda,* but it was not verified.

Ceroplastes longiseta Leonardi

Ceroplastes longiseta Leonardi, 1911: 268.
TYPE DATA. Syntypes female, ARGENTINA: Cacheuta, on *Fabiana denudata* (IEAP)
DISTRIBUTION. NEOTROPICAL REGION: Argentina.
HOST PLANTS. Solanaceae: *Fabiana denudata.*

Ceroplastes lucidus Hempel

Ceroplastes lucidus Hempel, 1900b: 465.

TYPE DATA. Syntypes female, BRAZIL: Sao Paulo, Ypiranga, on *Baccharis dracunculifolia* (MZSP).
DISTRIBUTION. NEOTROPICAL REGION: Argentina, Brazil Sao Paulo.
HOST PLANTS. Compositae: *Baccharis dracunculifolia.* **Polygonaceae:** *Muehlenbeckia.* **Verbenaceae:** *Lippia lycoides.*
REMARKS. Adult female redescribed by Hempel (1901b). Distribution and host plant records given by Lizer y Trelles (1939) and by Silva *et al.* (1968).

Ceroplastes macgregori Sampedro & Butze

Ceroplastes macgregori Sampedro & Butze, 1984: 143.
TYPE DATA. Holotype female, MEXICO: Tepetlixpita, Morelos, on *Spondias mombin* (IBEM).
DISTRIBUTION. NEOTROPICAL REGION: Mexico.
HOST PLANTS. Anacardiaceae: *Spondias mombin.* **Rutaceae:** *Citrus limon.*

Ceroplastes madagascariensis (Targioni Tozzetti)

Gascardia madagascariensis Targioni Tozzetti, 1893: 88.
Ceroplastes madagascariensis (Targioni Tozzetti); MacGillivray, 1921: 154.
TYPE DATA. Syntypes female, MADAGASCAR: on an undetermined tree of the Lauraceae (probably lost; G. Pellizzari Scaltriti, 1990, personal communication).
DISTRIBUTION. MADAGASIAN REGION: Madagascar.
HOST PLANTS. Lauraceae.
REMARKS. Targioni Tozzetti (1895) redescribed and illustrated the adult female, larva and the male cover. Adult female redescribed and illustrated by Newstead (1909).

Ceroplastes magnicauda Reyne

Ceroplastes magnicauda Reyne, 1964: 126.
TYPE DATA. Syntypes female, CURACAO: St. Christoffe, on *Croton flavens* (ZMAN).
DISTRIBUTION. NEOTROPICAL REGION: Curacao.
HOST PLANTS. Euphorbiaceae: *Croton flavens.*

Ceroplastes marmoreus Cockerell

Ceroplastes marmoreus Cockerell, 1903b: 158.
TYPE DATA. Syntypes female, MEXICO: Zapotlan, on sage, *Catalpa* and Compositae (USNM).
DISTRIBUTION. NEOTROPICAL REGION: Mexico.
HOST PLANTS. Bignoniaceae: *Catalpa.* **Compositae.**

Ceroplastes martinae Mosquera

Ceroplastes martinae Mosquera, 1979: 612.
TYPE DATA. Holotype female, COLOMBIA: Esmeralda (Caqueta), on *Mangifera indica* (CTBC).
DISTRIBUTION. NEOTROPICAL REGION: Colombia.
HOST PLANTS. Anacardiaceae: *Mangifera indica.*

Ceroplastes mierii (Targioni Tozzetti) n. comb.

Columnea mierii Targioni Tozzetti, 1866b: 145.
TYPE DATA. Syntypes female, BRAZIL: on unspecified shrub (probably lost; Pellizzari Scaltriti, 1990, personal communication).
DISTRIBUTION. NEOTROPICAL REGION: Brazil.
REMARKS. Targioni Tozzetti (1866b) gave this name to an unnamed species which has been described and illustrated by Westwood (1853b: 484). Both authors indicated its similarity to *Ceroplastes ceriferus*.

Ceroplastes milleri Takahashi

Ceroplastes milleri Takahashi, 1939e: 323.
TYPE DATA. Syntypes female, SARAWAK (BORNEO): Mt. Matang, on undetermined plant (IMZT).
DISTRIBUTION. AUSTRO ORIENTAL REGION: Sarawak.

Ceroplastes minutus Cockerell

Ceroplastes minutus Cockerell, 1898f: 434.
TYPE DATA. Syntypes female, MEXICO: Tabasco, Las Minas, on "escobillo" (USNM).
DISTRIBUTION. NEOTROPICAL REGION: Mexico.

Ceroplastes mosquerai n. name

Ceroplastes bicolor Mosquera, 1984: 126. [HOMONYM of *Ceroplastes bicolor* Hempel].
TYPE DATA. Holotype female, COLOMBIA: Suba, on *Schinus molle* (CTBC).
DISTRIBUTION. NEOTROPICAL REGION: Colombia.
HOST PLANTS. Anacardiaceae: *Schinus molle.*

Ceroplastes murrayi Froggatt

Ceroplastes murrayi Froggatt, 1919: 439.
TYPE DATA. Lectotype female designated by Williams & Watson (1990), PAPUA NEW GUINEA: Kikori River, Delta Division, on *Mangifera* sp. (BCRI).
DISTRIBUTION. AUSTRO ORIENTAL REGION: Papua New Guinea.
HOST PLANTS. Anacardiaceae: *Mangifera.*
REMARKS. Adult female redescribed and illustrated by Williams & Watson (1990).

Ceroplastes myricae (Linnaeus)

Coccus myricae Linnaeus, 1767: 741.
Columnea myricae (Linnaeus); Targioni Tozzetti, 1866b: 143.
Ceroplastes myricae (Linnaeus); Signoret, 1872b: 39.
TYPE DATA. Syntypes female, SOUTH AFRICA: Cape of Good Hope, on *Myrica quercifolia* (probably lost).
DISTRIBUTION. ETHIOPIAN REGION: South Africa.
HOST PLANTS. Myricaceae: *Myrica quercifolia.*

Ceroplastes nakaharai Gimpel

Ceroplastes nakaharai Gimpel in Gimpel *et al.*, 1974: 52.
TYPE DATA. Holotype female, U.S.A.: Florida, Dade County, on *Coccolobis diversifolia* (USNM).
DISTRIBUTION. NEOTROPICAL REGION: Cuba. **NEARCTIC REGION:** Florida.
HOST PLANTS. Loranthaceae: *Phoradendron flavescens.* **Myrtaceae:** *Eugenia myrtoides.* **Polygonaceae:** *Coccoloba diversifolia, C. floridana.* **Rubiaceae:** *Ixora acuminata.* **Tamaricaceae:** *Tamarix.*
REMARKS. Adult female redescribed and illustrated by Hamon & Williams (1984). Male test described and illustrated by Miller & Williams (1990). Colour photograph given by Hamon & Williams (1984).

Ceroplastes novaesi Hempel

Ceroplastes novaesi Hempel, 1900b: 457.
TYPE DATA. Syntypes female, BRAZIL: Capoeira Grande, Campinas, Sao Paulo and Cachoeira, on *Abutilon* sp., *Baccharis dracunculifoliae* and *Vernonia riedelii* (MZSP).
DISTRIBUTION. NEOTROPICAL REGION: Argentina, Brazil Rio Grande do Sul, Brazil Sao Paulo.
HOST PLANTS. Compositae: *Baccharis dracunculifolia, Tessaria absinthioides, Vernonia riedelii.* **Loranthaceae:** *Moquiniella polymorpha.* **Malvaceae:** *Abutilon.* **Verbenaceae:** *Lantana camara.*
REMARKS. Adult female redescribed by Hempel (1901a). Distribution and host plant records given by Lizer y Trelles (1939) and by Silva *et al.* (1968).

Ceroplastes novaesi mendozae Cockerell

Ceroplastes novaesi mendozae Cockerell, 1902e: 92.
TYPE DATA. Syntypes female, ARGENTINA: Mendoza, on pity stems of an herbaceous plant (BMNH, USNM).
DISTRIBUTION. NEOTROPICAL REGION: Argentina Mendoza.

Ceroplastes ocreus Mosquera

Ceroplastes ocreus Mosquera, 1984: 139.
TYPE DATA. Holotype female, COLOMBIA: Tocaima, on *Achatocarpus* aff. *nigricans* (CTBC).
DISTRIBUTION. NEOTROPICAL REGION: Colombia.
HOST PLANTS. Achatocarpaceae: *Achatocarpus nigricans.*

Ceroplastes parvus Green

Ceroplastes parvus Green, 1935b: 272.
TYPE DATA. Syntypes female, URUGUAY: Canelones Atlantida, on *Baccharis articulata* (BMNH).
DISTRIBUTION. NEOTROPICAL REGION: Uruguay.
HOST PLANTS. Compositae: *Baccharis articulata.*

Ceroplastes paucispinus De Lotto

Ceroplastes paucispinus De Lotto, 1970a: 143.
TYPE DATA. Holotype female, SOUTH AFRICA: Cape Province, Nossob Camp, on *Acacia giraffae* (SANC).
DISTRIBUTION. ETHIOPIAN REGION: South Africa.
HOST PLANTS. Leguminosae: *Acacia giraffae.*

Ceroplastes personatus Newstead

Ceroplastes personatus Newstead, 1898: 94.
TYPE DATA. Syntypes female, NIGERIA: Lagos, on unspecified host plant (BMNH).
DISTRIBUTION. ETHIOPIAN REGION: Ghana, Nigeria.
HOST PLANTS. Rubiaceae: *Coffea liberica.*

Ceroplastes pseudoceriferus Green

Ceroplastes ceriferus (Fabricius); Green, 1921b: 259 [MISIDENTIFICATION].
Ceroplastes pseudoceriferus Green, 1935a: 180.
TYPE DATA. Syntypes female, SRI LANKA: on undetermined plant, and INDIA: on *Azidarachta indica* and on *Diospyros montana* (BMNH).
DISTRIBUTION. NEW ZEALAND and PACIFIC REGION: Palaus. **ORIENTAL REGION:** India, Sri Lanka, Taiwan. **PALAEARCTIC REGION:** Japan, Korea, Tibet.
HOST PLANTS. Aceraceae: *Acer buergerianum, A. palmatum.* **Anacardiaceae:** *Mangifera indica, Rhus succedanea, R. verniciflua.* **Apocynaceae:** *Alstonia scholaris, Nerium oleander.* **Aquifoliaceae:** *Ilex crenata, I. integra, I. mutchagara, I. oldhami, I. rotunda.* **Araceae:** *Amorphophalus konjac.* **Araliaceae:** *Fatsia japonica, Gilibertia trifida.* **Berberidaceae:** *Berberis thunbergii, Mahonia fortunei, Nandina domestica.* **Caprifoliaceae:** *Viburnum awabuki, V. japonicum.* **Celastraceae:** *Euonymus alata, E. japonicus, E. sieboldiana.* **Commelinaceae:** *Commelina communis.* **Compositae:** *Artemisia capillaris, A. japonica, Blumea lacera, Chrysanthemum indicum, Erigeron annuus, E. canadensis, E. linifolius, Solidago vigra-aurea.* **Convolvulaceae:** *Ipomoea batatas.* **Cucurbitaceae:** *Cucurbita moschata.* **Ebenaceae:** *Diospyros kaki, D. montana.* **Elaeagnaceae:** *Elaeagnus pungens, E. multiflora.* **Euphorbiaceae:** *Croton, Mallotus japonicus.* **Fagaceae:** *Castanopsis acuta, Shiia cuspidata.* **Hippocastanaceae:** *Aesculus turbinata.* **Lauraceae:** *Actinodaphne lancifolia, Benzoin strychnifolium, Cinnamomum sericeum, Laurus nobilis, Machilus thunbergii, Persea americana.* **Leguminosae:** *Glycine max, Lespedeza bicolor.* **Magnoliaceae:** *Magnolia compressa, M. obovata, M. compressa.* **Malvaceae:** *Hibiscus rosa-sinensis.* **Melastomataceae:** *Melastoma candidum.* **Meliaceae:** *Melia indica, Toona sinensis.* **Menispermaceae:** *Sinomenium diversifolium.* **Moraceae:** *Artocarpus heterophyllus, Ficus benghalensis, F. vasculosa, Humulus lupulus, Morus alba, M. bombycis.* **Myrtaceae:** *Psidium guajava, Rhodomyrtus tomentosa.* **Oleaceae:** *Osmanthus ilicifolius.* **Oxalidaceae:** *Oxalis corniculata.* **Pittosporaceae:** *Pittosporum tobira.* **Platanaceae:** *Platanus occidenta-*

lis, P. orientalis. **Polygonaceae:** *Polygonum chinensis, P. conspicuum, P. cuspidatum.* **Punicaceae:** *Punica granatum.* **Rosaceae:** *Agrimonia eupatoria, Armeniaca vulgaris, Chaenomeles lagenaria, Crataegus cuneata, Cydonia oblonga, Eriobotrya japonica, Malus pumila, M. sieboldii, Mespilus germanica, Persica vulgaris, Photinia glabra, P. serrulata, Pourthiaea villosa, Prunus mume, P. preslii, P. salicina, P. yedoensis, P. zippeliana, Pyracantha angustifolia, Pyrus serotina, Spiraea cantoniensis, S. thunbergii.* **Rubiaceae:** *Gardenia jasminoides, Ixora chinensis.* **Rutaceae:** *Citrus natsudaidai, C. unshiu, Fortunella japonica, Poncirus trifoliata.* **Sapindaceae:** *Pometia pinnata, Sapindus mukurossi.* **Solanaceae:** *Solanum melongena, S. tuberosum.* **Tamaricaceae:** *Tamarix chinensis.* **Theaceae:** *Camellia japonica, C. sasanqua, C. sinensis, Cleyera ochnacea, Eurya emarginata, E. japonica, Gordonia axillaris, Stewartia pseudo-camellia, Ternstroemia japonica, Thea sinensis.* **Ulmaceae:** *Celtis sinensis.* **Urticaceae:** *Boehmeria nivea, Villebrunea frutescens.*

REMARKS. Adult female redescribed and illustrated by Sankaran (1959), Kawai & Tamaki (1967), De Lotto (1971c), Paik (1978), Wang (1980), Tao *et al.* (1983), Avasthi & Shafee (1986) and by Tang (1991). Adult male described and illustrated by Sankaran (1962) and by Paik (1978). Larval instars described and illustrated by Sankaran (1962) and by Kawai & Tamaki (1967). Kawai & Tamaki (1967) described the morphology of the wax cover of larval instars and adult female. Parida & Moharana (1982) and Moharana (1990) reported on chromosome number 2n=36 in India. Kajita (1965) described the changes in size and colour of parasitized and unparasitized scales. Distribution and host plant records given by Takahashi & Tachikawa (1956), Sankaran (1959), Kajita (1964), Beardsley (1966a), Ali (1971), De Lotto (1971c), Paik (1978), Wang (1980, 1981), Tao *et al.* (1983), Sinha &Denish (1984), Avasthi & Shafee (1986), Varshney & Moharana (1987), Shafee *et al.* (1989) and by Tang (1991). Amino acids in the honeydew were analyzed by Tamaki (1964a) and the carbohydrates by Tamaki (1964b). Tamaki (1966) analyzed the chemical composition of the wax secretion. Tamaki & Kawai (1966) studied the seasonal changes in the wax covering and its components. Tamaki & Kawai (1967) analyzed fatty acids, alcohols and hydrocarbons in the body lipids.

BIOLOGY. Sankaran (1959) reported this species as univoltine in India (Banaras) and described its life history.

Ceroplastes psidii (Chavannes)

Coccus psidii Chavannes, 1848: 139.

Columnea chavannesii Targioni Tozzetti, 1866b: 145 [UNJUSTIFIED REPLACEMENT NAME]

Ceroplastes chavannesii (Targioni Tozzetti); Targioni Tozzetti, 1868: 35.

Ceroplastes psidii (Chavannes); Signoret, 1869a: 867.

TYPE DATA. Syntypes female, BRAZIL: Rio de Janeiro, Bosafogo Bay, on *Psidium* sp. (MNHM).

DISTRIBUTION. NEOTROPICAL REGION: Brazil Paraiba, Brazil Pernambuco, Brazil Rio de Janeiro, Brazil Sao Paulo, .

HOST PLANTS. Myrtaceae: *Psidium.*

REMARKS. Adult female redescribed and illustrated by Westwood (1853b) and by Signoret (1872b). Adult female redescribed by Hempel (1920b)
ECONOMIC IMPORTANCE. Hempel (1920b) reported this wax scale as a serious pest of guavas in Northern Brazil (Paraiba and Pernambuco).

Ceroplastes purpurellus Cockerell

Ceroplastes purpurellus Cockerell, 1903b: 159.
TYPE DATA. Syntypes female, MEXICO: Tonila, Jalisco, on undetermined plant (BMNH, USNM).
DISTRIBUTION. NEOTROPICAL REGION: Mexico.

Ceroplastes purpureus Hempel

Ceroplastes purpureus Hempel, 1900b: 466.
TYPE DATA. Syntypes female, BRAZIL: Sao Paulo, Ypiranga, on *Miconia* sp. and other trees (MZSP).
DISTRIBUTION. NEOTROPICAL REGION: Brazil Sao Paulo.
HOST PLANTS. Melastomataceae: *Miconia.*
REMARKS. Adult female redescribed by Hempel (1901b).

Ceroplastes quadratus Green

Ceroplastes quadratus Green, 1935b: 274.
TYPE DATA. Syntypes female, BRAZIL: Goyaz, on undetermined plant (BMNH).
DISTRIBUTION. NEOTROPICAL REGION: Brazil.

Ceroplastes quadrilineatus Newstead

Ceroplastes quadrilineatus Newstead, 1910c: 193.
Gascardia quadrilineata (Newstead); De Lotto, 1965: 182.
TYPE DATA. Syntypes female, UGANDA: Kyetume near Kampala, on *Annona muricata*, and at Ndege, on *Ficus* sp. (BMNH)
DISTRIBUTION. ETHIOPIAN REGION: Ivory Coast, Uganda, Zimbabwe.
HOST PLANTS. Annonaceae: *Annona muricata.* **Ixonanthaceae:** *Ochtocosmus mirabilis.* **Labiatae:** *Salvia confertiflora.* **Meliaceae:** *Melia.* **Moraceae:** *Ficus, Morus.* **Rosaceae:** *Persica vulgaris.* **Sterculiaceae:** *Theobroma cacao.*
REMARKS. Adult female redescribed and illustrated by Hodgson (1969a). Intraspecific variation in taxonomic characters given by Hodgson (1969a). Distribution and host plant records given by Newstead (1917c), De Lotto (1965), Hodgson (1969a) and by Couturier *et al.* (1985).

Ceroplastes rarus Hempel

Ceroplastes rarus Hempel, 1900b: 469.
TYPE DATA. Syntypes female, BRAZIL: Sao Paulo, Ypiranga, on twigs of an indigenous tree (MZSP).
DISTRIBUTION. NEOTROPICAL REGION: Brazil Sao Paulo.
REMARKS. Adult female redescribed by Hempel (1901b).

Ceroplastes rhizophorae Hempel

Ceroplastes rhizophorae Hempel, 1918: 201.
TYPE DATA. Syntypes female, BRAZIL: Sao Paulo State, Santos, on *Rhizophora mangle* (MZSP).
DISTRIBUTION. NEOTROPICAL REGION: Brazil Sao Paulo.
HOST PLANTS. Rhizophoraceae: *Rhizophora mangle*.

Ceroplastes rotundus Hempel

Ceroplastes rotundus Hempel, 1900b: 473.
TYPE DATA. Syntypes female, BRAZIL: Sao Paulo, Ypiranga, on twigs of *Maytenus* sp. (MZSP).
DISTRIBUTION. NEOTROPICAL REGION: Brazil Sao Paulo.
HOST PLANTS. Celastraceae: *Maytenus*.
REMARKS. Adult female redescribed by Hempel (1901b).

Ceroplastes royenae Hall

Ceroplastes quadrilineatus royenae Hall, 1931: 297.
Gascardia quadrilineata royenae (Hall); De Lotto, 1965: 182.
Gascardia royenae (Hall); Hodgson, 1969a: 34.
TYPE DATA. Syntypes female, ZIMBABWE: Macheke, on *Royena pallens* (BMNH).
DISTRIBUTION. ETHIOPIAN REGION: Zimbabwe.
HOST PLANTS. Ebenaceae: *Diospyros lycioides sericea*. **Euphorbiaceae:** *Pseudolachnostylus maprounifolia*.
REMARKS. Adult female redescribed and illustrated by Hodgson (1969a).

Ceroplastes rubens Maskell

Ceroplastes rubens Maskell, 1893b: 214.
Ceroplastes rubens minor Maskell, 1897a: 309. Syntypes female, HONG KONG: on *Pinus sinensis* and *P. thunbergii* (NZAC, USNM). Syn. by Gimpel *et al.*, 1974: 57
Ceroplastes myricae (Linnaeus); Green, 1900c: 8 [MISIDENTIFICATION].
TYPE DATA. Syntypes female, AUSTRALIA: Queensland, Brisbane, on *Mangifera indica* and *Ficus* sp. (NZAC).
DISTRIBUTION. AUSTRO ORIENTAL REGION: Hong Kong, Java, Malaysia, Papua New Guinea, Philippines, Solomon Islands. **AUSTRALIAN REGION:** Queensland. **ETHIOPIAN REGION:** Kenya, South Africa, Zanzibar. **MADAGASIAN REGION:** Seychelles. **NEOTROPICAL REGION:** Puerto Rico, Trinidad. **NEW ZEALAND and PACIFIC REGION:** Cook Islands, Fiji, French Polynesia, Guam, Hawaii, Kiribati, Mariana Islands, New Caledonia, Norfolk Island, Palaus, Rota Island, Saipan Island, Samoa, Tahiti, Vanuatu, Western Samoa. **NEARCTIC REGION:** Florida. **ORIENTAL REGION:** India, Sri Lanka, Taiwan, Thailand, Vietnam. **PALAEARCTIC REGION:** China, Japan, Korea, Tibet.
HOST PLANTS. Acanthaceae: *Strobilanthes japonicus*. **Aceraceae:** *Acer ginnala, A. palmatum, A. trifidum*. **Amaranthaceae:** *Celosia cristata*. **Anacardiaceae:** *Anacardium occidentale, Mangifera indica, Rhus succedanea, Schinus terebinthifolius*. **Apocynaceae:**

Allamanda cathartica, Alstonia scholaris, Alyxia elliptica, A. gynopogon, A. olivaeformis, Melodinus baueri, Nerium oleander, Plumeria. **Aquifoliaceae:** *Ilex aquifolium, I. cornuta, I. latifolia, I. oldhami, I. othera, I. pedunculosa, I. rotunda, I. serrata.* **Araceae:** *Aglaonema commutatum, Aglaonema modestum, A. pictum, A. tricolor, Anthurium andraeanum, Dieffenbachia, Epipremnum pinnatum, Philodendron gigantium, Syngonium.* **Araliaceae:** *Aralia elegantissima, Brassaia actinophylla, Dizygotheca elegantissima, Fatsia japonica, Hedera helix, Meryta angustifolia, Polyscias guilfoylei, Schefflera.* **Arecaceae:** *Monstera deliciosa.* **Aspleniaceae:** *Asplenium nidum.* **Barringtoniaceae:** *Barringtonia butonica, B. racemosa.* **Berberidaceae:** *Nandina domestica.* **Bischofiaceae:** *Bischofia javanica.* **Bixaceae:** *Bixa orellana.* **Blechnaceae:** *Blechnum orientalis.* **Buxaceae:** *Buxus microphylla.* **Caprifoliaceae:** *Viburnum.* **Celastraceae:** *Euonymus alata, E. europaeus, E. japonicus.* **Cephalotaxaceae:** *Cephalotaxus.* **Clusiaceae:** *Montrouziera.* **Compositae:** *Artemisia vulgaris, Chrysanthemum sinense, Fitchia, Gerbera, Helianthus.* **Cucurbitaceae. Cunoniaceae:** *Weinmannia rarotongensis.* **Cycadaceae:** *Cycas.* **Davalliaceae:** *Arthropteris palisotii, Davallia.* **Dicksoniaceae:** *Cibotium.* **Ebenaceae:** *Diospyros kaki.* **Elaeocarpaceae:** *Elaeocarpus bifidus.* **Ericaceae:** *Rhododendron indicum.* **Euphorbiaceae:** *Euphorbia pulcherrima.* **Gleicheniaceae:** *Dicranopteris flexuosa, Gleichenia.* **Guttiferae:** *Calophyllum inophyllum, Garcinia subelliptica.* **Heliconiaceae:** *Heliconia.* **Hernandiaceae:** *Hernandia peltata.* **Hypoxidaceae:** *Molineria recurvata.* **Lauraceae:** *Cinnamomum pedunculatum, C. zeylanicum, Laurus nobilis, Lindera citriodora, Machilus thunbergii, Persea americana.* **Leguminosae:** *Acacia, Cytisus scoparius, Dioclea violacea, Inocarpus fagifer, Palaquium formosanum, Spartium junceum.* **Lomariopsidaceae:** *Elaphoglossum reticulatum.* **Loranthaceae:** *Loranthus.* **Magnoliaceae:** *Illicium anisatum, Magnolia salicifolia.* **Malvaceae:** *Hibiscus tiliaceus.* **Marantaceae. Melastomataceae:** *Astronidium robustum.* **Moraceae:** *Artocarpus altilis, A. integra, Cudrania javanesis, Ficus foveolata, F. prolixa, Morus alba.* **Musaceae:** *Musa paradisiaca.* **Myristicaceae:** *Myristica cagayanensis, M. fragrans.* **Myrsinaceae:** *Ardisia humilis, Ardisia japonica, Myrsine, Rapanea crassifolia.* **Myrtaceae:** *Eucalyptus globulus, Eugenia cumini, E. jambolana, E. jambos, Eugenia javanica, Feijoa sellowiana, Melaleuca, Metrosideros collina, Pimenta officinalis, Psidium guajava, Rhodomyrtus tomentosa.* **Nyctaginaceae:** *Bougainvillea.* **Oleaceae:** *Ligustrum japonicum, Olea verrucosa.* **Orchidaceae:** *Grammatophyllum, Stanhopea.* **Palmae:** *Cocos nucifera.* **Peperomiaceae:** *Peperomia.* **Pinaceae:** *Pinus caribaea, P. densiflora, P. parviflora, P. sinensis, P. thunbergii.* **Piperaceae:** *Macropiper excelsum, Piper.* **Pittosporaceae:** *Pittosporum bracteolatum, P. tobira.* **Polygonaceae:** *Coccoloba uvifera.* **Polypodiaceae:** *Belvisia, Phymatodes scolopendria, Polypodium.* **Pteridaceae:** *Acrostichum aureum.* **Rhizophoraceae:** *Rhizophora.* **Rosaceae:** *Chaenomeles, Eriobotrya japonica, Malus, Photinia glabra, Prunus domestica, P. mume, Pyrus serotina, Spiraea thunbergii.* **Rubiaceae:** *Gardenia jasminoides, G. taitensis, Gouldia, Ixora coccinea, Paederia tomentosa, Randia tahitensis, Straussia.* **Rutaceae:** *Citrus aurantifolia, C. deliciosa, C. limon, C.*

paradisi, *C. reticulata, C. sinensis, C. unshiu, Evodia littoralis, Pelea, Poncirus trifoliata.* **Santalaceae:** *Exocarpos phyllanthoides.* **Sapindaceae:** *Euphoria longana, Litchi, Nephelium lappaceum.* **Sapotaceae:** *Calocarpum.* **Schisandraceae:** *Kadsura japonica.* **Sinopteridaceae:** *Pellaea.* **Stilaginaceae:** *Antidesma.* **Symplocaceae:** *Symplocos japonica.* **Tamaricaceae:** *Tamarix chinensis.* **Taxaceae:** *Podocarpus nageia.* **Theaceae:** *Camellia japonica, C. rusticans, C. sasanqua, C. sinensis, Cleyera ochnacea, Eurya emarginata, E. japonica, E. ochnacea, Ternstroemia japonica.* **Thymelaeaceae:** *Daphne odora.* **Ulmaceae:** *Celtis.* **Verbenaceae:** *Premna.* **Zingiberaceae:** *Alpinia purpurata, Zingiber officinale.*
REMARKS. Common name - red wax scale. Adult female redescribed and illustrated by Green (1909), Froggatt (1915), Kuwana (1917a, 1923c), Morrison (1920), Zimmerman (1948), Borchsenius (1957), Gimpel *et al.* (1974), Paik (1978), Kawai (1980), Wang (1980), Tao *et al.* (1983), Hamon & Williams (1984), Williams & Watson (1990) and by Tang (1991). Adult male described and illustrated by Paik (1978). Distribution and host plant records given by Ramakrishna Ayyar (1919, 1930), Kuwana (1923c), Zimmerman (1948), Takahashi & Tachikawa (1956), De Lotto (1965), Ali (1971), Gimpel *et al.* (1974), Nakahara (1981b), Nakahara & Miller (1981), Wang (1981), Tao *et al.* (1983), Sinha &Denish (1984), Varshney & Moharana (1987), Shafee *et al.* (1989), Williams & Watson (1990), Danzig & Konstantinova (1990) and by Tang (1991). Colour photograph given by Kawai (1980) and by Hamon & Williams (1984). Chemistry of the wax cover discussed by Brown (1975). Tamaki & Kawai (1967) analyzed fatty acids, alcohols and hydrocarbons in the body lipids. Komura *et al.* (1982) identified ceroalbolinic acid, a body pigment of the female. Parida & Moharana (1982) and Moharana (1990) reported chromosome number 2n-36 in India.
BIOLOGY. Life history in Japan studied by Kuwana (1923a, 1923c). Tanaka (1953) evaluated the suitability of 15 cultivars of potatoes for laboratory rearing of this wax scale. Ohgushi & Nishino (1975) studied the life tables in Japan.
ECONOMIC IMPORTANCE. A major pest of citrus in Australia (Sabine, 1969), Hawaii and Japan (Ebeling, 1959). Mitsuhashi *et al.* (1956) described several methods for distinguishing between dead and live scale after treatment with hydrogen cyanide. Biology and pest status on coffee presented by Le Pelley (1968).

Ceroplastes rufus De Lotto
Ceroplastes rufus De Lotto, 1966b: 143.
Cerostegia rufa (De Lotto); De Lotto, 1969b: 215.
TYPE DATA. Holotype female, SOUTH AFRICA: Cape Province, Bitterfontein, on *Ruschia* sp. (SANC).
DISTRIBUTION. ETHIOPIAN REGION: South Africa.
HOST PLANTS. Aizoaceae: *Ruschia.*
REMARKS. Adult female redescribed and illustrated by De Lotto (1969b).

Ceroplastes rusci (Linnaeus)
Coccus rusci Linnaeus, 1758: 456.

Coccus artemisiae Rossi, 1794: 56. Syntypes female, ITALY: on twigs of *Artemisia* (probably lost). Syn. by Signoret, 1872b: 37.

Coccus caricae Fabricius, 1794: 225. Syntypes females, GALLIA [=FRANCE]: on *Ficus carica* (lost; Zimsen 1964). Syn. by Fronscolombe, 1834: 205.

Calypticus testudineus Costa, 1829: 12. Syntypes female, ITALY: on seven host plants, including myrtle, oleander and fig (probably lost; G. Pellizzari Scaltriti, 1990, personal communication). Syn. by Signoret, 1869a: 870.

Calypticus radiatus Costa, 1829: 12. Syntypes female, ITALY: Livedia, near Capo di Bova, on oleander (probably lost; G. Pellizzari Scaltriti, 1990, personal communication). Syn. by Signoret, 1869a: 871.

Coccus hydatis Costa, 1829: 14. Syntypes female, ITALY: on oleander (probably lost; G. Pellizzari Scaltriti, 1990, personal communication). Syn. by Signoret, 1869a: 871.

Lecanium rusci (Linnaeus); Walker, 1852: 1072.

Lecanium radiatum (Costa); Walker, 1852: 1078.

Lecanium testudineum (Costa); Walker, 1852: 1078.

Columnea testudiniformis Targioni Tozzetti; Targioni Tozzetti, 1866b: 142 [UNJUSTIFIED REPLACEMENT NAME].

Columnea caricae (Fabricius); Targioni Tozzetti, 1867: 12.

Chermes caricae (Bernard); Boisduval, 1867: 320 [ERRONEUS AUTHORSHIP].

Columnea testudinata Targioni Tozzetti; Targioni Tozzetti, 1868: 35 [UNJUSTIFIED REPLACEMENT NAME].

Calypticus hydatis (Costa); Signoret, 1869a: 871.

Ceroplastes rusci (Linnaeus); Signoret, 1872b: 35.

Lecanium artemisiae (Rossi); Signoret, 1872b: 37.

Ceroplastes denudatus Cockerell, 1893c: 82. Syntypes female, ANTIGUA: on *Annona muricata* (USNM) Syn. by Nakahara, 1978: 657.

Ceroplastes nerii Newstead, 1897: 101. Syntypes female, ALGERIA: Constantine, on *Nerium oleander* (BMNH). Syn. by Borchsenius, 1957: 455.

Coccus caricae Bernard; Fernald, 1903: 156 [ERRONEUS AUTHORSHIP].

Ceroplastes tenuitectus Green, 1907: 204. Syntypes female, ALDABRA ISLAND: on "Bois la fumee" (BMNH). Syn. by Green, 1923a: 94.

TYPE DATA. Syntypes female, ITALY: Apulia; according to Linnaeus (1767) on *Myrtus* and *Ruscus* (lost; D. J. Williams, 1990, Personal communication).

DISTRIBUTION. AUSTRO ORIENTAL REGION: Irian Jaya. **ETHIOPIAN REGION:** Angola, Cape Verde Island, Zimbabwe. **MADAGASIAN REGION:** Aldabra Island. **NEOTROPICAL REGION:** Antigua, Brazil, Guyana, Puerto Rico, Virgin Islands. **PALAEARCTIC REGION:** Afghanistan, Algeria, Azores Islands, Canary Islands, Corsica, Crete, Cyprus, Egypt, France, Greece, Iraq, Israel, Italy, Lebanon, Madeira, Portugal, Saudi Arabia, Spain, Turkey.

HOST PLANTS. Anacardiaceae: *Mangifera indica, Pistacia lentiscus, P. terebinthus, P. vera, Rhus coriaria, Schinus molle, S. terebinthifolius.* **Annonaceae:** *Annona cherimolia, A. muricata, A. squamo-*

sa. **Apocynaceae:** *Nerium oleander, Thevetia peruviana.* **Aquifoliaceae:** *Ilex aquifolium.* **Araliaceae:** *Hedera helix.* **Balsaminaceae:** *Impatiens sultani.* **Compositae:** *Argyranthemum frutescens, Artemisia* **Convolvulaceae:** *Convolvulus, Ipomoea batatas.* **Cyperaceae:** *Cyperus communis, C. flabelliformis.* **Ebenaceae:** *Euclea schimperi.* **Euphorbiaceae:** *Euphorbia longan.* **Guttiferae:** *Psorospermum febrifugum.* **Juncaceae:** *Juncus acutus.* **Lauraceae:** *Laurus nobilis, Persea americana.* **Malvaceae:** *Gossypium.* **Moraceae:** *Ficus benghalensis, F. benjamina, F. carica, F. elastica, F. indica, F. obliqua, F. retusa, F. rubiginosa, F. sycomorus, Morus alba, M. nigra.* **Musaceae:** *Musa cavendishi, M. sapientum.* **Myrtaceae:** *Myrtus communis, Psidium guajava.* **Palmae:** *Chamaerops humilis.* **Pittosporaceae:** *Pittosporum tobira.* **Platanaceae:** *Platanus orientalis.* **Proteaceae:** *Grevillea robusta.* **Rosaceae:** *Crataegus azarolus, Cydonia oblonga, C. vulgaris, Mespilus germanica, Prunus dulcis, Pyrus communis.* **Ruscaceae:** *Ruscus aculeatus.* **Rutaceae:** *Citrus aurantium, C. limon, C. paradisi.* **Salicaceae:** *Populus alba, P. deltoides.* **Sapindaceae:** *Litchi chinensis, Nephelium lappaceum, Sapindus saponaria.* **Sebestenaceae:** *Cordia myxa.* **Strelitziaceae:** *Strelitzia reginae.* **Vitidaceae:** *Vitis vinifera.*

REMARKS. Common name - fig wax scale. Adult female redescribed and illustrated by Paoli (1916), Leonardi (1920), Gómez-Menor Ortega (1937, 1958c), Borchsenius (1937, 1957), Vilar (1951, 1952), Hodgson (1969a) and by Fernandes (1973). First instar described and illustrated by Leonardi (1920), Gómez-Menor Ortega (1937) and by Vilar (1951, 1952). Colour photograph given by Chapot & Delucchi (1964), Delucchi (1975) and by Tranfaglia & Viggiani (1988). Hodgson (1969a) and De Lotto (1978) discussed the characters distinguishing this species from *C. eucleae* Brain and *C. eugeniae* Hall. Distribution and host plant records given by Newstead (1897), Hall (1922), Bodenheimer (1924, 1928, 1944a, 1953), Balachowsky (1926, 1932d, 1933a, 1935), Hodgson (1969a), Ben-Dov (1970d), Fernandes (1973), Georghiou (1977), Nakahara & Miller (1981), Danzig (1972d), Vieira *et al.* (1983), Podsiadlo (1983), Martin Mateo (1984), Matile-Ferrero (1984b), Carnero Hernanadez & Perez Guerra (1986) and by Fernandes (1989). Boisduval (1867) and Fernald (1903) erroneously credited the authorship of *Coccus caricae* to Bernard (1773). However, the latter did not describe a species in that name, but rather discussed a wax scale on figs, which he consistently named by the vernacular "Kermes du figuier". The synonymy of *C. tenuitectus* with *C. rusci*, is based on synonymy of the former, introduced by Green (1923a), with *C. denudatus*. The redescription and illustration of *C. rusci* by Ezzat & Hussein (1969) is a misidentification of *C. floridensis.* Matile-Ferrero (1988) discussed the taxonomic characters of this species as observed in material from Saudi Arabia. Foldi & Cassier (1985) described the ultrastructure of wax secereting gland of the adult female.

BIOLOGY. Khasawinah & Talhouk (1964) studied the biology and phenology in Lebanon. Life history and economic importance in Israel studied by Bodkin (1927). Inserra (1970) studied the phenology and natural enemies on citrus in Sicily, Italy. Develops two generations a year in Greece (Argyriou & Santorini, 1980). Benassy

& Franco (1974) observed one annual generation on fig trees in Southern France.

ECONOMIC IMPORTANCE. A pest of fig in the Mediterranean region (Bodkin, 1927; Bodenheimer, 1951b; Talhouk, 1969). Generally it is a minor pest of Citrus in Israel, however, in several periods became a serious pest (Ben-Dov, 1988).

Ceroplastes rusticus De Lotto

Ceroplastes rusticus De Lotto, 1961: 318.
Gascardia rustica (De Lotto); De Lotto, 1965: 204.
TYPE DATA. Holotype female, SOUTH AFRICA: Cape Province, Hartman's Kloof, on *Selago glutinosa* (BMNH).
DISTRIBUTION. ETHIOPIAN REGION: South Africa.
HOST PLANTS. Scrophulariaceae: *Selago corymbosa, S. glutinosa.*

Ceroplastes sanguineus Cockerell

Ceroplastes sanguineus Cockerell, 1905c: 162.
TYPE DATA. Syntypes female, PARAGUAY: Villa Encarnacion, on *Maytenus* sp. (USNM).
DISTRIBUTION. NEOTROPICAL REGION: Paraguay.
HOST PLANTS. Celastraceae: *Maytenus.*

Ceroplastes schrottkyi Cockerell

Ceroplastes schrottkyi Cockerell, 1905c: 162.
TYPE DATA. Syntypes female, PARAGUAY: Villa Encarnacion, on *Salix chilensis* (USNM).
DISTRIBUTION. NEOTROPICAL REGION: Paraguay.
HOST PLANTS. Salicaceae: *Salix chilensis.*

Ceroplastes scutigera Cockerell

Ceroplastes scutigera Cockerell, 1902e: 92.
TYPE DATA. Syntypes female, ARGENTINA: Ceres, on a shrub with small, entire oval-lanceolate leaves (USNM).
DISTRIBUTION. NEOTROPICAL REGION: Argentina.

Ceroplastes simplex Hempel

Ceroplastes simplex Hempel, 1900b: 475.
TYPE DATA. Syntypes female, BRAZIL: Sao Paulo, Ypiranga, on twigs of plant of Myrtaceae (MZSP).
DISTRIBUTION. NEOTROPICAL REGION: Brazil Sao Paulo.
HOST PLANTS. Myrtaceae.
REMARKS. Adult female redescribed by Hempel (1901b).

Ceroplastes sinensis Del Guercio

Ceroplastes sinensis Del Guercio, 1900: 3.
TYPE DATA. Syntypes female, ITALY: Liguria, on *Citrus sinensis* (probably lost; G. Pellizzari Scaltriti, 1990, personal communication).
DISTRIBUTION. AUSTRALIAN REGION: New South Wales. **NEOTROPICAL REGION:** Brazil Rio Grande do Sul, Chile, Mexico. **NEW ZEALAND and PACIFIC REGION:** New Zealand,

Norfolk Island. **NEARCTIC REGION:** California, North Carolina, Virginia. **PALAEARCTIC REGION:** Canary Islands, Corsica, France, Italy, Madeira, Morocco, Portugal, Spain, USSR Caucasus, USSR Georgia.

HOST PLANTS. Anacardiaceae: *Mangifera indica, Rhodosphaera rhodanthema, Schinus molle, S. terebinthifolius.* **Annonaceae:** *Annona cherimolia, A. squamosa.* **Aquifoliaceae:** *Ilex aquifolium, I. crenata, I. vomitoria.* **Araliaceae:** *Aralia.* **Asclepiadaceae:** *Araujia sericofera, Hoya carnosa.* **Bignoniaceae:** *Bignonia, B. unguiscati.* **Buxaceae:** *Buxus sempervirens.* **Celastraceae:** *Euonymus europaeus, E. japonicus.* **Compositae:** *Aster formosissima, Baccharis, Chrysanthemum frutescens, C. grandiflora, C. indicum, Dahlia variabilis, Felicia angustifolia, Olearia paniculata, Osteospermum moniliferum.* **Cucurbitaceae:** *Cucurbita pepo.* **Ebenaceae:** *Diospyros fasciculosa, D. ferrea, D. kaki.* **Ericaceae:** *Arbutus unedo, Erica arborea, Vaccinium myrtillus.* **Escalloniaceae:** *Escallonia rubra.* **Euphorbiaceae:** *Drypetes australasica, Euphorbia longan, E. pulcherrima, Mercurialis annua.* **Flacourtiaceae:** *Casearia sylvestris.* **Hippocastanaceae:** *Aesculus hippocastanum, A. pavia.* **Labiatae:** *Melissa officinalis, Rosmarinus officinalis, Salvia splendens.* **Lauraceae:** *Apollonias barbujana, Cinnamomum burmannii, C. zeylanicum, Laurus, Persea americana.* **Leguminosae:** *Caesalpinia sepiaria, Cytisus scoparius.* **Loranthaceae:** *Loranthus.* **Malvaceae:** *Lagunaria patersonii.* **Moraceae:** *Ficus carica, F. virens, F. watkinsiana.* **Musaceae:** *Musa nana.* **Myoporaceae:** *Myoporum acuminatum.* **Myrsinaceae:** *Myrsine africana.* **Myrtaceae:** *Acmena, Baeckea virgata, Callistemon citrinus, C. linearis, Eugenia greggii, Feijoa sellowiana, Melaleuca armillaris, M. decussata, M. diasmifolia, M. hypericifolia, M. pauciflora, Psidium guajava, Rhodomyrtus psidioides, Syncarpia glomulifera, Syzygium floribundum, S. francisii, S. luehmannii, S. moorei, Tristania conferta.* **Ochnaceae:** *Ochna mauritiana.* **Oleaceae:** *Syringa vulgaris.* **Onagraceae:** *Fuchsia.* **Philadelphaceae:** *Philadelphus coronaria.* **Pittosporaceae:** *Bursaria spinosa.* **Platanaceae:** *Platanus orientalis.* **Plumbaginaceae:** *Ceratostigma willmottianum.* **Polygonaceae:** *Muehlenbeckia platyclada.* **Portulacaceae:** *Portulaca oleracea.* **Punicaceae:** *Punica granatum.* **Rhamnaceae:** *Frangula alnus.* **Rosaceae:** *Chaenomeles lagenaria, Cotoneaster dammeri, Osteomeles schwerinae, Pyrus communis.* **Rubiaceae:** *Burchellia bubalina, Canthium ventosum, Gardenia florida, G. thunbergia, Rondeletia amoena.* **Rutaceae:** *Citrus limon, C. reticulata, C. sinensis, Coleonema pulchrum, Medicosma cunninghamii.* **Sapindaceae:** *Arytera divaricata, Dodonaea triquerta, D. viscosa, Harpullia pendula, Sapindus saponaria, Sarcopteryx stipitata.* **Sapotaceae:** *Achras, Martiusella imperialis, Planchonella australis, P. myrsinoides, Pouteria wakere.* **Scrophulariaceae:** *Halleria lucida, Veronica salicifolia.* **Solanaceae:** *Capsicum annuum, Cyphomandra betacea, Lycopersicon esculentum, Solanum nigrum, S. paniculatum, Streptosolen jamesonii.* **Styracaceae:** *Styrax officinalis.* **Tiliaceae:** *Grewia flava, G. oppositifolia.* **Ulmaceae:** *Trema micrantha.* **Urticaceae:** *Urtica dioica.* **Verbenaceae:** *Avicennia officinalis, Duranta plumieri.*

REMARKS. Adult female redescribed and illustrated by Leonardi

(1920), Silvestri (1920b), Gómez-Menor Ortega (1937, 1958c), Borchsenius (1937, 1957), Gomes Costa (1949), Vilar (1951, 1952),
Monastero & Zaami (1959), De Lotto (1971c), Williams & Kosztarab
(1972), Gimpel *et al.* (1974), Hamon & Williams (1984), Tremblay
(1988a), Gill (1988) and by Williams & Watson (1988). Adult male
described and illustrated by Silvestri (1920b). Larval instars described and illustrated by Silvestri (1920b), Vilar (1951, 1952),
Monastero & Zaami (1959) and by Snowball (1970). Colour photograph given by Chapot & Delucchi (1964), Delucchi (1975), Gill
(1988), Tranfaglia & Viggiani (1988), Johnson & Lyon (1988) and by
Gonzalez (1989). Intraspecific variation in taxonomic characters of
the adult female given by Gimpel *et al.* (1974) (Californian and Virgi-
nian populations) and by Tranfaglia (1980) (Italian populations).
Distribution and host plant records given by Balachowsky (1931b),
Gomes Costa (1949), Vilar (1951, 1952), Borchsenius (1957), Monas-
tero & Zaami (1959), Snowball (1970), De Lotto (1971c), Corseuil &
Barbosa (1971), Gimpel *et al.* (1974), Vieira *et al.* (1983), Martin
Mateo (1984), Carnero Hernandez & Perez Guerra (1986), Marotta
(1987), Gonzalez (1989) and by Williams & Watson (1990).
BIOLOGY. One annual generation was reported in Italy on citrus
(Monastero & Zaami, 1959; Frediani, 1960) and on pears in Italy
(Frediani, 1960). Snowball (1970) outlined the life history in New
South Wales.
ECONOMIC IMPORTANCE. A sporadic pest of citrus in Italy and
Spain. A pest of citrus in Australia (Snowball, 1970).

Ceroplastes singularis Newstead

Ceroplastes singularis Newstead, 1910c: 188.
TYPE DATA. Syntypes female, UGANDA: Entebbe, on *Psidium
    guajava* (BMNH).
DISTRIBUTION. ETHIOPIAN REGION: Uganda.
HOST PLANTS. Myrtaceae: *Psidium guajava.*

Ceroplastes sinoiae Hall

Ceroplastes helichrysi sinoiae Hall, 1931: 296.
Gascardia sinoiae (Hall); De Lotto, 1965: 204.
TYPE DATA. Syntypes female, ZIMBABWE: Sinoia, on *Ficus* sp.
    (BMNH)
DISTRIBUTION. ETHIOPIAN REGION: Angola, South Africa, Zim-
    babwe.
HOST PLANTS. Bignoniaceae: *Jacaranda mimosaefolia.* **Guttiferae:**
    Hypericum revolutum. **Labiatae:** *Coleus.* **Leguminosae:** *Acacia,
    Brachystegia spiciformis.* **Moraceae:** *Ficus burkei.* **Rubiaceae:**
    Gardenia.
REMARKS. Adult female redescribed and illustrated by De Lotto
(1965). The morphology of the wax covering of the adult female and
larval instars described by Bedford (1968). Distribution and host
plant records given by De Lotto (1965), Bedford (1968), Hodgson
(1969a) and by Almeida (1973).
BIOLOGY. A uniparental species, which develops one annual generation on Jacaranda in South Africa. Bedford (1968) presented a comprehensive study on various aspects of the biology of this wax scale.

ECONOMIC IMPORTANCE. An urban pest of Jacaranda trees in South Africa (Bedford, 1968).

Ceroplastes speciosus Hempel

Ceroplastes speciosus Hempel, 1900b: 464.
TYPE DATA. Syntypes female, BRAZIL: Sao Paulo, Ypiranga, on various plants of Myrtaceae (MZSP).
DISTRIBUTION. NEOTROPICAL REGION: Brazil Sao Paulo.
HOST PLANTS. Myrtaceae.

Ceroplastes spicatus Hall

Ceroplastes toddaliae spicatus Hall, 1937: 122.
Ceroplastes spicatus Mamet, 1954a: 12.
TYPE DATA. Syntypes female, ZIMBABWE: South Marendellas, on *Uapaca kirkiana* (BMNH).
DISTRIBUTION. ETHIOPIAN REGION: Zimbabwe. **MADAGASIAN REGION:** Madagascar.
HOST PLANTS. Euphorbiaceae: *Uapaca kirkiana*. **Guttiferae:** *Harungana madagascariensis*. **Ulmaceae:** *Ulmus parifolia*.
REMARKS. Adult female redescribed and illustrated by Hodgson (1969a). Distribution and host plant records given by Mamet (1954a) and by Hodgson (1969a).

Ceroplastes stenocephalus De Lotto

Ceroplastes stenocephalus De Lotto, 1961: 320.
Gascardia stenocephala (De Lotto); De Lotto, 1965: 206.
TYPE DATA. Holotype female, KENYA: Magadi, on *Acacia* sp. (BMNH).
DISTRIBUTION. ETHIOPIAN REGION: Kenya.
HOST PLANTS. Leguminosae: *Acacia*.
REMARKS. Adult female redescribed and illustrated by De Lotto (1965).

Ceroplastes subrotundus Leonardi

Ceroplastes subrotundus Leonardi, 1911: 266.
TYPE DATA. Syntypes female, ARGENTINA: Cacheuta, on *Cercidium andicolum* (IEAP)
DISTRIBUTION. NEOTROPICAL REGION: Argentina.
HOST PLANTS. Leguminosae: *Cercidium andicolum, C. praecox*.

Ceroplastes sumatrensis Reyne

Ceroplastes sumatrensis Reyne, 1965: 155.
TYPE DATA. Syntypes female, SUMATRA: at Buo, on a dicotyledone shrub or tree (ZMAN).
DISTRIBUTION. AUSTRO ORIENTAL REGION: Sumatra.

Ceroplastes tachardiaformis Brain

Ceroplastes tachardiaformis Brain, 1920b: 35.
Gascardia tachardiaformis (Brain); De Lotto, 1965: 182.
TYPE DATA. Syntypes female, SOUTH AFRICA: Cape Province, Aberdeen, on *Elytropappus rhinocerotis* (SANC).

DISTRIBUTION. ETHIOPIAN REGION: South Africa.
HOST PLANTS. Compositae: *Elytropappus rhinocerotis*, *Stoebe cinerea*.
REMARKS. Adult female redescribed and illustrated by De Lotto (1978).

Ceroplastes theobromae Newstead

Ceroplastes theobromae Newstead, 1906: 74 [NOMEN NUDUM].
Ceroplastes theobromae Newstead, 1908c: 38.
TYPE DATA. Syntypes female, CAMEROON: Soppo and Bamba, on cacao (BMNH).
DISTRIBUTION. ETHIOPIAN REGION: Cameroon.
HOST PLANTS. Sterculiaceae: *Theobroma cacao*.

Ceroplastes titschaki Lindinger

Ceroplastes titschaki Lindinger, 1942: 113.
TYPE DATA. Syntypes female, PERU: South Peru, near Apurimac, on *Salix humboldtiana* (lost; Weidner & Wagner, 1968).
DISTRIBUTION. NEOTROPICAL REGION: Peru.
HOST PLANTS. Salicaceae: *Salix humboldtiana*.

Ceroplastes toddaliae Hall

Ceroplastes toddaliae Hall, 1931: 299.
TYPE DATA. Syntypes female, ZIMBABWE: Embeza, on *Toddalia austriaca*, *Annona senegalensis*, *Psorospermum febrifugum* and *Rhus* sp. (BMNH).
DISTRIBUTION. ETHIOPIAN REGION: Ivory Coast, Malawi, Mozambique, Zambia, Zimbabwe.
HOST PLANTS. Anacardiaceae: *Rhus*. **Annonaceae:** *Annona cherimolia*, *A. senegalensis*, *Artobotrys brachypetalus*. **Celastraceae:** *Cassine aethiopica*. **Ebenaceae:** *Euclea*. **Guttiferae:** *Psorospermum febrifugum*. **Lauraceae:** *Persea gratissima*. **Loganiaceae:** *Strychnos innocua*. **Moraceae:** *Ficus capensis*. **Myricaceae:** *Myrica pululifera*. **Ochnaceae:** *Ochna lanceolata*. **Olacaceae:** *Coula edulis*. **Rosaceae:** *Chaenomeles lagenaria*, *Cliffortia nitidula*. **Rubiaceae:** *Craterispermum caudatum*. **Rutaceae:** *Toddalia austriaca*. **Santalaceae:** *Osyris lanceolata*. **Sapotaceae:** *Bequaertiodendron megalismontanum*. **Sterculiaceae:** *Theobroma cacao*.
REMARKS. Hodgson (1969a) redescribed and illustrated this species and compared it with *C. spicatus*. Distribution and host plant records given by Hodgson (1969a) and by Couturier *et al* (1985).

Ceroplastes trochezi Mosquera

Ceroplastes trochezi Mosquera, 1979: 618.
TYPE DATA. Holotype female, COLOMBIA: Buga, on mango (CTBC).
DISTRIBUTION. NEOTROPICAL REGION: Colombia.
HOST PLANTS. Anacardiaceae: *Mangifera indica*.

Ceroplastes uapacae Hall

Ceroplastes uapacae Hall, 1931: 300.
Ceroplastes uapacae chrysophylli Hall, 1931: 302. Syntypes female,

ZIMBABWE [=RHODESIA]: Mtoroshanga Pass, Umvukwes, on *Chrysophyllum argyrophyllum* (BMNH). Syn. by Hodgson, 1969a: 14.
TYPE DATA. Syntypes female, ZIMBABWE: Mazoe and Salisbury on *Uapaca kirkiana*, and at Umtali on *Eugenia maluccensis* (BMNH).
DISTRIBUTION. ETHIOPIAN REGION: Malawi, Mozambique, South Africa, Zambia, Zimbabwe.
HOST PLANTS. Euphorbiaceae: *Uapaca kirkiana*. **Meliaceae:** *Khaya*. **Myrtaceae:** *Eugenia, Syzygium cordatum, S. gerrardii, S. guineense*. **Rubiaceae:** *Aida micrantha*. **Sapotaceae:** *Bequaertiodendron, Chrysophyllum argyrophyllum*.
REMARKS. Adult female redescribed and illustrated by De Lotto (1967b) and by Hodgson (1969a). Intraspecific variation in taxonomic characters given by Hodgson (1969a). Distribution and host plant records given by De Lotto (1967b) and by Hodgson (1969a).

Ceroplastes utilis Cockerell

Ceroplastes utilis Cockerell; Riley & Howard, 1892: 139 [NOMEN NUDUM].
Ceroplastes utilis Cockerell, 1893c: 83.
Ceroplastes dozieri Cockerell & Bueker, 1930: 7. Syntypes female, HAITI: Sources Puantes, on *Maytenus buxifolia* (USNM). Syn. by Gimpel et al., 1974: 67.
TYPE DATA. Lectotype female designated by Gimpel et al. (1974), GRAND TURK ISLAND: on undetermined tree or bush (USNM).
DISTRIBUTION. NEOTROPICAL REGION: Grand Turk Island, Haiti. **NEARCTIC REGION:** Florida.
HOST PLANTS. Bignoniaceae: *Catalpa longissima*. **Celastraceae:** *Maytenus buxifolia*. **Myrtaceae:** *Eugenia myrtoides*. **Rosaceae:** *Crataegus*. **Verbenaceae:** *Avicennia marina*.
REMARKS. Adult female redescribed and illustrated by Gimpel et al. (1974) and by Hamon & Williams (1984). Distribution and host plant records given by Cockerell & Bueker (1930), Gimpel et al. (1974) and by Hamon & Williams (1984).

Ceroplastes variegatus Hempel

Ceroplastes variegatus Hempel, 1900b: 462.
TYPE DATA. Syntypes female, BRAZIL: Sao Paulo, Ypiranga, on *Miconia* sp. and various Myrtaceae (MZSP).
DISTRIBUTION. NEOTROPICAL REGION: Brazil Sao Paulo.
HOST PLANTS. Melastomataceae: *Miconia*. **Myrtaceae.**
REMARKS. Adult female redescribed by Hempel (1901a).

Ceroplastes vinsoni Signoret

Ceroplastes vinsonii Signoret, 1872b: 38.
TYPE DATA. Syntypes female, MAURITIUS: on *Eriobotrya japonica* (probably lost; curator of VMNH, 1990, personal communication).
DISTRIBUTION. MADAGASIAN REGION: Mauritius.
HOST PLANTS. Rosaceae: *Eriobotrya japonica*.
REMARKS. Green (1937) suggested that this species is very likely identical with *C. floridensis*, but since the type material of the former

is lost it cannot be verified. For the sake of stability it is advisable to retain *C. vinsoni*, which antedated *C. floridensis*, as a valid, but unrecognizable species. Reducing *C. floridensis*, a name well-established in applied and systematic literature, as a synonym of *C. vinsonii* will greatly harm stability.

Ceroplastes vinsonioides Newstead

Ceroplastes vinsonioides Newstead, 1911a: 96.
TYPE DATA. Syntypes female, UGANDA: Entebbe, Namukekera, on coffee (BMNH).
DISTRIBUTION. ETHIOPIAN REGION: Angola, Cameroon, Kenya, Nigeria, Uganda.
HOST PLANTS. Loganiaceae: *Strychnos*. **Rubiaceae:** *Coffea arabica*, *C. canephora*. **Rutaceae:** *Citrus*.
REMARKS. Adult female redescribed and illustrated by De Lotto (1965) and by Hodgson (1969a). Distribution and host plant records given by De Lotto (1965, 1968b), Hodgson (1969b), Boboye (1971), Almeida (1973) and by Matile-Ferrero & Nonveiller (1984).
ECONOMIC IMPORTANCE. Biology and pest status on coffee presented by Le Pelley (1968).

Ceroplastes xishuangensis Tang & Xie

Ceroplastes xishuangensis Tang & Xie in Tang, 1991: 301.
TYPE DATA. Holotype female, CHINA: Yunnan Province, Jinghong Town, on *Cycas siamensis* (EISC).
DISTRIBUTION. PALAEARCTIC REGION: China.
HOST PLANTS. Cycadaceae: *Cycas siamensis*. **Palmae:** *Elaeis guineensis*.

Ceroplastidia Cockerell

Ceroplastes (Ceroplastidia) Cockerell, 1910a: 76.
Ceroplastidia Cockerell; Morrison & Morrison, 1966: 32.
TYPE-SPECIES: *Ceroplastes bruneri* Cockerell & Cockerell, by original designation.
REMARKS. Morrison (1919) commented on this genus, noting its similarity to *Gascardia* and doubted its distinction from *Ceroplastes*. De Lotto (1971c) and Gimpel *et al.* (1974) regarded this genus as a subjective synonym of *Gascardia* and *Ceroplastes*, respectively. Here it is placed as a subjective synonym of *Ceroplastes*.

Ceroplastina Cockerell

Ceroplastes (Ceroplastina) Cockerell, 1910a: 76.
Ceroplastina Cockerell; Morrison & Morrison, 1966: 32.
TYPE-SPECIES: *Ceroplastes lahillei* Cockerell, by original designation.
REMARKS. This genus is here regarded as a subjective synonym of *Ceroplastes*.

Ceroplastodes Cockerell

Ceroplastodes Cockerell, 1893r: 350.

TYPE-SPECIES: *Fairmairia (Ceroplastodes) nivea* Cockerell, by original designation.
REMARKS. Generic characters discussed by Green (1909), Morrison (1920), Steinweden (1929), Hodgson (1971), Tao *et al.* (1983) and by Tang (1991). Key to species: ASIA - Tang (1991); ETHIOPIAN REGION - Hodgson (1971); SRI LANKA - Green (1909).

Ceroplastodes acaciae Cockerell

Ceroplastodes acaciae Cockerell, 1895o: 2.
TYPE DATA. Syntypes female, U.S.A.: Arizona, near Tucson, on *Acacia constricta* (USNM).
DISTRIBUTION. NEARCTIC REGION: Arizona, New Mexico.
HOST PLANTS. Leguminosae: *Acacia constricta.*
REMARKS. Male test described and illustrated by Miller & Williams (1990).

Ceroplastodes bahiensis Bondar

Ceroplastodes bahiensis Bondar, 1925: 58.
TYPE DATA. Syntypes female, BRAZIL: Bahia, Sul, on cacao (USNM).
DISTRIBUTION. NEOTROPICAL REGION: Brazil Bahia.
HOST PLANTS. Sterculiaceae: *Theobroma cacao.*

Ceroplastodes dugesii (Signoret)

Lecanopsis dugesii Signoret, 1886: xxxix.
Inglisia nivea Cockerell, 1893n: 160. Syntypes female, MEXICO: Chihuahua State, Montezuma, on *Acacia* sp. (USNM). Syn. by Cockerell, 1902h: 194.
Fairmairia (Ceroplastodes) nivea Cockerell; Cockerell, 1893r: 350 [UNJUSTIFIED REPLACEMENT NAME].
Ceroplastodes daleae Cockerell, 1894i: 13. Syntypes female, U. S. A.: New Mexico, Mesilla Valley, Tortugas Mountain, on *Dalea formosa* (USNM). Syn. by Avasthi & Shafee, 1991a: 1.
Ceroplastodes nivea (Cockerell); Cockerell, 1895c: 209.
Ceroplastodes dugesii (Lichtenstein & Signoret); Fernald, 1903: 164 [ERRONEOUS AUTHORSHIP].
Ceroplastodes deani Lawson, 1917: 203. Syntypes female, U. S. A.: Kansas, Phillips County, near Marvin, on *Petalostemon violaceus* (USNM). Syn. by Avasthi & Shafee, 1991a: 1.
Ceroplastodes dugesii (Lichtenstein & Signoret); Avasthi & Shafee, 1991a: 1 [ERRONEOUS AUTHORSHIP].
TYPE DATA. Syntypes female, MEXICO: on *Mimosa* sp. (VMNH).
DISTRIBUTION. NEOTROPICAL REGION: Mexico. **NEARCTIC REGION:** Kansas, New Mexico.
HOST PLANTS. Leguminosae: *Dalea formosa, Kuhnistera purpurea, Mimosa, Petalostemon violaceus.*
REMARKS. Adult female redescribed and illustrated by Avasthi & Shafee (1991a).

Ceroplastodes melaleucae (Green)

Eriochiton melaleucae Green, 1900b: 12.
Ceroplastodes melaleucae (Green); Fernald, 1903: 164.

TYPE DATA. Syntypes female, **AUSTRALIA:** Victoria, Myrniong, on *Melaleuca nodosa* (BMNH).
DISTRIBUTION. AUSTRALIAN REGION: Victoria.
HOST PLANTS. Myrtaceae: *Melaleuca nodosa.*
REMARKS. Adult female redescribed by Froggatt (1915).

Ceroplastodes melzeri Bondar

Ceroplastodes melzeri Bondar, 1925: 56.
TYPE DATA. Syntypes female, BRAZIL: Bahia, on cacao (USNM).
DISTRIBUTION. NEOTROPICAL REGION: Brazil Bahia.
HOST PLANTS. Sterculiaceae: *Theobroma cacao.*
REMARKS. Adult female redescribed and illustrated by Avasthi & Shafee (1991a).

Ceroplastodes misiones Morrison

Ceroplastodes misiones Morrison, 1919: 81.
TYPE DATA. Syntypes female, ARGENTINA: Bomplana, Misiones, on undetermined plant (USNM).
DISTRIBUTION. NEOTROPICAL REGION: Argentina.
HOST PLANTS. Compositae: *Baccharis oxyodonta.*

Ceroplastodes ritchiei Laing

Ceroplastodes ritchiei Laing, 1925: 55.
TYPE DATA. Syntypes female, TANZANIA: Morogoro, Government Experimental Farm, on *Annona* sp. (BMNH).
DISTRIBUTION. ETHIOPIAN REGION: Ivory Coast, Sierra Leone, Tanzania.
HOST PLANTS. Annonaceae: *Annona muricata,*
REMARKS. Adult female redescribed and illustrated by Hodgson (1971). Distribution and host plant records given by Hodgson (1971) and by Couturier *et al.* (1985).

Ceroplastodes theobromae Bondar

Ceroplastodes theobromae Bondar, 1925: 59.
TYPE DATA. Syntypes female, BRAZIL: Bahia, on cacao (USNM).
DISTRIBUTION. NEOTROPICAL REGION: Brazil Bahia.
HOST PLANTS. Sterculiaceae: *Theobroma cacao.*
REMARKS. Adult female redescribed and illustrated by Avasthi & Shafee (1991a).

Ceroplastodes wandoorensis Yousuf & Shafee

Ceroplastodes wandoorensis Yousuf & Shafee, 1988: 57.
TYPE DATA. Holotype female, INDIA: Aldaman Islands, Port Blair, Wandoor, on wild plant (AMUI).
DISTRIBUTION. ORIENTAL REGION: India Aldaman Islands.

Ceroplastodes zavattarii Bellio

Ceroplastodes zavattarii Bellio, 1939: 232.
TYPE DATA. Syntypes female, ETHIOPIA: Neghelli, on twigs of undetermined plant (IEAP).

DISTRIBUTION. ETHIOPIAN REGION: Ethiopia, Gambia, Guinea, Nigeria, Senegal.
HOST PLANTS. Gramineae: *Bambusa*. **Labiatae:** *Coleus*. **Malvaceae:** *Pavonia hirsuta*. **Tiliaceae:** *Triumfetta*.
REMARKS. Adult female redescribed and illustrated by Hodgson (1971). Distribution and host plant records given by Hodgson (1971) and by Fernandes (1990).

Cerostegia De Lotto

Cerostegia De Lotto, 1969b: 211.
TYPE-SPECIES: *Ceroplastes rufus* De Lotto, by original designation.
REMARKS. This genus is regarded here as a subjective synonym of *Ceroplastes*.

Chlamydolecanium Goux

Chlamydolecanium Goux, 1933: 119.
Chlamidolecanium Lindinger; Lindinger, 1935a: 142 [MIS-SPELLING].
TYPE-SPECIES: *Chlamydolecanium conchioides* Goux, by original designation and monotypy.
REMARKS. Generic characters discussed by Borchsenius (1957).

Chlamydolecanium conchioides Goux

Chlamydolecanium conchioides Goux, 1933: 120.
TYPE DATA. Syntypes female and larva, FRANCE: Corsica, Bastia, on *Lavandula stoechas* (GOUX).
DISTRIBUTION. PALAEARCTIC REGION: Corsica, Italy.
HOST PLANTS. Labiatae: *Lavandula stoechas*.
REMARKS. Adult female redescribed by Borchsenius (1957). Distribution and host plant records given by Marotta (1987).

Chloropulvinaria Borchsenius

Chloropulvinaria Borchsenius, 1952: 299.
TYPE-SPECIES: *Coccus floccifera* Westwood, by original designation.
REMARKS. Generic characters discussed by Borchsenius (1957), Wang (1980), Yang (1982), Hadzibejli (1983), Kosztarab & Kozár (1988) and by Tang (1991). Key to species: ASIA - Tang (1991). The genus is here regarded as a subjective synonym of *Pulvinaria*.

Cissococcus Cockerell

Cissococcus Cockerell, 1902a: 23.
TYPE-SPECIES: *Cissococcus fulleri* Cockerell, by original designation and monotypy.
REMARKS. Cockerell (1902a) assigned the genus to the Eriococcidae, Brain (1918) established for it the Cissococcinae. Ferris (1920a) assigned the genus to the Coccidae. Generic characters discussed by Ferris (1920a) and by Steinweden (1929).

Cissococcus fulleri Cockerell

Cissococcus fulleri Cockerell, 1902a: 23.

TYPE DATA. Syntypes female, SOUTH AFRICA: Umquahumbi Valley, on *Cissus cuneifolia* (SANC, BMNH, USNM).
DISTRIBUTION. ETHIOPIAN REGION: South Africa.
HOST PLANTS. Vitidaceae: *Cissus cuneifolia.*
REMARKS. Adult female redescribed and illustrated by Brain (1918) and by Ferris (1919b).
BIOLOGY. The insect induces formation of galls in which it develops.

Coccus Linnaeus

Coccus Linnaeus, 1758: 455.
Calymmata Costa, 1828: 6. Type-species: *Coccus hesperidum* Linnaeus. Synonymy by community of type species.
Coeus Lopez y Ramos; Lopez y Ramos, 1835: 15 [MIS-SPELLING of *Coccus*].
Cocus Watt & Mann; Watt & Mann, 1903: 300 [MIS-SPELLING of *Coccus*].
TYPE-SPECIES: *Coccus hesperidum* Linnaeus, by Opinion 1303 (1985), under the plenary powers of the International Commission on Zoological Nomenclature, Name Number 2244.
REMARKS. The generic name *Coccus* and its type species, by designation under the plenary powers, were placed on the Official Lists of Generic and Specific Names in Zoology, in Opinion 1303 of ICZN (1985). The generic name *Lecanium* Burmeister, 1835 (a junior objective synonym of *Coccus* Linnaeus) was placed on the official Index of Rejected and Invalid Generic Names in Zoology, in Opinion 1303 of ICZN (1985).

Generic characters discussed by Leonardi (1920), Steinweden (1929), Gómez-Menor Ortega (1937), Zimmerman (1948), Borchsenius (1957), De Lotto (1959, 1965), Hodgson (1967a), Williams & Kosztarab (1972), Gill *et al.* (1977), Paik (1978), Wang (1980), Kawai (1980), Yang (1982), Tao *et al.* (1983), Hadzibejli (1983), Hamon & Williams (1984), Gill (1988), Williams & Watson (1990) and by Tang (1991). Key to species: AFRICA - De Lotto (1957b, 1959, 1960, 1965). ASIA - Tang (1991). CALIFORNIA - Gill (1988). CHINA - Wang (1980), Tang (1991). FLORIDA - Hamon & Williams (1984). HAWAII - Zimmerman (1948). JAPAN - Kawai (1980). MICRONESIA - Beardsley (1966a). NORTH AMERICA - Gill *et al.* (1977). SINGAPORE - Morrison (1921). SOUTH AFRICA - De Lotto (1966A). TAIWAN - Tao *et al.* (1983). TROPICAL SOUTH PACIFIC - Williams & Watson (1990). USSR - Borchsenius (1957). ZIMBABWE - Hodgson (1967a, 1969b).

Coccus acaciae (Newstead)

Lecanium acaciae Newstead, 1917a: 355.
Coccus elongatus (Signoret); De Lotto, 1957b: 301 [MISIDENTIFICATION and ERRONEOUS SYNONYMIZATION].
Coccus acaciae (Newstead); Ben-Dov, 1977: 94 [RESTORED STATUS].
TYPE DATA. Lectotype female designated by Ben-Dov (1977), KENYA: Nairobi, on Acacia melanoxylon (BMNH).
DISTRIBUTION. ETHIOPIAN REGION: Kenya.
HOST PLANTS. Leguminosae: *Acacia melanoxylon, Albizia molucanna.*
REMARKS. Adult female redescribed by Ben-Dov (1977).

Coccus acrossus De Lotto

Coccus acrossus De Lotto, 1969a: 415.
Coccus atrichos De Lotto; Carvalho & Cardoso, 1970: 15 [MIS-SPELLING].
TYPE DATA. Holotype female, ANGOLA: Novo Redondo, on *Elaeis guineensis* (SANC).
DISTRIBUTION. ETHIOPIAN REGION: Angola.
HOST PLANTS. Palmae: *Elaeis guineensis*.

Coccus acutissimus (Green)

Lecanium acutissimum Green, 1896: 10.
Coccus acutissimus (Green); Fernald, 1903: 168.
TYPE DATA. Lectotype female designated by Williams & Watson (1990), SRI LANKA: Paradeniya, on *Areca catechu* (BMNH).
DISTRIBUTION. AUSTRO ORIENTAL REGION: Indonesia, Malaysia, Papua New Guinea, Singapore, Sumatra. **ETHIOPIAN REGION:** Kenya. **MADAGASIAN REGION:** Mauritius. **NEW ZEALAND and PACIFIC REGION:** French Polynesia, Hawaii, Mariana Islands, Palaus, Western Samoa. **NEARCTIC REGION:** Florida, Texas. **ORIENTAL REGION:** India, Ryukyu Islands, Sri Lanka, Taiwan, Thailand, Vietnam. **PALAEARCTIC REGION:** Japan.
HOST PLANTS. Anacardiaceae: *Anacardium occidentale, Mangifera indica*. **Annonaceae:** *Cananga odorata*. **Apocynaceae:** *Ochrosia, Plumeria acutifolia*. **Aquifoliaceae:** *Ilex aquifolium*. **Boraginaceae:** *Cordia alliodora*. **Cycadaceae:** *Cycas circinalis, C. revoluta*. **Euphorbiaceae:** *Glochidion*. **Flacourtiaceae:** *Hydnocarpus wightiana*. **Guttiferae:** *Calophyllum, Clusia rosea*. **Iridaceae:** *Gladiolus illyricus*. **Lauraceae:** *Licaria triandra, Persea americana*. **Leguminosae:** *Bauhinia*. **Linaceae:** *Durandea*. **Lythraceae:** *Lagerstroemia indica*. **Magnoliaceae:** *Magnolia grandiflora, Michelia alba*. **Moraceae:** *Artocarpus communis, A. gomezianus, A. heterophyllus, A. integrifolius, Ficus tinctoria*. **Myrsinaceae:** *Ardisia*. **Myrtaceae:** *Eugenia jambos, Metrosideros, Psidium guajava*. **Oleandraceae:** *Nephrolepis*. **Orchidaceae:** *Sobralia macrantha, Stanhopea*. **Palmae:** *Areca catechu, Chrysalidocarpus lutescens, Cocos nucifera*. **Rubiaceae:** *Canthium odorata, Gardenia*. **Sapindaceae:** *Euphoria longana, Litchi chinensis, Nephelium lappaceum*. **Sapotaceae:** *Palaquium formosanum*. **Smilacaceae:** *Smilax*. **Verbenaceae:** *Premna*. **Zingiberaceae:** *Hedychium*.
REMARKS. Adult female redescribed and illustrated by Green (1904d), Zimmerman (1948), Gill *et al.* (1977), Kawai (1980), Tao *et al.* (1983), Hamon & Williams (1984), Williams & Watson (1990) and by Tang (1991). Colour photograph given by Kawai (1980) and by Hamon & Williams (1984). Distribution and host plant records given by Takahashi (1942a, 1952), De Lotto (1957b), Beardsley (1966a), Ali (1971), Gill *et al.* (1977), Kawai (1980), Nakahara (1981b), Tao *et al.* (1983), Williams & Williams (1988), Williams & Watson (1990) and by Tang (1991).

Coccus aequale (Newstead) n. comb.

Lecanium aequale Newstead, 1917a: 354.

TYPE DATA. Syntypes female, and female second instar, GUYANA: Sea Shore, East Coast, on *Avicennia nitida* (BMNH).
DISTRIBUTION. NEOTROPICAL REGION: Guyana.
HOST PLANTS. Verbenaceae: *Avicennia nitida.*

Coccus africanus (Newstead)

Lecanium viride africanum Newstead, 1898: 95.
Coccus viridis africanus (Newstead); Fernald, 1903: 174.
Lecanium africanum Newstead; Newstead, 1917a: 357.
Coccus africanus (Newstead); De Lotto, 1960: 389.
TYPE DATA. Syntypes female, NIGERIA: Lagos, on coffee (BMNH).
DISTRIBUTION. ETHIOPIAN REGION: Angola, Eritrea, Kenya, Nigeria, Uganda.
HOST PLANTS. Apocynaceae: *Carissa edulis.* **Celastraceae:** *Gymnosporia.* **Ehretiaceae:** *Ehretia silvatica.* **Myrtaceae:** *Psidium guajava.* **Rubiaceae:** *Coffea arabica, C. robusta, Gardenia.* **Rutaceae:** *Citrus limon.*
REMARKS. Adult female redescribed and illustrated by Newstead (1917a) and by De Lotto (1957b, 1960). Distribution and host plant records given by Newstead (1917a), De Lotto (1957b, 1960) and by Almeida (1973).
ECONOMIC IMPORTANCE. Biology and pest status on coffee presented by Le Pelley (1968).

Coccus almoraensis Avasthi & Shafee

Coccus almoraensis Avasthi & Shafee, 1983: 389.
TYPE DATA. Holotype female, INDIA: Uttar Pradesh, Almora, on undetermined wild plant (AMUI).
DISTRIBUTION. ORIENTAL REGION: India.
HOST PLANTS. Anacardiaceae: *Mangifera indica.*

Coccus alpinus De Lotto

Lecanium africanum Newstead; Newstead, 1917a: 357 [MISIDENTIFICATION].
Coccus africanus (Newstead); De Lotto, 1957b: 296 [MISIDENTIFICATION].
Coccus alpinus De Lotto, 1960: 393.
TYPE DATA. Holotype female, KENYA: Ruiru, on *Coffea arabica* (BMNH).
DISTRIBUTION. ETHIOPIAN REGION: Angola, Eritrea, Ethiopia, Kenya, Tanzania, Uganda, Zaire, Zimbabwe.
HOST PLANTS. Apocynaceae: *Carissa edulis.* **Celastraceae:** *Gymnosporia.* **Ehretiaceae:** *Ehretia silvatica.* **Myrtaceae:** *Psidium guajava.* **Rubiaceae:** *Coffea arabica, C. canephora, C. robusta, Gardenia.* **Rutaceae:** *Citrus limon.*
REMARKS. Distribution and host plant records given by Newstead (1917a), De Lotto (1965, 1968b), Hodgson (1967a) and by Almeida (1973).
BIOLOGY. De Lotto (1960) noted that this species occurs in localities situated at altitudes above 1 - 1.3 km. Biology and pest status on coffee presented by Le Pelley (1968).

Coccus anneckei De Lotto

Coccus anneckei De Lotto, 1962: 263.
TYPE DATA. Holotype female, SOUTH AFRICA: Cape Province, Middelburg, on *Lycium* sp. (BMNH).
DISTRIBUTION. ETHIOPIAN REGION: South Africa.
HOST PLANTS. Solanaceae: *Lycium.*

Coccus antidesmae (Green)

Lecanium antidesmae Green, 1896: 10.
Coccus antidesmae (Green); Fernald, 1903: 168.
TYPE DATA. Syntypes female, SRI LANKA: Punduloya, on *Antidesma bunius* (BMNH).
DISTRIBUTION. ORIENTAL REGION: Sri Lanka.
HOST PLANTS. Stilaginaceae: *Antidesma bunius.*
REMARKS. Adult female redescribed and illustrated by Green (1904d). Adult female redescribed by Tang (1991).

Coccus asiaticus Lindinger

Lecanium caudatum Green, 1896: 10.
Coccus caudatus (Green); Fernald, 1903: 168 [SECONDARY HOMONYM of *Coccus caudatus* Walker].
Coccus asiaticus Lindinger, 1932f: 201 [REPLACEMENT NAME].
Coccus caudaus (Green); Tao, 1978: 80 [MIS-SPELLING].
TYPE DATA. Syntypes female, SRI LANKA: Pundaluoya and Kandy, on *Passiflora* sp., *Coffea arabica*, *Loranthus* sp., *Memecylon umbellatum*, *Pergularia odoratissima* and *Pavetta* sp. (BMNH).
DISTRIBUTION. ETHIOPIAN REGION: Cameroon, Kenya, Tanzania, Uganda. **ORIENTAL REGION:** Sri Lanka, Taiwan.
HOST PLANTS. Asclepiadaceae: *Pergularia odoratissima*. **Loranthaceae:** *Loranthus*. **Memecylaceae:** *Memecylon umbellatum*. **Passifloraceae:** *Passiflora*. **Rubiaceae:** *Coffea arabica, C. robusta, Pavetta*. **Rutaceae:** *Murraya exotica.*
REMARKS. Adult female redescribed and illustrated (as *C. caudatus*) by De Lotto (1957b). Adult female redescribed by Matile-Ferrero (1987) and by Tang (1991) (as *C. caudatus*). Tao *et al.* (1983) synonymized *Coccus caudatus* (Green) with *Parasaissetia nigra* (Nietner), however, the redescriptions of both species, by Tao *et al.*, strongly indicate that this is an error. Distribution and host plant records given by Green (1904d), Takahashi (1929a), De Lotto (1957b), Ali (1971), Matile-Ferrero (1987) and by Tang (1991).
ECONOMIC IMPORTANCE. Biology and pest status on coffee presented by Le Pelley (1968).

Coccus brasiliensis Fonseca

Coccus brasiliensis Fonseca, 1957: 128.
TYPE DATA. Syntypes female, BRAZIL: Sao Paulo State, Botucatu and Sao Paulo, on coffee (MZSP).
DISTRIBUTION. NEOTROPICAL REGION: Brazil Sao Paulo.
HOST PLANTS. Rubiaceae: *Coffea.*

Coccus bromeliae Bouche

Coccus bromeliae Bouche, 1833: 49.
TYPE DATA. Syntypes female, GERMANY: Berlin, on *Ananas* and
   Hibiscus (lost; Sachtleben, 1944).
DISTRIBUTION. PALAEARCTIC REGION: Germany.
HOST PLANTS. Bromeliaceae: *Ananas.* **Malvaceae:** *Hibiscus.*
REMARKS. Both Signoret (1875) and Fernald (1903) interpreted this
species as being a mealybug. However, as shown by Lindinger
(1932f) the original description is clearly that of a Coccidae. Lindinger (1934e) suggested that *C. bromeliae* may be a synonym of *Saissetia coffeae* (Walker). This may be correct, but this synonymy, if
accepted would change the name of a very well-known pest species
and harm stability. In the absence of definite evidence as to the
taxonomic identity of *C. bromeliae*, it seems more appropriate that
this nominal species should remain as an unrecognizable species of
Coccidae. See discussion in Ben-Dov & Cox (1990).

Coccus cajani (Newstead)

Lecanium cajani Newstead, 1917a: 359.
Coccus cajani (Newstead); De Lotto, 1962: 264.
TYPE DATA. Syntypes female, NIGERIA: on pigeon-pea (BMNH).
DISTRIBUTION. ETHIOPIAN REGION: Nigeria.
HOST PLANTS. Leguminosae: *Cajanus indicus.*
REMARKS. Adult female redescribed and illustrated by De Lotto
(1962).

Coccus cambodiensis Takahashi

Coccus cambodiensis Takahashi, 1942a: 17.
TYPE DATA. Syntypes female, CAMBODIA: Angkor, on *Ficus* sp.
   (IMZT).
DISTRIBUTION. AUSTRO ORIENTAL REGION: Cambodia.
HOST PLANTS. Moraceae: *Ficus, Ficus retusa.*
REMARKS. Adult female redescribed and illustrated by Tang (1991).

Coccus cameronensis Takahashi

Coccus cameronensis Takahashi, 1952: 16.
TYPE DATA. Syntypes female, MALAYSIA: Cameron Highlands, on
   undetermined tree (SMKM).
DISTRIBUTION. AUSTRO ORIENTAL REGION: Malaysia.
REMARKS. Adult female redescribed by Tang (1991).

Coccus capparidis (Green)

Lecanium capparidis Green, 1904d: 187.
Coccus capparidis (Green); Sanders, 1906: 8.
Lecanium (Coccus) capparidis Green; Green, 1937: 299.
Coccus arens Hodgson, 1968b: 114. Holotype female, EGYPT: Giza,
   on *Panax* sp. (BMNH). Syn. by Ben-Dov, 1980: 262.
TYPE DATA. Lectotype female designated by Williams & Watson
   (1990), SRI LANKA: Pundaluoya, on *Capparis moonii*, (BMNH).
DISTRIBUTION. NEARCTIC REGION: Florida. **NEOTROPICAL
   REGION:** Bahamas, Honduras. **NEW ZEALAND and PACIFIC**

REGION: Hawaii, Kiribati, Tonga, Western Samoa. **ORIENTAL REGION:** India, Sri Lanka. **PALAEARCTIC REGION:** Egypt, Israel.

HOST PLANTS. Apocynaceae: *Alyxia olivaeformis, Nerium oleander, Plumeria rubra.* **Araceae:** *Xanthosoma.* **Araliaceae:** *Aralia balfouriana, Meryta macrophylla, Panax.* **Asclepiadaceae:** *Asclepias curassavica.* **Boraginaceae:** *Cordia subcordata.* **Campanulaceae:** *Clermontia.* **Cannaceae:** *Canna indica.* **Capparidaceae:** *Capparis moonii.* **Compositae:** *Bidens pilosa.* **Ebenaceae:** *Diospyros virginiana.* **Euphorbiaceae:** *Codiaeum, Croton.* **Malvaceae:** *Hibiscus manihot.* **Moraceae:** *Artocarpus altilis.* **Musaceae:** *Musa paradisiaca.* **Myoporaceae:** *Myoporum acuminatum.* **Nyctaginaceae:** *Mirabilis jalapa.* **Orchidaceae:** *Cypripedium villosum, Dendrobium.* **Pteridaceae:** *Acrostichum aureum.* **Rubiaceae:** *Guettarda speciosa, Morinda citrifolia.* **Rutaceae:** *Citrus aurantium, C. paradisi, C. sinensis, Murraya paniculata.* **Verbenaceae:** *Lantana camara, Premna, Stachytarpheta.*

REMARKS. Adult female redescribed and illustrated by Hodgson (1968b) (as *C. arens*), Gill *et al.* (1977), Hamon & Williams (1984) Williams & Watson (1990). Adult female redescribed by Tang (1991). Colour photograph given by Hamon & Williams (1984). Distribution and host plant records given by Green (1937), Hodgson (1968b), Gill *et al.* (1977), Ben-Dov (1980), Nakahara (1981b), Hamon & Williams (1984) and by Williams & Watson (1990).

BIOLOGY. The female in Israel reproduces parthenogenetically. On citrus females infest the undersurface of leaves (Ben-Dov, 1980). Blumberg & Swirski (1984) studied the encapsulation response to parasitoids.

Coccus caviramicolus Morrison

Coccus caviramicolus Morrison, 1921: 659.

TYPE DATA. Holotype female, SINGAPORE: from hollow stems of *Macaranga* sp. (USNM).

DISTRIBUTION. AUSTRO ORIENTAL REGION: Indonesia, Malaysia, Singapore.

HOST PLANTS. Euphorbiaceae: *Macaranga griffithiana, M. triloba.*

REMARKS. Distribution and host plant records given by Takahashi (1950a).

Coccus celatus De Lotto

Lecanium viride Green; Newstead, 1910c: 187 [MISIDENTIFICATION].

Coccus celatus De Lotto, 1960: 395.

Coccus consimilis De Lotto, 1960: 397. Holotype female, UGANDA: Kampala, on *Coffea robusta* (BMNH). Syn. by Williams, 1982a: 108.

TYPE DATA. Holotype female, UGANDA: Kampala, on *Coffea robusta* (BMNH).

DISTRIBUTION. AUSTRO ORIENTAL REGION: Brunei, Malaysia, Papua New Guinea. **ETHIOPIAN REGION:** Cameroon, Ivory Coast, Kenya, Somalia, Sudan, Tanzania, Uganda, Zimbabwe. **NEW ZEALAND and PACIFIC REGION:** Irian Jaya. **ORIENTAL REGION:** Vietnam.

HOST PLANTS. Annonaceae: *Annona muricata.* **Apocynaceae:**

Carissa. **Casuarinaceae:** *Casuarina.* **Euphorbiaceae:** *Macaranga barteri.* **Guttiferae:** *Calophyllum soulattrii, C. spectabile.* **Leguminosae:** *Arachis hypogaea.* **Myrtaceae:** *Eugenia javanica, Syzygium.* **Rubiaceae:** *Bertiera racemosa, Coffea arabica, C. canephora, C. excelsa, C. robusta, Corynanthe pachyceras, Gardenia, Ixora coccinea, Psychotria.* **Rutaceae:** *Citrus.*

REMARKS. Williams (1982a) pointed out some diagnostic characters for adequate identification of this species. Adult female redescribed and illustrated by Williams & Watson (1990). Distribution and host plant records given by De Lotto (1960, 1969a), Hodgson (1969b), Williams (1982a), Matile-Ferrero & Nonveiller (1984), Couturier *et al.* (1985), Williams & Watson (1990) and by Danzig & Konstantinova (1990).

BIOLOGY. This soft scale is protected under shelters constructed by the ant *Macromischoides aculeatum* Mayer, in Uganda (De Lotto, 1960).

ECONOMIC IMPORTANCE. A coffee pest in Papua New Guinea (Williams, 1982a). Biology and pest status on coffee presented by Le Pelley (1968).

Coccus circularis Morrison

Coccus circularis Morrison, 1921: 665.
TYPE DATA. Holotype female, SINGAPORE: in hollow stems of *Macaranga* sp. (USNM).
DISTRIBUTION. AUSTRO ORIENTAL REGION: Singapore.
HOST PLANTS. Euphorbiaceae: *Macaranga.*

Coccus colemani Kannan

Coccus colemani Kannan, 1918: 135.
TYPE DATA. Syntypes female, INDIA: Mysore, host plant not indicated, probably on coffee and other host plants (depository unknown).
DISTRIBUTION. ORIENTAL REGION: India.
HOST PLANTS. Anacardiaceae: *Mangifera indica.* **Apocynaceae:** *Wrightia tinctoria.* **Leguminosae:** *Albizia.* **Moraceae:** *Artocarpus integrifolia, Ficus.* **Myrtaceae:** *Eugenia jambolana, Psidium guajava.* **Rutaceae:** *Aegle marmelos, Citrus* .
REMARKS. Adult female and larval instars redescribed and illustrated by Coleman & Kannan (1918). Green (1918) was doubtful whether there was sufficient evidence for the erection of this species, as distinct from *C. viridis.*
BIOLOGY. Life history in India presented by Coleman & Kannan (1918).

Coccus deformosum (Newstead) n. comb.

Lecanium (Eulecanium) deformosum Newstead, 1920: 190.
TYPE DATA. Syntypes female, GUYANA: "Cattle Trail Survey", on undetermined plant, protected under paper nests constructed by a species of *Acromyrmex* (BMNH).
DISTRIBUTION. NEOTROPICAL REGION: Guyana.

Coccus delottoi Matile-Ferrero & Le Ruyet

Coccus delottoi Matile-Ferrero & Le Ruyet, 1985: 259.
TYPE DATA. Holotype female, IVORY COAST: Tai, on *Diospyros subreana* (MNHN).
DISTRIBUTION. ETHIOPIAN REGION: Ivory Coast.
HOST PLANTS. Ebenaceae: *Diospyros soubreana.*

Coccus discrepans (Green)

Lecanium discrepans Green, 1904d: 204.
Saissetia discrepans (Green); Sanders, 1906: 9.
Coccus discrepans (Green); Morrison, 1920: 654.
TYPE DATA. Syntypes female, SRI LANKA: Punadluoya, Yatiyantota and Passara, on tea plant (BMNH).
DISTRIBUTION. AUSTRO ORIENTAL REGION: Singapore. **ORIENTAL REGION:** India, Pakistan, Sri Lanka, Taiwan. **PALAEARCTIC REGION:** Japan.
HOST PLANTS. Caricaceae: *Carica papaya.* **Euphorbiaceae:** *Glochidion callicarpa.* **Myrsinaceae:** *Maesa pedicellata.* **Palmae:** *Areca oleracea, Cocos nucifera.* **Rutaceae:** *Citrus, Murraya caloxylon.* **Theaceae:** *Thea.* **Verbenaceae:** *Callicarpa formosana.*
REMARKS. Adult female redescribed and illustrated by Tao *et al.* (1983) and by Tang (1991). Adult female redescribed by Morrison (1921) and by Kawai (1980). Distribution and host plant records given by Morrison (1921), Takahashi (1935), Green (1937), Ali (1971), Kawai (1980), Tao *et al.* (1983), Varshney (1985), Shafee *et al.* (1989) and by Tang (1991).
BIOLOGY. Attended by and enclosed in nests of *Crematogaster dohrni* (Green, 1937).

Coccus duartei (Almeida)

Lecanium duartei Almeida, 1969: 22.
Coccus duartei (Almeida); Almeida, 1973: 3.
TYPE DATA. Syntypes female, ANGOLA: Tchivinguiro, on trunk of undetermined tree (CZLP).
DISTRIBUTION. ETHIOPIAN REGION: Angola.

Coccus ehretiae (Brain)

Lecanium ehretiae Brain, 1920b: 4.
Coccus ehretiae (Brain); De Lotto, 1957b: 303.
TYPE DATA. Syntypes female, SOUTH AFRICA: Transvaal, Pretoria, on *Ehretia hottentotica* (SANC).
DISTRIBUTION. ETHIOPIAN REGION: South Africa. **MADAGASIAN REGION:** Madagascar.
HOST PLANTS. Ehretiaceae: *Ehretia hottentotica.*
REMARKS. Adult female redescribed and illustrated by De Lotto (1957b). Distribution and host plant records given by De Lotto (1957b) and by Mamet (1959b).

Coccus elatensis Ben-Dov

Coccus elatensis Ben-Dov, 1981: 649.

TYPE DATA. Holotype female, ISRAEL: Elot, on *Mangifera indica* (ICV).
DISTRIBUTION. PALAEARCTIC REGION: Israel.
HOST PLANTS. Anacardiaceae: *Mangifera indica.*
REMARKS. Adult female redescribed by Tang (1991).

Coccus erythrinae (Ihering)

Lecanium erythrinae Ihering, 1897: 407.
Coccus erythrinae (Ihering); Corseuil & Barbosa, 1971: 238.
TYPE DATA. Syntypes female, BRAZIL: Rio Grande do Sul, on *Erythrina cristagalli* (probably lost; Oliveira, M. G., 1990, personal communication).
DISTRIBUTION. NEOTROPICAL REGION: Brazil Rio Grande do Sul.
HOST PLANTS. Leguminosae: *Erythrina cristagalli.*
REMARKS. Adult female redescribed by Hempel (1900b). Distribution and host plant records given by Hempel (1900b) and by Corseuil & Barbosa (1971).

Coccus formicarii (Green)

Lecanium formicarii Green, 1896: 10.
Lecanium globulosum Maskell, 1897b: 243. Syntypes female, HONG KONG: on *Stillingia sebifera* (NZAC, UCD). Syn. by Ferris, 1936: 14.
Lecanium (Saissetia) formicarii Green; Cockerell & Parrott, 1899: 164.
Saissetia formicarii (Green); Fernald, 1903: 202.
Coccus formicarii (Green); Mamet, 1954a: 13.
Taiwansaissetia formicarii (Green); Tao *et al.*, 1983: 77.
TYPE DATA. Syntypes female, SRI LANKA: Punduloya, on tea and undetermined shrubs (BMNH).
DISTRIBUTION. AUSTRO ORIENTAL REGION: Indonesia, Malaysia. **MADAGASIAN REGION:** Madagascar. **ORIENTAL REGION:** Hong Kong, India, Sri Lanka, Taiwan, Thailand. **PALAEARCTIC REGION:** Nepal.
HOST PLANTS. Anacardiaceae: *Mangifera indica, Rhus succedanea.* **Bischofiaceae:** *Bischofia javanica.* **Buxaceae:** *Buxus microphylla.* **Capparidaceae:** *Crataeva religiosa.* **Ebenaceae:** *Diospyros discolor, D. kaki.* **Elaeocarpaceae:** *Elaeocarpus.* **Ericaceae:** *Philippia.* **Euphorbiaceae:** *Macaranga Sapium sebiferum.* **Fagaceae:** *Lithocarpus.* **Guttiferae:** *Garcinia subelliptica.* **Lauraceae:** *Cinnamomum camphora, C. camphora, C. zeylanicum, Machilus, Persea americana.* **Leguminosae:** *Palaquium formosanum.* **Lythraceae:** *Lagerstroemia speciosa.* **Magnoliaceae:** *Michelia alba.* **Meliaceae:** *Aglaia odorata.* **Moraceae:** *Artocarpus heterophyllus, Ficus septica, F. vasculosa, F. wightiana.* **Myristicaceae:** *Myristica cagayanensis, Ardisia sieboldii.* **Myrtaceae:** *Eucalyptus, Eugenia javanica, Psidium guajava.* **Oleaceae:** *Olea europaea.* **Palmae:** *Areca catechu, Ptychosperma macarthurii.* **Proteaceae:** *Grevillea robusta.* **Rosaceae:** *Prunus.* **Rubiaceae:** *Cinchona, Gardenia jasminoides.* **Rutaceae:** *Aegle marmelos.* **Sapindaceae:** *Euphoria longana.* **Theaceae:** *Camellia japonica, C. sinensis, Gordonia axillaris.* **Verbenaceae:** *Callicarpa formosana.*
REMARKS. Adult female redescribed and illustrated by Green

(1904d), Ferris (1936), Takagi (1975), Tao *et al.* (1983) and by Tang (1991). Distribution and host plant records given by Green (1904d, 1937), Ferris, (1936), Takahashi (1942a, 1950a), Mamet (1954a), Ali (1971), Takagi (1975), Tao *et al.* (1983) and by Tang (1991).
BIOLOGY. Recorded in India and Sri Lanka only from nests and shelters of the ant *Crematogaster dohrni* (Green, 1937).

Coccus guerinii (Signoret)

Lecanium guerinii Signoret, 1869b: 96.
Coccus guerinii (Signoret); Mamet, 1943: 151.
TYPE DATA. Syntypes female, MAURITIUS: on sugarcane (probably lost).
DISTRIBUTION. MADAGASIAN REGION: Mauritius.
HOST PLANTS. Gramineae: *Saccharum officinarum.*
REMARKS. Signoret (1877a) did not include this species in his Catalogue. Fernald (1903) retained it in *Lecanium* until further study will clarify its proper generic placement.

Coccus gymnospori (Green)

Lecanium gymnospori Green, 1908: 29.
Coccus gymnospori (Green); Sanders, 1909b: 45.
Lecanium gymnosporiae Green; Lindinger, 1932f: 197 [UNJUSTIFIED EMENDATION].
TYPE DATA. Lectotype female designated by Ben-Dov (1981), INDIA: Poona, on *Gymnosporia montana* (BMNH).
DISTRIBUTION. ORIENTAL REGION: India.
HOST PLANTS. Celastraceae: *Gymnosporia montana.*
REMARKS. Adult female redescribed and illustrated by Ben-Dov (1981). Adult female redescribed by Tang (1991).

Coccus hesperidum Linnaeus

Coccus hesperidum Linnaeus, 1758: 455.
Calypticus hesperidum (Linnaeus); Costa, 1829: 8.
Calypticus laevis Costa, 1829: 11. Syntypes female, ITALY: on peach (probably lost; G. Pellizzari Scaltriti, 1990, personal communication). Syn. by Fernald, 1903: 169.
Lecanium hesperidum (Linnaeus); Burmeister, 1835: 69.
Coccus patellaeformis Curtis, 1843b: 517. Syntypes female, ENG-LAND: Location and host plant not indicated, in greenhouse (probably lost). Syn. by Fernald, 1903: 169.
Chermes lauri Boisduval, 1867: 340. Syntypes female, FRANCE: locality not clearly indicated, on *Laurus nobilis* (lost; D. Matile-Ferrero, personal communication, 1991). Syn. by Maskell, 1893a: 103.
Lecanium angustatus Signoret, 1873b: 398. Syntypes female, FRANCE: in greenhouse, on *Cyperus rotundus* (probably lost). Syn. by Sanders, 1909a: 436.
Lecanium lauri (Boisduval); Signoret, 1873b: 400.
Lecanium maculatum Signoret, 1873b: 400. Syntypes female, FRANCE: on *Hedera helix* (VMNH). Syn. by Ben-Dov, 1976b: 115.
Kermes aurantj Alfonso, 1875: 431. Syntypes female, ITALY: Sicily,

locality not given, on orange (probably lost) Syn. by Fernald,
1903: 169.
Lecanium alienum Douglas, 1886c: 77. Syntypes female, ENGLAND:
on *Asplenium bulbiferum* (BMNH). Syn. by Leonardi, 1920: 320.
Lecanium depressum simulans Douglas; Douglas, 1887a: 28
[NOMEN NUDUM].
Lecanium minimum Newstead, 1892: 141. Syntypes female, ENG-
LAND: on *Areca* sp. and *Abutilon* sp. (BMNH). Syn. by Boratynski
& Williams, 1964: 108.
Lecanium terminaliae Cockerell, 1893f: 254. Syntypes female,
JAMAICA: Kingston, on *Terminalia catappa* (USNM). Syn. by
Sanders, 1909a: 436.
Lecanium assimile amaryllidis Cockerell; Cockerell, 1893f: 254
[NOMEN NUDUM].
Lecanium assimile amaryllis Cockerell, 1894e: 19. Syntypes female,
JAMAICA: on *Amaryllis* (USNM). Syn. by Sanders, 1909a: 436.
Lecanium ceratoniae Gennadius, 1895: cclxxvii. Syntypes female,
CYPRUS: on carob (probably lost). Syn. by Fernald, 1903: 170.
Lecanium hesperidum lauri (Boisduval); Cockerell, 1896b: 331.
Lecanium nanum Cockerell, 1896e: 19. Syntypes female, TRINIDAD:
on leaves of "Balata" (USNM). Syn. by Sanders, 1909a: 436.
Lecanium minimum pinicola Maskell, 1897a: 310. Syntypes female,
SOUTH AFRICA: Cape of Good Hope, on *Pinus insignis* (NZAC,
USNM). Syn. by Sanders, 1909a: 436.
Lecanium flaveolum Cockerell, 1897b: 52. Syntypes female, U.S.A.:
New Mexico, Mesilla Park, on *Pilea* sp. (BMNH, USNM). Syn. by
Sanders, 1909a: 436.
Lecanium ventrale Ehrhorn, 1898a: 245. Syntypes female, U.S.A.:
California, San Jose, Japanese Nursery, on tuberous plant
(USNM). Syn. by Sanders, 1909a: 436.
Lecanium hesperidum alienum Douglas; Cockerell, 1899b: 393.
Lecanium (Calymnatus) hesperidum pacificum Kuwana, 1902b: 30.
Syntypes female, ECUADOR: Galapagos Islands, on 12 species of
plants (ITLJ). Syn. by Sanders, 1909a: 436.
Coccus (Lecanium) minimus Newstead; Cockerell, 1903b: 162.
Coccus angustatus (Signoret); Fernald, 1903: 168.
Coccus flaveolus (Cockerell); Fernald, 1903: 168.
Coccus patelliformis Curtis; Fernald, 1903: 169 [MIS-SPELLING].
Chermes aurantii (Alfonso); Fernald, 1903: 169 [MIS-SPELLING of
Kermes aurantj Alfonso].
Coccus hesperidum alienus (Douglas); Fernald, 1903: 171.
Coccus hesperidum lauri (Boisduval); Fernald, 1903: 171.
Coccus hesperidum pacificus (Kuwana); Fernald, 1903: 171.
Coccus minimus (Newstead); Fernald, 1903: 172.
Coccus minimus pinicola (Maskell); Fernald, 1903: 172.
Coccus maculatus (Signoret); Fernald, 1903: 172.
Coccus nanus (Cockerell); Fernald, 1903: 172.
Coccus terminaliae (Cockerell); Fernald, 1903: 173.
Coccus ventralis (Ehrhorn); Fernald, 1903: 174.
Eulecanium assimile amaryllidis (Cockerell); Fernald, 1903: 181.
Lecanium hesperidum minimum Newstead; Newstead, 1903: 85.
Lecanium signiferum Green, 1904d: 197. Syntypes female, SRI
LANKA: Pundaluoya, on *Begonia* sp. (BMNH, USNM). Syn. by
Boratynski & Williams, 1964: 108.

Lecanium punctuliferum Green, 1904d: 205. Syntypes female, SRI LANKA: Paradeniya, on *Michelia champaca* (BMNH, USNM). Syn. by De Lotto, 1959: 160.

Saissetia punctulifera (Green); Sanders, 1906: 10.

Coccus signiferus (Green); Sanders, 1906: 8.

Lecanium (Coccus) hesperidum Linnaeus; Pettit & McDaniel, 1920: 16.

Coccus (Lecanium) hesperidum Linnaeus; Hall, 1922: 18.

Coccus jungi Chen, 1936: 218. Syntypes female, CHINA: Huangyan County, Chekiang Province, on *Citrus* sp. (probably lost; Tang, F. T., 1989, personal communication). Syn. by Tang, 1991: 75.

Lecanium mauritiense Mamet, 1936a: 96. Syntypes female, MAURI-TIUS: Rose Hill and Ebene, on *Furcraea gigantea* (MNHN). Syn. by De Lotto, 1959: 160.

Lecanium (Coccus) hesperidum Linnaeus; Green, 1937: 298.

Lecanium (Coccus) signiferum Green; Green, 1937: 298.

Coccus mauritiensis (Mamet); Mamet, 1949: 25.

TYPE DATA. Syntypes female, EUROPE: probably on *Citrus*, *Laurus* and other host plants (lost; D. J. Williams, 1989, personal communication).

DISTRIBUTION. AUSTRO ORIENTAL REGION: Indonesia, Malaysia, Papua New Guinea, Philippines, Solomon Islands. **AUSTRALIAN REGION:** Queensland, South Australia, Western Australia. **MADAGASIAN REGION:** Aldabra Island, Mauritius, Reunion, Seychelles. **NEW ZEALAND and PACIFIC REGION:** American Samoa, Caroline Islands, Cook Islands, Fiji, French Polynesia, Hawaii, Irian Jaya, Kiribati, Mariana Islands, Marshall Islands, New Caledonia, New Zealand, Norfolk Island, Palaus, Tonga, Tuvalu, Vanuatu, Western Samoa. **ORIENTAL REGION:** Afghanistan, India, Sri Lanka, Taiwan, Thailand, Vietnam. **PALAEARCTIC REGION:** Belgium, Bulgaria, Canary Islands, Crete, Cyprus, Denmark, Egypt, England, France, Germany, Greece, Hungary, Iran, Iraq, Israel, Italy, Japan, Korea, Lebanon, Madeira, Portugal, Romania, Saudi Arabia, Spain, Sweden, Tibet, Turkey, USSR Crimea, USSR Georgia, Yugoslavia. **ETHIOPIAN REGION:** Angola, Cameroon, Cape Verde Island, Eritrea, Ethiopia, Kenya, Malawi, Nigeria, Sierra Leone, South Africa, Sudan, Tanzania, Uganda, Zaire, Zambia, Zanzibar, Zimbabwe. **NEOTROPICAL REGION:** Argentina, Bermuda, Brazil Rio Grande do Sul, Chile, Colombia, Cuba, Ecuador, Galapagos Islands, Guadeloupe, Guyana, Jamaica, Mexico, Panama, Puerto Rico, Salvador, Surinam, Trinidad, Virgin Islands. **NEARCTIC REGION:** California, Canada, Florida, Iowa, Maryland, Missouri, Tennessee, Texas, Virginia, Wyoming.

HOST PLANTS. Acanthaceae: *Hemigraphis palmata, Pseuderanthemum.* **Agavaceae:** *Agave americana, A. rigida variegata, Cordyline terminalis, Dracaena, Furcraea gigantea, Yucca.* **Amaryllidaceae:** *Clivia miniata, Hippeastrum equestre.* **Anacardiaceae:** *Mangifera indica, Pistacia palestina, Rhus succedanea, Schinus molle.* **Annonaceae:** *Annona muricata.* **Apocynaceae:** *Melodinus baueri,* **Apocynaceae:** *Nerium indicum, Nerium oleander, Plumeria acutifolia, P. rubra, Thevetia peruviana, Vinca major.* **Araceae:** *Anthurium acaule, Cryptocoryne, Philodendron pertusum, Scindapsus aureus, Zantedeschia.* **Araliaceae:** *Brassaia actinophylla,*

Fatsia, Schefflera. **Aspleniaceae:** *Asplenium nidum.* **Balsaminaceae:** *Impatiens.* **Barringtoniaceae:** *Barringtonia asiatica.* **Begoniaceae:** *Begonia.* **Berberidaceae:** *Berberis vulgaris, Mahonia aquifolium.* **Bignoniaceae:** *Bignonia unguiscati, Tecomaria capensis.* **Bombacaceae:** *Ceiba pentandra.* **Boraginaceae:** *Cordia alliodora.* **Bromeliaceae:** *Bilbergia.* **Burseraceae:** *Canarium indicum.* **Cannaceae:** *Canna.* **Caprifoliaceae:** *Lonicera.* **Caricaceae:** *Carica papaya.* **Casuarinaceae:** *Casuarina.* **Celastraceae:** *Euonymus japonicus.* **Combretaceae:** *Conocarpus erecta, Lumnitzera coccinea, Terminalia bellerica, T. brassii, T. calamansanay, T. catappa, T. chebula.* **Compositae:** *Pluchea odorata, Solidago, Tridax procumbens, Vernonia.* **Convolvulaceae:** *Ipomoea.* **Costaceae:** *Costus speciosus.* **Dioscoreaceae:** *Dioscorea.* **Dipterocarpaceae:** *Anisoptera thrifera.* **Ebenaceae:** *Diospyros kaki.* **Ehretiaceae:** *Cordia myxa, Ehretia petiolaris, E. silvatica.* **Elaeocarpaceae:** *Elaeocarpus.* **Ericaceae:** *Rhododendron.* **Euphorbiaceae:** *Acalypha. Bischofia javanica, Codiaeum, Glochidion puberum, Macaranga tanarius, Mallotus philippinensis, Phyllanthus, Ricinus communis.* **Flacourtiaceae:** *Dovyalis caffra, Hydnocarpus wightiana.* **Geraniaceae:** *Pelargonium.* **Goodeniaceae:** *Scaevola taccada.* **Gramineae:** *Bambusa vulgaris.* **Guttiferae:** *Calophyllum inophyllum.* **Hippocrateaceae:** *Salacia.* **Iridaceae:** *Gladiolus,* **Iridaceae:** *Iris, Moraea bicolor.* **Labiatae:** *Coleus blumei,* **Labiatae:** *Micromeria teneriffae, Thymus.* **Lauraceae:** *Cinnamomum camphora, Laurus nobilis, Persea americana, P. borbonia.* **Leguminosae:** *Acacia cyclops, Bauhinia purpurea, B. variegata, Begonia radicans, Canavalia macrocarpa, Cassia nodosa, Ceratonia siliqua, Dalbergia, Erythrina indica, Gliricidia, Lespedeza cuneata, Milletia nitida, Platypodium, Robinia pseudacacia, Sesbania sesban, Sophora chrysophylla, Spartocytisus filipes, Wisteria.* **Liliaceae:** *Aloe ciliaris, A. neglectus, Haworthia fasciata.* **Loranthaceae:** *Viscum cruciatum.* **Lythraceae:** *Lawsonia inermis.* **Magnoliaceae:** *Elmerrillia papuana, Magnolia grandiflora, Michelia alba, M. champaca.* **Malvaceae:** *Abutilon grandiflorum, Althaea rosae, Gossypium, Hibiscus manihot, H. sarabdiffa, Malvaviscus arboreus, Sida rhombifolia.* **Melastomataceae:** *Melastoma.* **Meliaceae:** *Amoora, Toona ciliata.* **Moraceae:** *Artocarpus altilis, A. communis, Ficus benghalensis, F. carica, F. obliqua, F. retusa, F. septica, F. sycomorus, F. theophrastoides, F. verrucocarpa, Moringa oleifera, Morus alba.* **Musaceae:** *Musa paradisiaca, M. sapientum.* **Myoporaceae:** *Myoporum.* **Myrsinaceae:** *Maesa lanceolata, Myrsine africana.* **Myrtaceae:** *Decaspermum, Eucalyptus deglupta, Eugenia jambolana, E. paniculata, E. pendula, Metrosideros, Myrtus communis, Psidium guajava.* **Nyctaginaceae:** *Bougainvillea spectabilis, Mirabilis jalapa, Pisonia grandis.* **Ochnaceae:** *Schuurmansia henningsii.* **Oleaceae:** *Ligustrum, Olea europaea.* **Oleandraceae:** *Nephrolepis exaltata.* **Onagraceae:** *Fuchsia.* **Orchidaceae:** *Stanhopea, Vanda.* **Palmae:** *Areca oleracea, Chamaedorea, Cocos nucifera, Phoenix dactylifera.* **Peperomiaceae:** *Peperomia.* **Phytolaccaceae:** *Phytolacca dioica.* **Pinaceae:** *Cedrus, Pinus caribaea, P. elliotti, P. halepensis, P. radiata.* **Platanaceae:** *Platanus orientalis.* **Polygalaceae:** *Polygala virgata.* **Polypodiaceae:** *Adiantum capillus-veneris, Platycerium alcicorne.*

Primulaceae: *Cyclamen persicum.* **Proteaceae:** *Finschia.* **Pteridaceae:** *Acrostichum aureum.* **Rosaceae:** *Amygdalus communis, Cliffortia nitidula, Cotoneaster pannosa, Cydonia oblonga, Eriobotrya japonica, Malus sylvestris, Persica vulgaris, Prunus armeniaca, Prunus salicina, Pyracantha, Raphiolepis umbellata, Rosa, Rubus.* **Rubiaceae:** *Canthium, Coprosma montana, Gardenia jasminoides, Ixora, Oxyanthus speciosus, Platanocephalus morindaefolius, Timonius.* **Rutaceae:** *Citrus aurantium, C. grandis, C. limon, C. paradisi, C. reticulata, Fortunella margarita, Pelea.* **Salicaceae:** *Populus alba.* **Santalaceae:** *Santalum haleakale.* **Sapindaceae:** *Euphoria longana,* **Sapindaceae:** *Litchi chinensis.* **Sapotaceae:** *Calocarpum, Pometia pinnata.* **Saxifragaceae:** *Hydrangea integerrima.* **Selanginellaceae:** *Selanginella.* **Solanaceae:** *Capsicum frutescens, Datura metel, Lycium, Solanum villosum.* **Sterculiaceae:** *Dombeya, Theobroma cacao.* **Styracaceae:** *Styrax officinalis.* **Theaceae:** *Camellia japonica, Camellia sinensis.* **Tiliaceae:** *Grewia.* **Verbenaceae:** *Callicarpa formosana, Clerodendrum inerme, Duranta repens, Premna corymbosa.* **Vitidaceae:** *Rhoicissus tridentata, Vitis vinifera.* **Zingiberaceae:** *Alpinia mutica, Curcuma longa, Hedychium coronarium, Nicolaia speciosa.*

REMARKS. Common Name - Brown soft scale. Adult female redescribed and illustrated by Tyrrell (1896), Hempel (1900b), Thro (1903), Newstead (1903), Green (1904d), Kuwana (1917a), Leonardi (1920), Pettit & McDaniel (1920), Steinweden (1930), Gómez-Menor Ortega (1937, 1958c), Borchsenius (1937, 1957), Zimmerman (1948), Gomes Costa (1949), Fonseca (1953), De Lotto (1959), Hodgson (1967a), Ezzat & Hussein (1969), Williams & Kosztarab (1972), Fernandes (1973), Öncüer (1974), Gill *et al.* (1977), Paik (1978), Wang (1980), Kawai (1980), Tereznikova (1981), Tao *et al.* (1983), Hamon & Williams (1984), Gill (1988), Tang & Li (1988), Gonzalez (1989) and by Williams & Watson (1990).

Adult male described and illustrated by Giliomee (1967). Adult male described by Newstead (1917a). The redescription and illustration, as *Calymnatus hesperidum,* by Valemberg (1980), appears to be a misidentification of a *Coccus* sp., different from *C. hesperidum.* First instar larva described and illustrated by Gómez-Menor Ortega (1937, 1958c) and by Borchsenius (1957). Larval instars described and illustrated by Fonseca (1953). Colour photograph of the adult female given by Chapot & Delucchi (1964), Delucchi (1975), Kawai (1980), Hamon & Williams (1984), Stimmel (1987), Gill (1988) and by Tranfaglia & Viggiani (1988). Foldi (1978) and Waku & Foldi (1984) described the ultrastructure of wax secreting glands in the adult female. Nur (1979) studied the cytology and reported chromosome number 2n=14 in U. S. A. Parida & Ghosh (1984) and Moharana (1990) reported chromosome number 2n=14 in India. Intraspecific variation in 18 characters of the adult female studied by Blair *et al.* (1964). Annecke (1966) reported that the female moults only twice before reaching maturity. However, this might be an erroneous observation, because in other species of *Coccus* the female are known to develop through three larval instars. Distribution and host plant records given by Tyrrell (1896), Kuwana (1902b), Green (1904d, 1907), Lindinger (1911b), Bodkin (1914), Brain (1920b), Pettit & McDaniel (1920), Hall (1922), Bodenheimer (1924, 1926a,

1928, 1944a, 1944b, 1951), Balachowsky (1932b, 1935, 1957),
Mamet (1936a, 1943), Gomes Costa (1949), Takahashi (1950a,
1952), Schmutterer (1952a), Ossiannilsson (1959), De Lotto (1959,
1965, 1967a), Beardsley (1966a), Hodgson (1967a, 1969b), Almeida
(1969, 1973), Ali (1971, 1973), Ben-Dov (1971), Corseuil & Barbosa
(1971), Williams & Kosztarab (1972), Gill et al. (1977), Wise (1977),
Kozár et al. (1979), Lambdin & Watson (1980), Tereznikova (1981),
Nakahara & Miller (1981), Wang (1981), Nakahara (1983), Podsiadlo
(1983), Hadzibejli (1983), Tao et al. (1983), Vieira et al. (1983),
Martin Mateo (1984), Matile-Ferrero (1984b), Matile-Ferrero & Nonveiller (1984), Kozár (1985), Carnero Hernandez & Perez Guerra
(1986), Waite (1986), Varshney & Moharana (1987), Fernandes
(1987), Marotta (1987), Stimmel (1987), Tang & Li (1988), Williams &
Williams (1988), Williams & Watson (1990), Danzig & Konstantinova
(1990), Tang (1991) and by Hodgson & Hilburn (1991a, 1991b).
BIOLOGY. Female reproduces parthenogenetically. Males have not
been observed in California, Israel and South Africa, however they
were reported from England (Newstead, 1917a) and Russia (Saakyan-Baranova, 1964). Bodenheimer (1951b) presented in great
detail the biology, natural enemies and economic importance. The
encapsulation of parasitoid eggs was studied by Blumberg (1977)
and Blumberg & DeBach (1981). The encapsulation of eggs of the
parasitoids Encyrtus infelix (Embleton) and E. lecaniorum (Mayr) was
determined by Blumberg & Goldenberg (1992). Hart & Ingle (1971)
observed increased fecundity of the female after exposure to Methyl
Parathion.
ECONOMIC IMPORTANCE. This is one of the most cosmopolitan
species. Generally it is not a serious pest, except on ornamental or
greenhouse plants. It was reported at times, as a noxious pest of
citrus in Texas, California and South Africa, following the application
of non-selective insecticides which disrupted the activity of natural
enemies of the soft scale (Bartlett, 1978).

Coccus hesperidum javanensis (Newstead)

Lecanium hesperidum javanensis Newstead, 1908c: 38.
Coccus hesperidum javanensis (Newstead); Sanders, 1909b: 45.
TYPE DATA. Syntypes female, JAVA: East Java, Molio-ardjo, on
Liberian coffee (BMNH).
DISTRIBUTION. AUSTRO ORIENTAL REGION: Java.
HOST. Rubiaceae: Coffea liberica. .lm0"
REMARKS. Sanders (1909b) erroneously placed Lecanium (Trechoco-
rys) hesperidum africanum Newstead NOMEN NUDUM, as a
synonym of this species.

Coccus illuppalamae (Green)

Lecanium illuppalamae Green, 1922b: 1021.
Coccus illuppalamae (Green); Ali, 1971: 25.
TYPE DATA. Syntypes female, SRI LANKA: Maha Illuppalama, on
undetermined tree (BMNH).
DISTRIBUTION. ORIENTAL REGION: Sri Lanka.
REMARKS. Adult female redescribed by Tang (1991).

Coccus incisus (King)

Calymnatus incisus King in Cockerell, 1902i: 255.
Coccus incisus (King); Sanders, 1906: 8.
TYPE DATA. Syntypes female, SOUTH AMERICA (Country not indicated): on nutmeg [-*Torreya nucifera*] (depository not traced).
DISTRIBUTION. NEOTROPICAL REGION: South America.
HOST PLANTS. Taxaceae: *Torreya nucifera*.
REMARKS. Cockerell (1902i) indicated that this species is close to *Milviscutulus mangiferae*.

Coccus inquilinum (Newstead) n. comb.

Lecanium inquilinum Newstead, 1920: 189.
Myzolecanium inquilinum (Newstead); Lindinger, 1957: 544.
TYPE DATA. Syntypes female, GUYANA: "Cattle Trail Survey", host plant not recorded, enclosed in small paper nests constructed by ants of *Acromyrmex* sp. (BMNH).
DISTRIBUTION. NEOTROPICAL REGION: Guyana.

Coccus insolens (King)

Lecanium insolens King in Cockerell, 1902i: 255.
Coccus insolens (King); Silva et al., 1968: 143.
TYPE DATA. Syntypes female, BRAZIL: Locality not indicated, on *Philodendron* sp. (depository not traced).
DISTRIBUTION. NEOTROPICAL REGION: Brazil.
HOST PLANTS. Araceae: *Philodendron*.

Coccus inyangombae Hodgson

Coccus inyangombae Hodgson, 1967a: 5.
TYPE DATA. Holotype female, ZIMBABWE: Inyangombe Falls, Inyanga, on *Strychnos lugens* (SANC).
DISTRIBUTION. ETHIOPIAN REGION: Zimbabwe.
HOST PLANTS. Loganiaceae: *Strychnos lucens*.

Coccus jaculator (Green & Laing) n. comb.

Lecanium jaculator Green & Laing, 1924: 418.
TYPE DATA. Syntypes female, GUYANA: East Coast of Demerara, Belfield, on *Montrichardia aculeata* (BMNH).
DISTRIBUTION. NEOTROPICAL REGION: Guyana.
HOST PLANTS. Araceae: *Montrichardia aculeata*.

Coccus kosztarabi Avasthi & Shafee

Coccus kosztarabi Avasthi & Shafee, 1983: 389.
TYPE DATA. Holotype female, INDIA: Mysore, Tumkur, on *Mangifera indica* (AMUI).
DISTRIBUTION. ORIENTAL REGION: India.
HOST PLANTS. Anacardiaceae: *Mangifera indica*.

Coccus latioperculatum (Green)

Lecanium latioperculatum Green, 1922b: 1022.

Lecanium latioperculum Green; Ramakrishna Ayyar; 1930: 30 [MIS-SPELLING].
Coccus latioperculatum (Green); Ali, 1971: 26.
TYPE DATA. Syntypes female, SRI LANKA: Paradeniya, on undetermined shrub (BMNH).
DISTRIBUTION. ORIENTAL REGION: India, Sri Lanka.
HOST PLANTS. Anacardiaceae: *Anacardium occidentale*, *Mangifera indica*.
REMARKS. Distribution and host plant records given by Shafee *et al.* (1989).
BIOLOGY. Attended by the ant *Oecophylla smaragdina* (Green, 1922).

Coccus leurus De Lotto

Coccus leurus De Lotto, 1966a: 43.
TYPE DATA. Holotype female, SOUTH AFRICA: Natal, Umkomaas, on *Putterlickia verrucosa* (SANC).
DISTRIBUTION. ETHIOPIAN REGION: South Africa.
HOST PLANTS. Celastraceae: *Putterlickia verrucosa*.

Coccus lidgetti (Fernald) n. comb.

Lecanium australis Lidgett, 1901: 59 [HOMONYM of *Lecanium australe* Walker].
Lecanium lidgetti Fernald; Fernald, 1903: 212 [REPLACEMENT NAME].
TYPE DATA. Syntypes female, AUSTRALIA: Victoria, Myrniong, on *Acacia implexa* (BMNH).
DISTRIBUTION. AUSTRALIAN REGION: Victoria.
HOST PLANTS. Leguminosae: *Acacia implexa*.

Coccus litzeae Rutherford

Coccus litzeae Rutherford, 1915a: 111.
Lecanium litseae (Rutherford); Green, 1937: 305 [MIS-SPELLING].
Coccus litseae Rutherford; Ali, 1971: 26 [MIS-SPELLING].
TYPE DATA. Syntypes female and male, SRI LANKA: Paradeniya, on *Litsea longifolia* (BMNH).
DISTRIBUTION. ORIENTAL REGION: Sri Lanka.
HOST PLANTS. Lauraceae: *Litsea longifolia*.

Coccus lizeri (Fonseca) n. comb.

Lecanium lizeri Fonseca, 1957: 133.
TYPE DATA. Syntypes female, BRAZIL: Sao Paulo, Campinas, on coffee (MZSP).
DISTRIBUTION. NEOTROPICAL REGION: Brazil Sao Paulo.
HOST PLANTS. Rubiaceae: *Coffea*.

Coccus longulus (Douglas)

Lecanium angustatum Signoret; Douglas, 1887a: 25 [MISIDENTIFICATION].
Lecanium longulum Douglas, 1887b: 97.
Lecanium chirimoliae Maskell, 1890a: 137. Syntypes female, FIJI: on

Annona tripetala (probably lost; Deitz & Tocker, 1980). Syn. by Cockerell, 1893g: 50.
Lecanium ficus Maskell, 1897b: 243. Syntypes female, CHINA: Swatow, on *Ficus* sp. (NZAC). Syn. by Sanders, 1909a: 438.
Coccus longulum (Douglas); Kirkaldy, 1902: 106.
Coccus ficus (Maskell); Fernald, 1903: 168.
Coccus longulus (Douglas); Fernald, 1903: 171.
Lecanium frontale Green, 1904d: 192. Lectotype female designated by Ben-Dov (1977), SRI LANKA [=CEYLON]: Pundaluoya, on *Calophyllum* sp. (BMNH). Syn. by Sanders, 1909a: 438.
Coccus frontalis (Green); Sanders, 1906: 8.
Coccus elongatus (Signoret); Sanders, 1909a: 438 [ERRONEOUS SYNONYMIZATION].
Lecanium (Coccus) celtium Kuwana, 1909b: 162. Syntypes female, JAPAN: Ogasawara Islands [=Bonin Islands], on *Celtis sinensis* (ITLJ). Syn. by Takahashi, 1955b: 69.
Coccus celtium Kuwana; Sasscer, 1911: 66.
Lecanium wistariae Brain, 1920b: 8. Syntypes female, SOUTH AFRICA: Cape Province, Uitenhage, on *Wistaria* [=*Wisteria*] sp. (SANC). Syn. by De Lotto, 1957b: 301.
Lecanium (Coccus) longulus (Douglas); Pettit & McDaniel, 1920: 17.
Coccus (Lecanium) longulus Douglas; Hall, 1922: 19.
Lecanium kraunhiarum Lindinger, 1928: 107 [REPLACEMENT NAME for *Lecanium wistariae* Brain].
Lecanium (Coccus) frontale Green; Green, 1937: 299.
Coccus frontalis (Green); Mamet, 1943: 151.
Coccus celticum Kuwana; Takahashi, 1955b: 69 [MIS-SPELLING].
Parthenolecanium wistaricola Borchsenius, 1957: 349 [REPLACEMENT NAME for *Lecanium wistariae* Brain].
Coccus longulus (Douglas); Ben-Dov, 1977: 89 [RESTORED STATUS].
Coccus logulus (Douglas); Moharana, 1990: 48 [MIS-SPELLING].
TYPE DATA. Lectotype female designated by Ben-Dov (1977), ENGLAND: Harrow, on *Acacia catechu* (BMNH).
DISTRIBUTION. AUSTRO ORIENTAL REGION: Papua New Guinea, Philippines, Solomon Islands. **AUSTRALIAN REGION:** New South Wales, Northern Territory, Queensland, South Australia. **ETHIOPIAN REGION:** Kenya, South Africa, Uganda, Zimbabwe. **MADAGASIAN REGION:** Agalega Island, Madagascar, Mauritius, Seychelles. **NEOTROPICAL REGION:** Bermuda, Colombia, Cuba, Ecuador, Honduras, Mexico, Panama, Puerto Rico, Virgin Islands. **NEW ZEALAND and PACIFIC REGION:** Caroline Islands, Cook Islands, Fiji, French Polynesia, Gilbert Islands, Guam, Hawaii, Kiribati, Mariana Islands, Marshall Islands, New Caledonia, New Zealand, Palaus, Samoa, Tonga, Vanuatu, Western Samoa. **NEARCTIC REGION:** Alabama, California, Delaware, Florida, Georgia, Illinois, Louisiana, Maryland, Massachusetts, Missouri, New York, Ohio, Pennsylvania, Texas. **ORIENTAL REGION:** India, Ogasawara Islands, Sri Lanka, Taiwan, Thailand. **PALAEARCTIC REGION:** Canary Islands, China, Cyprus, Egypt, England, Germany, Israel, Lebanon, Saudi Arabia.
HOST PLANTS. Agavaceae: *Agave sisalana, Cordyline terminalis, Dracaena.* **Anacardiaceae:** *Mangifera indica.* **Annonaceae:** *Annona muricata, A. reticulata, A. squamosa, Cananga odorata.*

Apocynaceae: *Ervatamia orientalis, Plumeria.* **Araceae:** *Anthurium, Caladium, Colocasia esculenta, Cyrtosperma chamissonis, Dieffenbachia picta, Epipremnum pinnatum, Spathiphyllum lancefolium.* **Araliaceae:** *Schefflera.* **Begoniaceae:** *Begonia.* **Bignoniaceae:** *Tabebuia heterophylla.* **Boraginaceae:** *Cordia.* **Cannaceae:** *Canna.* **Caricaceae:** *Carica papaya.* **Casuarinaceae:** *Casuarina equisetifolia.* **Cobaeaceae:** *Cobaea scandens.* **Commelinaceae:** *Zebrina pendula.* **Compositae:** *Cosmos, Senecio, Wedelia biflora.* **Cucurbitaceae:** *Cucurbita pepo.* **Erythroxylaceae:** *Erythroxylum, E. coca.* **Euphorbiaceae:** *Acalypha tricolor, Aleurites fordii, A. moluccana, A. triloba, Codiaeum variegatum, Croton, Euphorbia pulcherrima, Excoecaria agallocha, Jatropha curcas, Pedilanthus.* **Geraniaceae:** *Pelargonium.* **Gramineae:** *Bambusa.* **Guttiferae:** *Calophyllum.* **Labiatae:** *Coleus.* **Lauraceae:** *Persea americana, Persea borbonia.* **Leguminosae:** *Acacia catechu, A. culturiformis, A. longifolia, A. melanoxylon, A. mellifera, A. simplex, Albizia falcata, Al. lebbeck, Arachis hypogaea, Bauhinia, Caesalpinia decapetala, C. pulcherrima, Cajanus cajan, C. flavus, Cassia grandis, Ceratonia siliqua, Dalbergia assamica, Delonix regia, Derris, Desmodium umbellatum, Dolichos, Gliricidia maculata, G. sepium, Glycine hispida, Inocarpus fagifer, Leucaena glauca, L. leucocephala, Mimosa pudica, Mucuna, Phaseolus puearia, Pithecellobium dulce, P. saman, Pterocarpus indicus, Pueraria thunbergiana, Tephrosia candida, Vigna unquiculata, Wisteria.* **Malvaceae:** *Hibiscus tiliaceus, Malvastrum tricuspidatum, Malvaviscus arboreus.* **Moraceae:** *Artocarpus heterophyllus, A. integra, Ficus benjamina, F. boninsimae, F. carica, F. lyrata, F. megapoda, F. retusa, F. rubiginosa, F. tinctoria, Morus alba.* **Musaceae:** *Musa.* **Myrtaceae:** *Metrosideros, Psidium guajava.* **Nyctaginaceae:** *Bougainvillea.* **Oleaceae:** *Ligustrum.* **Oxalidaceae:** *Averrhoa carambola.* **Palmae:** *Archontophoenix cunninghami, Areca catechu, Chamaerops humilis, Cocos nucifera, Ptychosperma sanderianum.* **Pinaceae:** *Pinus caribaea.* **Plumbaginaceae:** *Plumbago capensis.* **Proteaceae:** *Finschia.* **Rhizophoraceae:** *Bruguiera sexangula.* **Rosaceae:** *Rosa, Rubus.* **Rubiaceae:** *Coffea arabica, C. canephora, Gardenia, Ixora.* **Rutaceae:** *Boninia glabra, Citrus limon, C. reticulata.* **Sapindaceae:** *Litchi chinensis.* **Solanaceae:** *Cestrum.* **Sterculiaceae:** *Theobroma cacao.* **Strelitziaceae:** *Strelitzia.* **Theaceae:** *Camellia.* **Ulmaceae:** *Celtis boninensis, C. sinensis.* **Urticaceae:** *Boehmeria boninensis.* **Vitidaceae:** *Vitis vinifera.*

REMARKS. Common Name - Long brown scale. Ben-Dov (1977) showed that the earliest available name for this species is *Lecanium longulum* Douglas and not *Lecanium elongatum* Signoret. The erroneous senior synonymy of *Lecanium elongatum* Signoret over *Lecanium longulum* was introduced by Sanders (1909a). *Lecanium elongatum* Signoret is a junior synonym of *Parthenolecanium persicae* (Fabicius). Records of *Coccus elongatus* (Signoret) - by authors until 1977 - were misidentifications of this species. Adult female redescribed and illustrated by Thro (1903), Green (1904d), Froggatt (1915), Kuwana (1917a), Pettit & McDaniel (1920), Zimmerman (1948), Ezzat & Hussein (1969), Ben-Dov (1977), Gill *et al.* (1977), Kawai (1980), Tao *et al.* (1983), Hamon & Williams (1984), Gill (1988), Williams & Watson (1990) and by Tang (1991). Colour photograph given by

Kawai (1980), Hamon & Williams (1984) and by Gill (1988). Intraspecific variation in taxonomic characters given by Ben-Dov (1977). Parida & Moharana (1982) and Moharana (1990) reported chromosome number 2n=18 in India. Distribution and host plant records given by Green (1907, 1908), Kuwana (1909b), Pettit & McDaniel (1920), Brain (1920b), Hall (1922, 1935), Morrison (1929), Takahashi (1942a), Strickland (1947), Mamet (1954a, 1962, 1978), De Lotto (1957b), Brookes (1964), Beardsley (1966a), Hodgson (1967a), Ali (1971), Georghiou (1977), Ben-Dov (1977), Gill et al. (1977), Wise (1977), Nakahara & Miller (1981), Nakahara (1981b), Tao et al. (1983), Hamon & Williams (1984), Carnero Hernandez & Perez Guerra (1986), Varshney & Moharana (1987), Matile-Ferrero (1988), Williams & Williams (1988), Williams & Watson (1990), Tang (1991) and by Hodgson & Hilburn (1991a, 1991b).

BIOLOGY. Females reproduce parthenogenetically in Egypt, Israel and in California. El-Minshawy & Moursi (1976) studied the duration of development and fecundity on guava, in Egypt.

Coccus lumpurensis Takahashi

Coccus lumpurensis Takahashi, 1952: 12.
TYPE DATA. Syntypes female, MALAYSIA: Kuala Lumpur, on *Ficus* sp. (SMKM).
DISTRIBUTION. AUSTRO ORIENTAL REGION: Malaysia.
HOST PLANTS. Moraceae: *Ficus*.
REMARKS. Adult female redescribed by Tang (1991).

Coccus macarangae Morrison

Coccus macarangae Morrison, 1921: 663.
TYPE DATA. Holotype female, SINGAPORE: Selangor Forest, in hollow stems of *Macaranga* sp. (USNM).
DISTRIBUTION. AUSTRO ORIENTAL REGION: Singapore.
HOST PLANTS. Euphorbiaceae: *Macaranga*.
REMARKS. Adult female redescribed by Tang (1991).

Coccus macarangicolus Takahashi

Coccus macarangicolus Takahashi, 1952: 14.
TYPE DATA. Syntypes female, MALAYSIA: Kuala Lumpur, on *Macaranga triloba* (SMKM).
DISTRIBUTION. AUSTRO ORIENTAL REGION: Malaysia.
HOST PLANTS. Euphorbiaceae: *Macaranga triloba*.
REMARKS. Adult female redescribed by Tang (1991).
BIOLOGY. Found in hollow on the stem, associated with *Crematogaster* sp. (Takahashi, 1952).

Coccus malloti (Takahashi)

Pulvinaria malloti Takahashi, 1956: 25.
Coccus malloti (Takahashi); Kawai, 1980: 144.
TYPE DATA. Syntypes female, JAPAN: Kyoto, Tokyo and Yokohama, on *Euonymus oxyphilla*, *Aphananthe aspera*, *Mallotus japonicus*, *Pittosporum tobira*, *Illicium religiosum* and *Cornus controversa* (HUSJ).
DISTRIBUTION. PALAEARCTIC REGION: Japan.

HOST PLANTS. Celastraceae: *Euonymus oxyphilla.* **Cornaceae:** *Cornus controversa.* **Euphorbiaceae:** *Mallotus japonicus.* **Illiciaceae:** *Illicium religiosum.* **Pittosporaceae:** *Pittosporum tobira.* **Ulmaceae:** *Aphananthe aspera.*
REMARKS. Adult female redescribed by Kawai (1980) and by Tang (1991). Colour photograph given by Kawai (1980).

Coccus melaleucae (Maskell)

Lecanium melaleucae Maskell, 1898: 239.
Coccus melaleucae (Maskell); Fernald, 1903: 172.
TYPE DATA. Syntypes female, AUSTRALIA: New South Wales, Palmer Island, Clarence River, on *Melaleuca* sp. (NZAC).
DISTRIBUTION. AUSTRALIAN REGION: New South Wales.
HOST PLANTS. Myrtaceae: *Melaleuca.*
REMARKS. Adult female redescribed and illustrated by Froggatt (1915). King (1902b) reported this species as introduced into Massachusetts, but this needs verification.

Coccus milanjianus Hodgson

Coccus milanjianus Hodgson, 1968b: 114.
TYPE DATA. Holotype female, MALAWI: Mount Mlanje (7000 feet), on *Tarenna pavettoides* (BMNH).
DISTRIBUTION. ETHIOPIAN REGION: Malawi.
HOST PLANTS. Rubiaceae: *Tarenna pavettoides.*

Coccus moestus De Lotto

Coccus moestus De Lotto, 1959: 164.
TYPE DATA. Holotype female, ZANZIBAR: on clove tree (BMNH).
DISTRIBUTION. ETHIOPIAN REGION: Kenya, Zanzibar. **NEOTROPICAL REGION:** Costa Rica, Guyana, Haiti, Jamaica, Puerto Rico, Trinidad. **NEW ZEALAND and PACIFIC REGION:** Caroline Islands, Guam, Palaus, Truk Islands, Vanuatu. **ORIENTAL REGION:** Ogasawara Islands. **PALAEARCTIC REGION:** China, Japan.
HOST PLANTS. Anacardiaceae: *Anacardium occidentale, Mangifera indica.* **Aspleniaceae:** *Neottopteris nidus.* **Euphorbiaceae:** *Drypetes integerrima.* **Goodeniaceae:** *Scaevola.* **Lauraceae:** *Persea americana.* **Malvaceae:** *Hibiscus glabra.* **Moraceae:** *Artocarpus altilis.* **Myrtaceae:** *Eugenia aromatica.*
REMARKS. Adult female redescribed and illustrated by Beardsley (1966a) and by Williams & Watson (1990). Adult female redescribed by Kawai (1980) and by Tang (1991). Distribution and host plant records given by Beardsley (1966a), Kawai *et al.* (1971), Gill *et al.* (1977), Kawai (1980), Nakahara &Miller (1981), Williams & Watson (1990) and by Tang (1991).

Coccus muiri Kotinsky

Coccus tuberculatus Kotinsky, 1908a: 168 [HOMONYM of *Coccus tuberculatus* Bouche].
Coccus muiri Kotinsky, 1908b: 37 [REPLACEMENT NAME].
TYPE DATA. Syntypes female, SINGAPORE: on leaves of undetermined tree (BP).

DISTRIBUTION. AUSTRO ORIENTAL REGION: Singapore.
HOST PLANTS. Cruciferae: *Brassica actinophylla.* **Rubiaceae:** *Gardenia.*
REMARKS. Adult female redescribed by Takahashi (1952) and by Tang (1991). Distribution and host plant records given by Takahashi (1952) and by Tang (1991).

Coccus murex Hodgson

Coccus murex Hodgson, 1969b: 6.
TYPE DATA. Holotype female, ZIMBABWE: Chimanimani Mountains, near Outward Bound School, on *Brachystegia tamarinoides* (BMNH).
DISTRIBUTION. ETHIOPIAN REGION: Zimbabwe.
HOST PLANTS. Leguminosae: *Brachystegia tamarinoides.*

Coccus nyika Hodgson

Coccus nyika Hodgson, 1970a: 35.
TYPE DATA. Holotype female, MALAWI: Chilinda Bridge on the Nyika Plateau, on *Myrica salicifolia* (BMNH).
DISTRIBUTION. ETHIOPIAN REGION: Malawi.
HOST PLANTS. Myricaceae: *Myrica salicifolia.*

Coccus ophiorrhizae (Green)

Lecanium ophiorrhizae Green, 1896: 10.
Coccus ophiorrhizae (Green); Fernald, 1903: 173.
TYPE DATA. Syntypes female, SRI LANKA: Punduloya, on *Ophiorrhiza pectinata* (BMNH).
DISTRIBUTION. ORIENTAL REGION: India, Sri Lanka.
HOST PLANTS. Rubiaceae: *Ophiorrhiza pectinata.*
REMARKS. Adult female redescribed and illustrated by Green (1904d). Adult female redescribed by Tang (1991). Distribution and host plant records given by Ali (1971) and by Tang (1991).

Coccus opimum (Green) n. comb.

Lecanium opimum Green, 1913: 313.
TYPE DATA. Syntypes female, JAVA: Samarang, on *Cassia fistula* (BMNH).
DISTRIBUTION. AUSTRO ORIENTAL REGION: Java.
HOST PLANTS. Leguminosae: *Cassia fistulata.*

Coccus paradeformosum (Fonseca) n. comb.

Lecanium paradeformosum Fonseca, 1975: 79.
TYPE DATA. Syntypes female, BRAZIL: Rio Grande do Sul, Porto Alegre, on *Glycine hispida* (MZSP).
HOST PLANTS. Leguminosae: *Glycine hispida.*

Coccus penangensis Morrison

Coccus penangensis Morrison, 1921: 657.
TYPE DATA. Syntypes female and larva, MALAYSIA: Penang Island, in hollow stems of *Macaranga triloba* (USNM).
DISTRIBUTION. AUSTRO ORIENTAL REGION: Malaysia.

HOST PLANTS. Euphorbiaceae: *Macaranga triloba.*

Coccus perlatus (Cockerell)

Lecanium perlatum Cockerell, 1898b: 65.
Coccus perlatus (Cockerell); Fernald, 1903: 173.
TYPE DATA. Syntypes female, AZORES: Ponta Delgado, on orange trees (BMNH, USNM).
DISTRIBUTION. NEOTROPICAL REGION: Argentina. **PALAEARCTIC REGION:** Azores Islands.
HOST PLANTS. Rutaceae: *Citrus.*
REMARKS. Adult female redescribed by Borchsenius (1957) and by Tang (1991).

Coccus piperis namunakuli (Green)

Lecanium piperis namunakuli Green, 1922b: 1024.
Coccus namunakuli (Green); Ali, 1971: 27.
Coccus piperis namunakuli (Green); Varshney, 1985: 26.
TYPE DATA. Syntypes female, SRI LANKA: Namunakuli Hill, Badulla, on *Piper* sp. (BMNH).
DISTRIBUTION. ORIENTAL REGION: Sri Lanka.
HOST PLANTS. Piperaceae: *Piper.*

Coccus pseudelongatus (Brain)

Lecanium pseudelongatum Brain, 1920b: 6.
Coccus pseudelongatus (Brain); De Lotto, 1957b: 305.
TYPE DATA. Syntypes female, SOUTH AFRICA: Transvaal, Pretoria, on *Acacia caffra* (SANC).
DISTRIBUTION. ETHIOPIAN REGION: South Africa.
HOST PLANTS. Leguminosae: *Acacia caffra.*
REMARKS. Adult female redescribed and illustrated by De Lotto (1957b).

Coccus pseudohesperidum (Cockerell)

Lecanium pseudohesperidum Cockerell, 1895h: 381.
Coccus pseudohesperidum (Cockerell); Fernald, 1903: 173.
Coccus pseadohesperidum (Cockerell); Tang, 1991: 95 [MIS-SPELLING].
TYPE DATA. Syntypes female, CANADA: Ottawa (in greenhouse), on *Cattleya* sp. (USNM).
DISTRIBUTION. NEOTROPICAL REGION: Brazil, Guatemala. **NEW ZEALAND and PACIFIC REGION:** Hawaii. **NEARCTIC REGION:** California, Canada Ottawa, Florida, Indiana, Maryland, Massachusetts, Missouri, New Hampshire, New Jersey, New York, North Carolina, Pennsylvania, Washington, D. C. **PALAEARCTIC REGION:** England, Latvia.
HOST PLANTS. Iridaceae: *Iris.* **Orchidaceae:** *Cattleya mossiae, C. skinneri, C. trianca, Cymbidium, Epidendrum, Laelia anceps, L. purpurata, Octomeria, Oncidium, Phalaenopsis, Vanda teres, Vanilla.*
REMARKS. Adult female redescribed and illustrated by Zimmerman (1948), Borchsenius (1957), Gill *et al.* (1977), Hamon & Williams (1984) and by Gill (1988). Adult female redescribed by Tang (1991).

Colour photograph given by Gill (1988). Distribution and host plant records given by Zimmerman (1948), Borchsenius (1957), Gill *et al.* (1977), Nakahara (1981b), Hamon & Williams (1984) and by Gill (1988).

BIOLOGY. The species has been recorded almost exclusively on various species of the Orchidaceae. There is one exceptional record from the Iridaceae, in North Carolina.

ECONOMIC IMPORTANCE. Steinweden (1945) reported it as a pest of orchids in California, however, Gill *et al.* (1977) indicated that chemical insecticides effectively control it.

Coccus pseudomagnoliarum (Kuwana)

Lecanium (Eulecanium) pseudomagnoliarum Kuwana, 1914: 7.
Coccus citricola Campbell, 1914: 222. Syntypes female, U.S.A.: California, Claremont, on *Citrus* sp. (USNM). Syn. by Clausen, 1923: 225.
Lecanium pseudomagnoliarum Kuwana; Sasscer, 1915: 32.
Coccus aegaeus De Lotto, 1973: 291. Holotype female, TURKEY: Izmir, on *Citrus* sp. (SANC). Syn. by Tranfaglia, 1976: 129.

TYPE DATA. Syntypes female, JAPAN: Tokyo and Shizuoka, on *Citrus* (ITLJ).

DISTRIBUTION. NEARCTIC REGION: Arizona, California, Maryland. **NEOTROPICAL REGION:** Mexico. **PALAEARCTIC REGION:** France, Greece, Iran, Israel, Italy, Japan, Korea, Turkey, USSR Azerbaijan, USSR Georgia, USSR Krasnodar.

HOST PLANTS. Lauraceae: *Laurus nobilis.* **Rutaceae:** *Citrus aurantium, Citrus limon, C. paradisi, Evodia rutaecarpa, Poncirus trifoliata.* **Ulmaceae:** *Celtis australis, Zelkova serrata.*

REMARKS. Common Name - grey citrus scale. Adult female redescribed and illustrated by Kuwana (1917a), Steinweden (1930), Borchsenius (1957), De Lotto (1973) (as *C. aegaeus*), Tranfaglia (1974) (as *C. aegaeus*), Öncüer (1974), Gill *et al.* (1977), Paik (1978), Hadzibejli (1983) and by Gill (1988). Adult female redescribed by Tang (1991). First instar larva described and illustrated by Borchsenius (1957). Colour photograph given by Delucchi (1975), Kawai (1980), Gill (1988) and by Tranfaglia & Viggiani (1988). Distribution and host plant records given by Kuwana (1917a), Clausen (1923), Takahashi (1955c), Borchsenius (1957), De Lotto (1973), Öncüer (1974), Tranfaglia (1974, 1976), Gill *et al.* (1977), Paik (1978), Kawai (1980), Ben-Dov (1980), Hadzibejli (1983), Marotta (1987), Gill (1988), Marotta & Tranfaglia (1990) and by Tang (1991).

BIOLOGY. Develops one annual generation in California (Quayle, 1915), Israel (Ben-Dov, 1980) and in Greece (Argyriou & Ioanides, 1975). Females reproduce parthenogenetically.

ECONOMIC IMPORTANCE. Has been a serious pest of citrus in Arizona and California (Ebeling, 1959), but became less injurious in recent years (Gill *et al.*, 1977). A minor pest of citrus in Israel (Ben-Dov, 1980) and in Italy (Longo & Russo, 1986). Kennett (1988) reported on exploration for parasites in Japan.

Coccus pumilum (Brain) n. comb.

Lecanium pumilum Brain, 1920b: 5.

TYPE DATA. Syntypes female, SOUTH AFRICA: Cape Province, Robertson, on stem of a native shrub (SANC).
DISTRIBUTION. ETHIOPIAN REGION: South Africa.
HOST PLANTS. Bruniaceae: *Berzelia lanuginosa.*
REMARKS. Male described and illustrated by Durr (1954). Distribution and host plant records given by Durr (1954).
BIOLOGY. This coccid was found under a carton shelter constructed by ants (Durr, 1954).

Coccus ramakrishnai (Ramakrishna Ayyar)

Lecanium ramakrishnae Ramakrishna Ayyar, 1919: 35.
Coccus ramakrishnai (Ramakrishna Ayyar); Varshney, 1985: 26.
TYPE DATA. Syntypes female, INDIA: Kothapetta, Godavari District, on *Ficus benghalensis* (BMNH).
DISTRIBUTION. ORIENTAL REGION: India.
HOST PLANTS. Moraceae: *Ficus benghalensis.*
REMARKS. Ramakrishna Ayyar (1919) credited this species to Green, as a manuscript name, however, according to International Code (1985), the authorship must be credited to Ramakrishna Ayyar. The original description (Ramakrishna Ayyar, 1919), and the redescription (Ramakrishna Ayyar, 1930), were very brief, but included data and illustration which provided for the availability of this species.

Coccus resinatum (Kieffer & Herbst) n. comb.

Lecanium resinatum Kieffer & Herbst, 1909: 122.
TYPE DATA. Syntypes female and first instar larva, CHILE: on *Baccharis rosmarinifolius* (DEPOSITORY UNKNOWN).
DISTRIBUTION. NEOTROPICAL REGION: Chile.
HOST PLANTS. Compositae: *Baccharis rosmarinifolia.*
REMARKS. The original description, though available, is too general as to give enough characters for the correct generic placement of this species. Consequently, the species is here assigned to *Coccus* for technical reasons, in order to remove it from the officially-rejected genus *Lecanium.*
BIOLOGY. This insect causes the formation of galls on stems of the host plant. First instar crawlers appear in January. The galls are attended by ants, as well as by Sphegidae and Pompilidae (Kieffer & Herbst, 1909).

Coccus rhodesiensis (Hall)

Lecanium rhodesiensis Hall, 1935: 76.
Coccus rhodesiensis De Lotto, 1959: 170.
TYPE DATA. Syntypes female, ZIMBABWE [=RHODESIA]: El Dorado and Victoria Falls, on undetermined plant (BMNH).
DISTRIBUTION. ETHIOPIAN REGION: Mozambique, Zimbabwe.
HOST PLANTS. Olacaceae: *Ximenia americana.*
REMARKS. Adult female redescribed and illustrated by De Lotto (1959). Distribution and host plant records given by De Lotto (1959) and by Hodgson (1967a, 1969b).

Coccus rubellus (Cockerell)

Lecanium rubellum Cockerell, 1893h: 378.
Coccus rubellus (Cockerell); Fernald, 1903: 173.
TYPE DATA. Syntypes female, JAMAICA: Westmoreland, on unde-
termined plant (probably lost; Gill et al., 1979).
DISTRIBUTION. NEOTROPICAL REGION: Jamaica.
REMARKS. Gill et al. (1977) erroneously regarded this species as a
Nomen Dubium, because they were unable to locate type material.

Coccus saltuarius Hodgson

Coccus saltuarius Hodgson, 1968b: 116.
TYPE DATA. Holotype female, MALAWI: Mount Mlanje (7000 feet),
on Pterocelastrus echinatus (BMNH).
DISTRIBUTION. ETHIOPIAN REGION: Malawi.
HOST PLANTS. Celastraceae: Pterocelastrus echinatus.

Coccus schini (Cockerell)

Lecanium schini Cockerell, 1893i: 325.
Lecanium schini Cockerell; Cockerell, 1893j: 167.
Coccus schini (Cockerell); Fernald, 1903: 173.
TYPE DATA. Syntypes female, MEXICO: Guanajuato, on Schinus
molle (USNM).
DISTRIBUTION. NEOTROPICAL REGION: Mexico.
HOST PLANTS. Anacardiaceae: Schinus molle.
REMARKS. Gill et al. (1977) noted that this species does not belong
to Coccus and its generic placement is uncertain.

Coccus secretus Morrison

Coccus secretus Morrison, 1921: 662.
TYPE DATA. Holotype female, MALAYSIA: Penang Island, in hollow
stem of Macaranga triloba (USNM).
DISTRIBUTION. AUSTRO ORIENTAL REGION: Malaysia, Singa-
pore.
HOST PLANTS. Euphorbiaceae: Macaranga triloba.

Coccus sectilis De Lotto

Coccus sectilis De Lotto, 1966a: 46.
TYPE DATA. Holotype female, SOUTH AFRICA: Cape Province, Cape
of Good Hope, on Maytenus oleoides (SANC).
DISTRIBUTION. ETHIOPIAN REGION: South Africa.
HOST PLANTS. Celastraceae: Maytenus oleoides.

Coccus smaragdinus De Lotto

Coccus smaragdinus De Lotto, 1965: 193.
TYPE DATA. Holotype female, KENYA: Nairobi, on Strychnos sp.
(BMNH).
DISTRIBUTION. ETHIOPIAN REGION: Kenya.
HOST PLANTS. Loganiaceae: Strychnos.

Coccus sociabilis Hodgson

Coccus sociabilis Hodgson, 1969b: 8.
TYPE DATA. Holotype female, ZIMBABWE [=RHODESIA]: Juliasdale, Rodel Farm, on *Acacia* sp. (BMNH).
DISTRIBUTION. ETHIOPIAN REGION: Zimbabwe.
HOST PLANTS. Leguminosae: *Acacia.*

Coccus sordidus De Lotto

Coccus sordidus De Lotto, 1957b: 308.
TYPE DATA. Holotype female, KENYA: Nairobi, on *Afrormosia angolensis* (BMNH).
DISTRIBUTION. ETHIOPIAN REGION: Kenya.
HOST PLANTS. Leguminosae: *Afrormosia angolensis.*

Coccus stipulaeformis Haworth

Coccus stipulaeformis Haworth, 1812: 308.
TYPE DATA. Syntypes female, INDIA: taken from specimens of trees in the Herbarium of J. Banks (lost; See Williams, 1957).
DISTRIBUTION. ORIENTAL REGION: India.
REMARKS. De Lotto (1971c) discussed the original description, but gave no indication as to the generic placement or specific identity of this species.

Coccus subacutus (Newstead)

Lecanium subacutum Newstead, 1920: 187.
Coccus subacutus (Newstead); De Lotto, 1957b: 308.
TYPE DATA. Syntypes female, UGANDA: Lake Victoria, Jana and Sesse Islands, on *Coffea robusta,* and on undetermined plant in Bufumira and Sesse Islands (BMNH).
DISTRIBUTION. ETHIOPIAN REGION: Uganda.
HOST PLANTS. Rubiaceae: *Coffea robusta.*
REMARKS. Adult female redescribed and illustrated by De Lotto (1957b).

Coccus subhemisphaericus (Newstead)

Lecanium (Saissetia) subhemisphaericum Newstead, 1917a: 363.
Coccus subhemisphaericus (Newstead); De Lotto, 1957b: 310.
TYPE DATA. Syntypes female, UGANDA: Naguriga, Chagwe, on coffee, and GHANA: Aburi, on coffee (BMNH).
DISTRIBUTION. ETHIOPIAN REGION: Angola, Ghana, Ivory Coast, Kenya, Uganda, Zanzibar.
HOST PLANTS. Myrtaceae: *Eugenia aromatica.* **Rubiaceae:** *Bertiera racemosa, Coffea arabica, C. canephora, C. robusta.*
REMARKS. Adult female redescribed and illustrated by De Lotto (1957b). Distribution and host plant records given by De Lotto (1957b, 1967a), Almeida (1973) and by Couturier *et al.* (1985).

Coccus synapheae (Froggatt) n. comb.

Lecanium synapheae Froggatt, 1915: 613.
TYPE DATA. Syntypes female and male, AUSTRALIA: Western Australia, Boyanup, on *Synaphea petiolaris* (BCRI).

DISTRIBUTION. AUSTRALIAN REGION: Western Australia.
HOST PLANTS. Proteaceae: *Synaphea petiolaris*.
REMARKS. This species is here temporarily assigned to *Coccus* for technical reasons, in order to remove it from the officially-rejected genus *Lecanium*. Penny Gullan and Ting Kui Qin kindly informed me, on March 1992, that this species does not belong to *Coccus*.

Coccus takanoi Takahashi

Coccus takanoi Takahashi, 1932a: 45.
TYPE DATA. Syntypes female, TAIWAN: Shinka, on *Saccharum officinarum* (IMZT).
DISTRIBUTION. ORIENTAL REGION: Taiwan.
HOST PLANTS. Gramineae: *Saccharum officinarum*.
REMARKS. Takahashi (1932a) indicated the affinity of this species to *Saccharolecanium krugeri* (Zehntner).

Coccus tangandae Hodgson

Coccus tangandae Hodgson, 1967a: 6.
TYPE DATA. Holotype female, ZIMBABWE: Tanganda Halt, on *Markhamia acuminata* (SANC).
DISTRIBUTION. ETHIOPIAN REGION: Zimbabwe.
HOST PLANTS. Bignoniaceae: *Markhamia acuminata*. **Oleaceae:** *Schrebera alata*.

Coccus tenebricophilum (Green) n. comb.

Lecanium tenebricophilum Green, 1904b: 204.
TYPE DATA. Syntypes female, INDONESIA: Java, Bogor (Botanic Gardens), on *Erythrina lithosperma* (BMNH).
DISTRIBUTION. AUSTRO ORIENTAL REGION: Java.
HOST PLANTS. Leguminosae: *Erythrina lithosperma*.
REMARKS. This species is here assigned to *Coccus* for technical reasons, in order to remove it from the officially-rejected genus *Lecanium*.
BIOLOGY. Found within tunnels formed in branches of the tree by some boring insect. The coccids are entirely concealed, attached to the tunnel walls and always attended by ants (Green, 1904b).

Coccus tumuliferus Morrison

Coccus tumuliferus Morrison, 1921: 655.
TYPE DATA. Syntypes female and larva, SINGAPORE: in hollow stems of *Macaranga hypolema* (USNM).
DISTRIBUTION. AUSTRO ORIENTAL REGION: Singapore.
HOST PLANTS. Euphorbiaceae: *Macaranga hypolema*.

Coccus viridis (Green)

Lecanium viride Green, 1889: 248.
Coccus viridis (Green); Fernald, 1903: 174.
Lecanium (Trechocorys) hesperidum africanum Newstead; Newstead, 1906: 74 [NOMEN NUDUM].
Lecanium (Coccus) viride Green; Green, 1937: 299.
Coccus viridis viridis Köhler; Köhler, 1978: 564 [NOMEN NUDUM].

Coccus viridis bisexualis Köhler; Köhler, 1978: 564 [NOMEN NUDUM].
TYPE DATA. Lectotype female designated by Williams & Watson (1990), SRI LANKA: Pundaluoya, on coffee (BMNH).
DISTRIBUTION. AUSTRO ORIENTAL REGION: Indonesia, Papua New Guinea, Philippines, Solomon Islands. **ETHIOPIAN RE-GION:** Angola, Cameroon, Ghana, Guinea, Ivory Coast, Kenya, Nigeria, Principe, Sao Tome, Sierra Leone, South Africa, Tanzania, Uganda, Zanzibar. **MADAGASIAN REGION:** Agalega Island, Comoros, Madagascar, Mauritius, Reunion, Rodrigues, Seychelles. **NEOTROPICAL REGION:** Bermuda, Brazil Rio Grande do Sul, Colombia, Cuba, Dominican Republic, Guadeloupe, Guyana, Honduras, Jamaica, Panama, Puerto Rico, Virgin Islands. **NEW ZEALAND and PACIFIC REGION:** Cook Islands, Fiji, French Polynesia, Futuna Island, Guam, Hawaii, Irian Jaya, Kiribati, Mariana Islands, Nauru, New Caledonia, Palaus, Samoa, Tahiti, Tonga, Tuvalu, Vanuatu, Western Samoa. **NEARCTIC REGION:** Florida. **ORIENTAL REGION:** Cambodia, India, Ogasawara Islands, Sri Lanka, Taiwan, Thailand, Vietnam. **PALAEARCTIC REGION:** Madeira.
HOST PLANTS. Acanthaceae: *Odontonema , Sanchezia nobilis.* **Agavaceae:** *Cordyline terminalis, Dracaena.* **Amaranthaceae:** *Gomphrena globosa.* **Anacardiaceae:** *Campnosperma brevipetiolata, Mangifera indica, Schinus molle, S. terebinthifolius.* **Apocynaceae:** *Alstonia macrophylla, Alyxia olivaeformis, Carissa grandiflora, Nerium oleander, Ochrosia nakaiana, Plumeria acutifolia, P. obtusa, P. rubra, P. tricolor, Rauwolfia vomitoria.* **Aquifoliaceae:** *Ilex chinensis, I. macrothyrsa.* **Araceae:** *Caladium.* **Araliaceae:** *Meryta macrophylla, Polyscias guilfoylei.* **Barringtoniaceae:** *Barringtonia speciosa.* **Bignoniaceae:** *Tecomaria capensis.* **Boraginaceae:** *Cordia alliodora.* **Bromeliaceae:** *Ananas comosus.* **Celastraceae:** *Maytenus.* **Combretaceae:** *Terminalia catappa.* **Commelinaceae:** *Commelina.* **Compositae:** *Arctotis, Fitchia, Gerbera, Gerbera jamesonii, Pluchea indica, Senecio.* **Crassulaceae:** *Bryophyllum.* **Cucurbitaceae:** *Cucurbita pepo.* **Dioscoreaceae:** *Dioscorea.* **Ehretiaceae:** *Cordia.* **Euphorbiaceae:** *Carissa carandas, Codiaeum, Croton.* **Goodeniaceae:** *Scaevola taccada.* **Guttiferae:** *Mammea americana.* **Hydrangaceae:** *Hydrangea.* **Lauraceae:** *Persea americana.* **Leguminosae:** *Cassia, Gliricidia, Inocarpus fagifer, Tipuana.* **Melastomataceae:** **Meliaceae:** *Melia azedarach.* **Moraceae:** *Ficus elastica.* **Myristicaceae:** *Myristica.* **Myrsinaceae:** *Ardisia crispa.* **Myrtaceae:** *Eucalyptus, Eugenia, Melaleuca, Myrtella, Psidium guajava, P. littorale, P. pyriferum.* **Nyctaginaceae:** *Ceodes umbellifera.* **Orchidaceae:** *Broughtonia.* **Palmae:** *Areca catechu, Cocos nucifera.* **Pandanaceae:** *Pandanus.* **Periplocaceae:** *Cryptostegia grandiflora.* **Pittosporaceae:** *Pittosporum tobira.* **Polygonaceae:** *Coccoloba uvifera.* **Rubiaceae:** *Bobea mauaii, Borreria laevis, Canthium odoratum, Cinchona calisaya, Coffea arabica, C. canephora, C. liberica, C. robusta, Gardenia jasminoides, G. taitensis, Ixora macrothyrsa, Morinda citrifolia, Platanocephalus chinensis, P. morindaefolius, Psychotria boninensis, Randia tahitensis, Timonius.* **Rutaceae:** *Aegle marmelos, Boninia grisea, Citrus aurantifolia, C. aurantium, C. grandis, C. histrix, C. limon,*

*C. paradisi, Citrus reticulata, Citrus sinensis, Clausena excavata,
Poncirus trifoliata, Triphasia trifolia,* **Sapindaceae:** *Dodonaea
eriocarpa, Litchi chinensis,* **Sapotaceae:** *Manilkara zapota,
Palaquium formosanum, Planchonella, Pouteria obovata.* **Solanaceae:** *Cestrum.* **Sterculiaceae:** *Ileritiera littoralis, Theobroma
cacao.* **Theaceae:** *Camellia sinensis.* **Umbelliferae:** *Apium grave-
olans.* **Verbenaceae:** *Clerodendrum, Lantana camara.* **Zingiberaceae:** *Alpiniapurpurata, Zingiber officinale.*

REMARKS. Common name - Green scale or Green coffee scale. Adult
female redescribed and illustrated by Hempel (1900b), Green
(1904d), Morrison (1920), Zimmerman (1948), Gomes Costa (1949),
De Lotto (1960, 1978), Gill *et al.* (1977), Kawai (1980), Wang (1980),
Tao *et al.* (1983), Hamon & Williams (1984), Williams & Watson
(1990) and by Tang (1991). Colour photograph of adult female given
by Kawai (1980), Hamon & Williams (1984). Adult male described
and illustrated by Köhler (1976). De Lotto (1959) placed *Lecanium
(Trechocorys) hesperidum africanum* Newstead NOMEN NUDUM, as a
synonym of *C. viridis,* after examining the "types" in BMNH. On the
other hand Sanders (1909b) erroneously placed this NOMEN
NUDUM as a synonym of *Coccus hesperidum javanensis* (Newstead).
Distribution and host plant records given by Newstead (1917c),
Takahashi (1942a), Gomes Costa (1949), Balachowsky (1957),
Mamet (1959b, 1978), De Lotto (1957b, 1960), Castel-Branco (1963),
Beardsley (1966a), Ali (1971), Corseuil & Barbosa (1971), Kawai *et
al.* (1971), Almeida (1973), Köhler (1976, 1978), Takagi (1977), Gill *et
al.* (1977), Matile-Ferrero (1978), Nakahara & Miller (1981), Tao *et al.*
(1983), Nakahara (1983), Vieira *et al.* (1983), Matile-Ferrero & Non-
veiller (1984), Couturier *et al.* (1985), Varshney & Moharana (1987),
Fernandes (1987), Williams & Williams (1988), Shafee *et al.* (1989),
Williams & Watson (1990), Danzig & Konstantinova (1990), Tang
(1991) and by Hodgson & Hilburn (1991a, 1991b).

BIOLOGY. De Lotto (1960) observed in East Africa that this species
occurs in coastal areas and up to an altitude of 1 - 1.3 km. Köhler
(1976, 1978) studied the life history in Cuba, observing males at low
frequency that was correlated with population density. The females
are parthenogenetic and ovoviviparous. In Florida females mature in
50-70 days in late summer months (Fredrick, 1943). The scales
infest branches, shoots and leaves; on leaves usually on the undersurface.

ECONOMIC IMPORTANCE. A serious pest of coffee, citrus and other
crops in several regions in the tropics. A major citrus pest in Bolivia,
whereas a minor pest in other South American and South East Asian
countries (Talhouk, 1975). Biology and pest status on coffee presented by Le Pelley (1968). Natural enemies on coffee in Cuba were
studied by Köhler (1980).

Coccus viridulus De Lotto

Coccus viridulus De Lotto, 1960: 399.
TYPE DATA. Holotype female, KENYA: Nandi Hills (6100 feet), on
Coffea arabica (BMNH).
DISTRIBUTION. ETHIOPIAN REGION: Kenya, Uganda.
HOST PLANTS. Rubiaceae: *Coffea arabica.*

REMARKS. Distribution and host plant records given by De Lotto (1969a).
BIOLOGY. Biology and pest status on coffee presented by Le Pelley (1968).

Columnea Targioni Tozzetti

Columnea Targioni Tozzetti, 1866b: 142.
Columna Signoret; Signoret, 1877a: 658 [MIS-SPELLING].
Columella Sulc; Sulc, 1936: 65 [MIS-SPELLING].
TYPE-SPECIES: *Coccus caricae* Fabricius [=*Ceroplastes rusci* (L.)], by subsequent restriction of Targioni Tozzetti (1867).
REMARKS. This genus is here regarded as a subjective synonym of *Ceroplastes*. If *Ceroplastes* should be split into smaller generic units, *Columnea* might be an available name.

Conofilippia Brain

Conofilippia Brain, 1920b: 25.
TYPE-SPECIES: *Conofilippia subterranea* Brain, by original designation and monotypy.

Conofilippia subterranea Brain

Conofilippia subterranea Brain, 1920b: 25.
TYPE DATA. Syntypes female, SOUTH AFRICA: Transvaal, Pretoria District, De Wildt, on roots of a native shrub (SANC).
DISTRIBUTION. ETHIOPIAN REGION: South Africa.

Couturierina Matile-Ferrero & Le Ruyet

Couturierina Matile-Ferrero & Le Ruyet, 1985: 260.
TYPE-SPECIES: *Couturierina piptadeniastrae* Matile-Ferrero & Le Ruyet, by original designation and monotypy.

Couturierina piptadeniastrae

atile-Ferrero & Le Ruyet
Couturierina piptadeniastrae Matile-Ferrero & Le Ruyet, 1985: 261.
TYPE DATA. Holotype female, IVORY COAST: Tai, on *Piptadeniastrum africanum* (MNHN).
DISTRIBUTION. ETHIOPIAN REGION: Ivory Coast.
HOST PLANTS. Leguminosae: *Piptadeniastrum africanum*.
BIOLOGY. Attended by ants (Matile-Ferrero & Le Ruyet, 1985).

Cribrolecanium Green

Cribrolecanium Green, 1921c: 639.
Cribrolebanium Tang; Tang, 1991: 99 [MIS-SPELLING].
TYPE-SPECIES: *Cribrolecanium formicarum* Green, by original designation.
REMARKS. Generic characters discussed by Hodgson (1990) and by Tang (1991).

Cribrolecanium andersoni (Newstead)

Akermes andersoni Newstead, 1917a: 347.

Cribrolecanium andersoni (Newstead); De Lotto, 1968b: 83.
Parakermes andersoni (Newstead); Fonseca, 1973: 247.
TYPE DATA. Syntypes female, KENYA: Kabete, on orange leaves (BMNH).
DISTRIBUTION. ETHIOPIAN REGION: Angola, Cameroon, Kenya, Mozambique, South Africa, Swaziland, Zambia, Zimbabwe. **MADAGASIAN REGION:** Mauritius.
HOST PLANTS. Anacardiaceae: *Mangifera indica*. **Boraginaceae:** *Ehretia silvatica*. **Meliaceae:** *Toona ciliata*. **Moraceae:** *Ficus elastica, F. verrucocarpa*. **Myrtaceae:** *Callistemon, Psidium guajava*. **Passifloraceae:** *Passiflora edulis*. **Rubiaceae:** *Psychotria zombamontana*. **Rutaceae:** *Citrus*. **Sterculiaceae:** *Theobroma cacao*.
REMARKS. Adult female redescribed and illustrated by De Lotto (1965) and by Almeida (1969). Colour photograph given by Brink & Bruwer (1989). Distribution and host plant records given by De Lotto (1965, 1968b), Hodgson (1969b), Almeida (1969, 1973), Matile-Ferrero & Nonveiller (1984), Williams & Williams (1988), Brink & Bruwer (1989) and by Brink & Hewitt (1991).
BIOLOGY. Brink & Hewitt (1991) discussed the effect of various mortality factors on populations of the pest in South Africa.
ECONOMIC IMPORTANCE. A citrus pest in some citrus-producing areas of South Africa and Swaziland (Brink & Bruwer, 1989; Brink & Hewitt, 1991).

Cribrolecanium formicarum Green

Cribrolecanium formicarum Green, 1921c: 639.
TYPE DATA. Syntypes female, SRI LANKA: Paradeniya, on *Stereospermum chelonioides* (BMNH).
DISTRIBUTION. ORIENTAL REGION: Sri Lanka.
HOST PLANTS. Bignoniaceae: *Stereospermum chelonioides*.
REMARKS. Adult female redescribed and illustrated by Tang (1991).
BIOLOGY. Lives in hollow branches of its host plant and attended by *Crematogaster* sp. (Green, 1921c).

Cribrolecanium radicicola Green

Cribrolecanium radicicola Green, 1921c: 642.
Cribrolebanium radicicola Tang, 1991: 99 [MIS-SPELLING].
TYPE DATA. Syntypes female, INDIA: Coimbatore, on *Cassia* sp. (BMNH).
DISTRIBUTION. ORIENTAL REGION: India.
HOST PLANTS. Leguminosae: *Cassia*.
REMARKS. Adult female redescribed and illustrated by Tang (1991).
BIOLOGY. Living on roots of its host plant (Green, 1921c).

Cryptes Maskell

Cryptes Maskell, 1892a: 21.
Cryptes Crawford; Fernald, 1903: 209 [ERRONEOUS AUTHORSHIP].
Cryptes Crawford; Froggatt, 1915: 614 [ERRONEOUS AUTHORSHIP].
Cryptes Cockerell & Parrott; Morrison & Morrison, 1922: 80 [ERRONEOUS AUTHORSHIP].

TYPE-SPECIES: *Lecanium baccatum* Maskell, by subsequent designation of Cockerell & Parrott (1899).
REMARKS. Morrison & Morrison (1922) discussed the authorship of this genus, showed that it cannot be referred to Crawford and credited it to Cockerell & Parrott. Morrison & Morrison (1966), while discussing the issue again, were less decisive and proposed several alternatives. The genus is here credited to Maskell, because this case is similar to that of *Aspidoproctus* (Opinion 268, 1954) and to several other generic names in the Coccoidea which are credited to the first author who published it. Generic characters discussed by Froggatt (1915), Morrison & Morrison (1922) and by Steinweden (1929).

Cryptes baccatus (Maskell)

Lecanium baccatum Maskell, 1892a: 20.
Kermes maskelli Maskell, 1892a: 21. Syntypes female, AUSTRALIA: New South Wales, Melbourne and Adelaide, on *Acacia armata, A. calamifolia* and *A. longifolia* (NZAC). Syn. by community of type-series.
Kermes acaciae Maskell, 1894c: 81. Syntypes female, AUSTRALIA: New South Wales, Sydney, on *Acacia* sp. (NZAC) Syn. by Lindinger, 1943b: 148.
Cryptes baccatus (Maskell); Cockerell & Parrott, 1899: 161.
TYPE DATA. Syntypes female, AUSTRALIA: New South Wales, Melbourne and Adelaide, on *Acacia armata, A. calamifolia* and *A. longifolia* (NZAC).
DISTRIBUTION. AUSTRALIAN REGION: New South Wales, South Australia, Western Australia.
HOST PLANTS. Leguminosae: *Acacia armata, A. calamifolia, A. decurrens, A. linearis, A. longifolia, A. melanoxylon, A. pendula.*
REMARKS. Adult female redescribed and illustrated by Froggatt (1915) and by Morrison & Morrison (1922).

Cryptes baccatus marmoreus (Fuller)

Lecanium baccatum marmoreum Fuller, 1897b: 1345.
Cryptes baccatus marmoreus (Fuller); Fernald, 1903: 209.
TYPE DATA. Syntypes female, AUSTRALIA: Western Australia, Geraldton, on *Acacia* sp. (Probably lost; P. Gullan, 1990, personal communication).
DISTRIBUTION. AUSTRALIAN REGION: Western Australia.
HOST PLANTS. Leguminosae: *Acacia.*
REMARKS. Adult female redescribed by Fuller (1899).

Cryptinglisia Cockerell

Cryptinglisia Cockerell, 1900a: 173.
TYPE-SPECIES: *Cryptinglisia lounsburyi* Cockerell, by monotypy.
REMARKS. Generic characters discussed by De Lotto (1978).

Cryptinglisia lounsburyi Cockerell

Cryptinglisia lounsburyi Cockerell, 1900a: 173.
Inglisia geranii Brain, 1920b: 37. Syntypes female, SOUTH AFRICA: Cape Province, King Williamstown, on *Geranium* sp. (SANC) Syn. by De Lotto, 1978: 142

TYPE DATA. Syntypes female, SOUTH AFRICA: Cape Province, Constantia, on roots of *Vitis vinifera* (BMNH, USNM).
DISTRIBUTION. ETHIOPIAN REGION: South Africa, Zimbabwe, Italy.
HOST PLANTS. Geraniaceae: *Geranium, Pelargonium peltatum. Vitis vinifera.*
REMARKS. Adult female redescribed and illustrated by Brain (1920b) and by De Lotto (1970a) (as *I. geranii* Brain). Adult female redescribed by Tranfaglia & Marotta (1982). Colour photograph given by Tranfaglia & Marotta (1982), Tranfaglia & Viggiani (1988) and by Tremblay (1988b). Distribution and host plant records given by Brain (1920b), Hall (1935), Hodgson (1967d), De Lotto (1970a, 1978), Tranfaglia & Marotta (1982) and by Marotta (1987).

Cryptostigma Ferris

Cryptostigma Ferris, 1922: 160.
TYPE-SPECIES: *Cryptostigma ingae* Ferris [=*Cryptostigma inquilina* (Newstead)], by original designation and monotypy.
REMARKS. Generic characters discussed by Steinweden (1929), Qin & Gullan (1989), Williams & Watson (1990) and by Hodgson (1990). Key to species given by Qin & Gullan (1989).

Cryptostigma biorbiculus Morrison

Cryptostigma biorbiculus Morrison, 1929: 48.
TYPE DATA. Holotype female, PANAMA: Canal Zone, Ancon, on *Cordia alliodora* (USNM).
DISTRIBUTION. NEOTROPICAL REGION: Panama.
HOST PLANTS. Ehretiaceae: *Cordia alliodora.*
BIOLOGY. Attended by the ants *Pseudomyrma sericea* and *Azteca longiceps* (Morrison, 1929).

Cryptostigma endoeucalyptus Qin & Gullan

Cryptostigma endoeucalyptus Qin & Gullan, 1989: 226.
TYPE DATA. Holotype female, AUSTRALIA: New South Wales, Monga, On *Eucalyptus viminalis* (ANIC).
DISTRIBUTION. AUSTRALIAN REGION: Australian Capital Territory, New South Wales.
HOST PLANTS. Myrtaceae: *Eucalyptus mannifera, E. viminalis.*
REMARKS. First instar larva described by Qin & Gullan (1989).
BIOLOGY. Lives in galleries of *Crematogaster* sp. and *Iridomyrmex* sp., within trunks and branches of the trees (Qin & Gullan, 1989).

Cryptostigma inquilina (Newstead)

Pseudophilippia inquilina Newstead, 1920: 181.
Akermes secretus Morrison, 1922: 145. Holotype female, PUERTO RICO: on *Inga laurina* (USNM). Syn. by Qin & Gullan, 1989: 225.
Cryptostigma ingae Ferris, 1922: 160. Holotype female, PUERTO RICO: on *Inga laurina* (UCD). Syn. by Qin & Gullan, 1989: 225.
Cryptostigma secretus (Morrison); Morrison, 1929: 53.
Cryptostigma inquilina (Newstead); Qin & Gullan, 1989: 225.
Cryptostigma jamaicensis (Newstead); Qin & Gullan, 1989: 225 [NOMEN NUDUM].

Lecanopsis jamaicensis Newstead; Qin & Gullan, 1989: 225 [NOMEN NUDUM].

TYPE DATA. Lectotype female designated by Qin & Gullan (1989), JAMAICA: at banks of Great River, near Montpelier, on undetermined tree (BMNH).

DISTRIBUTION. NEOTROPICAL REGION: Jamaica, Panama, Puerto Rico, Virgin Islands.

HOST PLANTS. Ehretiaceae: *Cordia alliodora*. **Leguminosae:** *Inga laurina*.

REMARKS. Distribution and host plant records given by Morrison (1922, 1929), Ferris (1922), Nakahara & Miller (1981), Nakahara (1983) and by Qin & Gullan (1989).

BIOLOGY. Living on bark of the host plant beneath a blackish shelter constructed by the ant *Crematogaster brevispinosa* Mayr (Newstead, 1920). Attended by the ant *Azteca longiceps* in Panama (Morrison, 1929).

Cryptostigma magnetinsulae Qin & Gullan

Cryptostigma magnetinsulae Qin & Gullan, 1989: 231.

TYPE DATA. Holotype female, AUSTRALIA: Queensland, Magnetic Island, under loose bark of undetermined tree (BMNH).

DISTRIBUTION. AUSTRALIAN REGION: Queensland.

Cryptostigma quinquepori (Newstead)

Akermes quinquepori Newstead, 1917a: 349.
Cryptostigma quinquepori (Newstead); Morrison, 1929: 50.
Cryptostigma bunzlii Green, 1933: 57. Holotype female, SURINAM: on *Erythrina* sp. (BMNH). Syn. by Qin & Gullan, 1989.

TYPE DATA. Syntypes female, GUYANA: Georgetown, on *Microlobium acaciaefolium* (BMNH).

DISTRIBUTION. NEOTROPICAL REGION: Guyana, Surinam.

HOST PLANTS. Leguminosae: *Erythrina, Microlobium acaciaefolium*. **Polygonaceae:** *Triplaris surinamensis*. **Urticaceae:** *Cecropia*.

REMARKS. Distribution and host plant records given by Morrison (1929), Green (1933) and by Qin & Gullan (1989).

BIOLOGY. Attended by the ant *Azteca alfari cecropiae*, in Surinam (Morrison, 1929).

Cryptostigma reticulolaminae Morrison

Cryptostigma reticulolaminae Morrison, 1929: 51.

TYPE DATA. Holotype female, PANAMA: Canal Zone, Frijoles, on *Cordia alliodora* (USNM).

DISTRIBUTION. NEOTROPICAL REGION: Panama.

HOST PLANTS. Ehretiaceae: *Cordia alliodora*.

BIOLOGY. Attended by the ant *Azteca longiceps* in Panama (Morrison, 1929).

Cryptostigma robertsi Williams & Watson

Cryptostigma robertsi Williams & Watson, 1990: 101.

TYPE DATA. Holotype female, PAPUA NEW GUINEA: Madang P., Baku, on trunk of *Terminalia brassii* (BMNH).

DISTRIBUTION. AUSTRO ORIENTAL REGION: Papua New Guinea.

HOST PLANTS. Combretaceae: *Terminalia brassii.*

Cryptostigma saundersi Laing

Cryptostigma saundersi Laing, 1925: 59.
TYPE DATA. Syntypes female, BRAZIL: Rio de Janeiro, on trunk of undetermined tree (BMNH).
DISTRIBUTION. NEOTROPICAL REGION: Brazil Rio de Janeiro.

Ctenochiton Maskell

Ctenochiton Maskell, 1879: 208. .
Cnetochiton Balachowsky; Balachowsky, 1932b: 36 [MIS-SPELLING].
TYPE-SPECIES: *Ctenochiton viridis* Maskell, by subsequent designation of Fernald (1903).
REMARKS. Generic characters discussed by Froggatt (1915), Morrison & Morrison (1922), Steinweden (1929), Hodgson (1969b) and by Tang (1991). Key to species: ZIMBABWE - Hodgson (1969b).

Ctenochiton araucariae Green

Ctenochiton araucariae Green, 1900d: 449.
TYPE DATA. Syntypes female, AUSTRALIA: Victoria, on *Araucaria* sp. (BMNH).
DISTRIBUTION. AUSTRALIAN REGION: Victoria.
HOST PLANTS. Araucariaceae: *Araucaria.*
REMARKS. Adult female redescribed by Froggatt (1915).

Ctenochiton aztecus Townsend & Cockerell

Ctenochiton aztecus Townsend & Cockerell, 1898: 176.
TYPE DATA. Syntypes female, MEXICO: Arroyo San Isidoro, near Frontera, Tabasco, on trunk of a tree named "cafetilla cimarron" (USNM).
DISTRIBUTION. NEOTROPICAL REGION: Mexico.

Ctenochiton carinatus Takahashi

Ctenochiton carinatus Takahashi, 1951b: 107.
Platysaissetia carinata (Takahashi); Tang, 1991: 206.
TYPE DATA. Syntypes female, INDONESIA: Riau [-Riouw] Islands, Rempang, host plant not indicated (HUSJ).
DISTRIBUTION. AUSTRO ORIENTAL REGION: Indonesia.
REMARKS. Adult female redescribed by Tang (1991).

Ctenochiton cellulosus Cockerell

Ctenochiton cellulosa Cockerell, 1899e: 88.
TYPE DATA. Syntypes female, AUSTRALIA: Victoria, Myrniong, on *Melaleuca nodosa* (BMNH, USNM).
DISTRIBUTION. AUSTRALIAN REGION: Victoria.
HOST PLANTS. Myrtaceae: *Melaleuca nodosa.*
REMARKS. Adult female redescribed by Froggatt (1915).

Ctenochiton cinnamomi Green

Ctenochiton cinnamomi Green, 1922b: 1030.

Platysaissetia cinnamomi (Green); Tang, 1991: 207.
TYPE DATA. Syntypes female, SRI LANKA: Colombo, on *Cinnamomum* sp. and Chilaw, on undetermined tree (BMNH).
DISTRIBUTION. ORIENTAL REGION: Sri Lanka.
HOST PLANTS. Lauraceae: *Cinnamomum*.
REMARKS. Adult female redescribed by Tang (1991).

Ctenochiton crematogasteri Takahashi

Ctenochiton crematogasteri Takahashi, 1942a: 25.
Platysaissetia crematogasteri (Takahashi); Tang, 1991: 208.
TYPE DATA. Syntypes female, THAILAND: Mt. Sutep, on *Quercus* sp. (IMZT).
HOST PLANTS. Fagaceae: *Quercus*.
REMARKS. Adult female redescribed by Tang (1991).
BIOLOGY. Protected under shelter constructed by *Crematogaster* sp. (Takahashi, 1942a).

Ctenochiton dacrydii Maskell

Ctenochiton dacrydii Maskell, 1892a: 18.
TYPE DATA. Syntypes female, NEW ZEALAND: Reefton District, on *Dacrydium cupressinum* (NZAC).
DISTRIBUTION. NEW ZEALAND and PACIFIC REGION: New Zealand.
HOST PLANTS. Podocarpaceae: *Dacrydium cupressinum*.
BIOLOGY. Found under the bark of the host plant (Maskell, 1892a).

Ctenochiton depressus Maskell

Ctenochiton depressus Maskell, 1884: 132.
Ctenochiton depressum minor Maskell, 1895a: 19. Syntypes female, NEW ZEALAND: Reefton, on *Coprosma* sp. (NZAC). Syn. by Wise, 1977: 104.
TYPE DATA. Syntypes female, NEW ZEALAND: Hawke's Bay, on *Plagianthus* sp. and *Cyathea* sp. (NZAC).
DISTRIBUTION. NEW ZEALAND and PACIFIC REGION: New Zealand.
HOST PLANTS. Cyatheaceae: *Cyathea*. **Malvaceae:** *Plagianthus*. **Rubiaceae:** *Coprosma*.
REMARKS. Adult female redescribed and illustrated by Maskell (1887b).

Ctenochiton elaeocarpi Maskell

Ctenochiton elaeocarpi Maskell, 1885: 26.
TYPE DATA. Syntypes female, NEW ZEALAND: neighbourhood of Wellington, on *Elaeocarpus dentatus* (NZAC).
DISTRIBUTION. NEW ZEALAND and PACIFIC REGION: New Zealand.
HOST PLANTS. Elaeocarpaceae: *Elaeocarpus dentatus*.
REMARKS. Adult female redescribed and illustrated by Maskell (1887b).

Ctenochiton elongatus Maskell
Ctenochiton elongatus Maskell, 1879: 212.
TYPE DATA. Syntypes female, NEW ZEALAND: Auckland, on *Geniostoma ligustrifolium* (NZAC).
DISTRIBUTION. NEW ZEALAND and PACIFIC REGION: New Zealand.
HOST PLANTS. Loganiaceae: *Geniostoma ligustrifolium.*
REMARKS. Adult female redescribed and illustrated by Maskell (1887b).

Ctenochiton eucalypti Maskell
Ctenochiton eucalypti Maskell, 1895b: 52.

TYPE DATA. Syntypes female, AUSTRALIA: New South Wales, Maitland and Newcastle, on *Eucalyptus siderophloia* (NZAC, USNM).
DISTRIBUTION. AUSTRALIAN REGION: New South Wales.
HOST PLANTS. Myrtaceae: *Eucalyptus siderophloia.*
REMARKS. Adult female redescribed and illustrated by Froggatt (1915).

Ctenochiton flavus Maskell
Ctenochiton flavus Maskell, 1884: 130.
TYPE DATA. Syntypes female, NEW ZEALAND: North Island, on *Brachyglottis repanda* and *Panax arboreum* (CMNZ, BMNH, USNM).
DISTRIBUTION. NEW ZEALAND and PACIFIC REGION: New Zealand.
HOST PLANTS. Araliaceae: *Panax arboreum.* **Compositae:** *Brachyglottis repanda.*
REMARKS. Adult female redescribed and illustrated by Maskell (1887b).

Ctenochiton formicophilus Green
Ctenochiton formicophilus Green, 1930c: 289.
TYPE DATA. Syntypes female, INDONESIA: Sumatra, Kloof van Airpulih (West Coast), from nest of ants (BMNH).
DISTRIBUTION. AUSTRO ORIENTAL REGION: Sumatra.
BIOLOGY. Found in carton nests of the ant *Dlichoderus carbonarius-latisquama* (Green, 1930c).

Ctenochiton froggatti n. name
Ctenochiton serrata Froggatt, 1915: 512 [HOMONYM of *Ctenochiton-serratus* Green].
TYPE DATA. Syntypes female and male, AUSTRALIA: Western Australia, Geraldton, on *Acacia* sp. (BCRI).
DISTRIBUTION. AUSTRALIAN REGION: Western Australia.
HOST PLANTS. Leguminosae: *Acacia.*

Ctenochiton fryeri Green
Ctenochiton fryeri Green, 1922b: 1031.
Platysaissetia fryeri (Green); Tang, 1991: 209.

TYPE DATA. Syntypes female, SRI LANKA: Vavuniya, on bark of
undetermined tree (BMNH).
DISTRIBUTION. ORIENTAL REGION: Sri Lanka.
REMARKS. Adult female redescribed by Tang (1991).

Ctenochiton fuscus Maskell

Ctenochiton fuscus Maskell, 1884: 131.
TYPE DATA. Syntypes female, NEW ZEALAND: near Christchurch,
on *Brachyglottis repanda* (CMNZ, NZAC).
DISTRIBUTION. NEW ZEALAND and PACIFIC REGION: New Zeal-
and.
HOST PLANTS. Araliaceae: *Panax arboreum.* **Compositae:** *Brachyglottis repanda.*
REMARKS. Adult female redescribed and illustrated by Maskell
(1887b).

Ctenochiton hymenantherae Maskell

Ctenochiton hymenantherae Maskell, 1885: 25.
TYPE DATA. Syntypes female, NEW ZEALAND: on *Hymenanthera
crassifolia* (NZAC, USNM).
DISTRIBUTION. NEW ZEALAND and PACIFIC REGION: New Zeal-
and.
HOST PLANTS. Violaceae: *Hymenanthera crassifolia.*
REMARKS. Adult female redescribed and illustrated by Maskell
(1887b).

Ctenochiton inclusus Green

Ctenochiton inclusus Green, 1930c: 289.
TYPE DATA. Syntypes female, INDONESIA: Sumatra, in carton nest
of ants (BMNH).
DISTRIBUTION. AUSTRO ORIENTAL REGION: Sumatra.
BIOLOGY. This species was described from a single female found in
carton nest of ants (Green, 1930c).

Ctenochiton olivaceum Green

Ctenochiton olivaceum Green, 1922b: 1032.
TYPE DATA. Syntypes female, SRI LANKA: Matale, on *Pterospermum
suberifolium* (BMNH).
DISTRIBUTION. ORIENTAL REGION: India, Sri Lanka.
HOST PLANTS. Sterculiaceae: *Pterospermum suberifolium,*
REMARKS. Adult female redescribed and illustrated by Tang (1991).
Distribution and host plant records given by Green (1937), Ali
(1971), Varshney (1985) and by Tang (1991).

Ctenochiton perforatus Maskell

Ctenochiton perforatus Maskell, 1879: 208.
TYPE DATA. Syntypes female, NEW ZEALAND: near Christchurch,
on *Pittosporum* sp., *Drimys* sp., *Coprosma* sp., *Rubus* sp. and
Panax sp. (NZAC).
DISTRIBUTION. NEW ZEALAND and PACIFIC REGION: New Zeal-
and.

HOST PLANTS. Araliaceae: *Panax arboreum.* **Pittosporaceae:** *Pittosporum eugenioides.* **Rosaceae:** *Rubus.* **Rubiaceae:** *Coprosma lucida.* **Winteraceae:** *Drimys.*
REMARKS. Adult female redescribed and illustrated by Maskell (1887b).

Ctenochiton piperis Maskell

Ctenochiton piperis Maskell, 1882: 218.
TYPE DATA. Syntypes female, NEW ZEALAND: North Island, on *Piper excelsum* (CMNZ, NZAC, USNM).
DISTRIBUTION. NEW ZEALAND and PACIFIC REGION: New Zealand.
HOST PLANTS. Piperaceae: *Piper excelsum.*
REMARKS. Adult female redescribed and illustrated by Maskell (1887b).

Ctenochiton rhizophorae Maskell

Ctenochiton rhizophorae Maskell, 1895b: 54.
TYPE DATA. Syntypes female, AUSTRALIA: Queensland, Brisbane, on *Rhizophora mucronata* (NZAC, USNM).
DISTRIBUTION. AUSTRALIAN REGION: Queensland.
HOST PLANTS. Rhizophoraceae: *Rhizophora mucronata.*
REMARKS. Adult female redescribed by Froggatt (1915).

Ctenochiton serratus Green

Ctenochiton serratus Green, 1904a: 67.
TYPE DATA. Syntypes female, AUSTRALIA: Victoria, Warranambool, on *Styphelia* sp. (BMNH).
DISTRIBUTION. AUSTRALIAN REGION: Victoria.
HOST PLANTS. Epacridaceae: *Styphelia.*

Ctenochiton spinosus Maskell

Ctenochiton spinosus Maskell, 1879: 212.
Eriochiton spinosus (Maskell); Maskell, 1887a: 47.
Lecanium armatus Brittin, 1915: 152. Syntypes female, NEW ZEALAND: Oamaru, on *Muehlenbeckia australis* (NZAC). Syn. by Brittin, 1916: 425.
TYPE DATA. Syntypes female, NEW ZEALAND: on *Atherosperma novae-zealandiae* (NZAC).
DISTRIBUTION. NEW ZEALAND and PACIFIC REGION: New Zealand.
HOST PLANTS. Atherospermataceae: *Atherosperma novae-zealandiae.* **Elaeocarpaceae:** *Elaeocarpus dentatus.* **Polygonaceae:** *Muehlenbeckia australis.* **Rutaceae:** *Melicope ternata.*

Ctenochiton transparens Froggatt

Ctenochiton transparens Froggatt, 1915: 513.
TYPE DATA. Syntypes female and male, AUSTRALIA: Western Australia, Geraldton, on *Acacia* sp. (BCRI).
DISTRIBUTION. AUSTRALIAN REGION: Western Australia.
HOST PLANTS. Leguminosae: *Acacia.*

Ctenochiton viridis Maskell

Ctenochiton viridis Maskell, 1879: 211.

TYPE DATA. Syntypes female, NEW ZEALAND: in Riccarton Bush near Christchurch, on *Coprosma* sp., *Panax* sp. and *Rubus* sp. (NZAC).

DISTRIBUTION. NEW ZEALAND and PACIFIC REGION: New Zealand.

HOST PLANTS. Araliaceae: *Panax arboreum.* **Atherospermataceae:** *Atherosperma novae-zealandiae.* **Monimiaceae:** *Hedycaria dentata.* **Rosaceae:** *Rubus australis.* **Rubiaceae:** *Coprosma lucida.*

REMARKS. Adult female redescribed and illustrated by Maskell (1887b) and by Morrison & Morrison (1922).

Cyclolecanium Morrison

Cyclolecanium Morrison, 1929: 56.

TYPE-SPECIES: *Cyclolecanium hyperbaterum* Morrison, by original designation and monotypy.

Cyclolecanium hyperbaterum Morrison

Cyclolecanium hyperbaterum Morrison, 1929: 56.

TYPE DATA. Holotype female, PANAMA: Canal Zone, Ancon, on *Cordia alliodora* (USNM).

DISTRIBUTION. NEOTROPICAL REGION: Panama.

HOST PLANTS. Ehretiaceae: *Cordia alliodora.*

BIOLOGY. Morrison (1929) recorded this species from Panama in association with the ants *Pseudomyrma sericea, Camponotus* sp., *Azteca longiceps* and *Crematogaster* sp.

Cyphococcus Laing

Cyphococcus Laing, 1925: 56.

TYPE-SPECIES: *Cyphococcus caesalpiniae* Laing, by original designation and monotypy.

Cyphococcus caesalpiniae Laing

Cyphococcus caesalpiniae Laing, 1925: 56.

TYPE DATA. Syntypes female, UGANDA: Kampala, on *Caesalpinia dasyrachis* (BMNH).

DISTRIBUTION. ETHIOPIAN REGION: Uganda.

HOST PLANTS. Leguminosae: *Caesalpinia dasyrachis.*

Dermolecanium Zavattari

Dermolecanium Zavattari in Casazza, 1928: 410.

Dermatolecanium Lindinger; Lindinger, 1937: 183 [UNJUSTIFIED EMENDATION].

TYPE-SPECIES: *Dermolecanium migrans* Zavattari, by monotypy.

REMARKS. This genus was established for the type species which was based on the description of a Coccidae larvae. The larvae were reported to be responsible for serious skin injury to a human being. Silvestri (1939) and Morrison & Morrison (1966) critically commented on this unusual report. However, nomenclaturally the genus is valid.

Dermolecanium migrans Zavattari

Dermolecanium migrans Zavattari in Casazza, 1928: 410.
TYPE DATA. Syntypes first instar larva, ITALY: Garlasco, Pavia District, on face of a woman, and Pavia Hospital, on face of a man (probably lost).
DISTRIBUTION. PALAEARCTIC REGION: Italy.
REMARKS. Whereas the biological observations on this species appear to be peculiar, its nomenclatural status is sound.
BIOLOGY. Zavattari in Casazza (1928) reported that first instar crawlers of this species were taken on human skin, on which they caused itches and skin lesion. Silvestri (1939) and Morrison & Morrison (1966) were doubtful whether the above reported crawlers, of this plant-feeding group, actually attacked and damaged the human skin.

Dicyphococcus Borchsenius

Dicyphococcus Borchsenius, 1959: 165.
TYPE-SPECIES: Dicyphococcus bigibbus Borchsenius, by original designation.
REMARKS. Generic characters discussed by Yang (1982) and by Tang (1991). Key to species given by Borchsenius (1959) and by Tang (1991).

Dicyphococcus bigibbus Borchsenius

Dicyphococcus bigibbus Borchsenius, 1959: 168.
TYPE DATA. Syntypes female, CHINA: Yunnan Province, near Ching-Tung, on Moghania sp., Cinnamomum sp., Melastoma sp. and Wendlandia sp. (IEBC).
DISTRIBUTION. PALAEARCTIC REGION: China.
HOST PLANTS. Lauraceae: Cinnamomum. **Leguminosae:** Moghania. **Melastomataceae:** Melastoma. **Platanaceae:** Platanus orientalis. **Rubiaceae:** Wendlandia.
REMARKS. Adult female redescribed and illustrated by Yang (1982) and by Tang (1991). Distribution and host plant records given by Yang (1982) and by Tang (1991).

Dicyphococcus castilloae (Green)

Inglisia castilloae Green, 1911: 29.
Cardiococcus castilloae (Green); Cockerell, 1911: 327.
Dicyphococcus castilloae (Green); Borchsenius, 1959: 169.
TYPE DATA. Syntypes female, male and larva, SRI LANKA: Koslanda, on Hevea brasiliensis, Grewia microcos, Adenochlaena zeylanica, Solanum sp., Vernonia sp. and Thea sp. (BMNH).
DISTRIBUTION. ORIENTAL REGION: India, Sri Lanka.
HOST PLANTS. Compositae: Vernonia. **Euphorbiaceae:** Adenochlaena zeylanica, Hevea brasiliensis. **Moraceae:** Castilla elastica. **Solanaceae:** Solanum. **Theaceae:** Thea. **Tiliaceae:** Grewia microcos.

REMARKS. Distribution and host plant records given by Green (1937), Borchsenius (1959), Ali (1971) and by Varshney (1985).

Dicyphococcus ficicola Borchsenius

Dicyphococcus ficicola Borchsenius, 1959: 169.
TYPE DATA. Holotype female, CHINA: Yunnan Province, Ching-Dung District, near Pabyen-Chian, on *Ficus pyriformis* (IEBC).
DISTRIBUTION. PALAEARCTIC REGION: China.
HOST PLANTS. Moraceae: *Ficus pyriformis*.
REMARKS. Adult female redescribed by Tang (1991).

Didesmococcus Borchsenius

Didesmococcus Borchsenius, 1953: 281.
TYPE-SPECIES: *Didesmococcus megriensis* Borchsenius [=*D. unifasciatus* (Archangelskaya)], by original designation.
REMARKS. Generic characters discussed by Borchsenius (1957), Wang (1980), Yang (1982) and by Tang (1991). Key to species given by Borchsenius (1957) and by Tang (1991).

Didesmococcus koreanus Borchsenius

Didesmococcus koreanus Borchsenius, 1955b: 288.
TYPE DATA. Syntypes female, KOREA: Pyongyang, on *Prunus* sp., *Cerasus* sp. and *Armeniaca* sp. (ZIAS).
DISTRIBUTION. PALAEARCTIC REGION: China, Inner Mongolia, Korea.
HOST PLANTS. Rosaceae: *Armeniaca, Cerasus, Prunus*.
REMARKS. Adult female redescribed and illustrated by Borchsenius (1957), Wang (1980), Tang & Li (1988) and by Tang (1991). Adult male and first instar larva described and illustrated by Borchsenius (1957). Distribution and host plant records given by Borchsenius (1957, 1960), Wang (1980), Tang & Li (1988) and by Tang (1991).

Didesmococcus unifasciatus (Archangelskaya)

Physokermes unifasciatus Archangelskaya, 1923: 265.
Physokermes (Eulecanium) unifasciatus Archangelskaya; Archangelskaya, 1931: 79.
Sphaerolecanium unifasciatus (Archangelskaya); Kiritshenko, 1936: 70.
Lecanium unifsciatus (Archangelskaya); Borchsenius, 1937: 85.
Eriochiton amygdalae Rao, 1939: 59. Holotype female, PAKISTAN: Baluchistan, Loralai, on almond tree (PUSA). Syn. by Tang, 1991: 102.
Eulecanium unifasciatus (Archangelskaya); Borchsenius, 1949c: 172.
Didesmococcus megriensis Borchsenius, 1953: 282. Syntypes female, USSR: Armenia, Megri, on *Persica vulgaris, Amygdalus communis, Armeniaca* sp. and *Prunus* sp. (ZIAS). Syn. by Danzig, 1970: 1017.
Didesmococcus unifasciatus (Archangelskaya); Borchsenius, 1953: 282.
TYPE DATA. Syntypes female, USSR: Uzbekistan, Samarkand, on *Persica* sp. (ZIAS).
DISTRIBUTION. PALAEARCTIC REGION: Afghanistan, Inner

Mongolia, Mongolia, Pakistan, USSR Armenia, USSR Tadzhikistan, USSR Uzbekistan.

HOST PLANTS. Rosaceae: *Amygdalus communis, Amygdalus nana, Amygdalus pedunculata, Armeniaca, Persica vulgaris, Prunus prostrata.*

REMARKS. Archangelskaya (1931) again described this species as new, however, it is evident that she referred to the same species. Adult female redescribed and illustrated by Archangelskaya (1931, 1937) and by Borchsenius (1957). Adult female redescribed by Tang (1991). Adult male and first instar crawler described and illustrated by Borchsenius (1957). Distribution and host plant records given by Archangelskaya (1931, 1937), Kiritshenko (1936), Borchsenius (1937, 1949c, 1953, 1957), Danzig (1970, 1972d, 1974, 1977b) and by Tang (1991).

Drepanococcus Williams & Watson

Drepanococcus Williams & Watson, 1990: 102.
TYPE-SPECIES: *Eriochiton cajani* Maskell, by original designation.

Drepanococcus cajani (Maskell)

Eriochiton cajani Maskell, 1891b: 61.
Ceroplastodes cajani (Maskell); Cockerell, 1900c: 368.
Drepanococcus cajani (Maskell); Williams & Watson, 1990: 102.
TYPE DATA. Syntypes female, INDIA: Madras, on *Cajanus indicus* (NZAC).
DISTRIBUTION. AUSTRO ORIENTAL REGION: Malaysia, Philippines, Taiwan. **ORIENTAL REGION:** India, Sri Lanka.
HOST PLANTS. Convolvulaceae: *Ipomoea.* **Labiatae:** *Coleus, Ocimum sanctum.* **Leguminosae:** *Abrus precatorius, Atylosia candollei, Cajanus indicus.* **Myrtaceae:** *Psidium guajava.* **Rhamnaceae:** *Ziziphus jujuba.*
REMARKS. Maskell (1892a) also referred to this species as n. sp. Adult female redescribed and illustrated by Green (1909), Morrison (1920), Ramakrishna Ayyar (1930), Williams & Watson (1990) and by Tang (1991). Parida & Moharana (1982) and Moharana (1990) reported on chromosome number 2n-18 in India. Distribution and host plant records given by Green (1909, 1937), Newstead (1917c), Morrison (1920), Takahashi (1952), Ali (1971), Yang (1982), Varshney & Moharana (1987), Shafee *et al.* (1989), Williams & Watson (1990) and by Tang (1991).

Drepanococcus chiton (Green)

Ceroplastodes chiton Green, 1908: 32 [NOMEN NUDUM].
Ceroplastodes chiton Green, 1909: 287.
Drepanococcus chiton (Green); Williams & Watson, 1990: 102.
TYPE DATA. Lectotype female designated by Williams & Watson (1990), SRI LANKA: Maha Illuppalama, on *Cassia* sp. (BMNH).
DISTRIBUTION. AUSTRO ORIENTAL REGION: Malaysia, Papua New Guinea, Solomon Islands. **ORIENTAL REGION:** Andaman Islands, India, Sri Lanka, Taiwan, Thailand, Vietnam.
HOST PLANTS. Anacardiaceae: *Semecarpus magnifica.* **Annonaceae:** *Annona muricata.* **Caricaceae:** *Carica papaya.* **Clusiaceae:**

Calophyllum inophyllum. **Euphorbiaceae:** *Aleurites moluccana*.
Labiatae: *Coleus*. **Lauraceae:** *Litsea*. **Leguminosae:** *Bauhinia*,
Cajanus indicus, *Canavalia*, *Cassia*, *Dalbergia*, *Gliricidia sepium*.
Malvaceae: *Thespesia propulnea*. **Moraceae:** *Ficus*. **Proteaceae:**
Grevillea papuana. **Rhamnaceae:** *Colubrina*. **Rutaceae:** *Citrus
aurantifolia*. **Solanaceae:** *Solanum melongena*. **Sterculiaceae:**
Theobroma cacao.
REMARKS. Adult female redescribed and illustrated by Yang (1982),
Tao *et al.* (1983), Williams & Watson (1990) and by Tang (1991).
Distribution and host plant records given by Green (1937), Taka-
hashi (1952), Ali (1971), Yang (1982), Tao *et al.* (1983), Shafee *et al.*
(1989), Williams & Watson (1990), Danzig & Konstantinova (1990)
and by Tang (1991).

Drepanococcus magnospinus (Mamet)

Ceroplastodes magnospinus Mamet, 1959b: 417.
Drepanococcus magnospinus (Mamet); Williams & Watson, 1990:
102.
TYPE DATA. Holotype female, MADAGASCAR: Ifanadiana, Ranoma-
fana, on undetermined plant (MNHN).
DISTRIBUTION. MADAGASIAN REGION: Madagascar.
REMARKS. Williams & Watson (1990) placed the species in *Drepanococcus* and suggested that future study may show that it is a
synonym of *D. cajani* (Maskell).

Drepanococcus virescens (Green)

Ceroplastodes virescens Green, 1909: 288.
Drepanococcus virescens (Green); Williams & Watson, 1990: 102.
TYPE DATA. Syntypes female, SRI LANKA: Matale, on *Theobroma
cacao* (BMNH).
DISTRIBUTION. AUSTRO ORIENTAL REGION: Singapore. **ORIEN-
TAL REGION:** Sri Lanka.
HOST PLANTS. Moraceae: *Artocarpus*. **Sterculiaceae:** *Theobroma
cacao*.
REMARKS. Adult female redescribed by Morrison (1921). Williams &
Watson (1990) placed the species in *Drepanococcus* and suggested
that it may be a synonym of *D. chiton* (Green).

Edwallia Hempel

Edwallia Hempel, 1899: 131.
Edwalia Borchsenius; Borchsenius, 1957: 47 [MIS-SPELLING].
TYPE-SPECIES: *Edwallia rugosa* Hempel, by original designation
and monotypy.
REMARKS. Generic characters discussed by Hempel (1900b).

Edwallia rugosa Hempel

Edwallia rugosa Hempel, 1899: 131.
TYPE DATA. Syntypes female and male, BRAZIL: Sao Paulo, on
Eugenia jaboticaba (MZSP).
DISTRIBUTION. NEOTROPICAL REGION: Brazil Sao Paulo.
HOST PLANTS. Myrtaceae: *Eugenia jaboticaba*.

REMARKS. Adult female redescribed and illustrated by Hempel (1900b, 1920a).

Ericeroides Danzig

Ericeroides Danzig, 1990: 374.
TYPE-SPECIES: *Ericeroides zaitzevi* Danzig, by original designation and monotypy.

Ericeroides zaitzevi Danzig

Ericeroides zaitzevi Danzig, 1990: 376.
TYPE DATA. Holotype female, VIETNAM: Tamdao, Vinfu Province, on undetermined plant (ZIAS).
DISTRIBUTION. ORIENTAL REGION: Vietnam.

Ericerus Guerin-Meneville

Coccus (Ericerus) Guerin-Meneville, 1858: lxvii.
Pela Targioni Tozzetti, 1866: 140. TYPE-SPECIES: *Pela cerifera* Targioni Tozzetti, by original indication and monotypy. Synonym by synonymy of type species
Eurycerus Targioni Tozzetti; Targioni Tozzetti, 1867: 19 [MIS-SPELL-ING].
Curycerus Targioni Tozzetti; Targioni Tozzetti, 1867: 19 [MIS-SPELL-ING].
Ericerus Guerin-Meneville; Targioni Tozzetti, 1868: 729.
TYPE-SPECIES: *Coccus ceriferus* Fabricius (as a misidentification of *Coccus pela* Chavannes) by monotypy.
REMARKS. This genus constitutes a case of misidentified type species which, according to Article 70 (b) of the International Code, should be referred to ICZN. Generic characters discussed by Danzig (1967, 1980c), Paik (1978), Wang (1980), Yang (1982) and by Tang (1991).

Ericerus pela (Chavannes)

Coccus pela Chavannes, 1848: 144.
Coccus sinensis Walker, 1852: 1085. Syntypes female and male, CHINA, (BMNH). Syn. by Green, 1904c: 374.
Coccus cereus Walker, 1852: 1087 [UNJUSTIFIED REPLACEMENT NAME].
Coccus sinensis Westwood, 1853a: 95. Syntypes female and male, CHINA: probably Shanghai, host plant not indicated (probably lost). Syn. by Westwood, 1853c: 532.
Ericerus ceriferus (Anderson); Guerin-Meneville, 1858: lxvii [MISI-DENTIFICATION].
Pela cerifera Targioni Tozzetti; Targioni Tozzetti, 1866b: 140 [UNJUSTIFIED REPLACEMENT NAME].
Ericerus pela Westwood; Signoret, 1869c: 102 [ERRONEOUS AUTHORSHIP].
Eulecanium potanini Borchsenius, 1955b: 297. Syntypes female, CHINA: Sinan Province, on undetermined plant (ZIAS). Syn. by Danzig, 1967: 167.
TYPE DATA. Syntypes CHINA: location and host plant not indicated (probably lost).

DISTRIBUTION. PALAEARCTIC REGION: China, Japan, Korea, Tibet, USSR Primorye Territory.

HOST PLANTS. Oleaceae: *Chionanthus retusens, Fraxinus bungeana, F. longicuspis, F. mandshurica, F. rhynchophylla, Ligustrum ibota, L. japonicum, L. medium, Syringa amurensis.*

REMARKS. Common Name - China wax scale insect, Chinese scale or Pela scale. Adult female redescribed and illustrated by Kuwana (1917a, 1923b), Ferris (1950), Borchsenius (1955b, 1957) (as *E. potanini*), Danzig (1965, 1967, 1980c), Paik (1978), Wang (1980), Kawai (1980), Yang (1982) and by Tang (1991). Adult male described and illustrated by Kuwana (1917a), Danzig (1965) and by Gilliomee (1967). Larval instars described and illustrated by Kuwana (1923b) and by Ferris (1950). First and second instar larva described by Danzig (1965). De Lotto (1971c) discussed the identity of *Coccus sinensis* Westwood, and suggested that it might be distinct from *C. sinensis* Walker. However, because of the meagre original description of Westwood's species and inavailability of type material, it would be advisable to accept Westwood's (1853c) interpretation that both species are synonyms of *E. pela.* Colour photograph given by Kawai (1980). Distribution and host plant records given by Kuwana (1917a, 1923b), Ferris (1950), Takahashi & Tachikawa (1956), Danzig (1965, 1967, 1980c), Paik (1978), Kawai (1980), Wang (1980, 1981), Yang (1982) and by Tang (1991).

BIOLOGY. A biparental species, which develops one annual generation in Japan (Kuwana, 1923b), Russia, Vladivostok region (Danzig, 1965, 1967, 1980c) and in China (Li, 1985). A comprehensive account on the life history and commercial breeding was presented by Li (1985).

ECONOMIC IMPORTANCE. The wax secreted by the male nymphs is harvested in China and processed to produce the China wax, a commodity of considerable economic value in China (Li, 1985).

Eriochiton Maskell

Eriochiton Maskell, 1887a: 46.
Eriochitin MacGillivray; MacGillivray, 1921: 168 [MIS-SPELLING].

TYPE-SPECIES: *Eriochiton hispidus* Maskell, by subsequent designation of Fernald (1903).

REMARKS. Generic characters discussed by Morrison & Morrison (1922), Yang (1982) and by Tao *et al.* (1983).

Eriochiton hispidus Maskell

Eriochiton hispidus Maskell, 1887a: 47.

TYPE DATA. Syntypes female, NEW ZEALAND: Wellington, Botanic Gardens, on *Olearia haastii* (NZAC).

DISTRIBUTION. NEW ZEALAND and PACIFIC REGION: New Zealand.

HOST PLANTS. Compositae: *Olearia haastii.*

REMARKS. Adult female redescribed and illustrated by Maskell (1887b) and by Morrison & Morrison (1922).

Eriochiton theae Green

Eriochiton theae Green, 1900c: 10.

TYPE DATA. Syntypes female, BANGLADESH: Darjeeling, on tea plant (BMNH).
DISTRIBUTION. ORIENTAL REGION: Bangladesh, Taiwan.
HOST PLANTS. Theaceae: *Camellia sinensis.*
REMARKS. Borchsenius (1957) and Yang (1982) regarded this species as a synonym of *Metaceronema japonica* (Maskell). However, Tao *et al.* (1983) redescribed and illustrated it as a distinct species.

Eriopeltis Signoret

Eriopeltis Signoret, 1872a: 429.
Eriopettis Tang; Tang, 1991: 40 [MIS-SPELLING].
TYPE-SPECIES: *Coccus festucae* Fonscolombe, by original designation and monotypy.
REMARKS. Morrison & Morrison (1966) discussed and established that *Coccus festucae* Fonscolombe is the type-species. Generic characters discussed by Newstead (1903), Leonardi (1920), Steinweden (1929), Gómez-Menor Ortega (1937), Borchsenius (1956, 1957), Williams & Kosztarab (1972), Paik (1978), Danzig (1975b, 1980c), Kosztarab & Kozár (1978, 1988), Tereznikova (1981), Yang (1982), Manawadu (1986), Gill (1988) and by Tang (1991). Key to species: ASIA - Tang (1991); DENMARK - Kozárzhevskaya & Reitzel (1975); PALAEARCTIC REGION - Borchsenius (1956, 1957); - Danzig (1964, 1975b); Kosztarab & Kozár (1978, 1988); Manawadu (1986); UKRAINE - Tereznikova (1981); USSR (FAR EAST) - Danzig (1980c).

Eriopeltis coloradensis Cockerell

Eriopeltis coloradensis Cockerell, 1905b: 136.
TYPE DATA. Syntypes female, U.S.A.: Colorado, Boulder, on grass (BMNH, USNM).
DISTRIBUTION. NEARCTIC REGION: Colorado.
HOST PLANTS. Gramineae.
REMARKS. Male test described and illustrated by Miller & Williams (1990).

Eriopeltis festucae (Fonscolombe)

Coccus festucae Fonscolombe, 1834: 216.
Eriopeltis festucae (Fonscolombe); Signoret, 1872a: 430.
Eriopeltis brachypodii Giard; Giard, 1894: cxcix [NOMEN NUDUM].
Eriopeltis agropyri Borchsenius, 1956: 399. Syntypes female, USSR: Crimea, Simferopol, Odessa, Kiev, Stalingrad, on *Agropyrum* [=*Agropyron*] sp. (ZIAS). Syn. by Danzig, 1975b: 809.
Eriopeltis araxis Borchsenius, 1956: 401. Syntypes female, USSR: Armenia, Megri, and Azerbaijan, Ordubada, on *Agropyrum* sp. (ZIAS). Syn. by Danzig, 1975b: 809.
Eriopeltis caucasicus Borchsenius, 1956: 402. Syntypes female, USSR: Northern Caucasus, Krasnodarsk, on grass (ZIAS). Syn. by Danzig, 1975b: 809.
Eriopeltis desertus Borchsenius, 1956: 403. Syntypes female, USSR: Western Kazakhstan, on grass (ZIAS). Syn. by Danzig, 1975b: 809.
Eriopeltis eversmanni Borchsenius, 1956: 403. Syntypes female, USSR: Urals, on grass (ZIAS). Syn. by Danzig, 1975b: 809.

Eriopeltis ferganensis Borchsenius, 1956: 404. Syntypes female,
   USSR: Uzbekistan, Min-Bulak, on grass (ZIAS). Syn. by Danzig,
   1975b: 809.
Eriopeltis hamberdiensis Borchsenius, 1956: 405. Syntypes female,
   USSR: Armenia, Ashtaraksk Ridge, on *Agropyrum* sp. (ZIAS).
   Syn. by Danzig, 1975b: 809.
Eriopeltis maximus Borchsenius, 1956: 408. Syntypes female, USSR:
   Kazakhstan, Kustanaisk Region, on grass (ZIAS). Syn. by Danzig,
   1975b: 809.
Eriopeltis phragmitidis Borchsenius, 1956: 408. Syntypes female,
   USSR: Crimea, Yevpatoriya, on *Phragmites communis* (ZIAS).
   Syn. by Danzig, 1975b: 809.
Eriopeltis pratensis Borchsenius, 1956: 411. Syntypes female, USSR:
   Yaroslavsk Region, on grass (ZIAS). Syn. by Danzig, 1980c: 252.
Eriopeltis rasinae Borchsenius, 1956: 411. Syntypes female, USSR:
   Leningrad Region, and Latvia, Riga, on *Agrostis vulgaris* (ZIAS).
   Syn. by Danzig, 1975b: 809.
Eriopeltis zolotarevae Borchsenius, 1956: 416. Syntypes female and
   male, USSR: Primorye Territory, on grass (ZIAS). Syn. by Danzig,
   1975b: 809.
TYPE DATA. Syntypes female, FRANCE: Provence, Aix, on *Festuca
   phaenicoides* and *Festuca caespitosa* (lost; D. Matile-Ferrero,
   personal communication, 1989).
DISTRIBUTION. NEARCTIC REGION: California, Nebraska, Virgi-
   nia. **PALAEARCTIC REGION:** Austria, Bulgaria, Czechoslovakia,
   Denmark, England, France, Germany, Hungary, Inner Mongolia,
   Iran, Iraq, Israel, Italy, Korea, Mongolia, Netherlands, Poland,
   Romania, Spain, Sweden, Switzerland, Turkey, USSR Armenia,
   USSR Caucasus, USSR Crimea, USSR Georgia, USSR Kazakh-
   stan, USSR Latvia, USSR Moldavia, USSR Primorye Territory,
   USSR Ukraine, USSR Uzbekistan, Yugoslavia.
HOST PLANTS. Cyperaceae: *Carex.* **Gramineae:** *Agropyron repens,
   Agrostis alba, Agrostis perennans, Agrostis vulgaris, Aneurolepid-
   ium chinensis, Arundinella anomala, Brachypodium pinnatum,
   Brachypodium silvaticum, Bromus, Calamagrostis epigeios,
   Corynephorus canescens, Deschampsia flexuosa, Eragrostis,
   Festuca ovina, F. rubra, Phalaris tuberosum, Phleum pratense,
   Phragmites communis.*
REMARKS. Adult female redescribed and illustrated by Signoret
(1872a), Newstead (1903), Leonardi (1920), Green (1921a), Gómez-
Menor Ortega (1937), Borchsenius (1956, 1957), Kozár (1971), Williams & Kosztarab (1972), Hadzibejli (1973) (as *E. araxis*), Danzig
(1975b, 1980c), Paik (1978), Kosztarab & Kozár (1978, 1988), Tereznikova (1981), Gill (1988), Tang & Li (1988) and by Tang (1991).
Male described and illustrated by Green (1923b), Borchsenius (1956)
(as *E. zolotarevae*) and by Giliomee (1967). Male test described and
illustrated by Miller & Williams (1990). First instar crawler described
and illustrated by Leonardi (1920) and by Gómez-Menor Ortega
(1937). Gill (1988) discussed the records of this species from California and Nevada. Colour photograph given by Tranfaglia & Viggiani
(1988). Distribution and host plant records given by Leonardi (1920),
Green (1921a, 1923b), Bodenheimer (1926b, 1943, 1944b, 1953),
Borchsenius (1956), Ossiannilsson (1959), Tereznikova (1963a,
1981), Boratynski & Williams (1964), Giliomee (1967), Kozár (1971,

1983, 1985), Hadzibejli (1973), Williams & Kosztarab (1972), Kozárzhevskaya & Reitzel (1975), Danzig (1975b, 1977b, 1980c), Podsiadlo
& Komosinska (1976), Paik (1978), Koteja (1983), Martin Mateo
(1984), Kozár & Ostafichuk (1987), Marotta (1987), Kosztarab &
Kozár (1978, 1988), Gill (1988), Tang & Li (1988) and by Tang
(1991).

Eriopeltis lichtensteini Signoret

Eriopeltis lichtensteinii Signoret, 1877a: 607.
TYPE DATA. Syntypes female, FRANCE: Hyeres Islands, on grass
(VMNH).
DISTRIBUTION. PALAEARCTIC REGION: Bulgaria, Czechoslovakia,
Denmark, Finland, France, Germany, Hungary, Netherlands,
Poland, Sweden, USSR Latvia, USSR Leningrad Region, USSR
Ukraine.
HOST PLANTS. Gramineae: *Agrostis vulgaris, Brachypodium silvati-
cum, Calamagrostis arundinacea, C. epigeios, C. pseudophrag-
mites, Holcus lanatus, Melica nutans, Phalaris arundinacea,
Phragmites communis, Poa nemoralis.* **Juncaceae:** *Luzula pilosa.*
REMARKS. Green (1923b) critically separated this species from its
close congener *E. festucae.* Adult female redescribed and illustrated
by Green (1923b), Borchsenius (1957), Kosztarab & Kozár (1978,
1988) and by Tereznikova (1981). Distribution and host plant
records given by Herberg (1916), Tiensuu (1951), Borchsenius
(1957), Reyne (1957), Koteja (1971), Kosztarab & Kozár (1978, 1988),
Nuorteva (1974), Kozárzhevskaya & Reitzel, (1975), Danzig (1975b),
Kozár *et al.* (1977), Tereznikova (1981), Ossiannilsson (1985) Kozár &
Drozdjak (1990).
BIOLOGY. Herberg (1916) presented a monographic study on morphology, biology, anatomy and histology.

Eriopeltis sachalinensis Borchsenius

Eriopeltis festucae (Fonscolombe); Kuwana, 1917b: 171 [MISIDEN-
TIFICATION].
Eriopeltis sachalinensis Borchsenius, 1956: 413.
Eriopeltis koreanus Borchsenius, 1956: 406. Syntypes female,
NORTH KOREA: on grass (ZIAS). Syn. by Danzig, 1975b: 811.
Eriopeltis strelkovi Borchsenius, 1956: 414. Syntypes female and
male, USSR: Kuril Islands, on grass (ZIAS). Syn. by Danzig,
1975b: 811.
Eriopeltis japonensis Takahashi, 1957: 65. Syntypes female, JAPAN:
Mt. Rokko near Kobe, on *Festuca* sp. (HUSJ). Syn. by Danzig,
1975b: 811.
Eriopeltis zolotarevae Borchsenius; Danzig, 1967: 141 [MISIDEN-
TIFICATION].
TYPE DATA. Syntypes female, USSR: Sakhalin Island, Dolinsk, on
Calamagrostis sp. (ZIAS).
DISTRIBUTION. PALAEARCTIC REGION: Inner Mongolia, Japan,
Korea, USSR Kunashir Island, USSR Primorye Territory, USSR
Sakhalin Island, USSR Shikotan Island.
HOST PLANTS. Gramineae: *Calamagrostis purpurea, Festuca,
Pennisetum flaccidum.*
REMARKS. Adult female redescribed and illustrated by Borchsenius

(1957), Paik (1978) (as *E. koreanus*), Danzig (1980c) and by Tang & Li (1988). Adult female redescribed by Tang (1991). Adult male described and illustrated (as *E. strelkovi*) by Borchsenius (1957). Danzig (1967) misidentified this species as *E. zolotarevae* Borchsenius and erroneously placed *E. sachalinensis* as a junior synonym of the former. Distribution and host plant records given by Takahashi (1957), Borchsenius (1957), Danzig (1967, 1975b, 1978a, 1980c), Paik (1978), Tang & Li (1988) and by Tang (1991).

Eriopeltis stammeri Schmutterer

Eriopeltis festucae (Fonscolombe); Kiritshenko, 1931: 319 [MISIDEN-TIFICATION].
Eriopeltis stammeri Schmutterer, 1952a: 555.
Eriopeltis plumeus Borchsenius, 1956: 409. Syntypes female, USSR: Leningrad Region, near Lugi, on *Festuca* sp. (ZIAS). Syn. by Danzig, 1975b: 812.
TYPE DATA. Syntypes female, GERMANY: Becken, Bubenreuth and Adlitz, on *Festuca ovina* (SENC).
DISTRIBUTION. PALAEARCTIC REGION: Czechoslovakia, Denmark, Germany, Hungary, Poland, Sweden, USSR Crimea, USSR Kazakhstan, USSR Leningrad Region, USSR Ukraine, USSR Yakutsk.
HOST PLANTS. Gramineae: *Corynephorus*, *Deschampsia*, *Festuca capillata*, *F. kolymensis*, *F. ovina*, *F. rubra*, *Nardus stricta*.
REMARKS. Adult female redescribed and illustrated by Borchsenius (1957), Kosztarab & Kozár (1978, 1988) and by Tereznikova (1981). Distribution and host plant records given by Borchsenius (1956), Matesova (1968), Kozárzhevskaya & Reitzel (1975), Komosinska (1977), Danzig (1978b), Kozár (1980, 1986), Tereznikova (1981), Koteja & Zak-Ogaza (1983), Ossiannilsson (1985) and by Kosztarab & Kozár (1988).

Eriopeltis stipae Ishii

Eriopeltis stipae Ishii, 1935: 1.
TYPE DATA. Syntypes female, CHINA: Inner Mongolia, Jehol, Cheng-te, on grass (probably lost; S. Takagi, 1989, personal communication).
DISTRIBUTION. PALAEARCTIC REGION: Inner Mongolia.
HOST PLANTS. Gramineae: *Stipa*.
REMARKS. Adult female redescribed by Borchsenius (1957) and by Tang (1991).

Eriopeltis varleyi Manawadu

Eriopeltis varleyi Manawadu, 1986: 318.
TYPE DATA. Holotype female and paratype male, ENGLAND: Oxford, Wytham Woods, on *Brachypodium pinnatum* (BMNH).
DISTRIBUTION. PALAEARCTIC REGION: England.
HOST PLANTS. Gramineae: *Brachypodium pinnatum*.
REMARKS. Manawadu (1986) also described first and second instars of male nymphs and the adult male.

Etiennea Matile-Ferrero

Etiennea Matile-Ferrero, 1984a: 100.
TYPE-SPECIES: *Etiennea villiersi* Matile-Ferrero, by original designation.
REMARKS. Generic characters discussed by Hodgson (1991a). Hodgson (1991a) revised the genus and gave key to 19 species.

Etiennea cacao Hodgson

Etiennea cacao Hodgson, 1991a: 179.
TYPE DATA. Holotype female, NIGERIA: Indiayunre, on Cacao [=*Theobroma* sp.] (BMNH).
DISTRIBUTION. ETHIOPIAN REGION: Nigeria.
HOST PLANTS. Sterculiaceae: *Theobroma*.

Etiennea candelabra Hodgson

Etiennea candelabra Hodgson, 1991a: 181.
TYPE DATA. Holotype female, SOUTH AFRICA: Transvaal, Tshipise, on *Albizia anthelminthica* (SANC).
DISTRIBUTION. ETHIOPIAN REGION: South Africa.
HOST PLANTS. Leguminosae: *Albizia anthelminthica*.

Etiennea capensis Hodgson

Etiennea capensis Hodgson, 1991a: 185.
TYPE DATA. Holotype female, SOUTH AFRICA: Cape Province, Saldanha, on *Diosma*? (SANC).
DISTRIBUTION. ETHIOPIAN REGION: South Africa.
HOST PLANTS. Rutaceae: *Diosma*?

Etiennea carpenteri (Newstead)

Platysaissetia carpenteri Newstead, 1917a: 343.
Etiennea carpenteri (Newstead); Hodgson, 1991a: 187.
TYPE DATA. Syntypes female, UGANDA: Ngamba Island, on trunk of a fig tree (BMNH).
DISTRIBUTION. ETHIOPIAN REGION: Uganda.
HOST PLANTS. Moraceae: *Ficus*.
REMARKS. Adult female redescribed and illustrated by Hodgson (1991a).

Etiennea cephalomeatus Hodgson

Etiennea cephalomeatus Hodgson, 1991a: 188.
TYPE DATA. Holotype female, UGANDA: Kampala, on *Ficus* sp. (SANC).
DISTRIBUTION. ETHIOPIAN REGION: Uganda.
HOST PLANTS. Moraceae: *Ficus*.

Etiennea combreti (Hodgson)

Platysaissetia combreti Hodgson, 1969b: 25.
Etiennea combreti (Hodgson); Hodgson, 1991a: 190.
TYPE DATA. Holotype female, ZIMBABWE: Mazoe, on *Combretum* sp. (BMNH).

DISTRIBUTION. ETHIOPIAN REGION: Zimbabwe.
HOST PLANTS. Combretaceae: *Combretum*.
REMARKS. Adult female redescribed and illustrated by Hodgson (1991a).

Etiennea ferina (De Lotto)

Platysaissetia ferina De Lotto, 1978: 145.
Etiennea ferina (De Lotto); Hodgson, 1991a: 192.
TYPE DATA. Holotype female, SOUTH AFRICA: Keiskammahoek, on *Scutia myrtina* (SANC).
DISTRIBUTION. ETHIOPIAN REGION: South Africa.
HOST PLANTS. Rhamnaceae: *Scutia myrtina*.
REMARKS. Adult female redescribed and illustrated by Hodgson (1991a).

Etiennea ferox (Newstead)

Platysaissetia ferox Newstead, 1917a: 344.
Platysaissetia fouabii Matile-Ferrero & Le Ruyet, 1985: 263. Holotype female, IVORY COAST: Tai, on *Xylopia aethiopica* (MNHN). Syn. by Hodgson, 1991a: 194.
Etiennea ferox (Newstead); Hodgson, 1991a: 194.
TYPE DATA. Syntypes female, NIGERIA: Southern Nigeria, Calabar Botanical Gardens, on a hard-wooded scrub (BMNH).
DISTRIBUTION. ETHIOPIAN REGION: Ivory Coast, Nigeria.
HOST PLANTS. Annonaceae: *Xylopia aethiopica*.
REMARKS. Adult female redescribed and illustrated by Hodgson (1991a).

Etiennea gouligouli Hodgson

Etiennea gouligouli Hodgson, 1991a: 197.
TYPE DATA. Holotype female, CENTRAL AFRICAN REPUBLIC: Gouli-Gouli, on *Theobroma cacao* (MNHN).
DISTRIBUTION. ETHIOPIAN REGION: Central African Republic.
HOST PLANTS. Sterculiaceae: *Theobroma cacao*.

Etiennea halli Hodgson

Platysaissetia kellyi (Brain); Hall, 1935: 77 [MISIDENTIFICATION].
Platysaissetia kellyi (Brain); Hodgson, 1969b 27 [MISIDENTIFICATION].
Etiennea halli Hodgson, 1991a: 199.
TYPE DATA. Holotype female, ZIMBABWE [=RHODESIA]: Harare, The Kopje, on *Berlinia globiflora* (BMNH).
DISTRIBUTION. ETHIOPIAN REGION: Zimbabwe.
HOST PLANTS. Leguminosae: *Berlinia globiflora*.
REMARKS. This species was misidentified as *Platysaissetia kellyi* by Hall (1935) and by Hodgson (1969b).

Etiennea kellyi (Brain)

Saissetia kellyi Brain, 1920b: 12.
Etiennea kellyi (Brain); Hodgson, 1991a: 201.
TYPE DATA. Lectotype female designated by Hodgson (1991a),

SOUTH AFRICA: Natal, Pietermaritzburg, on *Acacia melanoxylon* (SANC).
DISTRIBUTION. ETHIOPIAN REGION: South Africa.
HOST PLANTS. Leguminosae: *Acacia melanoxylon.*
REMARKS. Adult female redescribed and illustrated by Hodgson (1991a). The redescription and illustration of *Platysaissetia kellyi* by Hodgson (1969b) is a misidentification of *Etiennea halli.* The records of *Platysaissetia kellyi* from Zimbabwe by Hall (1935) and by Hodgson (1969b) are misidentifications of *E. halli.*

Etiennea madagascariensis Hodgson

Etiennea madagascariensis Hodgson, 1991a: 203.
TYPE DATA. Holotype female, MADAGASCAR: on *Daniellia* aff. *similis* (MNHN).
DISTRIBUTION. MADAGASIAN REGION: Madagascar.
HOST PLANTS. Bignoniaceae: *Daniellia* aff. *similis.*

Etiennea montrichardiae (Newstead)

Platysaissetia montrichardiae Newstead, 1920: 192.
Etiennea montrichardiae (Newstead); Hodgson, 1991a: 205.
TYPE DATA. Lectotype female designated by Hodgson (1991a), GUYANA: Ikruaka Lake, Essequibo, on *Montrichardia aculeata* (BMNH).
DISTRIBUTION. NEOTROPICAL REGION: Guyana.
HOST PLANTS. Araceae: *Montrichardia aculeata.*
REMARKS. Adult female redescribed and illustrated by Hodgson (1991a).

Etiennea multituberculum Hodgson

Etiennea multituberculum Hodgson, 1991a: 208.
TYPE DATA. Holotype female, GABON: Cap Etrias, on *Ouratea* sp. (MNHN).
DISTRIBUTION. ETHIOPIAN REGION: Gabon.
HOST PLANTS. Ochnaceae: *Ouratea.*

Etiennea petasus Hodgson

Etiennea petasus Hodgson, 1991a: 210.
TYPE DATA. Holotype female, NIGERIA: Ibadan, International Institute of Tropical Agriculture, on undetermined tree (BMNH).
DISTRIBUTION. ETHIOPIAN REGION: Nigeria.

Etiennea sinetuberculum Hodgson

Etiennea sinetuberculum Hodgson, 1991a: 212.
TYPE DATA. Holotype female, CENTRAL AFRICAN REPUBLIC: La Moboke, on *Pancovia laurantii* (MNHN).
DISTRIBUTION. ETHIOPIAN REGION: Central African Republic.
HOST PLANTS. Sapindaceae: *Pancovia laurantii.*

Etiennea tafoensis Hodgson

Etiennea tafoensis Hodgson, 1991a: 214.
TYPE DATA. Holotype female, GHANA: Tafo, on *Ficus* sp. (BMNH).

DISTRIBUTION. ETHIOPIAN REGION: Ghana, Sierra Leone, Zaire.
HOST PLANTS. Moraceae: *Ficus*. **Rubiaceae:** *Morinda geminata*.

Etiennea ulcusculum Hodgson

Etiennea ulcusculum Hodgson, 1991a: 216.
TYPE DATA. Holotype female, ZAIRE: Eala, on *Macrolobium* sp. (MNHN).
DISTRIBUTION. ETHIOPIAN REGION: Zaire.
HOST PLANTS. Leguminosae: *Macrolobium*.

Etiennea villiersi Matile-Ferrero

Etiennea villiersi Matile-Ferrero, 1984a: 101.
TYPE DATA. Holotype female, SENEGAL: Djibelor, on *Aphania senegalensis* (MNHN).
DISTRIBUTION. ETHIOPIAN REGION: Senegal, Sierra Leone, Zaire.
HOST PLANTS. Sapindaceae: *Aphania senegalensis*. **Sapotaceae:** *Chrysophyllum cainito*. **Sterculiaceae:** *Cola*.
REMARKS. Adult female redescribed and illustrated by Hodgson (1991a). Distribution and host plant records given by Hodgson (1991a).

Eucalymnatus Cockerell

Eucalymnatus Cockerell in Cockerell & Parrott, 1901: 57.
Eucalymnatus Lindinger; Lindinger, 1943b: 147 [UNJUSTIFIED EMENDATION].
TYPE-SPECIES: *Lecanium tessellatum* Signoret, by original designation and monotypy.
REMARKS. Generic characters discussed by Steinweden (1929), Gómez-Menor Ortega (1937, 1958c), Zimmerman (1948), Borchsenius (1957), De Lotto (1965), Paik (1978), Hadzibejli (1983), Wang (1980), Yang (1982), Tao *et al.* (1983), Williams & Watson (1990) and by Tang (1991).

Eucalymnatus brunfelsiae (Hempel)

Lecanium brunfelsia Hempel, 1900b: 418.
Eucalymnatus brunfelsiae (Hempel); Cockerell, 1902f: 453 [JUSTIFIED EMENDATION].
TYPE DATA. Syntypes female, BRAZIL: Sao Paulo, Pilar and Alto da Serra, on *Brunfelsia* sp. (MZSP).
DISTRIBUTION. NEOTROPICAL REGION: Brazil Sao Paulo.
HOST PLANTS. Lauraceae: *Laurus*. **Solanaceae:** *Brunfelsia*.
REMARKS. Fernald (1903) explained the emendation of the specific name to *brunfelsiae*, which was adopted by Hempel (1912). Adult female redescribed by Hempel (1901a).

Eucalymnatus chelonioides Newstead

Lecanium (Eucalymnatus) chelonioides Newstead, 1917a: 369.
TYPE DATA. Syntypes female, and second instar, GUYANA: Georgetown, Botanic Gardens, on *Pachira insignis*, and Essequibo River near Agatask, on *Pachira aquatica* (BMNH).
DISTRIBUTION. NEOTROPICAL REGION: Guyana.

HOST PLANTS. Bombacaceae: *Pachira aquatica, Pachira insignis.*

Eucalymnatus decemplex Newstead

Lecanium (Eucalymnatus) decemplex Newstead, 1920: 188.
TYPE DATA. Syntypes female, GUYANA: Ayaria, Thuraka Lake, Ituribisci Creek, Essequebo, on *Lecythis* sp. (BMNH).
DISTRIBUTION. NEOTROPICAL REGION: Guyana.
HOST PLANTS. Lecythidaceae: *Lecythis.*

Eucalymnatus delicatus Hempel

Eucalymnatus delicatus Hempel, 1937: 12.
TYPE DATA. Syntypes female, BRAZIL: Sao Paulo, Bofete, on "arbusto sylvestre do campo" (MZSP).
DISTRIBUTION. NEOTROPICAL REGION: Brazil Sao Paulo.

Eucalymnatus gracilis (Hempel)

Lecanium gracile Hempel, 1900b: 419.
Eucalymnatus gracilis (Hempel); Cockerell, 1902f: 453.
TYPE DATA. Syntypes female, BRAZIL: Sao Paulo, Villa Americana, on Sapindaceae (MZSP).
DISTRIBUTION. NEOTROPICAL REGION: Brazil Sao Paulo.
HOST PLANTS. Sapindaceae.
REMARKS. Adult female redescribed by Hempel (1901a).

Eucalymnatus gracilis nictheroyensis Costa Lima

Eucalymnatus gracilis nictheroyensis Costa Lima, 1923: 36.
TYPE DATA. Syntypes female, BRAZIL: Minas Gerais, Nichteroy, Sao Francisco, on *Cestrum nocturnum* (MZSP).
DISTRIBUTION. NEOTROPICAL REGION: Brazil Minas Gerais.
HOST PLANTS. Solanaceae: *Cestrum nocturnum.*

Eucalymnatus hempeli Costa Lima

Eucalymnatus hempeli Costa Lima, 1923: 40.
TYPE DATA. Syntypes female, BRAZIL: Minas Gerais, Nictheroy on *Mangifera indica*, and Santa Thereza on *Lacuma caimito* (MZSP).
DISTRIBUTION. NEOTROPICAL REGION: Brazil Minas Gerais.
HOST PLANTS. Anacardiaceae: *Mangifera indica.*

Eucalymnatus hirsutus Hempel

Eucalymnatus hirsutus Hempel, 1937: 14.
TYPE DATA. Syntypes female, BRAZIL: Sao Paulo, Alto da Serra, on "arbusto silvestre" (MZSP).
DISTRIBUTION. NEOTROPICAL REGION: Brazil Sao Paulo.

Eucalymnatus itanhaensis Mendes

Eucalymnatus itanhaensis Mendes, 1931: 395.
TYPE DATA. Syntypes female, male and larva, BRAZIL: Sao Paulo, Linha Santos-Juquia, on Myrtaceae (MZSP).
DISTRIBUTION. NEOTROPICAL REGION: Brazil Sao Paulo.
HOST PLANTS. Myrtaceae.

Eucalymnatus magarinosi Costa Lima

Eucalymnatus magarinosi Costa Lima, 1923: 38.
TYPE DATA. Syntypes female, BRAZIL: Rio de Janeiro State, Alto de Therezopolis, on undetermined plant (MZSP).
DISTRIBUTION. NEOTROPICAL REGION: Brazil Rio de Janeiro.

Eucalymnatus rigidus Hempel

Eucalymnatus rigidus Hempel, 1937: 15.
TYPE DATA. Syntypes female, BRAZIL: Sao Paulo, Horto Florestal, on "arvore silvestre" (MZSP).
DISTRIBUTION. NEOTROPICAL REGION: Brazil Sao Paulo.

Eucalymnatus scutigerus Costa Lima

Eucalymnatus scutigerus Costa Lima, 1930b: 86.
TYPE DATA. Holotype female, BRAZIL: Bahia, on *Moquilea tomentosa* (FOCB).
DISTRIBUTION. NEOTROPICAL REGION: Brazil Bahia.
HOST PLANTS. Chrysobalanaceae: *Moquilea tomentosa*.

Eucalymnatus spinosus Costa Lima

Eucalymnatus spinosus Costa Lima, 1923: 37.
TYPE DATA. Syntypes female, BRAZIL: Minas Gerais, Nictheroy, Botanic Garden, on undetermined plant (MZSP).
DISTRIBUTION. NEOTROPICAL REGION: Brazil Minas Gerais.

Eucalymnatus tessellatus (Signoret)

Lecanium tessellatum Signoret, 1873b: 401.
Lecanium perforatum Newstead, 1894b: 233. Lectotype female designated by Williams & Watson (1990), ENGLAND: Kew, on *Caryota cumingii* (BMNH). Syn. by Green, 1922a: 461.
Lecanium tessellatum swainsonae Cockerell, 1897c: 109. Syntypes female, JAMAICA: on lignumvitae [=*Guaiacum sanctum*] (USNM). Syn. by Ray & Williams, 1981: 231.
Lecanium tessellatum perforatum (Newstead); Cockerell, 1897d: 90.
Lecanium (Eucalymnatus) tessellatum (Signoret); Cockerell & Parrott, 1901: 57.
Eucalymnatus tessellatus (Signoret); Cockerell, 1902f: 453.
Coccus tessellatum (Signoret); Kirkaldy, 1902: 106.
Eucalymnatus perforatus (Newstead); Fernald, 1903: 166.
Eucalymnatus tessellatus swainsonae Cockerell; Fernald, 1903: 167.
Lecanium subtessellatum Green, 1904d: 206. Lectotype female designated by Williams & Watson (1990), SRI LANKA: Kandy, on undetermined tree (BMNH). Syn. by Green, 1907: 205.
Eucalymnatus subtessellatus (Green); Sanders, 1906: 7.
Lecanium (Eucalymnatus) tessellatum prforatum Newstead; Newstead, 1914a: 306.
Lecanium (Eucalymnatus) perforatum Newstead; Pettit & McDaniel, 1920: 18.
Lecanium tessellatum obsoletum Green, 1922b: 1024. Lectotype female designated by Williams & Watson (1990), SRI LANKA: Matale, on *Myrtus communis* (BMNH). Syn. by Green, 1937: 300.

Lecanium tesselatum obsoletum Green; Ramakrishna Ayyar, 1926: 454 [MIS-SPELLING].
Lecanium tesselatum Signoret; Ramakrishna Ayyar, 1926: 454 [MIS-SPELLING].
Eucalymnatus tesselatus (Signoret); Balachowsky, 1927: 185 [MIS-SPELLING].
TYPE DATA. Lectotype female designated by Ray & Williams (1981), FRANCE: Montpellier (in greenhouse), on *Caryota ursus* (VMNH).
DISTRIBUTION. AUSTRO ORIENTAL REGION: Indonesia, Malaysia, Papua New Guinea, Solomon Islands. **AUSTRALIAN REGION:** New South Wales. **ETHIOPIAN REGION:** Kenya, Zanzibar. **MADAGASIAN REGION:** Comoros, Farquhar Island, Madagascar, Mauritius, Providence Island, Reunion, Seychelles. **NEOTROPICAL REGION:** Bermuda, Brazil Rio Grande do Sul, Brazil Sao Paulo, Cuba, Dominican Republic, Guyana, Puerto Rico, Venezuela, Virgin Islands. **NEW ZEALAND and PACIFIC REGION:** Cook Islands, Fiji, French Polynesia, Hawaii, Kiribati, New Caledonia, Norfolk Island, Ogasawara Islands, Tonga, Tuvalu, Vanuatu, Western Samoa. **NEARCTIC REGION:** Alabama, Arkansas, California, Florida, Illinois, Indiana, Kansas, Louisiana, Maryland, Massachusetts, Michigan, Missouri, New Jersey, New York, Ohio, Pennsylvania, South Carolina, Tennessee, Texas, Washington, D. C. **ORIENTAL REGION:** India, Sri Lanka, Taiwan, Thailand, Vietnam. **PALAEARCTIC REGION:** Canary Islands, Egypt, England, France, Germany, Italy, Korea, Madeira, Poland, Spain, Sweden, Tibet, Turkey, USSR Crimea, USSR Georgia.
HOST PLANTS. Acanthaceae: *Sanchezia nobilis.* **Anacardiaceae:** *Mangifera indica, Schinus terebinthifolius.* **Annonaceae:** *Annona muricata.* **Apocynaceae:** *Alyxia olivaeformis, Nerium oleander, Plumeria acutifolia, P. rubra.* **Aquifoliaceae:** *Ilex cassine.* **Araceae:** *Anthurium andraeanum, Scindapsus aureus.* **Araliaceae:** *Brassaia actinophylla, Meryta angustifolia.* **Arecaceae:** *Elaeis guineensis.* **Asclepiadaceae:** *Calotropis gigantea.* **Barringtoniaceae:** *Barringtonia asiatica.* **Bischofiaceae:** *Bischofia javanica.* **Caricaceae:** *Carica papaya.* **Compositae:** *Baccharis halimifolia, Fitchia.* **Cucurbitaceae: Elaeocarpaceae:** *Elaeocarpus.* **Euphorbiaceae:** *Drypetes integerrima.* **Flagellariaceae:** *Flagellaria.* **Gnetaceae:** *Gnetum gnemon.* **Goodeniaceae:** *Scaevola taccada.* **Guttiferae:** *Calophyllum indicum, C. inophyllum.* **Heliconiaceae:** *Heliconia.* **Hippocrateaceae:** *Salacia.* **Lauraceae:** *Cinnamomum elegans, C. zeylanicum, Machilus, Persea.* **Leguminosae:** *Cajanus cajan.* **Malvaceae:** *Lagunaria patersonii.* **Meliaceae:** *Aglaia formosana.* **Moraceae:** *Ficus aurea, F. boninsimae, F. elastica, F. tinctoria.* **Musaceae:** *Musa paradisiaca, M. sapientum.* **Myrsinaceae:** *Ardisia, A. sieboldii.* **Myrtaceae:** *Eucalyptus citriodora, Eugenia caryophyllata, E. cumini, E. jambolana, E. malaccensis, Metrosideros, M. boninensis, Psidium guajava, Syzygium buxiflorum.* **Oleaceae:** *Jasminum, Olea verrucosa.* **Orchidaceae:** *Calanthe, Vanilla.* **Palmae:** *Archontophoenix cunninghamii, Arenga engleri, Caryota rumphiana, C. urens, C. ursus, Chrysalidocarpus lutescens, Cocos nucifera, Howeia forsterana, Kentia, Livistona chinensis, Neodypsis decaryi, Nypa fruticans, Phoenix canariensis, P. roebelinii, Rhapis humilis, Sabal blackburniana.*

Pandanaceae: *Pandanus.* **Pittosporaceae:** *Pittosporum bracteo-latum.* **Rubiaceae:** *Gardenia jasminoides, Morinda citrifolia, Randia tahitensis, Timonius.* **Rutaceae:** *Citrus aurantifolia, Citrus limon, C. paradisi, C. reticulata, Evodia.* **Sapindaceae:** *Dodonaea viscosa, Euphoria longana, Litchi chinensis.* **Sapotaceae:** *Chrysophyllum cainito, Palaquium formosanum.* **Solanaceae:** *Solanum wendlandii.* **Sonneratiaceae:** *Sonneratia casseolaris.* **Sterculiaceae:** *Heritiera littoralis, Pterospermum acerifolium, Theobroma cacao.* **Theaceae:** *Schima mertensiana.* **Urticaceae:** *Boehmeria, Pilea urticifolia.* **Verbenaceae:** *Lantana camara, Premna.* **Zingiberaceae:** *Alpinia purpurata, Elettaria cardamomum, Zingiber officinale.* **Zygophyllaceae:** *Guaiacum sanctum.*

REMARKS. Common Name - Tessellated scale. Adult female redescribed and illustrated by Tyrell (1896), Thro (1903) (as *L. perforatum*), Newstead (1903) (as *E. perfortum*), Green (1907), Froggatt (1915), Dietz & Morrison (1916), Kuwana (1917a), Pettit & McDaniel (1920) (as *E. perforatum*), Leonardi (1920), Borchsenius (1937, 1957), Gómez-Menor Ortega (1937, 1958c), Zimmerman (1948), Ezzat & Hussein (1969), Paik (1978), Tereznikova (1981), Ray & Williams (1981), Wang (1980), Kawai (1980), Yang (1982), Tao *et al.* (1983), Hamon & Williams (1984), Gill (1988), by Williams & Watson (1990) and by Tang (1991). Colour photograph given by Kawai (1980) and by Gill (1988). Larval instars described and illustrated by Gómez-Menor Ortega (1937, 1958c) and by Ray & Williams (1981). Nur (1971, 1972) studied the cytology and reported chromosome number 2n=16 in California. Distribution and host plant records given by Tyrell (1896), Thro (1903), Green (1904d, 1907, 1937), Newstead (1914a), Leonardi (1917, 1920), Kuwana (1917a), Pettit & McDaniel (1920), Hall (1925), Borchsenius (1937, 1957), Takahashi (1942a, 1952), Mamet (1943, 1959b), Zimmerman (1948), Bodenheimer (1953), Gómez-Menor Ortega (1958c), Ossiannilsson (1959), De Lotto (1965), Beardsley (1966a), Ezzat & Hussein (1969), Ali (1971), Kawai *et al.* (1971), Koteja (1972), Kozárzhevskaya & Reitzel (1975), Panis & Martin (1976), Takagi (1977), Matile-Ferrero (1978), Wang (1980, 1981), Tereznikova (1981), Ray & Williams (1981), Nakahara (1981b, 1983), Nakahara & Miller (1981), Yang (1982), Tao *et al.* (1983), Vieira *et al.* (1983), Hamon & Williams (1984), Williams & Williams (1988), Gill (1988), Jhala *et al.* (1989), Williams & Watson (1990), Danzig & Konstantinova (1990), Tang (1991) and by Hodgson & Hilburn (1991a, 1991b).

Eulecanium Cockerell

Eulecanium Cockerell, 1893d: 54.

Lecanium (Globulicoccus) Lindinger, 1907b: 138. Type-species: *Coccus fuscus* Gmelin, by original indication. Syn. by synonymy of type-species.

Globulicoccus Lindinger; MacGillivray, 1921: 180.

Enlecanium Cockerell; Cockerell, 1929: 150 [MIS-SPELLING].

TYPE SPECIES: *Lecanium tiliae* Linnaeus, by original designation and by Opinion 1303 (1985).

REMARKS. Generic characters discussed by Leonardi (1920), Gómez-Menor Ortega (1937, 1958c), Borchsenius (1957), Danzig (1967, 1980c), Paik (1978), Kosztarab & Kozár (1978, 1988), Wang

(1980), Tereznikova (1981), Nakahara (1981a), Yang (1982), Hamon & Williams (1984), Gill (1988) and by Tang (1991). Key to species: ASIA - Tang (1991). CALIFORNIA - Gill (1988). EUROPE - Borchsenius (1957), Kosztarab & Kozár (1978, 1988). ITALY -Leonardi (1920). KOREA - Paik (1978). SPAIN - Gómez-Menor Ortega (1958c). UKRAINE - Tereznikova (1981). USSR (EUROPEAN) - Borchsenius (1957), Danzig (1964). USSR (FAR EAST) - Borchsenius (1957), Danzig (1967, 1980c).

Eulecanium albodermis Chen

Eulecanium albodermis Chen, 1962: 284.
TYPE DATA. Syntypes female, CHINA: Szechwan, Kingtang Hsien, on *Prunus salicina* and *Citrus* sp. (EISC).
DISTRIBUTION. PALAEARCTIC REGION: China.
HOST PLANTS. Rosaceae: *Prunus salicina.* **Rutaceae:** *Citrus.*

Eulecanium alnicola Chen

Eulecanium alnicola Chen, 1962: 283.
TYPE DATA. Syntypes female, CHINA: Szechwan, Chengtu and Kingtang Hsien, on *Alnus* sp., *Prunus salicina, Salix babylonica* and *Pyrus* sp. (EISC).
DISTRIBUTION. PALAEARCTIC REGION: China.
HOST PLANTS. Betulaceae: *Alnus.* **Rosaceae:** *Prunus salicina, Pyrus.* **Salicaceae:** *Salix babylonica,* .

Eulecanium berberidis major (Maskell)

Lecanium berberidis Schrank; Maskell, 1897a: 311 [MISIDENTIFICATION].
Lecanium berberidis major Maskell, 1898: 238.
Eulecanium berberidis major (Maskell); Fernald, 1903: 182.
TYPE DATA. Syntypes female, AUSTRALIA: Victoria, Melbourne, on *Vitis vinifera* (probably lost; Deitz & Tocker, 1980).
DISTRIBUTION. AUSTRALIAN REGION: Victoria.
HOST PLANTS. Vitidaceae: *Vitis vinifera.*

Eulecanium caraganae Borchsenius

Eulecanium caraganae Borchsenius, 1953: 285.
TYPE DATA. Syntypes female, USSR: Ukraine, near Odessa, and Kazakhstan, on *Caragana frutex* and on *C. arborescens* (ZIAS).
DISTRIBUTION. PALAEARCTIC REGION: Mongolia, USSR Kazakhstan, USSR Ukraine.
HOST PLANTS. Leguminosae: *Caragana arborescens, C. frutex, C. pygmaea, C. spinosa.*
REMARKS. Adult female redescribed and illustrated by Borchsenius (1957), Kosztarab & Kozár (1978, 1988) and by Tereznikova (1981). Adult female redescribed by Tang (1991). Larval instars described by Borchsenius (1957) and by Tereznikova (1981). Distribution and host plant records given by Borchsenius (1957), Hadzibejli (1967b), Tereznikova (1981), Kosztarab & Kozár (1978, 1988) and by Tang (1991).
BIOLOGY. Develops one annual generation in USSR, Georgia (Hadzibejli, 1967b).

Eulecanium caryae (Fitch)

Lecanium caryae Fitch, 1857d: 443.
Lecanium (Eulecanium) caryae Cockerell, 1896b: 332.
Lecanium cockerelli Hunter, 1899b: 70. Syntypes female, U.S.A.:
  Kansas, Lawrence, on *Ulmus americana* (USNM). Syn. by San-
  ders, 1909a: 442.
Eulecanium caryae (Fitch); King, 1902c: 160.
Eulecanium cockerelli (Hunter); Fernald, 1903: 185.
TYPE DATA. Syntypes female, U.S.A.: New York, Salem, west side of
  Jarvis Martin's woods, on *Carya* sp. (AMNY).
DISTRIBUTION. NEARCTIC REGION: California, Canada Ontario,
  Connecticut, Indiana, Kansas, Maine, Maryland, Massachusetts,
  Michigan, Mississippi, New Jersey, New York, Ohio, Pennsylva-
  nia, Virginia, West Virginia.
HOST PLANTS. Fagaceae: *Castanea dentata, Quercus rubra.* **Ju-
  glandaceae:** *Carya.* **Platanaceae:** *Platanus occidentalis.* **Rosa-
  ceae:** *Persica vulgaris, Pyrus.* **Ulmaceae:** *Ulmus americana.*
REMARKS. Adult female redescribed and illustrated by Dietz &
  Morrison (1916), Pettit & McDaniel (1920), Richards (1958), Williams
  & Kosztarab (1972), Hamon & Williams (1984) and by Gill (1988).
  Male test described and illustrated by Miller & Williams (1990).
  Distribution and host plant records given by Pettit & McDaniel
  (1920), Richards (1958), Williams & Kosztarab (1972), Hamon &
  Williams (1984) and by Gill (1988).

Eulecanium cerasorum (Cockerell)

Lecanium cerasorum Cockerell, 1900b: 71.
Eulecanium cerasorum (Cockerell); Fernald, 1903: 184.
Lecanium (Saissetia) cerasorum (Cockerell); Reh, 1903: 417.
Eulecanium cerosarum (Cockerell); Takahashi & Tachikawa, 1956: 5
  [MIS-SPELLING].
TYPE DATA. Syntypes female, JAPAN: Intercepted at San Francisco,
  California, U. S. A., on cherry tree (USNM).
DISTRIBUTION. NEARCTIC REGION: California. **PALAEARCTIC
  REGION:** China, Japan, Korea.
HOST PLANTS. Magnoliaceae: *Magnolia kobus.* **Rosaceae:** *Cerasus
  vulgaris.*
REMARKS. Adult female redescribed and illustrated by Takahashi
  (1955b), Paik (1978) and by Gill (1988). Adult female redescribed by
  Borchsenius (1957), Kawai (1980) and by Tang (1991). Colour photograph given by Kawai (1980), Johnson & Lyon (1988) and by Gill
  (1988). Distribution and host plant records given by Takahashi
  (1955b), Takahashi & Tachikawa (1956), Borchsenius (1957), Paik
  (1978), Kawai (1980), Nakahara (1981a), Gill (1988) and by Tang
  (1991).

Eulecanium ciliatum (Douglas)

Lecanium ciliatum Douglas, 1891a: 67.
Lecanium (Eulecanium) ciliatum Douglas; Cockerell, 1896b: 332.
Eulecanium ciliatum (Douglas); Fernald, 1903: 184.
Palaeolecanium ciliatum (Douglas); Lindinger, 1935a: 137.

TYPE DATA. Syntypes female, ENGLAND: Stonehouse, Devon and Delamere, on *Quercus robur* (BMNH).
DISTRIBUTION. PALAEARCTIC REGION: Austria, Bulgaria, Czechoslovakia, England, France, Germany, Hungary, Inner Mongolia, Netherlands, Poland, Romania, Sweden, USSR Caucasus, USSR Primorye Territory, USSR Ukraine, Yugoslavia.
HOST PLANTS. Aceraceae: *Acer ginnala.* **Betulaceae:** *Alnus hirsuta, Alnus incana, Betula verrucosa.* **Caprifoliaceae:** *Lonicera.* **Corylaceae:** *Corylus.* **Fagaceae:** *Fagus silvatica, Quercus mongolica, Q. robur, Q. sessilis.* **Juglandaceae:** *Juglans mandshurica.* **Rosaceae:** *Malus, Prunus vulgaris.* **Salicaceae:** *Populus balsamifera, P. euphratica, Salix arenaria.* **Ulmaceae:** *Ulmus.*
REMARKS. Adult female redescribed and illustrated by Newstead (1903), Sulc (1932), Borchsenius (1957), Savescu (1961), Danzig (1967, 1980c), Tereznikova (1981) and by Kosztarab & Kozár (1978, 1988). Adult female redescribed by Tang (1991). Distribution and host plant records given by Cockerell (1896), Newstead (1903), Green (1923b), Sulc (1932), Borchsenius (1957), Reyne (1957), Ossiannilsson (1959), Danzig (1967, 1980c), Koteja (1971, 1983), Kozár et al. (1977), Tereznikova (1981), Kosztarab & Kozár (1978, 1988), Koteja & Zak-Ogaza (1983) and by Tang (1991).

Eulecanium circumfluum Borchsenius

Eulecanium circumfluum Borchsenius, 1955b: 292.
TYPE DATA. Syntypes female, CHINA: Khyebye Province, Tian-Zin, on *Robinia* sp. (ZIAS).
DISTRIBUTION. PALAEARCTIC REGION: China, Inner Mongolia.
HOST PLANTS. Leguminosae: *Robinia.* **Salicaceae:** *Salix babylonica.*
REMARKS. Adult female redescribed and illustrated by Borchsenius (1957). Adult female redescribed by Tang (1991).

Eulecanium coryli cimbricus Wünn

Eulecanium coryli cimbricus Wünn, 1937: 47.
TYPE DATA. Syntypes female, GERMANY: Schleswig-Holstein, Husum, on *Myrica gale* (DEPOSITORY UNKNOWN).
DISTRIBUTION. PALAEARCTIC REGION: Germany.
HOST PLANTS. Myricaceae: *Myrica gale.*

Eulecanium distinguendum (Douglas)

Lecanium distinguendum Douglas, 1891b: 96.
Lecanium (Eulecanium) distinguendum Cockerell, 1896b: 332.
Eulecanium distinguendum (Douglas); Fernald, 1903: 186.
TYPE DATA. Syntypes female, ENGLAND: Delamere Forest, on *Vaccinium myrtillus* (BMNH).
DISTRIBUTION. PALAEARCTIC REGION: England.
HOST PLANTS. Ericaceae: *Vaccinium myrtillus.*

Eulecanium douglasi (Sulc)

Lecanium douglasi Sulc, 1895: 37.
Lecanium (Eulecanium) douglasi Sulc; Cockerell & Parrott, 1899: 237.
Eulecanium douglasi (Sulc); Fernald, 1903: 186.

Physokermes douglasi (Sulc); Lindinger, 1911a: 381.
Lecanium zebrinum Green, 1917: 203. Syntypes female, ENGLAND: Camberley, on *Betula alba* and *Populus tremula* (BMNH). Syn. by Green, 1934: 110.
Eulecanium longisetum Borchsenius, 1955b: 296. Syntypes female, USSR: Crimea, Simferopol, and Yukalovsk Region, on *Populus alba* (ZIAS). Syn. by Danzig, 1978a: 14.
Eulecanium trjapitzini Danzig, 1967: 164. Holotype female, USSR: Primorye Territory, near Tigrovi, on *Sorbaria sorbifolia* (ZIAS). Syn. by Danzig, 1978a: 14.
Eulecanium eoum Danzig, 1967: 165. Holotype female, USSR: Primorye Territory, Ussurinsk, on *Salix rorida* (ZIAS). Syn. by Danzig, 1978a: 14.
Eulecanium coangustum Danzig, 1967: 166. Holotype female, USSR: Primorye Territory, on *Rosa dahurica* (ZIAS). Syn. by Danzig, 1978a: 14.
Eulecanium coum Danzig; Danzig, 1980c: 280 [MIS-SPELLING].
TYPE DATA. Syntypes female and larva, CZECHOSLOVAKIA: Mosoly near Prague, on *Betula alba* (MMBC).
DISTRIBUTION. PALAEARCTIC REGION: Czechoslovakia, Germany, Mongolia, Poland, USSR Altai, USSR Crimea, USSR Kazakhstan, USSR Kunashir Island, USSR Primorye Territory, USSR Sakhalin Island, USSR Ukraine, USSR Ural, USSR Yakutsk.
HOST PLANTS. Betulaceae: *Alnaster fruticosa, Alnus hirsuta, Betula alba, B. middendorfii, B. nana, B. platyphylla, B. pubescens, B. verrucosa.* **Corylaceae:** *Corylus.* **Grossulariaceae:** *Grossularia, Ribes rubrum, R. vulgare.* **Rosaceae:** *Rosa dahurica, R. spinosissima, Sorbaria sorbifolia, Sorbus, Spiraea salicifolia.* **Salicaceae:** *Populus alba, P. tremula, Salix caprea, S. rorida, S. viminalis.*
REMARKS. Adult female redescribed and illustrated by Borchsenius (1957), Danzig (1967, 1980c), Tereznikova (1981), Kosztarab & Kozár (1979, 1988) and by Tang (1991). Distribution and host plant records given by Lindinger (1911a), Borchsenius (1955b, 1957), Matesova (1968), Danzig (1967, 1977b, 1978a, 1978b, 1980c), Tereznikova (1981), Koteja & Zak-Ogaza (1983), Kosztarab & Kozár (1988) and by Tang (1991).

Eulecanium elegans Leonardi

Eulecanium elegans Leonardi, 1911: 273.
Lecanium elegans (Leonardi); Sasscer, 1912: 89.
TYPE DATA. Syntypes female, ARGENTINA: Cacheuta, on *Larrea cuneata* (IEAP).
DISTRIBUTION. NEOTROPICAL REGION: Argentina.
HOST PLANTS. Zygophyllaceae: *Bulnesia retama, Bulnesia schickendanzii, Larrea cuneata, L. cuneifolia, L. divaricata.*
REMARKS. Adult female redescribed and illustrated by Teran (1973). Distribution and host plant records given by Lizer y Trelles (1939) and by Teran (1973).

Eulecanium emerici (Planchon)

Chermes emerici Planchon, 1864: 21.
Lecanium emerici (Planchon); Signoret, 1869a: 852.
Sphaerolecanium emerici (Planchon); Leonardi, 1908: 180.

Eulecanium emerici (Planchon); Kozár & Walter, 1985: 77.
TYPE DATA. Syntypes female, FRANCE: on *Quercus coccifera* and *Q. ilex* (PROBABLY LOST).
DISTRIBUTION. PALAEARCTIC REGION: Algeria, France.
HOST PLANTS. Fagaceae: *Quercus coccifera*, *Quercus ilex.*
REMARKS. Adult female redescribed and illustrated by Leonardi (1908, 1920). Lindinger (1912) regarded this species as a synonym of *Eulecanium tiliae* (Linnaeus).

Eulecanium eugeniae (Hempel)

Lecanium eugeniae Hempel, 1900b: 439.
Eulecanium eugeniae (Hempel); Cockerell, 1902f: 453.
TYPE DATA. Syntypes female, BRAZIL: Ypiranga, on *Eugenia* sp. (MZSP).
DISTRIBUTION. NEOTROPICAL REGION: Brazil.
HOST PLANTS. Myrtaceae: *Eugenia.*
REMARKS. Adult female redescribed by Hempel (1901a).

Eulecanium excrescens (Ferris)

Lecanium excrescens Ferris, 1920b: 37.
Eulecanium excrescens (Ferris); Lindinger, 1933b: 159.
TYPE DATA. Holotype female, U.S.A.: California, Palo Alto, on cultivated English walnut [=*Juglans regia*] (UCD).
DISTRIBUTION. NEARCTIC REGION: California, Connecticut, New York, Pennsylvania.
HOST PLANTS. Juglandaceae: *Juglans regia.*
REMARKS. Adult female redescribed and illustrated by Gill (1988). Colour photograph given by Gill (1988).

Eulecanium ficiphilum Borchsenius

Eulecanium ficiphilum Borchsenius, 1955b: 293.
TYPE DATA. Syntypes female, IRAN: Persepolis, on *Ficus carica* (ZIAS).
DISTRIBUTION. PALAEARCTIC REGION: Afghanistan, Iran.
HOST PLANTS. Moraceae: *Ficus carica.*
REMARKS. Adult female redescribed and illustrated by Borchsenius (1957). Adult female redescribed by Tang (1991). Distribution and host plant records given by Borchsenius (1957), Danzig (1972d) and by Tang (1991).

Eulecanium fradei Almeida

Eulecanium fradei Almeida, 1969: 23.
TYPE DATA. Syntypes female, ANGOLA: Belas (Luanda), on undetermined tree (CZLP).
DISTRIBUTION. ETHIOPIAN REGION: Angola.

Eulecanium franconicum (Lindinger)

Lecanium vaccinii macrocarpum Goethe, 1884: 125. Syntypes female, GERMANY: on cuttings of American Preiselbeere [=*Vaccinium vitis-idaea*] (DEPOSITORY UNKNOWN). Syn. by Lindinger, 1935a: 137.

Lecanium rubellum Lindinger, 1907b: 138 [HOMONYM of *Lecanium rubellum* Cockerell].
Lecanium franconicum Lindinger, 1908: 181 [REPLACEMENT NAME].
Lecanium (Eulecanium) franconicum Lindinger; Sulc, 1932: 95.
Palaeolecanium franconicum (Lindinger); Lindinger, 1935a: 137.
Eulecanium franconicum (Lindinger); Borchsenius, 1957: 406.
Lecanium slavum Kawecki, 1961: 66. Syntypes female, POLAND: Tatra Mountains, near Jedrzejow, on *Vaccinium* sp. and *Calluna vulgaris* (PASK). Syn. by Danzig, 1980c: 282.
Eulecanium franconicum calluneti Danzig, 1961: 573. Syntypes female and first instar larva, USSR: on *Calluna vulgaris* (ZIAS). Syn. by Danzig, 1980c: 282.
Eulecanium franconicum vaccinicola Danzig, 1961: 571. Syntypes female and first instar larva, USSR: on *Vaccinium uliginosum* and *V. myrtillus* (ZIAS). Syn. by Danzig, 1980c: 282.
Eulecanium rhododendri Danzig, 1967: 162. Holotype female, USSR: Primorye Territory, Pidan, near Tigrovoi, on *Rhododendron mucronulatum* (ZIAS). Syn. by Danzig, 1978a: 14.
Eulecanium slavum (Kawecki); Komosinska, 1977: 23.
Eulecanium fraconicum (Lindinger); Tang, 1991: 157 [MISSPELLING].
TYPE DATA. Syntypes female, GERMANY: near Grafenberg, in Steinau and in Dresden, on *Calluna vulgaris* (HMNH).
DISTRIBUTION. PALAEARCTIC REGION: Bulgaria, Czechoslovakia, France, Germany, Hungary, Italy, Latvia, Poland, Sweden, USSR Irkutsk Region, USSR Kirgizia, USSR Leningrad Region, USSR Primorye Territory, USSR Ukraine, USSR Yakutsk.
HOST PLANTS. Ericaceae: *Calluna arborea, C. vulgaris, Rhododendron mucronulatum, Vaccinium myrtillus, V. uliginosum, V. vittisidaea.*
REMARKS. Lindinger (1935a) placed *L. vaccinii macrocarpum* Goethe, 1884 as a junior synonym of *P. franconicum* (Lindinger, 1908). Although Lindinger's interpretation disagrees with the Principle of Priority, his act is here accepted rather than resurrect the forgotten and unused name by Goethe. Adult female redescribed and illustrated by Sulc (1932), Borchsenius (1957), Danzig (1961) (as *E. franconicum f. calluneti* and *E. franconicum f. vaccinicola*), Danzig (1967) (as *E. rhododendri*), Danzig (1980c), Tereznikova (1981) and by Kosztarab & Kozár (1979, 1988). Adult female redescribed by Tang (1991). Danzig (1961) described two food forms, namely form *calluneti* and form *vaccinicola*, which were characterized by differences in body colour, size, form and host plants. Later these were regarded by Danzig (1980c) as host-induced variations of *E. franconicum*. Distribution and host plant records given by Sulc (1932), Borchsenius (1957), Danzig (1967, 1980c), Tereznikova (1981), Komosinska (1977), Koteja (1983), Ossiannilsson (1985), Marotta (1987), Dziedzicka (1988), Kosztarab & Kozár (1988) and by Tang (1991).

Eulecanium giganteum (Shinji)

Lecanium gigantea Shinji, 1935a: 289.
Eulecanium diminutum Borchsenius, 1955b: 293. Syntypes female,

USSR: Primorye Territory, Lyanzikhe and Okeanskoe, on *Quercus* sp. (ZIAS). Syn. by Tang, 1981: 165.
Eulecanium gigantea (Shinji); Wang, 1980: 45.
TYPE DATA. Syntypes female, JAPAN: Morioka, on *Magnolia kobus* (lost; S. Takagi, 1989, personal communication).
DISTRIBUTION. PALAEARCTIC REGION: China, Inner Mongolia, Japan, USSR Primorye Territory.
HOST PLANTS. Aceraceae: *Acer negundo.* **Corylaceae:** *Corylus mandshurica.* **Fagaceae:** *Quercus mongolica.* **Juglandaceae:** *Juglans mandshurica.* **Leguminosae:** *Maackia amurensis.* **Magnoliaceae:** *Magnolia kobus.* **Salicaceae:** *Salix.* **Ulmaceae:** *Ulmus.*
REMARKS. Adult female redescribed and illustrated by Borchsenius (1957) (as *E. diminutum*), Danzig (1967, 1980c) (as *E. diminutum*) and by Tang & Li (1988). Adult female redescribed by Wang (1980) and by Tang (1991). Distribution and host plant records given by Borchsenius (1957), Danzig (1967, 1980c), Wang (1980), Tang & Li (1988) and by Tang (1991). Tang (1991: 165) regarded the record of *Parthenolecanium glandi* (Kuwana) by Takahashi (1955b: 72) as a MISIDENTIFICATION of this species.

Eulecanium hirsutum (Newstead) n. comb.
Lecanium hirsutum Newstead; Newstead, 1917a: 350.
TYPE DATA. Syntypes female, EAST AFRICA: Country not given, on undetermined plant (BMNH).
DISTRIBUTION. ETHIOPIAN REGION: East Africa.

Eulecanium hissaricum Borchsenius
Eulecanium hissaricum Borchsenius, 1955b: 294.
TYPE DATA. Syntypes female, USSR: Tadzhikistan, Gissarsk Range, Zid, on *Lonicera* sp. (ZIAS).
DISTRIBUTION. PALAEARCTIC REGION: USSR Tadzhikistan.
HOST PLANTS. Caprifoliaceae: *Lonicera.*
REMARKS. Adult female redescribed and illustrated by Borchsenius (1957). Adult female redescribed by Tang (1991).

Eulecanium juniperi Danzig
Eulecanium juniperi Danzig, 1972b: 272.
TYPE DATA. Holotype female, USSR: Kazakhstan, Kalbinsk Range, near Nikitink, on *Juniperus* sp. (ZIAS).
DISTRIBUTION. PALAEARCTIC REGION: USSR Kazakhstan.
HOST PLANTS. Cupressaceae: *Juniperus.*

Eulecanium kostylevi Borchsenius
Eulecanium kostylevi Borchsenius, 1955b: 295.
TYPE DATA. Syntypes female, KOREA: South Hamgen, near Gapsan, on *Ulmus* sp. (ZIAS).
DISTRIBUTION. PALAEARCTIC REGION: Inner Mongolia, Korea, Mongolia, USSR Primorye Territory.
HOST PLANTS. Corylaceae: *Corylus heterophylla.* **Juglandaceae:** *Juglans mandshurica.* **Leguminosae:** *Maackia amurensis.* **Rosaceae:** *Rosa dahurica.* **Ulmaceae:** *Ulmus propinqua, U. pumila.*

REMARKS. Adult female redescribed and illustrated by Borchsenius (1957) and by Danzig (1967, 1980c). Adult female redescribed by Tang (1991). Distribution and host plant records given by Borchsenius (1957), Danzig (1967, 1977b, 1980c), Tang & Li (1988) and by Tang (1991).

Eulecanium kunmingi (Ferris)

Lecanium kunmingi Ferris, 1950: 73.
Coccus kunming (Ferris); Ali, 1971: 26 [MIS-SPELLING].
Eulecanium kunmingi (Ferris); Yang, 1982: 179.
TYPE DATA. Holotype female, CHINA: Yunnan Province, Kunming north park, on *Rhamnus* sp. (UCD).
DISTRIBUTION. PALAEARCTIC REGION: China.
HOST PLANTS. Rhamnaceae: *Rhamnus.* **Rosaceae:** *Pyracantha crenata.*
REMARKS. Adult female redescribed and illustrated by Tang (1991). Distribution and host plant records given by Tang (1991).

Eulecanium kunoense (Kuwana)

Lecanium kunoensis Kuwana, 1907: 191.
Eulecanium kunoense (Kuwana); Lindinger, 1933b: 159.
TYPE DATA. Syntypes female, JAPAN: on *Rhamnus japonicus, Prunus mume* and *Pyrus sinensis* (ITLJ).
DISTRIBUTION. NEARCTIC REGION: California. **PALAEARCTIC REGION:** China, Japan, Korea.
HOST PLANTS. Grossulariaceae: *Grossularia, Ribes.* **Hippocastanaceae:** *Aesculus.* **Juglandaceae:** *Juglans regia.* **Rhamnaceae:** *Rhamnus japonicus.* **Rosaceae:** *Amygdalus communis, Cerasus vulgaris, Crataegus , Cydonia oblonga, Malus sylvestris, Photinia glabra, Photinia villosa, Prunus mume, Prunus salicina, Prunus triflora, Prunus yedoensis, Pyracantha, Pyrus baccata, Pyrus sinensis.*
REMARKS. Adult female redescribed and illustrated by Kuwana (1917a), McKenzie (1951), Takahashi (1955b), Borchsenius (1957), Husseiny & Madsen (1962), Paik (1978), Kawai (1980), Gill (1988) and by Tang (1991). Male test described and illustrated by Miller & Williams (1990). Adult male and larvae described by Husseiny & Madsen (1962). Colour photograph given by Kawai (1980) and by Gill (1988). Distribution and host plant records given by Takahashi (1955b), Takahashi & Tachikawa (1956), Borchsenius (1957), Husseiny & Madsen (1962), Paik (1978), Kawai (1980), Gill (1988) and by Tang (1991).
BIOLOGY. Develops one annual generation in California (McKenzie, 1951; Husseiny & Madsen, 1962).
ECONOMIC IMPORTANCE. A minor pest to ornamentals in California, but regarded a potential pest in this state (Gill, 1988).

Eulecanium kuwanai (Kanda)

Eulecanium kuwanai Kanda, 1934: 405.
Lecanium kuwanai (Kanda); Takahashi, 1955b: 69.
Eulecanium knwanai (Kanda); Tang, 1991: 158 [MIS-SPELLING].

TYPE DATA. Syntypes female, JAPAN: Yokohama, on *Hedera* sp. and *Viburnum dilatatum* (KAYJ).
DISTRIBUTION. PALAEARCTIC REGION: China, Inner Mongolia, Japan.
HOST PLANTS. Araliaceae: *Hedera helix.* **Caprifoliaceae:** *Viburnum dilatatum.* **Ulmaceae:** *Ulmus laciniana.*
REMARKS. Adult female redescribed and illustrated by Tang & Li (1988). Adult female redescribed by Takahashi (1955b), Borchsenius (1957), Kawai (1980), Wang (1980) and by Tang (1991). Colour photograph given by Kawai (1980). Distribution and host plant records given by Takahashi (1955b), Borchsenius (1957), Kawai (1980), Wang (1980), Tang & Li (1988) and by Tang (1991).

Eulecanium lespedezae Danzig

Eulecanium lespedezae Danzig, 1967: 160.
TYPE DATA. Holotype female, USSR: Primorye Territory, Chernyatino, in Razdolna River valley, near Pokrovka, on *Lespedeza bicolor* (ZIAS).
DISTRIBUTION. PALAEARCTIC REGION: USSR Primorye Territory.
HOST PLANTS. Leguminosae: *Lespedeza bicolor.*
REMARKS. Adult female redescribed and illustrated by Danzig (1980c). Adult female redescribed by Tang (1991).

Eulecanium lymani King

Eulecanium lymani King, 1901i: 335.
TYPE DATA. Syntypes female, CANADA: Quebec, on oak (CNOC).
DISTRIBUTION. NEARCTIC REGION: Canada Quebec.
HOST PLANTS. Fagaceae: *Quercus.*

Eulecanium melzeri Hempel

Eulecanium melzeri Hempel, 1920b: 351.
TYPE DATA. Syntypes female, BRAZIL: Sao Paulo, Bosque de Saude, on indigenous shrub (MZSP).
DISTRIBUTION. NEOTROPICAL REGION: Brazil Sao Paulo.

Eulecanium nigrivitta Borchsenius

Eulecanium nigrivitta Borchsenius, 1959: 172.
TYPE DATA. Holotype female, CHINA: Yunnan Province, near Kunming, on *Castanopsis* sp. (IEBC).
DISTRIBUTION. PALAEARCTIC REGION: China.
HOST PLANTS. Fagaceae: *Castanopsis.*
REMARKS. Adult female redescribed by Tang (1991).

Eulecanium nocivum Borchsenius

Eulecanium nocivum Borchsenius, 1953: 286.
TYPE DATA. Syntypes female, USSR: Georgia, Poti, on *Liquidambar* sp. (ZIAS).
DISTRIBUTION. PALAEARCTIC REGION: Turkey, USSR Georgia.
HOST PLANTS. Aceraceae: *Acer.* **Ebenaceae:** *Diospyros kaki.* **Hamamelidaceae:** *Liquidambar.* **Rosaceae:** *Cotoneaster, Cratae-*

gus, Cydonia vulgaris, Malus sylvestris. **Salicaceae:** *Populus deltoides.*
REMARKS. Adult female redescribed and illustrated by Borchsenius (1957). Adult female redescribed by Tang (1991). Distribution and host plant records given by Borchsenius (1957), Hadzibejli (1967b, 1983), Kozár *et al.* (1979) and by Tang (1991).
BIOLOGY. Develops one annual generation in USSR, Georgia (Hadzibejli, 1967b).

Eulecanium pallidior (Cockerell & King)

Lecanium pallidior Cockerell & King, 1899: 350.
Eulecanium pallidior (Cockerell & King); Fernald, 1903: 191.
TYPE DATA. Syntypes female, U.S.A.: Massachusetts, Methuen, on *Chamaecyparis thyoides* (BMNH, USNM).
DISTRIBUTION. NEARCTIC REGION: Massachusetts.
HOST PLANTS. Cupressaceae: *Chamaecyparis thyoides.*

Eulecanium patersoniae (Maskell) n. comb.

Lecanium patersoniae Maskell, 1895b: 57.
Lecanium pattersoniae Maskell; Froggatt, 1915: 611 [MIS-SPELLING].
TYPE DATA. Syntypes female, AUSTRALIA: New South Wales, Sydney, on *Patersonia glabrata* (NZAC).
DISTRIBUTION. AUSTRALIAN REGION: New South Wales.
HOST PLANTS. Iridaceae: *Patersonia glabrata.*
REMARKS. Adult female redescribed and illustrated by Froggatt (1915).

Eulecanium paucispinosum Danzig

Eulecanium paucispinosum Danzig, 1967: 157.
TYPE DATA. Holotype female, USSR: Vladivostok, Okeanskaya, on *Ulmus laciniata* (ZIAS).
DISTRIBUTION. PALAEARCTIC REGION: USSR Primorye Territory, USSR Vladivostok.
HOST PLANTS. Aceraceae: *Acer mandshuricum, A. ukurunduense.* **Caprifoliaceae:** *Lonicera maackii.* **Carpinaceae:** *Carpinus cordata.* **Corylaceae:** *Corylus heterophylla, Corylus mandshurica.* **Rosaceae:** *Cerasus maximoviczii, Rosa rugosa.* **Ulmaceae:** *Ulmus laciniata, U. propinqua.*
REMARKS. Adult female redescribed and illustrated by Danzig (1980c). Adult female redescribed by Tang (1991). Distribution and host plant records given by Danzig (1980c) and by Tang (1991).

Eulecanium perinflatum (Cockerell)

Lecanium perinflatum Cockerell, 1914: 332.
Eulecanium perinflatum (Cockerell); Lizer y Trelles, 1939: 191.
TYPE DATA. Syntypes female, ARGENTINA: Santa Ana, Missiones, on twigs of herbaceous plant (USNM).
DISTRIBUTION. NEOTROPICAL REGION: Argentina, Uruguay.
HOST PLANTS. Compositae: *Baccharis.* **Convolvulaceae:** *Ipomoea.* **Myrtaceae:** *Eugenia edulis, Psidium guajava.* **Rutaceae:** *Citrus.*

Solanaceae: *Cestrum parqui, Solanum nodiflorum, Solanum pseudoquina.*
REMARKS. Distribution and host plant records given by Morrison (1923) and by Lizer y Trelles (1939).

Eulecanium pistaciae Borchsenius

Eulecanium pistaciae Borchsenius, 1955b: 297.
TYPE DATA. Syntypes female, USSR: Armenia, near Megri, on *Pistacia mutica* (ZIAS).
DISTRIBUTION. PALAEARCTIC REGION: USSR Armenia.
HOST PLANTS. Anacardiaceae: *Pistacia mutica.*
REMARKS. Adult female redescribed and illustrated by Borchsenius (1957). Adult female redescribed by Tang (1991).

Eulecanium pseudotessellatum (Newstead) n. comb.

Lecanium pseudotessellatum Newstead, 1917a: 351.
TYPE DATA. Syntypes female, TRINIDAD: Aripo Savana, on *Chrysobalanus pellocarpus* (BMNH).
DISTRIBUTION. NEOTROPICAL REGION: Trinidad.
HOST PLANTS. Chrysobalanaceae: *Chrysobalanus pellocarpus.*

Eulecanium pubescens (Ehrhorn)

Lecanium pubescens Ehrhorn, 1898a: 244.
Lecanium (Eulecanium) pubescens Ehrhorn; Cockerell & Parrott, 1899: 237.
Eulecanium pubescens (Ehrhorn); Fernald, 1903: 194.
TYPE DATA. Syntypes female, U.S.A.: California, near Mountain View, on *Quercus* sp. (USNM).
DISTRIBUTION. NEARCTIC REGION: California.
HOST PLANTS. Fagaceae: *Quercus.*
REMARKS. Ferris (1920b) noted the great similarity of this species to *Parthenolecanium corni.*

Eulecanium rugulosum (Archangelskaya)

Lecanium ciliatum Douglas; Archangelskaya, 1923: 264 [MISIDENTIFICATION].
Lecanium rugulosum Archangelskaya, 1937: 46.
Eulecanium rugulosum (Archangelskaya); Borchsenius, 1957: 395.
TYPE DATA. Syntypes female, USSR: Uzbekistan, Tadzhikistan, Turkmenia, Kirgizia and Kazakhstan, on *Populus ontariensis* (ZIAS).
DISTRIBUTION. PALAEARCTIC REGION: USSR Armenia, USSR Kazakhstan, USSR Kirgizia, USSR Tadzhikistan, USSR Turkmenia, USSR Uzbekistan.
HOST PLANTS. Hippocastanaceae: *Aesculus hippocastanum.* **Juglandaceae:** *Juglans regia.* **Rosaceae:** *Armeniaca, Cerasus, Crataegus, Cydonia vulgaris, Malus, Persica vulgaris, Prunus, Pyrus.* **Salicaceae:** *Populus, Salix.* **Ulmaceae:** *Ulmus.*
REMARKS. Adult female redescribed and illustrated by Borchsenius (1957). Adult female redescribed by Tang (1991). Distribution and host plant records given by Borchsenius (1957) and by Tang (1991).

Eulecanium sachalinense Danzig

Eulecanium sachalinense Danzig, 1972b: 271.
TYPE DATA. Holotype female, USSR: South Sakhalin, on *Rosa amblyotis* (ZIAS).
DISTRIBUTION. PALAEARCTIC REGION: USSR Sakhalin Island, USSR Yakutsk.
HOST PLANTS. Rosaceae: *Rosa acicularis, Rosa amblyotis.*
REMARKS. Adult female redescribed and illustrated by Danzig (1980c). Adult female redescribed by Tang (1991). Distribution and host plant records given by Danzig (1978a, 1978b, 1980c) and by Tang (1991).

Eulecanium sansho (Shinji) n. comb.

Lecanium sansho Shinji, 1935b: 773.
TYPE DATA. Syntypes female, JAPAN: Morioka, on *Zanthoxylum piperitum* (lost; S. Takagi, 1989, personal communication).
DISTRIBUTION. PALAEARCTIC REGION: Japan.
HOST PLANTS. Rutaceae: *Zanthoxylum piperitum.*
REMARKS. Takahashi (1955b) and Kawai (1980) regarded this species as undeterminate and retained it in *Lecanium*. Here it is placed in *Eulecanium*, for technnical reason, since *Lecanium* was placed on the Official Index of Rejected Names (Opinion 1303, 1985).

Eulecanium secretum Borchsenius

Eulecanium secretum Borchsenius, 1955b: 299.
TYPE DATA. Syntypes female, KOREA: Hamgen Province, Zan-Bai-Shan, on *Dasiphora dahurica* (ZIAS).
DISTRIBUTION. PALAEARCTIC REGION: Korea, Mongolia, USSR Altai, USSR Irkutsk Region, USSR Yakutsk.
HOST PLANTS. Rosaceae: *Cotoneaster lucida, Dasiphora dahurica, Dasiphora fruticosa.*
REMARKS. Adult female redescribed and illustrated by Borchsenius (1957) and by Danzig (1980c). Adult female redescribed by Tang (1991). Distribution and host plant records given by Borchsenius (1957), Danzig (1972c, 1978b, 1980c) and by Tang (1991).

Eulecanium sericeum (Lindinger)

Lecanium sericeum Lindinger, 1906: 147.
Lecanium (Globulicoccus) sericeum Lindinger; Lindinger, 1907b: 138.
Physokermes sericeus (Lindinger); Lindinger, 1911a: 381.
Eulecanium sericeum (Lindinger); Leonardi, 1920: 310.
Physockermes sericeus (Lindinger); Leonardi, 1920: 310 [MIS-SPELL-ING].
Enlecanium sericeum (Lindinger); Tang, 1991: 175 [MIS-SPELLING].
TYPE DATA. Syntypes female, GERMANY: Erlangen, on *Abies pectinata* (HMNG).
DISTRIBUTION. PALAEARCTIC REGION: Corsica, Czechoslovakia, Germany, Italy, Poland, USSR Caucasus, USSR Ukraine.
HOST PLANTS. Pinaceae: *Abies alba, Picea pectinata.*
REMARKS. Adult female redescribed and illustrated by Leonardi (1920), Kawecki (1938, 1956), Borchsenius (1957), Kosztarab & Kozár (1979, 1988), Tereznikova (1981). Adult female redescribed by

Tang (1991). Larval instars described and illustrated by Kawecki (1956). Distribution and host plant records given by Lindinger (1911a), Leonardi (1917, 1920), Balachowsky (1934a), Kawecki (1938), Borchsenius (1957), Rehacek (1957), Żak-Ogaza & Koteja (1964), Hadzibejli (1967b), Tereznikova (1981), Kosztarab & Kozár (1978, 1988) and by Tang (1991).
BIOLOGY. Develops one annual generation in USSR, Georgia (Hadzibejli, 1967b).

Eulecanium sibiricum Borchsenius

Eulecanium sibiricum Borchsenius, 1955b: 300.
TYPE DATA. Syntypes female, USSR: Irkutsk Region, Kazugsk, near Biryulki, on *Picea* sp. (ZIAS).
DISTRIBUTION. PALAEARCTIC REGION: USSR Irkutsk Region.
HOST PLANTS. Pinaceae: *Picea.*
REMARKS. Adult female redescribed and illustrated by Borchsenius (1957). Adult female redescribed by Tang (1991). Adult male described and illustrated by Borchsenius (1957). Distribution and host plant records given by Borchsenius (1957) and by Tang (1991).

Eulecanium subaustrale Cockerell

Lecanium (Eulecanium) subaustrale Cockerell, 1898d: 131.
Eulecanium subaustrale Cockerell; Fernald, 1903: 197.
TYPE DATA. Syntypes female, MEXICO: Amecameca, on *Celtis occidentalis* (BMNH, USNM).
DISTRIBUTION. NEOTROPICAL REGION: Mexico.
HOST PLANTS. Ulmaceae: *Celtis occidentalis.*
REMARKS. Sanders (1909a) and Borchsenius (1957) regarded this species as a synonym of *Parthenolecanium persicae.*

Eulecanium takachihoi Kuwana

Lecanium (Eulecanium) takachihoi Kuwana, 1902a: 63.
Eulecanium takachihoi Kuwana; Fernald, 1903: 197.
Parthenolecanium takachihoi (Kuwana); Borchsenius, 1957: 376.
TYPE DATA. Syntypes female, JAPAN: Kyushu, Hikosan, on chestnut tree (ITLJ).
DISTRIBUTION. PALAEARCTIC REGION: Japan, Korea, USSR Primorye Territory.
HOST PLANTS. Fagaceae: *Castanea, Quercus mongolica.*
REMARKS. Adult female redescribed and illustrated by Kuwana (1917a), Takahashi (1955b), Borchsenius (1957), Danzig (1967, 1980c), Paik (1978) and by Kawai (1980). Adult female redescribed by Tang (1991). Colour photograph given by Kawai (1980). Distribution and host plant records given by Kuwana (1917a), Takahashi (1955b), Borchsenius (1957), Danzig (1967, 1980c), Paik (1978), Kawai (1980) and by Tang (1991).

Eulecanium tiliae (Linnaeus)

Coccus tiliae Linnaeus, 1758: 456.
Coccus capreae Linnaeus, 1767: 741. Syntypes female, EUROPE: on *Salix* (lost; D. J. Williams, 1990, personal communication). Syn. by Lindinger, 1912: 363.

Coccus alni Modeer, 1778: 23. Syntypes female, SWEDEN: near
  Göthenburg, on common alder [=*Alnus glutinosa*] (DEPOSITORY
  UNKNOWN). Syn. by Marchal, 1908a: 296.
Coccus mali Schrank, 1781: 295. Syntypes female, AUSTRIA: on Pyro
  malo (DEPOSITORY UNKNOWN). Syn. by Lindinger, 1912: 364.
Coccus salicum Fabricius, 1781: 394. Syntypes female, EUROPE: on
  Salix (lost; Zimsen, 1964). Syn. by Fernald, 1903: 333.
Coccus fuscus Gmelin, 1790: 2221. Syntypes female, EUROPE: on
  Quercus roboris (DEPOSITORY UNKNOWN). Syn. by Lindinger,
  1912: 364.
Coccus aceris Fabricius, 1794: 225. Syntypes female, EUROPE: on
  Acer (lost; Zimsen, 1964). Syn. by Marchal, 1908a: 296.
Coccus rubi Schrank, 1801: 144. Syntypes female, AUSTRIA: on nut
  tree (DEPOSITORY UNKNOWN). Syn. by Lindinger, 1912: 365.
Coccus pyri Schrank, 1801: 145. Syntypes female, AUSTRIA: on pear
  (DEPOSITORY UNKNOWN). Syn. by Lindinger, 1912: 365.
Coccus xylostei Schrank, 1801: 145. Syntypes female, AUSTRIA: on
  cherry (DEPOSITORY UNKNOWN). Syn. by Sulc, 1932: 119.
Coccus aceris campestris Schrank, 1801: 147. Syntypes female,
  AUSTRIA: Ingolstadt, on Massholders [=*Acer*] (DEPOSITORY
  UNKNOWN). Syn. by Marchal, 1908a: 296.
Calypticus fasciatus Costa, 1829: 14. Syntypes female, NETHER-
  LANDS: on *Ulmus* (probably lost; G. Pellizzari Scaltriti, 1990,
  personal communication). Syn. by Leonardi, 1920: 296.
Lecanium juglandis Bouche, 1844: 299. Syntypes female, and male,
  GERMANY: Berlin, on *Juglans regia* and *J. nigra* (lost; Sachtle-
  ben, 1944). Syn. by Borchsenius, 1957: 422.
Coccus aesculi Kollar; Kollar, 1848: 188 [NOMEN NUDUM]. See
  Newstead, 1903: 105.
Lecanium rubi (Schrank); Walker, 1852: 1073.
Lecanium fasciatum (Costa); Walker, 1852: 1078.
Lecanium genevense Targioni Tozzetti, 1868: 731. Syntypes female,
  EUROPE: on *Oxyacantha* sp. (probably lost; G. Pellizzari Scaltri-
  ti, 1990, personal communication). Syn. by Lindinger, 1912: 38.
Lecanium alni (Modeer); Signoret, 1869a: 843.
Lecanium pyri (Schrank); Signoret, 1869a: 868.
Lecanium berberidis (Schrank); Signoret, 1874a: 403 [MISIDENTIFI-
  CATION].
Lecanium aesculi (Kollar) [NOMEN NUDUM]. See Signoret, 1874a:
  412.
Lecanium cerasi Goethe; Goethe, 1884: 125. Syntypes female,
  GERMANY: on cherry and plum trees (DEPOSITORY
  UNKNOWN). Syn. by Lindinger, 1912: 370.
Lecanium variegatum Goethe, 1884: 125. Syntypes female, GERMA-
  NY: on plum and apple trees (DEPOSITORY UNKNOWN). Syn. by
  Lindinger, 1912: 123.
Lecanium fuscum (Gmelin); Douglas, 1887b: 98.
Lecanium (Eulecanium) aesculi (Kollar) [NOMEN NUDUM]. See Cock-
  erell, 1896b: 332
Lecanium (Eulecanium) capreae (Linnaeus); Cockerell, 1896b: 332.
Lecanium (Eulecanium) genevense Targioni Tozzetti; Cockerell,
  1896b: 332.
Lecanium (Eulecanium) juglandis Bouche; Cockerell, 1896b: 332.
Lecanium (Eulecanium) rubi (Schrank); Cockerell, 1896b: 332.

Lecanium (Eulecanium) cerasi Goethe; Cockerell, 1896b: 332.
Lecanium (Eulecanium) variegatum Goethe; Cockerell & Parrott, 1899: 235.
Eulecanium fuscum (Gmelin); Cockerell, 1901b: 91.
Lecanium (Eulecanium) hoferi King in Hofer, 1903: 478. Syntypes female, SWITZERLAND: on pear, apple and plum (USNM). Syn. by Lindinger, 1912: 370.
Lecanium websteri mirabilis King in Hofer, 1903: 482. Syntypes female, SWITZERLAND: on *Acer negundo* (USNM). Syn. by Lindinger, 1912: 371.
Eulecanium aceris (Curtis); Fernald, 1903: 180 [ERRONEOUS AUTHORSHIP].
Eulecanium aesculi (Kollar) [NOMEN NUDUM]. See Fernald, 1903: 180.
Eulecanium alni (Modeer); Fernald, 1903: 180.
Eulecanium capreae (Linnaeus); Fernald, 1903: 183.
Eulecanium cerasi (Goethe); Fernald, 1903: 184.
Eulecanium fasciatum (Costa); Fernald, 1903: 187.
Eulecanium pyri (Schrank); Fernald, 1903: 194.
Eulecanium rubi (Schrank); Fernald, 1903: 196.
Eulecanium tiliae (Linnaeus); Fernald, 1903: 197.
Eulecanium variegatum (Goethe); Fernald, 1903: 198.
Eulecanium websteri mirabile (King); Fernald, 1903: 198.
Lecanium (Saissetia) capreae (Linnaeus); Reh, 1903: 416.
Eulecanium curtisi Kirkaldy; Kirkaldy, 1904b: 257 [UNNECESSARY REPLACEMENT NAME for *Coccus aceris* Fabricius, as redescribed by Curtis, 1838]. Syn. by Lindinger, 1912: 363.
Lecanium (Globulicoccus) fuscum (Gmelin); Lindinger, 1907b: 138.
Eulecanium emerici (Planchon); Leonardi, 1908: 178 [MISIDENTIFICATION].
Globulicoccus fuscus (Gmelin); MacGillivray, 1921: 180.
Eulecanium mali (Schrank); Borchsenius, 1955a: 866.
Eulecanium ibericum Hadzibejli, 1960b: 316. Syntypes female, USSR: Georgia, on *Juglans regia* and *Quercus iberica* (ZIAS). Syn. by Danzig, 1977a: 100.
Eulecanium gyrcanicum Hadzibejli, 1967a: 719. Syntypes female, USSR: Caucasus, Astar, on *Parrotia persicae*, *Crataegus lagenaria* and *C. kyrtostyla* (ZIAS). Syn. by Danzig, 1977a: 100.
TYPE DATA. Syntypes female, EUROPE: on *Tilia* (lost; D. J. Williams, 1990, personal communication).
DISTRIBUTION. NEARCTIC REGION: California, Canada British Columbia, Canada Ontario. **ORIENTAL REGION:** India, Pakistan. **PALAEARCTIC REGION:** Austria, Bulgaria, Canary Islands, Corsica, Czechoslovakia, England, France, Germany, Greece, Hungary, Israel, Italy, Netherlands, Poland, Romania, Spain, Sweden, Switzerland, Turkey, USSR Caucasus, USSR Georgia, USSR Moldavia, USSR Ukraine, Yugoslavia.
HOST PLANTS. Aceraceae: *Acer campestre*, *A. circinatum*, *A. macrophyllum*, *A. negundo*, *A. platanoides*, *A. pseudoplatanus*. **Betulaceae:** *Betula alba*. **Carpinaceae:** *Carpinus betulus*. **Cornaceae:** *Cornus sanguinea*. **Corylaceae:** *Corylus avellana*. **Ericaceae:** *Vaccinium myrtillus*. **Fagaceae:** *Quercus coccifera*, *Q. iberica*, *Q. ilex*, *Q. robur*, *Q. suber*. **Grossulariaceae:** *Ribes uvacrispa*. **Hippocastanaceae:** *Aesculus hippocastanum*, *A. pavia*.

Juglandaceae: *Juglans regia.* **Moraceae:** *Ficus carica, Morus.* **Myricaceae:** *Myrica gale.* **Rosaceae:** *Armeniaca vulgaris, Cotoneaster, Crataegus azarolus, C. intricata, C. monogyna, C. oxyacantha, Cydonia oblonga, Fragaria vesca, Persica vulgaris, Prunus laurocerasus, Prunus spinosa, Pyracantha coccinea, Pyrus communis, Rosa, Rubus.* **Salicaceae:** *Populus canadensis, Salix alba, S. caprea, S. scoparius, S. viminalis.* **Tiliaceae:** *Tilia cordata.* **Ulmaceae:** *Ulmus foliacea, U. glabre, U. minor, U. scabra.* **Vitidaceae:** *Vitis vinifera.*

REMARKS. Danzig (1977a) and Danzig & Kerzhner (1981) have shown that this species was named by some students - *Eulecanium coryli* (Linnaeus), a name also used by others for *Parthenolecanium corni* (Bouche), thus creating confusion and instability. Opinion 1303 (1985) placed the name *Coccus coryli* Linnaeus on the Official Index of Rejected Names, and accepted *Coccus tiliae* Linnaeus as the senior synonym of this soft scale. Fernald (1903) erroneously credited the authorship of *Coccus aceris* Fabricius as redescribed by Curtis (1838: 717) to Curtis. Kirkaldy (1904b: 257) also erroneously referred to the redescription of *Coccus aceris* Fabricius by Curtis (1838) as *Coccus aceris* Curtis and replaced the latter by *Eulecanium curtisi* Kirkaldy. However, the latter was unnecessary and became a synonym of *E. tiliae.* Adult female redescribed and illustrated by Silvestri (1919, 1920b), Ferris (1925), Gómez-Menor Ortega (1937, 1958c), Borchsenius (1957), Kawecki (1958b), Phillips (1965a), Kosztarab & Kozár (1978, 1988), Tereznikova (1981), Hadzibejli (1983) and by Gill (1988). Adult female redescribed by Lellakova-Duskova (1966), Canard (1980) and by Tang (1991). Adult male described and illustrated by Silvestri (1919, 1920b) (as *E. coryli*). Kawecki (1958b) (as *L. coryli*), Giliomee (1967). Male test described and illustrated by Miller & Williams (1990). Larval instars described and illustrated by Silvestri (1919) and by Kawecki (1958b). Colour photograph of adult female, male and nymphs given by Gill (1988). Colour photograph of adult female and male given by Kosztarab & Kozár (1988). Savescu (1943, 1944) recognized, on the basis of morphology of the adult female, 5 ecological forms: *coryli coniformis* (on *Malus pumila*), *coryli hoferi* (on *Prunus domestica*), *coryli aesculi* (on *Aesculus hypocastanum*) and *coryli aceris* (on *Acer tataricum*). All these are here regarded as host-induced variations of *Eulecanium tiliae*. Distribution and host plant records given by Leonardi (1920), Silvestri (1919, 1920b), Borchsenius (1957), Richards (1958), Gómez-Menor Ortega (1958c), Ossiannilsson (1959), Hadzibejli (1960b, 1967a, 1967b), Phillips (1965a), Kozárzhevskaya & Reitzel (1975), Kozár et al. (1977), Danzig (1977a), Kozár & Kosztarab (1978, 1988), Canard (1980), Varshney (1985), Marotta (1987), Dziedzicka (1988), Gill (1988) and by Tang (1991).

BIOLOGY. Develops one annual generation in Europe (Borchsenius, 1957; Hadzibejli, 1967b; Kosztarab & Kozár, 1988). Adult female develops in South Germany by April-May, and oviposits at the beginning of May (Schmutterer, 1952b). An obligatory biparental species (Kawecki, 1961). Rubin & Beirne (1975a, 1975b) studied the life history and natural enemies in British Columbia.

Eulecanium transcaucasicum Borchsenius

Eulecanium transcaucasicum Borchsenius, 1955b: 301.
TYPE DATA. Syntypes female, USSR: Armenia, Yerevan, on *Cydonia vulgaris* and on *Ulmus* sp., (ZIAS).
DISTRIBUTION. PALAEARCTIC REGION: USSR Armenia.
HOST PLANTS. Rosaceae: *Cydonia vulgaris.* **Ulmaceae:** *Ulmus.*
REMARKS. Adult female redescribed and illustrated by Borchsenius (1957). Adult female redescribed by Tang (1991). Distribution and host plant records given by Borchsenius (1957) and by Tang (1991).

Eulecanium transvittatum (Green)

Lecanium transvittatum Green, 1917: 206.
Eulecanium transvittatum (Green); Borchsenius, 1957: 397.
TYPE DATA. Syntypes female, ENGLAND: Camberley, on *Betula alba* (BMNH).
DISTRIBUTION. PALAEARCTIC REGION: England.
HOST PLANTS. Betulaceae: *Betula alba.*
REMARKS. Adult female redescribed by Borchsenius (1957) and by Tang (1991). Lindinger (1932f) regarded this species as a synonym of *E. tiliae.*

Eulecanium zygophylli Danzig

Eulecanium zygophylli Danzig; Danzig, 1972c: 344.
Eulecanium zygophylli Danzig; Tang, 1991: 158 [MIS-SPELLING].
TYPE DATA. Holotype female, MONGOLIA: Gobi-Altai Aymak, near Begera, on *Zygophyllum xanthoxylon* (ZIAS).
DISTRIBUTION. PALAEARCTIC REGION: Mongolia.
HOST PLANTS. Zygophyllaceae: *Zygophyllum xanthoxylon.*
REMARKS. Adult female redescribed by Tang (1991).

Eumashona Hodgson

Mashona Hodgson, 1967a: 8. [HOMONYM of *Mashona* Pate].
Eumashona Hodgson, 1973: 63. [REPLACEMENT NAME].
TYPE-SPECIES: *Mashona tachardia* Hodgson, by original designation.
REMARKS. Key to species given by Hodgson (1967a, 1969b).

Eumashona msasae (Hall)

Lecanium msasae Hall, 1935: 75.
Coccus msasae (Hall); De Lotto, 1959: 165.
Mashona msasae (Hall); Hodgson, 1967a: 10.
Eumashona msasae (Hall); Hodgson, 1973: 63.
TYPE DATA. Syntypes female, ZIMBABWE: Salisbury, The Kopje, on *Brachystegia* sp. (BMNH).
DISTRIBUTION. ETHIOPIAN REGION: Zimbabwe.
HOST PLANTS. Leguminosae: *Brachystegia.*
REMARKS. Adult female redescribed and illustrated by De Lotto (1959).

Eumashona tachardia (Hodgson)

Mashona tachardia Hodgson, 1967a: 9.

Eumashona tachardia (Hodgson); Hodgson, 1973: 63.
TYPE DATA. Holotype female, ZIMBABWE: Marandellas, on *Brachystegia glaucescens* (SANC).
DISTRIBUTION. ETHIOPIAN REGION: Zimbabwe.
HOST PLANTS. Leguminosae: *Brachystegia glaucescens.*

Eupulvinaria Borchsenius

Eupulvinaria Borchsenius, 1953: 288.
Pulvinaria (Eupulvinaria) Borchsenius; Danzig, 1980c: 260.
TYPE-SPECIES: *Eupulvinaria peregrina* Borchsenius, by original designation.
REMARKS. Generic characters discussed by Borchsenius (1957), Danzig (1967a, 1980c), Hadzibejli (1983) and by Tang (1991). Danzig (1980c) amended the status of this taxon to a subgenus of *Pulvinaria.* The genus is here regarded as a subjective synonym of *Pulvinaria.* Key to species: ASIA - Tang (1991). USSR - Borchsenius (1957).

Eutaxia Green

Eutaxia Green, 1926a: 60.
TYPE-SPECIES: *Eutaxia moreirae* Green, by original designation and monotypy.

Eutaxia moreirai Green

Eutaxia moreirae Green, 1926a: 61.
Eutaxia moreirai Green; Corseuil & Barbosa, 1971: 239 [JUSTIFIE-DEMENDATION].
TYPE DATA. Syntypes female, BRAZIL: Rio Grande Do Sul, on undetermined shrub (BMNH).
DISTRIBUTION. NEOTROPICAL REGION: Brazil Rio Grande do Sul.

Exaeretopus Newstead

Exaeretopus Newstead, 1894a: 204.
Exoerctopus Cockerell; Cockerell, 1894m: 1051 [MIS-SPELLING].
Exeraetopus Bodenheimer; Bodenheimer, 1928: 192 [MIS-SPELLING].
TYPE-SPECIES: *Exaeretopus formiceticola* Newstead, by monotypy.
REMARKS. Koteja (1980b) revised the genus and gave key to species. Lindinger (1937) regarded it as a synonym of *Lecanopsis.* Goux (1937) regarded this genus as a subgenus of *Luzulaspis.* Generic characters discussed by Green (1921a), Steinweden (1929), Goux (1937), Borchsenius (1957), Koteja (1978), Kosztarab & Kozár (1978), Danzig (1980c), Koteja (1980b) and by Tang (1991). Key to species: ASIA - Tang (1991).

Exaeretopus agropyri (Hadzibejli)

Luzulaspis agropyri Hadzibejli, 1960b: 312.
Exaeretopus agropyri (Hadzibejli); Koteja, 1978: 319.
TYPE DATA. Syntypes female and male, USSR: Georgia, on *Agropyrum caucasicum* (ZIAS, GPPT).
DISTRIBUTION. PALAEARCTIC REGION: USSR Georgia.
HOST PLANTS. Gramineae: *Agropyron caucasicum.*

REMARKS. Adult female redescribed and illustrated by Hadzibejli (1973) and by Koteja (1980b).
BIOLOGY. Phenology in Georgia, USSR, given by Hadzibejli (1973).

Exaeretopus boonei Hollinger

Exaeretopus boonei Hollinger, 1923: 41.
TYPE DATA. Syntypes female, U.S.A.: Missouri, near McBaine, on *Ulmus* sp. (USNM).
DISTRIBUTION. NEARCTIC REGION: Missouri.
HOST PLANTS. Ulmaceae: *Ulmus.*
REMARKS. Koteja (1978) discussed the taxonomic features of this species and suggested that it does not belong to *Exaeretopus* or *Luzulaspis.*

Exaeretopus dianthi Koteja

Exaeretopus dianthi Koteja, 1980b: 357.
Exaeretopus dianthus Koteja; Tang, 1991: 48.
TYPE DATA. Holotype female, USSR: Tadzhikistan, southern slopes of Hissar Mts., Donbar, on *Dianthus* sp. (ZIAS).
DISTRIBUTION. PALAEARCTIC REGION: USSR Tadzhikistan.
HOST PLANTS. Caryophyllaceae: *Dianthus.*
REMARKS. Adult female redescribed and illustrated by Tang (1991).

Exaeretopus formiceticola Newstead

Spermococcus fallax Giard; Newstead, 1893b: 207 [MISIDENTIFICA-TION].
Exaeretopus formiceticola Newstead, 1894a: 204.
Luzulaspis luzulae (Dufour); Lindinger, 1912: 132 [ERRONEOUS SYNONYMY].
Luzulaspis (Exaeretopus) formiceticola Newstead; Goux, 1937: 96.
TYPE DATA. Syntypes female and larva, ENGLAND: Guernsey, in ants' nest (BMNH).
DISTRIBUTION. PALAEARCTIC REGION: Channel Islands, England, Yugoslavia.
HOST PLANTS. Gramineae: *Dactylis glomerata, Thymus.*
REMARKS. Adult female redescribed and illustrated by Green (1928b), Goux (1939), Kosztarab & Kozár (1978), Koteja (1980b) and by Kozár (1983). Adult female redescribed by Borchsenius (1957). Distribution and host plant records given by Green (1928b), Goux (1939), Kosztarab & Kozár (1978), Koteja (1980b) and by Kozár (1983).

Exaeretopus harpazi Ben-Dov

Exaeretopus harpazi Ben-Dov, 1987: 111.
TYPE DATA. Holotype female, ISRAEL: Rehovot, on *Aegilops* sp. (ICV).
DISTRIBUTION. PALAEARCTIC REGION: Israel.
HOST PLANTS. Gramineae: *Aegilops, Avena sterilis, Hordeum spontaneum, Lolium rigidum.*
BIOLOGY. Adult and ovipositing females occur in March - April, end of winter, on leaves and stems of annual grasses (Ben-Dov, 1987).

Exaeretopus mahunkai Kozár & Drozdják

Exaeretopus mahunkai Kozár & Drozdják, 1990: 364.
TYPE DATA. Holotype female, HUNGARY: Tarpa, on leaf of *Festuca*
sp. (HNHB).
DISTRIBUTION. PALAEARCTIC REGION: Hungary.
HOST PLANTS. Gramineae: *Festuca.*
BIOLOGY. The paratypes were collected from leaf litter of a *Quercus*-
forest (Kozár & Drozdják, 1990).

Exaeretopus orientalis Danzig

Exaeretopus orientalis Danzig, 1975a: 137.
TYPE DATA. Holotype female, USSR: Eastern Sayan, on *Carex pedi-
formis* (ZIAS).
DISTRIBUTION. PALAEARCTIC REGION: USSR Primorye Territory.
HOST PLANTS. Cyperaceae: *Carex nanella, C. pediformis.*
REMARKS. Adult female redescribed and illustrated by Koteja
(1980b) and by Danzig (1980c).

Exaeretopus pimpinellae Borchsenius

Exaeretopus pimpinellae Borchsenius, 1957: 117.
TYPE DATA. Syntypes female, USSR: Azerbaijan, near Vishnevsk
Station, on *Pimpinella* sp. (ZIAS)
DISTRIBUTION. PALAEARCTIC REGION: USSR Azerbaijan, USSR
Ukraine.
HOST PLANTS. Labiatae: *Thymus dimorphus.* **Umbelliferae:** *Pimpi-
nella.*
REMARKS. Adult female redescribed and illustrated by Koteja
(1980b) and by Tereznikova (1981). Distribution and host plant
records given by Koteja (1980b) and by Tereznikova (1981).

Exaeretopus tritici Williams

Exaeretopus tritici Williams, 1977: 281.
TYPE DATA. Holotype female, IRAQ: Naynawa, on *Triticum aestivum*
(BMNH).
DISTRIBUTION. PALAEARCTIC REGION: Iraq.
HOST PLANTS. Gramineae: *Triticum aestivum.*
REMARKS. Adult female redescribed and illustrated by Koteja
(1980b).

Filippia Targioni Tozzetti

Philippia Targioni Tozzetti, 1867: 23 [HOMONYM of *Philippia* Gray].
Filippia Targioni Tozzetti, 1868: 726 [REPLACEMENT NAME].
Defilippia Targioni Tozzetti; Targioni Tozzetti, 1868: 726 [UNJUSTI-
FIED REPLACEMENT NAME and HOMONYM of *Defilippia* Salva-
dori].
Euphilippia Berlese & Silvestri, 1906: 396. TYPE-SPECIES: *Euphilip-
pia olivina* Berlese & Silvestri, by monotypy. Syn. by community
of type species.
Phylippia Leonardi; Leonardi, 1920: 340 [MIS-SPELLING].
TYPE-SPECIES: *Philippia follicularis* Targioni Tozzetti, by subsequent
designation of Morrison & Morrison (1966).

REMARKS. Ben-Dov (1973, 1975) discussed the nomenclature and clarified its taxonomic identity. Generic characters discussed by Ben-Dov (1973, 1975) and by Tang (1991).

Filippia follicularis (Targioni Tozzetti)

Coccus oleae Olivier; Costa, 1857: 74 [MISIDENTIFICATION in part].
Philippia follicularis Targioni Tozzetti, 1867: 23.
Filippia follicularis (Targioni Tozzetti); Targioni Tozzetti, 1868: 726.
Philippia oleae (A. Costa); Signoret, 1872a: 433 [ERRONEOUS SYNONYMY and AUTHORSHIP].
Coccus oleae A. Costa; Costa, 1877: 117 [MISIDENTIFICATION and ERRONEOUS AUTHORSHIP].
Filippia oleae (O. G. Costa); Fernald, 1903: 146 [MISIDENTIFICATION and ERRONEOUS AUTHORSHIP].
Euphilippia olivina Berlese & Silvestri, 1906: 398. Syntypes female, ITALY: Toscana and Portici, on olive (IEAP). Syn. by Ben-Dov, 1975: 111.

TYPE DATA. Syntypes female, ITALY: on olive (probably lost; G. Pellizzari Scaltriti, 1990, personal communication).
DISTRIBUTION. PALAEARCTIC REGION: France, Greece, Israel, Italy, Spain, Turkey.
HOST PLANTS. Anacardiaceae: *Pistacia lentiscus*. **Araliaceae:** *Hedera helix*. **Caprifoliaceae:** *Viburnum tinus*. **Myrtaceae:** *Myrtus communis*. **Oleaceae:** *Olea europaea, Phyllirea latifolia, Phyllirea variabilis*.
REMARKS. Ben-Dov (1973, 1975) clarified the nomenclature and taxonomy of this species and showed that *Filippia oleae* (Costa) of authors, generally was a misidentification of *F. follicularis*. Adult female redescribed and illustrated by Ben-Dov (1973), Leonardi (1920) (as *E. olivina*) and by Tang (1991). Larval instars described and illustrated by Leonardi (1920), Gómez-Menor Ortega (1937) and by Ben-Dov (1973). Male described and illustrated by Leonardi (1920) and by Gómez-Menor Ortega (1937). Colour photograph given by Tranfaglia & Viggiani (1988). Distribution and host plant records given by Leonardi (1920), Gómez-Menor Ortega (1937), Bodenheimer (1953), Ben-Dov (1973, 1975) and by Marotta (1987).
BIOLOGY. Quaglia & Raspi (1979a) studied the ecology and life history on olive in Italy, Toscana. One annual generation was observed in Sicily, Italy (Longo, 1988), Verona, Italy (Pellizzari Scaltriti, 1981) and in Greece (Argyriou & Kourmadas, 1977).

Gascardia Targioni Tozzetti

Gascardia Targioni Tozzetti, 1893: 88.
TYPE-SPECIES: *Gascardia madagascariensis* Targioni Tozzetti, by monotypy.
REMARKS. This genus is regarded here as a subjective synonym of *Ceroplastes*. Newstead (1909) noted the affinity of this genus to *Ceroplastes*. De Lotto (1965) discussed this genus and separated it from *Ceroplastes*.

Hadzibejliaspis Koteja

Hadzibejliaspis Koteja, 1978: 317.

Hadzibejliaspis Tang; Tang, 1991: 51 [MIS-SPELLING].
TYPE-SPECIES: *Exaeretopus stipae* Hadzibejli, by original designation and monotypy.
REMARKS. Generic characters discussed by Tang (1991).

Hadzibejliaspis stipae (Hadzibejli)
Exaeretopus stipae Hadzibejli, 1960b: 310.
Hadzibejliaspis stipae (Hadzibejli); Koteja, 1978: 318.
Hadzibejliaspis stipae (Hadzibejli); Tang, 1991: 52 [MIS-SPELLING].
TYPE DATA. Syntypes female, USSR: Georgia, near Tbilisi, on *Stipa caucasica* (ZIAS, GPPI).
DISTRIBUTION. PALAEARCTIC REGION: USSR Georgia.
HOST PLANTS. Gramineae: *Stipa caucasica*.
REMARKS. Adult female redescribed and illustrated by Hadzibejli (1973). Adult female redescribed by Tang (1991).

Halococcus Takahashi
Halococcus Takahashi, 1951a: 1.
TYPE-SPECIES: *Halococcus formicarii* Takahashi, by original designation and monotypy.
REMARKS. Generic characters discussed by Hodgson (1990) and by Tang (1991).

Halococcus formicarii Takahashi
Halococcus formicarii Takahashi, 1951a: 1.
TYPE DATA. Syntypes female, MALAYSIA: Morib, Port Swettenham, Selangor, Port Dickson and Negri Sembilan, on *Sonneratia cascolaris*, *S. alba* and on *Avicennia* sp. (HUSJ).
DISTRIBUTION. AUSTRO ORIENTAL REGION: Malaysia.
HOST PLANTS. Celastraceae: *Sonneratia alba*, *S. casseolaris*. **Verbenaceae:** *Avicennia*.
REMARKS. Adult female redescribed and illustrated by Tang (1991).
BIOLOGY. Found in hollows made by *Crematogaster* sp. in branches of the host plants (Takahashi, 1951a)).

Hemilecanium Newstead
Hemilecanium Newstead, 1906: 74 [NOMEN NUDUM].
Hemilecanium Newstead, 1908c: 39.
TYPE-SPECIES: *Hemilecanium theobromae* Newstead, by monotypy.
REMARKS. Generic characters discussed Steinweden (1929) and by Hodgson (1969b, 1969c).

Hemilecanium coriaceum Hall
Hemilecanium coriaceum Hall, 1935: 73.
TYPE DATA. Syntypes female, ZIMBABWE: Victoria Falls, on undetermined plant (BMNH).
DISTRIBUTION. ETHIOPIAN REGION: Zimbabwe.

Hemilecanium imbricans (Green)
Lecanium imbricans Green, 1903: 94.
Hemilecanium imbricans (Green); Newstead, 1908c: 39.

TYPE DATA. Syntypes female, INDIA: Nilgiris, On *Ficus mysorensis* (BMNH).
DISTRIBUTION. ORIENTAL REGION: India.
HOST PLANTS. Euphorbiaceae: *Jatropha*. **Fagaceae:** *Quercus*. **Meliaceae:** *Cedrela toona*. **Moraceae:** *Ficus mysorensis*. **Rubiaceae:** *Coffea*.
REMARKS. Adult female redescribed and illustrated by Ramakrishna Ayyar (1919, 1930) and by Hodgson (1969c). Distribution and host plant records given by Newstead (1908c, 1917c), Hodgson (1969c), Ali (1971) and by Shafee *et al.* (1989).

Hemilecanium recurvatum Newstead
Hemilecanium recurvatum Newstead, 1910a: 18.
TYPE DATA. Syntypes female, ZAIRE: Romee near Stanleyville, on *Plectronia laurentii* (BMNH).
DISTRIBUTION. ETHIOPIAN REGION: Zaire.
HOST PLANTS. Rubiaceae: *Plectronia laurentii*.

Hemilecanium theobromae Newstead
Hemilecanium theobromae Newstead, 1906: 74 [NOMEN NUDUM].
Hemilecanium theobromae Newstead; Newstead, 1908c: 39.
Hemilecanium imbricans (Green); Hall, 1932: 195 [MISIDENTIFICATION].
TYPE DATA. Syntypes female, CAMEROON: Soppo, on *Theobroma cacao* (BMNH).
DISTRIBUTION. ETHIOPIAN REGION: Angola, Cameroon, South Africa, Zimbabwe.
HOST PLANTS. Anacardiaceae: *Harpephyllum caffrum*. **Apocynaceae:** *Nerium oleander*. **Euphorbiaceae:** *Acalypha wilkesiana, Euphorbia, Poinsettia*. **Sterculiaceae:** *Theobroma cacao*.
REMARKS. Adult female redescribed and illustrated by Hodgson (1969c). Intraspecific variation in taxonomic characters given by Hodgson (1969c). Distribution and host plant records given by Brain (1920b), Hall (1932), Hodgson (1969c) and by Almeida (1973).

Houardia Marchal
Houardia Marchal, 1909a: 586.
TYPE-SPECIES: *Houardia troglodytes* Marchal, by monotypy.
REMARKS. Hodgson (1990) revised the genus and provided a key to species. Generic characters discussed by Marchal (1909c) and by Hodgson (1990).

Houardia abdita De Lotto
Houardia abdita De Lotto, 1966b: 143.
TYPE DATA. Holotype female, SOUTH AFRICA: Transvaal, Pretoria, on *Burkea africana* (SANC).
DISTRIBUTION. ETHIOPIAN REGION: South Africa, Zimbabwe.
HOST PLANTS. Leguminosae: *Brachystegia spiciformis, Burkea africana*.
REMARKS. Adult female redescribed and illustrated by Hodgson (1990). Distribution and host plant records given by Hodgson (1969b, 1990).

BIOLOGY. De Lotto (1966b) found it in small coaxial galleries or cavities on branches of its host plant. Hodgson (1969b) collected it from the main stem of *Burkea spiciformis* where the bark had been damaged.

Houardia mozambiquensis Hodgson

Houardia mozambiquensis Hodgson, 1990: 222.
TYPE DATA. Holotype female, MOZAMBIQUE: Barra falsa, on *Casuarina cunninghamiana* (BMNH).
DISTRIBUTION. ETHIOPIAN REGION: Mozambique.
HOST PLANTS. Casuarinaceae: *Casuarina cunninghamiana.*

Houardia troglodytes Marchal

Houardia troglodytes Marchal, 1909a: 586.
TYPE DATA. Lectotype female designated by Hodgson (1990), SENEGAL: on *Balanites aegyptiaca* (MNHN).
DISTRIBUTION. ETHIOPIAN REGION: Senegal, Sudan, Zimbabwe.
HOST PLANTS. Balanitaceae: *Balanites.* **Leguminosae:** *Burkea africana.*
REMARKS. Adult female redescribed and illustrated by Marchal (1909c) and by Hodgson (1990). Larva and male described and illustrated by Marchal (1909c). Distribution and host plant records given by Hodgson (1990).
BIOLOGY. This scale insect occupies, together with ants of the genus *Crematogaster*, galls on twigs of the host plant (Marchal, 1909a, 1909c).

Idiosaissetia Brain

Idiosaissetia Brain, 1920b: 40.
TYPE-SPECIES: *Idiosaissetia peringueyi* Brain, by original designation and monotypy.

Idiosaissetia peringueyi Brain

Idiosaissetia peringueyi Brain, 1920b: 40.
TYPE DATA. Syntypes female, SOUTH AFRICA: Cape Province, on grass or thin reed (SANC).
DISTRIBUTION. ETHIOPIAN REGION: South Africa.
HOST PLANTS. Gramineae:

Inglisia Maskell

Inglisia Maskell, 1879: 213.
Inglesia Bruner *et al.* ; Bruner *et al.*, 1975: 361 [MIS-SPELLING].
TYPE-SPECIES: *Inglisia patella* Maskell, 1879, by monotypy.
REMARKS. Generic characters discussed by Green (1909), Froggatt (1915), Morrison & Morrison (1922), Steinweden (1929), Hodgson (1967d, 1969b), Yang (1982), Tao *et al.* (1983), Hamon & Williams (1984) and by Tang (1991). Key to species: ASIA - Tang (1991). ETHIOPIAN REGION - Hodgson (1967d). ZIMBABWE - Hodgson (1969b).

Inglisia australis Hempel

Inglisia australis Hempel, 1937: 10.
TYPE DATA. Syntypes female, BRAZIL: Sao Paulo, on *Galactia stenophylla* (MZSP).
DISTRIBUTION. NEOTROPICAL REGION: Brazil Sao Paulo.
HOST PLANTS. Leguminosae: *Galactia stenophylla.*

Inglisia chelonioides Green

Inglisia chelonioides Green, 1909: 283.
TYPE DATA. Syntypes female, SRI LANKA: Punadaluoya, on *Gelonium lanceolatum* (BMNH).
DISTRIBUTION. ORIENTAL REGION: India, Sri Lanka.
HOST PLANTS. Casuarinaceae: *Casuarina equisetifolia.* **Leguminosae:** *Acacia leucophloea, Caesalpinia, Parkinsonia aculeata, Pithecellobium dulce.* **Sapindaceae:** *Gelonium lanceolatum.*
REMARKS. Adult female redescribed and illustrated by Green (1922), Ramakrishna Ayyar (1919, 1930) and by Tang (1991). Distribution and host plant records given by Newstead (1917c), Ramakrishna Ayyar (1919, 1930), Green (1922, 1937), Ali (1971) and by Tang (1991).

Inglisia conchiformis Newstead

Inglisia conchiformis Newstead, 1910c: 185.
Onicoccus conchiformis Newstead, 1910c: 186 [NOMEN NUDUM].
Cardiococcus cenehiformis (Newstead); Green, 1922b: 1034 [MISSPELLING].
TYPE DATA. Syntypes female, GUINEA: Conakry, Jardin Botanique, on *Averrhoa carambola* (BMNH).
DISTRIBUTION. ETHIOPIAN REGION: Cameroon, Ghana, Guinea, Sierra Leone, Sudan, Tanzania, Uganda, Zaire.
HOST PLANTS. Euphorbiaceae: *Acalypha.* **Guttiferae:** *Harungana madagascariensis.* **Leguminosae:** *Acacia, Albizia, Cajanus indicus, Derris dalbergioides, Gliricidia maculata.* **Myrtaceae:** *Psidium guajava.* **Oxalidaceae:** *Averrhoa carambola.* **Sterculiaceae:** *Theobroma cacao.*
REMARKS. Adult female redescribed and illustrated by Hodgson (1967d). Intraspecific variation in taxonomic characters given by Hodgson (1967d). Distribution and host plant records given by Newstead (1911a, 1917c), Strickland (1947), Hodgson (1967d) and by Matile-Ferrero & Nonveiller (1984).

Inglisia elytropappi Brain

Inglisia elytropappi Brain, 1920b: 36.
Cryptinglisia elytropappi (Brain); De Lotto, 1978: 140.
TYPE DATA. Syntypes female, SOUTH AFRICA: Cape Province, Groot Drakenstein, Somerset West and Cape Flats, on *Elytropappus rhinocerotis* (SANC).
DISTRIBUTION. ETHIOPIAN REGION: South Africa.
HOST PLANTS. Compositae: *Elytropappus rhinocerotis.*
REMARKS. Adult female redescribed and illustrated by De Lotto (1967b).

Inglisia fagi Maskell

Inglisia fagi Maskell, 1890b: xv [NOMEN NUDUM].
Inglisia fagi Maskell, 1891a: 13.
TYPE DATA. Syntypes female, NEW ZEALAND: Reefton District, on *Fagus* sp. (NZAC, USNM).
DISTRIBUTION. NEW ZEALAND and PACIFIC REGION: New Zealand.
HOST PLANTS. Fagaceae: *Fagus*.

Inglisia foraminifer major Maskell

Inglisia foraminifer major Maskell, 1897a: 309.
TYPE DATA. Syntypes female, AUSTRALIA: South Australia, Murray River, Swan Hill, on *Muehlenbeckia adpressa* (NZAC, USNM).
DISTRIBUTION. AUSTRALIAN REGION: South Australia.
HOST PLANTS. Polygonaceae: *Muehlenbeckia adpressa*.

Inglisia grevilleae Hall

Inglisia grevilleae Hall, 1935: 83.
TYPE DATA. Holotype female, ZIMBABWE: Bulawayo, on *Grevillea robusta* (BMNH).
DISTRIBUTION. ETHIOPIAN REGION: Zimbabwe.
HOST PLANTS. Proteaceae: *Grevillea robusta*.
REMARKS. Adult female redescribed and illustrated by Hodgson (1967d).

Inglisia inconspicua Maskell

Inglisia inconspicua Maskell, 1892a: 19.
TYPE DATA. Syntypes female, NEW ZEALAND: Reefton District, on *Corokia cotoneaster* (NZAC, USNM).
DISTRIBUTION. NEW ZEALAND and PACIFIC REGION: New Zealand.
HOST PLANTS. Escalloniaceae: *Corokia cotoneaster*.

Inglisia leptospermi Maskell

Inglisia leptospermi Maskell, 1882: 220.
TYPE DATA. Syntypes female, NEW ZEALAND: Christchurch, Kaiapoi, Wellington and Auckland, on *Leptospermum scoparium* (CMNZ, NZAC, USNM).
DISTRIBUTION. NEW ZEALAND and PACIFIC REGION: New Zealand.
HOST PLANTS. Myrtaceae: *Leptospermum scoparium*.
REMARKS. Adult female redescribed and illustrated by Maskell (1887b).

Inglisia malvacearum Cockerell

Inglisia malvacearum Cockerell, 1898f: 432.
TYPE DATA. Syntypes female, MEXICO: Morelos, on *Malva* sp. (BMNH, USNM).
DISTRIBUTION. NEOTROPICAL REGION: Mexico. **NEARCTIC REGION:** Texas.
HOST PLANTS. Malvaceae: *Malva, Malvaviscus penduliflorus*.

REMARKS. Male test described and illustrated by Miller & Williams (1990). Distribution and host plant records given by Miller & Williams (1990).

Inglisia ornata Maskell

Inglisia ornata Maskell, 1885: 27.
TYPE DATA. Syntypes female, NEW ZEALAND: North Island, on *Elaeocarpus* sp. and *Leptospermum* sp. (NZAC, USNM).
DISTRIBUTION. NEW ZEALAND and PACIFIC REGION: New Zealand.
HOST PLANTS. Elaeocarpaceae: *Elaeocarpus dentatus.* **Myrtaceae:** *Leptospermum scoparium.*

Inglisia patella Maskell

Inglisia patella Maskell, 1879: 213.
TYPE DATA. Syntypes female, NEW ZEALAND: Christchurch, Riccarton Bush, on *Coprosma* sp. (NZAC).
DISTRIBUTION. NEW ZEALAND and PACIFIC REGION: New Zealand.
HOST PLANTS. Atherospermataceae: *Atherosperma.* **Rubiaceae:** *Coprosma lucida.*
REMARKS. Adult female redescribed and illustrated by Maskell (1887b) and by Morrison & Morrison (1922).

Inglisia pluvialis Hodgson

Inglisia pluvialis Hodgson, 1969b: 18.
TYPE DATA. Holotype female, ZIMBABWE: Victoria Falls, on *Ficus capensis* (BMNH).
DISTRIBUTION. ETHIOPIAN REGION: Zimbabwe.
HOST PLANTS. Moraceae: *Ficus capensis.*

Inglisia speciosa Takahashi

Inglisia speciosa Takahashi, 1951b: 108.
TYPE DATA. Syntypes female, INDONESIA: Riau [-Riouw] Islands, host plant not indicated (HUSJ).
DISTRIBUTION. AUSTRO ORIENTAL REGION: Indonesia.
REMARKS. Adult female redescribed and illustrated by Tang (1991).

Inglisia theobromae Newstead

Inglisia theobromae Newstead, 1917b: 33.
Inglisia castilloae theobromae Newstead; Gowdey, 1917: 188 [MISIDENTIFICATION].
TYPE DATA. Syntypes female, UGANDA: Nagunga, on stems of cacao pods and flowers (BMNH).
DISTRIBUTION. ETHIOPIAN REGION: Cameroon, Uganda, Zambia.
HOST PLANTS. Euphorbiaceae: *Euphorbia hirta.* **Sterculiaceae:** *Theobroma cacao.*
REMARKS. Adult female redescribed and illustrated by Hodgson (1967d). Adult male described and illustrated by Giliomee (1967). Intraspecific variation in taxonomic characters of the female given by

Hodgson (1967d). Distribution and host plant records given by
Hodgson (1967d) and by Matile-Ferrero & Nonveiller (1984).

Inglisia vitrea Cockerell

Inglisia vitrea Cockerell, 1894b: 308.
TYPE DATA. Syntypes female, TRINIDAD: Port-of-Spain, on *Acacia*
     sp. (USNM).
DISTRIBUTION. NEOTROPICAL REGION: Brazil Amazonas, Brazil
     Sao Paulo, Cuba, Puerto Rico, Trinidad. **NEARCTIC REGION:**
     Florida.
HOST PLANTS. Annonaceae: *Rollinia mucosa.* **Lauraceae:** *Laurus
     nobilis, Persea borbonia.* **Leguminosae:** *Acacia, Calliandra.*
Myricaceae: *Myrica cerifera.*
REMARKS. Adult female redescribed and illustrated by Hamon &
Williams (1984) and by Foldi (1988). Colour photograph given by
Hamon & Williams (1984). Distribution and host plant records given
by Ballou (1926), Silva *et al.* (1968), Nakahara & Miller (1981),
Hamon & Williams (1984) and by Foldi (1988).

Inglisia zizyphy Brain

Inglisia zizyphy Brain, 1920b: 37.
Cryptinglisia zizyphi (Brain); De Lotto, 1978: 140.
TYPE DATA. Syntypes female, SOUTH AFRICA: Transvaal, Pretoria,
     on *Zizyphus* sp. (SANC).
DISTRIBUTION. ETHIOPIAN REGION: South Africa, Zimbabwe.
HOST PLANTS. Ebenaceae: *Royena glabra.* **Labiatae:** *Rosmarinus.*
Rhamnaceae: *Ziziphus.*
REMARKS. Adult female redescribed and illustrated by Hodgson
(1967d). Intraspecific variation in taxonomic characters given by
Hodgson (1967d). Distribution and host plant records given by
Hodgson (1967d).

Kilifia De Lotto

Platycoccus Takahashi, 1959: 76 [HOMONYM of *Platycoccus* Stick-
     ney; Coccoidea].
Kilifia De Lotto, 1965: 206 [REPLACEMENT NAME].
Habibius Ezzat & Hussein, 1969: 402. TYPE-SPECIES: *Lecanium
     acuminatum* Signoret, by original designation. Syn. by Commun-
     ity of type species.
TYPE-SPECIES: *Lecanium acuminatum* Signoret, by original designa-
     tion.
REMARKS. Ben-Dov (1979) revised the genus and provided key to
species. Generic characters discussed by Tao *et al.* (1983), Hamon &
Williams (1984), Williams & Watson (1990) and by Tang (1991). Key
to species given by De Lotto (1965), Ben-Dov (1979) and by Tang
(1991).

Kilifia acuminata (Signoret)

Lecanium acuminatum Signoret, 1873b: 397.
Coccus acuminatum (Signoret); Kirkaldy, 1902: 105.
Coccus acuminatus (Signoret); Fernald, 1903: 167.
Lecanium acuminatum Signoret; Green, 1904d: 195.

Calymmata acuminatum (Signoret); Kirkaldy, 1904: 228.
Protopulvinaria acuminata (Signoret); Steinweden, 1929: 223.
Lecanium (Coccus) acuminatum Signoret; Green, 1937: 299.
Platycoccus acuminatus (Signoret); Takahashi, 1959: 76.
Kilifia acuminata (Signoret); De Lotto, 1965: 208.
Habibius acuminatus (Signoret); Ezzat & Hussein, 1969: 402.

TYPE DATA. Neotype female designated by Ben-Dov (1976a) and confirmed by Opinion 1192 (1981). SRI LANKA [-CEYLON]: Pundaluoya, on *Jasminum* sp. (BMNH).

DISTRIBUTION. AUSTRO ORIENTAL REGION: Malaysia, Papua New Guinea. **ETHIOPIAN REGION:** Cameroon. **NEOTROPICAL REGION:** Bermuda, Brazil, Cuba, Dominican Republic, Guatemala, Jamaica, Mexico, Panama, Puerto Rico, Surinam, Trinidad, Venezuela, Virgin Islands. **NEW ZEALAND and PACIFIC REGION:** Cook Islands, Hawaii, New Hebrides, Tonga, Vanuatu, Western Samoa. **NEARCTIC REGION:** Alabama, Georgia, Missouri, New York, Texas. **ORIENTAL REGION:** Sri Lanka, Taiwan. **PALAEARCTIC REGION:** China, Egypt, Japan.

HOST PLANTS. Agavaceae: *Cordyline terminalis*. **Anacardiaceae:** *Mangifera indica, Schinus terebinthifolius*. **Apocynaceae:** *Alyxia olivaeformis, Plumeria acuminata, P. acutifolia, P. rubra*. **Aquifoliaceae:** *Ilex vomitoria*. **Araceae:** *Anthurium, Monstera deliciosa, Zantedeschia*. **Aspleniaceae:** *Asplenium*. **Bischofiaceae:** *Bischofia*. **Bixaceae:** *Bixa orellana*. **Crassulaceae:** *Bryophyllum pinnatum*. **Flacourtiaceae:** *Casearia guianiensis*. **Guttiferae:** *Calophyllum inophyllum*. **Hernandiaceae:** *Gyrocarpus*. **Hydrangeaceae:** *Broussaisia*. **Lauraceae:** *Cinnamomum zeylanicum, Persea americana*. **Melastomataceae:** *Medinilla*. **Meliaceae:** *Aglaia odorata*. **Moraceae:** *Artocarpus altilis*. **Myrtaceae:** *Eugenia jambolana, Eugenia jambos, E. malaccensis, Jambosa vulgaris, Metrosideros, Pimenta officinalis, Psidium guajava, Syzygium sizygioides*. **Nyctaginaceae:** *Bougainvillea*. **Oleaceae:** *Jasminum, Osmanthus fragrans*. **Oxalidaceae:** *Averrhoa carambola*. **Passifloraceae:** *Passiflora*. **Potaliaceae:** *Fagraea berteriana*. **Rhizophoraceae:** *Rhizophora stylosa*. **Rubiaceae:** *Canthium odoratum, Cinchona ledgeriana, Coffea, Gardenia jasminoides, Ixora, Randia tahitensis, Straussia*. **Rutaceae:** *Citrus aurantifolia, C. limon, Pelea*. **Sapindaceae:** *Litchi chinensis*. **Sapotaceae:** *Achras sapota*. **Theaceae:** *Eurya nitida, Gordonia lasianthus*. **Zingiberaceae:** *Alpinia purpurata*.

REMARKS. Common Name - Acuminate scale. The nomenclature and identity of the species clarified and established by Ben-Dov (1976a) and in Opinion 1192 (1981). Adult female redescribed and illustrated by Green (1904d), Zimmerman (1948), Ezzat & Hussein (1969), Ben-Dov (1979), Kawai (1980), Tao *et al.* (1983), Hamon & Williams (1984), Williams & Watson (1990) and by Tang (1991). Colour photograph of adult female given by Kawai (1980), Hamon & Williams (1984). First instar larva and pre adult female described and illustrated by Ezzat & Hussein (1969). Distribution and host plant records given by Green (1904d), Ballou (1926), Zimmerman (1948), Ferris (1950), Ezzat & Hussein (1969), Panis & Martin (1976), Ben-Dov (1979), Kawai (1980), Nakahara (1981b), Nakahara (1983),

Tao *et al.* (1983), Hamon & Williams (1984), Matile-Ferrero & Nonveiller (1984), Williams & Watson (1990), Tang (1991) and by Hodgson & Hilburn (1991a, 1991b).
BIOLOGY. Hafez *et al.* (1971) studied the duration of development
and fecundity on mango in Egypt.

Kilifia americana Ben-Dov

Kilifia americana Ben-Dov, 1979: 316.
TYPE DATA. Holotype female, U.S.A.: Texas, Brownsville, on *Citrus*
   sp. (USNM).
DISTRIBUTION. NEOTROPICAL REGION: Mexico. **NEARCTIC
   REGION:** Texas.
HOST PLANTS. Anacardiaceae: *Mangifera indica.* **Apocynaceae:**
   Ervatamia coronaria. **Araceae:** *Dieffenbachia.* **Palmae:** *Chamae-
   dorea.* **Rubiaceae:** *Coffea, Gardenia.* **Rutaceae:** *Citrus limon.*
   Sapotaceae: *Achras.*
REMARKS. Distribution and host plant records given by Gonzalez
Hernandez & Atkinson (1984).

Kilifia deltoides De Lotto

Kilifia deltoides De Lotto, 1965: 208.
TYPE DATA. Holotype female, KENYA: Kilifi, on *Mangifera indica*
   (BMNH).
DISTRIBUTION. AUSTRO ORIENTAL REGION: Java, Malaysia,
   Sarawak. **ETHIOPIAN REGION:** Kenya, Zanzibar. **MADAGASIAN
   REGION:** Comoros.
HOST PLANTS. Anacardiaceae: *Mangifera indica, M. laurina.*
   Lauraceae: *Cinnamomum.* **Leguminosae:** *Adianthum fergonosi.*
   Myrtaceae: *Callistemon, Psidium guajava.*
REMARKS. Adult female redescribed and illustrated by Ben-Dov
(1979). Adult female redescribed by Tang (1991). Distribution and
host plant records given by Matile-Ferrero (1978), Ben-Dov (1979)
and by Tang (1991).

Kilifia diversipes (Cockerell)

Coccus diversipes Cockerell, 1905e: 130.
Kilifia diversipes (Cockerell); De Lotto, 1965: 208.
TYPE DATA. Lectotype female designated by Ben-Dov (1979), PHIL-
   IPPINES: Lucena, on cultivated fern (USNM).
DISTRIBUTION. AUSTRO ORIENTAL REGION: Philippines.
HOST PLANTS. Aspleniaceae: *Asplenium.*
REMARKS. Adult female redescribed and illustrated by Morrison
(1920) and by Ben-Dov (1979). Adult female redescribed by Tang
(1991). Distribution and host plant records given by Robinson
(1917), Morrison (1920), Ben-Dov (1979) and by Tang (1991).

Kilifia guizhouensis Qin & Gullan

Kilifia guizhouensis Qin & Gullan, 1991: 21.

TYPE DATA. Holotype female, CHINA: Huaxi, Guizhou, on *Myrsine semiserrata* (GAC).
DISTRIBUTION. PALAEARCTIC REGION: China.
HOST PLANTS. Myrsinaceae: *Myrsine semiserrata*.

Kilifia sinensis Ben-Dov

Kilifia sinensis Ben-Dov, 1979: 321.
TYPE DATA. Holotype female, CHINA: Si-Shan, Kunming, on *Eurya nitida* (USNM).
DISTRIBUTION. PALAEARCTIC REGION: China.
HOST PLANTS. Theaceae: *Eurya nitida*.
REMARKS. Adult female redescribed by Tang (1991).

Kozáricoccus Avasthi & Shafee

Kozáricoccus Avasthi & Shafee, 1984b: 31.
TYPE-SPECIES: *Ceroplastodes bituberculatus* Brain, by original designation and monotypy.

Kozáricoccus bituberculatus (Brain)

Ceroplastodes bituberculatus Brain, 1920b: 40.
Ceronema bituberculatus (Brain); Matile-Ferrero, 1978: 44.
Kozáricoccus bituberculatus (Brain); Avasthi & Shafee, 1984b: 32.
TYPE DATA. Syntypes female, SOUTH AFRICA: Cape Province, Somerset West and Stellenbosch, on stems of native shrub (SANC).
DISTRIBUTION. ETHIOPIAN REGION: South Africa.
REMARKS. Adult female redescribed and illustrated by Hodgson (1971) and by Avasthi & Shafee (1984b). First instar larva described by Hodgson (1971).

Lacca Signoret NOMEN NUDUM

Lacca Signoret, 1869a: 848 [NOMEN NUDUM].
REMARKS. The generic name *Lacca* Signoret NOMEN NUDUM was linked (Signoret, 1869a; Morrison & Morrison, 1966) with *Ceroplastes ceriferus*. Signoret (1869a) first published it in the binomen *Lacca alba* NOMEN NUDUM, which he credited to Pearson (1794). However, this binomen was never used by Pearson (1794), who just presented a chemical study of wax, which he named "white wax", produced by scale insects in India. It is evident that Signoret introduced the binomen as a translatinization of "white wax", while erroneously crediting it to Pearson. Both the generic and specific names are NOMINA NUDA.

Lagosinia Cockerell

Lagosinia Cockerell, 1899f: 332.
TYPE-SPECIES: *Lecanium strachani* Cockerell, by original designation and monotypy.
REMARKS. Generic characters discussed by Hodgson (1968a, 1969b). Key to species: ZIMBABWE - Hodgson (1968a, 1969b).

Lagosinia aristolochiae (Newstead)

Pulvinaria aristolochiae Newstead, 1917b: 19.
Lagosinia aristolochiae (Newstead); Hodgson, 1968a: 144.
TYPE DATA. Syntypes female, GHANA: Aburi, on *Aristolochia* sp. (BMNH).
DISTRIBUTION. ETHIOPIAN REGION: Ghana.
HOST PLANTS. Aristolochiaceae: *Aristolochia*. **Sterculiaceae:** *Theobroma cacao*.
REMARKS. Adult female redescribed and illustrated by Hodgson (1968a). Distribution and host plant records given by Strickland (1947) and by Hodgson (1968a).

Lagosinia mazoeensis Hodgson

Pulvinaria aristolochiae Newstead; Hall, 1932: 187 [MISIDENTIFICATION].
Lagosinia mazoeensis Hodgson, 1968a: 145.
TYPE DATA. Holotype female, ZIMBABWE: Mazoe, on *Chlorocodon* sp. (BMNH).
DISTRIBUTION. ETHIOPIAN REGION: Zimbabwe.
HOST PLANTS. Ericaceae: *Chlorocodon*.

Lagosinia rhodesiensis Hodgson

Pulvinaria aristolochiae Newstead; Hall, 1937: 123 [MISIDENTIFICATION].
Lagosinia rhodesiensis Hodgson, 1968a: 149.
TYPE DATA. Holotype female, ZIMBABWE: South Marandellas, on *Uapaca kirkiana* (BMNH).
DISTRIBUTION. ETHIOPIAN REGION: Zimbabwe.
HOST PLANTS. Uapacaceae: *Uapaca kirkiana*.

Lagosinia strachani (Cockerell)

Lecanium strachani Cockerell, 1898l: 259.
Lagosinia strachani (Cockerell); Cockerell, 1899f: 332.
Pulvinaria aristolochiae Newstead; Hall, 1937: 123 [MISIDENTIFICATION].
TYPE DATA. Syntypes female, NIGERIA: Lagos, on *Annona squamosa* (USNM, BMNH).
DISTRIBUTION. ETHIOPIAN REGION: Eritrea, Ghana, Malawi, Mauritania, Nigeria, Senegal, Sudan, Uganda, Zimbabwe.
HOST PLANTS. Acanthaceae: *Acanthus*. **Anacardiaceae:** *Mangifera indica*. **Annonaceae:** *Annona reticulata, A. squamosa*. **Asclepiadaceae:** *Calotropis procera*. **Compositae:** *Ageratum, Vernonia*. **Solanaceae:** *Datura fastuosa*. **Vitidaceae:** *Cissus*.
REMARKS. Adult female redescribed and illustrated by Hodgson (1968a). Intraspecific variation in taxonomic characters given by Hodgson (1968a). Distribution and host plant records given by Hodgson (1968a, 1969b) and by Balachowsky & Matile-Ferrero (1970).

Lagosinia vayssierei (Castel-Branco)

Pulvinaria vayssierei Castel-Branco, 1952: 33.

Lagosinia montana Hodgson, 1968a: 147. Holotype female, MALAWI: Mlanje, on *Acalypha colorata* (BMNH). Syn. by Hodgson, 1970b: 279.
Lagosinia vayssieret (Castel-Branco); Hodgson, 1970b: 179.
TYPE DATA. Syntypes female and first instar larva, MOZAMBIQUE: Nampula, Experimental Station of C. I. C. A., on cotton plant (CZLP).
DISTRIBUTION. ETHIOPIAN REGION: Malawi, Mozambique.
HOST PLANTS. Euphorbiaceae: *Acalypha colorata*. **Malvaceae:** *Gossypium*.
REMARKS. Adult female redescribed and illustrated by Hodgson (1968a) (as *L. montana*) and by Hodgson (1970b). First instar larva described by Hodgson (1970b). Distribution and host plant records given by Hodgson (1968a, 1970b).

Lecaniococcus Danzig

Lecaniococcus Danzig, 1967: 148.
TYPE-SPECIES: *Lecaniococcus ditispinosus* Danzig, by original designation and monotypy.
REMARKS. Generic characters discussed by Danzig (1980c) and by Tang (1991).

Lecaniococcus ditispinosus Danzig

Lecaniococcus ditispinosus Danzig, 1967: 149.
TYPE DATA. Holotype female, USSR: Primorye Territory, near Vladivostok, on *Quercus mongolica* (ZIAS).
DISTRIBUTION. PALAEARCTIC REGION: USSR Primorye Territory.
HOST PLANTS. Fagaceae: *Quercus mongolica*. **Tiliaceae:** *Tilia*.
REMARKS. Adult female redescribed and illustrated by Danzig (1980c) and by Tang (1991).

Lecanochiton Maskell

Lecanochiton Maskell, 1882: 221.
Lecaniochiton Lindinger; Lindinger, 1932f: 197 [UNJUSTIFIED EMENDATION].
TYPE-SPECIES: *Lecanochiton metrosideri* Maskell, by monotypy.
REMARKS. Generic characters discussed by Morrison & Morrison (1922).

Lecanochiton metrosideri Maskell

Lecanochiton metrosideri Maskell, 1882: 222.
TYPE DATA. Syntypes female, NEW ZEALAND: Milford Sound, on *Metrosideros* sp. (NZAC).
DISTRIBUTION. NEW ZEALAND and PACIFIC REGION: New Zealand.
HOST PLANTS. Myrtaceae: *Metrosideros*.
REMARKS. Adult female redescribed and illustrated by Maskell (1887b) and by Morrison & Morrison (1922).

Lecanochiton minor Maskell

Lecanochiton minor Maskell, 1891a: 12.

TYPE DATA. Syntypes female, NEW ZEALAND: Reefton District, on *Metrosideros robusta* (NZAC, BMNH, USNM).
DISTRIBUTION. NEW ZEALAND and PACIFIC REGION: New Zealand.
HOST PLANTS. Myrtaceae: *Metrosideros robusta.*

Lecanopsis Targioni Tozzetti

Rhizobium Targioni Tozzetti; Targioni Tozzetti, 1867: 23 [NOMEN NUDUM].
Lecanopsis Targioni Tozzetti, 1868: 729.
Lecaniopsis Lindinger; Lindinger, 1923: 148 [UNJUSTIFIED EMENDATION].
TYPE-SPECIES: *Lecanopsis rhyzophila* Targioni Tozzetti, by monotypy.
REMARKS. Morrison & Morrison (1966) discussed the nomenclature of the genus. Generic characters discussed by Signoret (1874b), Leonardi (1920), Steinweden (1929), Goux (1937), Gómez-Menor Ortega (1937), Borchsenius (1952, 1957), Kosztarab & Kozár (1978, 1988), Danzig (1980c), Yang (1982), Tao *et al.* (1983) and by Tang (1991). Key to species: ASIA - Tang (1991). EUROPE - Kosztarab & Kozár (1988). UKRAINE - Tereznikova (1981). USSR - Borchsenius (1952, 1957), Danzig (1964).

Lecanopsis aphenogastrorum Gómez-Menor Ortega

Lecanopsis aphenogastrorum Gómez-Menor Ortega, 1928: 350.
TYPE DATA. Syntypes female, SPAIN: Colldejou (Tarragona), in nest of the ant *Aphenogaster barbara* (MNCN).
DISTRIBUTION. PALAEARCTIC REGION: Spain.
REMARKS. Adult female redescribed and illustrated by Gómez-Menor Ortega (1937).
BIOLOGY. Found in nest of the ant *Aphenogaster barbara* (Gómez-Menor Ortega, 1928).

Lecanopsis ceylonica Green

Lecanopsis ceylonica Green, 1922b: 1026.
TYPE DATA. Syntypes female, SRI LANKA: Pattipola, on grass (BMNH).
DISTRIBUTION. ORIENTAL REGION: India, Sri Lanka.
HOST PLANTS. Gramineae:
REMARKS. Adult female and male redescribed and illustrated by Venkatraman (1941). Adult female redescribed by Tang (1991). Distribution and host plant records given by Green (1937), Varshney (1985) and by Shafee *et al.* (1989).

Lecanopsis fallax (Giard)

Spermococcus fallax Giard; Giard, 1894: cxcix.
Lecanopsis fallax (Giard); Borchsenius, 1957: 92.
TYPE DATA. Syntypes female, FRANCE: Wimereux, on roots of various plants mainly on Gramineae (probably lost; D. Matile-Ferrero, 1989, personal communication).
DISTRIBUTION. PALAEARCTIC REGION: France.
HOST PLANTS. Gramineae:

REMARKS. Lindinger (1935a) synonymized this species with *Coccus radicum graminis* Fonscolombe, however, Ben-Dov & Matile-Ferrero (1989) have shown that the latter is a mealybug. Borchsenius (1957) accepted *L. fallax* as a valid species in the Coccidae.

Lecanopsis festucae Borchsenius

Lecanopsis formicarum Newstead; Kiritshenko, 1931: 318 [MISIDEN-TIFICATION in part].
Lecanopsis festucae Borchsenius, 1952: 288.
TYPE DATA. Syntypes female, USSR: Ukraine, Odessa, Crimea and Caucasus, on *Festuca* sp. (ZIAS).
DISTRIBUTION. PALAEARCTIC REGION: Bulgaria, Hungary, Mongolia, Poland, Romania, USSR Caucasus, USSR Crimea, USSR Moldavia, USSR Ukraine.
HOST PLANTS. Gramineae: *Agrostis vulgaris, Dactylis glomerata, Elytrigia repens, Festuca sulcata, Lolium perenne, Poa compressa, P. nemoralis.*
REMARKS. Adult female redescribed and illustrated by Borchsenius (1957), Kosztarab & Kozár (1978, 1988), Tereznikova (1981) and by Tang (1991). Larval instars described by Borchsenius (1957) and by Tereznikova (1981). Distribution and host plant records given by Borchsenius (1957), Koteja & Zak-Ogaza (1969), Podsiadlo & Komosinska (1976), Danzig (1977b), Tereznikova (1981), Kozár (1985), Kozár & Ostafichuk (1987), Kosztarab & Kozár (1988) and by Tang (1991).

Lecanopsis formicarum Newstead

Lecanopsis formicarum Newstead, 1893a: 138.
Lecanopsis brevicornis Newstead, 1896: 59. Syntypes nymphal stages, ENGLAND: Norfolk, Snettisham Beach, near King's Lynn, on grass roots (BMNH). Syn. by Green, 1921a: 193.
Lecanopsis butleri Green, 1917: 208. Syntypes female and larva, ENGLAND: Herts, Royson Heath on grass (BMNH). Syn. by Green, 1921a: 193.
Lecaniopsis formicarum Newstead; Lindinger, 1935a: 139 [UNJUSTIFIED EMENDATION].
Lecaniopsis butleri Green; Lindinger, 1936: 158 [UNJUSTIFIED EMENDATION].
Filippia subterranea Gómez-Menor Ortega, 1948: 94. Lectotype female designated by A. Blay (Museo Nacional de Ciencias Naturales, Madrid), SPAIN: Monte Araca, de Vitoria, on roots of Gramineae (MNCN). Syn. by NEW SYNONYMY introduced here.
TYPE DATA. Syntypes female, ENGLAND: Chesil Beach, in nest of the ant *Formica nigra* (BMNH).
DISTRIBUTION. PALAEARCTIC REGION: Czechoslovakia, Denmark, England, France, Germany, Guernsey Islands, Hungary, Italy, Mongolia, Netherlands, Poland, Spain, Sweden, Switzerland.
HOST PLANTS. Gramineae: *Agrostis vulgaris, Festuca rubra, Poa compressa, P. pratensis.*
REMARKS. Adult female redescribed and illustrated by Leonardi (1920) (as *L. brevicornis*), Green (1921a), Gómez-Menor Ortega (1937), Schmutterer (1952a), Borchsenius (1957), Boratynski *et al.*

(1982) and by Kosztarab & Kozár (1978, 1988). Adult female redescribed by Tang (1991). Male and female larval instars described and illustrated by Boratynski *et al.* (1982). First instar crawler described and illustrated by Schmutterer (1952a). Distribution and host plant records given by Leonardi (1908, 1920), Balachowsky (1936c), Green (1925b), Gómez-Menor Ortega (1937, 1948), Schmutterer (1952a), Reyne (1957), Kozárzhevskaya & Reitzel (1975), Danzig (1980b), Koteja & Zak-Ogaza (1983), Ossiannilsson (1985) and by Kosztarab & Kozár (1988).

BIOLOGY. A biparental species that develops one annual generation in Poland (Boratynski *et al.*, 1982). Annual life cycle presented in details by Boratynski *et al.* (1982). Found in England in nest of the ant *Formica nigra* (Newstead, 1893a).

Lecanopsis iridis Borchsenius

Lecanopsis iridis Borchsenius, 1952: 291.
TYPE DATA. Syntypes larva of first, before last and last instar, USSR: Primorye Territory, near Grigoryevka, on *Iris uniflora* (ZIAS).
DISTRIBUTION. PALAEARCTIC REGION: USSR Primorye Territory.
HOST PLANTS. Iridaceae: *Iris uniflora.*
REMARKS. The female of this species has not been observed. Larvae redescribed and illustrated by Borchsenius (1957) and by Danzig (1980c).

Lecanopsis lineolatae King & Cockerell

Lecanopsis lineolatae King & Cockerell, 1897: 90.
TYPE DATA. Syntypes female, U.S.A.: Massachusetts, Lawrence, in nest of *Crematogaster lineolata* (USNM).
DISTRIBUTION. NEARCTIC REGION: Massachusetts.
BIOLOGY. Found in nest of the ant *Crematogaster lineolata* (King & Cockerell, 1897).

Lecanopsis myrmecophila Leonardi

Lecanopsis mirmecophila Leonardi; Leonardi, 1908: 181 [MIS-SPELLING].
Lecanopsis myrmecophila Leonardi, 1908: 181.
Lecaniopsis myrmecophila Leonardi; Lindinger, 1935a: 139 [UNJUSTIFIED EMENDATION].
TYPE DATA. Syntypes female, ITALY: Sardinia, Tempio, in nest of *Tetramorium coespitum* (IEAP).
DISTRIBUTION. PALAEARCTIC REGION: Italy.
REMARKS. In the title of the original description (Leonardi, 1908) the specific name was mis-spelled, however it was corrected further in the text, and by Leonardi (1920). Adult female redescribed and illustrated by Leonardi (1920). Adult female redescribed by Borchsenius (1957).
BIOLOGY. Found in nest of the ant *Tetramorium coespitum.*

Lecanopsis nevesi Gómez-Menor Ortega

Lecanopsis nevesi Gómez-Menor Ortega, 1946: 88.

TYPE DATA. Syntypes female, SPAIN: Toledo, on *Santolina chamae-cyparis* (MNCN).
DISTRIBUTION. PALAEARCTIC REGION: Spain.
HOST PLANTS. Compositae: *Santolina chamaecyparis*.

Lecanopsis porifera Borchsenius

Lecanopsis formicarum Newstead; Kiritshenko, 1931: 318 [MISIDEN-TIFICATION].
Lecanopsis formicarum Newstead; Borchsenius, 1936: 113 [MIS-IDENTIFICATION].
Lecanopsis porifera Borchsenius, 1952: 282.
TYPE DATA. Syntypes female and larva, USSR: Crimea, Ukraine, Georgia and Armenia, on *Agropyron repens*, *Bromus* sp. and *Festuca* sp. (ZIAS).
DISTRIBUTION. PALAEARCTIC REGION: Hungary, Romania, USSR Armenia, USSR Crimea, USSR Georgia, USSR Ukraine, Yugoslavia.
HOST PLANTS. Gramineae: *Agropyron repens*, *Bromus*, *Dactylis glomerata*, *Festuca*, *Lolium perenne*.
REMARKS. Adult female redescribed and illustrated by Borchsenius (1957), Kosztarab & Kozár (1978, 1988) and by Tereznikova (1981). Larval instars described and illustrated by Borchsenius (1957) and by Tereznikova (1981).

Lecanopsis radicumgraminis (Fonscolombe)

Coccus radicumgraminis Fonscolombe, 1834: 212.
TYPE DATA. Syntypes female, FRANCE: southern France, Saint-Canadet, on *Festuca caespitosa* (probably lost; D. Matile-Ferrero, 1989, personal communication).
DISTRIBUTION. PALAEARCTIC REGION: France.
HOST PLANTS. Gramineae: *Festuca caespitosa*.
BIOLOGY. Found on roots of the host plant.
REMARKS. Ben-Dov & Matile-Ferrero (1989a, 1989b) discussed the identity of this species, rejected its interpretation, by some earlier students, as a mealybug and established it as a member of the Coccidae.

Lecanopsis rhizophila Targioni Tozzetti

Lecanopsis rhyzophila Targioni Tozzetti, 1868: 729.
Lecanopsis rhizophila Targioni Tozzetti; Signoret, 1874b: 93 [JUSTI-FIED EMENDATION OF SPELLING].
TYPE DATA. Syntypes female, ITALY: Florence, on roots of *Asperula* sp. (probably lost; G. Pellizzari Scaltriti, 1990, personal communication).
DISTRIBUTION. PALAEARCTIC REGION: Italy.
HOST PLANTS. Rubiaceae: *Asperula*.
REMARKS. Adult female redescribed and illustrated by Signoret (1874b).

Lecanopsis sacchari Takahashi

Lecanopsis sacchari Takahashi, 1928a: 345.

TYPE DATA. Syntypes female. TAIWAN: Kori, Taichu, on *Saccharum officinarum* and *Miscanthus* sp. (IMZT).
DISTRIBUTION. ORIENTAL REGION: Taiwan.
HOST PLANTS. Gramineae: *Miscanthus.* **Gramineae:** *Saccharum officinarum.*
REMARKS. Adult female redescribed and illustrated by Tao *et al.* (1983). Adult female redescribed by Tang (1991).

Lecanopsis shutovae Borchsenius

Lecanopsis shutovae Borchsenius, 1952: 293.
TYPE DATA. Syntypes female and larva, USSR: Primorye Territory, on *Festuca* sp. (ZIAS).
DISTRIBUTION. PALAEARCTIC REGION: USSR Primorye Territory.
HOST PLANTS. Gramineae: *Festuca.*
REMARKS. Adult female redescribed and illustrated by Borchsenius (1957) and by Danzig (1980c). Adult female redescribed by Tang (1991). Larvae described by Borchsenius (1957).

Lecanopsis taurica Borchsenius

Lecanopsis formicarum Newstead; Borchsenius, 1950: 142 [MISIDENTIFICATION].
Lecanopsis taurica Borchsenius, 1952: 285.
TYPE DATA. Syntypes female and larva, USSR: Crimea, near Simferopol, on *Bromus erectus* (ZIAS).
DISTRIBUTION. PALAEARCTIC REGION: USSR Crimea.
HOST PLANTS. Gramineae: *Bromus erectus.*
REMARKS. Adult female redescribed and illustrated by Borchsenius (1957). Larval instars described and illustrated by Borchsenius (1957).

Lecanopsis terrestris Borchsenius

Lecanopsis formicarum Newstead; Kiritshenko, 1931: 318 [MISIDENTIFICATION in part].
Lecanopsis terrestris Borchsenius, 1952: 287.
TYPE DATA. Syntypes female, USSR: Ukraine, near Poltava and Chkalovsk Region, Orsk, on grass (ZIAS).
DISTRIBUTION. PALAEARCTIC REGION: Hungary, USSR Chkalovsk Region, USSR Ukraine.
HOST PLANTS. Gramineae: *Agropyron repens, Agrostis vulgaris.*
REMARKS. Adult female redescribed and illustrated by Borchsenius (1957). Adult female redescribed by Kosztarab & Kozár (1988). Distribution and host plant records given by Kozár & Drozdjak (1990).

Lecanopsis turcica (Bodenheimer)

Paralecanopsis turcica Bodenheimer, 1951a: 329.
Paralecanium turcica Bodenheimer; Bodenheimer, 1953: 110 [MISSPELLING].
Lecanopsis turcica (Bodenheimer); Ben-Dov, 1980: 263.
TYPE DATA. Syntypes larval instars, TURKEY: Kaya-ardi, near Nigde, on grass (BCR).
DISTRIBUTION. PALAEARCTIC REGION: Turkey.
HOST PLANTS. Gramineae:

REMARKS. Bodenheimer (1953) presented this species again as new, however, the 1951a description fully satisfies the Code. The nomenclature, type-series and close affinity of this species with *Lecanopsis porifera* Borchsenius were discussed by Ben-Dov (1980).

Leptopulvinaria Kanda

Leptopulvinaria Kanda, 1960: 118.
TYPE-SPECIES: *Leptopulvinaria elaeocarpi* Kanda, by original designation and monotypy.
REMARKS. Generic characters discussed by Tang (1991).

Leptopulvinaria elaeocarpi Kanda

Leptopulvinaria elaeocarpi Kanda, 1960: 118.
TYPE DATA. Syntypes female, JAPAN: Honshu, Wakayama Prefecture, Goshiki, Kanaya-machi, Avita-gun, on *Elaeocarpus sylvestris var. ellipticus* (KAYJ).
DISTRIBUTION. PALAEARCTIC REGION: Japan.
HOST PLANTS. Elaeocarpaceae: *Elaeocarpus sylvestris ellipticus*.
REMARKS. Adult female redescribed by Kawai (1980) and by Tang (1991).

Lichtensia Signoret

Lichtensia Signoret, 1873a: 27.
Lichstensia Fuller; Fuller, 1899: 457 [MIS-SPELLING].
TYPE-SPECIES: *Lichtensia viburni* Signoret, by monotypy.
REMARKS. Generic characters discussed by Newstead (1903), Steinweden (1929), Gómez-Menor Ortega (1937), Ben-Dov (1975a), Kosztarab & Kozár (1988) and by Tang (1991).

Lichtensia argentata Hempel

Lichtensia argentata Hempel, 1900b: 492.
Filippia argentata (Hempel); Silva *et al.*, 1968: 146.
TYPE DATA. Syntypes female, BRAZIL: Sao Paulo, Ypiranga, on a tree of the order Ilicineae (MZSP).
DISTRIBUTION. NEOTROPICAL REGION: Brazil Sao Paulo.
REMARKS. The syntypes were probably collected from a tree of the Aquifoliaceae. Adult female redescribed by Hempel (1901b).

Lichtensia carissae (Brain)

Filippia carissae Brain, 1920b: 24.
Lichtensia carissae (Brain); Matile-Ferrero, 1978: 44.
TYPE DATA. Syntypes female, SOUTH AFRICA: Natal Coast, on *Carissa grandiflora* (SANC).
DISTRIBUTION. ETHIOPIAN REGION: Namibia, South Africa, Zimbabwe.
HOST PLANTS. Apocynaceae: *Carissa grandiflora*. **Crassulaceae:** *Cotyledon orbiculata*. **Rubiaceae:** *Psychotria zombamontana*.
REMARKS. Adult female redescribed and illustrated by De Lotto (1967b). Distribution and host plant records given by De Lotto (1967b, 1968b) and by Hodgson (1969b).

Lichtensia chilianthi (Brain)

Filippia chilianthi Brain, 1920b: 23.
Filippia oleae rhusae Hall, 1932: 186. Syntypes female, ZIMBABWE
 [=RHODESIA]: Embeza, on *Rhus* sp. (BMNH). Syn. by Hodgson,
 1969b: 13.
Pulvinaria syzygii Hall, 1941: 235. Syntypes female, ZIMBABWE
 [=RHODESIA]: Inyanga, Inyangombi Falls, on *Syzygium corda-
 tum* (BMNH). Syn. by Hodgson, 1969b: 13.
Lichtensia chilianthi (Brain); Matile-Ferrero, 1978: 44.
TYPE DATA. Syntypes female, SOUTH AFRICA: Natal, Illovo River,
 on native shrub, and Orange Free State, Bloemfontein, on *Chi-
 lianthus oleaceus* (SANC).
DISTRIBUTION. ETHIOPIAN REGION: Mozambique, South Africa,
 Zimbabwe.
HOST PLANTS. Anacardiaceae: *Rhus chirindensis.* **Compositae:**
 Brachylaena discolor. **Loganiaceae:** *Chilianthus oleaceus.*
 Myrtaceae: *Syzygium cordatum.* **Oleaceae:** *Olea africana.*
 Rubiaceae: *Canthium obovatum.* **Solanaceae:** *Solanum quadran-
 gulare.* **Urticaceae:** *Girandinia eylesii.*
REMARKS. Adult female redescribed and illustrated by De Lotto
(1967b). Intraspecific variation in taxonomic characters given by
Hodgson (1969b). Distribution and host plant records given by Hall
(1941), De Lotto (1967b) and by Hodgson (1969b).

Lichtensia gemina (De Lotto) n. comb.

Filippia gemina De Lotto, 1974: 209.
TYPE DATA. Holotype female, SOUTH AFRICA: Natal, St. Lucia
 Lake, on *Pterocelastrus echinatus* (SANC).
DISTRIBUTION. ETHIOPIAN REGION: South Africa.
HOST PLANTS. Celastraceae: *Pterocelastrus echinatus.*

Lichtensia madagascariensis (Mamet) n. comb.

Filippia madagascariensis Mamet, 1950: 28.
TYPE DATA. Holotype female, MADAGASCAR: Antsingy, on twigs of
 undetermined plant (MNHN).
DISTRIBUTION. MADAGASIAN REGION: Madagascar.

Lichtensia peringueyi Joubert

Lichtensia peringueyi Joubert, 1925: 120.
TYPE DATA. Syntypes female, SOUTH AFRICA: Cape Province, Stel-
 lenbosch, on *Pennisetum macrorum* (SANC).
DISTRIBUTION. ETHIOPIAN REGION: South Africa.
HOST PLANTS. Gramineae: *Pennisetum macrorum.*

Lichtensia polychaeta (De Lotto) n. comb.

Filippia polychaeta De Lotto, 1974: 209.
TYPE DATA. Holotype female, SOUTH AFRICA: Cape Province, Cape
 Town, on *Berzelia lanuginosa* (SANC).
DISTRIBUTION. ETHIOPIAN REGION: South Africa.
HOST PLANTS. Bruniaceae: *Berzelia lanuginosa.*

Lichtensia schini Hempel

Lichtensia schini Hempel, 1932: 320.
Filippia schini (Hempel); Corseuil & Barbosa, 1971: 239.
TYPE DATA. Syntypes female, BRAZIL: Rio Grande do Sul, Lavras,
  on *Schinus dependens* (MZSP).
DISTRIBUTION. NEOTROPICAL REGION: Brazil Rio Grande do Sul.
HOST PLANTS. Anacardiaceae: *Schinus dependens*.
REMARKS. Adult female redescribed by Gomes Costa (1949).

Lichtensia simillima Cockerell

Lichtensia simillima Cockerell, 1902e: 90.
TYPE DATA. Syntypes female, ARGENTINA: General Acha, on a
  shrubby plant (BMNH, USNM).
DISTRIBUTION. NEOTROPICAL REGION: Argentina.

Lichtensia spanochaeta (De Lotto) n. comb.

Filippia spanochaeta De Lotto, 1974: 211.
TYPE DATA. Holotype female, SOUTH AFRICA: Orange Free State,
  Ficksburg, on *Aster filifolius* (SANC).
DISTRIBUTION. ETHIOPIAN REGION: South Africa
HOST PLANTS. Compositae: *Aster filifolius*.

Lichtensia strigosa (De Lotto) n. comb.

Filippia strigosa De Lotto, 1974: 214.
TYPE DATA. Holotype female, SOUTH AFRICA: Transvaal, Rusten-
  burg, on unidentified grass (SANC).
DISTRIBUTION. ETHIOPIAN REGION: South Africa.
HOST PLANTS. Gramineae.

Lichtensia viburni Signoret

Lichtensia viburni Signoret, 1873a: 28.
Philippia viburni (Signoret); Lichtenstein, 1881: cxv.
Philippia hederae Lichtenstein; Lichtenstein, 1881: cxv [NOMEN
  NUDUM].
Lichtensia eatoni Newstead, 1895: 166. Syntypes female, ALGERIA:
  Constantine, on olive (BMNH). Syn. by Lindinger, 1912: 232.
Filippia oleae (Costa); Lindinger, 1912: 336 [MISIDENTIFICATION
  and ERRONEOUS AUTHORSHIP].
Philippia oleae (Costa); Leonardi, 1920: 340 [MISIDENTIFICATION
  and ERRONEOUS AUTHORSHIP].
Lichtensia rifana Balachowsky, 1931c: 215. Syntypes female,
  MOROCCO: Beni-Hosmar Mountains, on *Silene gibraltarica*
  (MNHN). Syn. by Matile-Ferrero, 1978: 45.
Filippia rosmarini Goux, 1934: 321. Syntypes female and larva,
  FRANCE: Var, near Cuges, on *Rosmarinus officinalis* (GOUX).
  Syn. by Matile-Ferrero, 1978: 45.
Filippia viburni (Signoret); Balachowsky, 1935: 264.
TYPE DATA. Syntypes female, FRANCE: Montpellier, on "laurier-tin"
  (VMNH).
DISTRIBUTION. PALAEARCTIC REGION: Algeria, Czechoslovakia,
  England, France, Germany, Greece, Ireland, Israel, Italy, Major-

ca, Malta, Morocco, Spain, USSR Azerbaijan, USSR Crimea,
USSR Georgia, USSR Ukraine, Wales, Yugoslavia.
HOST PLANTS. Anacardiaceae: *Pistacia lentiscus.* **Aquifoliaceae:**
Ilex aquifolium. **Araliaceae:** *Fatsia japonica, Hedera helix.* **Capri-
foliaceae:** *Viburnum tinus.* **Caryophyllaceae:** *Silene gibraltarica.*
Ericaceae: *Arbutus unedo.* **Labiatae:** *Rosmarinus officinalis.*
Myrtaceae: *Myrtus communis.* **Oleaceae:** *Jasminum, Olea*
europaea, Phillyrea media. **Rhamnaceae:** *Rhamnus palestina.*
Rosaceae: *Prunus laurocerasus, Spiraea salicifolia.* **Ulmaceae:**
Celtis sinensis
REMARKS. Adult female redescribed and illustrated by Newstead
(1903), Gómez-Menor Ortega (1937), Rehácek (1955), Borchsenius
(1957), Ben-Dov (1975), Tereznikova (1981), Tremblay (1988a) and
by Tang (1991). Male described and illustrated by Leonardi (1920),
Borchsenius (1957), Giliomee (1967), Tereznikova (1981) and by
Kosztarab & Kozár (1988). First instar described by Leonardi (1920).
First and last instars female described and illustrated by Ben-Dov
(1975). Intraspecific variation of taxonomic characters given by Ben-
Dov (1975). The redescription of *Philippia oleae* (Costa) by Leonardi
(1920), very likely refers to *L. viburni.* Colour photograph of female
and male given by Tranfaglia & Viggiani (1988). Distribution and
host plant records given by Newstead (1895), Leonardi (1920), Goux
(1934), Gómez-Menor Ortega (1937), Rehacek (1955), Borchsenius
(1957), Ben-Dov (1975), Tereznikova (1981), Marotta (1987) and by
Kosztarab & Kozár (1988).
BIOLOGY. Quaglia & Raspi (1979b) studied (as *Philippia oleae*) the
ecology and life history on olive in Italy, Toscana. In Sicily, Italy, it
develops two annual generations (Longo, 1988).

Loemica Laing
Loemica Laing, 1929: 476.
TYPE-SPECIES: *Loemica ghesquierei* Laing, by original designation
and monotypy.

Loemica ghesquierei Laing
Loemica ghesquierei Laing, 1929: 477.
TYPE DATA. Syntypes female, ZAIRE: Kasai, on roots of grass
(BMNH).
DISTRIBUTION. ETHIOPIAN REGION: Zaire.
HOST PLANTS. Gramineae.

Luzulaspis Cockerell
Signoretia Targioni Tozzetti, 1868: 727. [HOMONYM of *Signoretia*
Stal].
Signorettia Targioni Tozzetti; Targioni Tozzetti, 1869: 699 [MIS-
SPELLING].
Luzulaspis Cockerell, 1902a: 25 [REPLACEMENT NAME].
TYPE-SPECIES: *Aspidiotus luzulae* Dufour, by monotypy.
REMARKS. The genus was revised by Koteja (1979b). Generic char-
acters discussed by Newstead (1903), Steinweden (1929), Goux
(1937, 1939), Borchsenius (1952, 1957), Koteja (1978), Kosztarab &
Kozár (1978), Koteja & Howell (1979), Koteja (1979b), Danzig

(1980c), Kawai (1980), Tereznikova (1981), Yang (1982), Hamon & Williams (1984), Gill (1988), Kosztarab & Kozár (1988) and by Tang (1991). Key to species: AMERICA - Koteja & Howell (1979). ASIA - Tang (1991). CALIFORNIA - Gill (1988). EUROPE - Kosztarab & Kozár (1988). JAPAN - Kawai (1980). USSR - Borchsenius (1952, 1957); Danzig (1964). USSR (FAR EAST) - Danzig (1980c). UKRAINE - Tereznikova (1981). WORLD - Koteja (1979b).

Luzulaspis americana Koteja & Howell

Luzulaspis americana Koteja & Howell, 1979: 335.
TYPE DATA. Holotype female, USA: Georgia, Echols County, on grass (USNM).
DISTRIBUTION. NEARCTIC REGION: Georgia.
HOST PLANTS. Gramineae.
REMARKS. Adult female redescribed and illustrated by Koteja (1979b) and by Hamon & Williams (1984).

Luzulaspis bisetosa Borchsenius

Luzulaspis bisetosa Borchsenius, 1952: 277.
Luzulaspis amabilis Kanda, 1960: 116. Syntypes female, JAPAN: Sagamihara, Kanagawa Prefecture, on *Carex rugata* (KAYJ). Syn. by Koteja, 1979b: 625.
TYPE DATA. Syntypes female, USSR: Primorye Territory, near Ussurisk, on *Carex* sp. (ZIAS).
DISTRIBUTION. PALAEARCTIC REGION: Japan, Korea, USSR Kunashir Island, USSR Primorye Territory.
HOST PLANTS. Cyperaceae: *Carex nanella, C. rugata.* **Juncaceae:** *Luzula.*
REMARKS. Adult female redescribed and illustrated by Takahashi (1955a), Borchsenius (1957), Koteja (1979b) and by Danzig (1980c). Adult female redescribed by Tang (1991). Intraspecific variation in taxonomic characters given by Koteja (1979b). Distribution and host plant records given by Takahashi (1955a), Borchsenius (1957), Danzig (1967, 1978a, 1980c) and by Koteja (1979b).

Luzulaspis borealis Koteja & Howell

Luzulaspis borealis Koteja & Howell, 1979: 339.
TYPE DATA. Holotype female, USA: Alaska, Mouth of Beluga River, Cook Inlet, on *Carex* sp. (USNM).
DISTRIBUTION. NEARCTIC REGION: Alaska.
HOST PLANTS. Cyperaceae: *Carex.*
REMARKS. Adult female redescribed and illustrated by Koteja (1979b).

Luzulaspis caricicola (Lindinger) n. comb.

Luzulaspis caricis Takahashi, 1955a: 37 [HOMONYM of *Luzulaspis caricis* (Ehrhorn)].
Lecaniopsis caricicola Lindinger, 1957: 550 [REPLACEMENT NAME].
Luzulaspis takahashii Koteja; Koteja, 1979b: 623 [UNJUSTIFIED REPLACEMENT NAME].
TYPE DATA. Syntypes female, JAPAN: Hikawa near Tokyo, on *Carex brunea* (HUSJ).

DISTRIBUTION. PALAEARCTIC REGION: Japan.
HOST PLANTS. Cyperaceae: *Carex brunnea*.
REMARKS. Adult female redescribed and illustrated by Koteja
(1979b). Adult female redescribed by Borchsenius (1957), Kawai
(1980) and by Tang (1991) (as *L. takahashii*). Colour photograph
given by Kawai (1980).

Luzulaspis caricis (Ehrhorn)

Exaeretopus caricis Ehrhorn, 1902: 193.
Luzulaspis caricis (Ehrhorn); Koteja, 1978: 323.
TYPE DATA. Syntypes female, USA: California, Mt. Shasta, on *Carex
breweri* (USNM).
DISTRIBUTION. NEARCTIC REGION: California.
HOST PLANTS. Cyperaceae: *Carex breweri*.
REMARKS. Adult female redescribed and illustrated by Koteja &
Howell (1979), Koteja (1979b) and by Gill (1988).

Luzulaspis crassispina Borchsenius

Luzulaspis crassispina Borchsenius, 1959: 171.
TYPE DATA. Holotype female, CHINA: Yunnan Province, Kingtung,
on *Carex* sp. (IEBC).
DISTRIBUTION. PALAEARCTIC REGION: China.
HOST PLANTS. Cyperaceae: *Carex*.
REMARKS. Adult female redescribed and illustrated by Koteja
(1979b). Adult female redescribed by Tang (1991). Intraspecific varia-
tion in taxonomic characters given by Koteja (1979b).

Luzulaspis dactylis Green

Luzulaspis dactylis Green, 1928b: 24.
Luzulaspis montana Schmutterer, 1955a: 163. Holotype female,
GERMANY: Maria Gern near Berchtesgaden, host plant not
indicated (SENC). Syn. by Koteja, 1979b: 611.
TYPE DATA. Syntypes female, ENGLAND: Swanage, Dorset, on
Carex sp. and grass (BMNH).
DISTRIBUTION. PALAEARCTIC REGION: Czechoslovakia, England,
Germany, Greece, Italy, Poland, USSR Leningrad Region.
HOST PLANTS. Cyperaceae: *Carex praecox*. **Gramineae:** *Dactylis
glomerata, Festuca pratensis*.
REMARKS. Adult female redescribed and illustrated by Koteja
(1979b). Adult female redescribed by Borchsenius (1957) and by
Kosztarab & Kozár (1988). Distribution and host plant records given
by Schmutterer (1955a), Borchsenius (1957), Podsiadlo & Komosins-
ka (1976), Koteja (1979b), Kozár (1985), Marotta (1987) and by
Kosztarab & Kozár (1988).

Luzulaspis frontalis Green

Luzulaspis frontalis Green, 1928b: 25.
TYPE DATA. Syntypes female, ENGLAND: Kent, Bearstead near
Maidstone, on *Carex remota* (BMNH).
DISTRIBUTION. PALAEARCTIC REGION: Austria, Czechoslovakia,
England, Germany, Poland.
HOST PLANTS. Cyperaceae: *Carex brizoides, C. remota*.

REMARKS. Adult female redescribed and illustrated by Kosztarab & Kozár (1978) and by Koteja (1964, 1966a, 1979b). Adult female redescribed by Borchsenius (1957) and by Kosztarab & Kozár (1988). Koteja (1966a) described and illustrated larval instars and the adult male. Distribution and host plant records given by Rehacek (1957), Borchsenius (1957), Koteja (1964, 1966a, 1979b) and by Kosztarab & Kozár (1988).
BIOLOGY. The biology in Poland was studied by Koteja (1966a).

Luzulaspis grandis Borchsenius

Luzulaspis frontalis Green; Borchsenius, 1939: 44 [MISIDENTIFICA-TION].
Luzulaspis grandis Borchsenius, 1952: 279.
Luzulaspis caucasica Borchsenius, 1952: 278. Syntypes female, USSR: Georgia, Abkhazia, on grass (ZIAS). Syn. by Danzig, 1985: 122.
Luzulaspis pieninica Koteja & Zak-Ogaza, 1966: 322. Holotype female, POLAND: Cisowa Skala in the Nowy Targ Valley, on *Carex ornithopoda* (PASK). Syn. by Danzig, 1980c: 248
TYPE DATA. Syntypes female, USSR: Primorye Territory, Khasanski Region, on *Carex* sp. (ZIAS).
DISTRIBUTION. PALAEARCTIC REGION: Czechoslovakia, Germany, Poland, USSR Georgia, USSR Primorye Territory.
HOST PLANTS. Cyperaceae: *Carex ornithopoda*. **Gramineae**.
REMARKS. Adult female redescribed and illustrated by Borchsenius (1957), Koteja & Zak-Ogaza (1966) (as *L. pieninica*), Koteja (1979b) and by Danzig (1980c). Adult female redescribed by Kosztarab & Kozár (1988) and by Tang (1991) (as *L. pieninica*). Distribution and host plant records given by Borchsenius (1957), Koteja & Zak-Ogaza (1966), Danzig (1967, 1980c), Koteja (1979b) and by Kosztarab & Kozár (1988).

Luzulaspis kosztarabi Koteja & Kozár

Luzulaspis kosztarabi Koteja & Kozár, 1979: 121.
TYPE DATA. Holotype female, HUNGARY: Tornanadaska, on *Carex* sp. (HNHB).
DISTRIBUTION. PALAEARCTIC REGION: Hungary.
HOST PLANTS. Cyperaceae: *Carex*.
REMARKS. Adult female redescribed and illustrated by Koteja (1979b) and by Kosztarab & Kozár (1988).

Luzulaspis luzulae (Dufour)

Aspidiotus luzulae Dufour, 1864: 208.
Signoretia clypeata Targioni Tozzetti; Targioni Tozzetti, 1868: 727 [UNJUSTIFIED REPLACEMENT NAME].
Signoretia luzulae (Dufour); Targioni Tozzetti, 1868: 727.
Luzulaspis luzulae (Dufour); Cockerell, 1902a: 25.
Luzulaspis (Luzulaspis) luzulae (Dufour); Goux, 1937: 96.
TYPE DATA. Syntypes female and first instar larva, FRANCE: Saint Sever, on *Luzula maxima* (lost; Koteja, 1979b).
DISTRIBUTION. PALAEARCTIC REGION: Bulgaria, Czechoslovakia, England, France, Germany, Hungary, Italy, Poland, Scotland,

USSR Karelia, USSR Kunashirlsland, USSR Leningrad Region, USSR Sakhalin Island, USSR Ukraine.
HOST PLANTS. Compositae: *Bellis perennis.* **Cyperaceae:** *Carex, Scirpus caespitosus.* **Gramineae:** *Agrostis, Nardus stricta, Sieglingia decumbens.* **Juncaceae:** *Juncus silvatica, J. squarrosus, Luzula campestris, L. capitata, L. maxima, L. multiflora, L. nemorosa, L. pilosa.* **Rosaceae:** *Potentilla erecta.*
REMARKS. Adult female redescribed and illustrated by Newstead (1903), Rehacek (1955), Borchsenius (1957), Koteja (1964, 1966b, 1979b), Kosztarab& Kozár (1978, 1988), Danzig (1980c), Tereznikova (1981) and by Tang (1991). Male described and illustrated by Giliomee (1967). Distribution and host plant records given by Rehacek (1955), Borchsenius (1957), Koteja (1964, 1966b, 1979b), Kozár *et al.* (1977), Danzig (1978c, 1980c), Tereznikova (1981), Marotta (1987), Dziedzicka (1988), Kosztarab & Kozár (1988), Marotta & Tranfaglia (1990) and by Tang (1991). Koteja (1966b) supposed that *Lecanium insulae* by Rondani (1876) was a mis-spelling for *L. luzulae.* However, Rondani (1876) listed "*Lecanium insulae* Prrs. Cicaditi", which cannot be regarded a mis-spelling of this species.

Luzulaspis macrospinus Savescu
Luzulaspis macrospinus Savescu, 1985: 123.
TYPE DATA. Syntypes female, ROMANIA: Rupea, Brasov District, on oak (ASAR).
DISTRIBUTION. PALAEARCTIC REGION: Romania.
HOST PLANTS. Fagaceae: *Quercus.*

Luzulaspis minima Koteja & Howell
Luzulaspis minima Koteja & Howell, 1979: 339.
TYPE DATA. Holotype female, USA: California, Tulare County, Eagle Lake, Mineral King, on *Carex* sp. (USNM).
DISTRIBUTION. NEARCTIC REGION: California.
HOST PLANTS. Cyperaceae: *Carex.*
REMARKS. Adult female redescribed and illustrated by Koteja (1979b) and by Gill (1988).

Luzulaspis nemorosa Koteja
Luzulaspis nemorosa Koteja, 1966b: 45.
TYPE DATA. Holotype female, POLAND: Przegorzaly (Cracow), on *Luzula nemorosa* (PASK).
DISTRIBUTION. PALAEARCTIC REGION: Bulgaria, Czechoslovakia, France, Hungary, Poland.
HOST PLANTS. Cyperaceae: *Carex nemorosa.* **Gramineae:** *Deschampsia caespitosa.*
REMARKS. Adult female redescribed and illustrated by Koteja (1979b) and by Kosztarab & Kozár (1978, 1988). Koteja (1979b) listed many earlier records by European workers, of *Luzulaspis luzulae,* which are misidentifications of this species. Distribution and host plant records given by Kozár *et al.* (1979), Koteja (1979b), Kozár (1985, 1986) and by Kosztarab & Kozár (1988).

Luzulaspis rajae Kozár

Luzulaspis rajae Kozár, 1981b: 318.
TYPE DATA. Holotype female, HUNGARY: Szar, Vertes Mountains (Sozo), on *Carex montana* (HNHB).
DISTRIBUTION. PALAEARCTIC REGION: Hungary.
HOST PLANTS. Cyperaceae: *Carex montana*.
REMARKS. Adult female redescribed and illustrated by Kosztarab & Kozár (1988).

Luzulaspis saueri Lepage & Giannotti

Luzulaspis saueri Lepage & Giannotti, 1944: 302.
TYPE DATA. Syntypes female, BRAZIL: Sao Paulo State, Cunha, on wild grass (IBSP).
DISTRIBUTION. NEOTROPICAL REGION: Brazil Sao Paulo.
HOST PLANTS. Gramineae.
REMARKS. Koteja (1978) suggested that this species does not belong to *Luzulaspis*, nor to the Eriopeltini, but probably represents a distinct genus in the *Coccus* group of genera.

Luzulaspis scotica Green

Luzulaspis scotica Green, 1926b: 179.
Luzulaspis (Luzulaspis) scoticae Green; Goux, 1937: 96 [MIS-SPELL-ING].
Luzulaspis borchsenii Rehacek, 1959: 176. Syntypes female, CZECHOSLOVAKIA: Slovakia, near River Morava, between Kuty and Devin, on *Carex* sp. (NMPC). Syn. by Koteja, 1979b: 609.
TYPE DATA. Syntypes female, SCOTLAND: Aberlady, Haddington, on grass (BMNH).
DISTRIBUTION. PALAEARCTIC REGION: Czechoslovakia, France, Hungary, Poland, Romania, Scotland, Sweden, USSR Leningrad Region, Wales.
HOST PLANTS. Cyperaceae: *Carex fusca, C. gracilis, C. stellulata, Eriophorum vaginatum.*
REMARKS. Adult female redescribed and illustrated by Koteja (1979b), Tereznikova (1981) (as *L. borchsenii*) and by Kosztarab & Kozár (1988). Adult female redescribed by Borchsenius (1957). Distribution and host plant records given by Goux (1937), Borchsenius (1957), Ossiannilsson (1959), Rehacek (1959), Podsiadlo & Komosinska (1976), Koteja (1979b, 1986), Koteja & Zak-Ogaza (1983), and by Kosztarab & Kozár (1988).

Luzulaspis spinulosa Leonardi

Luzulaspis spinulosa Leonardi, 1911: 262.
TYPE DATA. Syntypes female, ARGENTINA: Cacheuta, on *Atriplex lampa* (IEAP).
DISTRIBUTION. NEOTROPICAL REGION: Argentina.
HOST PLANTS. Chenopodiaceae: *Atriplex lampa*.
REMARKS. Koteja (1978) indicated that this species has no connection with *Luzulaspis*, but he did not assign it to another genus.

Maacoccus Tao, Wong & Chang

Maacoccus Tao, Wong & Chang, 1983: 71.
TYPE-SPECIES: *Lecanium bicruciatus* Green, by original designation.
REMARKS. Generic characters discussed by Tang (1991). Key to species given by Tao *et al.* (1983) and by Tang (1991).

Maacoccus arundinariae (Green)

Lecanium arundinariae Green, 1904d: 220.
Coccus arundinariae (Green); Sanders, 1906: 8.
Maacoccus arundinariae (Green); Tang, 1991: 111.
TYPE DATA. Holotype female, SRI LANKA: Pundaluoya, on *Arundinaria* sp. (BMNH).
DISTRIBUTION. ORIENTAL REGION: Sri Lanka.
HOST PLANTS. Gramineae: *Arundinaria*.
REMARKS. Adult female redescribed by Tang (1991).

Maacoccus bicruciatus (Green)

Lecanium bicruciatus Green, 1904d: 214.
Coccus bicruciatus (Green); Sanders, 1906: 8.
Coccus bicurciatus (Green); Tao, 1978: 80 [MIS-SPELLING].
Maacoccus bicruciatus (Green); Tao *et al.*, 1983: 71.
TYPE DATA. Syntypes female, SRI LANKA: Paradeniya, Pundaluoya, Watawella and Nuwera Eliya, on *Memecylon umbellatum, Nothopegia colebrookiana, Elaeagnus latifolia, Calophyllum* sp. and *Eugenia* sp. (BMNH).
DISTRIBUTION. ETHIOPIAN REGION: Kenya, Zanzibar. **ORIENTAL REGION**: India, Sri Lanka, Taiwan.
HOST PLANTS. Anacardiaceae: *Mangifera indica, Nothopegia colebrookiana*. **Elaeagnaceae**: *Elaeagnus latifolia*. **Guttiferae**: *Calophyllum*. **Memecylaceae**: *Memecylon umbellatum*. **Myrtaceae**: *Eugenia*. **Rutaceae**: *Citrus. Murraya exotica*.
REMARKS. Adult female redescribed and illustrated by Mamet (1956a), De Lotto (1957b) and by Tao *et al.* (1983). Adult female redescribed by Tang (1991). Distribution and host plant records given by Ferris (1921b), Takahashi (1928), Mamet (1956a), De Lotto (1957b), Ali (1971), Tao (1978), Tao *et al.* (1983) and by Tang (1991).

Maacoccus cinnamomicolus (Takahashi)

Coccus cinnamomicolus Takahashi, 1952: 13.
Maacoccus cinnamomicolus (Takahashi); Tang, 1991: 113.
TYPE DATA. Syntypes female, MALAYSIA: Kuala Lumpur, on *Cinnamomum* sp. (SMKM).
DISTRIBUTION. AUSTRO ORIENTAL REGION: Malaysia.
HOST PLANTS. Lauraceae: *Cinnamomum*.
REMARKS. Adult female redescribed by Tang (1991).

Maacoccus piperis (Green)

Lecanium piperis Green, 1896: 10.
Coccus piperis (Green); Fernald, 1903: 173.
Coccus piperis piperis (Green); Varshney, 1985: 26.
Maacoccus piperis (Green); Tang, 1991: 113.

TYPE DATA. Syntypes female, SRI LANKA: Punduloya, on wild pepper (BMNH).
DISTRIBUTION. ORIENTAL REGION: Sri Lanka.
HOST PLANTS. Piperaceae: *Piper.*
REMARKS. Adult female redescribed and illustrated by Green (1904d). Adult female redescribed by Tang (1991).

Maacoccus scolopiae (Takahashi)

Coccus scolopiae Takahashi, 1933: 35.
Maacoccus scolopiae (Takahashi); Tao *et al.*, 1983: 72.
TYPE DATA. Syntypes female, TAIWAN: Garambi near Koshun, on *Scolopia oldhami* (IMZT).
DISTRIBUTION. ORIENTAL REGION: Taiwan.
HOST PLANTS. Flacourtiaceae: *Scolopia oldhami.*
REMARKS. Adult female redescribed and illustrated by Tao *et al.* (1983) and by Tang (1991).

Maacoccus watti (Green)

Lecanium watti Green, 1900c: 6.
Coccus watti (Green); Fernald, 1903: 174.
Saissetia watti (Green); Ali, 1971: 45.
Maacoccus watti (Green); Tang, 1991: 115.
TYPE DATA. Syntypes female, INDIA: Assam, on tea plant (BMNH).
DISTRIBUTION. ORIENTAL REGION: India.
HOST PLANTS. Rutaceae: *Citrus.* **Theaceae:** *Thea.*
REMARKS. Adult female redescribed by Tang (1991). Distribution and host plant records given by Ali (1971), Shafee *et al.* (1989) and by Tang (1991).

Macropulvinaria Hodgson

Macropulvinaria Hodgson, 1968a: 155.
TYPE-SPECIES: *Pulvinaria jacksoni* Newstead, by original designation.
REMARKS. Generic characters discussed by Tao *et al.* (1983), Williams & Watson (1990) and by Tang (1991). Key to species: ETHIOPIAN REGION - Hodgson (1968a); TAIWAN - Tao *et al.* (1983); ZIMBABWE - Hodgson (1969b).

Macropulvinaria acaciae Hodgson

Macropulvinaria acaciae Hodgson, 1968a: 155.
TYPE DATA. Syntypes female, SUDAN: Medani, on *Acacia arabica* (BMNH).
DISTRIBUTION. ETHIOPIAN REGION: Sudan.
HOST PLANTS. Leguminosae: *Acacia arabica.*

Macropulvinaria filamentosa (Newstead)

Lecanium (Eulecanium) filamentosum Newstead, 1913: 74.
Lecanium filamentosum Newstead; Sasscer, 1915: 31.
Eulecanium filamentosum Newstead; Gowdey, 1917: 188.
Pulvinaria filamentosum (Newstead); De Lotto, 1959: 152.

Macropulvinaria filamentosa (Newstead); Hodgson, 1968a: 157.
TYPE DATA. Syntypes female, UGANDA: Tero Forest, on undetermined shrub (BMNH)
DISTRIBUTION. ETHIOPIAN REGION: Uganda.
REMARKS. Adult female redescribed and illustrated by Hodgson (1968a). Distribution and host plant records given by De Lotto (1959) and by Hodgson (1968a).

Macropulvinaria inopheron (Laing)

Lecanium inopheron Laing, 1925: 57.
Pulvinaria jacksoni Newstead; Hall, 1932: 191 [MISIDENTIFICATION].
Pulvinaria inopheron (Laing); Hodgson, 1967b: 201.
Macropulvinaria inopheron (Laing); Hodgson, 1968a: 159.
TYPE DATA. Syntypes female, UGANDA: Kampala, on *Erythrina* sp. (BMNH).
DISTRIBUTION. ETHIOPIAN REGION: Uganda, Zimbabwe.
HOST PLANTS. Annonaceae: *Annona.* **Araliaceae:** *Cussonia spicata.* **Euphorbiaceae:** *Croton sylvaticus.* **Labiatae:** *Salvia.* **Leguminosae:** *Erythrina.* **Malvaceae:** *Hibiscus.* **Moraceae:** *Ficus.* **Rhamnaceae:** *Ziziphus.*
REMARKS. Adult female redescribed and illustrated by Hodgson (1967b, 1968a). Distribution and host plant records given by Hall (1932) and by Hodgson (1967a, 1968a, 1969b).

Macropulvinaria jacksoni (Newstead)

Pulvinaria jacksoni Newstead, 1908b: 155.
Lecanium nyassae Newstead, 1911b: 162. Syntypes female, MALAWI: North Nyassa, Rungwa, Utengule, on undetermined plant (BMNH). Syn. by Hodgson, 1968a: 159.
Coccus nyassae (Newstead); Sasscer, 1912: 88.
Macropulvinaria jacksoni (Newstead); Hodgson, 1968a: 159.
TYPE DATA. Syntypes female, SENEGAL: Dakar, Botanic Gardens, on *Ficus* sp. (BMNH).
DISTRIBUTION. ETHIOPIAN REGION: Cameroon, Ghana, Ivory Coast, Kenya, Malawi, Nigeria, Senegal, Zimbabwe.
HOST PLANTS. Euphorbiaceae: *Acalypha.* **Leguminosae:** *Cassia dydimobotrya, Erythrina.* **Moraceae:** *Ficus.* **Passifloraceae:** *Passiflora.* **Sterculiaceae:** *Theobroma cacao.* **Strelitziaceae:**
REMARKS. Adult female redescribed and illustrated by Hodgson (1967b, 1968a). Distribution and host plant records given by Strickland (1947), Hodgson (1967b, 1968a), Matile-Ferrero & Nonveiller (1984) and by Couturier *et al.* (1985).

Mallococcus Maskell

Mallophora Maskell, 1897a: 314 [HOMONYM of *Mallophora* Macquart, Diptera: Asylidae].
Mallococcus Maskell, 1898: 242 [REPLACEMENT NAME].
TYPE-SPECIES: *Mallophora sinensis* Maskell, by monotypy.
REMARKS. Generic characters discussed by Morrison & Morrison

(1922), Steinweden (1929), Ferris (1955), Williams & Kosztarab (1973) and by Tang (1991).

Mallococcus lanigerus (Hempel)

Lecanium lanigerum Hempel, 1900b: 446.
Mallococcus lanigerus (Hempel); Cockerell, 1902f: 452.
Pseudophilippia lanigera (Hempel); Lizer y Trelles, 1942b: 233.
TYPE DATA. Syntypes female, BRAZIL: Rio Grande Do Sul, near Itapira, at banks of Mogy-guassu River, on undetermined tree (USNM).
DISTRIBUTION. NEOTROPICAL REGION: Argentina, Brazil Rio Grande do Sul.
HOST PLANTS. Rutaceae: *Citrus*. **Sterculiaceae:** *Astrapaea*.
REMARKS. Adult female redescribed by Hempel (1901a). Distribution and host plant records given by Lizer y Trelles (1942b), Silva *et al.* (1968) and by Corseuil & Barbosa (1971).

Mallococcus sinensis (Maskell)

Mallophora sinensis Maskell, 1897a: 314.
Mallococcus sinensis (Maskell); Maskell, 1898: 242.
TYPE DATA. Lectotype female designated by Lambdin & Kosztarab (1973), HONG KONG: on *Callicarpa tomentosa* (USNM).
DISTRIBUTION. PALAEARCTIC REGION: China, Hong Kong.
HOST PLANTS. Verbenaceae: *Callicarpa tomentosa*.
REMARKS. Adult female redescribed and illustrated by Morrison & Morrison (1922), Ferris (1955), Lambdin & Kosztarab (1973) and by Tang (1991). First and second larval instars described and illustrated by Lambdin & Kosztarab (1973). Distribution and host plant records given by Ferris (1921b, 1950) and by Lambdin & Kosztarab (1973).

Mallococcus vitecicola Young

Mallococcus vitecicola Young in Wan *et al.*, 1985: 267.
TYPE DATA. Holotype female, CHINA: Peng Ze, Jiangxi, on *Vitex rotundifolia* (SIEC).
DISTRIBUTION. PALAEARCTIC REGION: China.
HOST PLANTS. Begoniaceae: *Begonia*. **Compositae:** *Artemisia capillaris, Tagetes erecta.* **Gramineae:** *Bromus japonicus.* **Malvaceae:** *Hibiscus syriacus.* **Rosaceae:** *Rosa chinensis.* **Verbenaceae:** *Vitex rotundifolia*.
REMARKS. Wan *et al.* (1985) erroneously assigned this species to the Lecanodiaspididae. Tang (1991) placed it in the Coccidae. Adult female redescribed by Tang (1991).
BIOLOGY. Wan *et al.* (1985) presented data on the biology, phenology, natural enemies and control.
ECONOMIC IMPORTANCE. Regarded a pest of the medical herb, *Vitex rotundifolia* in China (Wan *et al.*, 1985).

Mametia Matile-Ferrero

Mametia Matile-Ferrero, 1978: 44.
TYPE-SPECIES: *Mametia louisieae* Matile-Ferrero, by original designation.
REMARKS. Key to separation from four related genera given by

Matile-Ferrero (1978). Generic characters discussed by Tang (1991).

Mametia grandis (Green & Laing)

Ceronema grandis Green & Laing, 1924: 417.
Mametia grandis (Green & Laing); Matile-Ferrero, 1978: 44.
TYPE DATA. Syntypes female, KENYA: on native plant (BMNH).
DISTRIBUTION. ETHIOPIAN REGION: Angola, Kenya.
HOST PLANTS. Leguminosae: *Cassia.*
REMARKS. Adult female redescribed and illustrated by Hodgson (1967c). Distribution and host plant records given by Hodgson (1967c) and by Almeida (1969).

Mametia louisieae Matile-Ferrero

Mametia louisieae Matile-Ferrero, 1978: 45.
TYPE DATA. Holotype female, COMOROS: Grande Comore, route de M'Vouni, on *Eugenia caryophyllata* (MNHN).
DISTRIBUTION. MADAGASIAN REGION: Comoros.
HOST PLANTS. Myrtaceae: *Eugenia caryophyllata.*

Marsipococcus Cockerell & Bueker

Marsipococcus Cockerell & Bueker, 1930: 7.
TYPE-SPECIES: *Lecanium marsupiale* Green, by monotypy.
REMARKS. Generic characters discussed by De Lotto (1965) and by Tang (1991). Key to species given by Tang (1991).

Marsipococcus durbanensis (Brain)

Lecanium durbanense; Brain, 1920b: 8.
Coccus durbanensis (Brain); De Lotto, 1959: 158.
Marsipococcus durbanensis (Brain); De Lotto, 1967b: 785.
TYPE DATA. Syntypes female, SOUTH AFRICA: Natal, Durban, on leaves of undetermined plant (SANC).
DISTRIBUTION. ETHIOPIAN REGION: South Africa.
HOST PLANTS. Zamiaceae: *Encephalartos horridus.*
REMARKS. Adult female redescribed and illustrated by De Lotto (1959, 1967b). Distribution and host plant records given by De Lotto (1959, 1967b).

Marsipococcus marsupialis (Green)

Lecanium marsupiale Green, 1904d: 212.
Coccus marsupialis (Green); Sanders, 1906: 8.
Marsipococcus marsupialis (Green); Cockerell & Bueker, 1930: 7.
TYPE DATA. Syntypes female, SRI LANKA: Paradeniya, Rambukkana and Matale, on *Piper nigrum, Piper* sp., *Pothos scandens* and *Annona* sp. (BMNH).
DISTRIBUTION. ETHIOPIAN REGION: Tanzania. **ORIENTAL REGION:** India, Sri Lanka.
HOST PLANTS. Annonaceae: *Annona.* **Araceae:** *Pothos scandens.* **Euphorbiaceae:** *Manihot glaziout.* **Piperaceae:** *Piper nigrum.*
REMARKS. Adult female redescribed and illustrated by Ramakrishna Ayyar (1919, 1930), De Lotto (1965) and by Tang (1991). Distribution and host plant records given by Lindinger (1913a), Ramakrishna

Ayyar (1919, 1930), Green (1937), De Lotto (1965), Ali (1971), Shafee
et al. (1989) and by Tang (1991).

Marsipococcus proteae (Brain)

Lecanium proteae Brain, 1920b: 7.
Coccus proteae (Brain); De Lotto, 1959: 168.
Marsipococcus proteae (Brain); De Lotto, 1967b: 786.
TYPE DATA. Syntypes female, SOUTH AFRICA: Transvaal, Pretoria,
  on *Protea* sp. (BMNH).
DISTRIBUTION. ETHIOPIAN REGION: South Africa.
HOST PLANTS. Proteaceae: *Protea cafra, P. cynaroides, P. nerifolia.*
REMARKS. Adult female redescribed and illustrated by De Lotto
(1959, 1967b). Distribution and host plant records given by De Lotto
(1959, 1967b).

Megalecanium Hempel

Megalecanium Hempel, 1920b: 352.
Megalolecanium Lindinger; Lindinger, 1937: 189 [UNJUSTIFIED
  EMENDATION]
TYPE-SPECIES: *Megalecanium testudinis* Hempel, by original desig-
  nation and monotypy.

Megalecanium testudinis Hempel

Megalecanium testudinis Hempel, 1920b: 352.
TYPE DATA. Syntypes female, BRAZIL: Sao Paulo, Cantareira near
  Sao Paulo, on bark of "cambara preto" and "cambara branco"
  (MZSP).
DISTRIBUTION. NEOTROPICAL REGION: Brazil Sao Paulo.
HOST PLANTS. Verbenaceae: *Lantana.*
REMARKS. Distribution and host plant records given by Silva *et al.*
(1968).

Megalocryptes Takahashi

Megalocryptes Takahashi, 1942a: 20.
TYPE-SPECIES: *Megalocryptes buteae* Takahashi, by original desig-
  nation and monotypy.
REMARKS. Generic characters discussed by Tang (1991).

Megalocryptes bambusicola (Green)

Lecanium bambusicola Green, 1930c: 285.
Megalocryptes bambusicola (Green); Ali, 1971: 34.
TYPE DATA. Syntypes female, INDONESIA: Sumatra, Fort de Kock,
  on *Bambusa nana* (BMNH).
DISTRIBUTION. AUSTRO ORIENTAL REGION: Sumatra.
HOST PLANTS. Gramineae: *Bambusa nana.*

Megalocryptes buteae Takahashi

Megalocryptes buteae Takahashi, 1942a: 20.
TYPE DATA. Syntypes female, THAILAND: Bangkok, Chiengmai,
  Lampoon and Mt. Sutep, on *Butea frondosa, Ziziphus jujuba,
  Sesbania glandiflora,* mango, *Quercus* sp. and bamboo (IMZT).

DISTRIBUTION. ORIENTAL REGION: Thailand. **PALAEARCTIC REGION:** China.
HOST PLANTS. Anacardiaceae: *Mangifera indica*, **Fagaceae:** *Quercus*. **Gramineae:** *Bambusa*. **Leguminosae:** *Butea frondosa*, *Sesbania glandiflora*. **Rhamnaceae:** *Ziziphus jujuba*.
REMARKS. Adult female redescribed and illustrated by Tang (1991). Distribution and host plant records given by Tang (1991).

Megapulvinaria Yang

Megapulvinaria Yang, 1982: 162.
TYPE-SPECIES: *Pulvinaria maxima* Green, by original designation and monotypy.
REMARKS. Tang (1991) regarded this genus as a subjective synonym of *Macropulvinaria*.

Megapulvinaria burkilli (Green)

Pulvinaria burkilli Green, 1908: 31.
Megapulvinaria burkilli (Green); Avasthi & Shafee, 1991b: 23.
TYPE DATA. Syntypes female, INDIA: Siugaing near Calcutta, on *Croton tiglium* (BMNH).
DISTRIBUTION. ORIENTAL REGION: India.
HOST PLANTS. Euphorbiaceae: *Croton tiglius*. **Rhamnaceae:** *Ziziphus*.
REMARKS. Newstead (1917c) supplemented the taxonomic characters. Distribution and host plant records given by Newstead (1917c).

Megapulvinaria maxima (Green)

Pulvinaria maxima Green, 1904b: 206.
Pulvinaria thespesiae Green, 1909: 259. Syntypes female, SRI LANKA: Jaffna, on *Thespesia populnea* (BMNH). Syn. by Takahashi, 1935: 10.
Eriochiton formosae Takahashi, 1929a: 64. Syntypes female, TAIWAN: Taito, on undetermined plant (IMZT). Syn. by Takahashi, 1935: 10.
Megapulvinaria maxima (Green); Yang, 1982: 162.
Macropulvinaria maxima (Green); Tao *et al.*, 1983: 87.
Pulvinaria maxima maxima Green; Varshney, 1985: 25.
Pulvinaria maxima thespesiae Green; Varshney, 1985: 25.
TYPE DATA. Lectotype female designated by Williams & Watson (1990), JAVA: Buitenzorg [Bogor], on *Erythrina* (BMNH).
DISTRIBUTION. AUSTRO ORIENTAL REGION: Java, Papua New Guinea, Philippines. **NEW ZEALAND and PACIFIC REGION:** Truk Islands. **ORIENTAL REGION:** India, Sri Lanka, Taiwan, Thailand, Vietnam.
HOST PLANTS. Annonaceae: *Annona*. **Euphorbiaceae:** *Croton tiglium*, *Glochidion callicarpa*, *Homonoia riparia*, *Jatropha curcas*, *J. pendurifolia*, *Phyllanthus takaoensis*. **Leguminosae:** *Erythrina*, *Samanea saman*. **Malvaceae:** *Thespesia populnea*. **Melastomataceae:** *Medinilla*. **Moraceae:** *Morus acidosa*, *M. alba*. **Rhamnaceae:** *Ziziphus jujuba*. **Urticaceae:** *Boehmeria frutescens*.
REMARKS. Adult female redescribed and illustrated by Ramakrishna Ayyar (1919, 1930), Morrison (1920) (as *P. thespesiae*), Tao *et al.*

(1983), Williams & Watson (1990) and by Tang (1991). Distribution and host plant records given by Green (1909, 1937), Ramakrishna Ayyar (1919, 1930), Morrison (1920), Takahashi (1935, 1942a), Beardsley (1966a), Ali (1971), Yang (1982), Tao *et al.* (1983), Varshney (1984), Varshney & Moharana (1987), Shafee *et al.* (1989), Williams & Watson (1990), Danzig & Konstantinova (1990) and by Tang (1991).

Megapulvinaria orientalis (Reyne) n. comb.
Filippia orientalis Reyne, 1963a: 30.
Lichtensia orientalis (Reyne); Tang, 1991: 54.
TYPE DATA. Syntypes female, male and larva, THAILAND: Pi-Mai, on undetermined plant (ZMAN).
DISTRIBUTION. ORIENTAL REGION: Thailand.
REMARKS. The assignment of this species to *Megapulvinaria* is based on a study of the type-series which became available from ZMAN.

Megasaissetia Cockerell
Saissetia (Megasaissetia) Cockerell, 1901d: 32.
Megasaissetia Cockerell; Fernald, 1903: 207.
Megalosaissetia Lindinger; Lindinger, 1937: 189 [UNJUSTIFIED EMENDATION].
TYPE-SPECIES: *Lecanium (Saissetia) inflatum* Cockerell & Parrott, by original designation and monotypy.

Megasaissetia brasiliensis Hempel
Megasaissetia brasiliensis Hempel, 1912: 68.
TYPE DATA. Syntypes female, BRAZIL: Sao Paulo, Alto da Serra, on undetermined plant (MZSP).
DISTRIBUTION. NEOTROPICAL REGION: Brazil Sao Paulo.

Megasaissetia inflata (Cockerell & Parrott)
Lecanium (Saissetia) inflatum Cockerell & Parrott in Cockerell, 1899g: 13.
Saissetia inflata Cockerell & Parrott; Cockerell, 1901d: 32.
Saissetia (Megasaissetia) inflata Cockerell & Parrott; Cockerell, 1901d: 32.
Lecanium inflatum Cockerell & Parrott; Cockerell, 1901d: 32.
Megasaissetia inflata (Cockerell & Parrott); Fernald, 1903: 207.
TYPE DATA. Syntypes female, MEXICO: Coatzocoalcos in Vera Cruz, on a tree named "laurel" (BMNH, USNM).
DISTRIBUTION. NEOTROPICAL REGION: Mexico.

Megasaissetia nectandrae Hempel
Megasaissetia nectandrae Hempel, 1918: 203.
TYPE DATA. Syntypes female, BRAZIL: Sao Paulo, Piracicaba, on *Nectandra* sp. (MZSP).
DISTRIBUTION. NEOTROPICAL REGION: Brazil Sao Paulo.
HOST PLANTS. Lauraceae: *Nectandra*.

Melanesicoccus Williams & Watson

Melanesicoccus Williams & Watson, 1990: 113.
TYPE-SPECIES: *Melanesicoccus kleinhoviae* Williams & Watson, by original designation.
REMARKS. Key to species given by Williams & Watson (1990).

Melanesicoccus kleinhoviae Williams & Watson

Melanesicoccus kleinhoviae Williams & Watson, 1990: 116.
TYPE DATA. Holotype female, PAPUA NEW GUINEA: Milne Bay Province, on *Kleinhovia hospita* (BMNH).
DISTRIBUTION. AUSTRO ORIENTAL REGION: Papua New Guinea.
HOST PLANTS. Sterculiaceae: *Kleinhovia hospita.*

Melanesicoccus myrmecariae Williams & Watson

Melanesicoccus myrmecariae Williams & Watson, 1990: 116.
TYPE DATA. Holotype female, SOLOMON ISLANDS: Malaita Province, Malaita, Baunani, on bark of a plant named "be-be-ro", in galleries of *Iridomyrmex* sp. (BMNH).
DISTRIBUTION. AUSTRO ORIENTAL REGION: Solomon Islands.
BIOLOGY. Found in galleries of *Iridomyrmex* sp. (Williams & Watson, 1990).

Melanesicoccus solomonensis Williams & Watson

Melanesicoccus solomonensis Williams & Watson, 1990: 118.
TYPE DATA. Holotype female, SOLOMON ISLANDS: Guadalcanal Province, Guadalcanal, on *Acalypha* sp. (BMNH).
DISTRIBUTION. AUSTRO ORIENTAL REGION: Solomon Islands.
HOST PLANTS. Euphorbiaceae: *Acalypha.*

Membranaria Brain

Membranaria Brain, 1920b: 41.
TYPE-SPECIES: *Membranaria pretoriae* Brain, by original designation and monotypy.
REMARKS. Brain (1920b) supposed that this genus is related to *Pulvinaria*, whereas De Lotto (1970a) indicated its affinity to *Ceronema*.

Membranaria pretoriae Brain

Membranaria pretoriae Brain, 1920b: 41.
TYPE DATA. Syntypes female, SOUTH AFRICA: Transvaal, Pretoria, on grass (SANC).
DISTRIBUTION. ETHIOPIAN REGION: South Africa.
HOST PLANTS. Gramineae:
REMARKS. Adult female redescribed and illustrated by De Lotto (1970a).

Mesembryna De Lotto

Mesembryna De Lotto, 1979: 245.
TYPE-SPECIES: *Mesembryna fasciata* De Lotto, by original designation and monotypy.

Mesembryna fasciata De Lotto

Mesembryna fasciata De Lotto, 1979: 246.
TYPE DATA. Holotype female, SOUTH AFRICA: Cape Province, Bitterfontein, on *Mesembryanthemum* sp. (SANC).
DISTRIBUTION. ETHIOPIAN REGION: South Africa.
HOST PLANTS. Aizoaceae: *Mesembryanthemum.*

Mesolecanium Cockerell

Mesolecanium Cockerell, 1902f: 451.
TYPE-SPECIES: *Lecanium nocturnum* Cockerell & Parrott, by original designation.
REMARKS. Generic characters discussed by Morrison (1929) and by Hamon & Williams (1984).

Mesolecanium argaformis Hempel

Mesolecanium argaformis Hempel, 1920b: 346.
TYPE DATA. Syntypes female, BRAZIL: Sao Paulo, Cantareira near Sao Paulo, on bark of a plant named "canella poca" (MZSP).
DISTRIBUTION. NEOTROPICAL REGION: Brazil Sao Paulo.

Mesolecanium baccharidis (Cockerell)

Lecanium baccharidis Cockerell, 1895d: 174.
Mesolecanium baccharidis (Cockerell); Cockerell, 1902f: 452.
TYPE DATA. Syntypes female, BRAZIL: Sao Paulo, on *Baccharis* sp. (USNM).
DISTRIBUTION. NEOTROPICAL REGION: Brazil Rio Grande do Sul, Brazil Sao Paulo.
HOST PLANTS. Compositae: *Baccharis dracunculifolia.*
REMARKS. Adult female redescribed and illustrated by Hempel (1900b). Distribution and host plant records given by Hempel (1900b), Silva *et al.* (1968) and by Corseuil & Barbosa (1971).

Mesolecanium batatae (Cockerell)

Lecanium batatae Cockerell, 1895n: 61.
Mesolecanium batatae (Cockerell); Cockerell, 1902f: 452.
TYPE DATA. Syntypes female, ANTIGUA: on tuberous roots of *Ipomoea batatas* (USNM).
DISTRIBUTION. NEOTROPICAL REGION: Antigua.
HOST PLANTS. Convolvulaceae: *Ipomoea batatas.*

Mesolecanium campomanesiae (Hempel)

Lecanium campomanesiae Hempel, 1900b: 447.
Mesolecanium campomanesiae (Hempel); Cockerell, 1902f: 452.
TYPE DATA. Syntypes female, BRAZIL: Sao Paulo, Ypiranga, on *Campomanesia* sp. (MZSP).
DISTRIBUTION. NEOTROPICAL REGION: Brazil Sao Paulo.
HOST PLANTS. Myrtaceae: *Campomanesia.*
REMARKS. Adult female redescribed by Hempel (1901a).

Mesolecanium deltae Lizer y Trelles

Mesolecanium deltae Lizer y Trelles, 1917: 103.
Lecanium deltae (Lizer y Trelles); Lizer y Trelles, 1939: 190.
Coccus deltae (Lizer y Trelles); Corseuil & Barbosa, 1971: 238.
TYPE DATA. Syntypes female and larva, ARGENTINA: Buenos Aires
  and region of the Parana delta, on *Citrus* (DPBA).
DISTRIBUTION. NEOTROPICAL REGION: Argentina, Brazil Rio
  Grande do Sul, Paraguay, Uruguay.
HOST PLANTS. Rutaceae: *Citrus sinensis*.
REMARKS. Adult female redescribed and illustrated by Gomes
  Costa (1949) and by Fonseca (1972a). Distribution and host plant
  records given by Lizer y Trelles (1939), Gomes Costa (1949), Teran &
  Guyot (1969), Corseuil & Barbosa (1971) and by Fonseca (1972a).
BIOLOGY. Lizer y Trelles (1939) indicated the exclusiveness of this
  species to *Citrus*. Teran & Guyot (1969) observed one annual genera-
  tion on citrus in Argentina (Tucuman), discussed the life history,
  mortality factors and natural enemies.
ECONOMIC IMPORTANCE. A citrus pest in Argentina (Teran &
  Guyot, 1969).

Mesolecanium ferum Hempel

Mesolecanium ferum Hempel, 1920b: 350.
TYPE DATA. Syntypes female, BRAZIL: Sao Paulo, Campinas, on
  Croton floribundus (MZSP).
DISTRIBUTION. NEOTROPICAL REGION: Brazil Sao Paulo.
HOST PLANTS. Euphorbiaceae: *Croton floribundus*.

Mesolecanium impar (Cockerell)

Lecanium (Calymnatus) impar Cockerell, 1898d: 131.
Mesolecanium impar (Cockerell); Fernald, 1903: 175.
Coccus impar (Cockerell); Bodkin, 1917: 109.
TYPE DATA. Syntypes female, MEXICO: Las Minas, Tabasco, on
  leaves of a plant named "Escobillo" (USNM).
DISTRIBUTION. NEOTROPICAL REGION: Guyana, Mexico.

Mesolecanium inflatum Hempel

Mesolecanium inflatum Hempel, 1904: 316.
TYPE DATA. Syntypes female, BRAZIL: Rio de Janeiro, Maua, on
  Myrtaceae (MZSP).
DISTRIBUTION. NEOTROPICAL REGION: Brazil Sao Paulo.
HOST PLANTS. Myrtaceae:

Mesolecanium inquilinum Morrison

Mesolecanium inquilinum Morrison, 1929: 40.
TYPE DATA. Syntypes female, PANAMA: Canal Zone, Las Cascadus
  Cacao Plantation, on undetermined plant, in tents of an ant
  Azteca sp. (USNM).
DISTRIBUTION. NEOTROPICAL REGION: Panama.
BIOLOGY. Attended and protected by an *Azteca* ant (Morrison,
  1929).

Mesolecanium jaboticabae (Hempel)

Lecanium jaboticabae Hempel, 1900b: 443.
Mesolecanium jaboticabae (Hempel); Cockerell, 1902f: 452.
TYPE DATA. Syntypes female, BRAZIL: Sao Paulo, Ypiranga, on *Eugenia jaboticabu* (MZSP).
DISTRIBUTION. NEOTROPICAL REGION: Brazil Sao Paulo.
HOST PLANTS. Myrtaceae: *Eugenia jaboticaba.*
REMARKS. Adult female redescribed by Hempel (1901a, 1920a).

Mesolecanium lucidum Hempel

Mesolecanium lucidum Hempel, 1912: 67.
TYPE DATA. Syntypes female, BRAZIL: Rio Grande do Sul, Porto Alegre, on a plant of Solanaceae (MZSP).
DISTRIBUTION. NEOTROPICAL REGION: Brazil Rio Grande do Sul.
HOST PLANTS. Rutaceae: *Citrus aurantium.* **Solanaceae:**
REMARKS. Distribution and host plant records given by Corseuil & Barbosa (1912).

Mesolecanium marmoratum Hempel

Mesolecanium marmoratum Hempel, 1920b: 348.
TYPE DATA. Syntypes female, BRAZIL: Sao Paulo, Cantareira near Sao Paulo, on bark of plants "canella branca e canella poca" (MZSP).
DISTRIBUTION. NEOTROPICAL REGION: Brazil Sao Paulo.

Mesolecanium mayteni (Hempel)

Lecanium mayteni Hempel, 1900b: 438.
Mesolecanium mayteni (Hempel); Cockerell, 1902f: 452.
TYPE DATA. Syntypes female, BRAZIL: Sao Paulo, Ypiranga, on *Maytenus* sp. (MZSP).
DISTRIBUTION. NEOTROPICAL REGION: Brazil Sao Paulo.
HOST PLANTS. Celastraceae: *Maytenus.*
REMARKS. Adult female redescribed by Hempel (1901a).

Mesolecanium nigrofasciatum (Pergande)

Lecanium nigrofasciatum Pergande, 1898: 26.
Lecanium (Eulecanium) nigrofasciatum Pergande; Cockerell & Parrott, 1899: 234.
Eulecanium nigrofasciatum (Pergande); King, 1902c: 160.
Mesolecanium nigrofasciatum (Pergande); Nakahara, 1981a: 284.
TYPE DATA. Syntypes female, U.S.A.: From 11 Eastern states, on several hosts mainly peach and plums (USNM).
DISTRIBUTION. NEARCTIC REGION: Alabama, Arkansas, Canada Ontario, Florida, Georgia, Indiana, Kentucky, Louisiana, Maryland, Michigan, Mississippi, New Jersey, New York, North Carolina, Ohio, Oklahoma, Pennsylvania, South Carolina, Tennessee, Texas, Virginia, Washington, D. C., West Virginia, Wisconsin.
HOST PLANTS. Aceraceae: *Acer dasycarpum, A. platanoides, A. rubrum, A. saccharinum.* **Altingiaceae:** *Liquidambar styraciflua.* **Ericaceae:** *Vaccinium darrowi.* **Lauraceae:** *Lindera benzoin, Persea borbonia, Sassafras albidum.* **Leguminosae:** *Mimosa.*

Loranthaceae: *Phoradendron.* **Magnoliaceae:** *Magnolia virginiana.* **Platanaceae:** *Platanus occidentalis.* **Rosaceae:** *Persica, Prunus, Prunus cerasifera.*
REMARKS. Adult female redescribed and illustrated by Thro (1903), Dietz & Morrison (1916), Pettit & McDaniel (1920), Richards (1958), Williams & Kosztarab (1972) and by Hamon & Williams (1984). Male test described and illustrated by Miller & Williams (1990). Colour photograph given by Hamon & Williams (1984). Distribution and host plant records given by Dietz & Morrison (1916), Pettit & McDaniel (1920), Richards (1958), Williams & Kosztarab (1972), Lambdin & Watson (1980), Nakahara (1981a), Hamon & Williams (1984) and by Miller & Williams (1990).

Mesolecanium nocturnum (Cockerell & Parrott)

Lecanium nocturnum Cockerell & Parrott in Cockerell, 1899g: 13.
Mesolecanium nocturnum (Cockerell & Parrott); Cockerell, 1902f: 451.
Lecanium nocturnium Cockerell & Parrott; Thro, 1903: 210 [MIS-SPELLING].
TYPE DATA. Syntypes female, MEXICO: Vera Cruz, Alvarado, on a bush named "huele de noche" (BMNH, USNM).
DISTRIBUTION. NEOTROPICAL REGION: Mexico.

Mesolecanium obscurum (Hempel)

Lecanium obscurum Hempel, 1900a: 5.
Mesolecanium obscurum (Hempel); Cockerell, 1902f: 452.
TYPE DATA. Syntypes female, male and larva, BRAZIL: Sao Paulo, Ypiranga, on *Maytenus* sp. (MZSP).
DISTRIBUTION. NEOTROPICAL REGION: Brazil Sao Paulo.
HOST PLANTS. Celastraceae: *Maytenus.*
REMARKS. Adult female redescribed and illustrated by Hempel (1900b).

Mesolecanium perditulum Cockerell & Robbins

Mesolecanium perditulum Cockerell & Robbins, 1909: 150.
TYPE DATA. Syntypes female, NICARAGUA: Quesalquaque, on bark of undetermined tree (USNM).
DISTRIBUTION. NEOTROPICAL REGION: Nicaragua.

Mesolecanium perditum (Cockerell)

Lecanium (Eulecanium) perditum Cockerell; Cockerell, 1897j: 267.
Mesolecanium perditum (Cockerell); Fernald, 1903: 175.
TYPE DATA. Syntypes female, MEXICO: Yucatan, Xcloak near Izamal, on undetermined plant (USNM).
DISTRIBUTION. NEOTROPICAL REGION: Mexico.

Mesolecanium phoradendri (Cockerell)

Lecanium phoradendri Cockerell, 1894i: 14.
Mesolecanium phoradendri (Cockerell); Cockerell, 1902f: 452.
Mesolecanium phoridendri (Cockerell); Miller & Williams, 1990: 348 [MIS-SPELLING].

TYPE DATA. Syntypes female, U.S.A.: Arizona, Tucson, on *Phoradendron* sp. (BMNH, USNM).
DISTRIBUTION. NEARCTIC REGION: Arizona.
HOST PLANTS. Loranthaceae: *Phoradendron.*
REMARKS. Male test described and illustrated by Miller & Williams (1990).

Mesolecanium planum Hempel

Mesolecanium planum Hempel, 1932: 327.
TYPE DATA. Syntypes female, BRAZIL: Minas Gerais, Cajury, on undetermined tree (MZSP).
DISTRIBUTION. NEOTROPICAL REGION: Brazil Minas Gerais.

Mesolecanium pseudosemen (Cockerell)

Lecanium pseudosemen Cockerell, 1895p: 202.
Mesolecanium pseudosemen (Cockerell); Cockerell, 1902f: 452.
TYPE DATA. Syntypes female, BRAZIL: Sao Paulo, on undetermined plant (USNM).
DISTRIBUTION. NEOTROPICAL REGION: Brazil Rio Grande do Sul, Brazil Sao Paulo.
HOST PLANTS. Nyctaginaceae: *Bougainvillea.* **Rutaceae:** *Citrus.* **Solanaceae:** *Capsicum, Solanum paniculatum.*
REMARKS. Adult female redescribed and illustrated by Hempel (1900b). Distribution and host plant records given by Hempel (1900b), Silva *et al.* (1968) and by Corseuil & Barbosa (1971).

Mesolecanium rhizophorae (Cockerell)

Lecanium (Calymnatus) rhizophorae Cockerell, 1898k: 501.
Mesolecanium rhizophorae (Cockerell); Fernald, 1903: 176.
TYPE DATA. Syntypes female, BRAZIL: Sao Paulo State, Cubatao near Santos, on *Rhizophora mangle* (USNM).
DISTRIBUTION. NEOTROPICAL REGION: Brazil Sao Paulo.
HOST PLANTS. Rhizophoraceae: *Rhizophora mangle.*
REMARKS. Adult female redescribed and illustrated by Hempel (1900b).

Mesolecanium uvicola Hempel

Mesolecanium uvicola Hempel, 1920b: 349.
TYPE DATA. Syntypes female, BRAZIL: Minas Gerais, Taboas, on grapevine imported from Chile (MZSP).
DISTRIBUTION. NEOTROPICAL REGION: Brazil Minas Gerais.
HOST PLANTS. Vitidaceae: *Vitis.*

Mesolecanium viticis (Morrison)

Lecanium viticis Morrison, 1923: 123.
Mesolecanium viticis (Morrison); Lizer y Trelles, 1939: 190.
TYPE DATA. Syntypes female, ARGENTINA: Missiones, on *Vitex montevidensis* (USNM).
DISTRIBUTION. NEOTROPICAL REGION: Argentina.
HOST PLANTS. Verbenaceae: *Vitex montevidensis.*

Messinea De Lotto

Messinea De Lotto, 1966b: 147.
TYPE-SPECIES: *Messinea conica* De Lotto, by original designation and monotypy.
REMARKS. Generic characters discussed by Hodgson (1969b).

Messinea conica De Lotto

Messinea conica De Lotto, 1966b: 147.
TYPE DATA. Holotype female, SOUTH AFRICA: Transvaal, Messina, on *Colophospermum mopane* (SANC).
DISTRIBUTION. ETHIOPIAN REGION: Eritrea, Nigeria, South Africa, Saudi Arabia.
HOST PLANTS. Leguminosae: *Acacia abyssinica, Caesalpinia pulcherrima, Colophospermum mopane.*
REMARKS. Intraspecific variation in taxonomic characters given by Hodgson (1969b). Distribution and host plant records given by Hodgson (1969b) and by Matile-Ferrero (1988).

Messinea loisa Hodgson

Messinea loisa Hodgson, 1969b: 22.
TYPE DATA. Holotype female, ZIMBABWE: Tanganda Halt, on *Acacia nigrescens* (BMNH).
DISTRIBUTION. ETHIOPIAN REGION: Zimbabwe.
HOST PLANTS. Leguminosae: *Acacia nigrescens.*

Messinea plana De Lotto

Messinea plana De Lotto, 1967b: 787.
TYPE DATA. Holotype female, SOUTH AFRICA: Natal, Durban, on *Chaetachme aristata* (SANC).
DISTRIBUTION. ETHIOPIAN REGION: South Africa.
HOST PLANTS. Ulmaceae: *Chaetachme aristata.*

Metaceronema Takahashi

Metaceronema Takahashi, 1955d: 27.
TYPE-SPECIES: *Ceronema japonica* Maskell, by original designation and monotypy.
REMARKS. Generic characters discussed by Borchsenius (1957), Paik (1978), Wang (1980), Yang (1982) and by Tang (1991).

Metaceronema japonica (Maskell)

Ceronema japonicum Maskell, 1897b: 243.
Lichtensia japonica Kuwana, 1909a: 152. Syntypes female, JAPAN: Tokyo and Kanagawa, on *Thea japonica* (ITLJ). Syn. by Takahashi, 1955d: 27.
Euphilippia aquifoliae Chen, 1937: 383. Syntypes female, CHINA: Hwangyen, Chekiang, on *Osmanthus aquifolia* (probably lost; F. T. Tang, 1990, personal communication). Syn. by Yang, 1982: 156.
Metaceronema japonica (Maskell); Takahashi, 1955d: 27.
Euphilippia monticola Wang, 1976: 342. Holotype female, CHINA: Chekiang, on *Thea drupifera* (IEBC). Syn. by Yang, 1982: 156.

TYPE DATA. Syntypes female, JAPAN: Miyanoshita, on *Ilex crenata*, and INDIA: on tea (NZAC, UCD, USNM).
DISTRIBUTION. ORIENTAL REGION: India. **PALAEARCTIC REGION:** China, Japan, Korea.
HOST PLANTS. Aquifoliaceae: *Ilex crenata*, *I. integra*. **Buxaceae:** *Buxus microphylla*. **Oleaceae:** *Osmanthus aquifolia*. **Theaceae:** *Camellia japonica*, *Thea drupifera*, *T. japonica*.
REMARKS. Adult female redescribed and illustrated by Kuwana (1917a), Wang (1976) (as *E. monticola*), Paik (1978), Kawai (1980) and by Tang (1991). Adult female redescribed by Takahashi (1955d), Borchsenius (1957) and by Wang (1980). Colour photograph given by Kawai (1980). *Eriochiton theae* Green, 1900 was synonymized with this species by Borchsenius (1957) and by Yang (1982), whereas Tao et al., (1983) regarded it as a distinct species; see under *E. theae*. Distribution and host plant records given by Chen (1937), Takahashi (1955d), Takahashi & Tachikawa (1956), Borchsenius (1957), Ali (1971), Wang (1976, 1980), Kawai (1980), Yang (1982), Varshney (1985), Shafee et al. (1989) and by Tang (1991).

Metapulvinaria Nakahara & Gill

Metapulvinaria Nakahara & Gill, 1985: 36.
TYPE-SPECIES: *Lichtensia lycii* Cockerell, by original designation and monotypy.
REMARKS. Generic characters discussed by Gill (1988).

Metapulvinaria lycii (Cockerell)

Lichtensia lycii Cockerell, 1895l: 254.
Ctenochiton lycii (Cockerell); Steinweden, 1929: 234.
Filippia lycii (Cockerell); Lindinger, 1932f: 203.
Metapulvinaria lycii (Cockerell); Nakahara & Gill, 1985: 37.
TYPE DATA. Lectotype female designated by Nakahara & Gill (1985), U.S.A.: New Mexico, Las Cruces, on *Lycium torreyi* (USNM).
DISTRIBUTION. NEOTROPICAL REGION: Mexico. **NEARCTIC REGION:** Arizona, California, New Mexico, Texas.
HOST PLANTS. Malvaceae: *Hibiscus denudatus*. **Solanaceae:** *Lycium cooperi*, *L. torreyi*, *Solanum*.
REMARKS. Adult female redescribed and illustrated by Nakahara & Gill (1985) and by Gill (1988). Distribution and host plant records given by Ferris (1921a), Ferris & Kelley (1923), Nakahara & Gill (1985) and by Gill (1988).

Millericoccus Avasthi & Shafee

Millericoccus Avasthi & Shafee, 1984a: 7.
TYPE-SPECIES: *Ceroplastodes costalimae* Bondar, by original designation and monotypy.

Millericoccus costalimai (Bondar)

Ceroplastodes costalimae Bondar, 1925: 60.
Ceroplastes costalimai Bondar; Silva et al., 1968: 141 [JUSTIFIED EMENDATION].
Millericoccus costalimae (Bondar); Avasthi & Shafee, 1984a: 8.
TYPE DATA. Syntypes female, BRAZIL: Bahia, on cacao (USNM).

DISTRIBUTION. NEOTROPICAL REGION: Brazil Bahia.
HOST PLANTS. Sterculiaceae: *Theobroma cacao*.
REMARKS. Adult female redescribed and illustrated by Avasthi &
Shafee (1984a). Distribution and host plant records given by Silva *et
al.* (1968).

Milviscutulus Williams & Watson

Milviscutulus Williams & Watson, 1990: 119.
TYPE-SPECIES: *Lecanium mangiferae* Green, by original designa-
tion.
REMARKS. Key to species given by Williams & Watson (1990).

Milviscutulus ciliatus Williams & Watson

Milviscutulus ciliatus Williams & Watson, 1990: 121.
TYPE DATA. Holotype female, PAPUA NEW GUINEA: Morobe Prov-
ince, Buso, on *Gmelina mlouccana* (BMNH).
DISTRIBUTION. AUSTRO ORIENTAL REGION: Papua New Guinea.
NEW ZEALAND and PACIFIC REGION: Western Samoa.
HOST PLANTS. Apocynaceae: *Cerbera manghas*. **Connaraceae:**
Connarus. **Euphorbiaceae:** *Macaranga*. **Myrtaceae:** *Decasper-
mum*, *Eugenia*, *Melaleuca*, *Psidium guajava*. **Verbenaceae:**
Gmelina moluccana.

Milviscutulus mangiferae (Green)

Lecanium mangiferae Green, 1889: 249.
Coccus mangiferae (Green); Fernald, 1903: 172.
Lecanium psidii Green, 1904d: 225. Lectotype female designated by
Williams & Watson (1990), SRI LANKA: Colombo, on *Mangifera
indica* (BMNH). Syn. by Williams & Watson, 1990: 122.
Saissetia psidii (Green); Sanders, 1906: 10.
Lecanium wardi Newstead, 1917a: 353. Lectotype female designated
by Williams & Watson (1990), GUYANA: Georgetown, on Molucca
Apple (BMNH). Syn. by Williams, 1963: 100.
Coccus wardi (Newstead); Bodkin, 1917: 108.
Lecanium desolatum Green, 1922b: 1020. Lectotype female designat-
ed by Williams & Watson (1990), SRI LANKA: Paradeniya, on
Ficus gibbosa (BMNH). Syn. by Williams & Watson, 1990: 122.
Lecanium ixorae Green, 1922b: 1022. Lectotype female designated by
Williams & Watson (1990), SRI LANKA: Heneratgoda, on *Ixora
coccinea* (BMNH). Syn. by Williams & Watson, 1990: 122.
Protopulvinaria mangiferae (Green); Steinweden, 1929: 223.
Coccus ixorae (Green); Green, 1937: 303.
Coccus kuraruensis Takahashi, 1939d: 314. Syntypes female,
TAIWAN: Kuraru near Koshun, on lemon (IMZT). Syn. by Tao *et
al.*, 1983: 81.
Protopulvinaria ixorae (Green); Takahashi, 1955a: 37.
Coccus desolatum (Green); Ali, 1971: 23.
Milviscutulus mangiferae (Green); Williams & Watson, 1990: 122.
Ptoropulvinaria mangiferae (Green); Danzig & Konstantinova, 1990:
44 [MIS-SPELLING].
Udinia psidii (Green); Tang, 1991: 222.

TYPE DATA. Lectotype female designated by Ben-Dov *et al.* (1975).
SRI LANKA: Punduloya, on *Mangifera indica* (BMNH).
DISTRIBUTION. AUSTRO ORIENTAL REGION: Malaysia, Papua
New Guinea, Philippines, Solomon Islands. **ETHIOPIAN RE-
GION:** Ghana, Kenya, South Africa, Zanzibar. **MADAGASIAN
REGION:** Agalega Island, Comoros, Madagascar, Mauritius,
Reunion, Rodrigues, Seychelles. **NEOTROPICAL REGION:**
Barbados, Brazil, Colombia, Costa Rica, Cuba, Dominican
Republic, Ecuador, Guyana, Honduras, Martinique, Mexico,
Nicaragua, Panama, Puerto Rico, Salvador, Venezuela, Virgin
Islands. **NEW ZEALAND and PACIFIC REGION:** Fiji, Hawaii,
Irian Jaya, Palaus, Tonga, Western Samoa. **NEARCTIC REGION:**
Florida, Texas. **ORIENTAL REGION:** Hong Kong, India, Pakis-
tan, Singapore, Sri Lanka, Taiwan, Thailand, Vietnam. **PAL-
AEARCTIC REGION:** Israel, Japan.
HOST PLANTS. Agavaceae: *Cordyline terminalis*. **Anacardiaceae:**
Campnosperma brevipetiolata, Gluta turtur, Mangifera indica.
Apocynaceae: *Alstonia spectabilis, Plumeria, Thevetia peruviana.*
Araceae: *Caladium, Epipremnum.* **Araliaceae:** *Meryta macrophyl-
la, Schefflera.* **Arecaceae:** *Monstera deliciosa.* **Bischofiaceae:**
Bischofia javanica. **Bromeliaceae:** *Ananas.* **Caricaceae:** *Carica
papaya.* **Combretaceae:** *Terminalia brassii, Terminalia catappa,
T. complanata.* **Compositae:** *Wedelia biflora.* **Convolvulaceae:**
Merremia. **Ehretiaceae:** *Cordia myxa.* **Elaeocarpaceae:** *Elaeocarpus.* **Euphorbiaceae:** *Breynia cernua, Codiaeum variegatum,
Pimelodendron amboinicum.* **Flagellariaceae:** *Flagellaria.* **Gnetaceae:** *Gnetum gnemon.* **Lauraceae:** *Cinnamomum, Litsea zeylanica, Persea americana.* **Leguminosae:** *Gliricidia, Palaquium
formosanum.* **Loranthaceae: Malvaceae:** *Hibiscus.* **Moraceae:**
*Artocarpus altilis, A. heterophyllus, A. integrifolia, Ficus glanduli-
fera, F. septica, F. theophrastoides, F. tinctoria.* **Myrsinaceae:**
Gymnacranthera, Rapanea quianensis. **Myrtaceae:** *Decaspermum, Eucalyptus citriodora, E. deglupta, Eugenia aquea, E.
caryophyllata, E. jambos, E. malaccensis, E. parkeri, Psidium
guajava, Rhodomyrtus tomentosa.* **Opiliaceae:** *Champereia
manillana.* **Orchidaceae:** *Dendrobium spectabile, Vanilla.* **Palmae:** *Cocos nucifera.* **Rhizophoraceae:** *Gynotroches axilaris, Rhizophora apiculata, R. mucronata.* **Rubiaceae:** *Morinda citrifolia,
Platanocephalus chinensis, P. morindaefolius, Psychotria rubra,
Timonius.* **Rutaceae:** *Citrus sinensis.* **Sapindaceae:** *Guioa.*
Sapotaceae: *Pometia pinnata.* **Strelitziaceae:** *Strelitzia.*
REMARKS. Common Name - The Mango shield scale. Adult female
and larval instars redescribed and illustrated by Ben-Dov *et al.*,
(1975). Adult female redescribed and illustrated by Green (1904d),
Morrison (1920), Zimmerman (1948), De Lotto (1957b), Tao *et al.*
(1983), Hamon & Williams (1984) and by Williams & Watson (1990).
Adult female redescribed by Takahashi (1955a) (as *P. ixorae*), Kawai
(1980) (as *P. ixorae*) and by Tang (1991) (as *Protopulvinaria mangiferae*, as well as *Coccus desolatus* and *Udinia psidii*). Male test described and illustrated by Miller & Williams (1990). Colour photograph given by Hamon & Williams (1984) and by Kamburov (1987).
Intraspecific variation of taxonomic characters given and discussed
by Ben-Dov *et al.* (1975) and by Williams & Watson (1990). Distribution and host plant records given by Houser (1918), Takahashi

(1955a), Mamet (1959b, 1962), De Lotto (1957b), Beardsley (1966a), Ali (1971), Ben-Dov *et al.* (1975), Panis & Martin (1976), Takagi (1977), Matile-Ferrero (1978), Nakahara & Miller (1981), Nakahara (1981b, 1983), Tao *et al.* (1983), Kamburov (1987), Williams & Watson (1990), Miller & Williams (1990), Danzig & Konstantinova (1990) and by Tang (1991).

BIOLOGY. Reproduction is parthenogenetic, however Otanes (1936) and Avidov & Zaitzov (1960) reported on the occurrence of males at a very low rate. Blumberg & Swirski (1984) studied the encapsulation response to parasitoids.

ECONOMIC IMPORTANCE. Reported as a mango pest in Israel (Avidov & Harpaz, 1969) and South Africa (Kamburov, 1987).

Milviscutulus pilosus Williams & Watson

Milviscutulus pilosus Williams & Watson, 1990: 128.

TYPE DATA. Holotype female, SOLOMON ISLANDS: Western Province, New Georgia Islands, New Georgia, Munda, on *Ficus* sp. (BMNH).

DISTRIBUTION. AUSTRO ORIENTAL REGION: Papua New Guinea, Solomon Islands.

HOST PLANTS. Barringtoniaceae: *Barringtonia.* **Boraginaceae:** *Mertensia.* **Combretaceae:** *Terminalia brassii.* **Convolvulaceae:** *Merremia pacifica.* **Moraceae:** *Ficus.* **Palmae:** *Cocos nucifera.*

Milviscutulus spiculatus Williams & Watson

Milviscutulus spiculatus Williams & Watson, 1990: 130.

TYPE DATA. Holotype female, PAPUA NEW GUINEA: Morobe Province, Buso, on *Morinda citrifolia* (BMNH).

DISTRIBUTION. AUSTRO ORIENTAL REGION: Papua New Guinea, Solomon Islands. **NEW ZEALAND and PACIFIC REGION:** Irian Jaya.

HOST PLANTS. Anacardiaceae: *Mangifera indica.* **Araceae:** *Epipremnum.* **Hernandiaceae:** *Gyrocarpus.* **Lauraceae:** *Persea americana.* **Moraceae:** *Ficus septica, F. theophrastoides.* **Rhizophoraceae:** *Rhizophora apiculata.* **Rubiaceae:** *Morinda citrifolia, Timonius.* **Verbenaceae:** *Faradaya splendida.*

Mitrococcus Borchsenius

Mitrococcus Borchsenius, 1959: 170.

TYPE-SPECIES: *Mitrococcus celsus* Borchsenius, by original designation and monotypy.

REMARKS. Generic characters discussed by Yang (1982) and by Tang (1991).

Mitrococcus celsus Borchsenius

Mitrococcus celsus Borchsenius, 1959: 171.

TYPE DATA. Holotype female, CHINA: Sichwan Province, Omyeshan, on undetermined plant (IEBC).

DISTRIBUTION. PALAEARCTIC REGION: China.

REMARKS. Adult female redescribed and illustrated by Yang (1982) and by Tang (1991). Distribution and host plant records given by Yang (1982) and by Tang (1991).

Mohelnia Sulc

Mohelnia Sulc, 1941: 1.
TYPE-SPECIES: *Mohelnia festuceti* Sulc, by monotypy.
REMARKS. Borchsenius (1957) regarded this genus as a subjective
synonym of *Scythia* Kiritshenko, an interpretation adopted in this
catalogue.

Myzolecanium Beccari

Myzolecanium Beccari, 1877: 191.
Myxolecanium Targioni Tozzetti; Targioni Tozzetti, 1877: 317 [MIS-
 SPELLING].
Myxilecanium MacGillivray; MacGillivray, 1921: 178, 492 [MIS-
 SPELLING].
TYPE-SPECIES: *Myzolecanium kibarae* Beccari, by monotypy.
REMARKS. Beccari (1877) credited this genus to Targioni Tozzetti,
however, in accordance with Article 50 of ICZN, it should be credited
to Beccari.

Myzolecanium kibarae Beccari

Myzolecanium kibarae Beccari, 1877: 191.
Myxolecanium kibarae Beccari; Targioni Tozzetti, 1877: 317 [MIS-
 SPELLING].
Myxilecanium kibarae Beccari; MacGillivray, 1921: 178 [MIS-SPELL-
 ING].
Myzolecanium kibarae Targioni Tozzetti; Lindinger, 1935b: 123
 [ERRONEOUS AUTHORSHIP].
TYPE DATA. Syntypes female, PAPUA NEW GUINEA: in cavities in
 twigs of *Kibara hospitans* (DEPOSITORY UNKNOWN).
DISTRIBUTION. AUSTRO ORIENTAL REGION: Papua New Guinea.
HOST PLANTS. Monimiaceae: *Kibara hospitans*.
REMARKS. Adult female redescribed and illustrated by Targioni
Tozzetti (1877). Lindinger (1935b) discussed the nomenclature and
erroneously credited the authorship to Targioni Tozzetti. Williams &
Watson (1990) discussed the species.
BIOLOGY. Beccari (1877) observed this species within cavities in
twigs of the host plant, where it lives in association with the ant
Hypoclinea scrutator.

Nemolecanium Borchsenius

Nemolecanium Borchsenius, 1955b: 290.
TYPE-SPECIES: *Nemolecanium abietis* Borchsenius, by original
 designation
REMARKS. Generic characters discussed by Borchsenius (1957),
Tereznikova (1981), Kosztarab & Kozár (1988) and by Tang (1991).
Key to species given by Borchsenius (1957), Danzig (1964) and by
Tang (1991).

Nemolecanium abietis Borchsenius

Nemolecanium abietis Borchsenius, 1955b: 290.
TYPE DATA. Syntypes female, USSR: Crimea, Yalta, on *Abies* sp.
 (ZIAS).

DISTRIBUTION. PALAEARCTIC REGION: USSR Crimea.
HOST PLANTS. Pinaceae: *Abies*.
REMARKS. Adult female redescribed and illustrated by Borchsenius (1957), Tereznikova (1981) and by Tang (1991). Male described and illustrated by Borchsenius (1957) and by Giliomee (1967). Distribution and host plant records given by Borchsenius (1957), Tereznikova (1981) and by Tang (1991).

Nemolecanium adventicium Borchsenius

Nemolecanium adventicium Borchsenius, 1955b: 291.
TYPE DATA. Syntypes female, USSR: Moscow, Botanical Gardens, on *Abies* sp. introduced from Germany (ZIAS).
DISTRIBUTION. PALAEARCTIC REGION: USSR Moscow.
HOST PLANTS. Pinaceae: *Abies*.
REMARKS. Adult female redescribed and illustrated by Borchsenius (1957). Adult female redescribed by Tang (1991). Distribution and host plant records given by Borchsenius (1957) and by Tang (1991).

Nemolecanium aptii (Bodenheimer)

Eulecanium aptii Bodenheimer, 1941: 74.
Nemolecanium aptii (Bodenheimer); Borchsenius, 1957: 379.
TYPE DATA. Syntypes female, TURKEY: Carsamba near Bolu, on *Abies nordmanniana* (probably lost; see Ben-Dov & Harpaz, 1985).
DISTRIBUTION. PALAEARCTIC REGION: Turkey.
HOST PLANTS. Pinaceae: *Abies nordmanniana*.
REMARKS. Adult female redescribed by Bodenheimer (1953), Borchsenius (1957) and by Tang (1991).
BIOLOGY. Occurs on the needles of the host plant.

Nemolecanium graniformis (Wünn)

Physokermes graniformis Wünn, 1921: 29.
Nemolecanium graniformis (Wünn); Borchsenius, 1955b: 289.
Lecanium graniforme (Wünn); Kawecki, 1957: 195.
TYPE DATA. Syntypes female, GERMANY: Elsassichen Jura, Grafschaft, on *Abies pectinata* (VMNH).
DISTRIBUTION. PALAEARCTIC REGION: Czechoslovakia, France, Germany, Italy, Poland, Switzerland.
HOST PLANTS. Pinaceae: *Abies alba, A. nebredensis, A. pectinata*.
REMARKS. Adult female redescribed and illustrated by Borchsenius (1957) and by Kosztarab & Kozár (1978, 1988). Adult female redescribed by Tang (1991). Distribution and host plant records given by Borchsenius (1957), Kawecki (1957), Rehacek (1957), Zak-Ogaza & Koteja (1964), Kosztarab & Kozár (1978), Marotta (1987), Dziedzicka (1988) and by Kosztarab & Kozár (1988).

Neolecanium Parrott

Lecanium (Neolecanium) Parrott in Cockerell & Parrott, 1901: 58.
Neolecanium Parrott; Cockerell, 1902f: 451.
TYPE-SPECIES: *Lecanium imbricatum* Cockerell, by original designation.
REMARKS. Generic characters discussed by Green (1909), Morrison

(1929), Williams & Kosztarab (1972) and by Hamon & Williams (1984).

Neolecanium amazonensis Foldi

Neolecanium amazonensis Foldi, 1988: 80.
TYPE DATA. Holotype female, BRAZIL: Amazonas, near Manaus, on *Pourouma cecropiaefoliae* (INPA).
DISTRIBUTION. NEOTROPICAL REGION: Brazil Amazonas.
HOST PLANTS. Urticaceae: *Pourouma cecropiaefoliae.*
BIOLOGY. Protected by a carton shelter constructed by ants (Foldi, 1988).

Neolecanium chilaspidis (Cockerell)

Lecanium chilaspidis Cockerell, 1897j: 268.
Neolecanium chilaspidis (Cockerell); Cockerell, 1902f: 451.
TYPE DATA. Syntypes female, MEXICO: Tehuantepe City, on *Chilaspis [=Chilopsis] linearis* (USNM).
DISTRIBUTION. NEOTROPICAL REGION: Mexico.
HOST PLANTS. Bignoniaceae: *Chilopsis linearis.*

Neolecanium cinnamomi Rutherford

Neolecanium cinnamomi Rutherford, 1914: 265.
TYPE DATA. Syntypes female, SRI LANKA: Paradeniya, on bark of Cinnamon (BMNH).
DISTRIBUTION. ORIENTAL REGION: Sri Lanka.
HOST PLANTS. Lauraceae: *Cinnamomum.*
REMARKS. Green (1937) suggested that this species might be identical with *Ctenochiton cinnamomi* Green, 1922.

Neolecanium cornuparvum (Thro)

Lecanium cornuparvum Thro, 1903: 216.
Neolecanium cornuparvum (Thro); Fernald, 1903: 176.
Lecanium (Neolecanium) cornuparvum Thro; Pettit & McDaniel, 1920: 9.
TYPE DATA. Syntypes female, U.S.A.: New York, Trumansburg, on *Magnolia* sp. (USNM).
DISTRIBUTION. NEARCTIC REGION: Alabama, Connecticut, Florida, Georgia, Indiana, Kentucky, Louisiana, Maine, Maryland, Mississippi, New York, North Carolina, Ohio, Pennsylvania, South Carolina, Tennessee, Virginia, West Virginia, Wisconsin.
HOST PLANTS. Magnoliaceae: *Magnolia acuminata, M. grandiflora, M. soulangeana, M. stellata, M. tripetala.*
REMARKS. Adult female redescribed and illustrated by Dietz & Morrison (1916), Pettit & McDaniel (1920), Williams & Kosztarab (1972) and by Hamon & Williams (1984). Adult male described and illustrated by Ray & Williams (1983). Larval instars described and illustrated by Ray & Williams (1983). Colour photograph given by Johnson & Lyon (1988). Distribution and host plant records given by Dietz & Morrison (1916), Pettit & McDaniel (1920), Williams & Kosztarab (1972), Lambdin & Watson (1980), Ray & Williams (1983) and by Hamon & Williams (1984).

Neolecanium craspeditae Morrison

Neolecanium craspeditae Morrison, 1929: 40.
TYPE DATA. Holotype female, PANAMA: Canal Zone, Rio Agua Salud, on mango (USNM).
DISTRIBUTION. NEOTROPICAL REGION: Panama, Trinidad.
HOST PLANTS. Anacardiaceae: *Mangifera indica.* **Ehretiaceae:** *Cordia.*
BIOLOGY. Protected under shelter made by the ant *Azteca trigona* in Trinidad (Morrison, 1929).

Neolecanium derameliae Morrison

Neolecanium derameliae Morrison, 1929: 43.
TYPE DATA. Syntypes female and larva, PANAMA: Canal Zone, Cooper's Place, Rio Aojeta, on undetermined plant, beneath sheds of *Azteca* sp. (USNM).
DISTRIBUTION. NEOTROPICAL REGION: Panama.
BIOLOGY. Protected beneath tents of an *Azteca* ant (Morrison, 1929).

Neolecanium herrerae Cockerell

Neolecanium herrerae Cockerell, 1902b: 143.
TYPE DATA. Syntypes female, MEXICO: Publa, on *Agave* sp. (BMNH, USNM).
DISTRIBUTION. NEOTROPICAL REGION: Mexico.
HOST PLANTS. Agavaceae: *Agave.*

Neolecanium imbricatum (Cockerell)

Lecanium imbricatum Cockerell, 1896d: 38.
Lecanium (Neolecanium) imbricatum Cockerell; Cockerell & Parrott, 1901: 58.
Neolecanium imbricatum (Cockerell); Cockerell, 1902f: 451.
TYPE DATA. Syntypes female, MEXICO: Alta Mira, Tamaulipas, on *Mimosa* sp. (USNM).
DISTRIBUTION. NEOTROPICAL REGION: Mexico.
HOST PLANTS. Leguminosae: *Mimosa.*

Neolecanium leucaenae Cockerell

Neolecanium leucaenae Cockerell, 1903c: 46.
TYPE DATA. Syntypes female, MEXICO: Zapotlan, on *Leucaena* sp. (BMNH, USNM).
DISTRIBUTION. NEOTROPICAL REGION: Mexico.
HOST PLANTS. Leguminosae: *Leucaena.*

Neolecanium manzanillense Cockerell

Neolecanium manzanillense Cockerell, 1903b: 161.
TYPE DATA. Syntypes female, MEXICO: Manzanillo, on Leguminosae shrub (USNM).
DISTRIBUTION. NEOTROPICAL REGION: Mexico.
HOST PLANTS. Leguminosae.

Neolecanium perconvexum (Cockerell)

Lecanium perconvexum Cockerell, 1898d: 132.
Neolecanium perconvexum (Cockerell); Cockerell, 1902f: 451.
TYPE DATA. Syntypes female, BRAZIL: Sao Paulo, Campinas, on *Nectandra* sp. (USNM).
DISTRIBUTION. NEOTROPICAL REGION: Brazil Minas Gerais, Brazil Sao Paulo.
HOST PLANTS. Lauraceae: *Nectandra*.
REMARKS. Adult female redescribed and illustrated by Hempel (1900b). Distribution and host plant records given by Cockerell (1902f) and by Hempel (1900b, 1912).

Neolecanium plebeium Cockerell

Neolecanium plebeium Cockerell, 1903b: 161.
TYPE DATA. Syntypes female, MEXICO: Colima, on *Ficus* sp. (BMNH, USNM).
DISTRIBUTION. NEOTROPICAL REGION: Mexico.
HOST PLANTS. Moraceae: *Ficus*.

Neolecanium pseudoleae Rutherford

Neolecanium pseudoleae Rutherford, 1915a: 112.
TYPE DATA. Syntypes female, SRI LANKA: Paradeniya, on cinnamon (BMNH).
DISTRIBUTION. ORIENTAL REGION: Sri Lanka.
HOST PLANTS. Lauraceae: *Cinnamomum*.
REMARKS. Green (1937) supposed that this species belongs to *Saissetia*.

Neolecanium sallei (Signoret)

Lecanium sallei Signoret, 1874a: 410.
Neolecanium sallei (Signoret); Cockerell, 1902f: 451.
TYPE DATA. Syntypes female, MEXICO: on undetermined tree (VMNH).
DISTRIBUTION. NEOTROPICAL REGION: Mexico.

Neolecanium silveirai (Hempel)

Lecanium silveirai Hempel, 1900a: 5.
Neolecanium silveirai (Hempel); Cockerell, 1902f: 451.
TYPE DATA. Syntypes female, BRAZIL: Minas Gerais, Sete Lagoas and Diamantina, on roots of grapevine (MZSP).
DISTRIBUTION. NEOTROPICAL REGION: Brazil Minas Gerais.
HOST PLANTS. Vitidaceae: *Vitis*.
REMARKS. Adult female redescribed and illustrated by Hempel (1900b) and by Lepage & Piza (1941). First instar larva described and illustrated by Lepage & Piza (1941).
ECONOMIC IMPORTANCE. Reported as a serious grapevine pest in Brazil (Lepage & Piza, 1941).

Neolecanium subterraneum Hempel

Neolecanium subterraneum Hempel; Hempel, 1918: 202.

TYPE DATA. Syntypes female, BRAZIL: Sao Paulo, Ypiranga, on roots of undetermined plant (MZSP).
DISTRIBUTION. NEOTROPICAL REGION: Brazil Sao Paulo.

Neolecanium tuberculatum (Townsend & Cockerell)

Lecanium tuberculatum Townsend & Cockerell, 1898: 177.
Neolecanium tuberculatum (Townsend & Cockerell); Cockerell, 1902f: 451.
Neolecanium tuberosum Lindinger; Lindinger, 1957: 550 [UNJUSTIFIED REPLACEMENT NAME].
TYPE DATA. Syntypes female, MEXICO: San Antonio del Sapotal, near Frontera, Tabasco, on a plant named "cafetillo" (BMNH, USNM).
DISTRIBUTION. NEOTROPICAL REGION: Mexico.

Neolecanium urichi (Cockerell)

Lecanium urichi Cockerell, 1894g: 203.
Neolecanium urichi (Cockerell); Cockerell, 1902f: 451.
TYPE DATA. Syntypes female, TRINIDAD: in nest of *Crematogaster brevispinosa* (USNM).
DISTRIBUTION. NEOTROPICAL REGION: Brazil Rio Grande do Sul, Grenada, Trinidad.
HOST PLANTS. Smilacaceae: *Smilax campestris.*
REMARKS. Redescribed by Hempel (1900b). Distribution and host plant records given by Cockerell (1896c, 1902f), Hempel (1900b) and by Corseuil & Barbosa (1971).
BIOLOGY. Attended by the ant *Crematogaster brevispinosa* Mayr (Hempel, 1900b).

Neolecanochiton Hempel

Neolecanochiton Hempel, 1932: 324.
Neolecaniochiton Lindinger; Lindinger, 1937: 190 [UNJUSTIFIED EMENDATION].
TYPE-SPECIES: *Neolecanochiton grevilleae* Hempel, by original designation and monotypy.

Neolecanochiton grevilleae Hempel

Neolecanochiton grevilleae Hempel, 1932: 324.
TYPE DATA. Syntypes female, BRAZIL: Sao Paulo, Capital, on a plant named "arvore de sombra cidade" (MZSP).
DISTRIBUTION. NEOTROPICAL REGION: Brazil Sao Paulo.
HOST PLANTS. Proteaceae: *Grevillea robusta.*
REMARKS. Distribution and host plant records given by Silva *et al.* (1968).

Neoplatylecanium Takahashi

Neoplatylecanium Takahashi, 1929a: 53.
TYPE-SPECIES: *Neoplatylecanium cinnamomi* Takahashi, by original designation and monotypy.
REMARKS. Generic characters discussed by Yang (1982), Tao *et al.* (1983), Williams & Watson (1990) and by Tang (1991).

Neoplatylecanium adersi (Newstead)

Lecanium adersi Newstead, 1917a: 357.
Coccus adersi (Newstead); De Lotto, 1959: 154.
Neoplatylecanium adersi (Newstead); Tang, 1991: 121.
TYPE DATA. Syntypes female, ZANZIBAR: Marahubi, on mango (BMNH).
DISTRIBUTION. ETHIOPIAN REGION: Zanzibar, India.
HOST PLANTS. Anacardiaceae: *Mangifera indica.*
REMARKS. Adult female redescribed and illustrated by De Lotto (1959) and by Tang (1991).

Neoplatylecanium cinnamomi Takahashi

Neoplatylecanium cinnamomi Takahashi, 1929a: 54.
TYPE DATA. Syntypes female, TAIWAN: Kuaru, on *Cinnamomum ceylanicum* (IMZT).
DISTRIBUTION. ORIENTAL REGION: Taiwan.
HOST PLANTS. Lauraceae: *Cinnamomum camphora, C. zeylanicum.*
REMARKS. Adult female redescribed and illustrated by Tao *et al.* (1983). Distribution and host plant records given by Tao *et al.* (1983).

Neoplatylecanium tripartitum (Green)

Lecanium tripartitum Green, 1922b: 1025.
Neoplatylecanium tripartitum (Green); Takahashi, 1929a: 54.
Marsipococcus tripartitus (Green); Tang, 1991: 118.
TYPE DATA. Syntypes female, SRI LANKA: Namunakuli Hill, Badulla, on *Calophyllum walkeri* (BMNH).
DISTRIBUTION. ORIENTAL REGION: Sri Lanka.
HOST PLANTS. Guttiferae: *Calophyllum walkeri.*
REMARKS. Adult female redescribed by Tang (1991).

Neopulvinaria Hadzibejli

Neopulvinaria Hadzibejli, 1955: 232.
TYPE-SPECIES: *Neopulvinaria imeretina* Hadzibejli, [=*Neopulvinaria innumerabilis* (Rathvon)] by original designation and monotypy.
REMARKS. Generic characters discussed by Borchsenius (1957) and by Tang (1991).

Neopulvinaria innumerabilis (Rathvon)

Coccus innumerabilis Rathvon, 1854: 256.
Lecanium acericorticis Fitch, 1861: 775. Syntypes female, U.S.A.: New York, Salem, west border of Jarvis Martin's woods, on maple (probably lost; see Barnes, 1988). Syn. by Fernald, 1903: 134.
Lecanium acerella Rathvon, 1876: 101. Syntypes female, U.S.A.: Pennsylvania, on *Acer* (DEPOSITORY UNKNOWN). Syn. by Fernald, 1903: 134.
Pulvinaria innumerabilis (Rathvon); Putnam, 1880: 293.
Pulvinaria tinsleyi King, 1900: 360. Syntypes female, U.S.A.: New Mexico, near Roswell, on *Celtis* sp. (USNM). Syn. by Steinweden, 1946: 7.
Neopulvinaria imeretina Hadzibejli, 1955: 233. Syntypes female, male

and larva, USSR: Georgia, on grapevine (ZIAS). Syn. by Danzig & Matile-Ferrero, 1990: 131.

TYPE DATA. Syntypes female, U.S.A.: Pennsylvania, Lancaster, on silver leaved maple, *Acer dasycarpum* (depository unknown).

DISTRIBUTION. NEARCTIC REGION: Arizona, California, Canada British Columbia, Canada Ontario, Canada Quebec, Florida, Illinois, Indiana, Louisiana, Mississippi, New Jersey, New Mexico, New York, Oregon, Pennsylvania, Tennessee, Texas, Utah, Virginia, Wisconsin. **PALAEARCTIC REGION:** France, Italy, USSR Armenia, USSR Georgia.

HOST PLANTS. Aceraceae: *Acer dasycarpum, A. negundo, A. nigrum, A. platanoides, A. rubrum, A. saccharinum.* **Anacardiaceae:** *Rhus.* **Aquifoliaceae:** *Ilex glabra.* **Betulaceae:** *Alnus rubra.* **Carpinaceae:** *Carpinus caroliniana.* **Compositae:** *Solidago.* **Cornaceae:** *Cornus florida, Cornus sanguinea.* **Ebenaceae:** *Diospyros kaki, D. lotus, D. virginiana.* **Empetraceae:** *Ceratiola ericoides.* **Fagaceae:** *Fagus grandifolia, Quercus laurifolia, Quercus nigra, Q. palustris, Q. rubra.* **Grossulariaceae:** *Ribes grossularia, R. nigrum.* **Juglandaceae:** *Carya ovata, Juglans nigra, J. regia.* **Lauraceae:** *Lindera benzoin.* **Leguminosae:** *Acacia, Amorpha fruticosa, Gleditsia triacanthos, Mimosa, Phaseolus vulgaris, Robinia pseudoacacia.* **Magnoliaceae:** *Magnolia grandiflora.* **Myricaceae:** *Myrica cerifera.* **Philadelphaceae:** *Philadelphus coronarius.* **Punicaceae:** *Punica granatum.* **Rosaceae:** *Cerasus vulgaris, Crataegus monogyna, Cydonia vulgaris, Malus, Persica vulgaris, Prunus divaricata, Pyrus.* **Rutaceae:** *Poncirus trifoliata.* **Salicaceae:** *Populus, Salix nigra.* **Tiliaceae:** *Tilia americana, T. caucasica.* **Ulmaceae:** *Celtis occidentalis, Ulmus alata, U. thomasii.* **Vitidaceae:** *Ampelopsis quinquefolia, A. tricuspidata, A. veitchii, Parthenocissus quinquefolia, Vitis riparia, V. rupestris, V. vinifera.*

REMARKS. Common name - cottony Maple scale. Adult female redescribed and illustrated by Steinweden (1946), Borchsenius (1957), Hadzibejli (1960a, 1983), Canard (1966a), Pellizzari Scaltriti (1977), Williams & Kosztarab (1972), Hamon & Williams (1984) and by Gill (1988). Adult female redescribed by Tang (1991). Adult male described and illustrated by Borchsenius (1957) and by Hadzibejli (1960a, 1983). Colour photograph given by Hamon & Williams (1984) and by Johnson & Lyon (1988). Larval instars described by Phillips (1962). Distribution and host plant records given by Takahashi (1942a), Steinweden (1946), Philipps (1962), Canard (1966a), Williams & Kosztarab (1972), Pellizzari Scaltriti (1977), Lambdin & Watson (1980), Hadzibejli (1983), Hamon & Williams (1984), Gill (1988), Danzig & Matile-Ferrero (1990) and by Tang (1991).

BIOLOGY. One annual generation was observed in France (Canard, 1966a) and in Italy (Pellizzari Scaltriti, 1977). Phenology in Georgia, USSR presented by Hadzibejli (1983).

ECONOMIC IMPORTANCE. A pest of grapevine in several countries of Europe and North America (Danzig & Matile-Ferrero, 1990).

Neosaissetia Tao, Wong & Chang

Neosaissetia Tao, Wong & Chang, 1983: 79.

TYPE-SPECIES: *Neosaissetia tropicalis* Tao & Wong, by original designation and monotypy.
REMARKS. Generic characters discussed by Williams & Watson (1990) and by Tang (1991). Key to species given by Tang (1991).

Neosaissetia keravatae Williams & Watson

Neosaissetia keravatae Williams & Watson, 1990: 133.
TYPE DATA. Holotype female, PAPUA NEW GUINEA: East New Britain Province, Keravat, on *Piper nigrum* (BMNH).
DISTRIBUTION. AUSTRO ORIENTAL REGION: Papua New Guinea.
HOST PLANTS. Piperaceae: *Piper nigrum.*

Neosaissetia laos (Takahashi)

Saissetia triangularum laos Takahashi, 1942a: 22.
Saissetia laos (Takahashi); Ali, 1971: 45.
Neosaissetia laos (Takahashi); Tang, 1991: 123.
TYPE DATA. Syntypes female, THAILAND: Mt. Sutep, on *Machilus* sp. (IMZT).
DISTRIBUTION. AUSTRO ORIENTAL REGION: Malaysia, Singapore. **ORIENTAL REGION:** Thailand.
HOST PLANTS. Lauraceae: *Cinnamomum, Machilus.* **Moraceae:** *Artocarpus.*
REMARKS. Adult female redescribed by Tang (1991). Distribution and host plant records given by Takahashi (1952) and by Tang (1991).

Neosaissetia triangularum (Morrison)

Saissetia triangularum Morrison, 1920: 197.
Neosaissetia triangularum (Morrison); Tang, 1991: 124.
Neosaissetia triangulaxum (Morrison); Tang, 1991: 124 [MIS-SPELLING].
TYPE DATA. Syntypes female, PHILIPPINES: Laguna Province, Paete, Luzon, on *Cocos nucifera* (USNM).
DISTRIBUTION. AUSTRO ORIENTAL REGION: Philippines.
HOST PLANTS. Palmae: *Cocos nucifera.*
REMARKS. Adult female redescribed and illustrated by Tang (1991).

Neosaissetia tropicalis Tao & Wong

Neosaissetia tropicalis Tao & Wong in Tao, Wong & Chang, 1983: 79.
TYPE DATA. Syntypes female, TAIWAN: Kenting, Pingtung Hsien, on *Palaquium formosanum* (IMZT).
DISTRIBUTION. ORIENTAL REGION: Taiwan.
HOST PLANTS. Sapotaceae: *Palaquium formosanum.*
REMARKS. Adult female redescribed and illustrated by Tang (1991).

Palaeolecanium Sulc

Palaeolecanium Sulc, 1908: 36.
Palacolecanium Lindinger; Lindinger, 1957: 548 [MIS-SPELLING].
Paloelecanium Lindinger; Lindinger, 1958: 368 [MIS-SPELLING].
TYPE-SPECIES: *Lecanium bituberculatum* Targioni Tozzetti, by original designation.

REMARKS. Generic characters discussed by Borchsenius (1957), Kosztarab & Kozár (1978, 1988) and by Tang (1991).

Palaeolecanium bituberculatum (Signoret)

Lecanium bituberculatum Targioni Tozzetti; Targioni Tozzetti, 1868: 731 [NOMEN NUDUM].
Lecanium bituberculatum Signoret, 1874a: 414.
Lecanium bituberculatum (Targioni Tozzetti); Reh, 1903: 407 [ERRONEOUS AUTHORSHIP].
Eulecanium bituberculatum (Targioni Tozzetti); Fernald, 1903: 182 [ERRONEOUS AUTHORSHIP].
Palaeolecanium bituberculatum (Targioni Tozzetti); Sulc, 1908: 36 [ERRONEOUS AUTHORSHIP].
Lecanium (Palaeolecanium) bituberculatum (Targioni Tozzetti); Sulc, 1932: 82 [ERRONEOUS AUTHORSHIP].
Palaeolecanium bituberculatum (Signoret); De Lotto, 1971c: 148.
Palaeolecanium bituberculatum (Targioni Tozzetti); Kosztarab & Kozár, 1988: 210 [ERRONEOUS AUTHORSHIP].
Palaeolecanium bituberculatum (Targioni Tozzetti); Kozár & Drozdják, 1990: 363 [ERRONEOUS AUTHORSHIP].
Palaeolecanium bituberculatum (Targioni Tozzetti); Tang, 1991: 185 [ERRONEOUS AUTHORSHIP].
TYPE DATA. Syntypes female, FRANCE: Cannes and Hyeres, ITALY: Florence, on "epine blanche" (VMNH).
DISTRIBUTION. PALAEARCTIC REGION: Afghanistan, Bulgaria, Cyprus, Czechoslovakia, Denmark, England, France, Germany, Hungary, Iraq, Israel, Italy, Netherlands, Poland, Romania, Spain, Sweden, Switzerland, Turkey, USSR Armenia, USSR Azerbaijan, USSR Caucasus, USSR Georgia, USSR Kazakhstan, USSR Kirgizia, USSR Moldavia, USSR Tadzhikistan, USSR Turkmenia, USSR Ukraine, USSR Uzbekistan, Yugoslavia.
HOST PLANTS. Betulaceae: *Corylus.* **Juglandaceae:** *Juglans regia.* **Rosaceae:** *Crataegus aronia, C. monogyna, C. oxyacantha, Cydonia oblonga, Malus sylvestris, Mespilus germanica, Prunus communis, P. divaricata, P. domestica, P. laurocerasus, P. spinosa.*
REMARKS. De Lotto (1971c) discussed the authorship of this species and verified that it should be credited to Signoret, not to Targioni Tozzetti as erroneously practiced until 1971. Adult female redescribed and illustrated by Newstead (1903), Leonardi (1920), Sulc (1932), Borchsenius (1957), Gómez-Menor Ortega (1960), Savescu (1961), Kosztarab & Kozár (1978, 1988), Tereznikova (1981) and by Tang (1991). Male described and illustrated by Giliomee (1967). Distribution and host plant records given by Sulc (1932), Borchsenius (1957), Reyne (1957), Gómez-Menor Órtega (1965), Giliomee (1967), Ben-Dov (1971), Danzig (1972d), Kozárzhevskaya & Reitzel (1975), Georghiou (1977), Kozár et al. (1977, 1979), Aziz (1977), Tereznikova (1981), Hadzibejli (1983), Koteja & Zak-Ogaza (1983), Kozár (1983), Kozár & Ostafichuk (1987), Dziedzicka (1988), Kosztarab & Kozár (1988), Kozár & Drozdjak (1990) and by Tang (1991).

Palaeolecanium kosswigi (Bodenheimer)

Eulecanium kosswigi Bodenheimer, 1953: 100.

Palaeolecanium kosswigi (Bodenheimer); Borchsenius, 1957: 347.
TYPE DATA. Syntypes female, TURKEY: Between Mardin and Diyar-
  bakir, on *Pyrus elaeagnifolia* (probably lost; Ben-Dov & Harpaz,
  1985).
DISTRIBUTION. PALAEARCTIC REGION: Turkey.
HOST PLANTS. Rosaceae: *Pyrus elaeagnifolia.*

Paracardiococcus Takahashi

Paracardiococcus Takahashi, 1935: 6.
Paracardicoccus Tao, 1978: 82 [MIS-SPELLING].
TYPE-SPECIES: *Paracardiococcus actinodaphnis* Takahashi, by
  original designation and monotypy.
REMARKS. Generic characters discussed by Yang (1982), Tao *et al.*
(1983) and by Tang (1991).

Paracardiococcus actinodaphnis Takahashi

Paracardiococcus actinodaphnis Takahashi, 1935: 6.
TYPE DATA. Syntypes female, TAIWAN: Kyanrawa, Saukan (Suo-
  Gun, Taihoku Prefecture), on *Actinodaphne mushaensis* (IMZT).
DISTRIBUTION. ORIENTAL REGION: Taiwan.
HOST PLANTS. Lauraceae: *Actinodaphne mushaensis.*
REMARKS. Adult female redescribed and illustrated by Tao *et al.*
(1983) and by Tang (1991).

Paracerostegia Tang

Paracerostegia Tang, 1991: 303.
TYPE-SPECIES: *Ceroplastes floridensis* Comstock, by original desig-
  nation.
REMARKS. This genus is here regarded as a subjective synonym of
Ceroplastes.

Paractenochiton Takahashi

Paractenochiton Takahashi, 1942a: 28.
TYPE-SPECIES: *Paractenochiton sutepensis* Takahashi, by original
  designation and monotypy.
REMARKS. Generic characters discussed by Tang (1991).

Paractenochiton sutepensis Takahashi

Paractenochiton sutepensis Takahashi, 1942a: 29.
TYPE DATA. Syntypes female, THAILAND: Mt. Sutep, on *Quercus* sp.
  (IMZT).
DISTRIBUTION. ORIENTAL REGION: Thailand.
HOST PLANTS. Fagaceae: *Quercus.*
REMARKS. Adult female redescribed by Tang (1991).

Parafairmairia Cockerell

Fairmairia Signoret, 1874b: 98. [HOMONYM of *Fairmairia* Robineau-
  Desvoidy].
Parafairmairia Cockerell, 1899f: 332 [REPLACEMENT NAME].
Farrmairia Hempel; Hempel, 1899: 131 [MIS-SPELLING]].

Fairmairea Lindinger; Lindinger, 1907a: 20 [UNJUSTIFIED EMEN-
DATION].
Parafairmairea Lindinger; Lindinger, 1907a: 20 [UNJUSTIFIED
ÉMENDATION].
TYPE-SPECIES: *Fairmairia bipartita* Signoret, by monotypy.
REMARKS. The genus was revised by Koteja & Rosciszewska (1970).
Generic characters discussed by Steinweden (1929), Gómez-Menor
Ortega (1948), Borchsenius (1957), Kosztarab & Kozár (1978, 1988),
Danzig (1980c), Tereznikova (1981) and by Tang (1991). Key to
species: Borchsenius (1957), Danzig (1964), Koteja & Rosciszewska
(1970), Tereznikova (1981), Kosztarab & Kozár (1988) and by Tang
(1991).

Parafairmairia bipartita (Signoret)
Fairmairia bipartita Signoret, 1874b: 99.
Parafairmairia bipartita (Signoret); Fernald, 1903: 163.
Parafairmairia delicata Borchsenius, 1952: 272. Syntypes female,
USSR: Latvia, on *Carex* sp. (ZIAS) Syn. by Koteja & Rosciszews-
ka, 1970: 252.
TYPE DATA. Syntypes female, FRANCE: Hyeres Islands, on *Agro-
pyrum campestre*, and Cannes on *Mesembryanthemum* (VMNH).
DISTRIBUTION. PALAEARCTIC REGION: Bulgaria, France,
Germany, Hungary, Italy, Poland, Spain, USSR Kunashir Island,
USSR Latvia, USSR Leningrad Region, USSR Sakhalin Island,
USSR Ukraine.
HOST PLANTS. Aizoaceae: *Mesembryanthemum.* **Cyperaceae:**
Carex arenaria, C. brizoides, C. hirta, C. leporina, Scirpus.
Gramineae: *Agropyron campestre, Agrostis vulgaris, Festuca
pseudovina, Stipa capillata.* **Juncaceae:** *Juncus effusa.*
REMARKS. Adult female redescribed and illustrated by Gómez-
Menor Ortega (1948), Borchsenius (1957), Koteja & Rosciszewska
(1970), Kosztarab & Kozár (1978, 1988), Danzig (1980c), Tereznikova
(1981) (as *P. delicata*) and by Tang (1991). Male described and illustrated by Koteja & Rosciszewska (1970). Larvae described and
illustrated by Gómez-Menor Ortega (1948). Distribution and host
plant records given by Gómez-Menor Ortega (1948), Borchsenius
(1952, 1957), Koteja & Rosciszewska (1970), Koteja (1971), Danzig
(1978a, 1980c), Kozár *et al.* (1979), Tereznikova (1981), Martin Mateo
(1984), Marotta (1987), Kosztarab & Kozár (1988), Kozár & Drozdjak
(1990) and by Tang (1991).

Parafairmairia ejaculatoria Mamet
Parafairmairia ejaculatoria Mamet, 1959b: 419.
TYPE DATA. Holotype female, MADAGASCAR: Manakambahiny, on
a plant named "Fotsy ambo" (MNHN).
DISTRIBUTION. MADAGASIAN REGION: Madagascar.

Parafairmairia elongata Matesova
Parafairmairia elongata Matesova, 1979: 49.
TYPE DATA. Holotype female, USSR: Kazakhstan, East shore of
Zaysan Lake, near influx of Cherny Irtysh River, on *Carex* sp.
(ZIAS).

DISTRIBUTION. PALAEARCTIC REGION: USSR Kazakhstan.
HOST PLANTS. Cyperaceae: *Carex*.
REMARKS. Adult female redescribed by Tang (1991).

Parafairmairia gracilis Green

Parafairmairia gracilis Green, 1916a: 24.
TYPE DATA. Syntypes female, ENGLAND: Camberley, Surrey, on grasses and sedges (BMNH).
DISTRIBUTION. PALAEARCTIC REGION: Austria, Czechoslovakia, Denmark, England, France, Germany, Hungary, Italy, Poland, Sweden, USSR Moldavia, USSR Ukraine, Yugoslavia.
HOST PLANTS. Cyperaceae: *Carex brizoides, C. digitata, C. elata, C. fusca, C. goodenoughii, C. gracilis, C. melanostachya, C. mellett, C. vesicaria, Eriophorum vaginatum, Scirpus silvaticus.* **Gramineae: Juncaceae:** *Juncus bufonius.* **Oleaceae:** *Fraxinus excelsior.* **Salicaceae:** *Salix.* **Typhaceae:** *Typha latifolia.*
REMARKS. Adult female redescribed and illustrated by Schmutterer (1952a), Borchsenius (1957), Tereznikova (1963a), Koteja & Rosciszewska (1970), Kosztarab & Kozár (1978, 1988), Tereznikova (1981). Adult male described and illustrated by Koteja & Rosciszewska (1970). Larval instars described by Schmutterer (1952a) and by Koteja & Rosciszewska (1970). Distribution and host plant records given by Schmutterer (1952a), Borchsenius (1957), Ossiannilsson (1959), Koteja & Rosciszewska (1970), Kozárzhevskaya & Reitzel (1975), Podsiadlo & Komosinska (1976), Kozár *et al.* (1977), Tereznikova (1963a, 1981), Koteja (1983), Koteja & Zak-Ogaza (1983), Kozár & Ostafichuk (1987), Kosztarab & Kozár (1988), Kozár & Pellizzari Scaltriti (1989) and by Kozár & Drozdjak (1990).

Parafairmairia hissarica Borchsenius

Parafairmairia hissarica Borchsenius, 1952: 273.
TYPE DATA. Syntypes female, USSR: Tadzhikistan, Gissarsk Range, on bank of Varzob, near Kondar, on grass (ZIAS).
DISTRIBUTION. PALAEARCTIC REGION: USSR Tadzhikistan.
HOST PLANTS. Gramineae.
REMARKS. Adult female redescribed and illustrated by Borchsenius (1957). Adult female redescribed by Koteja & Rosciszewska (1970).

Parafairmairia patellaeformis Brain

Parafairmatrea patellaeformis Brain, 1920b: 39 [MIS-SPELLING].
TYPE DATA. Syntypes female, SOUTH AFRICA: Cape Province, Port Alfred, on *Acacia karroo* (SANC).
DISTRIBUTION. ETHIOPIAN REGION: South Africa.
HOST PLANTS. Leguminosae: *Acacia karroo*.
BIOLOGY. Living on stems of the host plant.

Parakermes Fonseca

Parakermes Fonseca, 1973: 247.
TYPE-SPECIES: *Parakermes brasiliensis* Fonseca, by present designation.

Parakermes brasiliensis Fonseca

Parakermes brasiliensis Fonseca, 1973: 247.
TYPE DATA. Syntypes female and larva, BRAZIL: Sao Paulo, Parque Siqueira Campos, on *Follornia ilicifolia* (MZSP).
DISTRIBUTION. NEOTROPICAL REGION: Brazil Sao Paulo.
HOST PLANTS. Leguminosae: *Follornia ilicifolia.*

Paralecanium Cockerell

Lecanium (Paralecanium) Cockerell in Cockerell & Parrott, 1899: 227.
Paralecanium Cockerell; Fernald, 1903: 199.
Paralecanium Cockerell & Parrott; Tang, 1991: 126 [ERRONEOUS AUTHORSHIP].
TYPE-SPECIES: *Lecanium frenchii* Maskell, by original designation.
REMARKS. Generic characters discussed by Morrison (1920, 1921), Morrison & Morrison (1922), Steinweden (1929), Kawai (1980), Yang (1982), Tao *et al.* (1983), Williams & Watson (1990) and by Tang (1991). Key to species: ASIA - Tang (1991). JAPAN - Kawai (1980). MALAYSIA - Takahashi (1950b). PHILIPPINES - Morrison (1920). TAIWAN - Tao *et al.* (1983).

Paralecanium album Takahashi

Paralecanium vacuum Morrison; Takahashi, 1939a: 115 [MISIDEN-TIFICATION].
Paralecanium album Takahashi; Takahashi, 1950b: 53.
TYPE DATA. Syntypes female, MALAYSIA: Kuantan, Bukit Chera-kah, Dusun Tua, Selangor, on *Durio zibethinus* (SMKM).
DISTRIBUTION. AUSTRO ORIENTAL REGION: Malaysia.
HOST PLANTS. Bombacaceae: *Durio zibethinus.*
REMARKS. The redescription and illustration of *P. vacuum* Morrison, by Takahashi (1939a) actually refers to *P. album* Takahashi.

Paralecanium angkorense Takahashi

Paralecanium angkorense Takahashi, 1942a: 24.
TYPE DATA. Syntypes female, CAMBODIA: Angkor, on undetermined tree (IMZT).
DISTRIBUTION. ORIENTAL REGION: Cambodia.

Paralecanium calophylli Green

Lecanium (Paralecanium) calophylli Green, 1904d: 240.
Paralecanium calophylli Green; Sanders, 1906: 9.
TYPE DATA. Syntypes female, SRI LANKA: Newara Eliya, on *Calophyllum* sp. (BMNH).
DISTRIBUTION. ORIENTAL REGION: Sri Lanka.
HOST PLANTS. Guttiferae: *Calophyllum.*

Paralecanium carolinensis Beardsley

Paralecanium carolinensis Beardsley, 1966a: 488.
TYPE DATA. Holotype female, CAROLINE ISLANDS: Losap Island, Losap, on *Pandanus* sp. (BP).
DISTRIBUTION. NEW ZEALAND and PACIFIC REGION: Caroline Islands.

HOST PLANTS. Pandanaceae: *Pandanus, Freycinetia.*

Paralecanium cocophyllae Banks

Paralecanium cocophyllae Banks, 1906: 235.
TYPE DATA. Syntypes female, PHILIPPINES: Manila, on under sides of leaves of *Cocos nucifera* (ECMP).
DISTRIBUTION. AUSTRO ORIENTAL REGION: Malaysia, Philippines.
HOST PLANTS. Palmae: *Calamus, Cocos nucifera, Sabal adamsoni.*
REMARKS. Adult female redescribed by Takahashi (1950b). Distribution and host plant records given by Takahashi (1950b).

Paralecanium expansum (Green)

Lecanium expansum Green, 1896: 9.
Lecanium (Paralecanium) expansum Green; Cockerell & Parrott, 1899: 227.
Paralecanium expansum (Green); Fernald, 1903: 199.
TYPE DATA. Syntypes female, SRI LANKA: Punduloya, on *Litsea* sp. and *Dalbergia* sp. (BMNH).
DISTRIBUTION. AUSTRO ORIENTAL REGION: Singapore. **AUSTRALIAN REGION:** Queensland. **ORIENTAL REGION:** India, Sri Lanka, Taiwan. **PALAEARCTIC REGION:** Japan.
HOST PLANTS. Lauraceae: *Litsea, Machilus thunbergii.* **Leguminosae:** *Dalbergia.* **Moraceae:** *Ficus macrophylla, F. pumila, F. retusa.* **Myricaceae:** *Myrica rubra.*
REMARKS. Adult female redescribed and illustrated by Green (1904d), Froggatt (1915), Kawai (1980), Tao *et al.* (1983) and by Tang (1991). Distribution and host plant records given by Kotinsky (1908a), Froggatt (1915), Takahashi (1928a), Green (1937), Ali (1971), Kawai (1980), Tao *et al.* (1983), Shafee *et al.* (1989) and by Tang (1991).

Paralecanium expansum javanicum (Green)

Lecanium expansum javanicum Green, 1904b: 205.
Paralecanium expansum javanicum (Green); Sanders, 1906: 9.
TYPE DATA. Syntypes female, INDONESIA: Java, on *Anomianthus heterocarpus* (BMNH).
DISTRIBUTION. AUSTRO ORIENTAL REGION: Java.
HOST PLANTS. Annonaceae: *Anomianthus heterocarpus.*

Paralecanium expansum metallicum (Green)

Lecanium expansum metallicum Green, 1904b: 205.
Lecanium (Paralecanium) expansum metallicum Green; Green, 1904c: 377.
Paralecanium expansum metallicum (Green); Sanders, 1906: 9.
Paralecanium metallicum (Green); Ali, 1971: 38.
TYPE DATA. Syntypes female, INDONESIA: Java, Bogor (Botanic Gardens), on *Myristica fragrans* (BMNH).
DISTRIBUTION. AUSTRO ORIENTAL REGION: Java, Singapore.
HOST PLANTS. Myristicaceae: *Myristica fragrans.*
REMARKS. Adult female redescribed by Green (1904c).

Paralecanium expansum rotundum (Green)

Lecanium expansum rotundum Green, 1904b: 206.
Paralecanium expansum rotundum (Green); Sanders, 1906: 9.
TYPE DATA. Syntypes female, INDONESIA: Java, on *Rhizophora mucronata* (BMNH).
DISTRIBUTION. AUSTRO ORIENTAL REGION: Java.
HOST PLANTS. Rhizophoraceae: *Rhizophora mucronata.*

Paralecanium frenchi (Maskell)

Lecanium frenchi Maskell, 1891a: 17.
Lecanium (Paralecanium) frenchii Maskell; Cockerell & Parrott, 1899: 227.
Paralecanium frenchii (Maskell); Fernald, 1903: 199.
TYPE DATA. Syntypes female, AUSTRALIA: Melbourne, on *Banksia australis* (NZAC).
DISTRIBUTION. AUSTRALIAN REGION: Victoria.
HOST PLANTS. Proteaceae: *Banksia australis.*
REMARKS. Adult female redescribed and illustrated by Froggatt (1915), Morrison & Morrison (1922) and by Yang (1982).

Paralecanium frenchi macrozamiae (Fuller)

Lecanium macrozamiae Fuller, 1897b: 1345.
Lecanium (Paralecanium) frenchii macrozamiae (Fuller); Cockerell & Parrott, 1899: 227.
Paralecanium frenchii macrozamiae (Fuller); Fernald, 1903: 199.
TYPE DATA. Syntypes female, AUSTRALIA: Western Australia, on *Macrozamia frazeri* (probably lost; P. Gullan, 1990, personal communication).
DISTRIBUTION. AUSTRALIAN REGION: Western Australia.
HOST PLANTS. Zamiaceae: *Macrozamia frazeri.*
REMARKS. Adult female redescribed and illustrated by Fuller (1899).

Paralecanium geometricum (Green)

Lecanium geometricum Green, 1896: 9.
Lecanium (Paralecanium) geometricum Green; Cockerell & Parrott, 1899: 227.
Paralecanium geometricum (Green); Fernald, 1903: 199.
TYPE DATA. Syntypes female, SRI LANKA: Punduloya, on undetermined shrub (BMNH).
DISTRIBUTION. ORIENTAL REGION: Sri Lanka.
HOST PLANTS. Rutaceae: *Glycosmis pentaphylla.*
REMARKS. Adult female redescribed and illustrated by Green (1904d). Adult female redescribed by Tang (1991). Distribution and host plant records given by Green (1904d, 1937) and by Tang (1991).

Paralecanium hainanense Takahashi

Paralecanium hainanense Takahashi, 1942b: 500.
Paralecanium hainanensis Takahashi; Tang, 1991: 129.
TYPE DATA. Syntypes female, CHINA: Hainan Island, host plant not indicated (IMZT).

DISTRIBUTION. PALAEARCTIC REGION: China.
REMARKS. Adult female redescribed and illustrated by Tang (1991).

Paralecanium limbatum Green

Lecanium (Paralecanium) limbatum Green, 1922b: 1023.
Paralecanium limbatum Green; Green, 1937: 306.
TYPE DATA. Syntypes female, SRI LANKA: Batticaloa, on *Ixora coccinea* (BMNH).
DISTRIBUTION. ORIENTAL REGION: Sri Lanka.
HOST PLANTS. Rubiaceae: *Ixora coccinea.*

Paralecanium maculatum Takahashi

Paralecanium maculatum Takahashi, 1950a: 71.
Platylecanium maculatum (Takahashi); Ali, 1971: 41.
TYPE DATA. Syntypes female, INDONESIA: Riau [=Riouw] Islands, host plant not indicated (HUSJ).
DISTRIBUTION. AUSTRO ORIENTAL REGION: Indonesia.

Paralecanium malainum Takahashi

Paralecanium malainum Takahashi, 1950b: 54.
TYPE DATA. Syntypes female, MALAYSIA: Kuala Lumpur, Batu Arang, Selangor and SINGAPORE: on mangosteen and on *Palaquium* sp. (SMKM).
DISTRIBUTION. AUSTRO ORIENTAL REGION: Malaysia, Singapore.
HOST PLANTS. Rubiaceae: *Garcinia mangostana.* **Sapotaceae:** *Palaquium.*

Paralecanium mancum (Green)

Lecanium (Paralecanium) mancum Green, 1922b: 1023.
Paralecanium mancum (Green); Green, 1937: 305.
TYPE DATA. Syntypes female, SRI LANKA: Namunakuli Hill, Badulla, on *Calophyllum walkeri* (BMNH).
DISTRIBUTION. ORIENTAL REGION: Sri Lanka.
HOST PLANTS. Guttiferae: *Calophyllum walkeri.*

Paralecanium marginatum (Green)

Lecanium marginatum Green, 1896: 9.
Lecanium (Paralecanium) marginatum Green; Cockerell & Parrott, 1899: 227.
Paralecanium marginatum (Green); Fernald, 1903: 199.
TYPE DATA. Syntypes female, SRI LANKA: Punduloya, on *Psychotria thwaitesii* (BMNH).
DISTRIBUTION. ORIENTAL REGION: Sri Lanka.
HOST PLANTS. Rubiaceae: *Psychotria thwaitesii.*
REMARKS. Adult female redescribed and illustrated by Green (1904d).

Paralecanium marianum Cockerell

Paralecanium marianum Cockerell, 1902f: 455.

TYPE DATA. Syntypes female, BRAZIL: Rio de Janeiro State, Maria, on undetermined tree (USNM).
DISTRIBUTION. NEOTROPICAL REGION: Brazil Sao Paulo.
HOST PLANTS. Celastraceae: *Maytenus*.
REMARKS. Adult female redescribed by Hempel (1904). Distribution and host plant records given by Hempel (1904) and by Silva *et al.* (1968).

Paralecanium maritimum (Green)

Lecanium planum maritimum Green, 1896: 9.
Lecanium (Paralecanium) maritimum (Green); Cockerell & Parrott, 1899: 228.
Paralecanium maritimum (Green); Fernald, 1903: 199.
TYPE DATA. Syntypes female, SRI LANKA: Bentota, on leaves of a thorny bush growing on the sea-shore (BMNH).
DISTRIBUTION. ORIENTAL REGION: India, Sri Lanka.
HOST PLANTS. Apocynaceae: *Carissa*. **Rubiaceae:** *Ixora coccinea*.
REMARKS. Adult female redescribed and illustrated by Green (1904d). Distribution and host plant records given by Green (1937) and by Varshney (1985).

Paralecanium milleri Takahashi

Paralecanium milleri Takahashi, 1939a: 114.
Paralecanium milleti Takahashi; Tang, 1991: 131 [MIS-SPELLING].
TYPE DATA. Syntypes female, MALAYSIA: Kuala Lumpur, on *Annona muricata* (SMKM).
DISTRIBUTION. AUSTRO ORIENTAL REGION: Malaysia.
HOST PLANTS. Anacardiaceae: *Mangifera indica*. **Annonaceae:** *Annona muricata*. **Moraceae:** *Ficus religiosa*. **Palmae:** *Cocos nucifera, Elaeis gumlusis*.
REMARKS. Adult female redescribed by Takahashi (1950b) and by Tang (1991). Distribution and host plant records given by Takahashi (1950b) and by Tang (1991).
BIOLOGY. Attended by *Oecophylla* sp. (Takahashi, 1950b).

Paralecanium minutum Takahashi

Paralecanium minutum Takahashi, 1951b: 103.
TYPE DATA. Syntypes female, INDONESIA: Riau [-Riouw] Islands, Rempang, host plant not indicated (HUSJ).
DISTRIBUTION. AUSTRO ORIENTAL REGION: Indonesia.

Paralecanium neomaritimum Takahashi

Paralecanium neomaritimum Takahashi, 1950b: 49.
TYPE DATA. Syntypes female, MALAYSIA: Morib, Selangor, on *Avicennia* sp. (SMKM).
DISTRIBUTION. AUSTRO ORIENTAL REGION: Malaysia.
HOST PLANTS. Verbenaceae: *Avicennia*.

Paralecanium ovatum Morrison

Paralecanium ovatum Morrison, 1921: 671.

TYPE DATA. Syntypes female, male and larva, SINGAPORE: Botanic
Gardens, on *Pandanus* sp. (USNM).
DISTRIBUTION. AUSTRO ORIENTAL REGION: Indonesia, Malaysia, Papua New Guinea, Singapore.
HOST PLANTS. Pandanaceae: *Pandanus*.
REMARKS. Adult female redescribed by Williams & Watson (1990).
Distribution and host plant records given by Takahashi (1950b) and
by Williams & Watson (1990).

Paralecanium pahanense Takahashi
Paralecanium pahanense Takahashi, 1950b: 55.
TYPE DATA. Syntypes female, MALAYSIA: Fraser's Hills, Pahan, on
undetermined tree (SMKM).
DISTRIBUTION. AUSTRO ORIENTAL REGION: Malaysia.

Paralecanium paradeniyense Green
Lecanium (Paralecanium) paradeniyense Green, 1904d: 241.
Paralecanium paradeniyense Green; Sanders, 1906: 9.
TYPE DATA. Syntypes female, SRI LANKA: Paradeniya, on *Piper
nigrum* (BMNH).
DISTRIBUTION. ORIENTAL REGION: Sri Lanka.
HOST PLANTS. Piperaceae: *Piper nigrum*. **Smilacaceae:** *Smilax
ovalifolia*.
REMARKS. Distribution and host plant records given by Green
(1937).

Paralecanium planum (Green)
Lecanium planum Green, 1896: 9.
Lecanium (Paralecanium) planum Green; Cockerell & Parrott, 1899:
228.
Paralecanium planum (Green); Fernald, 1903: 200.
TYPE DATA. Syntypes female, SRI LANKA: Punduloya, on undetermined tree (BMNH).
DISTRIBUTION. ORIENTAL REGION: Sri Lanka.
HOST PLANTS. Anacardiaceae: *Nothopegia colebrookiana*.
REMARKS. Adult female redescribed and illustrated by Green
(1904d). Distribution and host plant records given by Green (1904d,
1937).

Paralecanium pseudexpansum (Green) n. comb.
Lecanium pseudexpansum Green, 1914: 233.
TYPE DATA. Syntypes female, AUSTRALIA: Northern Territory,
Koolpinyah near Darwin, on *Pandanus odoratissimus* (BMNH).
DISTRIBUTION. AUSTRALIAN REGION: Northern Territory.
HOST PLANTS. Pandanaceae: *Pandanus odoratissimus*.

Paralecanium quadratum (Green)
Lecanium expansum quadratum Green, 1904d: 236.
Paralecanium expansum quadratum (Green); Sanders, 1906: 9.
Paralecanium quadratum (Green); Takahashi, 1955f: 239.
TYPE DATA. Syntypes female, SRI LANKA: Balangoda on cultivated

nutmeg [=*Myristica fragrans*] and Kalutara on undetermined tree (BMNH).
DISTRIBUTION. AUSTRO ORIENTAL REGION: Java, Philippines, Sri Lanka, Taiwan. **PALAEARCTIC REGION:** Japan.
HOST PLANTS. Flacourtiaceae: *Scolopia oldhami.* **Guttiferae:** *Calophyllum inophyllum.* **Myristicaceae:** *Myristica fragrans.* **Tiliaceae:** *Diplodiscus paniculatus.*
REMARKS. Adult female redescribed and illustrated by Morrison (1920), Kawai (1980), Tao *et al.* (1983) and by Tang (1991). Distribution and host plant records given by Morrison (1920), Takahashi (1935, 1955f), Green (1937), Ali (1971), Kawai (1980), Tao *et al.* (1983) and by Tang (1991).

Paralecanium trifasciatum Green

Lecanium (Paralecanium) trifasciatum Green, 1922b: 1024.
Paralecanium trifasciatum Green, 1937: 307.
TYPE DATA. Syntypes female, SRI LANKA: Batticaloa, on *Hemicyclea* [=*Hemicyclia*] (BMNH).
DISTRIBUTION. ORIENTAL REGION: Sri Lanka.
HOST PLANTS. Euphorbiaceae: *Hemicyclia lanceolata.*

Paralecanium vacuum Morrison

Paralecanium vacuum Morrison, 1921: 674.
TYPE DATA. Syntypes female, SINGAPORE: on *Ficus* sp. (USNM).
DISTRIBUTION. AUSTRO ORIENTAL REGION: Indonesia, Malaysia, Singapore.
HOST PLANTS. Moraceae: *Ficus.* **Rhizophoraceae:** *Rhizophora.*
REMARKS. Adult female redescribed by Takahashi (1950b). Distribution and host plant records given by Takahashi (1950b).

Paralecanium zonatum Green

Lecanium (Paralecanium) zonatum Green, 1904d: 245.
Paralecanium zonatum Green; Sanders, 1906: 9.
TYPE DATA. Syntypes female, SRI LANKA: Paradeniya, on *Garcinia spicata* (BMNH).
DISTRIBUTION. ORIENTAL REGION: Sri Lanka.
HOST PLANTS. Guttiferae: *Garcinia spicata.*

Paralecanopsis Bodenheimer

Paralecanopsis Bodenheimer, 1951a: 329.
TYPE-SPECIES: *Paralecanopsis turcica* Bodenheimer, by original designation.
REMARKS. Bodenheimer (1953) presented this genus again as new, however the 1951a description made the genus available. Borchsenius (1957) discussed the characters of the genus. The genus is accepted here as a subjective synonym of *Lecanopsis* as suggested by Ben-Dov (1980).

Parapulvinaria Fonseca

Parapulvinaria Fonseca, 1969: 9.

TYPE-SPECIES: *Parapulvinaria cassariae* Fonseca, by original designation and monotypy.

Parapulvinaria cassariae Fonseca

Parapulvinaria cassariae Fonseca, 1969: 9.
TYPE DATA. Syntypes female, BRAZIL: Sao Paulo, Jardim da Luz, on *Cassaria [=Casearia] sylvestris* (MZSP).
DISTRIBUTION. NEOTROPICAL REGION: Brazil.
HOST PLANTS. Flacourtiaceae: *Casearia sylvestris*.

Parasaissetia Takahashi

Parasaissetia Takahashi, 1955d: 26.
TYPE-SPECIES: *Lecanium nigrum* Nietner, by original designation.
REMARKS. Generic characters discussed by De Lotto (1965), Yang (1982), Hamon & Williams (1984), Gill (1988), Williams & Watson (1990) and by Tang (1991). Key to species given by De Lotto (1965).

Parasaissetia ficicola De Lotto

Parasaissetia ficicola De Lotto, 1965: 214.
TYPE DATA. Holotype female, KENYA: Nairobi, on *Ficus mallatocarpa* (BMNH).
DISTRIBUTION. ETHIOPIAN REGION: Kenya, Tanzania, Uganda.
HOST PLANTS. Moraceae: *Ficus mallatocarpa, F. populifolia*.

Parasaissetia litorea De Lotto

Parasaissetia litorea De Lotto, 1967b: 788.
TYPE DATA. Holotype female, SOUTH AFRICA: Cape Province, Addo, on *Loranthus* sp. (SANC).
DISTRIBUTION. ETHIOPIAN REGION: South Africa.
HOST PLANTS. Loranthaceae: *Loranthus*. **Moraceae:** *Ficus*. **Sapotaceae:** *Sideroxylon inerme*. **Smilacaceae:** *Smilax*.

Parasaissetia nairobica (De Lotto)

Saissetia nairobica De Lotto, 1957a: 173.
Parasaissetia nairobica (De Lotto); De Lotto, 1965: 216.
TYPE DATA. Holotype female, KENYA: Nairobi, on *Cordia holstii* (BMNH).
DISTRIBUTION. ETHIOPIAN REGION: Ivory Coast, Kenya, Tanzania.
HOST PLANTS. Ehretiaceae: *Cordia holstii*. **Euphorbiaceae:** *Cleistanthus polystachyus, Mareya spicata*. **Flacourtiaceae:** *Caloncoba brevipes*. **Ixonanthaceae:** *Ochtocosmus africanus*. **Moraceae:** *Ficus sycomorus*. **Pandaceae:** *Microdesmis puberula*.
REMARKS. Distribution and host plant records given by De Lotto (1965) and by Couturier *et al.* (1985).

Parasaissetia nigra (Nietner)

Lecanium nigrum Nietner, 1861: 9.
Lecanium depressum Targioni Tozzetti, 1867: 29. Syntypes female, ITALY: Florence, Royal Botanic Gardens, on *Ficus* sp. and other

plants (probably lost; G. Pellizzari Scaltriti, 1990, personal
communication). Syn. by Maskell, 1894a: 166.
Lecanium depressum simulans Douglas, 1887a: 28. Syntypes female,
ENGLAND: on *Ficus elastica* and camellias (probably lost; Wil-
liams & Watson, 1990). Syn. by Maskell, 1894a: 166.
Lecanium begoniae Douglas, 1892b: 209. Syntypes female, GUYANA:
Demerara, on *Begonia* (probably lost; Williams & Watson, 1990).
Syn. by Maskell, 1894a: 166.
Lecanium caudatum Green, 1896: 10. Syntypes female, SRI LANKA:
Punduloya, on *Passiflora* and coffee (BMNH). Syn. by Tao *et al.*,
1983: 76.
Lecanium nigrum depressum Targioni Tozzetti; Cockerell, 1896b:
332.
Lecanium nigrum begoniae Douglas; Cockerell, 1896b: 332.
Lecanium (Saissetia) nigrum begoniae Nietner; Cockerell & Parrott,
1899: 163.
Saissetia nigra (Nietner); King, 1902a: 296.
Coccus nigrum (Nietner); Kirkaldy, 1902: 106.
Saissetia depressa (Targioni Tozzetti); Fernald, 1903: 201.
Lecanium (Saissetia) sideroxyllum Kuwana; Kuwana, 1909b: 162.
Syntypes female, JAPAN: Ogasawara Islands [=Bonin Islands],
on *Sideroxylon ferrugineum* (ITLJ). Syn. by Takahashi, 1955b:
70.
Lecanium (Saissetia) pseudonigrum Kuwana, 1909b: 162. Syntypes
female, JAPAN: Ogasawara Islands [=Bonin Islands], on a plant
named "Shirotsugi" (ITLJ). Syn. by Takahashi, 1955b: 70.
Saissetia pseudonigrum (Kuwana); Sasscer, 1911: 67.
Saissetia sideroxyllum Kuwana; Sasscer, 1911: 67.
Saissetia cuneiformis Leonardi, 1913: 33. Syntypes female and first
instar larva, ERITREA: on *Rhus aztechesan* (IEAP). Syn. by De
Lotto, 1956a: 240.
Lecanium (Saissetia) signatum Newstead, 1917a: 363. Lectotype
female designated by Williams & Watson (1990), UGANDA:
Entebbe, on guava (BMNH). Syn. by De Lotto, 1957a: 175.
Coccus signatus (Newstead); Gowdey, 1917: 188.
Lecanium (Saissetia) nigrum nitidum Newstead, 1920: 191. Lectotype
female designated by Williams & Watson (1990), UGANDA: Lake
Victoria, Bukeke and Sesse Islands, on "Luzibarziba" [=*Alchornea
cordifolia*], (BMNH). Syn. by De Lotto, 1957a: 175.
Saissetia perseae Brain, 1920b: 11. Syntypes female, SOUTH
AFRICA: Natal, Durban, on *Persea gratissima* (SANC). Syn. by De
Lotto, 1957a: 175.
Saissetia (Lecanium) nigra (Nietner); Hall, 1922: 22.
Lecanium (Saissetia) crassum Green, 1930c: 287. Syntypes female,
SUMATRA: on *Broussonetia papyrifera*, and from INDIA, WEST
BENGAL: on *Psidium guajava*, and from INDIA, Calcutta, on
Ficus religiosa (BMNH). Syn. by Williams, 1963: 100.
Saissetia nigrum depressum (Cockerell); Ramakrishna Ayyar, 1930:
45 [ERRONEOUS AUTHORSHIP].
Saissetia nigra depressa (Douglas); Balachowsky, 1939a: 257
[ERRONEOUS AUTHORSHIP].
Lecanium nigrum depressum Douglas; Balachowsky, 1939a: 257
[ERRONEOUS AUTHORSHIP].
Parasaissetia nigra (Nietner); Takahashi, 1955d: 26.

Saissetia crassum Green; Ali, 1971: 44.
TYPE DATA. Syntypes female, SRI LANKA [=CEYLON]: on coffee (probably lost).
DISTRIBUTION. AUSTRO ORIENTAL REGION: Indonesia, Java, Malaysia, Papua New Guinea, Philippines, Sabah, Sarawak, Singapore. **AUSTRALIAN REGION:** New South Wales, Northern Territory. **MADAGASIAN REGION:** Agalega Island, Aldabra Island, Comoros, Madagascar, Mauritius, Reunion, Rodrigues Island, Seychelles. **NEW ZEALAND and PACIFIC REGION:** American Samoa, Caroline Islands, Cook Islands, Fiji, French Polynesia, Futuna Island, Gilbert Islands, Guam, Hawaii, Johnston Island, Kiribati, Mariana Islands, Nauru, New Caledonia, New Hebrides, New Zealand, Ogasawara Islands, Palau, Palmyra Island, Phoenix Island, Swains' Islands, Tahiti, Tonga, Tuvalu, Vanuatu, Western Samoa. **ORIENTAL REGION:** India, Pakistan, Sri Lanka, Taiwan, Thailand, Vietnam. **ETHIOPIAN REGION:** Angola, Cameroon, Cape Verde Island, Chad, Eritrea, Ghana, Guinea, Ivory Coast, Kenya, Mozambique, Namibia, Nigeria, Sao Tome, Sierra Leone, South Africa, St. Helena, Sudan, Tanzania, Uganda, Upper Volta, Zambia, Zanzibar, Zimbabwe. **NEOTROPICAL REGION:** Argentina, Barbados, Bermuda, Brazil Rio Grande do Sul, Colombia, Costa Rica, Dominican Republic, Ecuador, Guadeloupe, Guatemala, Guyana, Honduras, Jamaica, Martinique, Mexico, Nicaragua, Panama, Puerto Rico, Trinidad, Virgin Islands. **NEARCTIC REGION:** Alabama, California, Canada, Florida, Louisiana, Maryland, Missouri, New Mexico, New York, Ohio, Oklahoma, Pennsylvania, Texas, Virginia. **PALAEARCTIC REGION:** Azores Islands, Belgium, Canary Islands, Denmark, Egypt, England, Inner Mongolia, Israel, Italy, Japan, Madeira, Saudi Arabia, Spain.
HOST PLANTS. Acanthaceae: *Adhatoda visica, Eranthemum, Graptophyllum pictum, Odontonema, Pseuderanthemum, Ruttya ovata.* **Agavaceae:** *Agave sisalana, Cordyline terminalis, Dracaena, Furcraea gigantea, Polyanthes.* **Alliaceae:** *Agapanthus.* **Amaranthaceae:** *Amaranthus.* **Anacardiaceae:** *Anacardium occidentale, Mangifera indica, Rhus abyssinica, R. culmenum, R. simarubaefolia, Schinus molle, S. terebinthifolius, Spondias dulcis.* **Annonaceae:** *Annona cherimolia, A. chrysophylla, A. macrocarpa, A. muricata, A. reticulata, Artabotrys odoratissimus, Brieya fasciculata, Xylopia quintasii.* **Apocynaceae:** *Adenium obesum, Carissa bispinosa acuminata, C. edulis, Nerium indicum, N. oleander, Plumeria acutifolia, P. rubra, P. tricolor.* **Aquifoliaceae:** *Ilex mitis.* **Araceae:** *Anthurium digitatum, Colocasia esculenta, Dieffenbachia regina, Monstera deliciosa, Philodendron, Zantedeschia aethiopica.* **Araliaceae:** *Aralia elegantissima, Brassaia actinophylla, Cussonia spicata, Hedera helix, Schefflera.* **Barringtoniaceae:** *Barringtonia.* **Begoniaceae:** *Begonia.* **Bignoniaceae:** *Jacaranda, Tecoma.* **Bischofiaceae:** *Bischofia javanica.* **Bombacaceae:** *Ceiba pentandra, Ochroma.* **Boraginaceae:** *Cordia subcordata.* **Bromeliaceae:** *Ananas comosus, Bromelia, Neoregelia spectabilis.* **Burseraceae:** *Garuga pacifica.* **Cactaceae: Cannaceae:** *Canna indica, C. speciosa.* **Caricaceae:** *Carica papaya.* **Celastraceae:** *Euonymus, Maytenus cymosa, M. senegalensis.* **Cobaeaceae:** *Cobaea scandens.* **Combretaceae:** *Terminalia*

brassii, Terminalia catappa. **Compositae:** *Artemisia, Bidens pilosa, Chrysanthemum, Elephantopus scaber, Helianthus annuus, Pluchea odorata, Solidago, Vernonia amygdalina.* **Convolvulaceae:** *Ipomoea tuberosa.* **Crassulaceae:** *Bryophyllum daigrimontianum, Kalanchoe pinnata.* **Cucurbitaceae:** *Cucurbita moschata, C. pepo, Sechium edule.* **Ebenaceae:** *Euclea kellau.* **Ehretiaceae:** *Cordia abyssinica, C. myxa, Ehretia silvatica.* **Euphorbiaceae:** *Alchornea cordifolia, Aleurites fordii, Antidesma venosum, Bridelia, Codiaeum variegatum, Croton macrostachys, Croton tiglium, Discogypremna caloneura, Erythrococca, Euphorbia fulgens, E. pulcherrima, Glochidion, Hevea brasiliensis, Jatropha acontifolia, Manihot esculentus, Mareya spicata, Phyllanthus, Poinsettia, Ricinus communis, Sapium sebiferum.* **Flacourtiaceae:** *Caloncoba brevipes, Dovyalis caffra.* **Geraniaceae:** *Geranium.* **Goodeniaceae:** *Scaevola frutescens, S. koenigii.* **Gramineae:** *Bambusa, Cyperus rotundus, Panicum distachyum, P. maximum.* **Guttiferae:** *Calophyllum inophyllum, Clusia grandiflora, Harungana, Mammea americana.* **Heliconiaceae:** *Heliconia humilis, H. metallica.* **Iridaceae:** *Gladiolus illyricus, Moraea bicolor.* **Labiatae:** *Coleus blumei.* **Lauraceae:** *Cinnamomum zeylanicum, Ocotea usambarensis, Persea americana.* **Leguminosae:** *Cassia grandis, Crotalaria, Delonix regia, Erythrina corallodendron, E. indica, Parkinsonia aculeata.* **Liliaceae:** *Aloe vera, Asparagus.* **Loganiaceae:** *Geniostoma.* **Loranthaceae:** *Loranthus quequensis.* **Malpighiaceae:** *Malpighia coccigera.* **Malvaceae:** *Abutilon molle, Gossypium arboreum, G. barbadense, G. hirsutum, G. tomentosum, Hibiscus esculentum, H. sabdariffa, H. sinensis, H. syriacus, H. tiliaceus, Kokia drynarioides, Malvaviscus arboreus, Sida acuta, S. rhombifolia, Thespesia populnea.* **Meliaceae:** *Melia azedarach.* **Monimiaceae:** *Tambourissa.* **Moraceae:** *Artocarpus altilis, A. heterophyllus, Castilla elastica, Ficus capensis, F. carica, F. dekdekena, F. ingens, F. lutea, F. mallatocarpa, F. megapoda, F. melleri, F. microcarpa, F. nekbuda, F. nitida, F. palmata, F. scabra, F. sycomorus, F. tinctoria, F. vasculosa, F. vasta, F. wightiana, F. wrightii, Maclura africana, Morus alba.* **Musaceae:** *Musa paradisiaca, M. sapientum.* **Myrsinaceae:** *Ardisia quinquegona, Myrsine.* **Myrtaceae:** *Eucalyptus deglupta, Eugenia javanica, Feijoa sellowiana, Metrosideros, Psidium cattleyanum, P. guajava, P. littorale, Syzygium cordatum, S. guiniense.* **Naucleaceae:** *Uncaria africana.* **Nyctaginaceae:** *Bougainvillea.* **Orchidaceae:** *Cymbidium, Piperia, Vanda.* **Palmae:** *Areca catechu, Cocos nucifera, Elaeis guineensis, Erythea armata.* **Passifloraceae:** *Adenia glauca, P. edulis, P. quadrangularis.* **Piperaceae:** *Piper methysticum.* **Pittosporaceae:** *Pittosporum arborescens, P. undulatum.* **Polygonaceae:** *Emex spinosa, Homalocladium platycladum.* **Proteaceae:** *Hakea.* **Rosaceae:** *Crataegus mexicana, Malus sylvestris, Prunus capuli, Rubus hawaiiensis.* **Rubiaceae:** *Canthium obovatum, Coffea arabica, C. canephora, Gardenia jasminoides, Guettarda speciosa, Ixora coccinea, I. macrothyrsa, Morinda citrifolia.* **Rutaceae:** *Casimiroa edulis, C. aurantifolia, C. limon, C. reticulata, Evodia hortensis, Fagara macrophylla, Pelea.* **Salicaceae:** *Salix warburgii.* **Santalaceae:** *Santalum paniculatum.* **Sapindaceae:** *Litchi chinensis.* **Sapotaceae:** *Chrysophyllum cainito, Sideroxylon oxyacantha.*

Smilacaceae: *Smilax.* **Solanaceae:** *Capsicum frutescens, Cestrum, Solanum jasminoides, S. melongena, Solanum nigrum, Solanum wendlandii.* **Sterculiaceae:** *Theobroma cacao.* **Strelitziaceae:** *Strelitzia nicolai, S. reginae.* **Strychnaceae:** *Strychnos lucens.* **Tamaricaceae:** *Tamarix.* **Theaceae:** *Thea.* **Ulmaceae:** *Ulmus.* **Umbelliferae:** *Foeniculum.* **Urticaceae:** *Pilea cadierei.* **Verbenaceae:** *Clerodendrum, Duranta plumieri, Premna taitensis.* **Vitidaceae:** *Cissus, Vitis vinifera.* **Zingiberaceae:** *Alpinia purpurata, Hedychium flavum, Zingiber officinale.*

REMARKS. Common Name - The nigra scale. Adult female redescribed and illustrated by Newstead (1903), Green (1904d), Froggatt (1915), Steinweden (1929), Smith (1944), Zimmerman (1948), Borchsenius (1957), De Lotto (1967b), Ezzat & Hussein (1969), Ben-Dov (1978), Kawai (1980), Hamon & Williams (1984), Gill (1988), Tang & Li (1988), Williams & Watson (1990) and by Tang (1991). Male test (based on material from U. S. A., Alabama) described and illustrated by Miller & Williams (1990). Larval instars described and illustrated by Smith (1944) and by Ben-Dov (1978). Colour photograph given by Kawai (1980), Hamon & Williams (1984) and by Johnson & Lyon (1988). De Lotto (1957a, 1967b) discussed morphological and habit differences between populations in East Africa. Distribution and host plant records given by Kuwana (1909b), Leonardi (1913), Froggatt (1915), Green (1916e), Morrison (1920), Newstead (1920), Hall (1922), Lizer y Trelles (1942b), Zimmerman (1948), Balachowsky (1957), Gómez-Menor Ortega (1958c), Mamet (1951, 1954a, 1959b, 1962, 1978), Ossiannilsson (1959), Beardsley (1966a), De Lotto (1956a, 1967b, 1968b), Hodgson (1969b), Corseuil & Barbosa (1971), Kawai *et al.* (1971), Ali (1971), Almeida (1973), Kozárzhevskaya & Reitzel (1975), Takagi (1977), Wise (1977), Matile-Ferrero (1970, 1976, 1978, 1984b), Ben-Dov (1978), Tao *et al.* (1983), Nakahara (1981b), Vieira *et al.* (1983), Hamon & Williams (1984), Matile-Ferrero & Nonveiller (1984), Williams (1985a), Couturier *et al.* (1985), Carnero Hernandez & Perez Guerra (1986), Fernandes (1987), Marotta (1987), Tang & Li (1988), Shafee *et al.* (1989), Williams & Watson (1990), Danzig & Konstantinova (1990), Tang (1991) and by Hodgson & Hilburn (1991a, 1991b).

BIOLOGY. Smith (1944) studied the biology, ecology and natural enemies in California. The scale successfully develops in the laboratory on *Cucurbita* squashes of the Butternut strain and on sprouting potato tubers; on squashes at 24°C a generation is completed within 45-60 days (Ben-Dov, 1978). The encapsulation of eggs of the parasitoid *Encyrtus infelix* (Embleton) was determined by Blumberg & Goldenberg (1992).

ECONOMIC IMPORTANCE. A minor pest of coffee in Africa (Le Pelley, 1968). A minor pest of ornamentals in Hawaii (Bartlett, 1978). Formerly reported as a moderate pest of ornamentals and citrus in California, but presently of no economic importance (Gill, 1988).

Parasaissetia tsaratananae (Mamet)

Platysaissetia tsaratananae Mamet, 1951: 238.
Parasaissetia tsaratananae (Mamet); Matile-Ferrero, 1978: 48.

TYPE DATA. Syntypes female, MADAGASCAR: Tsaratanana, on a plant named "Sarvyonga" and on *Panax* sp. (MNHN).

DISTRIBUTION. MADAGASIAN REGION: Comoros, Madagascar, Reunion.
HOST PLANTS. Araliaceae: *Cussonia bojeri, Panax.* **Compositae:** *Psiadia.* **Moraceae:** *Ficus.* **Myrtaceae:** *Eugenia vaccinifolia.* **Rutaceae:** *Citrus.*
REMARKS. Distribution and host plant records given by Mamet (1957, 1959b, 1960), Matile-Ferrero (1978) and by Williams & Williams (1988).

Parthenolecanium Sulc

Parthenolecanium Sulc, 1908: 36.
TYPE-SPECIES: *Lecanium corni* Bouche, by designation of Opinion 1303 (1985).
REMARKS. The nomenclature of the genus discussed by Morrison & Morrison (1966) and by Danzig (1967). Generic characters discussed by Borchsenius (1957), Kosztarab & Kozár (1978, 1988), Wang (1980), Danzig (1980c), Danzig & Kerzhner (1981), Tereznikova (1981), Hadzibejli (1983), Hamon & Williams (1984), Gill (1988) and by Tang (1991). Key to species: ASIA - Tang (1991). CALIFORNIA - Gill (1988). EUROPE - Kosztarab & Kozár (1988). FLORIDA -Hamon & Williams (1984). HUNGARY - Kosztarab & Kozár (1978). UKRAINE -Tereznikova (1981). USSR (EUROPEAN) - Borchsenius (1957), Danzig (1964). USSR (FAR EAST) - Danzig (1980c).

Parthenolecanium cerasifex (Fitch)

Lecanium cerasifex Fitch, 1857a: 368.
Lecanium (Eulecanium) cerasifex Fitch; Cockerell, 1896b: 332.
Eulecanium cerasifex (Fitch); King, 1902b: 60.
Parthenolecanium cerasifera (Fitch); Boratynski & Davies, 1971: 58 [MIS-SPELLING].
TYPE DATA. Syntypes female, U.S.A.: New York, Salem, on twigs of wild black cherry (USNM).
DISTRIBUTION. NEARCTIC REGION: Canada British Columbia, Canada Ontario, Canada Quebec, New York.
HOST PLANTS. Aceraceae: *Acer.* **Grossulariaceae:** *Ribes.* **Rosaceae:** *Cerasus, Malus sylvestris, Persica vulgaris, Rosa.* **Taxaceae:** *Taxus.* **Tiliaceae:** *Tilia.*
REMARKS. Adult female redescribed and illustrated by Richards (1958) and by Phillips (1965a). First and second instar larvae described by Richards (1958). Male scale described and illustrated by Richards (1958). Nur (1971, 1972) studied the cytology and reported chromosome number 2n-16 in Canada.
BIOLOGY. Phillips (1965b) separated this species from *P. putmani* on the basis of biological and behavioural attributes.

Parthenolecanium corni (Bouche)

Lecanium corni Bouche, 1844: 298.
Lecanium vini Bouche, 1851: 112. Syntypes female, GERMANY: Berlin, on grapevine (lost; Sachtleben, 1944). Syn. by Lindinger, 1912: 371.
Coccus tiliae Fitch, 1851: 69. Syntypes female, U.S.A.: New York, on

Tilia (probably lost; see Barnes, 1988). Syn. by Barnes, 1988: 104.

Lecanium ribis Fitch, 1857b: 427. Syntypes female, U.S.A.: New York, Albany, on *Ribes* sp. (probably lost; see Barnes, 1988). Syn. by Sanders, 1909a: 443.

Lecanium cynosbati Fitch, 1857c: 436. Syntypes female, U.S.A.: New York, Salem, on stalks of wild gooseberry, *Ribes cynosbati* (USNM). Syn. by Sanders, 1909a: 443.

Lecanium corylifex Fitch, 1857e: 473. Syntypes female, U.S.A.: New York, Salem, on Hazelnut, *Corylus americana* (USNM). Syn. by Sanders, 1909a: 443.

Lecanium juglandifex Fitch, 1857e: 463. Syntypes female, U.S.A.: New York, Salem, Martin's meadow, on butternut (probably lost; see Barnes, 1988). Syn. by Sanders, 1909a: 443.

Coccus rosarum Snellen van Volenhoven in De Graaf *et al.*, 1862: 94. Syntypes female, HOLLAND: on rose (probably lost). Syn. by Marchal, 1908a: 264.

Lecanium fitchii Signoret, 1874a: 404. Syntypes female, U.S.A.: Washington, D. C., on "ronce" [=raspberry], (VMNH). Syn. by Sanders, 1909a: 443.

Lecanium rugosum Signoret, 1874a: 429. Syntypes female, FRANCE: on peach (VMNH). Syn. by Lindinger, 1912: 371.

Lecanium tarsalis Signoret, 1874a: 430. Syntypes female, U.S.A.: New York, on *Cornus sanguineus* (VMNH). Syn. by Sanders, 1909a: 443.

Lecanium wistariae Signoret, 1874a: 433. Lectotype female designated by Ben-Dov (1976b), FRANCE: Clamart, on *Glycine* sp. (VMNH). Syn. by Danzig, 1980a: 594.

Lecanium robiniarum Douglas, 1890b: 318. Syntypes female, HUNGARY: Budapest, on *Robinia pseudacacia* (BMNH). Syn. by Lindinger, 1912: 294.

Lecanium armeniacum Craw, 1891: 12. Syntypes female, U.S.A.: California, on apricot, prune, plum, cherry and pear (USNM). Syn. by Sanders, 1909a: 443.

Lecanium assimile Newstead, 1892: 141. Syntypes female, ENGLAND: Colwyn Bay, N. W., on *Grindelia hirsuta* (BMNH). Syn. by Leonardi, 1920: 287.

Lecanium caryae canadense Cockerell, 1895r: 253. Syntypes female, CANADA: Ontario, Stittsville near Ottawa, on *Ulmus racemosa* (USNM). Syn. by Sanders, 1909a: 443.

Lecanium lintneri Cockerell & Bennett in Cockerell, 1895h: 381. Syntypes female, U.S.A.: New York, Greene County, Lake Mohonk, on *Sassafras* (USNM). Syn. by Sanders, 1909a: 443.

Lecanium pruinosum armeniacum Craw; Tyrrell, 1896: 268.

Lecanium (Eulecanium) corylifex Fitch; Cockerell, 1896b: 332.

Lecanium (Eulecanium) rugosum Signoret; Cockerell, 1896b: 332.

Lecanium (Eulecanium) lintneri Cockerell & Bennett; Cockerell, 1896b: 333.

Lecanium canadense Cockerell; Cockerell, 1898j: 294.

Lecanium (Eulecanium) caryarum Cockerell, 1898j: 293. Syntypes female, CANADA: Ontario, Niagara, on *Carya alba* (BMNH, USNM). Syn. by Sanders, 1909a: 443.

Lecanium (Eulecanium) kingii Cockerell, 1898i: 322. Syntypes female,
    U.S.A.: Massachusetts, Lawrence, on *Vaccinium corymbosum*
    (USNM). Syn. by Sanders, 1909a: 443.

Lecanium (Eulecanium) maclurarum Cockerell, 1898j: 294. Syntypes
    female, CANADA: Ontario, Niagara, on osage orange [=*Maclura
    aurantiaca*] (USNM). Syn. by Sanders, 1909a: 443.

Lecanium crawii Ehrhorn, 1898a: 245. Syntypes female, U.S.A.:
    California, near Mountain View, on *Acer macrophyllum* (USNM).
    Syn. by Sanders, 1909a: 443.

Lecanium maclurae Hunter, 1899b: 67 [HOMONYM of *Lecanium
    maclurae* Fitch] Syntypes female, U.S.A.: Kansas, Claflin, Barber
    County, on osage orange [=*Maclura aurantiaca*] (USNM). Syn. by
    Sanders, 1909a: 443.

Lecanium kansasense Hunter, 1899b: 69. Syntypes female, U.S.A.:
    Kansas, Lawrence, University Campus, on *Cersis canadensis*
    (USNM). Syn. by Sanders, 1909a: 443.

Lecanium (Eulecanium) canadense Cockerell; Cockerell & Parrott,
    1899: 232.

Lecanium (Eulecanium) armeniacum Craw; Cockerell & Parrott, 1899:
    233.

Lecanium (Eulecanium) cynosbati Fitch; Cockerell & Parrott, 1899:
    233.

Lecanium (Eulecanium) tarsale Signoret; Cockerell & Parrott, 1899:
    234.

Lecanium (Eulecanium) ribis Fitch; Cockerell & Parrott, 1899: 234.

Lecanium (Eulecanium) fitchii Signoret; Cockerell & Parrott, 1899:
    236.

Lecanium (Eulecanium) crawii Ehrhorn; Cockerell & Parrott, 1899:
    237.

Lecanium (Eulecanium) aurantiacum Hunter, 1900: 107 [REPLACE-
    MENT NAME for *Lecanium maclurae* Hunter].

Lecanium websteri King, 1901b: 106. Syntypes female, U.S.A.: Ohio,
    on mulberry (USNM). Syn. by Sanders, 1909a: 443.

Eulecanium maclurarum (Cockerell); King, 1901i: 335.

Eulecanium rosae King, 1901i: 336. Syntypes female, CANADA:
    Quebec, Sherbrooke, on rose bush (CNOC). Syn. by Sanders,
    1909a: 443.

Lecanium rehi King in King & Reh, 1901: 61. Syntypes female,
    GERMANY: Hamburg and Berlin, on *Ribes grossularia, R. rubrum*
    and *R. nigrum* (HMNG). Syn. by Lindinger, 1912: 371.

Lecanium adenostomae Kuwana, 1901: 402. Syntypes female, U.S.A.:
    California, Santa Clara County, Stanford University, Siera
    Morena, on *Adenostoma fasciculata* (USNM) Syn. by Sanders,
    1909a: 443.

Eulecanium corylifex (Fitch); King, 1901h: 314.

Eulecanium fitchii (Signoret); King, 1901i: 333.

Eulecanium guignardi King, 1901i: 334. Syntypes female, CANADA:
    Ontario, Niagara, on plum tree (USNM). Syn. by Sanders, 1909a:
    443.

Lecanium (Eulecanium) vini Bouche; King & Reh, 1901: 6.

Eulecanium vini (Bouche); Cockerell, 1901b: 92.

Eulecanium kansasense (Hunter); King, 1902b: 60.

Eulecanium websteri (King); King, 1902b: 60.

Eulecanium fraxini King, 1902c: 158. Syntypes female, CANADA:

Ottawa, on *Fraxinus americana* (USNM) Syn. by Sanders, 1909a: 443.
Eulecanium cynosbati (Fitch); King, 1902c: 159.
Eulecanium robiniarum (Douglas); Cockerell, 1902g: 178.
Eulecanium adenostomae (Kuwana); Fernald, 1903: 180.
Eulecanium armeniacum (Craw); Fernald, 1903: 181.
Eulecanium assimile (Newstead); Fernald, 1903: 181.
Eulecanium aurantiacum Hunter; Fernald, 1903: 182.
Eulecanium canadense (Cockerell); Fernald, 1903: 182.
Eulecanium caryarum (Cockerell); Fernald, 1903: 184.
Eulecanium corni (Bouche); Fernald, 1903: 185.
Eulecanium crawii (Ehrhorn); Fernald, 1903: 186.
Eulecanium kingii (Cockerell); Fernald, 1903: 189.
Eulecanium lintneri (Cockerell & Bennett); Fernald, 1903: 189.
Lecanium obtusum Thro, 1903: 212. Syntypes female, U.S.A.: New York, on *Rubus villosus* (USNM). Syn. by Sanders, 1909a: 443.
Eulecanium obtusum (Thro); Fernald, 1903: 191.
Eulecanium rehi (King); Fernald, 1903: 195.
Eulecanium ribis (Fitch); Fernald, 1903: 195.
Eulecanium rugosum (Signoret); Fernald, 1903: 196.
Eulecanium tarsale (Signoret); Fernald, 1903: 197.
Lecanium folsomi King, 1903c: 193. Syntypes female, U.S.A.: Illinois, Urbana, on paw-paw [=*Asimina triloba*] (USNM). Syn. by Sanders, 1909a: 443.
Lecanium (Eulecanium) assimile Newstead; Reh, 1903: 408.
Lecanium (Eulecanium) ribis Fitch; Reh, 1903: 412.
Lecanium (Eulecanium) robiniarum Douglas; Reh, 1903: 408.
Lecanium (Eulecanium) rosarum (Snellen van Vollenhoven); Reh, 1903: 408.
Eulecanium folsomi (King); Cockerell, 1905d: 129.
Lecanium corni robiniarum Marchal, 1908a: 278. Syntypes female, FRANCE: on *Robinia pseudo-acacia* (MNHN). Syn. by Lindinger, 1912: 294.
Lecanium coryli (Linnaeus); Sulc, 1908: 36 [MISIDENTIFICATION].
Lecanium persicae crudum Green, 1917: 202. Syntypes female, ENGLAND: on *Aralia* sp. (BMNH). Syn. by Habib, 1955a: 70.
Parthenolecanium coryli (Linnaeus); Sulc, 1932: 64 [MISIDENTIFICATION].
Eulecanium corni corni (Bouche); Schmutterer, 1952a: 546.
Parthenolecanium corni (Bouche); Borchsenius, 1957: 356.
TYPE DATA. Syntypes female, GERMANY: Berlin, on lower side of twigs of *Crunus sanguineum*, *Pyrus* sp., *Tilia* sp., *Corylus* sp., and *Ribes rubrum* (Lost; Sachtleben, 1944).
DISTRIBUTION. NEOTROPICAL REGION: Argentina, Brazil, Chile. **NEW ZEALAND and PACIFIC REGION:** New Zealand. **NEARCTIC REGION:** California, Canada British Columbia, Canada Ontario, Canada Saskatchewan, Florida, Tennessee. **ORIENTAL REGION:** Pakistan. **PALAEARCTIC REGION:** Bulgaria, Canary Islands, Czechoslovakia, Denmark, France, Germany, Greece, Hungary, Inner Mongolia, Iran, Italy, Japan, Korea, Netherlands, Poland, Romania, Spain, Sweden, Turkey, USSR Altai, USSR Armenia, USSR Azerbaijan, USSR Caucasus, USSR Georgia, USSR Kazakhstan, USSR Kirgizia, USSR Leningrad Region, USSR Moldavia, USSR Primorye Territory, USSR Sakhalin Is-

land, USSR Tadzhikistan, USSR Turkmenia, USSR Ukraine, USSR Uzbekistan, Yugoslavia.
HOST PLANTS. Aceraceae: *Acer negundo, A. rubrum, A. saccharinum.* **Altingiaceae:** *Liquidambar styraciflua.* **Annonaceae:** *Asimina triloba.* **Aquifoliaceae:** *Ilex.* **Araliaceae:** *Aralia.* **Aspleniaceae:** *Asplenium.* **Berberidaceae:** *Mahonia.* **Buxaceae:** *Buxus sempervirens.* **Cannabidaceae:** *Cannabis.* **Capparidaceae:** *Viburnum.* **Caprifoliaceae:** *Lonicera.* **Carpinaceae:** *Carpinus betulus, C. caroliniana.* **Celastraceae:** *Euonymus europaeus.* **Cornaceae:** *Cornus florida, C. sanguinea.* **Corylaceae:** *Corylus avellana.* **Ebenaceae:** *Diospyros.* **Elaeagnaceae:** *Elaeagnus.* **Ericaceae:** *Vaccinium myrtillus.* **Euphorbiaceae:** *Ricinus communis.* **Fagaceae:** *Quercus nigra.* **Geraniaceae:** *Pelargonium.* **Grossulariaceae:** *Ribes aureum, R. nigrum, R. rubrum.* **Hippocastanaceae:** *Aesculus hippocastanum.* **Juglandaceae:** *Carya illinoiensis, C. tomentosa, Juglans regia.* **Labiatae:** *Rosmarinus officinalis, Thymus.* **Leguminosae:** *Acacia, Caragana arborescens, Cercis siliquastrum, Colutea arborescens, Glycine, Robinia pseudoacacia, Sophora japonica, Wisteria sinensis.* **Magnoliaceae:** *Liriodendron tulipifera, Magnolia grandiflora.* **Moraceae:** *Maclura pomifera, Morus alba.* **Myricaceae:** *Myrica cerifera.* **Nyssaceae:** *Nyssa sylvatica.* **Oleaceae:** *Fraxinus americana, F. chinensis, F. excelsior, Jasminum, Syringa vulgaris.* **Pinaceae:** *Pinus.* **Platanaceae:** *Platanus orientalis.* **Ranunculaceae:** *Clematis vitalba, Thalictrum minus.* **Rosaceae:** *Amygdalus communis, Cotoneaster, Crataegus monogyna, C. oxyacantha, Cydonia vulgaris, Malus sylvestris, Mespilus germanica, Persica vulgaris, Potentilla fructicosa, Prunus cerasifera, P. laurocerasus, P. pritchardi, P. spinosa, Pyracantha, Pyrus communis, Rosa acicularis, R. canina, Rubus ideus, Sorbaria, Sorbus, Spiraea.* **Salicaceae:** *Populus, Salix nigra, S. pentandra.* **Taxaceae:** *Taxus baccata.* **Tiliaceae:** *Tilia americana.* **Ulmaceae:** *Celtis occidentalis, Ulmus americana, U. parvifolia, U. thomasii.* **Vitidaceae:** *Vitis vinifera.*

REMARKS. Common Name - European Fruit Lecanium. Opinion 1303 (1985) placed *Lecanium corni* Bouche, the senior synonym, on the Official List of Names. Kawecki (1958a) and Danzig & Kerzhner (1981) discussed the great confusion regarding the nomenclature and synonyms. Lindinger (1932f, 1934a, 1934b, 1935a) insisted that *Coccus pyri* Schrank, 1781 and *C. xylostei* Schrank, 1801 are synonyms of *P. corni* (Bouche, 1844). In order to stabilize the latter, Lindinger interpretation cannot be accepted. Both species of Schrank are here regarded as synonyms of *Eulecanium tiliae* (L.). Adult female redescribed and illustrated by Marchal (1908a) (as *L. coryli*), Pettit & McDaniel (1920), Gómez-Menor Ortega (1937, 1958c), Schmutterer (1954), Borchsenius (1957), Kawecki (1958a), Savescu (1960), Lellakova-Duskova (1966), Dziedzicka (1968), Williams & Kosztarab (1972), Paik (1978), Kosztarab & Kozár (1978, 1988), Kawai (1980), Wang (1980), Danzig (1980c), Tereznikova (1981), Hamon & Williams (1984), Gill (1988), Tang &Li (1988), Gonzalez (1989) and by Tang (1991). Adult male described and illustrated by Schmutterer (1954), Habib (1956a), Borchsenius (1957), Giliomee (1967) and by Tereznikova (1981). Male test described and illustrated by Miller & Williams (1990). Intraspecific variation in taxonomic characters of the male given by Giliomee (1967). Larval instars described by Schmutterer (1954), Borchsenius (1957), Kawecki (1958a), Dziedzicka (1968), Tereznikova (1981) and by Gonzalez (1989). Colour photograph given by Kawai (1980), Hamon & Williams (1984), Gill (1988), Tranfaglia & Viggiani (1988), Johnson & Lyon (1988) and by Gonzalez (1989). The effect of host plant on intraspecific variation of morphological characters studied by Ebeling (1938) and by Habib (1957). Bielenin (1958) and Dziedzicka & Sermak (1967) studied the frequency variation of dorso-marginal tubercles in second instar larvae. Dziedzicka & Sermak (1967) pointed out the variability in number and position of dorso-marginal tubercles in the adult female. Savescu (1943, 1944) distinguished, based on the adult female, the following 15 ecological forms, each specific to a certain host plant: *corni juglandis* (Juglans regia), *corni savesculi* (Ulmus scabra), *corni mori* (Morus alba), *corni ribis* (Ribes aureum), *corni knechteli* (Cydonia oblonga), *corni pyri* (Pyrus pyraster), *corni rugosum* (Prunus persica), *corni assimile* (Prunus domestica), *corni mali* (Malus pumila), *corni rosarum* (Rosa centifolia), *corni robiniarum* (Robinia pseudo-acacia), *corni sojae* (Soja hyspida), *corni corni* (Cornus sanguineus), *corni vini* (Vitis vinifera), *corni aceris* (Acer tataricum). All these are here regarded as host-induced variations of *P. corni*. Distribution and host plant records given by Signoret (1874a), Leonardi (1920), Pettit & McDaniel (1920), Balachowsky (1935), Takahashi & Tachikawa (1956), Reyne (1957), Borchsenius (1957), Richards (1958), Ossiannilsson (1959), Tereznikova (1963a), Komosinska (1977), Wise (1977), Kozár *et al.* (1977, 1979), Lambdin & Watson (1980), Danzig (1980c), Komeili-Birjandi (1981), Hadzibejli (1983), Kozár (1980, 1983, 1985), Koteja (1983), Martin Mateo (1984), Hamon & Williams (1984), Santas (1985), Carnero Hernandez & Perez Guerra (1986), Kozár & Ostafichuk (1987), Marotta (1987), Gill (1988), Tang & Li (1988), Dziedzicka (1988), Kosztarab & Kozár (1988), Gonzalez (1989), Foldi & Soria (1989), Kozár & Drozdjak (1990) and by Tang (1991).

BIOLOGY. Life history discussed and outlined by Schmutterer (1954) (in Germany), Habib (1955b) (in England), Peterson (1960) (as *L. coryli*, in Saskatchewan, Canada), Dziedzicka (1968) (in Poland), Gonzalez (1989) (in Chile), Santas (1985) (in Greece). Behaviour of nymphal stages studied by Habib (1955c). Komarek (1946) described the physiological damage to ash-tree in Czechoslovakia. Blahutiak (1972) studied the inter-relationships with the parasite *Blastothrix confusa*.

ECONOMIC IMPORTANCE. Widely distributed in the temperate and subtropical regions, where often it becomes a serious pest of deciduous orchards, vine and ornamentals (Bartlett, 1978; Kosztarab & Kozár, 1988; Gill, 1988).

Parthenolecanium corni apuliae (Nuzzaci) n. comb.

Eulecanium corni apuliae Nuzzaci, 1969: 9.

TYPE DATA. Syntypes female, male and larva, ITALY: Foggia, Stornara on grapevine (IEAB).

DISTRIBUTION. PALAEARCTIC REGION: Italy.

HOST PLANTS. Vitidaceae: *Vitis vinifera*.

BIOLOGY. Develops two generation per year in Italy (Nuzzaci, 1969).

Parthenolecanium fletcheri (Cockerell)

Lecanium fletcheri Cockerell, 1893p: 221.
Lecanium (Eulecanium) fletcheri (Cockerell); Cockerell, 1896b: 332.
Eulecanium fletcheri (Cockerell); King, 1902c: 159.
Lecanium arion Lindinger; Lindinger, 1911a: 379 [NOMEN NUDUM].
Lecanium arion Lindinger, 1912: 323. Syntypes female, GERMANY and NETHERLANDS, on *Thuja occidentalis* (DEPOSITORY UNKNOWN; Weidner & Wagner, 1968). Syn. by Borchsenius, 1957: 370.
Lecanium areon Lindinger; Borchsenius, 1936: 116 [MIS-SPELLING].
Eulecanium arion (Lindinger); Schmutterer, 1954: 75.
Parthenolecanium fletcheri (Cockerell); Borchsenius, 1957: 350.
TYPE DATA. Syntypes female, CANADA: Ottawa, on *Thuja* sp. and cedar (USNM).
DISTRIBUTION. NEARCTIC REGION: Alabama, Arkansas, California, Canada Ontario, Colorado, Connecticut, Idaho, Illinois, Indiana, Iowa, Kansas, Kentucky, Louisiana, Maryland, Massachusetts, Michigan, Minnesota, Missouri, New Jersey, New York, Ohio, Oklahoma, Oregon, Pennsylvania, Rhode Island, South Carolina, Tennessee, Texas, Virginia, Washington, West Virginia, Wisconsin. **PALAEARCTIC REGION:** Austria, Bulgaria, Czechoslovakia, France, Germany, Hungary, Latvia, Netherlands, Poland, Romania, Sweden, Switzerland, USSR Armenia, USSR Georgia, USSR Ukraine, USSR Uzbekistan.
HOST PLANTS. Cupressaceae: *Juniperus virginiana, Thuja occidentalis.* **Taxaceae:** *Taxus.*
REMARKS. Adult female redescribed and illustrated by Dietz & Morrison (1916), Schmutterer (1954) (as *E. arion*), Borchsenius (1957), Savescu (1961), Dziedzicka (1968), Kosztarab & Kozár (1979, 1988), Tereznikova (1981), Hamon & Williams (1984), Gill (1988) and by Kosztarab & Kozár (1988). Adult female redescribed by Tang (1991). Larval instars described and illustrated by Schmutterer (1954) and by Dziedzicka (1968). Colour photograph given by Kosztarab & Kozár (1988) and by Gill (1988). Phillips (1965) erroneously synonymized this species with *P. corni.* Williams & Kosztarab were unable to separate *P. fletcheri* from *P. corni.* Hamon & Williams (1984) clearly distinguished them from each other. Distribution and host plant records given by Lindinger (1912), Rehacek (1957), Borchsenius (1957), Ossiannilsson (1959), Phillips (1965a), Lambdin & Watson (1980), Tereznikova (1981), Koteja & Zak-Ogaza (1983), Hamon & Williams (1984), Gill (1988), Dziedzicka (1988), Kozár (1988), Kosztarab & Kozár (1978, 1988) and by Tang (1991).
BIOLOGY. Life history presented by Schmutterer (1954) (in Germany); Dziedzicka (1968) (in Poland); Kosztarab & Kozár (1988) (in Europe).

Parthenolecanium glandi (Kuwana)

Lecanium glandi Kuwana, 1907: 191.
Eulecanium glandi (Kuwana); Kuwana, 1917a: 19.
Parthenolecanium glandi (Kuwana); Borchsenius, 1957: 377.
TYPE DATA. Syntypes female, JAPAN: on apple, pear and other trees (ITLJ).
DISTRIBUTION. PALAEARCTIC REGION: Inner Mongolia, Japan,

Korea.
HOST PLANTS. Rosaceae: *Malus sylvestris, Pyrus communis.* **Salicaceae:** *Populus berolinensis.* **Ulmaceae:** *Zelkova serrata.*
REMARKS. Adult female redescribed and illustrated by Kuwana (1917a), Paik (1978), and by Tang & Li (1988). Adult female redescribed by Takahashi (1955b), Borchsenius (1957) and by Kawai (1980). Colour photograph given by Kawai (1980). Distribution and host plant records given by Takahashi (1955b), Borchsenius (1957), Paik (1978) and by Tang & Li (1988).

Parthenolecanium orientalis Borchsenius

Parthenolecanium corni orientalis Borchsenius, 1957: 369.
Parthenolecanium orientalis Borchsenius; Borchsenius, 1960: 215.
TYPE DATA. Syntypes female, CHINA (North East) and KOREA (North): on *Wisteria* sp., *Prunus* sp., *Salix* sp. and *Ribes* sp. (ZIAS).
DISTRIBUTION. PALAEARCTIC REGION: China, Korea.
HOST PLANTS. Grossulariaceae: *Ribes.* **Leguminosae:** *Wisteria chinensis.* **Rosaceae:** *Prunus.* **Salicaceae:** *Salix.*
REMARKS. Distribution and host plant records given by Borchsenius (1960).

Parthenolecanium persicae (Fabricius)

Chermes persicae Fabricius, 1776: 304.
Coccus persicorum Sulzer, 1776: 112. Syntypes female, GERMANY: on peach (probably lost). Syn. by Fernald, 1903: 191.
Coccus costatus Schrank, 1781: 589. Syntypes female, AUSTRIA: on twigs of *Amygdalus persicae* (DEPOSITORY UNKNOWN). Syn. by Fernald, 1903: 191.
Coccus clematidis Gmelin, 1790: 2220. Syntypes female, EUROPE: on *Clematis* (probably lost). Syn. by Lindinger, 1912: 363.
Coccus berberidis Schrank, 1801: 146. Syntypes female, AUSTRIA: on Sauerdorns [=*Berberis* sp.] (DEPOSITORY UNKNOWN). Syn. by Lindinger, 1912: 370.
Coccus persicae (Fabricius); Fonscolombe, 1834: 207.
Lecanium persicae (Fabricius); Bouche, 1844: 296.
Lecanium berberidis (Schrank); Walker, 1852: 1073.
Lecanium cymbiformis Targioni Tozzetti; Targioni Tozzetti, 1868: 730 [UNJUSTIFIED REPLACEMENT NAME].
Lecanium persicochilense Targioni Tozzetti; Targioni Tozzetti, 1868: 731 [UNJUSTIFIED REPLACEMENT NAME].
Lecanium elongatum Signoret, 1874a: 404. Syntypes female, FRANCE: Landes, Mont-de-Marsan, on laurier-cerise [=*Prunus laurocerasus*], (VMNH). Syn. by Ben-Dov, 1977: 91.
Lecanium genistae Signoret, 1874a: 405. Syntypes female, FRANCE: Alpes-Maritimes, on "genet epineux" (VMNH). Syn. by Lindinger, 1912: 370.
Lecanium mori Signoret, 1874a: 407. Syntypes female, FRANCE: Albertville and Savoie, on *Morus alba* (VMNH). Syn. by Sulc, 1932: 101.
Lecanium rosarum Snellen von Vollenhoven; Signoret, 1874a: 427 [MISIDENTIFICATION].
Lecanium sarothamni Douglas, 1891a: 65. Syntypes female, ENG-

LAND: Hereford, on *Sarothamnus scoparius* (BMNH). Syn. by Fernald, 1903: 192.
Lecanium (Eulecanium) mori Signoret; Cockerell, 1896b: 332.
Lecanium magnoliarum Cockerell, 1898e: 145. Syntypes female, U.S.A.: California, San Jose, on *Magnolia* sp. (USNM). Syn. by Sanders, 1909a: 441.
Lecanium (Eulecanium) magnoliarum Cockerell; Cockerell & Parrott, 1899: 236.
Lecanium (Eulecanium) berberidis (Schrank); Cockerell & Parrott, 1899: 237.
Coccus mori (Signoret); Kirkaldy, 1902: 106.
Eulecanium magnoliarum hortensiae Cockerell, 1903a: 19. Syntypes female, FRANCE: Nice, on *Hortensia* sp. (USNM). Syn. by Borchsenius, 1957: 351.
Coccus elongatus (Signoret); Fernald, 1903: 168.
Coccus genistae (Signoret); Fernald, 1903: 168.
Lecanium (Eulecanium) persicae (Fabricius); Reh, 1903: 409.
Eulecanium cecconi Leonardi, 1908: 178. Syntypes female, ITALY: Vallombrosa, on *Menispermum canadense* (IEAP). Syn. by Lindinger, 1912: 211.
Lecanium cecconi (Leonardi); Sanders, 1909b: 46.
Lecanium cymbiformis Targioni Tozzetti; Leonardi, 1920: 306 [MISSPELLING].
Lecanium (Parthenolecanium) persicae (Fabricius); Sulc, 1932: 75.
Palaeolecanium costatum (Schrank); Lindinger, 1935a: 136.
Palaeolecanium persicae (Fabricius); Lindinger, 1935a: 138.
Parthenolecanium persicae (Fabricius); Borchsenius, 1957: 350.
Lecanium persicae goidanichi Kawecki, 1962: 17. Syntypes adult female, first and second instar larva, ITALY: Western Alps, Cueno, on *Viscum album* and on *Pinus silvestris* (PASK). Syn. by Kosztarab & Kozár, 1988: 223.
Parthenolecanium thymi Danzig, 1967: 152. Holotype female, USSR: Primorye Territory, Zmeinaya Hill in Artemovka River valley, near Lesnoi Kordon, on *Thymus serphyllus* (ZIAS). Syn. by Danzig, 1980c: 272.
Lecanium persicae persicae (Fabricius); Kawecki, 1971: 258.
TYPE DATA. Syntypes female, EUROPE: on *Amygdalus persicae* (lost; Zimsen, 1964).
DISTRIBUTION. AUSTRALIAN REGION: New South Wales, Queensland, Victoria. **NEOTROPICAL REGION:** Argentina, Brazil Rio Grande do Sul, Chile. **NEW ZEALAND and PACIFIC REGION:** New Zealand. **NEARCTIC REGION:** Alabama, California, Canada Ontario, Idaho, Indiana, Maryland, Massachusetts, Mississippi, Missouri, New Jersey, New Mexico, New York, North Carolina, Ohio, Oregon, Pennsylvania, Rhode Island, South Carolina, Tennessee, Texas, Utah, Virginia. **ORIENTAL REGION:** India, Pakistan, Sri Lanka. **PALAEARCTIC REGION:** Afghanistan, Algeria, Austria, Bulgaria, Canary Islands, Corsica, Cyprus, Czechoslovakia, Denmark, Egypt, England, France, Greece, Hungary, Israel, Italy, Japan, Korea, Madeira, Morocco, Portugal, Romania, Spain, Switzerland, Turkey, USSR Caucasus, USSR Kazakhstan, USSR Moldavia, USSR Primorye Territory, USSR Sakhalin Island, USSR Tadzhikistan, USSR Ukraine, Yugoslavia.
HOST PLANTS. Anacardiaceae: *Mangifera indica*. **Araliaceae:**

Fatsia japonica. **Aspleniaceae:** *Asplenium nidum*. **Berberidaceae:** *Berberis canadensis*, *B. thunbergii*, *B. vulgaris*, *Mahonia aquifolium*. **Celastraceae:** *Euonymus japonicus*. **Ebenaceae:** *Diospyros kaki*, *D. lotus*. **Elaeagnaceae:** *Elaeagnus pungens*, *Hippophae rhamnoides*. **Euphorbiaceae:** *Ricinus communis*. **Hydrangeaceae:** *Hydrangea hortensis*. **Labiatae:** *Thymus serphyllum*. **Lauraceae:** *Persea americana*. **Leguminosae:** *Albizia julibrissin*, *Caragana*, *Cytisus hirsutus*, *Sophora*, *Wisteria sinensis*. **Menispermaceae:** *Menispermum canadense*. **Moraceae:** *Ficus carica*, *Morus alba*, *M. nigra*. **Oleaceae:** *Forsythia*, *Fraxinus excelsior*. **Ranunculaceae:** *Clematis vitalba*. **Rosaceae:** *Armeniaca vulgaris*, *Eriobotrya japonica*, *Persica vulgaris*, *Prunus domestica*, *P. laurocerasus*, *Pyracantha coccinea*, *Rosa*. **Rutaceae:** *Citrus aurantium*. **Tamaricaceae:** *Tamarix*. **Thymelaeaceae:** *Daphne*. **Ulmaceae:** *Ulmus thomasii*. **Vitidaceae:** *Ampelopsis*, *Vitis vinifera*.

REMARKS. Common Name - European Peach Scale. Adult female redescribed and illustrated by Signoret (1873b), Marchal (1908a), Froggatt (1915), Kuwana (1917a), Leonardi (1920) (as *E. cecconii*), Sulc (1932), Gómez-Menor Ortega (1937, 1958c), Gomes Costa (1949), Borchsenius (1957), Paik (1978), Kosztarab & Kozár (1978, 1988), Danzig (1967) (as *P. thymi*), Ezzat &Hussein (1969), Williams & Kosztarab (1972), Danzig (1980c), Kawai (1980), Tereznikova (1981), Hamon & Williams (1984), Gill (1988) and by Tang (1991). Larval instars described and illustrated by Green (1928b) and by Boratynski (1970). Colour photograph given by Kawai (1980), Kosztarab & Kozár (1988), Gonzalez (1989) and by Ben-Dov *et al.* (1991). Lindinger (1935a) regarded *Coccus costatus* a synonym of *Palaeolecanium bituberculatum*, but it is not accepted. Distribution and host plant records given by Signoret (1873b), Hofer (1903), Marchal (1908a), Leonardi (1920), Sulc (1932), Gomes Costa (1949), Takahashi & Tachikawa (1956), Boratynski & Williams (1964), Phillips (1965a), Ezzat & Hussein (1969), Corseuil & Barbosa (1971), Danzig (1967, 1972d, 1980c), Williams & Kosztarab (1972), Kozárzhevskaya & Reitzel (1975), Wise (1977), Georghiou (1977), Lambdin & Watson (1980), Tereznikova (1981), Vieira *et al.* (1983), Martin Mateo (1984), Hamon & Williams (1984), Kozár (1980, 1983, 1985), Varshney (1985), Carnero Hernandez & Perez Guerra (1986), Marotta (1987), Kozár & Ostafichuk (1987), Gill (1988), Kosztarab & Kozár (1988), Gonzalez (1989), Shafee *et al.* (1989), Nada *et al.* (1990), Tang (1991) and by Ben-Dov *et al.* (1991).

BIOLOGY. The female reproduces parthenogenetically in Europe (Kosztarab & Kozár, 1988), Virginia (Williams & Kosztarab, 1972), Chile (Gonzalez, 1989) and in Israel (Ben-Dov *et al.*, 1991). Develops one annual generation in Central Europe (Kosztarab & Kozár, 1988), Chile (Gonzalez, 1989) and in Israel (Ben-Dov *et al.*, 1991). Borchsenius (1957) reported two generations per year in Central Asia. Bazarov *et al.* (1975) studied the parasites in Tadzhikistan.

ECONOMIC IMPORTANCE. A minor pest in deciduous orchards in California (Gill, 1988), Europe (Kosztarab & Kozár, 1988) and in Israel (Ben-Dov *et al.*, 1991). However, upon its introduction into Western Australia about 1901 became a serious pest of vines and plums (Bartlett, 1978).

Parthenolecanium persicae spinosum (Brittin) n. comb.

Lecanium (Eulecanium) spinosum Brittin, 1940b: 420.
TYPE DATA. Syntypes female, NEW ZEALAND: Ngatea, on *Wisteria* sp. (NZAC).
DISTRIBUTION. NEW ZEALAND and PACIFIC REGION: New Zealand.
HOST PLANTS. Leguminosae: *Wisteria.*

Parthenolecanium pomeranicum (Kawecki)

Lecanium corni crudum Green; Green, 1930a: 14 [MISIDENTIFICATION in part].
Lecanium corni crudum Green; Gimmingham, 1934: 41 [MISIDENTIFICATION].
Lecanium corni crudum Green; Green, 1934: 109 [MISIDENTIFICATION].
Coccus corni crudum (Green); Ossiannilsson, 1951: 5 [MISIDENTIFICATION].
Eulecanium crudum (Green); Schmutterer, 1952b: 20 [MISIDENTIFICATION].
Lecanium pomeranicum Kawecki, 1954: 15.
Eulecanium taxi Habib, 1955a: 71. Holotype female, ENGLAND: Silwood Park, Sunninghill, Berks, on *Taxus baccata* (BMNH). Syn. by Boratynski & Williams, 1964: 107.
Parthenolecanium pomeranicum (Kawecki); Borchsenius, 1957: 372.
Eulecanium crudum (Green); Reyne, 1957: 14 [MISIDENTIFICATION].
TYPE DATA. Syntypes female, male and larva, POLAND: Bucze, Urle near Warsaw, Wierzchlas, Krakow, Dziwnow, and ENGLAND: all on *Taxus* spp. (PASK).
DISTRIBUTION. PALAEARCTIC REGION: Austria, Bulgaria, Czechoslovakia, England, France, Germany, Hungary, Italy, Netherlands, Poland, Romania, Sweden, Switzerland, USSR Ukraine.
HOST PLANTS. Cupressaceae: *Juniperus.* **Taxaceae:** *Taxus baccata, T. cuspidata.*
REMARKS. Common Name - Yew scale. Kawecki (1954) and Habib (1955a) presented a detailed account on pre-1954 misidentifications, by various authors, of this species. Adult female redescribed and illustrated by Schmutterer (1954) (as *E. crudum*), Habib (1956b) (as *E. taxi*), Borchsenius (1957), Dziedzicka (1968), Kosztarab & Kozár (1978, 1988) and by Tereznikova (1981). Male described and illustrated by Schmutterer (1954), Theron (1958) (as *E. taxi*) and by Giliomee (1967). Larval instars described and illustrated by Dziedzicka (1968). Habib (1956b) again described *E. taxi* as a new species, however the 1955a description takes priority. Bielenin (1962) described and illustrated the anatomy and histology of the female reproductive organs. Bielenin (1963) described and illustrated the alimentary canal of the male and female. Distribution and host plant records given by Reyne (1957), Borchsenius (1957), Ossiannilsson (1959), Giliomee (1967), Kozárzhevskaya & Reitzel (1975), Kozár *et*

al. (1977), Tereznikova (1981), Kozár (1985), Dziedzicka (1988), Kosztarab & Kozár (1988) and by Pellizzari Scaltriti (1991).
BIOLOGY. Life history presented by Schmutterer (1954), Habib (1956b), Dziedzicka (1968), and by Kosztarab & Kozár (1988).

Parthenolecanium pruinosum (Coquillett)

Lecanium pruinosum Coquillett, 1891: 382.
Lecanium robiniae Townsend, 1892a: 11. Syntypes female, U.S.A.: New Mexico, Las Cruces, on imported trees of black locust [=*Robinia pseudacacia*] (USNM). Syn. by Sanders, 1909a: 442.
Lecanium (Eulecanium) pruinosum Cocquillett; Cockerell, 1896b: 333.
Lecanium pruinosum pruinosum Coquillett; Tyrrell, 1896: 268.
Eulecanium robiniae (Townsend); Cockerell, 1902g: 178.
Eulecanium robiniae subsimile Cockerell, 1902g: 178. Syntypes female, MEXICO: Cerro Chilicote, on ash and *Rhus* sp. (USNM). Syn. by Sanders, 1909a: 442.
Parthenolecanium pruinosum (Coquillett); Nakahara, 1981a: 285.
TYPE DATA. Syntypes female, U.S.A.: California, Los Angeles, on apricot (USNM).
DISTRIBUTION. NEOTROPICAL REGION: Mexico. **NEARCTIC REGION:** California, Canada British Columbia, New Mexico.
HOST PLANTS. Leguminosae: *Robinia pseudacacia.* **Scrophularia-ceae:** *Veronica.*
REMARKS. Common Name - the frosted scale. Adult female redescribed and illustrated by Craw (1891), Tyrrell (1896), Gill (1988). Colour photograph given by Johnson & Lyon (1988). Distribution and host plant records given by Tyrrell (1896), Gill (1988) and by Kozár *et al.* (1989).

Parthenolecanium putmani (Phillips)

Lecanium putmani Phillips, 1965a: 234.
Parthenolecanium putmami (Phillips); Boratynski & Davies, 1971: 58 [MIS-SPELLING].
Parthenolecanium putmani (Phillips); Nakahara, 1981a: 285.
TYPE DATA. Holotype female, CANADA: Ontario, Jordan, on Japanese plum (CNOC).
DISTRIBUTION. NEARCTIC REGION: Canada Ontario.
HOST PLANTS. Carpinaceae: *Ostrya virginiana.* **Juglandaceae:** *Carya ovata, Juglans nigra.* **Lauraceae:** *Sassafras albidum.* **Leguminosae:** *Gleditsia triacanthos.* **Rosaceae:** *Prunus.* **Salica-ceae:** *Salix.*
REMARKS. Nur (1971, 1972) studied the cytology and reported chromosome number 2n=16 in Canada.
BIOLOGY. Phillips (1965a) separated this species from *P. cerasifex* on the basis of biological and behavioural attributes.

Parthenolecanium quercifex (Fitch)

Lecanium quercifex Fitch, 1859: 805.
Lecanium quercitronis Fitch, 1859: 805. Syntypes female, U.S.A.:

New York, Salem, on twigs of black oak (AMNY). Syn. by Sanders, 1909a: 445.

Lecanium antennatum Signoret, 1874a: 413. Syntypes female, U.S.A.: New York, on oak (VMNH). Syn. by Sanders, 1909a: 445.

Lecanium (Eulecanium) antennatum Signoret; Cockerell, 1896b: 332.

Lecanium pruinosum kermoides Tyrrell, 1896: 268. Syntypes female, and male, U.S.A.: California, on orange, oak and locust (USNM). Syn. by Sanders, 1909a: 445.

Lecanium (Eulecanium) quercitronis Fitch; Cockerell & Parrott, 1899: 232.

Lecanium (Eulecanium) quercifex Fitch; Cockerell & Parrott, 1899: 235.

Eulecanium quercifex (Fitch); King, 1901h: 315.

Eulecanium quercitronis (Fitch); King, 1901h: 315.

Eulecanium antennatum (Signoret); Fernald, 1903: 181.

Eulecanium quercitronis kermoides (Tyrrell); Fernald, 1903: 195.

Parthenolecanium quercifex (Fitch); Nakahara, 1981a: 285.

TYPE DATA. Syntypes female, U.S.A.: New York, Salem, on white oak (AMNY).

DISTRIBUTION. NEARCTIC REGION: California, Canada British Columbia, Florida, Indiana, New York, Tennessee, Virginia.

HOST PLANTS. Ebenaceae: *Diospyros virginiana.* **Fagaceae:** *Chrysolepis, Quercus alba, Q. coccinea, Q. nigra, Q. palustris, Q. phellos, Q. prinus, Q. rubra, Q. stellata, Q. velutina.* **Juglandaceae:** *Carya.* **Platanaceae:** *Platanus occidentalis.*

REMARKS. Adult female redescribed and illustrated by Thro (1903), Dietz & Morrison (1916), Hamon & Williams (1984) and by Gill (1988). Colour photograph given by Gill (1988). Distribution and host plant records given by Dietz & Morrison (1916), Williams & Kosztarab (1972), Lambdin & Watson (1980), Hamon & Williams (1984) and by Gill (1988).

Parthenolecanium rufulum (Cockerell)

Lecanium quercus (Linnaeus); Signoret, 1874a: 427 [MISIDENTIFICATION].

Eulecanium alni rufulum Cockerell, 1903a: 21.

Lecanium pulchrum Reh, 1903: 410. Syntypes female, FRANCE: Haute-Garonne, Terne-et-Garonne, Toulouse and Poitiers, on *Castanea vesca* (DEPOSITORY UNKNOWN). Syn. by Cockerell, 1922: 309.

Eulecanium pulchrum (King); Sanders, 1906: 9 [ERRONEOUS AUTHORSHIP].

Lecanium pulchrum King; Marchal, 1908a: 304 [ERRONEOUS AUTHORSHIP].

Palaeolecanium rufulum (Cockerell); Lindinger, 1932f: 184.

Lecanium (Pterolecanium) pulchrum King; Sulc, 1932: 53 [ERRONEOUS AUTHORSHIP].

Eulecanium pulchrum (Marchal); Schmutterer, 1952a: 548 [ERRONEOUS AUTHORSHIP].

Eulecanium pulchrum (King); Schmutterer, 1954: 72 [ERRONEOUS AUTHORSHIP].

Parthenolecanium rufulum (Cockerell); Borchsenius, 1957: 373.

TYPE DATA. Syntypes female, FRANCE: La Vienne, on *Carpinus* sp. (BMNH, USNM).
DISTRIBUTION. PALAEARCTIC REGION: Austria, Bulgaria, Czechoslovakia, England, France, Germany, Hungary, Italy, Poland, Romania, Spain, USSR Moldavia, USSR Ukraine, Yugoslavia.
HOST PLANTS. Carpinaceae: *Carpinus betulus.* **Corylaceae:** *Corylus avellana.* **Ebenaceae:** *Diospyros kaki.* **Ericaceae:** *Vaccinium myrtillus.* **Fagaceae:** *Castanea sativa, C. vesca, Quercus cerris, Q. lusitanica, Q. petraea, Q. pubescens, Q. robur, Q. sessilis, Q. toza.* **Leguminosae:** *Robinia pseudacacia.* **Rosaceae:** *Crataegus oxyacantha, Rosa, Rubus.* **Ulmaceae:** *Ulmus.*
REMARKS. Ben-Dov (1976b) erroneously synonymized this species with *Lecanium wistariae* Signoret, however Danzig (1980a) re-examined the type-series of *L. wistariae,* found that it is a synonym of *P. corni* and resurrected *P. rufulum* as a distinct species. Adult female redescribed and illustrated (as *P. pulchrum*) by Marchal (1908a), Sulc (1932), Schmutterer (1954), Borchsenius (1957), Gómez-Menor Ortega (1960) and by Savescu (1961). Adult female redescribed and illustrated (as *P. rufulum*) by Boratynski & Williams (1964), Tereznikova (1981), Kosztarab & Kozár (1978, 1988). Adult femaleredescribed by Tang (1991). Colour photograph given by Kosztarab & Kozár (1988). Larval instars described and illustrated by Schmutterer (1954) and by Dziedzicka (1968). Distribution and host plant records given by Sulc (1932), Schmutterer (1952a, 1954), Borchsenius (1957), Gómez-Menor Ortega (1960), Boratynski & Williams (1964), Komosinska (1977), Kozár et al. (1977), Tereznikova (1981), Koteja (1983), Martin Mateo (1984), Kozár (1985), Kozár & Ostafichuk (1987), Marotta (1987), Dziedzicka (1988), Kosztarab & Kozár (1988), Longo & Russo (1990), Kozár & Drozdják (1990) and by Tang (1991).
BIOLOGY. Life history presented by Schmutterer (1954), Dziedzicka (1968), and by Kosztarab & Kozár (1988).

Parthenolecanium smreczynskii (Kawecki)

Lecanium smreczynskii Kawecki, 1967: 687.
Parthenolecanium smreczynskii (Kawecki); Kosztarab & Kozár, 1988: 227.
TYPE DATA. Syntypes female, POLAND: Poronin near Zakopane, on *Thymus* sp. (ZAWP).
DISTRIBUTION. PALAEARCTIC REGION: Poland.
HOST PLANTS. Labiatae: *Thymus.*
REMARKS. Adult female redescribed by Kosztarab & Kozár (1988).

Parthenolecanium tamaricis (Bodenheimer)

Eulecanium tamaricis Bodenheimer, 1953: 103.
Parthenolecanium tamaricis (Bodenheimer); Borchsenius, 1957: 370.
TYPE DATA. Syntypes female, TURKEY: at the banks of Kizil Irmak River near Kalecik, on *Tamarix pallasii* (lost; Ben-Dov & Harpaz, 1985).
DISTRIBUTION. PALAEARCTIC REGION: Turkey.
HOST PLANTS. Tamaricaceae: *Tamarix pallasii.*

Pendularia Fonseca

Pendularia Fonseca, 1927: 268.
TYPE-SPECIES: *Pendularia pendens* Fonseca, by original designation and monotypy.
REMARKS. The original description was reprinted in Fonseca (1929). MacGregor (1981) discussed the genus and distinguished it from *Takahashia*.

Pendularia jaliscensis (Cockerell & Cockerell)

Takahashia jaliscensis Cockerell & Cockerell in Cockerell, 1902j: 466.
Pendularia jaliscensis (Cockerell); MacGregor, 1981: 305 [ERRONEOUS AUTHORSHIP].
TYPE DATA. Syntypes female, MEXICO: Jalisco, Barranca de Atenquique, on "copal" (? *Rhus* sp.) (BMNH, USNM).
DISTRIBUTION. NEOTROPICAL REGION: Mexico.
HOST PLANTS. Anacardiaceae: *Rhus*.
REMARKS. Adult female redescribed and illustrated by MacGregor (1981).

Pendularia pendens Fonseca

Pendularia pendens Fonseca, 1927: 268.
Takahashia pendens (Fonseca); Costa Lima, 1942: 245.
TYPE DATA. Syntypes female, BRAZIL: Sao Paulo, on *Eugenia jaboticaba* (IBSP).
DISTRIBUTION. NEOTROPICAL REGION: Brazil Sao Paulo.
HOST PLANTS. Myrtaceae: *Eugenia jaboticaba*.
REMARKS. The original description was reprinted in Fonseca (1929) including slightly different photographs. Adult female redescribed and illustrated by MacGregor (1981).

Perilecanium Fonseca

Perilecanium Fonseca, 1962: 13.
TYPE-SPECIES: *Lecanium transparens* Hempel, by original designation.

Perilecanium ocultus Fonseca

Perilecanium ocultus Fonseca, 1962: 15.
TYPE DATA. Syntypes female, BRAZIL: Sao Paulo, Parque Siqueira, Campos, on *Metrodorea stipulata* (MZSP).
DISTRIBUTION. NEOTROPICAL REGION: Brazil Sao Paulo.
HOST PLANTS. Flacourtiaceae: *Casearia sylvestris*. **Rutaceae:** *Metrodorea stipulata*.

Perilecanium transparens (Hempel)

Lecanium transparens Hempel, 1937: 12.
Perilecanium transparens (Hempel); Fonseca, 1962: 13.
TYPE DATA. Syntypes female, BRAZIL: Sao Paulo, Capital, jardim da Luz, on leaves of undetermined tree (MZSP).
DISTRIBUTION. NEOTROPICAL REGION: Brazil Sao Paulo.

REMARKS. Adult female redescribed and illustrated by Fonseca (1962).

Perilecanium urbanus Fonseca

Perilecanium urbanus Fonseca, 1969: 14.
TYPE DATA. Syntypes female, BRAZIL: Sao Paulo, Jardim da Luz, on *Cassaria [=Casearia] sylvestris* (MZSP).
DISTRIBUTION. NEOTROPICAL REGION: Brazil Sao Paulo.
HOST PLANTS. Flacourtiaceae: *Casearia sylvestris*.

Philephedra Cockerell

Pulvinaria (Philephedra) Cockerell, 1898h: 24.
Philephedra Cockerell; Cockerell, 1902f: 451.
TYPE-SPECIES: *Pulvinaria (Philephedra) ephedrae* Cockerell, by original designation.
REMARKS. Generic characters discussed by Steinweden (1929), Nakahara & Gill (1985) and by Gill (1988). Key to species given by Nakahara & Gill (1985).

Philephedra broadwayi (Cockerell)

Pulvinaria broadwayi Cockerell, 1896c: 306.
Philephedra theobromae Green; Green, 1916c: 377. Lectotype female designated by Nakahara & Gill (1985). TRINIDAD: on cacao pods (BMNH). Syn. by Nakahara & Gill, 1985: 7.
Philephedra broadwayi (Cockerell); Ballou, 1922: 74.
TYPE DATA. Lectotype female designated by Nakahara & Gill (1985). GRENADA: Broadway, on undetermined plant (USNM).
DISTRIBUTION. NEOTROPICAL REGION: Colombia, Dominica, Grenada, Nevis, Tobago, Trinidad.
HOST PLANTS. Anacardiaceae: *Mangifera indica, Spondias*. **Annonaceae:** *Annona muricata*. **Euphorbiaceae:** *Aleurites triloba*. **Solanaceae:** *Solanum melongena*. **Sterculiaceae:** *Theobroma bicolor, T. cacao*.
REMARKS. Adult female redescribed and illustrated by Nakahara & Gill (1985). Distribution and host plant records given by Green (1916c), Ballou (1922) and by Nakahara & Gill (1985).

Philephedra broadwayi echinopsidis (Newstead)

. comb.
Pulvinaria broadwayi echinopsidis Newstead, 1920: 187.
TYPE DATA. Syntypes female, GUYANA: Georgetown, Botanic Gardens, on *Echinopsis latiflora* (BMNH).
DISTRIBUTION. NEOTROPICAL REGION: Guyana.
HOST PLANTS. Cactaceae: *Echinopsis latiflora*.

Philephedra colimensis (Cockerell)

Lichtensia colimensis Cockerell, 1902j: 467.
Philephedra colimensis (Cockerell); Nakahara & Gill, 1985: 10.
TYPE DATA. Lectotype female designated by Nakahara & Gill (1985). MEXICO: Colima, on *Celtis?* (USNM).
DISTRIBUTION. NEOTROPICAL REGION: Mexico.

HOST PLANTS. Ulmaceae: *Celtis?*.
REMARKS. Adult female redescribed and illustrated by Nakahara & Gill (1985).

Philephedra crescentiae (Cockerell)

Lichtensia crescentiae Cockerell, 1898f: 435.
Philephedra crescentiae (Cockerell); Nakahara & Gill, 1985: 13.
TYPE DATA. Lectotype female designated by Nakahara & Gill (1985),
    MEXICO: Frontera, Tabasco, on *Crescentia* sp. (USNM).
DISTRIBUTION. NEOTROPICAL REGION: Mexico.
HOST PLANTS. Bignoniaceae: *Crescentia*.
REMARKS. Adult female redescribed and illustrated by Nakahara & Gill (1985).

Philephedra ephedrae (Cockerell)

Pulvinaria ephedrae Cockerell, 1898h: 24.
Pulvinaria (Philephedra) ephedrae Cockerell; Cockerell, 1898h: 24.
Philephedra ephedrae (Cockerell); Cockerell, 1902f: 451.
TYPE DATA. Lectotype female designated by Nakahara & Gill (1985),
    U.S.A.: New Mexico, Mesilla Park, on *Ephedra* sp. (USNM).
DISTRIBUTION. NEARCTIC REGION: California, New Mexico.
HOST PLANTS. Ephedraceae: *Ephedra*.
REMARKS. Adult female redescribed and illustrated by Nakahara & Gill (1985) and by Gill (1988).

Philephedra floridana Nakahara & Gill

Philephedra floridana Nakahara & Gill, 1985: 19.
TYPE DATA. Holotype female, U.S.A.: Florida, Ft. Pierce, on *Conocarpus erectus* (USNM).
DISTRIBUTION. NEARCTIC REGION: Florida.
HOST PLANTS. Combretaceae: *Conocarpus erectus*.

Philephedra lutea (Cockerell)

Pulvinaria lutea Cockerell, 1893m: 51.
Lichtensia lutea (Cockerell); Cockerell, 1895l: 255.
Philephedra lutea (Cockerell); Nakahara & Gill, 1985: 22.
TYPE DATA. Lectotype female designated by Nakahara & Gill (1985),
    MEXICO: Vera Cruz, on croton (USNM).
DISTRIBUTION. NEOTROPICAL REGION: Guatemala, Mexico,
    Texas.
HOST PLANTS. Euphorbiaceae: *Codiaeum variegatum*. **Lauraceae:**
    Persea americana. **Magnoliaceae:** *Magnolia*. **Moraceae:** *Ficus*.
REMARKS. Adult female redescribed and illustrated by Nakahara & Gill (1985). Distribution and host plant records given by Nakahara & Gill (1985).

Philephedra mimosae (Townsend & Cockerell)

Lichtensia mimosae Townsend & Cockerell, 1898: 175.
Philephedra mimosae (Townsend & Cockerell); Nakahara & Gill, 1985: 25.

TYPE DATA. Lectotype female designated by Nakahara & Gill (1985), MEXICO: Las Minas, Tab., on *Mimosa apotala* (USNM).
DISTRIBUTION. NEOTROPICAL REGION: Mexico.
HOST PLANTS. Leguminosae: *Mimosa apotala.*
REMARKS. Adult female redescribed and illustrated by Nakahara & Gill (1985).

Philephedra parvula (Cockerell)

Pulvinaria parvula Cockerell, 1899g: 19.
Lichtensia parvula (Cockerell); Cockerell, 1902j: 467.
Lichtensia zapotlana Cockerell; Cockerell, 1902j: 467. Lectotype female designated by Nakahara & Gill (1985), MEXICO: Zapotlan, Jalisco, on leguminous shrub (USNM). Syn. by Nakahara & Gill, 1985: 28.
Lichtensia zapotlana townsendi Cockerell, 1903b: 162. Lectotype female designated by Nakahara & Gill (1985), MEXICO: Ameria, Colima, on copal [=*Bursera jorullensis*] (USNM). Syn. by Nakahara & Gill, 1985: 28.
Filippia parvula (Cockerell); Steinweden, 1929: 238.
Filippia zapotlana (Cockerell); Steinweden, 1929: 238.
Philephedra parvula (Cockerell); Nakahara & Gill, 1985: 28.
TYPE DATA. Lectotype female designated by Nakahara & Gill (1985), MEXICO: Cuautla, on *Mimosa* sp. (USNM).
DISTRIBUTION. NEOTROPICAL REGION: Mexico. **NEARCTIC REGION:** Arizona.
HOST PLANTS. Burseraceae: *Bursera jorullensis.* **Leguminosae:** *Acacia constricta, Caesalpinia gilliesi, Mimosa, Prosopis.*
REMARKS. Adult female redescribed and illustrated by Nakahara & Gill (1985). Distribution and host plant records given by Nakahara & Gill (1985).

Philephedra tuberculosa Nakahara & Gill

Philephedra tuberculosa Nakahara & Gill, 1985: 31.
TYPE DATA. Holotype female, U.S.A.: Texas, La Feria, on *Schinus terebinthifolius* (USNM).
DISTRIBUTION. NEOTROPICAL REGION: Colombia, Costa Rica, Guatemala, Mexico, Nicaragua, Venezuela. **NEARCTIC REGION:** Florida, Texas.
HOST PLANTS. Anacardiaceae: *Mangifera indica, Schinus terebinthifolius.* **Annonaceae:** *Annona reticulata, A. squamosa.* **Bignoniaceae:** *Tabebuia pallida.* **Bombacaceae:** *Chorisia speciosa.* **Burseraceae:** *Bursera simaruba.* **Caricaceae:** *Carica papaya.* **Combretaceae:** *Bucida buceras, B. spinosa, Conocarpus erectus.* **Compositae:** *Ambrosia artemisifolia, Dahlia, Helianthus, Solidago chapmanii.* **Convolvulaceae:** *Ipomoea prescarpae.* **Euphorbiaceae:** *Acalypha wilkesiana, Codiaeum variegatum, Euphorbia heterophylla.* **Flacourtiaceae:** *Dovyalis hebecarpa.* **Guttiferae:** *Calophyllum brasiliense, C. inophyllum.* **Labiatae:** *Rosmarinus officinalis.* **Leguminosae:** *Bauhinia purpurea, Cassia fistulata, Gliricidia sepium, Lysiloma bahamensis, L. latisiliqua, Parkinsonia aculeata, Pithecellobium guadalupense, Pongamia pinnata.* **Malvaceae:** *Gossypium, Hibiscus acetosella, Malvaviscus arboreus.* **Moraceae:** *Ficus carica, F. rubiginosa, F. triangularis.*

Myrtaceae: *Psidium guajava.* **Polygonaceae:** *Coccoloba diversifo-lia, C. uvifera.* **Portulacaceae:** *Portulaca.* **Rutaceae:** *Amyris elemifera, Citrus.* **Sapotaceae:** *Chrysophyllum oliviforme, Mastichodendron foetidissimum, Mimusops elengi, Mimusops roxburghiana, Pouteria sapota.* **Zingiberaceae:** *Zingiber.*
REMARKS. Male test described and illustrated by Miller & Williams (1990).
BIOLOGY. Life history, behaviour and natural enemies in Florida studied by Pena *et al.* (1987).
ECONOMIC IMPORTANCE. A pest of papaya and Annona in Florida (Pena *et al.*, 1987).

Phyllostroma Sulc

Phyllostroma Sulc, 1942a: 5.
TYPE-SPECIES: *Pulvinaria ericae* Löw, by original designation and monotypy.
REMARKS. Generic characters discussed by Borchsenius (1957), Kosztarab & Kozár (1978, 1988) and by Tang (1991).

Phyllostroma myrtilli (Kaltenbach)

Lecanium myrtilli Kaltenbach, 1874: 420.
Pulvinaria ericae Löw, 1883: 115. Syntypes female, AUSTRIA: Weissenbach, Rauheneck, on *Erica carnea* (DEPOSITORY UNKNOWN). Syn. by Borchsenius, 1957: 281.
Phyllostroma ericae (Löw); Sulc, 1942a: 5.
TYPE DATA. Syntypes female and larva, GERMANY: on *Vaccinium myrtillus* (DEPOSITORY UNKNOWN).
DISTRIBUTION. PALAEARCTIC REGION: Austria, Czechoslovakia, Germany, Hungary, Italy, Poland, Romania, USSR Latvia, USSR Leningrad Region, USSR Ukraine.
HOST PLANTS. Ericaceae: *Erica arborea, E. carnea, Vaccinium myrtillus, V. vitisidea.*
REMARKS. Adult female redescribed and illustrated by Sulc (1942a, 1942b), Borchsenius (1957), Kosztarab & Kozár (1978, 1988), Tereznikova (1963a, 1981) and by Tang (1991). Male described and illustrated by Giliomee (1967). Fernald (1903: 224) placed *L. myrtilli* as a junior synonym of *Chionaspis salicis* (Linnaeus) in the Diaspididae, but it was rejected by subsequent workers. Distribution and host plant records given by Sulc (1942a), Borchsenius (1957), Giliomee (1967), Kozár *et al.* (1977), Tereznikova (1981), Koteja (1983), Marotta (1987), Kosztarab & Kozár (1988), Marotta & Tranfaglia (1990) and by Tang (1991).

Physokermes Targioni Tozzetti

Physokermes Targioni Tozzetti, 1868: 734.
Physochermes Targioni Tozzetti; Targioni Tozzetti, 1869: 259 [MIS-SPELLING].
Physockermes Leonardi; Leonardi, 1920: 310 [MIS-SPELLING].
Physocermes Atanasov; Atanasov, 1959: 429 [MIS-SPELLING].
TYPE-SPECIES: *Coccus hemicryphus* Dalman, by subsequent restriction of the genus to this species by Signoret (1874b).
REMARKS. Generic characters discussed by Ferris (1920b), Leonar-

di (1920), Steinweden (1929), Borchsenius (1957), Danzig (1967, 1972b, 1980c), Williams & Kosztarab (1972), Kosztarab & Kozár (1978, 1988), Tereznikova (1981), Yang (1982), Hadzibejli (1983), Gill (1988) and by Tang (1991). Key to species: ASIA - Tang (1991). EUROPE - Kosztarab & Kozár (1988). CALIFORNIA - Gill (1988). HUNGARY - Kosztarab & Kozár (1978). UKRAINE - Tereznikova (1981). USSR - Borchsenius (1957), Danzig (1964).

Physokermes coloradensis Cockerell

Physokermes coloradensis Cockerell, 1895f: 101.
TYPE DATA. Syntypes female, U.S.A.: Colorado, Manitou, on *Pinus edulis* (USNM).
DISTRIBUTION. NEARCTIC REGION: Colorado.
HOST PLANTS. Pinaceae: *Pinus edulis.*

Physokermes concolor Coleman

Physokermes concolor Coleman, 1903: 73.
TYPE DATA. Syntypes female, U.S.A.: California, Salmon Mts., Siskiyon County, on *Abies concolor* (UCD).
DISTRIBUTION. NEARCTIC REGION: California, Canada British Columbia, Colorado.
HOST PLANTS. Pinaceae: *Abies concolor, Abies lasiocarpa, Picea sitchensis, Pinus ponderosa.*
REMARKS. Adult female redescribed and illustrated by Gill (1988). Colour photograph given by Gill (1988). Distribution and host plant records given by Cockerell (1910b), Gill (1988) and by Kozár *et al.* (1989).

Physokermes fasciatus Borchsenius

Physokermes fasciatus Borchsenius, 1957: 445.
TYPE DATA. Syntypes female, and first instar larva, USSR: Kazakhstan, Dzhungarsk Altay, Mali Boskan gorge, on *Picea* sp. (ZIAS).
DISTRIBUTION. PALAEARCTIC REGION: USSR Tadzhikistan.
HOST PLANTS. Pinaceae: *Picea.*
REMARKS. Adult female redescribed by Tang (1991).

Physokermes hemicryphus (Dalman)

Coccus hemicryphus Dalman, 1826: 369.
Physokermes hemicryphus (Dalman); Targioni Tozzetti, 1868: 734.
Physokermes hemicriphus (Dalman); Danzig, 1967: 169 [MIS-SPELLING].
TYPE DATA. Syntypes female, SWEDEN: on *Abies* sp. (DEPOSITORY UNKNOWN).
DISTRIBUTION. NEARCTIC REGION: California, Canada British Columbia, Virginia. **PALAEARCTIC REGION:** Austria, Bulgaria, Czechoslovakia, Germany, Greece, Hungary, Italy, Mongolia, Poland, Romania, Sweden, USSR Crimea, USSR Moldavia, USSR Ukraine, Yugoslavia.
HOST PLANTS. Pinaceae: *Abies alba, Abies pectinata, Picea abies, P. engelmanni, P. excelsa, P. glauca, P. obovata, P. pungens, P. sitchensis.*
REMARKS. Adult female redescribed and illustrated by Schmutterer

(1956), Kosztarab & Kozár (1978, 1988), Tereznikova (1981) and by Gill (1988). Adult female redescribed by Tang (1991). Larval instars described by Schmutterer (1956) and by Tereznikova (1981). Colour photograph given by Williams & Kosztarab (1972) and by Gill (1988), indicated that this species was misidentified in the U. S. A. for many years as *P. piceae*. Nur (1979) studied the cytology and reported chromosome number 2n-18 and 27. Distribution and host plant records given by Signoret (1874b), Schmutterer (1956), Borchsenius (1957), Danzig (1967, 1972c), Williams & Kosztarab (1972), Kozár *et al.* (1977), Tereznikova (1981), Koteja (1983), Dziedzicka (1988), Kozár (1985), Gill (1988), Santas (1988) and by Kosztarab & Kozár (1988).

BIOLOGY. The female reproduces parthenogenetically (Schmutterer, 1956), developing one annual generation. Life history presented by Schmutterer (1956), Santas (1988) and by Kosztarab & Kozár (1988). Peachhacker (1976, 1977) attempted to forecast the honeydew flow, as a source for bee honey production, by sampling populations of the overwintering first instar larvae. Santas (1988) evaluated its role in apiculture in Greece.

Physokermes inopinatus Danzig & Kozár

Physokermes inopinatus Danzig & Kozár, 1973: 832.
TYPE DATA. Holotype female, HUNGARY: Csopak, on northern shores of Lake Balaton, on *Picea abies* (HNHB).
DISTRIBUTION. PALAEARCTIC REGION: Hungary.
HOST PLANTS. Pinaceae: *Picea abies*.
REMARKS. Adult female redescribed and illustrated by Tereznikova (1981) and by Kosztarab & Kozár (1978, 1988). Colour photograph given by Kosztarab & Kozár (1988).

Physokermes insignicola (Craw)

Lecanium insignicola Craw, 1894: 14.
Physokermes insignicola (Craw); Cockerell, 1895r: 258.
TYPE DATA. Syntypes female, U.S.A.: California, San Francisco (Golden Gate Park) and Oakland, on *Pinus insignis* (USNM).
DISTRIBUTION. NEARCTIC REGION: California.
HOST PLANTS. Pinaceae: *Pinus insignicola, P. muricata, P. radiata*.
REMARKS. Adult female redescribed and illustrated by Tyrrell (1896) and by Gill (1988). Colour photograph given by Gill (1988). Distribution and host plant records given by Tyrrell (1896) and by Gill (1988).

Physokermes jezoensis Siraiwa

Physokermes jezoensis Siraiwa, 1939: 64.
TYPE DATA. Syntypes female, USSR: Southern Sakhalin, from Odomari, Toyohara, Esutori, Tomarioru, Maoka, Honto and Naihoro, on *Picea jezoensis* (probably lost; S. Takagi, 1989, personal communication).
DISTRIBUTION. PALAEARCTIC REGION: USSR Kunashir Island, USSR Primorye Territory, USSR Sakhalin Island.
HOST PLANTS. Pinaceae: *Picea ajanensis, P. glehnii, P. jezoensis, P. korajensis*.

REMARKS. Adult female redescribed and illustrated by Danzig (1967, 1980c). Adult female redescribed by Borchsenius (1957), Kawai (1980) and by Tang (1991). Colour photograph given by Kawai (1980). Distribution and host plant records given by Borchsenius (1957), Danzig (1967, 1978c) and by Kawai (1980).

Physokermes piceae (Schrank)

Coccus piceae Schrank, 1801: 146.
Coccus racemosus Ratzeburg, 1843: 204. Syntypes female, GERMA-NY: Harz, Thuringia and Sclesiens, on pine (DEPOSITORY UNKNOWN). Syn. by Signoret, 1877a: 673.
Lecanium racemosum (Ratzeburg); Walker, 1852: 1072.
Lecanium piceae (Schrank); Walker, 1852: 1073.
Physokermes racemosus (Ratzeburg); Targioni Tozzetti, 1868: 41.
Lecanium piceae (Schrank); Signoret, 1874a: 409.
Physokermes piceae (Schrank); Fernald, 1903: 208.
Physokermes latipes Borchsenius, 1957: 445. Syntypes female, USSR: White Russia, Byelovezh, on *Picea excelsa* (ZIAS) Syn. by Danzig, 1967: 169.
Physocermes piceae (Schrank); Atanasov, 1959: 429 [MIS-SPELLING].
TYPE DATA. Syntypes female, AUSTRIA: on spruce (=Fichte) (VMNH).
DISTRIBUTION. PALAEARCTIC REGION: Austria, Bulgaria, Czechoslovakia, Denmark, France, Germany, Hungary, Italy, Mongolia, Netherlands, Poland, Romania, Switzerland, Turkey, USSR Byelorussia, USSR Kazakhstan, USSR Ukraine, Yugoslavia.
HOST PLANTS. Pinaceae: *Picea abies*, *P. excelsa*, *P. obovata*, *P. orientalis*, *P. pungens*.
REMARKS. In several publications before 1954, this species was named *Physokermes abietis* (Geoffroy, 1762). Opinion 228 (1954) ruled that binomens published by Geoffroy (1762) are not available for nomenclatorial purposes, consequently *P. piceae* (Schrank) became the senior synonym. Adult female redescribed and illustrated by Leonardi (1920) (as *P. abietis* Geoffroy), Schmutterer (1956), Borchsenius (1957), Savescu (1961), Kozár (1972a) and by Kosztarab & Kozár (1978, 1988). Adult female redescribed by Tang (1991). Male described and illustrated by Jancke (1955), Schmutterer (1956) and by Gillomee (1967). Larval instars described and illustrated by Schmutterer (1956). Distribution and host plant records given by Leonardi (1920), Schmutterer (1956), Borchsenius (1957), Reync (1957), Atanasov (1959), Gillomee (1967), Matesova (1968), Kozár (1972a, 1980, 1985), Kozárzhevskaya & Reitzel (1975), Danzig (1977b), Tereznikova (1981), Marotta (1987), Dziedzicka (1988) and by Kosztarab & Kozár (1988).
BIOLOGY. Life history presented by Schmutterer (1956) and by Kosztarab & Kozár (1988).

Physokermes shanxiensis Tang

Physokermes shanxiensis Tang, 1991: 202.
TYPE DATA. Holotype female, CHINA: Shanxi, Tai-yuan city, on *Picea asparata* (EISC).
DISTRIBUTION. PALAEARCTIC REGION: China.

HOST PLANTS. Pinaceae: *Picea asparata.*

Physokermes sugonjaevi Danzig

Physokermes sugonjaevi Danzig, 1972b: 274.

TYPE DATA. Holotype female, USSR: East Sayan, Tunkinsk Range, Arshan, on *Picea obovata* (ZIAS).

DISTRIBUTION. PALAEARCTIC REGION: Mongolia, USSR Altai, USSR East Sayan, USSR Irkutsk Region, USSR Kazakhstan, USSR Yakutsk.

HOST PLANTS. Pinaceae: *Picea obovata.*

REMARKS. Adult female redescribed by Tang (1991). Distribution and host plant records given by Danzig (1977b, 1978b) and by Tang (1991).

Physokermes taxifoliae Coleman

Physokermes taxifoliae Coleman, 1903: 72.

TYPE DATA. Syntypes female, U.S.A.: California, Santa Ana County, Stevens Creek, on *Pseudotsuga taxifolia* (UCD).

DISTRIBUTION. NEARCTIC REGION: California, Canada British Columbia.

HOST PLANTS. Pinaceae: *Pseudotsuga menziesii, P. taxifolia.*

REMARKS. Adult female redescribed and illustrated by Gill (1988). Colour photograph given by Gill (1988).

Platinglisia Cockerell

Platinglisia Cockerell, 1899a: 12.

TYPE-SPECIES: *Platinglisia noacki* Cockerell, by original designation and monotypy.

Platinglisia noacki Cockerell

Platinglisia noacki Cockerell, 1899a: 12.

TYPE DATA. Syntypes female, BRAZIL: Campinas, on leaves of a myrtaceous tree or shrub (BMNH, USNM).

DISTRIBUTION. NEOTROPICAL REGION: Brazil Sao Paulo.

HOST PLANTS. Aquifoliaceae: *Ilex.* **Begoniaceae:** *Begonia.* **Euphorbiaceae: Lauraceae:** *Laurus.* **Leguminosae:** *Inga edulis.* **Magnoliaceae:** *Magnolia pumila.* **Myrtaceae:** *Eugenia.* **Proteaceae:** *Grevillea robusta.* **Thymelaeaceae:**

REMARKS. Adult female redescribed by Hempel (1900b). Distribution and host plant records given by Hempel (1900b) and by Silva *et al.* (1968).

Platylecanium Cockerell & Robinson

Platylecanium Cockerell & Robinson, 1915b: 427.

Platilecanium Danzig & Konstantinova; Danzig & Konstantinova, 1990: 44 [MIS-SPELLING].

Platylecarium Tang; Tang, 1991: 135 [MIS-SPELLING].

TYPE-SPECIES: *Neolecanium cribrigerum* Cockerell & Robinson, by original designation.

REMARKS. Generic characters discussed by Morrison (1920), Takagi

(1975), Williams & Watson (1990) and by Tang (1991). Key to species: ASIA - Tang (1991). MALAYSIA - Takahashi (1950b).

Platylecanium asymmetricum Morrison

Platylecanium asymmetricum Morrison, 1921: 667.
Platylecanium assymmetricum Morrison; Takahashi, 1950b: 57 [MISSPELLING].
TYPE DATA. Syntypes female, male and larva, SINGAPORE: Government Hill, on *Pinanga* sp. (USNM).
DISTRIBUTION. AUSTRO ORIENTAL REGION: Malaysia, Singapore.
HOST PLANTS. Lauraceae: *Cinnamomum.* **Palmae:** *Pinanga.*
REMARKS. Adult female redescribed and illustrated by Tang (1991). Distribution and host plant records given by Takahashi (1950b) and by Tang (1991).

Platylecanium cappari (Froggatt) n. comb.

Lecanium cappari Froggatt, 1915: 604.
TYPE DATA. Syntypes female, AUSTRALIA: New South Wales, on *Capparis mitchelli* (BCRI).
DISTRIBUTION. AUSTRALIAN REGION: New South Wales.
HOST PLANTS. Capparidaceae: *Capparis mitchelli.*
REMARKS. This new combination introduced here is based on information kindly forwarded to me, on March 1992, by Penny Gullan and Ting Kui Qin (Canberra, Australia) who examined the syntypes.

Platylecanium citri Takahashi

Platylecanium citri Takahashi, 1942a: 23.
TYPE DATA. Syntypes female, THAILAND: Bangkok Noi, on *Citrus* sp. (IMZT).
DISTRIBUTION. ORIENTAL REGION: Thailand.
HOST PLANTS. Rutaceae: *Citrus.*
REMARKS. Adult female redescribed by Tang (1991).

Platylecanium cocotis Laing

Platylecanium cocotis Laing, 1925: 59.
TYPE DATA. Lectotype female designated by Williams & Watson (1990), VANUATU [-New Hebrides]: Efaate, on coconut (BMNH).
DISTRIBUTION. AUSTRO ORIENTAL REGION: Papua New Guinea, Solomon Islands, Vanuatu.
HOST PLANTS. Palmae: *Cocos nucifera.* **Zingiberaceae:** *Alpinia purpurata.*
REMARKS. Adult female redescribed and illustrated by Williams & Watson (1990). Adult female redescribed by Tang (1991). Distribution and host plant records given by Williams & Watson (1990) and by Tang (1991).

Platylecanium cribrigerum (Cockerell & Robinson)

Neolecanium cribrigerum Cockerell & Robinson, 1915a: 110.
Platylecanium cribrigerum (Cockerell & Robinson); Cockerell &

Robinson, 1915b: 427.
TYPE DATA. Syntypes female, PHILIPPINES: Los Banos, on *Piper loheri* (BMNH, USNM).
DISTRIBUTION. AUSTRO ORIENTAL REGION: Philippines.
HOST PLANTS. Piperaceae: *Piper loheri.*

Platylecanium cyperi Takahashi

Platylecanium cyperi Takahashi, 1950b: 59.
Platylecarium cyperi Takahashi; Tang, 1991: 135 [MIS-SPELLING].
TYPE DATA. Syntypes female, MALAYSIA: Kuala Lumpur, on plant of Cyperaceae (SMKM).
DISTRIBUTION. AUSTRO ORIENTAL REGION: Malaysia.
HOST PLANTS. Cyperaceae:
REMARKS. Adult female redescribed by Tang (1991).

Platylecanium elongatum Takahashi

Platylecanium elongatum Takahashi, 1951b: 104.
TYPE DATA. Syntypes female, INDONESIA: Riau [=Riouw] Islands, Rempang, on wild palm (HUSJ).
DISTRIBUTION. AUSTRO ORIENTAL REGION: Indonesia.
HOST PLANTS. Palmae.
REMARKS. Adult female redescribed by Tang (1991).

Platylecanium fusiforme Green

Lecanium (Platylecanium) fusiforme Green; Green, 1922b: 1020.
Lecanium fusiforme Green; Green, 1937: 303.
Coccus fusiforme (Green); Ali, 1971: 24.
TYPE DATA. Syntypes female, SRI LANKA: Ambalangoda, on leaves of undetermined shrub (BMNH).
DISTRIBUTION. ORIENTAL REGION: Sri Lanka.
REMARKS. Adult female redescribed by Tang (1991).

Platylecanium mesuae Takahashi

Platylecanium mesuae Takahashi, 1950b: 58.
TYPE DATA. Syntypes female, MALAYSIA: Kuala Lumpur, on *Mesua* sp. (SMKM).
HOST PLANTS. Guttiferae: *Mesua.*
REMARKS. Adult female redescribed by Tang (1991).

Platylecanium nepalense Takagi

Platylecanium nepalense Takagi, 1975: 7.
Platylecanium hepalense Takagi; Tang, 1991: 133 [MIS-SPELLING].
TYPE DATA. Holotype female, NEPAL: Balaju, on plant of the Anacardiaceae (HUSJ).
DISTRIBUTION. PALAEARCTIC REGION: Nepal.
HOST PLANTS. Anacardiaceae:
REMARKS. Adult female redescribed by Tang (1991).

Platylecanium riouwense Takahashi

Platylecanium riouwense Takahashi, 1951b: 104.
TYPE DATA. Syntypes female, INDONESIA: Riau [=Riouw] Islands,

Rempang, host plant not indicated (HUSJ)
DISTRIBUTION. AUSTRO ORIENTAL REGION: Indonesia.
REMARKS. Adult female redescribed by Tang (1991).

Platysaissetia Cockerell

Saissetia (Platysaissetia) Cockerell, 1901d: 32.
Platysaissetia Cockerell; Fernald, 1903: 207.
TYPE-SPECIES: *Lecanium (Saissetia) castiloae* Cockerell, by original designation.
REMARKS. Hodgson (1991a) revised the genus, redefined its characters and retained in it only the type-species. Generic characters discussed by Hodgson (1969b, 1991), De Lotto (1978) and by Tang (1991). Tang (1991) regarded *Taiwansaissetia* Tao, Wong & Chang, as a subjective synonym of this genus.

Platysaissetia castilloae (Cockerell)

Lecanium (Saissetia) castilloae Cockerell, 1898f: 436.
Lecanium castilloae Cockerell; Cockerell, 1899h: 271.
Saissetia (Platysaissetia) castilloae Cockerell; Cockerell, 1901d: 32.
Platysaissetia castilloae (Cockerell); Fernald, 1903: 208.
Cardiococcus castilloae (Cockerell); Cockerell, 1911: 327.
TYPE DATA. Syntypes female, MEXICO: Frontera, Tabasco, on *Castilloa [=Castilla] elastica* (BMNH, USNM).
DISTRIBUTION. NEOTROPICAL REGION: Mexico.
HOST PLANTS. Moraceae: *Castilla elastica*.
REMARKS. Adult female redescribed and illustrated by Hodgson (1991a).

Platysaissetia crustuliforme (Green)

Neolecanium crustuliforme Green, 1909: 252.
Lecanium (Platysaissetia) crustuliforme (Green); Green, 1937: 304.
Platysaissetia crustuliforme (Green); Varshney, 1985: 27.
TYPE DATA. Syntypes female, SRI LANKA: Chilaw, on undetermined tree (BMNH).
DISTRIBUTION. ORIENTAL REGION: Sri Lanka.
REMARKS. Adult female redescribed by Tang (1991).

Poaspis Koteja

Poaspis Koteja, 1978: 320.
TYPE-SPECIES: *Luzulaspis jahandiezi* Balachowsky, by original designation.
REMARKS. Koteja (1979a) revised the genus and gave key to species. Generic characters discussed by Kosztarab & Kozár (1988) and by Tang (1991).

Poaspis cunhii (Balachowsky)

Luzulaspis cunhii Balachowsky, 1937a: 116.
Poaspis cunhii (Balachowsky); Koteja, 1978: 321.
TYPE DATA. Syntypes female, MADEIRA ISLAND: pentes du Pico Ariero, on *Avena sulcata* (MNHN).
DISTRIBUTION. PALAEARCTIC REGION: Madeira.

HOST PLANTS. Gramineae: *Avena sulcata.*
REMARKS. Adult female redescribed and illustrated by Koteja (1979a). Adult female redescribed by Borchsenius (1957). Distribution and host plant records given by Borchsenius (1957), Koteja (1978, 1979b) and by Vieira *et al* (1983).

Poaspis intermedia (Goux)

Luzulaspis (Exaeretopus) intermedius Goux, 1939: 63.
Poaspis intermedia (Goux); Koteja, 1978: 321.
TYPE DATA. Holotype female, FRANCE: Savoie, Thuile, on bark of *Juniperus* (GOUX).
DISTRIBUTION. PALAEARCTIC REGION: France, Hungary.
REMARKS. The type-series was collected on the bark of *Juniperus*, however it is assumed that these insects wandered from neighbouring grasses. Adult female redescribed by Koteja (1979a). Distribution and host plant records given by Kozár (1986).

Poaspis jahandiezi (Balachowsky)

Luzulaspis jahandiezi Balachowsky, 1932a: 197.
Luzulaspis (Exaeretopus) jahandiezi Balachowsky; Goux, 1937: 96.
Luzulaspis jahadiezi Balachowsky; Kosztarab & Kozár, 1978: 107 [MIS-SPELLING].
Poaspis jahandiezi (Balachowsky); Koteja, 1978: 321.
Poaspis jahadiezi (Balachowsky); Kozár, 1986: 176 [MIS-SPELLING].
TYPE DATA. Syntypes female, FRANCE: Var, Ile de Port-Cros, on *Phragmites communis* (MNHN).
DISTRIBUTION. PALAEARCTIC REGION: Bulgaria, Cyprus, Czechoslovakia, France, Greece, Hungary.
HOST PLANTS. Gramineae: *Agropyron repens, Agrostis vulgaris, Calamagrostis epigeios, C. varia, Phragmites communis, Piptaterum multiflorum.*
REMARKS. Adult female redescribed and illustrated by Rehacek (1957), Kosztarab & Kozár (1978) and by Koteja (1979a). Adult female redescribed by Borchsenius (1957) and by Kosztarab & Kozár (1988). Koteja (1979a) confirmed that the specimens from Greece, which were named *Exaeretopus hellenicus* Green [NOMEN NUDUM] by Bodenheimer (1928), belong to this species. Distribution and host plant records given by Rehacek (1957), Koteja (1979a, 1979b), Kozár (1986) and by Kosztarab & Kozár (1988).

Poaspis kondarensis (Borchsenius)

Luzulaspis kondarensis Borchsenius, 1952: 276.
Poaspis kondarensis (Borchsenius); Koteja, 1978: 321.
TYPE DATA. Syntypes female, USSR: Tadzhikistan, Mt. Hissar, Kondara Gorge, on grass (ZIAS).
DISTRIBUTION. PALAEARCTIC REGION: USSR Kazakhstan, USSR Tadzhikistan, USSR Uzbekistan.
HOST PLANTS. Gramineae: *Secale silvestris.*
REMARKS. Adult female redescribed and illustrated by Borchsenius (1957) and by Koteja (1979a). Adult female redescribed by Tang (1991). Distribution and host plant records given by Borchsenius (1957) and by Koteja (1979a).

Poaspis kurilensis (Danzig)

Luzulaspis kurilensis Danzig, 1975a: 137.
Poaspis kurilensis (Danzig); Koteja, 1978: 321.
TYPE DATA. Holotype female. USSR: Kuril Islands. Kunashir. on
  Calamagrostis sp. (ZIAS).
DISTRIBUTION. PALAEARCTIC REGION: USSR Kunashir Island.
HOST PLANTS. Gramineae: *Calamagrostis*.
REMARKS. Adult female redescribed and illustrated by Koteja
(1979a), Danzig (1980c) and by Tang (1991).

Poaspis lata (Goux)

Luzulaspis (Exaeretopus) latus Goux, 1939: 68.
Poaspis lata (Goux); Koteja, 1978: 321.
TYPE DATA. Holotype female. FRANCE: Var, Carqueiranne, on
  Avena sp. (GOUX).
DISTRIBUTION. PALAEARCTIC REGION: France.
HOST PLANTS. Gramineae: *Avena*.
REMARKS. Adult female redescribed by Koteja (1979a).

Poaspis taurica (Borchsenius)

Luzulaspis taurica Borchsenius, 1952: 274.
Poaspis taurica (Borchsenius); Koteja, 1978: 321.
TYPE DATA. Syntypes female, USSR: Crimea, Yaltinsk Region,, Ayu-
  Dag, on *Luzula* sp. (ZIAS).
DISTRIBUTION. PALAEARCTIC REGION: USSR Crimea.
HOST PLANTS. Juncaceae: *Luzula*.
REMARKS. Adult female redescribed and illustrated by Borchsenius
(1957), Koteja (1979a) and by Tereznikova (1981). Distribution and
host plant records given by Borchsenius (1957), Koteja (1978, 1979a)
and by Tereznikova (1981).

Podoparalecanium Tao, Wong & Chang

Podoparalecanium Tao, Wong & Chang, 1983: 100.
TYPE-SPECIES: *Paralecanium machili* Takahashi, by original desig-
  nation and monotypy.
REMARKS. Tang (1991) regarded this genus a subjective synonym of
Paralecanium.

Podoparalecanium luzonicum (Cockerell)

Paralecanium luzonicum Cockerell, 1914: 333.
Podoparalecanium luzonicum (Cockerell); Tao *et al.*, 1983: 100.
TYPE DATA. Syntypes female, PHILIPPINES: Los Banos, on *Alectro-
  nia [=Alectryon] viridis* (BMNH, USNM).
DISTRIBUTION. AUSTRO ORIENTAL REGION: Philippines.
HOST PLANTS. Sapindaceae: *Alectryon viridis*.
REMARKS. Adult female redescribed and illustrated by Morrison
(1920).

Podoparalecanium machili (Takahashi)

Paralecanium machili Takahashi, 1933: 37.
Podoparalecanium machili (Takahashi); Tao *et al.*, 1983: 100.

TYPE DATA. Syntypes female, TAIWAN: Kuaru, Kankau, on *Machilus* sp. and *Cinnamomum zeylanicum* (IMZT).
DISTRIBUTION. ORIENTAL REGION: Taiwan.
HOST PLANTS. Lauraceae: *Cinnamomum zeylanicum, Machilus*.
REMARKS. Adult female redescribed and illustrated by Tao *et al.* (1983) and by Tang (1991). Distribution and host plant records given by Tao *et al.* (1983) and by Tang (1991).

Protopulvinaria Cockerell

Pulvinaria (Protopulvinaria) Cockerell, 1894b: 310.
Protopulvinaria Cockerell; Green, 1909: 253.
Ptoropulvinaria Danzig & Konstantinova; Danzig & Konstantinova, 1990: 44 [MIS-SPELLING].
Protopulvinarir Tang; Tang, 1991: 250 [MIS-SPELLING].
TYPE-SPECIES: *Pulvinaria (Protopulvinaria) pyriformis* Cockerell, by original designation and monotypy.
REMARKS. Fernald (1903) indicated *Pulvinaria convexa* Hempel as type-species, an erroneous designation which was repeated by Gómez-Menor Ortega (1937, 1958c), Borchsenius (1957) and by Yang (1982). Generic characters discussed by Green (1909), Morrison (1920), Steinweden (1929), Gómez-Menor Ortega (1948, 1958c), Borchsenius (1957), Williams & Kosztarab (1972), Kawai (1980), Yang (1982), Tao *et al.* (1983), Hamon & Williams (1984), Gill (1988), Williams & Watson (1990) and by Tang (1991). Key to species: ASIA - Tang (1991). FLORIDA - Hamon & Williams (1984). JAPAN - Kawai (1980). TAIWAN - Tao *et al.* (1983).

Protopulvinaria fukayai (Kuwana)

Lecanium (Coccus) fukayai Kuwana, 1909a: 154.
Coccus fukayai Kuwana; Sasscer, 1911: 67.
Protopulvinaria japonica Kuwana; Kuwana, 1916: 145. Syntypes female, JAPAN: Nagasaki, on *Fatsia japonica* (ITLJ) Syn. by Takahashi, 1955a: 35.
Protopulvinaria fukayai (Cockerell); Takahashi, 1955a: 35 [ERRONEOUS AUTHORSHIP].
TYPE DATA. Syntypes female, JAPAN: Ibaraki, on vine (ITLJ).
DISTRIBUTION. PALAEARCTIC REGION: China, Japan.
HOST PLANTS. Apocynaceae: *Trachelospermum asiaticum*. **Araliaceae:** *Fatsia japonica, Hedera japonica*. **Lauraceae:** *Cinnamomum chekiangensis, C. japonicum, Laurus nobilis, Machilus thunbergii*. **Rubiaceae:** *Gardenia florida, G. grandiflora, G. jasminoides*. **Vitidaceae:** *Vitis*.
REMARKS. Adult female redescribed and illustrated by Kuwana (1917a), Kawai (1980) and by Tang (1991). Adult female redescribed by Takahashi (1955a), Borchsenius (1957). Colour photograph given by Kawai (1980). Distribution and host plant records given by Kuwana (1916, 1917a), Takahashi (1955a), Takahashi & Tachikawa (1956), Borchsenius (1957), Kawai (1980) and by Tang (1991).

Protopulvinaria longivalvata Green

Protopulvinaria longivalvata Green, 1909: 254.
Protopulvinaria longivalvata bakeri Cockerell & Robinson, 1914: 332.

Syntypes female, PHILIPPINES: Los Banos, on leaves of "bocanga" (BMNH, USNM) Syn. by Morrison, 1920: 186.

TYPE DATA. Lectotype female designated by Williams & Watson (1990), SRI LANKA: Heneratgoda, on *Piper nigrum* (BMNH).

DISTRIBUTION. AUSTRO ORIENTAL REGION: Malaysia, Philippines. **MADAGASIAN REGION:** Reunion. **NEOTROPICAL REGION:** Brazil Rio Grande do Sul, Puerto Rico, Virgin Islands. **NEW ZEALAND and PACIFIC REGION:** French Polynesia. **ORIENTAL REGION:** Sri Lanka.

HOST PLANTS. Anacardiaceae: *Mangifera indica.* **Apocynaceae:** *Plumeria rubra.* **Euphorbiaceae:** *Claoxylon.* **Flacourtiaceae:** *Samyda spanulosa.* **Lauraceae:** *Laurus nobilis, Nectandra, Persea gratissima.* **Loranthaceae:** *Loranthus.* **Meliaceae:** *Cabralea cangerana.* **Myrtaceae:** *Eugenia jambos, Psidium guajava.* **Piperaceae:** *Piper betle, P. nigrum.* **Rubiaceae:** *Gardenia florida, G. jasminoides, G. latifolia.* **Rutaceae:** *Citrus.* **Theaceae:** *Camellia.*

REMARKS. Adult female redescribed and illustrated by Morrison (1920), Lepage & Giannotti (1943) and by Williams & Watson (1990). Adult female redescribed by Rutherford (1915b) and by Tang (1991). Distribution and host plant records given by Cockerell & Robinson (1914), Morrison (1920), Green (1937), Lepage & Giannotti (1943), Takahashi (1952), Corseuil & Barbosa (1971), Nakahara & Miller (1981), Nakahara (1983), Williams & Williams (1988), Williams & Watson (1990) and by Tang (1991).

Protopulvinaria pyriformis Cockerell

Pulvinaria (Protopulvinaria) pyriformis Cockerell, 1894b: 309.

Pulvinaria newsteadi Leonardi, 1898: 279. Syntypes female, MADEIRA ISLANDS: Funchal, on Caprifoglio [=*Caprifolium*] (IEAP) Syn. by Cockerell, 1899k: 311.

Pulvinaria plana Lindinger, 1911b: 34. Syntypes female, CANARY ISLANDS: Tenerife, on *Laurus canariensis* (HMNH). Syn. by Lindinger, 1912: 199.

Protopulvinaria piriformis Cockerell; Lindinger, 1912: 199 [MIS-SPELLING].

Protopulvinaria agalmae Takahashi, 1933: 39. Syntypes female, TAIWAN: Taihoku, Shinten, on *Agalma lutchuense* (IMZT). Syn. by Takahashi, 1955a: 36.

Protopulvinaria pyformis Cockerell; Tao, 1978: 82 [MIS-SPELLING].

Pulvinaria phriformis Cockerell; Pollard & Alleyne, 1986: 39 [MIS-SPELLING].

TYPE DATA. Syntypes female, TRINIDAD: St. Anns, on guava (USNM).

DISTRIBUTION. ETHIOPIAN REGION: South Africa, Zimbabwe. **MADAGASIAN REGION:** Comoros, Mauritius. **NEOTROPICAL REGION:** Bermuda, Chile, Cuba, Dominican Republic, Grenada, Guyana, Puerto Rico, Trinidad, Virgin Islands. **NEARCTIC REGION:** Alabama, California, Florida, Georgia, Louisiana, Mississippi, New Hampshire, New Mexico, New York, South Carolina, Texas, Virginia. **ORIENTAL REGION:** Taiwan, Vietnam. **PALAEARCTIC REGION:** Canary Islands, Israel, Japan, Madeira, Spain.

HOST PLANTS. Acanthaceae: *Adhatoda vasica.* **Agavaceae:** *Dracaena duranti.* **Anacardiaceae:** *Mangifera indica.* **Apocynaceae:** *Carissa grandiflora, Plumeria tricolor, Trachelospermum jasminoides.* **Aquifoliaceae:** *Ilex canariensis, I. perado.* **Araceae:** *Dizygotheca.* **Araliaceae:** *Agalma lutchuense, Aralia, Brassaia actinophylla, Fatsia japonica, Hedera canariensis, H. helix, Schefflera octophylla, Tetrapanax papyriferum.* **Asclepiadaceae:** *Araujia sericofera.* **Cannaceae:** *Canna indica.* **Caprifoliaceae:** *Caprifolium, Lonicera etrusca, Viburnum tinus.* **Caricaceae:** *Carica papaya.* **Convolvulaceae:** *Ipomoea.* **Elaeocarpaceae:** *Elaeocarpus elliptica, E. serratus.* **Euphorbiaceae:** *Antidesma bunius.* **Lauraceae:** *Apollonias barbujana, Cinnamomum camphora, C. zeylanicum, Laurus azorica, L. canariensis, L. nobilis, Ocotea foetens, Persea americana, P. borbonia, P. gratissima.* **Leguminosae:** *Bauhinia chamioni, B. vahlii.* **Malvaceae:** *Hibiscus sinensis.* **Moraceae:** *Ficus.* **Musaceae:** *Musa cavendishi,* **Myrtaceae:** *Eucalyptus, Eugenia jambolana, Myrtus communis, Psidium guajava.* **Orchidaceae:** *Cymbidium, Epidendrum.* **Passifloraceae:** *Passiflora.* **Pittosporaceae:** *Pittosporum tobira.* **Rubiaceae:** *Gardenia jasminoides.* **Rutaceae:** *Choisya ternata, Citrus.* **Scrophulariaceae:** *Veronica.* **Stilaginaceae:** *Antidesma.*

REMARKS. Common Name - Pyriform scale. Adult female redescribed and illustrated by Gómez-Menor Ortega (1948, 1958c), De Lotto (1967b), Williams & Kosztarab (1972), Kawai (1980), Tao *et al.* (1983), Hamon & Williams (1984), Gill (1988) and by Gonzalez (1989). Adult female redescribed by Tang (1991). Larval instars described and illustrated by Gómez-Menor Ortega (1948, 1958c) and by Ray & Williams (1982). Colour photograph given by Hamon & Williams (1984) and by Gill (1988). Wysoki (1987) compiled a bibliography of publications up to 1986. Distribution and host plant records given by Lindinger (1912), Mamet (1954b), Gómez-Menor Ortega (1958c), Del Rivero (1966), De Lotto (1967b), Hodgson (1969b), Williams & Kosztarab (1972), Panis (1976), Matile-Ferrero (1978), Kawai (1980), Nakahara & Miller (1981), Ray & Williams (1982), Nakahara (1983), Tao *et al.* (1983), Vieira *et al.* (1983), Martin Mateo (1984), Hamon & Williams (1984), Ben-Dov (1985), Williams & Williams (1988), Gill (1988), Gonzalez (1989), Danzig & Konstantinova (1990), Tang (1991) and by Hodgson & Hilburn (1991a, 1991b).

BIOLOGY. Females reproduce parthenogenetically. Males may occur in populations at low proportion. The entire life cycle is spent on the lower leaf surface. Develops several overlapping generations per year in California (Gill, 1988). In Israel, two annual generations were recorded on avocado, as compared to three on *Hedera helix* (Blumberg & Blumberg, 1991). Blumberg & Swirski (1984) and Blumberg & Blumberg (1991) studied the encapsulation response to parasitoids. The encapsulation of eggs of the parasitoid *Encyrtus infelix* (Embleton) was determined by Blumberg & Goldenberg (1992).

ECONOMIC IMPORTANCE. A serious pest of fruit trees and ornamentals in many parts of the world (Del Rivero, 1966; Hamon & Williams, 1984; Ben-Dov, 1985; Gill, 1988; De Meijer *et al.*, 1989). De Meijer *et al.* (1989) evaluated and showed differences in susceptibility of 9 avocado cultivars to this soft scale in Israel.

Pseudalichtensia Hempel

Pseudalichtensia Hempel, 1928: 237.
TYPE-SPECIES: *Pseudalichtensia brasiliae* Hempel, by original designation and monotypy.

Pseudalichtensia brasiliae Hempel

Pseudalichtensia brasiliae Hempel, 1928: 237.
TYPE DATA. Syntypes female, BRAZIL: Sao Paulo, Itarare, on *Phoradendron dipterum* and *Nectandra* sp. (IBSP).
DISTRIBUTION. NEOTROPICAL REGION: Brazil Sao Paulo.
HOST PLANTS. Lauraceae: *Nectandra*. **Viscaceae:** *Phoradendron dipterum*.
REMARKS. Adult female redescribed and illustrated by Hempel (1929).

Pseudokermes Cockerell

Lecanium (Pseudokermes) Cockerell, 1895p: 203.
Pseudokermes Cockerell; Cockerell, 1899f: 333.
TYPE-SPECIES: *Lecanium (Pseudokermes) nitens* Cockerell, by monotypy.

Pseudokermes armatus Cockerell

Lecanium (Pseudokermes) armatus Cockerell, 1898f: 436.
Pseudokermes armatus Cockerell; Cockerell, 1899h: 270.
TYPE DATA. Syntypes female, MEXICO: Tabasco, S. Francisco de Peal, on a plant named "Palo de Gusano" (BMNH, USNM).
DISTRIBUTION. NEOTROPICAL REGION: Mexico.

Pseudokermes cooleyi King

Pseudokermes cooleyi King, 1914: 246.
TYPE DATA. Syntypes female, U.S.A.: Montana, Corvallis, Ravalli County, on *Picea englemanni* (USNM).
DISTRIBUTION. NEARCTIC REGION: Montana.
HOST PLANTS. Pinaceae: *Picea engelmanni*.

Pseudokermes marginatus Newstead

Pseudokermes marginatus Newstead, 1920: 185.
TYPE DATA. Syntypes female, GUYANA: Ituni Savannah, on *Nectandra* sp. (BMNH).
DISTRIBUTION. NEOTROPICAL REGION: Guyana.
HOST PLANTS. Lauraceae: *Nectandra*.

Pseudokermes nitens Cockerell

Lecanium (Pseudokermes) nitens Cockerell; Cockerell, 1895p: 203.
Pseudokermes nitens Cockerell; Hempel, 1900b: 448.
Pseudokermes nitens (Hempel); Fernald, 1903: 165 [ERRONEOUS AUTHORSHIP].
TYPE DATA. Syntypes female, BRAZIL: Rio Grande do Sul, on *Myrtus (Blepharocalyx) tweedii* (USNM).

DISTRIBUTION. NEOTROPICAL REGION: Argentina, Brazil Rio Grande do Sul, Brazil Sao Paulo.
HOST PLANTS. Leguminosae: *Mimosa scabrella, Phyllocalyx*. **Malpighiaceae:** *Stenocalyx michelii, Stenocalyx pitanga*. **Myrtaceae:** *Campomanesia xanthocarpa, Eugenia pungeris, E. uniflora, Myrtus tweedii, Psidium guajava*. **Sapindaceae:** *Cupania vernalis*.
REMARKS. Adult female redescribed and illustrated by Hempel (1900b, 1901a, 1920a) and by Gomes Costa (1949). Distribution and host plant records given by Hempel (1900b, 1901a, 1920a, 1920b), Lizer y Trelles (1939), Gomes Costa (1949) and by Corseuil & Barbosa (1971).

Pseudokermes palmae Hempel
Pseudokermes palmae Hempel, 1937: 11.
TYPE DATA. Syntypes female, BRAZIL: Sao Paulo, on a cultivated palm (IBSP).
DISTRIBUTION. NEOTROPICAL REGION: Brazil Sao Paulo.
HOST PLANTS. Palmae.

Pseudophilippia Cockerell
Pseudophilippia Cockerell, 1897d: 89.
TYPE-SPECIES: *Pseudophilippia quaintancii* Cockerell, by monotypy.
REMARKS. Generic characters discussed by Williams & Kosztarab (1972) and by Hamon & Williams (1984).

Pseudophilippia quaintancii Cockerell
Pseudophilippia quaintancii Cockerell, 1897d: 90.
TYPE DATA. Syntypes female, U.S.A.: Florida, Lake City, on *Pinus australis* (USNM).
DISTRIBUTION. NEARCTIC REGION: Alabama, Florida, Georgia, Louisiana, Maryland, Mississippi, New Jersey, New York, North Carolina, Pennsylvania, South Carolina, Tennessee, Virginia.
HOST PLANTS. Pinaceae: *Pinus elliotti, P. taeda*.
REMARKS. Adult female redescribed and illustrated by Williams & Kosztarab (1972) and by Hamon & Williams (1984). Adult male described by Ray & Williams (1980) and by Hamon & Williams (1984). Male test described and illustrated by Miller & Williams (1990). Larval instars described by Ray & Williams (1980) and by Hamon & Williams (1984). Distribution and host plant records given by Williams & Kosztarab (1972), Ray & Williams (1980), Hamon & Williams (1984) and by Clarke *et al.* (1989a).
BIOLOGY. Clarke *et al.*, (1989a) studied the life history in U. S. A., Georgia coastal plain, on *Pinus taeda*, and observed two generations per year.

Pseudopulvinaria Atkinson
Pseudopulvinaria Atkinson, 1889: 4.
Lefroyia Green, 1908: 21. TYPE-SPECIES: *Lefroyia castaneae* Green, by original designation and monotypy. Syn. by synonymy of type-species; see Green, 1922b: 345.

TYPE-SPECIES: *Pseudopulvinaria sikkimensis* Atkinson, by original designation and monotypy.
REMARKS. Ferris (1950b) and Hoy (1963) assigned the genus to the Eriococcidae. However, Hodgson (1991b) following a detailed study of the type-species concluded that it is an aberrant member of the Coccidae. Generic characters discussed by Steinweden (1929), Tang (1991) and by Hodgson (1991b).

Pseudopulvinaria sikkimensis Atkinson

Pseudopulvinaria sikkimensis Atkinson, 1889: 4.
Lefroyia castaneae Green, 1908: 21. Lectotype female designated by Hodgson (1991b), INDIA: Assam, Shilong, on *Castanea* sp. (BMNH). Syn. by Green, 1922b: 345.
TYPE DATA. Syntypes female, INDIA: Sikkim, Mungphu, on *Quercus incana, Castanea indica* and *C. tribuloides* (DEPOSITORY UNKNOWN; Hodgson, 1991b).
DISTRIBUTION. ORIENTAL REGION: India. **PALAEARCTIC REGION:** China.
HOST PLANTS. Fagaceae: *Castanea indica, C. tribuloides, Quercus incana.*
REMARKS. Hodgson (1991b) redescribed and illustrated the adult female, adult male and pupa, second and third instar female, second instar male, and the crawler. Adult female redescribed and illustrated by Tang (1991). Distribution and host plant records given by Tang (1991) and by Hodgson (1991b).

Psilococcus Borchsenius

Psilococcus Borchsenius, 1952: 269.
TYPE-SPECIES: *Psilococcus ruber* Borchsenius, by original designation and monotypy.
REMARKS. Generic characters discussed by Borchsenius (1957), Tereznikova (1967), Kosztarab & Kozár (1978, 1988), Danzig (1980c) and by Tang (1991).

Psilococcus ruber Borchsenius

Psilococcus ruber Borchsenius, 1952: 270.
Psilococcus parvus Borchsenius, 1957: 115. Syntypes female, USSR: Latvia, on undetermined plant (ZIAS). Syn. by Danzig, 1980c: 255.
TYPE DATA. Syntypes female, USSR: Primorye Territory, Khasansk, on *Carex* sp. (ZIAS).
DISTRIBUTION. PALAEARCTIC REGION: Hungary, Korea, Poland, USSR Irkutsk Region, USSR Karelia, USSR Latvia, USSR Leningrad Region, USSR Primorye Territory, USSR Ukraine.
HOST PLANTS. Cyperaceae: *Carex brizoides, C. campylorhina, C. canescens, C. dispalata, C. divulsa, C. duriuscula, C. hirta, C. pallescens.*
REMARKS. Adult female redescribed and illustrated by Borchsenius (1957), Tereznikova (1967), Koteja (1969a) (as *P. parvus*), Kosztarab & Kozár (1978, 1988), Danzig (1980c), Tereznikova (1967, 1981) (as *P. parvus*) and by Tang (1991). Adult male described and illustrated by Koteja (1969a) (as *P. parvus*). Larval instars described and il-

lustrated by Koteja (1969a) (as *P. parvus*). Distribution and host plant records given by Borchsenius (1957), Danzig (1967, 1980c), Tereznikova (1967, 1981), Koteja (1969a, 1971), Kozár & Sugonyaev (1979), Kosztarab & Kozár (1988) and by Tang (1991).

Pterolecanium Sulc

Lecanium (Pterolecanium) Sulc, 1932: 53.
Pterolecanium Sulc; Morrison & Morrison, 1966: 169.
TYPE-SPECIES: *Lecanium pulchrum* Reh, by monotypy.
REMARKS. The type species is a synonym of *Parthenolecanium rufulum* (Cockerell). The genus is here accepted as a subjective synonym of *Parthenolecanium*, as suggested by Borchsenius (1957).

Pulvinaria Targioni Tozzetti

Pulvinaria Targioni Tozzetti, 1867: 13.
Pluvinaria Shinji; Shinji, 1935b: 771 [MIS-SPELLING].
Pulviferia Atanasov; Atanasov, 1959: 429 [MIS-SPELLING].
TYPE-SPECIES: *Coccus vitis* Linnaeus, by original designation and monotypy.
REMARKS. Generic characters discussed by Newstead (1903), Green (1909), Froggatt (1915), Dietz & Morrison (1916), Morrison (1920), Leonardi (1920), Steinweden (1929, 1946), Gómez-Menor Ortega (1937, 1958c), Sulc (1942a), Zimmerman (1948), Takahashi (1955e), Borchsenius (1957), De Lotto (1965), Hodgson (1967b, 1968a), Williams & Kosztarab (1972), Paik (1978), Kosztarab& Kozár (1978, 1988), Danzig (1980c), Wang (1980), Kawai (1980), Yang (1982), Tao *et al.* (1983), Hadzibejli (1983), Hamon & Williams (1984), Gill (1988), Williams & Watson (1990), Tang (1991) and by Qin & Gullan (1992). Key to species: ASIA - Tang (1991). AUSTRALIA - Qin & Gullan (1992). CALIFORNIA - Gill (1988). CHINA - Yang (1982), Tang (1991). ETHIOPIAN REGION - Hodgson (1967b, 1968a). FLORIDA - Hamon & Williams (1984). EUROPE- Kosztarab & Kozár (1978, 1988). HAWAII - Zimmerman (1948). ICERYI group -Williams (1982b). INDIANA - Dietz & Morrison (1916). JAPAN - Takahashi (1955e), Kawai (1980). MICRONESIA - Beardsley (1966a). SPAIN - Gómez-Menor Ortega (1937). PHILIPPINES - Morrison (1920). SRI LANKA - Green (1909). TAIWAN - Tao *et al.* (1983). TROPICAL SOUTH PACIFIC - Williams & Watson (1990). U. S. A. - Steinweden (1946). USSR (EUROPEAN) - Borchsenius (1957), Danzig (1964). USSR (FAR EAST) - Danzig (1967, 1980c). VIRGINIA - Williams & Kosztarab (1972). ZIMBABWE - Hodgson (1969b). Canard (1965c) suggested the use of attributes of the mouth parts of the first instar crawlers to separate between species of the genus.

Pulvinaria acericola (Walsh & Riley)

Lecanium acericola Walsh & Riley, 1868: 14.
Pulvinaria innumerabilis acericola (Walsh & Riley); Cockerell, 1896b: 329.
Pulvinaria acericola (Walsh & Riley); Fernald, 1903: 128.
TYPE DATA. Syntypes female, U.S.A.: Indiana and Iowa, on silver maple (USNM).
DISTRIBUTION. NEARCTIC REGION: Alabama, Arkansas, Canada,

Connecticut, Florida, Georgia, Indiana, Iowa, Louisiana, Maryland, Michigan, Mississippi, New York, North Carolina, Oklahoma, Pennsylvania, South Carolina, Tennessee, Texas, Virginia, West Virginia.
HOST PLANTS. Aceraceae: *Acer negundo, A. rubrum, A. saccharinum.* **Aquifoliaceae:** *Ilex cornuta, I. crenata, I. opaca.* **Cornaceae:** *Cornus florida.* **Ericaceae:** *Pieris japonicus.* **Lauraceae:** *Persea borbonia, Sassafras variifolium.* **Nyssaceae:** *Nyssa sylvatica.*
REMARKS. Adult female redescribed and illustrated by Dietz & Morrison (1916), Steinweden (1946), Williams & Kosztarab (1972) and by Hamon & Williams (1984). Colour photograph given by Hamon & Williams (1984) and by Johnson & Lyon (1988). Adult male described and illustrated by Gilliomee (1967). Male test described and illustrated by Miller & Williams (1990). Distribution and host plant records given by Dietz & Morrison (1916), Steinweden (1946), Baerg (1947), Gilliomee (1967), Williams & Kosztarab (1972), Lambdin & Watson (1980), Hamon & Williams (1984) and by Miller & Williams (1990).
BIOLOGY. Baerg (1947) studied the life history in Arkansas (U. S. A.) and observed one annual generation.

Pulvinaria aestivalis Danzig

Pulvinaria aestivalis Danzig, 1967: 146.
Pulvinaria (Pulvinaria) aestivalis Danzig; Danzig, 1980c: 267.
TYPE DATA. Holotype female, USSR: Primorye Territory, near Ussuri, on *Salix viminalis* (ZIAS).
DISTRIBUTION. PALAEARCTIC REGION: USSR Primorye Territory.
HOST PLANTS. Salicaceae: *Salix viminalis.*
REMARKS. Adult female redescribed and illustrated by Danzig (1980c).

Pulvinaria aethiopica (De Lotto)

Lecanium (Coccus) viride Green; Newstead, 1917c: 130 [MISIDENTIFICATION, in part].
Lecanium africanum Newstead; Brain, 1920b: 4 [MISIDENTIFICATION, in part].
Coccus aethiopicus De Lotto, 1959: 156.
Pulvinaria aethiopica (De Lotto); De Lotto, 1967a: 111.
TYPE DATA. Holotype female, SOUTH AFRICA: Transvaal, Buffelspoots, on *Citrus* sp. (BMNH).
DISTRIBUTION. ETHIOPIAN REGION: Angola, Cape Verde Island, South Africa, Zambia, Zimbabwe.
HOST PLANTS. Rubiaceae: *Coffea arabica, C. canephora.* **Rutaceae:** *Citrus aurantium.*
REMARKS. Adult female redescribed and illustrated by Fernandes (1972) and by De Lotto (1979). Distribution and host plant records given by De Lotto (1960, 1965, 1967a, 1979), Hodgson (1967a), Fernandes (1972) and by Almeida (1973).

Pulvinaria alboinducta Fonseca

Pulvinaria alboinducta Fonseca, 1962: 21.

TYPE DATA. Syntypes female, BRAZIL: Sao Paulo, Parque Siqueira Campos, on *Metrodorea stipulata* (MZSP).
DISTRIBUTION. NEOTROPICAL REGION: Brazil Sao Paulo.
HOST PLANTS. Rutaceae: *Metrodorea stipulata.*

Pulvinaria aligarhensis Avasthi & Shafee
Pulvinaria aligarhensis Avasthi & Shafee, 1985: 1289.
TYPE DATA. Holotype female, INDIA: Uttar Pradesh, Aligarh, Naurangabad, on *Azadirachta indica* (AMUI).
DISTRIBUTION. ORIENTAL REGION: India.
HOST PLANTS. Meliaceae: *Melia indica.*

Pulvinaria ampelopsidis Savescu
Pulvinaria ampelopsidis Savescu, 1983: 43.
TYPE DATA. Syntypes female, male and larva, ROMANIA: Bucarest, on *Ampelopsis quinquefolia* (ASAR).
DISTRIBUTION. PALAEARCTIC REGION: Romania.
HOST PLANTS. Vitidaceae: *Ampelopsis quinquefolia.*

Pulvinaria amygdali Cockerell
Pulvinaria amygdali Cockerell, 1896g: 225.
TYPE DATA. Syntypes female, U.S.A.: New Mexico, Pinos Altos, on peach (BMNH, USNM).
DISTRIBUTION. NEARCTIC REGION: New Mexico.
HOST PLANTS. Rosaceae: *Persica vulgaris.*
REMARKS. Adult female redescribed and illustrated by Steinweden (1946). Steinweden (1946) indicated that many records under this name from various US states were doubtful or erroneous.

Pulvinaria araliae Shinji
Pulvinaria araliae Shinji, 1935b: 771.
TYPE DATA. Syntypes female, JAPAN: Morioka, on *Aralia chinensis* and *Acanthopanax spinosum* (lost; S. Takagi, 1990, personal communication).
DISTRIBUTION. PALAEARCTIC REGION: Japan.
HOST PLANTS. Araliaceae: *Acanthopanax spinosum, Aralia chinensis.*

Pulvinaria areolata Fonseca
Pulvinaria areolata Fonseca, 1969: 11.
TYPE DATA. Syntypes female, BRAZIL: Sao Paulo, Parque Siqueira Campos, on undetermined plant (IBSP).
DISTRIBUTION. NEOTROPICAL REGION: Brazil Sao Paulo.

Pulvinaria argentina Leonardi
Pulvinaria argentina Leonardi, 1911: 260.
TYPE DATA. Syntypes female, ARGENTINA: Cacheuta, on *Lycium cilense* (IEAP).
DISTRIBUTION. NEOTROPICAL REGION: Argentina.
HOST PLANTS. Compositae: *Baccharis.* **Solanaceae:** *Cestrum*

parqui, Fabiana denudata, Lycium chilense, Nicotiana cavanille-sii.
REMARKS. Distribution and host plant records given by Lizer y Trelles (1939).

Pulvinaria aurantii Cockerell

Pulvinaria aurantii Cockerell, 1896f: 48.
Lecanium notatum Maskell, 1897b: 243. Syntypes female, JAPAN: Atami, on *Thea* sp., *Ilex crenata* and *Pittosporum* sp. (NZAC, USNM). Syn. by Takahashi, 1955b: 70.
Coccus notatus (Maskell); Fernald, 1903: 173.
Chloropulvinaria aurantii (Cockerell); Borchsenius, 1952: 300.
Pulvinaria notatum (Maskell); Takahashi, 1955b: 70.
TYPE DATA. Syntypes female, JAPAN: Tokyo, on orange (USNM).
DISTRIBUTION. NEW ZEALAND and PACIFIC REGION: Ogasawara Islands. **ORIENTAL REGION:** Vietnam. **PALAEARCTIC REGION:** China, Iraq, Japan, USSR Georgia.
HOST PLANTS. Apocynaceae: *Nerium oleander.* **Araliaceae:** *Fatsia japonica, Hedera rhombea.* **Ebenaceae:** *Diospyros kaki.* **Eucommiaceae:** *Eucommia ulmoides.* **Lauraceae:** *Laurus nobilis.* **Musaceae:** *Musa.* **Oleaceae:** *Osmanthus fragans.* **Pittosporaceae:** *Pittosporum tobira.* **Rosaceae:** *Eriobotrya japonica.* **Rutaceae:** *Citrus aurantium, C. limon, C. maxima, C. paradisi, C. sinensis, C. unshiu, Poncirus trifoliata.*
REMARKS. Adult female redescribed and illustrated by Kuwana (1917a), Borchsenius (1937, 1957), Takahashi (1955e), Kawai (1980), Wang (1980) and by Hadzibejli (1983). Adult female redescribed by Tang (1991). Adult male described and illustrated by Borchsenius (1957). Colour photograph given by Kawai (1980). Distribution and host plant records given by Maskell (1897b), Kuwana (1917a), Borchsenius (1937, 1952, 1957), Takahashi (1955e), Beardsley (1966a), Abdul-Rassoul (1976), Wang (1980), Kawai (1980), Yang (1982), Hadzibejli (1983), Danzig & Konstantinova (1990) and by Tang (1991).

Pulvinaria avasthii Yousuf & Shafee

Pulvinaria avasthii Yousuf & Shafee, 1988: 60.
TYPE DATA. Holotype female, INDIA: Andaman Islands, Port Blair, Sippighat, on *Mangifera indica* (AMUI).
DISTRIBUTION. ORIENTAL REGION: India Aldaman Islands.
HOST PLANTS. Anacardiaceae: *Mangifera indica.*

Pulvinaria bambusicola (Tang) n. comb.

Saccharipulvinaria bambusicola Tang, 1991: 269.
TYPE DATA. Holotype female, CHINA: Zheijiang Province, Hangzhou, on *Bambusa* sp. (EISC).
DISTRIBUTION. PALAEARCTIC REGION: China.
HOST PLANTS. Gramineae: *Bambusa.*

Pulvinaria bigeloviae Cockerell

Pulvinaria bigeloviae Cockerell, 1893u: 366.

TYPE DATA. Syntypes female, U.S.A.: Colorado, Custer County, West Cliff, on *Chrysothamnus [=Bigelovia]* sp. (USNM).
DISTRIBUTION. NEARCTIC REGION: Arizona, California, Colorado.
HOST PLANTS. Compositae: *Chrysothamnus, Franseria, Haplopappus, Hymenoclea*.
REMARKS. Adult female redescribed and illustrated by Gill (1988). Male test described and illustrated by Miller & Williams (1990). Colour photograph given by Gill (1988). Distribution and host plant records given by Cockerell (1910b), Gill (1988) and by Miller & Williams (1990).

Pulvinaria borchsenii Danzig

Pulvinaria borchsenii Danzig, 1967: 148.
Pulvinaria (Pulvinaria) borchsenii Danzig; Danzig, 1980c: 267.
TYPE DATA. Holotype female, USSR: Primorye Territory, Tigrovoi, on *Aralia mandshurica* (ZIAS).
DISTRIBUTION. PALAEARCTIC REGION: USSR Primorye Territory, USSR Sakhalin Island.
HOST PLANTS. Araliaceae: *Aralia mandshurica, Eleutherococcus senticosus*. **Betulaceae:** *Alnus hirsuta, Betula*. **Celastraceae:** *Euonymus*. **Compositae:** *Tanacetum sibiricum*. **Rosaceae:** *Crataegus, Sorbaria sorbifolia, Spiraea salicifolia*. **Salicaceae:** *Populus, Salix*.
REMARKS. Adult female redescribed and illustrated by Danzig (1980c). Adult female redescribed by Tang (1991). Distribution and host plant records given by Danzig (1980c) and by Tang (1991).

Pulvinaria brachiungualis Savescu

Pulvinaria brachiungualis Savescu, 1985: 126.
TYPE DATA. Syntypes female, ROMANIA: Bucarest, on *Acer pseudoplatanus* (ASAR).
DISTRIBUTION. PALAEARCTIC REGION: Romania.
HOST PLANTS. Aceraceae: *Acer pseudoplatanus*.

Pulvinaria brevicornis Newstead

Pulvinaria brevicornis Newstead, 1920: 186.
TYPE DATA. Syntypes female, GUYANA: Turkeyn, East Coast, on *Avicennia nitida* (BMNH).
DISTRIBUTION. NEOTROPICAL REGION: Guyana.
HOST PLANTS. Verbenaceae: *Avicennia nitida*.

Pulvinaria cacao Williams & Watson

Pulvinaria cacao Williams & Watson, 1990: 148.
TYPE DATA. Holotype female, PAPUA NEW GUINEA: Northern Province, Arehe, on *Theobroma cacao* (BMNH).
DISTRIBUTION. AUSTRO ORIENTAL REGION: Papua New Guinea.
HOST PLANTS. Sterculiaceae: *Theobroma cacao*.

Pulvinaria callosa (De Lotto)

Coccus callosus De Lotto, 1966a: 43.
Pulvinaria callosa (De Lotto); De Lotto, 1979: 250.

TYPE DATA. Holotype female, SOUTH AFRICA: Natal, Umkomaas, on *Ochna natalizia* (SANC).
DISTRIBUTION. ETHIOPIAN REGION: South Africa.
HOST PLANTS. Ochnaceae: *Ochna natalizia*. **Rubiaceae:** *Gardenia spatulifolia*.
REMARKS. Distribution and host plant records given by De Lotto (1979).

Pulvinaria carieri Grandpre & Charmoy

Pulvinaria carieri Grandpre & Charmoy, 1899: 41.
Pulvinaria carieri de Charmoy; Fernald, 1903: 130 [ERRONEOUS AUTHORSHIP].
Pulvinaria carieri de Charmoy; Mamet, 1941a: 25 [ERRONEOUS AUTHORSHIP].
TYPE DATA. Syntypes female, MAURITIUS: on the roots of various plants (lost; Mamet, 1941a).
DISTRIBUTION. MADAGASIAN REGION: Mauritius.
REMARKS. Mamet (1941a) could not locate type material of this species and strongly indicated that it may be identical with *P. grabhami* Cockerell, 1903. For the sake of stability I prefer to keep *P. carieri* as an unrecognizable species rather than to reduce the recognized and established *P. grabhami* as its junior synonym.
BIOLOGY. Lives subterraneously on roots of the host plants (Mamet, 1941a).

Pulvinaria cestri (Bouché)

Coccus cestri Bouché, 1833: 50.
Chermes cestri (Bouché); Boisduval, 1867: 336.
Pulvinaria cestri (Bouché); Signoret, 1873a: 35.
TYPE DATA. Syntypes female, GERMANY: Berlin, in greenhouse, on *Cestrum* sp. (lost; Sachtleben, 1944).
DISTRIBUTION. PALAEARCTIC REGION: Germany.
HOST PLANTS. Solanaceae: *Cestrum*.
REMARKS. The taxonomic identity of this species cannot be recognized from the original description. However, evidently it is a soft scale, because Bouché (1833) indicates that it resembles *C. hesperidum* and *C. bromeliae*, soft scale species with which he was familiar. For the sake of stability this species is retained as an unrecognizable one.

Pulvinaria chrysanthemi Hall

Pulvinaria chrysanthemi Hall, 1923: 15.
TYPE DATA. Syntypes female, EGYPT: Heliopolis, on roots of *Chrysanthemum coronarium* (BMNH)
DISTRIBUTION. PALAEARCTIC REGION: Egypt.
HOST PLANTS. Chenopodiaceae: *Beta vulgaris maritima*. **Compositae:** *Chrysanthemum coronarium*.
REMARKS. Adult female redescribed and illustrated by Ezzat & Hussein (1969). Adult female redescribed by Hosny (1939) and by Borchsenius (1957). Distribution and host plant records given by Hosny (1939), Borchsenius (1957) and by Ezzat & Hussein (1969).

Pulvinaria citricola Kuwana

Pulvinaria citricola Kuwana, 1914: 3.
Pulvinaria nipponica Lindinger; Lindinger, 1933a: 50 [UNJUSTIFIED REPLACEMENT NAME].
Eupulvinaria citricola (Kuwana); Borchsenius, 1953: 288.
TYPE DATA. Syntypes female, JAPAN: Okayama, Shizuoka on *Citrus* sp., and Tokyo on *Diospyros kaki*, *Hibiscus syriacus* and *Citrus* sp. (ITLJ).
DISTRIBUTION. NEARCTIC REGION: California, Maryland, Virginia. **PALAEARCTIC REGION:** China, Japan, Tibet.
HOST PLANTS. Ebenaceae: *Diospyros kaki*. **Malvaceae:** *Hibiscus syriacus*. **Rosaceae:** *Pyracantha coccinea*. **Rutaceae:** *Citrus*. **Ulmaceae:** *Zelkova serrata*.
REMARKS. Ferris (1922) transferred *Takahashia citricola* Kuwana, 1909 to *Pulvinaria*, making *P. citricola* Kuwana, 1914 a secondary homonym. Therefore, Lindinger (1933a) proposed *P. nipponica* as a replacement name for *P. citricola* Kuwana, 1909. However, the latter belongs to *Saissetia*, and *P. nipponica* is a synonym of *P. citricola* Kuwana, 1914. Adult female redescribed and illustrated by Kuwana (1917a), Steinweden (1946), Williams & Kosztarab (1972), Hamon & Williams (1984) and by Gill (1988). Adult female redescribed by Borchsenius (1957), Kawai (1980) and by Tang (1991). Distribution and host plant records given by Steinweden (1946), Takahashi & Tachikawa (1956), Williams & Kosztarab (1972), Wang (1980, 1981), Hamon & Williams (1984), Gill (1988) and by Tang (1991).

Pulvinaria claviseta De Lotto

Pulvinaria claviseta De Lotto, 1970a: 149.
TYPE DATA. Holotype female, SOUTH AFRICA: Transvaal, Pretoria, on *Maytenus polyacanthus* (SANC).
DISTRIBUTION. ETHIOPIAN REGION: South Africa.
HOST PLANTS. Celastraceae: *Maytenus polyacanthus*.

Pulvinaria cockerelli King

Pulvinaria cockerelli King, 1899c: 417.
TYPE DATA. Syntypes female, U.S.A.: Massachusetts, Methuen, on *Spiraea salicifolia* (USNM).
DISTRIBUTION. NEARCTIC REGION: Massachusetts.
HOST PLANTS. Rosaceae: *Spiraea*.

Pulvinaria cocolobae (Borchsenius) n. comb.

Chloropulvinaria cocolobae Borchsenius, 1957: 210.
TYPE DATA. Syntypes female, USSR: Leningrad, in greenhouse of the Botanical Institute, on *Coccoloba peltata* (ZIAS).
DISTRIBUTION. PALAEARCTIC REGION: USSR Leningrad Region.
HOST PLANTS. Polygonaceae: *Coccoloba peltata*.
REMARKS. Borchsenius (1957) described this species from specimens taken in a greenhouse in Leningrad, indicating that its country of origin was unknown. Adult female redescribed by Tang (1991).

Pulvinaria convexa (Hempel)

Protopulvinaria convexa Hempel, 1900b: 485.
Pulvinaria convexa (Hempel); Fernald, 1903: 131.
TYPE DATA. Syntypes female, BRAZIL: Sao Paulo, on *Smilax* sp. (MZSP).
DISTRIBUTION. NEOTROPICAL REGION: Argentina, Brazil Sao Paulo.
HOST PLANTS. Aquifoliaceae: *Ilex*. **Rosaceae:** *Rosa*. **Smilacaceae:** *Smilax assumptionis*, *Smilax campestris*.
REMARKS. Adult female redescribed and illustrated by Hempel (1901b, 1929). Distribution and host plant records given by Hempel (1901b, 1929), Lizer y Trelles (1939), and by Silva *et al.* (1968).

Pulvinaria corni Savescu

Pulvinaria corni Savescu, 1985: 124.
TYPE DATA. Syntypes female, ROMANIA: Bucarest, on *Cornus*, *Tilia*, *Vitis* and *Philadelphus* (ASAR).
DISTRIBUTION. PALAEARCTIC REGION: Romania.
HOST PLANTS. Cornaceae: *Cornus*. **Philadelphaceae:** *Philadelphus*. **Tiliaceae:** *Tilia*. **Vitidaceae:** *Vitis*.

Pulvinaria costata Borchsenius

Pulvinaria costata Borchsenius, 1952: 297.
Pulvinaria (Pulvinaria) costata Borchsenius; Danzig, 1980c: 267.
TYPE DATA. Syntypes female, USSR: Primorye Territory, Okeanskaya near Vladivostok, on *Alnus* sp. (ZIAS).
DISTRIBUTION. PALAEARCTIC REGION: Inner Mongolia, USSR Primorye Territory.
HOST PLANTS. Betulaceae: *Alnus*. **Salicaceae:** *Populus berolinensis*, *P. bolleana*, *P. cathayana*, *P. pekinensis*, *P. simonii*.
REMARKS. Adult female redescribed and illustrated by Borchsenius (1957), Danzig (1980c) and by Tang & Li (1988). Adult female redescribed by Tang (1991). Distribution and host plant records given by Tang & Li (1988) and by Tang (1991).

Pulvinaria crassispina Danzig

Pulvinaria crassispina Danzig; Danzig, 1966: 1491 [NOMEN NUDUM].
Pulvinaria crassispina Danzig, 1967: 145.
Pulvinaria (Pulvinaria) crassispina Danzig; Danzig, 1980c: 265.
TYPE DATA. Syntypes female, USSR: Primorye Territory, Ussurisk, on *Spiraea* sp. and on *Sorbaria sorbifolia* (ZIAS).
DISTRIBUTION. PALAEARCTIC REGION: USSR Primorye Territory.
HOST PLANTS. Rosaceae: *Sorbaria sorbifolia*, *Spiraea*.
REMARKS. Adult female redescribed and illustrated by Danzig (1980c). Adult female redescribed by Tang (1991). Distribution and host plant records given by Danzig (1980c) and by Tang (1991).

Pulvinaria crotonis De Lotto

Pulvinaria crotonis De Lotto, 1954: 213.

TYPE DATA. Holotype female, ERITREA: Ghescinascim, on *Croton macrostachys* (BMNH).
DISTRIBUTION. ETHIOPIAN REGION: Eritrea.
HOST PLANTS. Euphorbiaceae: *Croton macrostachys.*
REMARKS. Hodgson (1967b) gave supplementary notes for distinguishing this species from *P. inopheron* and *P. jacksoni.*

Pulvinaria decorata Borchsenius

Pulvinaria ornata Froggatt, 1921a: 427 [HOMONYM of *Pulvinaria ornata* Hempel].
Pulvinaria decorata Borchsenius, 1957: 228 [REPLACEMENT NAME].
TYPE DATA. Lectotype female designated by Qin & Gullan (1992), AUSTRALIA: New South Wales, Sydney, on lemon tree (BCRI).
DISTRIBUTION. AUSTRALIAN REGION: New South Wales.
HOST PLANTS. Rutaceae: *Citrus limon.*
REMARKS. Adult female redescribed and illustrated by Qin & Gullan (1992).

Pulvinaria delottoi Gill

Pulvinaria delottoi Gill, 1979: 241.
TYPE DATA. Holotype female, U.S.A.: California, Alameda County, Oakland, on *Mesembryanthemum* sp. (USNM).
DISTRIBUTION. ETHIOPIAN REGION: South Africa. **NEARCTIC REGION:** California.
HOST PLANTS. Aizoaceae: *Carpobrotus edulis, Cheiridopsis inaequalis, Lampranthus, Mesembryanthemum.* **Crassulaceae:** *Crassula lycopopioides, Sedum.*
REMARKS. Gill (1979) described the female larval instars. Adult female redescribed and illustrated by Gill (1988). Colour photograph given by Gill (1988).

Pulvinaria dendrophthorae Cockerell

Lecanium dendrophthorae Cockerell; Cockerell, 1892b: 333 [NOMEN NUDUM].
Pulvinaria dendrophthorae Cockerell, 1893l: 162.
TYPE DATA. Syntypes female, JAMAICA: on *Dendrophthora* sp. (USNM).
DISTRIBUTION. NEOTROPICAL REGION: Jamaica.
HOST PLANTS. Lranthaceae: *Dendrophthora.*

Pulvinaria depressa Hempel

Pulvinaria depressa Hempel, 1900b: 490.
TYPE DATA. Syntypes female, BRAZIL: Sao Paulo, Ypiranga, on *Miconia* sp. (MZSP).
DISTRIBUTION. NEOTROPICAL REGION: Brazil Sao Paulo.
HOST PLANTS. Melastomataceae: *Miconia.*
REMARKS. Adult female redescribed by Hempel (1901b).

Pulvinaria dicrostachys Leonardi

Pulvinaria dicrostachys Leonardi, 1913: 30.

TYPE DATA. Syntypes female and first instar larva, ERITREA: Agordad, on *Dicrostachys [=Dichrostachys] nutas* (IEAP).
DISTRIBUTION. ETHIOPIAN REGION: Eritrea.
HOST PLANTS. Leguminosae: *Dichrostachys nutans*.

Pulvinaria dodonaeae Maskell

Pulvinaria dodonaeae Maskell, 1893b: 222.
Pulvinaria greeni Froggatt, 1915: 415. Lectotype female designated by Qin & Gullan (1992), AUSTRALIA: New South Wales, Condobolin, on *Myoporum deserti* (BCRI). Syn. by Qin & Gullan, 1992: 115.
TYPE DATA. Lectotype female designated by Qin & Gullan (1992), AUSTRALIA: host plant not indicated (NZAC).
DISTRIBUTION. AUSTRALIAN REGION: New South Wales, Northern Territory, Queensland, South Australia, Victoria.
HOST PLANTS. Leguminosae: *Acacia*. **Myoporaceae:** *Eremophila gilesii, E. longifolia, E. mitchellii, Myoporum deserti*. **Myrsinaceae:** *Ardisia crispa*. **Sapindaceae:** *Dodonaea attenuata, D. bursarifolia*.
REMARKS. Adult female redescribed and illustrated by Qin & Gullan (1992).

Pulvinaria durantae Takahashi

Pulvinaria durantae Takahashi, 1931: 1.
Eupulvinaria durantae (Takahashi); Borchsenius, 1953: 288.
TYPE DATA. Syntypes female, TAIWAN: Taihoku, Kagi, on *Duranta repens* (IMZT).
DISTRIBUTION. ORIENTAL REGION: India, Taiwan.
HOST PLANTS. Verbenaceae: *Duranta repens*.
REMARKS. Adult female redescribed and illustrated by Tao *et al.* (1983). Adult female redescribed by Borchsenius (1957) and by Tang (1991). Distribution and host plant records given by Borchsenius (1957), Tao *et al.* (1983), Varshney (1985) and by Tang (1991).

Pulvinaria ellesmerensis Richards

Pulvinaria ellesmerensis Richards, 1964: 1457.
TYPE DATA. Holotype female, CANADA: Northwestern Territories, Ellesmere Island, Lake Hazen, on *Salix arctica* (CNOC).
DISTRIBUTION. NEARCTIC REGION: Canada Northwestern Territories.
HOST PLANTS. Salicaceae: *Salix arctica*.

Pulvinaria elongata Newstead

Pulvinaria elongata Newstead, 1917b: 20.
Coccus (Lecanium) elongatus (Signoret); Gómez-Menor Ortega, 1958a: 22 [MISIDENTIFICATION].
Coccus elongatus (Signoret); Gómez-Menor Ortega, 1958c: 70 [MISIDENTIFICATION].
Pulvinaria longisqua De Lotto, 1966c: 467. Holotype female, KENYA: Nairobi, on *Saccharum officinarum* (SANC). Syn. by Williams, 1982b: 113.
TYPE DATA. Lectotype female designated by Williams (1982b),

GUYANA: Demarara, Plantation La Bonne Intention, on sugar-cane (BMNH).
DISTRIBUTION. AUSTRO ORIENTAL REGION: Papua New Guinea. **AUSTRALIAN REGION:** Queensland. **ETHIOPIAN REGION:** Cameroon, Kenya, Nigeria. **NEOTROPICAL REGION:** Bahamas, Barbados, Colombia, Cuba, Dominican Republic, Grenada, Guyana, Jamaica, Mexico, Puerto Rico, Trinidad, Venezuela. **NEARCTIC REGION:** Florida, Georgia, Louisiana. **PALAEARCTIC REGION:** Morocco, Spain.
HOST PLANTS. Gramineae: *Andropogon gayanus, Oryza sativa, Panicum rhizomatum, Paspalum notatum, Saccharum officinarum.*
REMARKS. Adult female redescribed and illustrated by Gómez-Menor Ortega (1958a, 1958c) (as *Coccus elongatus*), De Lotto (1966c) (as *P. longisqua*), Mamet (1958), Hamon & Williams (1984), Williams & Watson (1990) and by Qin & Gullan (1992). Colour photograph given by Hamon & Williams (1984). Adult female redescribed by Williams (1982b) and by Gómez-Menor Ortega (1960). Distribution and host plant records given by Mamet (1958), Gómez-Menor Ortega (1958a, 1958c, 1960), De Lotto (1966), Panis (1975), Nakahara & Miller (1981), Williams (1982b), Hamon & Williams (1984), Williams & Watson (1990) and by Qin & Gullan (1992).

Pulvinaria enkianthi Takahashi

Pulvinaria enkianthi Takahashi, 1955e: 151.
TYPE DATA. Syntypes female, JAPAN: Tokyo, on *Enkianthus perulatus*, (HUSJ).
DISTRIBUTION. PALAEARCTIC REGION: Japan.
HOST PLANTS. Ericaceae: *Enkianthus.*
REMARKS. Adult female redescribed by Kawai (1980) and by Tang (1991). Colour photograph given by Kawai (1980).

Pulvinaria ericicola McConnell

Pulvinaria ericicola McConnell, 1949: 29.
TYPE DATA. Holotype female, U.S.A.: Maryland, College Park, on *Rhododendron nudiflorum* (USNM).
DISTRIBUTION. NEARCTIC REGION: Alabama, Florida, Maryland, New Hampshire, New York, Virginia.
HOST PLANTS. Ericaceae: *Lyonia ferruginea, Rhododendron nudiflorum, Vaccinium arboreum, V. vacillans.*
REMARKS. Adult female redescribed and illustrated by Williams & Kosztarab (1972) and by Hamon & Williams (1984). Male test described and illustrated by Miller & Williams (1990). Colour photograph given by Hamon & Williams (1984). Distribution and host plant records given by Williams & Kosztarab (1972), Hamon & Williams (1984) and by Miller & Williams (1990).

Pulvinaria eryngii Fonseca

Pulvinaria minuta Fonseca, 1969: 13 [HOMONYM of *Pulvinaria minuta* Brethes].
Pulvinaria eryngii Fonseca, 1973: 247 [REPLACEMENT NAME].
TYPE DATA. Syntypes female, BRAZIL: Sao Paulo State, Eldorado, on *Eryngium aloifolium* (MZSP).

DISTRIBUTION. NEOTROPICAL REGION: Brazil Sao Paulo.
HOST PLANTS. Umbelliferae: *Eryngium aloifolium*.

Pulvinaria eugeniae Hempel

Pulvinaria eugeniae Hempel, 1900b: 488.
TYPE DATA. Syntypes female, BRAZIL: Sao Paulo, Ypiranga and Sao Paulo, on *Eugenia jaboticaba* and other trees of the Myrtaceae (MZSP).
DISTRIBUTION. NEOTROPICAL REGION: Brazil Sao Paulo.
HOST PLANTS. Myrtaceae: *Eugenia jaboticaba*.
REMARKS. Adult female redescribed by Hempel (1901b, 1920a).

Pulvinaria euonymi Shinji

Pulvinaria euonymi Shinji, 1935b: 771.
Pulvinaria euonymicola Lindinger, 1957: 551 [UNJUSTIFIED RE-PLACEMENT NAME].
TYPE DATA. Syntypes female, JAPAN: Morioka, on *Euonymus japonicus var. radicans* (lost; S. Takagi, 1990, personal communication).
DISTRIBUTION. PALAEARCTIC REGION: Japan.
HOST PLANTS. Celastraceae: *Euonymus japonicus radicans*.
REMARKS. Adult female redescribed by Kawai (1980). Colour photographgiven by Kawai (1980). Lindinger (1957) introduced *P. euonymicola* as a New Name to replace *P. euonymi* Shinji, which he supposed to be a homonym of *P. evonymi* Goureau, 1869. However, *euonymi* and *evonymi* are not homonymous.

Pulvinaria ferrisi Ali

Pulvinaria marginata Ferris, 1950: 74 [HOMONYM of *Pulvinaria marginata* Targioni Tozzetti].
Pulvinaria ferrisi Ali, 1971: 50 [REPLACEMENT NAME].
TYPE DATA. Syntypes female, CHINA: Yunnan Province, Kunming, on shrub of Magnoliaceae (UCD).
DISTRIBUTION. PALAEARCTIC REGION: China.
HOST PLANTS. Magnoliaceae.

Pulvinaria ficus Hempel

Pulvinaria ficus Hempel, 1900b: 486.
Pulvinaria ficius Hempel; Pollard & Alleyne, 1986: 39 [MIS-SPELL-ING].
TYPE DATA. Syntypes female, BRAZIL: Sao Paulo, on *Ficus* sp., *Psidium* sp., *Mangifera* sp. and *Ixora coccinea* (MZSP).
DISTRIBUTION. NEOTROPICAL REGION: Brazil Rio Grande do Sul, Brazil Sao Paulo, Caribbean.
HOST PLANTS. Anacardiaceae: *Mangifera indica*. **Aquifoliaceae:** *Ilex coccinea*. **Compositae:** *Artemisia*. **Euphorbiaceae:** *Acalypha*. **Lauraceae:** *Persea americana*. **Meliaceae:** *Melia azedarach*. **Moraceae:** *Ficus benjamina, F. nitida*. **Myrtaceae:** *Psidium guajava*. **Polygonaceae:** *Muehlenbeckia platyclada*. **Rubiaceae:** *Gardenia florida, G. jasminoides, Ixora coccinea*. **Rutaceae:** *Citrus*.
REMARKS. Adult female redescribed by Hempel (1901b, 1920a) and

by Gomes Costa (1949). Distribution and host plant records given by
Hempel (1901b, 1920a), Gomes Costa (1949), Silva *et al.* (1968),
Corseuil & Barbosa (1971) and by Pollard & Alleyne (1986).

Pulvinaria flava Takahashi

Pulvinaria flavida Takahashi; Takahashi, 1955e: 150 [MIS-SPELL-
    ING].
Pulvinaria flava Takahashi, 1955e: 152.
TYPE DATA. Syntypes female, JAPAN: Habiki Hill near Kuroyama,
    Minami-Kawachi-gun, Osaka-fu, on *Symplocos* sp. (HUSJ).
DISTRIBUTION. PALAEARCTIC REGION: Japan.
HOST PLANTS. Symplocaceae: *Symplocos.*
REMARKS. Adult female redescribed by Kawai (1980) and by Tang
(1991) (as *P. flavida*).

Pulvinaria flavescens Brethes

Pulvinaria flavescens Brethes in Massini & Brethes, 1918: 150.
TYPE DATA. Syntypes female, ARGENTINA: Buenos Aires, on lemon
    (MLPA).
DISTRIBUTION. NEOTROPICAL REGION: Argentina, Brazil Rio
    Grande do Sul.
HOST PLANTS. Aquifoliaceae: *Ilex.* **Rutaceae:** *Citrus aurantium, C.
    deliciosa, C. limon, C. sinensis.*
REMARKS. Adult female redescribed and illustrated by Gomes
Costa (1949). First instar larva described and illustrated by Ringue-
let (1924). Adult male described and illustrated by Quintana (1956a).
Distribution and host plant records given by Gomes Costa (1949),
Silva *et al.* (1968) and by Corseuil & Barbosa (1971).

Pulvinaria flavicans Maskell

Pulvinaria flavicans Maskell, 1889: 103.
Pulvinaria tecta Maskell, 1894c: 79. Lectotype female designated by
    Qin & Gullan (1992), AUSTRALIA: New South Wales, Sydney,
    host plant not indicated (NZAC). Syn. by Qin & Gullan, 1992:
    122.
Pulvinaria contexta Froggatt, 1915: 413. Lectotype female (the buff
    variety) designated by Qin & Gullan (1992), AUSTRALIA: New
    South Wales, Mittagong, host plant not indicated (BCRI). Syn. by
    Qin & Gullan, 1992: 122. Lectotype female designated by Qin &
    Gullan (1992), AUSTRALIA: South Australia, host plant not
    indicated (NZAC).
DISTRIBUTION. AUSTRALIAN REGION: Australian Capital Terri-
    tory, New South Wales, South Australia, Victoria.
HOST PLANTS. Leguminosae: *Acacia longifolia, Bossiaea buxifolia,
    Daviesia corymbosa, Dillwynia juniperina, Oxylobium, Templeto-
    nia retusa.*
REMARKS. Adult female redescribed and illustrated by Qin &
Gullan (1992). Distribution and host plant records given Maskell
(1894c), Froggatt (1915) and by Qin & Gullan (1992).

Pulvinaria flavicans formicicola Newstead

Pulvinaria flavicans Maskell; Newstead, 1917b: 21 [MISIDENTIFICA-TION].
Pulvinaria flavicans formicicola Newstead in Bodkin, 1922: 61.
TYPE DATA. Syntypes female, GUYANA: Rockstone, on Bloodwood plant (BMNH).
DISTRIBUTION. NEOTROPICAL REGION: Guyana.
REMARKS. Newstead (1917b) presented a detailed description of what he apparently misidentified as *P. flavicans* Maskell. I could not trace any further description by Newstead under the above trinomen. The syntype specimens of this trinomen, deposited in BMNH, are labelled with the data as given by Newstead (1917b). It is concluded that the trinomen was first published by Bodkin (1922), who credited it to Newstead and that it refers to the material and description misidentified by Newstead (1917b) as *Pulvinaria flavicans* Maskell.

Pulvinaria floccifera (Westwood)

Coccus flocciferus Westwood, 1870: 308.
Pulvinaria camelicola Signoret, 1873a: 32. Syntypes female, FRANCE: Paris, Luxembourg Gardens, on *Camellia* (VMNH). Syn. by Lindinger, 1912: 93.
Pulvinaria linearis Targioni Tozzetti, 1884: 398. Syntypes male, female and larva, ITALY: on *Camellia japonica* (probably lost; G. Pellizzari Scaltriti, 1990, personal communication). Syn. by Lindinger, 1912: 93.
Pulvinaria brassiae Cockerell, 1895i: 135. Syntypes female, CANADA: Ontario, Ottawa (in greenhouse), on *Brassia verrucosa* (USNM). Syn. by Fernald, 1903: 132.
Pulvinaria floccifera (Westwood); Green, 1897: 72.
Pulvinaria floccosa (Westwood); Newstead, 1900: 26 [MIS-SPELLING].
Pulvinaria brassicae Cockerell; King, 1902c: 160 [MIS-SPELLING of *P. brassiae*].
Pulvinaria theae Froggatt, 1915: 418. Lectotype female designated by Qin & Gullan (1992), AUSTRALIA: New South Wales, Richmond, on *Thea viridis* (BMNH). Syn. by Qin & Gullan, 1992: 127.
Chloropulvinaria floccifera (Westwood); Borchsenius, 1952: 300.
TYPE DATA. Syntypes female, ENGLAND: on *Camellia* sp. and NETHERLANDS: Utrecht, on *Camellia* sp. (BMNH).
DISTRIBUTION. AUSTRALIAN REGION: New South Wales, South Australia, Victoria. **ETHIOPIAN REGION:** South Africa. **NEARCTIC REGION:** Alabama, California, Canada British Columbia, Connecticut, Georgia, Illinois, Indiana, Massachusetts, Missouri, New Jersey, North Carolina, Oregon, Rhode Island, South Carolina, Texas, Virginia, Washington, Washington D. C. . **ORIENTAL REGION:** Vietnam. **PALAEARCTIC REGION:** Canary Islands, Czechoslovakia, Denmark, Egypt, England, France, Germany, Greece, Hungary, Italy, Japan, Madeira, Netherlands, Poland, Romania, Saudi Arabia, Spain, Sweden, Switzerland, USSR Crimea, USSR Georgia, Yugoslavia.
HOST PLANTS. Anacardiaceae: *Rhus radicans*. **Apocynaceae:** *Thevetia peruviana*. **Aquifoliaceae:** *Ilex aquifolium, I. cornuta, I. integra, I. oldhami, I. pyramidalis*. **Araceae:** *Anthurium*. **Berberidaceae:** *Mahonia*. **Bignoniaceae:** *Phaedranthus buccinatorius*.

Celastraceae: *Euonymus japonicus.* **Chenopodiaceae:** *Chenopodium allum.* **Compositae:** *Montanoa bipinnatifidia.* **Cupressaceae:** *Retinospora.* **Ericaceae:** *Rhododendron.* **Euphorbiaceae:** *Acalypha, Ricinus communis.* **Hydrangeaceae:** *Hydrangea hortensis.* **Illiciaceae:** *Illicium anisatum.* **Magnoliaceae:** *Magnolia japonica.* **Malvaceae:** *Hibiscus.* **Moraceae:** *Ficus.* **Myrsinaceae:** *Ardisia crispa.* **Myrtaceae:** *Psidium guajava.* **Oleaceae:** *Jasminum.* **Orchidaceae: Pittosporaceae:** *Pittosporum tobira.* **Podocarpaceae:** *Podocarpus elongatus.* **Polypodiaceae:** *Adiantum capillus-veneris.* **Rutaceae:** *Citrus aurantium, C. deliciosa.* **Solanaceae:** *Capsicum, Nicotiana glauca.* **Taxaceae:** *Taxus baccata.* **Theaceae:** *Camellia japonica, Camellia sasanqua, Cleyera ochnacea, Eurya japonica, Ternstroemia japonica, Thea japonica.*

REMARKS. Common Name - Cottony Camellia scale. Adult female redescribed and illustrated by Targioni Tozzetti (1885) (as *P. linearis*), Newstead (1903), Dietz & Morrison (1916), Kuwana (1917a) (as *P. camelicola*), Leonardi (1920), Gómez-Menor Ortega (1937, 1958c), Steinweden (1946), Takahashi (1955e), Borchsenius (1957), Canard (1965b), Ezzat & Hussein (1969), Williams & Kosztarab (1972), Hadzibejli (1977, 1983), Kawai (1980), Wang (1980), Tereznikova (1981), Hamon & Williams (1984), Gill (1988), Kosztarab & Kozár (1988) and by Qin & Gullan (1992). Adult female redescribed by Tang (1991). Colour photograph given by Kawai (1980), Gill (1988) and by Johnson & Lyon (1988). Lindinger (1932f) indicated that *Lecanium cestri* Bouche, 1833 is identical with this species. If Lindinger's interpretation would be accepted, the well-established binomen *P. floccifera* would be removed and replaced by the forgotten *P. cestri.* Therefore, for the sake of stability, *P. cestri* is retained as an unrecognizable species. Distribution and host plant records given by Signoret (1873a), Newstead (1903), Froggatt (1915), Dietz & Morrison (1916), Kuwana (1917a), Leonardi (1920), Brain (1920b), Hall (1922), Bodenheimer (1928), Balachowsky (1935), Gómez-Menor Ortega (1937, 1958c), Steinweden (1946), Takahashi (1955e), Borchsenius (1957), Rehacek (1957), Ossiannilsson (1959), Canard (1965b), Ezzat & Hussein (1969), Williams & Kosztarab (1972), Kozárzhevskaya & Reitzel (1975), Hadzibejli (1977, 1983), Tereznikova (1981), Vieira *et al.* (1983), Kozár (1983, 1986), Martin Mateo (1984), Hamon & Williams (1984), Wang (1980), Kawai (1980), Matile-Ferrero (1984b), Gill (1988), Kosztarab & Kozár (1988), Danzig & Konstantinova (1990), Tang (1991) and by Qin & Gullan (1992).

BIOLOGY. Develops one annual generation in Virginia, U. S. A. (Williams & Kosztarab, 1972). Takahashi (1955e) noted that in Japan generally one annual generation is developed, however two generations develop on *Eurya* at Tokyo. El-Minshawy & Moursi (1976) studied the duration of development and fecundity on guava, in Egypt.

Pulvinaria fraxini Signoret

Pulvinaria fraxini Signoret, 1873a: 36.
TYPE-DATA: Syntypes female, FRANCE: Montpellier, on "frene" [-*Fraxinus*] (VMNH).
DISTRIBUTION. PALAEARCTIC REGION: France.

HOST PLANTS. Oleaceae: *Fraxinus.*
REMARKS. Fernald (1903) regraded it as a distinct species, Lindinger (1912) synonymized it with *P. betulae,* whereas Borchsenius (1957) treated it as an unrecognizable species.

Pulvinaria fujisana Kanda

Pulvinaria fujisana Kanda, 1960: 121.
TYPE DATA. Syntypes female, JAPAN: Honshu, Shizuoka Prefecture, at foot of Mt. Fuji, on *Prunus indica, P. donarium* and *P. donarium* var. *spontanea* (KAYJ).
DISTRIBUTION. PALAEARCTIC REGION: Japan.
HOST PLANTS. Rosaceae: *Prunus donarium, P. indica.*
REMARKS. Adult female redescribed by Kawai (1980) and by Tang (1991).

Pulvinaria gamazumii Kanda

Pulvinaria gamazumii Kanda, 1960: 119.
TYPE DATA. Syntypes female, JAPAN: Honshu, Aomori Prefecture, Mt. Towada, on *Viburnum wrightii* (KAYJ).
DISTRIBUTION. PALAEARCTIC REGION: Japan.
HOST PLANTS. Caprifoliaceae: *Viburnum wrightii.*
REMARKS. Adult female redescribed by Kawai (1980) and by Tang (1991).

Pulvinaria globosa Fonseca

Pulvinaria globosa Fonseca, 1962: 18.
TYPE DATA. Syntypes female, BRAZIL: Sao Paulo, Parque Siqueira Campos, on *Stizophyllum peforatum* (MZSP).
DISTRIBUTION. NEOTROPICAL REGION: Brazil Sao Paulo.
HOST PLANTS. Bignoniaceae: *Stizophyllum perforatum.*

Pulvinaria gossypii (Bodenheimer) n. comb.

Filippia gossypii Bodenheimer, 1944b: 89.
TYPE DATA. Syntypes female, IRAN: Chabahar, on branches of cotton (BCR).
DISTRIBUTION. PALAEARCTIC REGION: Iran.
HOST PLANTS. Malvaceae: *Gossypium.*
REMARKS. The new assignment of this species to *Pulvinaria* is based on a study of the syntypes. Adult female redescribed by Borchsenius (1957).

Pulvinaria grabhami Cockerell

Pulvinaria grabhami Cockerell, 1903d: 261.
Pulvinaria antigoni Green, 1907: 204. Syntypes female, SEYCHELLES: on *Antigonum* sp. (BMNH). Syn. by Green, 1923a: 88.
Lecanium nicotianae Newstead, 1908a: 8. Syntypes female, MADAGASCAR: Fenerivo, on *Nicotiana tabacum* (BMNH). Syn. by Hodgson, 1968a: 163.
Coccus nicotianae (Newstead); Sanders, 1909b: 45.
Pulvinaria subterranea Newstead, 1917b: 22. Syntypes female,

UGANDA: Entebbe, on roots of *Chrysanthemum* (BMNH). Syn. by Hodgson, 1968a: 163.

Pulvinaria africana Newstead, 1917b: 23. Syntypes female, GHANA: Accra, on guava (BMNH). Syn. by Hodgson, 1968a: 163.

Pulvinaria floccifera (Westwood); Brain, 1920b: 19 [MISIDENTIFICATION].

Coccus nicotianae (Newstead); Mamet, 1943: 152.

TYPE DATA. Syntypes female, MADEIRA: Funchal, Public Garden, on *Jossinia tinifolia* (USNM).

DISTRIBUTION. ETHIOPIAN REGION: Angola, Ghana, South Africa, Tanzania, Uganda, Zanzibar, Zimbabwe. **MADAGASIAN REGION:** Agalega Island, Madagascar, Mauritius, Seychelles.

HOST PLANTS. Bignoniaceae: *Jacaranda.* **Compositae:** *Chrysanthemum, Cynara scolymus.* **Crassulaceae:** *Bryophyllum pinnatum.* **Cucurbitaceae:** *Luffa acutangula.* **Euphorbiaceae:** *Ricinus communis.* **Myrtaceae:** *Psidium guajava, Jossinia tinifolia.* **Polygonaceae:** *Antigonon.* **Solanaceae:** *Capsicum annuum, Lycopersicon esculentum, Nicotiana glauca, N. tabacum, Solanum indicum.* **Verbenaceae:** *Lantana camara.*

REMARKS. Adult female redescribed and illustrated by Hodgson (1968a). Intraspecific variation in taxonomic characters given by Hodgson (1968a). Newstead (1909) again referred to *Lecanium nicotianae* Newstead, 1908, as n. sp. Distribution and host plant records given by Green (1907), Newstead (1908b, 1909, 1917b), Brain (1920b), Mamet (1943, 1978), Hodgson (1968a), Almeida (1973) and by Williams & Williams (1988).

Pulvinaria grandis Hempel

Pulvinaria grandis Hempel, 1900b: 491.

TYPE DATA. Syntypes female, BRAZIL: Sao Paulo, Ypiranga, on *Myrcia* sp. and other plants of the Myrtaceae (MZSP).

DISTRIBUTION. NEOTROPICAL REGION: Brazil Sao Paulo.

HOST PLANTS. Myrtaceae: *Myrcia.*

REMARKS. Adult female redescribed by Hempel (1901b).

Pulvinaria hazeae Kuwana

Pulvinaria hazeae Kuwana, 1902a: 61.

TYPE DATA. Syntypes female, JAPAN: Kyushu, Koishiwara, Chiku-jo-gun, on *Rhus succedane* (ITLJ).

DISTRIBUTION. PALAEARCTIC REGION: Japan.

HOST PLANTS. Anacardiaceae: *Rhus succedanea.*

REMARKS. Adult female redescribed and illustrated by Kuwana (1917a). Adult female redescribed by Takahashi (1956), Borchsenius (1957), Kawai (1980) and by Tang (1991). Distribution and host plant records given by Takahashi (1956), Borchsenius (1957), Kawai (1980) and by Tang (1991).

Pulvinaria hemiacantha (De Lotto) n. comb.

Pulvinariella hemiacantha De Lotto, 1979: 254.

TYPE DATA. Holotype female, SOUTH AFRICA: Natal, Durban, on *Mesembryanthemum* sp. (SANC).

DISTRIBUTION. ETHIOPIAN REGION: South Africa.

HOST PLANTS. Aizoaceae: *Mesembryanthemum.*

Pulvinaria horii Kuwana

Pulvinaria horii Kuwana, 1902a: 59.
Lecanium lichenoides Green, 1921b: 257. Syntypes female, ENG-
LAND: Herts, St. Albans, on *Quercus glandulifera* (imported from
Japan) (BMNH). Syn. by Takahashi, 1955b: 73.
Eupulvinaria horii (Kuwana); Borchsenius, 1953: 288.
Lecanium horii (Kuwana); Takahashi, 1955b: 73.
TYPE DATA. Syntypes female, and first instar larva, JAPAN: Tokyo,
Nishigahara Agricultural Experiment Station, on *Acer trifidum,*
Aesculus turbinata, Koelreuteria paniculata (ITLJ).
DISTRIBUTION. PALAEARCTIC REGION: England, Japan.
HOST PLANTS. Aceraceae: *Acer trifidum.* **Fagaceae:** *Quercus glan-*
dulifera, Shiia sieboldii. **Hippocastanaceae:** *Aesculus turbinata.*
Rosaceae: *Pyrus simonii.* **Sapindaceae:** *Koelreuteria paniculata.*
Ulmaceae: *Zelkovaserrata.*
REMARKS. Adult female redescribed and illustrated by Kuwana
(1917a). Adult female redescribed by Takahashi (1955b), Borchse-
nius (1957), Kawai (1980) and by Tang (1991). Colour photograph
given by Kawai (1980). Distribution and host plant records given by
Kuwana (1917a), Green (1921b), Borchsenius (1953), Takahashi
(1955b), Takahashi & Tachikawa (1956), Borchsenius (1957), Kawai
(1980) and by Tang (1991).

Pulvinaria hydrangeae Steinweden

Pulvinaria hydrangeae Steinweden, 1946: 7.
Eupulvinaria hydrangeae (Steinweden); Canard, 1965b: 411.
TYPE DATA. Syntypes female, U.S.A.: California, San Mateo, on
Hydrangea hortensis (UCD, USNM).
DISTRIBUTION. AUSTRALIAN REGION: New South Wales. **NEW**
ZEALAND and PACIFIC REGION: New Zealand. **NEARCTIC**
REGION: California, Florida, Massachusetts, New York, Virginia.
PALAEARCTIC REGION: Belgium, France, Italy, Japan.
HOST PLANTS. Aceraceae: *Acer negundo, Acer platanoides.* **Corna-**
ceae: *Cornus.* **Ebenaceae:** *Diospyros kaki.* **Hydrangeaceae:**
Hydrangea hortensis, Hydrangea macrophylla. **Philadelphaceae:**
Deutzia. **Rosaceae:** *Crataegus.* **Tiliaceae:** *Tilia platyphyllos.*
REMARKS. Adult female redescribed and illustrated by Canard
(1965b), Williams & Kosztarab (1972), Pellizzari Scaltriti (1976),
Hamon & Williams (1984), Gill (1988) and by Qin & Gullan (1992).
Adult female redescribed by Takahashi (1956), Borchsenius (1957),
Kawai (1980) and by Tang (1991). Adult male described and il-
lustrated by Canard (1969). Nur (1963) reported chromosome
number 2n=16 in California. Distribution and host plant records
given by Takahashi & Tachikawa (1956), Takahashi (1956), Borch-
senius (1957), Brookes (1964), Canard (1965b, 1969), Williams &
Kosztarab (1972), Pellizzari Scaltriti (1976), Deitz (1979), Hamon &
Williams (1984), Marotta (1987), Gill (1988), Tondeur *et al.* (1990),
Tang (1991) and by Qin & Gullan (1992).

BIOLOGY. Pellizzari Scaltriti (1976) observed one annual generation in Italy. Nur (1963) studied the meiotic parthenogenesis.
ECONOMIC IMPORTANCE. Tondeur *et al.* (1990) reported this species as a noxious urban pest in Region Bruxelloise, Belgium.

Pulvinaria iceryi (Signoret)

Lecanium iceryi Guerin-Meneville; Guerin-Meneville, 1869: 92 [NOMEN NUDUM].
Lecanium iceryi Signoret; Signoret, 1869a: 857 [NOMEN NUDUM].
Lecanium iceryi Signoret, 1869b: 95.
Lecanium gasteralphe Signoret; Signoret, 1869c: 101 [NOMEN NUDUM].
Pulvinaria gasteralphe Signoret; Signoret, 1869c: 101 [NOMEN NUDUM].
Pulvinaria gasteralpha Signoret, 1873a: 37. Syntypes female, MAURITIUS: on sugar-cane (probably lost). Syn. by Mamet, 1958: 69.
Pulvinaria iceryi (Guerin-Meneville); Fernald, 1903: 133 [ERRONEOUS AUTHORSHIP].
Pulvinaria lepida Brain, 1920b: 20. Syntypes female, SOUTH AFRICA: Transvaal, Standerton and Pretoria, on common veld grass (BMNH). Syn. by Mamet, 1958: 69.
Pulvinaria elongata durbanensis Munro & Fouche; Munro & Fouche, 1936: 94 [NOMEN NUDUM].
Pulvinaria elongata Newstead; Mamet, 1949: 27 [MISIDENTIFICATION].
Pulvinaria lepida Brain; Mamet, 1949: 28 [MISIDENTIFICATION].
Pulvinaria iceryi (Signoret); Mamet, 1958: 65.
Coccus iceryi (Signoret); Tao *et al.*, 1983: 87.
Saccharipulvinaria iceryi (Signoret); Tao *et al.*, 1983: 87.
TYPE DATA. Neotype female designated by Mamet (1958), MAURITIUS: Reduit, on sugar-cane (BMNH).
DISTRIBUTION. ETHIOPIAN REGION: Kenya, South Africa, Tanzania, Uganda, Zambia, Zimbabwe. **MADAGASIAN REGION:** Agalega Island, Madagascar, Mauritius, Reunion.
HOST PLANTS. Gramineae: *Agropyron repens, A. schinzii, Cymbopogon giganteus, Cynodon dactylon, Digitaria didactyla, Digitaria scalarum, Eleusine coracana, Heteropogon contortus, Panicum maximum, Paspalidium geminatum, Saccharum officinarum.*
REMARKS. Mamet (1958) clarified and established the nomenclature of this species. Adult female redescribed and illustrated by Mamet (1958). Adult female redescribed by Tang (1991). Guerin-Meneville (1868) named this species *Gasteralphes iceryi* Signoret [NOME-N NUDUM]. However, Mamet (1958) showed that *Gasteralphes* has no standing in nomenclature. Tao *et al.* (1983) redescription and illustration [(as *Saccharipulvinaria iceryi* (Signoret)] very likely refers to *P. elongata* Newstead. Distribution and host plant records given by Hall (1932), Munro & Fouche (1936), De Lotto (1966c), Mamet (1958, 1978), Hodgson (1969b), Williams (1982b) and by Tang (1991).

Pulvinaria idesiae Kuwana

Pulvinaria idesiae Kuwana, 1914: 6.

Eupulvinaria idesiae (Kuwana); Borchsenius, 1953: 288.
TYPE DATA. Syntypes female, JAPAN: Nishigahara, Tokyo, on *Idesia polycarpa* and *Phellodendron amurensis* (ITLJ).
DISTRIBUTION. PALAEARCTIC REGION: Japan.
HOST PLANTS. Betulaceae: *Alnus hirsuta*. **Cornaceae:** *Cornus*.
   Ebenaceae: *Diospyros kaki*. **Flacourtiaceae:** *Idesia polycarpa*.
   Hippocastanaceae: *Aesculus turbinata*. **Rutaceae:** *Phellodendron amurensis*. **Salicaceae:** *Salix glandulosa*.
REMARKS. Adult female redescribed and illustrated by Kuwana (1917a). Adult female redescribed by Takahashi (1956), Borchsenius (1957), Kawai (1980) and by Tang (1991). Colour photograph given by Kawai (1980). Distribution and host plant records given by Takahashi (1956), Takahashi & Tachikawa (1956), Borchsenius (1957) and by Kawai (1980).

Pulvinaria inconspiqua Danzig

Pulvinaria inconspiqua Danzig, 1967: 145.
Pulvinaria (Pulvinaria) inconspiqua Danzig; Danzig, 1980c: 266.
Pulvinaria inconspigua Danzig; Tang, 1991: 260 [MIS-SPELLING].
TYPE DATA. Holotype female, USSR: Primorye Territory, Khasansk Region, "Kedrovaya Pad" Reserve, on *Alnus hirsuta* (ZIAS).
DISTRIBUTION. PALAEARCTIC REGION: USSR Primorye Territory.
HOST PLANTS. Betulaceae: *Alnus hirsuta*.
REMARKS. Adult female redescribed and illustrated by Danzig (1980c).

Pulvinaria indica Avasthi & Shafee

Pulvinaria indica Avasthi & Shafee, 1985: 1290.
TYPE DATA. Holotype female, INDIA: Andhra Pradesh, Vishakhapatnam, Simahachalam, on *Duranta repens* (AMUI).
DISTRIBUTION. ORIENTAL REGION: India.
HOST PLANTS. Verbenaceae: *Duranta repens*.

Pulvinaria ixorae Green

Pulvinaria ixorae Green, 1909: 266.
TYPE DATA. Syntypes female, SRI LANKA: Batticaloa, on *Ixora coccinea* (BMNH).
DISTRIBUTION. ORIENTAL REGION: India, Sri Lanka.
HOST PLANTS. Rubiaceae: *Ixora coccinea*.
REMARKS. Distribution and host plant records given by Varshney (1985).

Pulvinaria juglandii Hadzibejli

Pulvinaria juglandii Hadzibejli, 1971: 449.
TYPE DATA. Syntypes female, male and first instar larva, USSR: Georgia, Mukhrani, Kavtiskhevi, and Azerbaijan, Zakatala, on *Juglans regia* (GPPI).
DISTRIBUTION. PALAEARCTIC REGION: USSR Azerbaijan, USSR Georgia.
HOST PLANTS. Juglandaceae: *Juglans regia*.
REMARKS. Adult female redescribed and illustrated by Hadzibejli (1983).

Pulvinaria justaserpentina Fonseca

Pulvinaria justaserpentina Fonseca, 1973: 251.
TYPE DATA. Syntypes female, BRAZIL: Sao Paulo, Parque Siqueira
  Campos, on indigenous plant (MZSP).
DISTRIBUTION. NEOTROPICAL REGION: Brazil Sao Paulo.

Pulvinaria katsurae Shinji

Pulvinaria katsurae Shinji, 1935b: 772.
TYPE DATA. Syntypes female, JAPAN: Morioka, on *Cercidiphyllum*
  japonicum (lost; S. Takagi, 1990, personal communication).
DISTRIBUTION. PALAEARCTIC REGION: Japan.
HOST PLANTS. Cercidiphyllaceae: *Cercidiphyllum japonicum.*

Pulvinaria kirgisica Borchsenius

Pulvinaria kirgisica Borchsenius, 1952: 298.
TYPE DATA. Syntypes female, USSR: Kazakhstan, bank of Karakol
  river near Pishpeka, on *Betula* sp. (ZIAS).
DISTRIBUTION. PALAEARCTIC REGION: USSR Kazakhstan.
HOST PLANTS. Betulaceae: *Betula.*
REMARKS. Adult female redescribed and illustrated by Borchsenius
(1957). Adult female redescribed by Tang (1991).

Pulvinaria kuwacola Kuwana

Pulvinaria kuwacola Kuwana, 1907: 188.
TYPE DATA. Syntypes female, JAPAN: on mulberry (ITLJ).
DISTRIBUTION. PALAEARCTIC REGION: Japan.
HOST PLANTS. Caprifoliaceae: *Viburnum odoratissimum.* **Euphor-**
  biaceae: *Mallotus japonicus.* **Magnoliaceae:** *Magnolia.* **Mora-**
  ceae: *Morus alba.* **Rosaceae:** *Prunus yedoensis.*
REMARKS. Adult female redescribed and illustrated by Kuwana
(1917a). Adult female redescribed by Takahashi (1956), Borchsenius
(1957), Kawai (1980) and by Tang (1991). Colour photograph given
by Kawai (1980). Distribution and host plant records given by
Takahashi & Tachikawa (1956), Takahashi (1956), Kawai (1980) and
by Tang (1991).

Pulvinaria loralaiensis Rao

Pulvinaria loralaiensis Rao, 1939: 60.
Anapulvinaria loralaiensis (Rao); Tang, 1991: 273.
TYPE DATA. Holotype female, PAKISTAN: Baluchistan, Loralai, on
  pistachio tree (PUSA).
DISTRIBUTION. ORIENTAL REGION: Pakistan.
HOST PLANTS. Anacardiaceae: *Pistacia.*
REMARKS. Adult female redescribed and illustrated by Tang (1991).

Pulvinaria mammeae Maskell

Pulvinaria mammeae Maskell, 1895b: 59.
TYPE DATA. Syntypes female, SANDWICH ISLANDS [=HAWAII]: on
  Mammea americana (NZAC).
DISTRIBUTION. NEW ZEALAND and PACIFIC REGION: Hawaii.
HOST PLANTS. Anacardiaceae: *Mangifera indica.* **Guttiferae:**

Mammea americana. **Lauraceae:** *Persea americana.* **Malvaceae:** *Hibiscus.* **Moraceae:** *Ficus palawanensis, F. variegata.* **Punica-ceae:** *Punica granatum.* **Rosaceae:** *Prunus.* **Rubiaceae:** *Coffea.* **Rutaceae:** *Citrus aurantium.* **Sapindaceae:** *Litchi chinensis.*
REMARKS. Adult female redescribed and illustrated by Zimmerman (1948). Distribution and host plant records given by Zimmerman (1948) and by Nakahara (1981b).

Pulvinaria marmorata Cockerell

Pulvinaria marmorata Cockerell, 1898d: 130.
Pulvinaria bigeloviae marmorata Cockerell; Cockerell, 1899b: 394.
TYPE DATA. Syntypes female, U.S.A.: New Mexico, Organ Mountains, Dripping Spring, on undetermined herbaceous plant (BMNH, USNM).
DISTRIBUTION. NEARCTIC REGION: New Mexico.

Pulvinaria maskelli Olliff

Pulvinaria maskelli Olliff, 1891: 667.
Signoretia atriplicis Maskell, 1892a: 23. Lectotype female designated by Qin & Gullan (1992), AUSTRALIA: New South Wales, Wentworth, on *Atriplex* sp. (ANIC). Syn. by Fernald, 1903: 136.
Pulvinaria maskelli spinosior Maskell, 1894c: 78. Lectotype female designated by Qin & Gullan (1992), AUSTRALIA: on *Frenela robusta* (BMNH). Syn. by Froggatt, 1915: 415 and by Qin & Gullan, 1992: 132.
Pulvinaria nuytsiae Maskell, 1897a: 313. Lectotype female designated by Qin & Gullan (1992), AUSTRALIA: Western Australia, Perth, on *Nuytsia floribunda* (ANIC). Syn. by Qin & Gullan, 1992: 132.
Pulvinaria maskelli viminariae Fuller, 1897b: 1345. Syntypes female, AUSTRALIA: Western Australia, Perth, on *Viminaria denudata* and *Hakea ilicifolia* (probably lost, Qin & Gullan, 1992). Syn. by Fernald, 1903: 136 and by Qin & Gullan, 1992: 132.
Ctenochiton nuytsiae Fuller, 1897b: 1345. Syntypes male, AUSTRALIA: Western Australia, on *Nuytsia floribunda* (ANIC). Syn. by Qin & Gullan, 1992: 132.
Ctenochiton nuytsiae minor Fuller; Fuller, 1897b: 1345 [NOMEN NUDUM].
Pulvinaria maskelli nuytsiae (Fuller); Cockerell, 1899b: 394.
Pulvinaria maskelli nuytsiae Maskell; Fernald, 1903: 136 [ERRONEOUS AUTHORSHIP].
Pulvinaria maskelli novemarticulata Green, 1915d: 48. Lectotype female designated by Qin & Gullan (1992), AUSTRALIA: Victoria, Mallee, on *Hymenanthera dentata* (BMNH). Syn. by Qin & Gullan, 1992: 132.
Pulvinaria newmani Froggatt, 1915: 417. Lectotype female designated by Qin & Gullan (1992), AUSTRALIA: Western Australia, Harvey District, on *Jacksonia* sp. (BCRI). Syn. by Qin & Gullan, 1992: 132.
Pulvinaria daveyi Froggatt, 1923: 162. Syntypes female, AUSTRALIA: Victoria, Bright, on *Callitris* sp. (probably lost; Qin & Gullan, 1992). Syn. by Qin & Gullan, 1992: 132.

TYPE DATA. Lectotype female designated by Qin & Gullan (1992) female, AUSTRALIA: New South Wales, host plant not indicated (ANIC).
DISTRIBUTION. AUSTRALIAN REGION: New South Wales, South Australia, Victoria, Western Australia.
HOST PLANTS. Chenopodiaceae: *Atriplex nummularia, A. rhagodioides, A. stipitata, A. vesicaria, Rhagodia hastata, R. obovata, R. spinesens, Salicornia.* **Cupressaceae:** *Callitris canescens, C. robusta.* **Euphorbiaceae:** *Beyeria viscosa.* **Leguminosae:** *Gastrolobium bilobum, Jacksonia, Oxylobium lanceolatum, Viminaria denudata.* **Loranthaceae:** *Nuytsia floribunda.* **Myoporaceae:** *Myoporum.* **Myrtaceae:** *Eucalyptus.* **Proteaceae:** *Hakea ilicifolia.* **Santalaceae:** *Santalum spicatum, Dodonaea.* **Solanaceae:** *Lycium australe.* **Violaceae:** *Hymenanthera dentata.*
REMARKS. Adult female redescribed and illustrated by Olliff (1892), Froggatt (1915) and by Qin & Gullan (1992). Adult male described and illustrated by Olliff (1892). Distribution and host plant records given by Olliff (1892), Maskell (1892a, 1897a), Fuller (1897b, 1899), Green (1915d), Froggatt (1915, 1923) and by Qin & Gullan (1992).

Pulvinaria merwei Joubert

Pulvinaria merwei Joubert, 1925: 121.
TYPE DATA. Syntypes female, SOUTH AFRICA: Durban, on sweet potato (SANC).
DISTRIBUTION. ETHIOPIAN REGION: South Africa.
HOST PLANTS. Convolvulaceae: *Ipomoea batatas.*
REMARKS. Adult female redescribed and illustrated by Hodgson (1968a).

Pulvinaria mesembryanthemi (Vallot)

Coccus mesembryanthemi Vallot, 1829: 31.
Calypticus mesembrianthemi Costa, 1844: 273. Syntypes female, ITALY: Napoli, Posilipo beach, on *Mesembryanthemum acinaciforme* (probably lost; Pellizzari Scaltriti, personal communication, 1990). Syn. by Fernald, 1903: 136.
Pulvinaria biplicata Targioni Tozzetti; Targioni Tozzetti, 1868: 34 [UNJUSTIFIED REPLACEMENT NAME].
Pulvinaria mesembrianthemi (Vallot); Signoret, 1873a: 39 [MISSPELLING].
Icerya mesembryanthemi Peringuey, 1893: 52. Syntypes larvae, SOUTH AFRICA: Cape Peninsula, on *Mesembryanthemum edule* (probably lost). Syn. by Lindinger, 1935a: 145.
Calypticus mesembryanthemi Peringuey; Gill, 1988: 89 [MIS-SPELLING].
Pulvirariella mesembrianthemi (Vallot); Tang, 1991: 281 [MIS-SPELLING].
TYPE DATA. Syntypes female, FRANCE: Dijon, Botanic Gardens, on *Mesembryanthemum coccineum* and *M. acinaciforme* (probably lost).
DISTRIBUTION. AUSTRALIAN REGION: Australian Capital Territory, New South Wales, Queensland, South Australia, Tasmania, Western Australia. **ETHIOPIAN REGION:** South Africa, Swaziland, Zimbabwe. **NEOTROPICAL REGION:** Argentina. **NEARC-**

TIC REGION: California. PALAEARCTIC REGION: Canary Islands, Corsica, Egypt, England Scilly Isles, France, Hungary, Israel, Italy, Madeira, Portugal, Spain.
HOST PLANTS. Aizoaceae: *Carpobrotus acinaciformis*, *C. chilensis*, *C. edulis*, *Disphyma crassifolium*, *Mesembryanthemum acutiforme*, *M. crystallinum*, *M. rigidicaule*, *Sesuvium portulacastrum*. Chenopodiaceae: *Atriplex vesicaria*, *Lampranthus glaucus*.
REMARKS. Common Name - Iceplant Scale. Adult female redescribed and illustrated by Newstead (1903), Leonardi (1920), Gómez-Menor Ortega (1937), Quintana (1956b), Borchsenius (1957), De Lotto (1967b), Hodgson (1967b, 1968a), Gill (1988) and by Qin & Gullan (1992). Adult female redescribed by Tang (1991). Adult male described and illustrated by Quintana (1956b). Male test described and illustrated by Miller & Williams (1990). First Instar crawler described and illustrated by Gómez-Menor Ortega (1937) and by Quintana (1956b). Colour photograph of adult female given by Tranfaglia & Viggiani (1988). Distribution and host plant records given by Fernald (1903), Leonardi (1920), Brain (1920b), Hall (1922, 1932), Lindinger (1935a), Gómez-Menor Ortega (1937), Borchsenius (1957), Bytinski-Salz (1966), De Lotto (1967b), Hodgson (1967b, 1968a), Ezzat & Hussein (1969), Donaldson *et al.* (1978), Vieira *et al.* (1983), Martin Mateo (1984), Williams (1985c), Marotta (1987), Kozár (1986), Carnero Hernandez & Perez Guerra (1986), Gill (1988), Miller & Williams (1990), Tang (1991) and by Qin & Gullan (1992).
BIOLOGY. Appears to be restricted to plants of the Aizoaceae, with preference to *Mesembryanthemum edulis*, however, it was recorded from plants of the Chenopodiaceae. Washburn & Frankie (1985) studied the life history in California. Washburn & Washburn (1984) showed that first instar crawlers exhibit active aerial dispersal behaviour.
ECONOMIC IMPORTANCE. A potential pest to aizoaceous ground covers in California (Donaldson *et al.*, 1978).

Pulvinaria minuscula Danzig

Eupulvinaria minuscula Danzig, 1967: 142.
Pulvinaria (Eupulvinaria) minuscula Danzig; Danzig, 1980c: 263.
Eupulvinaria minuscula Danzig; Tang, 1991: 238.
TYPE DATA. Holotype female, USSR: Primorye Territory, Shkotovsk Region, Zmeinaya Hill, in Artemovka River Valley, near Lesnoi Kordon, on *Crataegus maximoviczii* (ZIAS).
DISTRIBUTION. PALAEARCTIC REGION: USSR Primorye Territory.
HOST PLANTS. Rosaceae: *Crataegus maximoviczii*.
REMARKS. Adult female redescribed and illustrated by Danzig (1980c). Adult female redescribed by Tang (1991).

Pulvinaria minuta Brethes

Pulvinaria minuta Brethes in Massini & Brethes, 1918: 153.
TYPE DATA. Syntypes female, ARGENTINA: Buenos Aires, on *Schinus dependens* (MLPA).
DISTRIBUTION. NEOTROPICAL REGION: Argentina.
HOST PLANTS. Anacardiaceae: *Schinus dependens*.

Pulvinaria mkuzei Hodgson

Pulvinaria mkuzei Hodgson, 1968a: 166.
TYPE DATA. Syntypes female, SOUTH AFRICA: Mkuze Game Reserve, on *Hermbstaedtia caffra* (SANC).
DISTRIBUTION. ETHIOPIAN REGION: South Africa.
HOST PLANTS. Amaranthaceae: *Hermbstaedtia caffra*. **Santalaceae:** *Thesium virens*.

Pulvinaria myricariae Bazarov

Pulvinaria myricariae Bazarov, 1971a: 64.
TYPE DATA. Holotype female, USSR: Tadzhikistan, Pamir, Bodom, on *Myricaria* sp. (ZIAS).
DISTRIBUTION. PALAEARCTIC REGION: USSR Tadzhikistan.
HOST PLANTS. Tamaricaceae: *Myricaria*.
BIOLOGY. Lives on roots of the host plant (Bazarov, 1971a).

Pulvinaria neocellulosa Takahashi

Pulvinaria neocellulosa Takahashi, 1940a: 24.
Eupulvinaria neocellulosa (Takahashi); Tang, 1991: 238.
TYPE DATA. Syntypes female, TAIWAN: Taihoku, on *Evodia merrillii* and *Murraya exotica* (IMZT).
DISTRIBUTION. ORIENTAL REGION: Taiwan.
HOST PLANTS. Rutaceae: *Evodia confusa, Evodia merrillii, Murraya exotica*.
REMARKS. Adult female redescribed and illustrated by Tao *et al.* (1983). Adult female redescribed by Tang (1991). Distribution and host plant records given by Tao *et al.* (1983) and by Tang (1991).

Pulvinaria nishigaharae (Kuwana)

Lecanium nishigaharae Kuwana, 1907: 192.
Pulvinaria nishigaharae (Kuwana); Takahashi, 1955b: 70.
TYPE DATA. Syntypes female, JAPAN: on mulberry (ITLJ).
DISTRIBUTION. PALAEARCTIC REGION: Japan, Korea.
HOST PLANTS. Moraceae: *Morus*. **Ulmaceae:** *Zelkova serrata*.
REMARKS. Adult female redescribed and illustrated by Kuwana (1917a) and by Paik (1978). Adult female redescribed by Takahashi (1956), Borchsenius (1957), Kawai (1980) and by Tang (1991). Colour photograph given by Kawai (1980). Distribution and host plant records given by Takahashi (1955b, 1956), Borchsenius (1957), Paik (1978), Kawai (1980) and by Tang (1991).

Pulvinaria obscura Newstead

Pulvinaria obscura Newstead, 1894c: 23.
TYPE DATA. Syntypes female, INDIA: Madras, Nungumbaukum, on *Hygrophila spinosa* (BMNH).
DISTRIBUTION. ORIENTAL REGION: India.
HOST PLANTS. Acanthaceae: *Hygrophila spinosa*.

Pulvinaria occidentalis Cockerell

Pulvinaria innumerabilis occidentalis Cockerell, 1897a: 13.
Pulvinaria occidentalis Cockerell; King, 1901g: 197.

Pulvinaria ehrhorni King, 1901d: 145. Syntypes female, U.S.A.: California, at Mountain View, on alder and willow (USNM). Syn. by Steinweden, 1946: 9.

Pulvinaria coulteri Cockerell, 1905f: 514. Syntypes female, U.S.A.: Colorado, Coulter, Middle Park, on wild rose (USNM). Syn. by Steinweden, 1946: 9.

Pulvinaria occidentalis subalpina Cockerell, 1910b: 428. Syntypes female, U.S.A.: Colorado, Tolland, on *Betula glandulosa* (USNM) Syn. by Steinweden, 1946: 9.

TYPE DATA. Syntypes female, U.S.A.: Washington State, on currant, hawthorn, plum, pear, mountain ash, poplar, gooseberry and alder (USNM).

DISTRIBUTION. NEARCTIC REGION: California, Canada British Columbia, Colorado, Oregon, Washington.

HOST PLANTS. Betulaceae: *Alnus, Betula glandulosa*. **Cornaceae:** *Cornus*. **Grossulariaceae:** *Grossularia divaricata, Ribes lacustre*. **Philadelphaceae:** *Deutzia*. **Rosaceae:** *Crataegus, Prunus, Rosa*. **Salicaceae:** *Populus alba*.

REMARKS. Sanders (1909a) synonymized this species with *P. vitis*, but Steinweden (1946) regarded it as a distinct species. Adult female redescribed and illustrated by Steinweden (1946). Williams & Kosztarab (1972) described and illustrated the adult female of a species, which was named *Pulvinaria* sp. near *occidentalis* Cockerell, based on specimens from *Ilex cassine* collected at Richmond, Virginia, U. S. A. Until the above specimens will be named according to the ICZN Code, the above has no nomenclatural status. Distribution and host plant records given by Cockerell (1905f, 1910b), Steinweden (1946) and by Williams & Kosztarab (1972).

Pulvinaria ochnaceae (Kuwana)

Lecanium (Coccus) ochnaceae Kuwana, 1909a: 154.
Coccus ochnaceae Kuwana; Sasscer, 1911: 67.
Lecanium ochnaceae Kuwana; Kuwana, 1917a: 172.
Pulvinaria ochnaceae (Kuwana); Takahashi, 1955b: 70.
Chloropulvinaria ochnaceae (Kuwana); Borchsenius, 1957: 220.

TYPE DATA. Syntypes female, JAPAN: Tokyo, on *Eurya ochnacea* (ITLJ).

DISTRIBUTION. PALAEARCTIC REGION: Japan.

HOST PLANTS. Theaceae: *Eurya ochnacea*.

REMARKS. Adult female redescribed by Borchsenius (1957). Kawai (1980) synonymized this species with *P. okitsuensis* Kuwana, 1914, but this contradicts the Law of Priority. Until further study these species are regarded as separate.

Pulvinaria okitsuensis Kuwana

Pulvinaria okitsuensis Kuwana, 1914: 5.
Chloropulvinaria okitsuensis (Kuwana); Borchsenius, 1957: 216.

TYPE DATA. Syntypes female, JAPAN: Okitsu, Shizuoka-ken, on orange (ITLJ).

DISTRIBUTION. PALAEARCTIC REGION: China, Japan.

HOST PLANTS. Buxaceae: *Buxus microphylla*. **Rutaceae:** *Citrus, Poncirus trifoliata*. **Theaceae:** *Eurya japonica, Thea japonica, T. sinensis*.

REMARKS. Adult female redescribed and illustrated by Kuwana (1917a). Adult female redescribed by Takahashi (1956), Borchsenius (1957), Kawai (1980) and by Tang (1991). Colour photograph given by Kawai (1980). Distribution and host plant records given by Takahashi (1955b, 1956), Takahashi & Tachikawa (1956), Borchsenius (1957), Kawai (1980), Yang (1982) and by Tang (1991).

Pulvinaria ornata Hempel

Pulvinaria ornata Hempel, 1912: 61.
TYPE DATA. Syntypes female, BRAZIL: Sao Paulo, Campinas, on *Arabidaea* sp. (MZSP).
DISTRIBUTION. NEOTROPICAL REGION: Brazil Sao Paulo.
HOST PLANTS. Bignoniaceae: *Arabidaea.*

Pulvinaria oyamae Kuwana

Pulvinaria oyamae Kuwana, 1902a: 60.
TYPE DATA. Syntypes female, JAPAN: Nagano-ken, on undetermined plant (ITLJ).
DISTRIBUTION. PALAEARCTIC REGION: Japan.
HOST PLANTS. Salicaceae: *Salix.*
REMARKS. Adult female redescribed and illustrated by Kuwana (1917a). Adult female redescribed by Takahashi (1956), Borchsenius (1957), Kawai (1980) and by Tang (1991). Colour photograph given by Kawai (1980). Distribution and host plant records given by Kuwana (1917a), Takahashi (1956), Borchsenius (1957), Kawai (1980) and by Tang (1991).

Pulvinaria paranaensis Hempel

Pulvinaria paranaensis Hempel, 1928: 235.
TYPE DATA. Syntypes female, BRAZIL: Rio Grande do Sul and Parana, on "herva-mate" and "conghonha" [=*Ilex paraguaiensis*] (IBSP).
DISTRIBUTION. NEOTROPICAL REGION: Brazil Parana, Brazil Rio Grande do Sul.
HOST PLANTS. Aquifoliaceae: *Ilex paraguariensis.*

Pulvinaria peninsularis Ferris

Pulvinaria peninsularis Ferris, 1921a: 88.
TYPE DATA. Holotype female, MEXICO: Lower California, San Barton, on undetermined shrub (UCD).
DISTRIBUTION. NEOTROPICAL REGION: Mexico Lower California.
NEARCTIC REGION: Texas.
HOST PLANTS. Amaranthaceae: *Celosia floribunda.* **Asclepiadaceae:** *Philibertia tomentella.* **Lythraceae:** *Nesaea salicifolia.* **Rhamnaceae:** *Karwinskia humboldtiana.* **Rutaceae:** *Citrus.* **Solanaceae:** *Solanum.* **Zygophyllaceae:** *Porlieria angustifolia.*
REMARKS. Adult female redescribed and illustrated by Steinweden (1946). Distribution and host plant records given by Steinweden (1946). Qin & Gullan (1992) indicated that this species is probably a synonym of *P. urbicula* Cockerell.

Pulvinaria peregrina (Borchsenius) n. comb.

Pulvinaria horii Kuwana; Borchsenius, 1950: 151 [MISIDENTIFICA-TION].
Eupulvinaria peregrina Borchsenius, 1953: 288.
TYPE DATA. Syntypes female, USSR: Georgia, Adzhar, on *Hibiscus* sp. and *Diospyros kaki* (ZIAS).
DISTRIBUTION. PALAEARCTIC REGION: USSR Azerbaijan, USSR Georgia.
HOST PLANTS. Aceraceae: *Acer japonica.* **Ebenaceae:** *Diospyros kaki, D. lotus.* **Ericaceae:** *Vaccinium myrtillus.* **Malvaceae:** *Hibiscus japonica.* **Rosaceae:** *Cydonia vulgaris, Pyrus caucasica, Rosa.* **Rutaceae:** *Poncirus trifoliata.* **Ulmaceae:** *Celtis sinensis.*
REMARKS. Adult female redescribed and illustrated by Borchsenius (1957) and by Hadzibejli (1977, 1983). Adult female redescribed by Tang (1991). Distribution and host plant records given by Borchsenius (1957), Hadzibejli (1977, 1983) and by Tang (1991).

Pulvinaria persicae Newstead

Pulvinaria persicae Newstead, 1892: 142.
TYPE DATA. Syntypes female, ENGLAND: Cheshire, Knutsford, High Legh, on peach under glass (BMNH).
DISTRIBUTION. PALAEARCTIC REGION: England.
HOST PLANTS. Rosaceae: *Persica vulgaris.*

Pulvinaria phaiae Lull

Pulvinaria phaiae Lull, 1899: 237.
TYPE DATA. Syntypes female, U.S.A.: Massachusetts, Amherst (in greenhouse), on *Phaius maculatus* and *Phaius* sp. (USNM).
DISTRIBUTION. NEARCTIC REGION: California, Mississippi, New York.
HOST PLANTS. Orchidaceae: *Odontoglossum grande, Phaius maculatus, Phalaenopsis .*
REMARKS. Common Name - Cottony Orchid Scale. Fernald (1903) synonymized this species with *P. floccifera*, however Steinweden (1946) concluded that these are distinct species. Adult female redescribed and illustrated by Steinweden (1946) and by Gill (1988). Distribution and host plant records given by Steinweden (1946) and by Gill (1988).
BIOLOGY. This species appears to be restricted to orchids.

Pulvinaria photiniae Kuwana

Pulvinaria photiniae Kuwana, 1914: 4.
Eupulvinaria photiniae (Kuwana); Borchsenius, 1953: 288.
TYPE DATA. Syntypes female, JAPAN: Tokyo, Nishigahara, on *Photinia villosa* and *Celtis sinensis* (ITLJ).
DISTRIBUTION. PALAEARCTIC REGION: Japan.
HOST PLANTS. Rosaceae: *Photinia villosa.* **Ulmaceae:** *Celtis sinensis.*
REMARKS. Adult female redescribed and illustrated by Kuwana (1917a) and by Tang (1991). Adult female redescribed by Takahashi (1956), Borchsenius (1957) and by Kawai (1980). Colour photograph given by Kawai (1980). Distribution and host plant records given by

Kuwana (1917a), Takahashi (1956), Borchsenius (1957), Kawai (1980) and by Tang (1991).

Pulvinaria pistaciae Bodenheimer

Pulvinaria pistaciae Bodenheimer, 1926b: 189.
Anapulvinaria pistaciae (Bodenheimer); Borchsenius, 1952: 301.
Anapulvinaria pistaceae (Bodenheimer); Hadzibejli, 1977: 550 [MIS-
    SPELLING].
Anapulvinaria pistacia (Bodenheimer); Hadzibejli, 1983: 120 [MIS-
    SPELLING].
TYPE DATA. Lectotype female designated by Ben-Dov & Harpaz
    (1985), ISRAEL: Mount Kana'an, Safad, on *Pistacia palestina*
    (BMNH).
DISTRIBUTION. PALAEARCTIC REGION: Afghanistan, Cyprus,
    Iran, Iraq, Israel, Syria, Turkey, USSR Armenia, USSR Azerbai-
    jan, USSR Crimea, USSR Georgia, USSR Kirgizia, USSR Tadzhi-
    kistan, USSR Turkmenia, USSR Uzbekistan.
HOST PLANTS. Anacardiaceae: *Pistacia khinjuk, P. palestina, P.
    vera, Rhus coriaria.*
REMARKS. Adult female redescribed and illustrated by Borchsenius
    (1937, 1957), Davatchi (1958), Hadzibejli (1977, 1983), Tereznikova
    (1981) and by Tang (1991). Distribution and host plant records given
    by Bodenheimer (1943, 1944b, 1953), Borchsenius (1937, 1952,
    1957), Davatchi (1958), Abu-Yaman (1970), Danzig (1972d), Geor-
    ghiou (1977), Hadzibejli (1977, 1983), Tereznikova (1981) and by
    Tang (1991).
BIOLOGY. In Israel it develops on *P. palestina* one annual genera-
    tion. Larval instars develop on twigs during summer and winter
    (June until April). At April-May the young females migrate to the new
    flush of leaves on which they reproduce (Ben-Dov, 1971). Life cycle
    in Iraq studied by Abu-Yaman (1970).
ECONOMIC IMPORTANCE. Abu-Yaman (1970) evaluated the
    chemical control of this pest of pistachio in Iraq.

Pulvinaria platensis Brethes

Pulvinaria platensis Brethes in Massini & Brethes, 1918: 149.
TYPE DATA. Syntypes female, ARGENTINA: Buenos Aires, on *Euge-
    nia* sp. (MLPA).
DISTRIBUTION. NEOTROPICAL REGION: Argentina Buenos Aires.
HOST PLANTS. Myrtaceae: *Eugenia.*

Pulvinaria plucheae Ehrhorn

Pulvinaria plucheae Ehrhorn, 1906: 334.
TYPE DATA. Syntypes female, U.S.A.: California, San Diego, on
    Pluchea sericea (UCD).
DISTRIBUTION. NEARCTIC REGION: California.
HOST PLANTS. Compositae: *Pluchea sericea.*
REMARKS. Steinweden (1946) and Gill (1988) did not present a
redescription of this species.

Pulvinaria polygonata Cockerell

Pulvinaria polygonata Cockerell, 1905e: 131.

Pulvinaria cellulosa Green, 1909: 262. Lectotype female designated by Williams & Watson (1990), SRI LANKA: Pundaluoya, on *Citrus* sp. (BMNH). Syn. by Borchsenius, 1957: 219.
Pulvinaria nerii Kanda, 1950: 35. Syntypes female, CHINA: North China, Shanxi, on *Nerium odorum* (KAYJ). Syn. by Tang, 1991: 230.
Chloropulvinaria polygonata (Cockerell); Borchsenius, 1957: 219.
Chloropulvinaria polygonata (Green); Yang, 1982: 158 [ERRONEOUS AUTHORSHIP].
Macropulvinaria polygonata (Cockerell); Tao *et al.*, 1983: 89.
Chloropulvinaria nerii (Kanda); Tang & Li, 1988: 103.
TYPE DATA. Syntypes female, PHILIPPINES: Manila, on undetermined tree (BMNH, USNM).
DISTRIBUTION. AUSTRO ORIENTAL REGION: Philippines. **AUSTRALIAN REGION:** Queensland. **NEW ZEALAND and PACIFIC REGION:** Cook Islands. **ORIENTAL REGION:** India, Sri Lanka, Taiwan, Vietnam. **PALAEARCTIC REGION:** China, Inner Mongolia, Japan.
HOST PLANTS. Amaryllidaceae: *Clivia miniata.* **Anacardiaceae:** *Mangifera indica.* **Apocynaceae:** *Nerium odorum, Plumeria rubra,* **Hydrangeaceae:** *Hydrangea.* **Magnoliaceae:** *Michelia yunnanensis.* **Rutaceae:** *Citrus aurantifolia, C. limon, C. nobilis, C. reticulata, C. sinensis, Murraya exotica.*
REMARKS. Adult female redescribed and illustrated by Morrison (1920), Tao *et al.* (1983), Tang & Li (1988), Williams & Watson (1990), Tang (1991) and by Qin & Gullan (1992). Adult female redescribed by Borchsenius (1957) and by Kawai (1980). Colour photograph given by Kawai (1980). Parida & Ghosh (1984) and Moharana (1990) reported chromosome number 2n=18 in India. Distribution and host plant records given by Green (1909, 1937), Morrison (1920), Takahashi (1939b), Borchsenius (1957), Ali (1971), Kawai (1980), Yang (1982), Tao *et al.* (1983), Varshney (1984), Sinha & Dinesh (1984), Varshney & Moharana (1987), Tang & Li (1988), Shafee *et al.* (1989), Williams & Watson (1990), Danzig & Konstantinova (1990), Tang (1991) and by Qin & Gullan (1992).
BIOLOGY. Life history and control studied in Taiwan (Takahashi, 1939b) and in India (Chatterji & Datta, 1974). Ali (1964) observed (as *P. cellulosa*) five generations per year, on mango in Bihar, India. The insect overwinters in the fifth generation.

Pulvinaria populeti Borchsenius

Pulvinaria populeti Borchsenius, 1953: 289.
TYPE DATA. Syntypes female, USSR: Kazakhstan, Semipalatinsk, on *Populus* sp. (ZIAS).
DISTRIBUTION. PALAEARCTIC REGION: USSR Kazakhstan.
HOST PLANTS. Salicaceae: *Populus.*
REMARKS. Adult female redescribed and illustrated by Borchsenius (1957). Adult female redescribed by Tang (1991). Distribution and host plant records given by Borchsenius (1957), Matesova (1968) and by Tang (1991).

Pulvinaria portblairensis Yousuf & Shafee

Pulvinaria portblairensis Yousuf & Shafee, 1988: 58.

TYPE DATA. Holotype female, INDIA: Andaman Islands, Port Blair, Wright Myo, on wild plant (AMUI).
DISTRIBUTION. ORIENTAL REGION: India Aldaman Islands.

Pulvinaria pruni Hunter

Pulvinaria pruni Hunter, 1900: 104.
TYPE DATA. Syntypes female, U.S.A.: Kansas, Wichita, on *Prunus* sp. (USNM).
DISTRIBUTION. NEARCTIC REGION: Kansas, Oklahoma, Texas.
HOST PLANTS. Rosaceae: *Prunus.*
REMARKS. Adult female redescribed and illustrated by Steinweden (1946). Distribution and host plant records given by Steinweden (1946).

Pulvinaria psidii Maskell

Pulvinaria psidii Maskell, 1893b: 223.
Pulvinaria cupaniae Cockerell, 1893i: 159. Syntypes female, JAMAICA: Kingston, on *Cupania edulis* (BMNH, USNM). Syn. by Green, 1909: 265.
Pulvinaria psidii philippina Cockerell, 1905e: 132. Syntypes female, PHILIPPINES: Lucena, Tayabas, on *Ficus* sp. (USNM). Syn. by Morrison, 1920: 185.
Pulvinaria darwiniensis Froggatt, 1915: 414. Lectotype female designated by Qin & Gullan (1992), AUSTRALIA: Northern Territory, Palmerson, on palm (BCRI). Syn. by Qin & Gullan, 1992: 144.
Lecanium vacuolatum Green; Dash, 1916: 42 [NOMEN NUDUM].
Pulvinaria cussoniae Hall, 1932: 188. Lectotype female designated by Williams & Watson (1990), ZIMBABWE: Macheke, on *Cussonia arborea* (BMNH). Syn. by Hodgson, 1968a: 168.
Pulvinaria gymnosporiae Hall, 1932: 189. Lectotype female designated by Williams & Watson (1990), ZIMBABWE: Salisbury, on *Gymnosporia* sp. (BMNH). Syn. by Hodgson, 1968a: 168.
Chloropulvinaria psidii (Maskell); Borchsenius, 1957: 217.
TYPE DATA. Syntypes female, HAWAII [=SANDWICH ISLANDS]: on *Psidium* sp. (NZAC, USNM).
DISTRIBUTION. AUSTRO ORIENTAL REGION: Indonesia, Malaysia, New Britain, Papua New Guinea, Philippines, Solomon Islands. **AUSTRALIAN REGION:** Australian Capital Territory, New South Wales, Northern Territory, Queensland. **ETHIOPIAN REGION:** Angola, Ivory Coast, Kenya, Malawi, South Africa, St. Helena, Uganda, Zaire, Zimbabwe. **MADAGASIAN REGION:** Aldabra Island, Farquhar Island, Madagascar, Mauritius, Providence Island, Reunion, Seychelles. **NEOTROPICAL REGION:** Bermuda, Brazil, Cuba, Dominican Republic, Guadeloupe, Jamaica, Puerto Rico, Venezuela, Virgin Islands. **NEW ZEALAND and PACIFIC REGION:** Cook Islands, Fiji, French Polynesia, Hawaii, Irian Jaya, Kiribati, Mariana Islands, New Caledonia, New Zealand, Niue, Ogasawara Islands, Palaus, Tahiti, Tonga, Truk Islands, Vanuatu, Western Samoa. **NEARCTIC REGION:** Alabama, Florida, Georgia, Missouri, New York, Washington, D. C. **ORIENTAL REGION:** Cambodia, India, Sri Lanka, Taiwan, Thailand. **PALAEARCTIC REGION:** Canary Islands, Egypt.
HOST PLANTS. Agavaceae: *Cordyline terminalis.* **Aizoaceae:**

Amaranthaceae: *Alternathera versicolor.* **Amaryllidaceae:** *Crinum moorei.* **Anacardiaceae:** *Comocladia, Mangifera indica, Pistacia atlantica, Schinus molle, S. terebinthifolius, Spondias dulcis.* **Apocynaceae:** *Alstonia scholaris, Carissa, Plumeria acutifolia, P. lambertiana, P. rubra, Pteralyxia macrocarpa.* **Araceae:** *Anthurium, Anthurium triumphorus, Colocasia antiquorum, C. esculenta, Monstera deliciosa, Zantedeschia aethiopica.* **Araliaceae:** *Aralia, Brassaia actinophylla, Cussonia arborea, Hedera helix, Meryta macrophylla, Schefflera.* **Aspleniaceae:** *Asplenium.* **Barringtoniaceae:** *Barringtonia.* **Bignoniaceae:** *Bignonia, Stenolobium, Tecoma stans.* **Boraginaceae:** *Cordia alliodora.* **Cannaceae:** *Canna indica.* **Caryophyllaceae:** *Dianthus.* **Celastraceae:** *Gymnosporia.* **Chenopodiaceae:** *Chenopodium pumilio.* **Combretaceae:** *Terminalia brassii.* **Compositae:** *Bidens pilosa, Chrysanthemum indicum, Dahlia variabilis, Eupatorium, Gerbera, Wedelia biflora.* **Convolvulaceae:** *Ipomoea grandiflora.* **Crassulaceae:** *Bryophyllum, Kalanchoe.* **Dicksoniaceae:** *Cibotium.* **Euphorbiaceae:** *Antidesma membranaceum, Bischofia javanica, Codiaeum, Euphorbia, Macaranga, Uapaca kirkiana.* **Geraniaceae:** *Pelargonium.* **Goodeniaceae:** *Scaevola floribunda, S. gaudichaudiana.* **Guttiferae:** *Clusia rosea, Garcinia mangostana.* **Heliconiaceae:** *Heliconia humilis.* **Lauraceae:** *Persea.* **Lythraceae:** *Lagerstroemia indica.* **Malvaceae:** *Hibiscus rosa-sinensis, H. syriacus, Malvaviscus arboreus, Thespesia populnea.* **Melastomataceae:** **Moraceae:** *Artocarpus heterophyllus, Ficus benghalensis, F. capensis, F. elastica, F. indica, Morus alba.* **Myristicaceae:** *Myristica castanaefolia.* **Myrtaceae:** *Callistemon, Eucalyptus deglupta, Eugenia jambolana, E. jambos, E. malaccensis, Metrosideros, Psidium guajava, P. pomiferum.* **Oleaceae:** *Jasminum.* **Onagraceae:** *Ludwigia capitata.* **Orchidaceae:** *Vanilla.* **Palmae:** *Livistona chinensis.* **Pandanaceae:** *Pandanus.* **Pinaceae:** *Pinus caribaea.* **Piperaceae:** *Piper methysticum.* **Pittosporaceae:** *Pittosporum macrophyllum, P. tobira.* **Polemoniaceae:** *Phlox.* **Polygonaceae:** *Muehlenbeckia platyclados.* **Punicaceae:** *Punica granatum.* **Rubiaceae:** *Bouvardia, Coffea arabica, C. canephora, Gardenia jasminoides, Ixora chinensis, I. coccinea, I. macrothyrsa, Morinda citrifolia, Psychotria rubra, Randia tahitensis, Straussia, Tarenna sambucina.* **Rutaceae:** *Boronia serrulata, Citrus aurantifolia, C. aurantium, C. limon, C. sinensis, Poncirus trifoliata.* **Sapindaceae:** *Dodonaea, Dodonaea triquetra, Euphoria longana, Litchi chinensis.* **Sapotaceae:** *Chrysophyllum cainito, C. oliviforme, Pometia pinnata.* **Solanaceae:** *Capsicum frutescens.* **Tamaricaceae:** *Tamarix gallica.* **Theaceae:** *Camellia sinensis, Thea.* **Verbenaceae:** *Clerodendrum, Duranta, Stachytarpheta.* **Violaceae: Zingiberaceae:** *Alpinia purpurata, Hedychium, Phaeomeria, Zingiber officinale.*

REMARKS. Common Name - Green shield scale. Adult female redescribed and illustrated by Green (1909), Kuwana (1917a), Morrison (1920), Steinweden (1946), Zimmerman (1948), Borchsenius (1957), Hodgson (1968a), Ezzat & Hussein (1969), Wang (1980), Kawai (1980), Tao *et al.* (1983), Hamon &Williams (1984), Williams & Watson (1990), Tang (1991) and by Qin & Gullan (1992). Colour photograph given by Kawai (1980) and by Hamon & Williams (1984). Intraspecific variation in taxonomic characters given by Hodgson

(1968a). Parida & Ghosh (1984) and Moharana (1990) reported chromosome number 2n–14 in India. Distribution and host plant records given by Cockerell (1905e), Newstead (1911a, 1917c), Green (1907, 1916e), Brain (1920b), Ballou (1926), Hall (1926, 1932), Takahashi (1942a, 1950a), Mamet (1943, 1954a), Steinweden (1946), Balachowsky (1957), Beardsley (1966a), Hodgson (1968a, 1969b), Ali (1971), Kawai *et al.* (1971), Almeida (1973), Matile-Ferrero (1976), Panis & Martin (1976), Wise (1977), Nakahara & Miller (1981), Tao *et al.* (1983), Hamon & Williams (1984), Couturier *et al.* (1985), Carnero Hernandez & Perez Guerra (1986), Waite (1986), Varshney & Moharana (1987), Williams & Williams (1988), Shafee *et al.* (1989), Williams & Watson (1990), Tang (1991), Hodgson & Hilburn (1991a, 1991b) and by Qin & Gullan (1992).

BIOLOGY. Green (1909) reported on overlapping generations in Sri Lanka, three generations were indicated in Taiwan (Takahashi, 1939b), whereas only two in Egypt (Salama & Saleh, 1970). Salama & Saleh (1970) studied the effect of environmental factors on differential distribution of the scale on guava trees in Egypt. El-Minshawy & Moursi (1976) studied the duration of development and fecundity on guava, in Egypt.

ECONOMIC IMPORTANCE. Regarded a mango pest in Egypt (Nada *et al.*, 1990).

Pulvinaria pulchra Danzig

Eupulvinaria pulchra Danzig, 1967: 144.
Pulvinaria (Eupulvinaria) pulchra Danzig; Danzig, 1980c: 262.
TYPE DATA. Holotype female, USSR: Primorye Territory, Suchansk, on *Acer barbinerve* (ZIAS).
DISTRIBUTION. PALAEARCTIC REGION: USSR Primorye Territory.
HOST PLANTS. Aceraceae: *Acer barbinerve*, *Acer mono*, *A. ukurunduense*. **Araliaceae:** *Aralia mandshurica*, *Eleutherococcus senticossus*. **Rosaceae:** *Rosa multiflora*, *R. rugosa*. **Rutaceae:** *Phellodendron amurense*.
REMARKS. Adult female redescribed and illustrated by Danzig (1980c). Adult female redescribed by Tang (1991). Distribution and host plant records given by Danzig (1980c) and by Tang (1991).

Pulvinaria randiae Hall

Pulvinaria randiae Hall, 1932: 193.
TYPE DATA. Syntypes female, ZIMBABWE: Inyazura, on *Randia vestita* (BMNH).
DISTRIBUTION. ETHIOPIAN REGION: Zimbabwe.
HOST PLANTS. Rubiaceae: *Randia vestita*.
REMARKS. Adult female redescribed and illustrated by Hodgson (1967b, 1968a).

Pulvinaria regalis Canard

Pulvinaria regalis Canard, 1968b: 951.

TYPE DATA. Syntypes female, FRANCE: Versailles, Chateau de Versailles, on *Tilia vulgaris* (MNHN).
DISTRIBUTION. PALAEARCTIC REGION: England, France.
HOST PLANTS. Aceraceae: *Acer pseudoplatanus.* **Aquifoliaceae:** *Ilex aquifolium.* **Hippocastanaceae:** *Aesculus hippocastanum.* **Lauraceae:** *Laurus nobilis.* **Magnoliaceae:** *Magnolia.* **Rutaceae:** *Skimmia japonica.* **Tiliaceae:** *Tilia vulgaris.* **Ulmaceae:** *Ulmus.*
REMARKS. Distribution and host plant records given by Halstead (1982) and by Speight & Nicol (1984). The fine structure of wax glands and wax morphology in the adult female described and micrographed by Foldi & Pearce (1985).
BIOLOGY. A univoltine species in France (Canard, 1968b) and in Southern England (Halstead, 1982; Speight & Nicol, 1984).
ECONOMIC IMPORTANCE. Reported as an urban pest in London (Speight & Nicol, 1984).

Pulvinaria rhizophila Borchsenius

Pulvinaria artemisiae Lichtenstein; Archangelskaya, 1937: 134 [MISIDENTIFICATION and ERRONEOUS AUTHORSHIP].
Pulvinaria rhizophila Borchsenius, 1952: 298.
TYPE DATA. Syntypes female, USSR: Kirgizia, Makbal crossing, on roots of *Artemisia* sp. (ZIAS).
DISTRIBUTION. PALAEARCTIC REGION: Mongolia, USSR Kirgizia.
HOST PLANTS. Compositae: *Artemisia santolinifolia.*
REMARKS. Adult female redescribed and illustrated by Borchsenius (1957). Adult female redescribed by Tang (1991).

Pulvinaria rhoicina De Lotto

Pulvinaria rhoicina De Lotto, 1979: 250.
TYPE DATA. Holotype female, SOUTH AFRICA: Cape Province, Noll's Halt, on *Rhus* sp. (SANC).
DISTRIBUTION. ETHIOPIAN REGION: South Africa.
HOST PLANTS. Anacardiaceae: *Rhus.*

Pulvinaria rhois Ehrhorn

Pulvinaria rhois Ehrhorn, 1898b: 186.
TYPE DATA. Syntypes female, U.S.A.: California, Santa Clara County, Mountain View, on *Rhus diversiloba* (UCD).
DISTRIBUTION. NEARCTIC REGION: California.
HOST PLANTS. Anacardiaceae: *Rhus diversiloba.* **Grossulariaceae:** *Ribes.* **Rosaceae:** *Persica vulgaris, Prunus malus.*
REMARKS. Adult female redescribed and illustrated by Ferris (1920b), Steinweden (1946) and by Gill (1988). Distribution and host plant records given by Ferris (1920b), Steinweden (1946) and by Gill (1988).

Pulvinaria saccharia De Lotto

Pulvinaria saccharia De Lotto, 1964: 863.
Pulvinaria sorghicola De Lotto; Williams, 1982b: 115 [as TYPING ERROR for *P. saccharia*; line 23].
TYPE DATA. Holotype female, SOUTH AFRICA: Natal, Durban, on *Saccharum officinarum* (SANC).

DISTRIBUTION. ETHIOPIAN REGION: Kenya, Mali, Senegal, Sierra Leone, South Africa, Tanzania, Zimbabwe.
HOST PLANTS. Gramineae: *Bracharia brizantha, Oryza sativa, Saccharum officinarum.*
REMARKS. Adult female redescribed and illustrated by De Lotto (1966c). Adult female redescribed by Williams (1982b). The record of *Coccus elongatus* by Hall (1935: 74) from *Bracharia brizantha* is a misidentification of this species. Distribution and host plant records given by De Lotto (1966c), Hodgson (1967b, 1969b) and by Williams (1982b).

Pulvinaria salicicola Borchsenius

Pulvinaria betulae Signoret; Archangelskaya, 1923: 265 [MISIDEN-TIFICATION].
Pulvinaria salicicola Borchsenius, 1953: 289.
TYPE DATA. Syntypes female, USSR: Tadzikistan, Leninabad, on poplar (ZIAS).
DISTRIBUTION. PALAEARCTIC REGION: Inner Mongolia, USSR Tadzhikistan.
HOST PLANTS. Salicaceae: *Populus, Salix matsudana.*
REMARKS. Adult female redescribed and illustrated by Borchsenius (1957) and by Tang & Li (1988). Adult female redescribed by Tang (1991). Shmelev (1975) described and illustrated the adult female and male, and larval instars. Distribution and host plant records given by Borchsenius (1957), Shmelev (1975), Tang & Li (1988) and by Tang (1991).

Pulvinaria salicis (Bouche)

Lecanium salicis Bouche, 1851: 112.
Pulvinaria salicis (Bouche); Signoret, 1873a: 44.
TYPE DATA. Syntypes female, GERMANY: Berlin, on Weiden [=willow] and Papplen [=poplar] (lost; Sachtleben, 1944).
DISTRIBUTION. PALAEARCTIC REGION: Germany.
HOST PLANTS. Salicaceae: *Populus, Salix.*
REMARKS. Signoret (1873a) suggested that this species might be identical with *Lecanium capreae* (L.) and Fernald (1903) introduced the synonymy. However, Borchsenius (1957) rejected that synonymy and presented a redescription which was based on material from France. Bouche (1851) clearly described an ovisac, placed behind and under the female's body. The latter feature evidently refers this species to *Pulvinaria.*

Pulvinaria salicorniae Froggatt

Pulvinaria salicorniae Froggatt, 1915: 417.
TYPE DATA. Lectotype female designated by Qin & Gullan (1992), AUSTRALIA: Victoria, Little River, on *Salicornia* sp. (BCRI).
DISTRIBUTION. AUSTRALIAN REGION: Victoria.
HOST PLANTS. Chenopodiaceae: *Salicornia.*
REMARKS. Although Froggatt (1915) did not indicate this species as n. sp., but referred to it as a manuscript name by E. E. Green, the authorship should be credited to Froggatt, under Article 50 of the ICZN. Adult female redescribed and illustrated by Qin & Gullan

(1992). Distribution and host plant records given by Qin & Gullan (1992).

Pulvinaria savescui n. name

Pulvinaria euonymicola Savescu, 1983: 47. [HOMONYM of *Pulvinaria euonymicola* Lindinger].
TYPE DATA. Syntypes female, male and larva, ROMANIA: Mogosoaia, Otopeni, Valenii de Munte and Miercurea, on *Euonymus europaea* and on *E. verrucosa* (ASAR).
DISTRIBUTION. PALAEARCTIC REGION: Romania.
HOST PLANTS. Celastraceae: *Euonymus europaeus, E. verrucosa.*

Pulvinaria sericea (Fourcroy)

Chermes sericeus Fourcroy, 1785: 230.
Coccus lanatus Gmelin, 1790: 2221. Syntypes female, EUROPE: on *Quercus roboris* (lost). Syn. by Signoret, 1869a: 859.
Pulvinaria marginata Targioni Tozzetti; Targioni Tozzetti, 1868: 34 [UNJUSTIFIED REPLACEMENT NAME].
Pulvinaria lanatus (Gmelin); Signoret, 1873a: 38.
Pulvinaria sericea (Fourcroy); Cockerell, 1901c: 90.
TYPE DATA. Syntypes female, FRANCE: Paris, on *Quercus* sp. (FOUR).
DISTRIBUTION. PALAEARCTIC REGION: France.
HOST PLANTS. Fagaceae: *Quercus.*
REMARKS. Adult female redescribed by Borchsenius (1957).

Pulvinaria shinjii n. name

Pulvinaria acericola Shinji, 1935b: 772 [HOMONYM of *Pulvinaria acericola* (Walsh & Riley)].
TYPE DATA. Syntypes female, JAPAN: Morioka, on *Acer* sp., *Magnolia kobus* and *M. phypoleuca* (lost; S. Takagi, 1990, personal communication).
DISTRIBUTION. PALAEARCTIC REGION: Japan.
HOST PLANTS. Aceraceae: *Acer.* **Magnoliaceae:** *Magnolia kobus, M. phypoleuca.*

Pulvinaria simplex King

Pulvinaria simplex King in Hofer, 1903: 475.
TYPE DATA. Syntypes female, SWITZERLAND: on twigs of grapevine (USNM).
DISTRIBUTION. PALAEARCTIC REGION: Switzerland.
HOST PLANTS. Vitidaceae: *Vitis.*
REMARKS. Adult female redescribed by Borchsenius (1957).

Pulvinaria simulans Cockerell

Pulvinaria simulans Cockerell, 1894b: 310.
TYPE DATA. Syntypes female, TRINIDAD: Port-of-Spain, on undetermined tree (USNM).
DISTRIBUTION. NEOTROPICAL REGION: Guyana, Trinidad.
REMARKS. Distribution and host plant records given by Bodkin (1914).

Pulvinaria sorghicola De Lotto

Pulvinaria sorghicola De Lotto, 1979: 252.
TYPE DATA. Holotype female, SOUTH AFRICA: Transvaal, Pretoria,
on *Sorghum vulgare* (SANC).
DISTRIBUTION. ETHIOPIAN REGION: South Africa.
HOST PLANTS. Gramineae: *Sorghum vulgare.*

Pulvinaria taiwana Takahashi

Pulvinaria taiwana Takahashi, 1929a: 61.
Chloropulvinaria taiwana (Takahashi); Yang, 1982: 158.
TYPE DATA. Syntypes female, TAIWAN: Kagi, on *Mangifera Indica*
(IMZT).
DISTRIBUTION. ORIENTAL REGION: Taiwan.
HOST PLANTS. Anacardiaceae: *Mangifera indica.*
REMARKS. Adult female redescribed and illustrated by Tao *et al.*
(1983). Adult female redescribed by Tang (1991). Distribution and
host plantrecords given by Yang (1982), Tao *et al.* (1983) and by
Tang (1991).

Pulvinaria tapiae Mamet

Pulvinaria tapiae Mamet, 1951: 240.
TYPE DATA. Syntypes female, MADAGASCAR: Arivonimamo, on
"Tapia" (MNHN).
DISTRIBUTION. MADAGASIAN REGION: Madagascar.

Pulvinaria tenuivalvata (Newstead)

Lecanium tenuivalvatum Newstead, 1911a: 92.
Coccus tenuivalvatus (Newstead); Gowdey, 1917: 188.
Pulvinaria tenuivalvata (Newstead); De Lotto, 1965: 217.
TYPE DATA. Lectotype female immature designated by Williams
(1982b), UGANDA: Entebbe, on Citronella grass (BMNH).
DISTRIBUTION. ETHIOPIAN REGION: Uganda.
HOST PLANTS. Gramineae: *Cymbopogon citratus, Pennisetum
purpureum.*
REMARKS. Adult female redescribed and illustrated by De Lotto
(1965). Adult female redescribed by Williams (1982b). Distribution
and host plant records given by Gowdey (1917), De Lotto (1965) and
by Williams (1982b).

Pulvinaria terrestris Borchsenius

Pulvinaria betulae (Linnaeus); Borchsenius, 1949c: 175 [MISIDEN-
TIFICATION].
Pulvinaria terrestris Borchsenius; Borchsenius, 1953: 290.
TYPE DATA. Syntypes female, USSR: Armenia, Vagravar Megrinsk
Ridge, Ayrum, on *Crataegus* sp. and *Carpinus betulus* (ZIAS).
DISTRIBUTION. PALAEARCTIC REGION: USSR Armenia.
HOST PLANTS. Carpinaceae: *Carpinus betulus.* **Rosaceae:** *Cratae-
gus.*
REMARKS. Adult female redescribed and illustrated by Borchsenius
(1957). Adult female redescribed by Tang (1991).
BIOLOGY. Lives on roots of the host plant.

Pulvinaria tessellata Green

Pulvinaria tessellata Green, 1896: 8.
Propulvinarir tessellata (Green); Tang, 1991: 250 [MIS-SPELLING].
TYPE DATA. Syntypes female, SRI LANKA: Punduloya, on *Ophiorrhiza pectinata* (BMNH).
DISTRIBUTION. ORIENTAL REGION: Sri Lanka.
HOST PLANTS. Acanthaceae: *Strobilanthes.* **Rubiaceae:** *Ophiorrhiza pectinata.*
REMARKS. Adult female redescribed and illustrated by Green (1909). Adult female redescribed by Tang (1991). Distribution and host plant records given by Green (1909, 1937).

Pulvinaria thompsoni Maskell

Pulvinaria thompsoni Maskell, 1896b: 393.
Pulvinaria paradelpha Cockerell & Lidgett in Cockerell, 1899c: 15. Lectotype female designated by Qin & Gullan (1992), AUSTRALIA: Victoria, Mount Difficult, on *Acacia melanoxylon* (BMNH). Syn. by Qin & Gullan, 1992: 151.
TYPE DATA. Lectotype female designated by Qin & Gullan (1992), AUSTRALIA: locality and host plant not indicated (ANIC).
DISTRIBUTION. AUSTRALIAN REGION: Australian Capital Territory, Tasmania, Victoria.
HOST PLANTS. Leguminosae: *Acacia melanoxylon, Daviesia mimosoides.* **Sapindaceae:** *Dodonaea viscosa.*
REMARKS. Adult female redescribed and illustrated by Qin & Gullan (1992). Adult female redescribed by Froggatt (1915).

Pulvinaria tomentosa Green

Pulvinaria tomentosa Green, 1896: 8.
Eupulvinaria tomentosa (Green); Tang, 1991: 241.
TYPE DATA. Syntypes female, SRI LANKA: Pundaluoya, on leaves and small branches of undetermined shrub (BMNH).
DISTRIBUTION. ORIENTAL REGION: Sri Lanka.
REMARKS. Adult female redescribed and illustrated by Green (1909). Adult female redescribed by Tang (1991).

Pulvinaria torreyae Takahashi

Pulvinaria torreyae Takahashi; Takahashi, 1955e: 152 [NOMEN NUDUM].
Pulvinaria torreyae Takahashi, 1956: 29.
TYPE DATA. Syntypes female, JAPAN: Yokohama, on *Torreya nucifera, Taxus cuspidata* and *Cephalotaxus* sp. (HUSJ).
DISTRIBUTION. PALAEARCTIC REGION: Japan, Korea.
HOST PLANTS. Taxaceae: *Cephalotaxus, Taxus cuspidata, Torreya nucifera.*
REMARKS. Adult female redescribed and illustrated by Paik (1978). Adult female redescribed by Kawai (1980) and by Tang (1991). Colour photograph given by Kawai (1980). Distribution and host plant records given by Paik (1978), Kawai (1980) and by Tang (1991).

Pulvinaria tremulae Signoret

Pulvinaria tremulae Signoret, 1873a: 45.
TYPE DATA. Syntypes female, FRANCE: on "tremble" [=*Populus tremulae*] (VMNH).
DISTRIBUTION. PALAEARCTIC REGION: France.
HOST PLANTS. Salicaceae: *Populus tremulae.*
REMARKS. Lindinger (1912) synonymized this species with *Pulvinaria betulae*, whereas Borchsenius (1957) regarded it as a distinct species. Adult female redescribed by Tang (1991).

Pulvinaria tromelini Mamet

Pulvinaria tromelini Mamet, 1956a: 140.
TYPE DATA. Syntypes female, MADAGASCAR: Tromelin Island, on *Achyranthes aspera* (MNHN).
DISTRIBUTION. MADAGASIAN REGION: Tromelin Island.
HOST PLANTS. Amaranthaceae: *Achyranthes aspera.*

Pulvinaria tuberculata (Bouche)

Coccus tuberculatus Bouche, 1834: 18.
Pulvinaria tuberculatus (Bouche); Fernald, 1903: 139.
TYPE DATA. Syntypes female and male, GERMANY: Berlin, in greenhouse, on Malvaceae and *Cestrum* sp. imported from South America (lost; Sachtleben, 1944).
HOST PLANTS. Malvaceae: Solanaceae: *Cestrum.*
REMARKS. Adult female redescribed by Signoret (1877a).

Pulvinaria tyleri Cockerell

Pulvinaria tyleri Cockerell, 1905e: 132.
TYPE DATA. Syntypes female, PHILIPPINES: Batangas, on "cadena de amor" (USNM).
DISTRIBUTION. AUSTRO ORIENTAL REGION: Philippines.
HOST PLANTS. Polygonaceae: *Antigonon leptopus.* **Verbenaceae:** *Lantana camara.*
REMARKS. Adult female redescribed and illustrated by Morrison (1920) who suggested that it is doubtfully different from *P. psidii.* Distribution and host plant records given by Morrison (1920).

Pulvinaria uapacae Hodgson

Pulvinaria uapacae Hodgson, 1967b: 207.
TYPE DATA. Holotype female, ZIMBABWE: Inyanga, Honde Valley, on *Uapaca kirkiana* (SANC).
DISTRIBUTION. ETHIOPIAN REGION: Zimbabwe.
HOST PLANTS. Euphorbiaceae: *Uapaca kirkiana.*
REMARKS. Hodgson (1968a) presented additional notes on taxonomic characters.

Pulvinaria urbicola Cockerell

Pulvinaria urbicola Cockerell, 1893l: 160.
TYPE DATA. Syntypes female, JAMAICA: Kingston, on *Capsicum* sp. (USNM).
DISTRIBUTION. AUSTRO ORIENTAL REGION: Papua New Guinea,

Solomon Islands. **AUSTRALIAN REGION:** Northern Territory, Queensland. **NEOTROPICAL REGION:** Barbados, Bermuda, Cuba, Dominican Republic, Galapagos Islands, Jamaica, Puerto Rico, Trinidad. **NEW ZEALAND and PACIFIC REGION:** Caroline Islands, Cook Islands, Fiji, Hawaii, Kiribati, Mariana Islands, New Caledonia, Tuvalu, Vanuatu, Western Samoa. **NEARCTIC REGION:** Alabama, Florida, Louisiana, Texas. **PALAEARCTIC REGION:** Israel.

HOST PLANTS. Acanthaceae: *Pseuderanthemum, Ruellia.* **Adiantaceae:** *Adiantum.* **Amaranthaceae:** *Alternathera.* **Apocynaceae:** *Melochia tomentosa, Plumeria.* **Araceae:** *Alocasia macrorhiza, Monstera deliciosa.* **Bignoniaceae:** *Catalpa speciosa, Tecoma stans.* **Boraginaceae:** *Cordia subcordata.* **Bromeliaceae:** *Ananas.* **Caryophyllaceae:** *Dianthus caryophyllus.* **Compositae:** *Bidens pilosa, Dahlia pinnata, Eupatorium, Sonchus oleraceus.* **Convolvulaceae:** *Ipomoea batatas, I. horsfalliae.* **Cucurbitaceae:** *Momordica balsamina.* **Euphorbiaceae:** *Euphorbia heterophylla, Ricinus communis.* **Gentianaceae:** *Eustoma exaltatum.* **Hernandiaceae:** *Hernandia sonora.* **Malpighiaceae:** *Malpighia glabra.* **Myrtaceae:** *Psidium guajava.* **Nyctaginaceae:** *Mirabilis jalapa.* **Piperaceae:** *Piper methysticum.* **Polygonaceae:** *Coccoloba diversifolia, Polygonella fimbriata.* **Polypodiaceae: Sapindaceae:** *Litchi chinensis.* **Scrophulariaceae:** *Penstemon multiflorus.* **Solanaceae:** *Capsicum frutescens, C. grossum, Datura metel, Lycium carolinianum, Lycopersicon esculentum, Physalis peruviana, Solanum melongena, S. nigrum.* **Umbelliferae:** *Pteroselinum crispum.* **Verbenaceae:** *Citharexylum spinosum, Clerodendrum inerme, Lantana camara, Premna.* **Zingiberaceae:** *Alpinia purpurata.*

REMARKS. Adult female redescribed and illustrated by Zimmerman (1948), Hamon & Williams (1984), Williams & Watson (1990) and by Qin & Gullan (1992). Qin & Gullan (1992) indicated that *Pulvinaria peninsularis* Ferris is probably a synonym of this species. Colour photograph given by Hamon & Williams (1984). Distribution and host plant records given by Zimmerman (1948), Beardsley (1966a), Panis & Martin (1976), Nakahara & Miller (1981), Nakahara (1981b), Hamon & Williams (1984), Ben-Dov (1987), Williams & Watson (1990), Hodgson & Hilburn (1991a, 1991b) and by Qin & Gullan (1992).

BIOLOGY. Panis & Marro (1977) described a method for mass rearing this soft scale. The encapsulation of eggs of the parasitoid *Encyrtus infelix* (Embleton) was determined by Blumberg & Goldenberg (1992).

Pulvinaria vangueriae Hall

Pulvinaria vangueriae Hall, 1932: 194.
TYPE DATA. Syntypes female, ZIMBABWE: Mazoe, on *Ficus* sp. and *Vangueria* sp. (BMNH)
DISTRIBUTION. ETHIOPIAN REGION: Zimbabwe.
HOST PLANTS. Loganiaceae: *Strychnos spinosa.* **Moraceae:** *Ficus.* **Rubiaceae:** *Vanguera.*
REMARKS. Adult female redescribed and illustrated by Hodgson

(1967b, 1968a). Distribution and host plant records given by Hodgson (1967b, 1968a).

Pulvinaria viburni King

Pulvinaria viburni King, 1901i: 333.
TYPE DATA. Syntypes female, CANADA: Quebec, near Ottawa, in woods at Aylmer, on *Viburnum pubescens* (USNM).
DISTRIBUTION. NEARCTIC REGION: Canada Quebec.
HOST PLANTS. Capparidaceae: *Viburnum pubescens*.

Pulvinaria vini Hadzibejli

Pulvinaria vini Hadzibejli, 1960a: 54.
TYPE DATA. Syntypes female, male and larva, USSR: Georgia, on grapevine (GPPT, ZIAS).
DISTRIBUTION. PALAEARCTIC REGION: USSR Georgia.
HOST PLANTS. Vitidaceae: *Vitis vinifera*.
REMARKS. Adult female redescribed and illustrated by Hadzibejli (1983).

Pulvinaria vinifera King

Pulvinaria vinifera King in Hofer, 1903: 481.
TYPE DATA. Syntypes female, SWITZERLAND: on grapevine (USNM).
DISTRIBUTION. PALAEARCTIC REGION: Switzerland.
HOST PLANTS. Vitidaceae: *Vitis vinifera*.
REMARKS. Lindinger (1912) synonymized this species with *Pulvinaria betulae*, whereas Borchsenius (1957) regarded it as a distinct species.

Pulvinaria vitis (Linnaeus)

Coccus vitis Linnaeus, 1758: 456.
Coccus betulae Linnaeus, 1758: 455. Syntypes female, EUROPE: on *Betula alba* (lost; D. J. Williams, 1989, personal communication). Syn. by Newstead, 1903: 51.
Coccus carpini Linnaeus, 1758: 455. Syntypes female, EUROPE: on *Carpinus* (lost; D. J. Williams, 1989, personal communication). Syn. by Newstead, 1903: 51.
Coccus oxyacanthae Linnaeus, 1758: 456. Syntypes female, EUROPE: on *Crataegus oxyacantha* (lost; D. J. Williams, 1989, personal communication). Syn. by Newstead, 1903: 51.
Coccus crataegi Linnaeus; Linnaeus, 1766: 742 [UNJUSTIFIED REPLACEMENT NAME to *Coccus oxyacanthae*].
Chermes carpini (Linnaeus); Olivier, 1792: 441.
Chermes crataegi (Linnaeus); Olivier, 1792: 492.
Calypticus spumosus Costa, 1829: 10. Syntypes female, ITALY: San Jorio, on *Quercus* sp. (probably lost; G. Pellizzari Scaltriti, 1990, personal communication). Syn. by Signoret, 1869a: 875.
Coccus salicis Fitch, 1851: 69. Syntypes female, U.S.A.: New York, Salem, at the footbridge (AMNY). Syn. by Sanders, 1910: 60.
Lecanium crataegi (Linnaeus); Walker, 1852: 1073.
Lecanium maclurae Fitch, 1855: 38. Syntypes female, U.S.A.: New

York, on *Maclura aurantiaca* (AMNY). Syn. by Sanders, 1909a: 433.

Lecanium americanum Targioni Tozzetti; Targioni Tozzetti, 1868: 731 [UNJUSTIFIED REPLACEMENT NAME to *Coccus salicis* Fitch].

Pulvinaria betulae (Linnaeus); Signoret, 1873a: 31.

Pulvinaria carpini (Linnaeus); Signoret, 1873a: 34.

Pulvinaria oxyacanthae (Linnaeus); Signoret, 1873a: 40.

Pulvinaria populi Signoret, 1873a: 42. Syntypes female, FRANCE: Montpellier, on *Populus* (VMNH). Syn. by Newstead, 1903: 55.

Pulvinaria ribesiae Signoret, 1873a: 43. Syntypes female, FRANCE: Clamart, on grosseillier sanguin [=currant] (VMNH). Syn. by Newstead, 1903: 55.

Pulvinaria vitis (Linnaeus); Signoret, 1873a: 45.

Pulvinaria betulae alni Douglas 1891b: 100. Syntypes female, ENGLAND: Lewisham, on *Alnus glutinosa* (BMNH). Syn. by Lindinger, 1912: 375.

Pulvinaria maclurae (Fitch); Cockerell, 1894a: 32.

Pulvinaria innumerabilis tiliae King & Cockerell, 1898: 286. Syntypes female, U.S.A.: Massachusetts, Methuen, Lawrence and Andover, on *Tilia americana* (USNM). Syn. by Sanders, 1909a: 433.

Pulvinaria hunteri King, 1901d: 144. Syntypes female, U.S.A.: Kansas, Kansas City, on Maple, and Lawrence on honey locust [=*Gleditsia triacanthos*] (USNM). Syn. by Sanders, 1909a: 433.

Pulvinaria tiliae King & Cockerell; King, 1901i: 314.

Pulvinaria (Pulvinaria) betulae (Linnaeus); Danzig, 1980c: 263.

TYPE DATA. Syntypes female, EUROPE: on *Vitis vinifera* (lost; D. J. Williams, 1989, personal communication).

DISTRIBUTION. NEOTROPICAL REGION: Brazil. **NEW ZEALAND and PACIFIC REGION:** New Zealand. **NEARCTIC REGION:** California, Canada Ontario, Kansas, Massachusetts, New York. **PALAEARCTIC REGION:** Algeria, Austria, Bulgaria, China, Czechoslovakia, Denmark, England, Finland, France, Germany, Greece, Hungary, Iran, Ireland, Israel, Italy, Japan, Jordan, Luxemburg, Malta, Mongolia, Morocco, Netherlands, Norway, Poland, Portugal, Romania, Spain, Sweden, Switzerland, Turkey, USSR Altai, USSR Georgia, USSR Irkutsk Region, USSR Kazakhstan, USSR Moldavia, USSR Primorye Territory, USSR Sakhalin Island, USSR Ukraine, USSR Yakutsk, Yugoslavia.

HOST PLANTS. Aceraceae: *Acer.* **Betulaceae:** *Alnaster fruticosa, Alnus glutinosa, A. maximowiczii, Betula ermanii, B. exilis, B. mandshurica, B. platyphylla, B. pubescens, B. tauschii, B. verrucosa.* **Carpinaceae:** *Carpinus.* **Celastraceae:** *Euonymus europaeus.* **Corylaceae:** *Corylus.* **Fagaceae:** *Fagus.* **Grossulariaceae:** *Ribes grossularia, R. hispidulum, R. mejeri, R. nigrum, R. rubrum, R. sachalinense.* **Hippocastanaceae:** *Aesculus.* **Juglandaceae:** *Juglans.* **Oleaceae:** *Fraxinus mandshurica.* **Rosaceae:** *Crataegus chlorosarca, C. oxyacantha, Cydonia, Malus, Mespilus, Prunus, Pyrus, Rosa, Sorbus aucuparia, S. commixta, Spiraea hypericifolia, S. media, S. salicifolia.* **Salicaceae:** *Populus tremula, Salix arenaria, S. caprea, S. hultenii, S. pentandra, S. sachalinensis.* **Tiliaceae:** *Tilia.* **Vitidaceae:** *Vitis vinifera.*

REMARKS. Common Name - Cottony Grape Scale, or Cottony Vine Scale. Two interpretations are recognized as to the taxonomic and biological identity of this nominal species. East European students

(e. g. Borchsenius, 1957; Kosztarab & Kozár, 1988) regard *P. vitis, P. betulae, P. populi* and *P. ribesiae* as distinct nominal species. On the other hand Malumphy (1991), following morphological and experimental studies, concluded that *P. vitis* is a single plastic species, which exhibits considerable morphological, ecological, cytological, enzyme-isoenzyme and genetic variation. Malumphy's (1991) is adopted in this catalogue. If in the future, new species will have to be segregated from *P. vitis*, new names may be introduced, that will be based on adequate type-series. This approach appears to be more appropriate, rather than perpetuate the old nominal species, together with their burden of considerable confusion as to their taxonomy, biology and records. Adult female redescribed and illustrated by Douglas (1890a) (as *P. ribesiae*), Newstead (1903), Gómez-Menor Ortega (1937, 1958c), Steinweden (1946), Borchsenius (1957), Savescu (1960), Koteja (1964) (as *P. ribesiae*), Danzig (1980c), Tereznikova (1981), Hadzibejli (1983), Kosztarab & Kozár (1978, 1988), Tang (1991) (as *P. betulae*) and by Malumphy (1991). Adult male described and illustrated by Borchsenius (1957) (as *P. populi*). Phillips (1962) described the larval instars, based on material from Canada, Ontario. Danzig (1980c) discussed the differences in phenology and sex composition between populations of this species, but did not indicate whether these represent distinct species or merely host-induced variations. Distribution and host plant records given by Signoret (1873a), Douglas (1890a), Newstead (1903), Borchsenius (1957), Savescu (1960), Koteja (1964), Matesova (1968), Danzig (1977b, 1978b, 1980c), Kosztarab & Kozár (1978, 1988), Tereznikova (1981), Hadzibejli (1983), Koteja & Zak-Ogaza (1983), Kozár (1986), Gill (1988), Tang (1991) and by Malumphy (1991).
BIOLOGY. Develops one annual generation in Central Europe (Kosztarab & Kozár, 1988). Malumphy (1991) reported on one generation in England, but indicated that a second generation is possible in a single year, under suitable environmental conditions.
ECONOMIC IMPORTANCE. A pest of grapevine in Europe (Kosztarab & Kozár, 1988) and of peaches in Canada (Phillips, 1963).

Pulvinariella Borchsenius

Calittico Costa, 1844: 273. TYPE-SPECIES: *Calittico mesembrianthemi* Costa [=*Calypticus mesembrianthemi* Costa = *P. mesembryanthemi* (Vallot)], by present designation. Syn. by synonymy of type-species.
Pulvinariella Borchsenius, 1953: 287.
TYPE-SPECIES: *Coccus mesembryanthemi* Vallot, by original designation.
REMARKS. Generic characters discussed by Borchsenius (1957), De Lotto (1979) and by Tang (1991). The genus is here regarded as a subjective synonym of *Pulvinaria*. Costa (1844) prsented observations on a soft scale insect, which he found on *Mesembryanthemum acanaciforme* at Posilipo beach near Napoli, to which he referred as *Calittico* species. Signoret (1869a: 861) named this species *Calypticus mesembrianthemi* Costa. Although *Calittico* antedated *Pulvinariella* it is here regarded as an objective synonym of the latter, rather than resurrecting this unused, forgotten generic name.

Pulvinarisca Borchsenius

Pulvinarisca Borchsenius, 1953: 288.
TYPE-SPECIES: *Pulvinaria serpentina* Balachowsky, by original designation and monotypy.

Pulvinarisca serpentina (Balachowsky)

Pulvinaria serpentina Balachowsky, 1929: 310.
Pulvinarisca serpentina (Balachowsky); Borchsenius, 1953: 288.
TYPE DATA. Syntypes female, ALGERIA: Hoggar, Oued Aguelil (Tifedest), on *Acacia tortilis* (MNHN).
DISTRIBUTION. PALAEARCTIC REGION: Algeria.
HOST PLANTS. Leguminosae: *Acacia tortilis*.

Pulvinella Hempel

Pulvinaria (Pulvinella) Hempel, 1899: 132.
Pulvinella Hempel; Hempel, 1900: 416.
TYPE-SPECIES: *Pulvinaria (Pulvinella) pulchella* Hempel, by original-designation and monotypy.
REMARKS. Generic characters discussed by Hempel (1900b).

Pulvinella pulchella Hempel

Pulvinaria (Pulvinella) pulchella Hempel, 1899: 132.
Pulvinella pulchella Hempel; Hempel, 1900b: 481.
TYPE DATA. Syntypes female and larva, BRAZIL: Sao Paulo, Ypiranga, on *Baccharis* sp. (MZSP).
DISTRIBUTION. NEOTROPICAL REGION: Argentina, Brazil Sao Paulo.
HOST PLANTS. Anacardiaceae: *Schinus molle*. **Compositae:** *Baccharis*.
REMARKS. Adult female redescribed and illustrated by Hempel (1900b). Distribution and host plant records given by Lizer y Trelles (1942b).

Rhizopulvinaria Borchsenius

Rhizopulvinaria Borchsenius, 1952: 301.
Rhizopulvinara Tang; Tang, 1991: 282 [MIS-SPELLING].
TYPE-SPECIES: *Rhizopulvinaria virgulata* Borchsenius, by original designation.
REMARKS. Generic characters discussed by Borchsenius (1957), Canard (1966b, 1968a, 1971), Kosztarab & Kozár (1978, 1988), Tereznikova (1981) and by Tang (1991). Key to species: ARMENIA - Ter-Grigorian (1980). ASIA - Tang (1991). MEDITERRANEAN BASIN - Canard (1968a). MIDDLE ASIA - Matesova (1960). USSR - Borchsenius (1952, 1957), Danzig (1964). UKRAINE -Tereznikova (1981).

Rhizopulvinaria arenaria Canard

Rhizopulvinaria arenaria Canard, 1967b: 170.
TYPE DATA. Syntypes female, FRANCE: Saint-Cyprien-Plage (P. O.), on *Dianthus pyrenaicus* (MNHN).
DISTRIBUTION. PALAEARCTIC REGION: France.
HOST PLANTS. Caryophyllaceae: *Dianthus pyrenaicus*.

Rhizopulvinaria armeniaca Borchsenius

Pulvinaria artemisiae Signoret; Borchsenius, 1949c: 176 [MISIDEN-TIFICATION].
Rhizopulvinaria armeniaca Borchsenius, 1952: 306.
TYPE DATA. Syntypes female, USSR: Armenia, Legvaz Megrinsk Region, on Chenopodiaceae (ZIAS).
DISTRIBUTION. PALAEARCTIC REGION: USSR Armenia.
HOST PLANTS. Chenopodiaceae:
REMARKS. Adult female redescribed and illustrated by Borchsenius (1957).

Rhizopulvinaria artemisiae (Signoret)

Pulvinaria artemisiae Signoret, 1873a: 31.
Rhizopulvinaria artemisiae (Signoret); Borchsenius, 1952: 302.
TYPE DATA. Syntypes female, FRANCE: Montpellier, apparently on *Artemisia* (VMNH).
DISTRIBUTION. PALAEARCTIC REGION: Czechoslovakia, France, Germany, Hungary, Italy, Romania, Spain.
HOST PLANTS. Compositae: *Artemisia campestris, A. maritima.*
Cruciferae: *Alyssum.*
REMARKS. Adult female redescribed and illustrated by Borchsenius (1937), Savescu (1961), Canard (1968a) and by Kosztarab & Kozár (1978, 1988). Adult female redescribed by Borchsenius (1957). Bodenheimer (1953) recorded this species from Turkey and presented a detailed redescription, based on this record.

Rhizopulvinaria dianthi (Bodenheimer)

Pulvinaria dianthi Bodenheimer, 1943: 12.
Rhizopulvinaria dianthi (Bodenheimer); Borchsenius, 1952: 302.
TYPE DATA. Syntypes female, IRAQ: Ruwanduz gorge, on roots of *Dianthus* sp. (BCR).
DISTRIBUTION. PALAEARCTIC REGION: Iraq.
HOST PLANTS. Caryophyllaceae: *Dianthus.*
REMARKS. Adult female redescribed by Borchsenius (1957).

Rhizopulvinaria ericae (Balachowsky)

Ctenochiton ericae Balachowsky, 1936b: 122.
Rhizopulvinaria ericae (Balachowsky); Borchsenius, 1957: 270.
TYPE DATA. Syntypes female, FRANCE: Massif de l'Esterel, near Agay, on *Erica arborea* (MNHN).
DISTRIBUTION. PALAEARCTIC REGION: France.
HOST PLANTS. Ericaceae: *Erica arborea.*
REMARKS. Adult female redescribed by Borchsenius (1957).

Rhizopulvinaria gracilis Canard

Rhizopulvinaria gracilis Canard, 1967b: 179.
TYPE DATA. Syntypes female, FRANCE: Digne (B. A.), on *Dianthus virgineus* (MNHN).
DISTRIBUTION. PALAEARCTIC REGION: France.
HOST PLANTS. Caryophyllaceae: *Dianthus virgineus.*

Rhizopulvinaria grandicula Borchsenius

Rhizopulvinaria grandicula Borchsenius, 1952: 314.
TYPE DATA. Syntypes female, USSR: Armenia, Yerevan, on roots of *Artemisia* sp. and *Kochia prostrata* (ZIAS).
DISTRIBUTION. PALAEARCTIC REGION: USSR Armenia.
HOST PLANTS. Chenopodiaceae: *Kochia prostrata*. **Compositae:** *Artemisia*.
REMARKS. Adult female redescribed and illustrated by Borchsenius (1957).

Rhizopulvinaria grassei (Balachowsky)

Eulecanium grassei Balachowsky, 1936a: 56.
Rhizopulvinaria grassei (Balachowsky); Borchsenius, 1957: 280.
TYPE DATA. Syntypes female, FRANCE: Pyrenees-Orientales, plage de Pierrefitte, between Banyuls-s. -Mer and Cerbere, on *Anethum foeniculum* (MNHN).
DISTRIBUTION. PALAEARCTIC REGION: France, Italy.
HOST PLANTS. Caryophyllaceae: *Cerastium tomentosum*. **Umbelliferae:** *Anethum foeniculum*, *Foeniculum vulgare piperitum*.
REMARKS. Adult female redescribed and illustrated by Canard (1966b). Adult female redescribed by Borchsenius (1957). Distribution and host plant records given by Borchsenius (1957), Canard (1966b), Marotta (1987) and by Longo *et al.* (1989).
BIOLOGY. Develops one annual generation in France (Canard, 1966b).

Rhizopulvinaria halli Borchsenius

Ctenochiton artemisiae Hall, 1926: 15.
Rhizopulvinaria artemisiae (Hall); Borchsenius, 1952: 302 [SECONDARY HOMONYM of *Rhizopulvinaria artemisiae* (Signoret)].
Rhizopulvinaria halli Borchsenius, 1957: 279 [REPLACEMENT NAME].
TYPE DATA. Syntypes female, EGYPT: in the wadis east of Beni Suef, on roots of *Artemisia judaica* (BMNH).
DISTRIBUTION. PALAEARCTIC REGION: Egypt.
HOST PLANTS. Compositae: *Artemisia judaica*.
REMARKS. Adult female redescribed and illustrated by Borchsenius (1957) and by Ezzat & Hussein (1969) (as *C. artemisiae*). Adult female redescribed by Hosny (1939).

Rhizopulvinaria hissarica Borchsenius

Rhizopulvinaria hissarica Borchsenius, 1952: 314.
TYPE DATA. Syntypes female, USSR: Tadzhikistan, Gissarsk Ridge, on roots of *Dianthus* sp. (ZIAS).
DISTRIBUTION. PALAEARCTIC REGION: USSR Tadzhikistan.
HOST PLANTS. Caryophyllaceae: *Dianthus*.
REMARKS. Adult female redescribed and illustrated by Borchsenius (1957). Adult female redescribed by Tang (1991).

Rhizopulvinaria maritima Canard

Rhizopulvinaria maritima Canard, 1967a: 159.

TYPE DATA. Syntypes female, FRANCE: Plage du Racou, on *Helichrysum stoechas* (MNHN).
DISTRIBUTION. PALAEARCTIC REGION: France.
HOST PLANTS. Compositae: *Helichrysum stoechas.*

Rhizopulvinaria megriensis Borchsenius

Rhizopulvinaria megriensis Borchsenius, 1952: 307.
TYPE DATA. Syntypes female, USSR: Armenia, Nakhichevan-Dara gorge, Megri, on Chenopodiaceae (ZIAS).
DISTRIBUTION. PALAEARCTIC REGION: USSR Armenia.
HOST PLANTS. Chenopodiaceae:
REMARKS. Adult female redescribed and illustrated by Borchsenius (1957).

Rhizopulvinaria minima Borchsenius

Pulvinaria artemisiae Lichtenstein; Archangelskaya, 1937: 134 [MISIDENTIFICATION (in part) and ERRONEOUS AUTHORSHIP].
Rhizopulvinaria minima Borchsenius, 1952: 310.
TYPE DATA. Syntypes female, USSR: Kirgizia, Kirgizsk Ridge, Tald-Bulak and Tadzhikistan, Isfarinsk Region on *Astragalus* sp., *Rosa* sp. and on *Acantholimon* sp. (ZIAS).
DISTRIBUTION. PALAEARCTIC REGION: USSR Kirgizia, USSR Tadzhikistan.
HOST PLANTS. Leguminosae: *Astragalus.* **Plumbaginaceae:** *Acantholimon.* **Rosaceae:** *Rosa.*
REMARKS. Adult female redescribed and illustrated by Borchsenius (1957). Adult female redescribed by Tang (1991).

Rhizopulvinaria narzykulovi Bazarov & Shmelev

Rhizopulvinaria narzykulovi Bazarov & Shmelev, 1975: 107.
TYPE DATA. Holotype female, USSR: Tadzhikistan, Eastern Pamir near Kzhilrabat, on wormwood (ZIAS).
DISTRIBUTION. PALAEARCTIC REGION: USSR Tadzhikistan.
HOST PLANTS. Compositae: *Artemisia.*
BIOLOGY. Lives on roots and root neck of the plants (Bazarov & Shmelev, 1975).

Rhizopulvinaria polispina Matesova

Rhizopulvinaria polispina Matesova, 1960: 197.
TYPE DATA. Syntypes female, USSR: Kazakhstan, Kurchumskom Region, Irtsha, and Kokchetavsko Region, on *Artemisia* sp. and *Silene* sp. (ZIAS).
DISTRIBUTION. PALAEARCTIC REGION: USSR Kazakhstan.
HOST PLANTS. Caryophyllaceae: *Silene.* **Compositae:** *Artemisia, Seriphidium compactum.*
REMARKS. Adult female redescribed by Tang (1991). Distribution and host plant records given by Matesova (1968).

Rhizopulvinaria pyrethri Borchsenius

Rhizopulvinaria pyrethri Borchsenius, 1952: 313.

TYPE DATA. Syntypes female, USSR: Kazakhstan, near Uralska, on *Pyrethrum* sp. (ZIAS).
DISTRIBUTION. PALAEARCTIC REGION: USSR Kazakhstan.
HOST PLANTS. Compositae: *Pyrethrum.*
REMARKS. Adult female redescribed and illustrated by Borchsenius (1957). Adult female redescribed by Tang (1991).
BIOLOGY. Lives on roots of the host plant.

Rhizopulvinaria quadrispina Matesova

Rhizopulvinaria quadrispina Matesova, 1960: 201.
TYPE DATA. Syntypes female, USSR: Kazakhstan, Alma-Atinsk Region, on *Salsola* sp. (ZIAS).
DISTRIBUTION. PALAEARCTIC REGION: USSR Kazakhstan.
HOST PLANTS. Chenopodiaceae: *Salsola.*
REMARKS. Adult female redescribed and illustrated by Tang (1991).
BIOLOGY. Lives on roots of the host plant.

Rhizopulvinaria retamae (Hall)

Pulvinaria retamae Hall, 1923: 17.
Rhizopulvinaria retamae (Hall); Borchsenius, 1952: 302.
TYPE DATA. Syntypes female, EGYPT: Suez Road (at the 7th Tower), on *Retama roetam* (BMNH)
DISTRIBUTION. PALAEARCTIC REGION: Egypt, Israel.
HOST PLANTS. Compositae: *Artemisia herba-alba.* **Leguminosae:** *Retama roetam.* **Scrophulariaceae:** *Linaria.* **Umbelliferae:** *Pituranthos tortuosus.*
REMARKS. Adult female redescribed and illustrated by Borchsenius (1957) and by Ezzat & Hussein (1969). Distribution and host plant records given by Bodenheimer (1929), Borchsenius (1957), Ezzat & Hussein (1969) and by Ben-Dov (1971, 1980).

Rhizopulvinaria rhizophila Bazarov

Rhizopulvinaria rhizophila Bazarov, 1963: 41.
TYPE DATA. Holotype female, USSR: Tadzhikistan, at the road to Pamir near Faizabad, on *Artemisia* sp. (ZIAS).
DISTRIBUTION. PALAEARCTIC REGION: USSR Tadzhikistan.
HOST PLANTS. Compositae: *Artemisia.*
BIOLOGY. Lives on roots of the host plant (Bazarov, 1963).

Rhizopulvinaria saxatilis Canard

Rhizopulvinaria saxatilis Canard, 1967b: 176.
TYPE DATA. Syntypes female, FRANCE: Portel (Aude), on *Dianthus virgineus* (MNHN).
DISTRIBUTION. PALAEARCTIC REGION: France.
HOST PLANTS. Caryophyllaceae: *Dianthus virgineus.*

Rhizopulvinaria saxosa Shmelev

Rhizopulvinaria saxosa Shmelev, 1971: 61.
TYPE DATA. Holotype female, USSR: Tadzhikistan, Western Pamir, Khorog (Botanic Gardens), on *Dianthus* sp. (ZIAS).
DISTRIBUTION. PALAEARCTIC REGION: USSR Tadzhikistan.

HOST PLANTS. Caryophyllaceae: *Dianthus*.
BIOLOGY. Lives on roots and root neck of the host plant (Shmelev, 1971).

Rhizopulvinaria solitudina Matesova

Rhizopulvinaria solitudina Matesova, 1960: 202.
TYPE DATA. Syntypes female, USSR: Kazakhstan, Karagandinsk Region, Betpak-Dala Desert, on *Artemisia* sp. (ZIAS).
DISTRIBUTION. PALAEARCTIC REGION: USSR Kazakhstan.
HOST PLANTS. Compositae: *Artemisia*.
REMARKS. Adult female redescribed and illustrated by Tang (1991).
BIOLOGY. Lives on roots of the host plant.

Rhizopulvinaria spinifera Borchsenius

Pulvinaria artemisiae Signoret; Kiritshenko, 1931: 316 [MISIDEN-TIFICATION].
Rhizopulvinaria spinifera Borchsenius, 1952: 305.
Rhizopulvinaria spinifera (Bodenheimer); Matesova, 1968: 121 [ERRONEOUS AUTHORSHIP].
TYPE DATA. Syntypes female, USSR: Ukraine (near Odessa) and Georgia (near Borshomi) on *Dianthus orientalis*, *Gypsophila* sp. and *Achillea* sp. (ZIAS).
DISTRIBUTION. PALAEARCTIC REGION: Hungary, USSR Crimea, USSR Georgia, USSR Kazakhstan, USSR Moldavia, USSR Ukraine.
HOST PLANTS. Caryophyllaceae: *Dianthus orientalis*, *D. ponteder-ae*, *D. serotinus*, *Gypsophila*, *Minuartia setacea*, *Silene nutans*, *S. otites*. **Compositae:** *Achillea*, *Helichrysum arenarium*. **Legumi-nosae:** *Genista ovata*. **Rubiaceae:** *Galium verum*.
REMARKS. Adult female redescribed and illustrated by Borchsenius (1957), Kosztarab & Kozár (1978, 1988), Tereznikova (1981). Distribution and host plant records given by Borchsenius (1957), Matesova (1968), Tereznikova (1981), Kozár (1986), Kozár & Ostafichuk (1987) and by Kosztarab & Kozár (1988).
BIOLOGY. Lives on roots of the host plant.

Rhizopulvinaria transcaspica Borchsenius

Pulvinaria artemisiae Lichtenstein; Archangelskaya, 1937: 134 [MISIDENTIFICATION (in part) and ERRONEOUS AUTHORSHIP].
Rhizopulvinaria transcaspica Borchsenius, 1952: 316.
TYPE DATA. Syntypes female, USSR: Uzbekistan, near Samarkand and Turkmenia, near Krasnovodsk, on *Artemisia* sp. (ZIAS).
DISTRIBUTION. PALAEARCTIC REGION: USSR Turkmenia, USSR Uzbekistan.
HOST PLANTS. Compositae: *Artemisia*.
REMARKS. Adult female redescribed and illustrated by Borchsenius (1957). Adult female redescribed by Tang (1991).

Rhizopulvinaria turkestanica (Archangelskaya)

Pulvinaria artemisiae turkestanica Archangelskaya, 1931: 81.
Rhizopulvinaria turkestanica (Archangelskaya); Borchsenius, 1952: 302.

TYPE DATA. Syntypes female, USSR: Uzbekistan, Samarkand and Ferghana Districts, and Tadzhikistan, Kanibadam, on *Artemisia, Acanthophyllum spinosum, Scutellaria, Scrophularia, Silene* and on *Teucrium polium* (ZIAS).
DISTRIBUTION. PALAEARCTIC REGION: Mongolia, USSR Tadzhikistan, USSR Uzbekistan.
HOST PLANTS. Caryophyllaceae: *Acanthophyllum spinosum, Silene.* **Compositae:** *Artemisia.* **Labiatae:** *Scutellaria, Teucrium polium.* **Scrophulariaceae:** *Scrophularia.*
REMARKS. Adult female redescribed and illustrated by Borchsenius (1957). Adult female redescribed by Tang (1991).

Rhizopulvinaria turkmenica Borchsenius

Rhizopulvinaria turkmenica Borchsenius, 1952: 304.
TYPE DATA. Syntypes female, USSR: Turkmenia, near Ashkhabad, on *Aster* sp. (ZIAS).
DISTRIBUTION. PALAEARCTIC REGION: USSR Turkmenia.
HOST PLANTS. Compositae: *Aster.*
REMARKS. Adult female redescribed and illustrated by Borchsenius (1957). Adult female redescribed by Tang (1991).

Rhizopulvinaria ucrainica Tereznikova

Rhizopulvinaria ucrainica Tereznikova, 1981: 147.
TYPE DATA. Holotype female, USSR: Crimea, Karadag, on plant of Cruciferae (ZIAS).
DISTRIBUTION. PALAEARCTIC REGION: USSR Crimea.
HOST PLANTS. Cruciferae.
BIOLOGY. Lives on root neck of the host plant.

Rhizopulvinaria variabilis Borchsenius

Rhizopulvinaria variabilis Borchsenius, 1952: 311.
TYPE DATA. Syntypes female, USSR: Stalingrad Region, on *Artemisia* sp. (ZIAS).
DISTRIBUTION. PALAEARCTIC REGION: USSR Kazakhstan, USSR Stalingrad region.
HOST PLANTS. Compositae: *Artemisia scoparia.* **Labiatae:** *Scutellaria supina.*
REMARKS. Adult female redescribed and illustrated by Borchsenius (1957). Adult female redescribed by Tang (1991). Distribution and host plant records given by Borchsenius (1957), Matesova (1968) and by Tang (1991).

Rhizopulvinaria virgulata Borchsenius

Pulvinaria artemisiae Lichtenstein; Archangelskaya, 1937: 134 [MISIDENTIFICATION (in part) and ERRONEOUS AUTHORSHIP].
Rhizopulvinaria virgulata Borchsenius, 1952: 309.
TYPE DATA. Syntypes female, USSR: Turkmenia, Firyuzinsk gorge, and IRAN: Shakhrud, on roots of *Artemisia* sp. (ZIAS).
DISTRIBUTION. PALAEARCTIC REGION: Iran, Mongolia, USSR Turkmenia.
HOST PLANTS. Chenopodiaceae: *Sympegma regelii.* **Compositae:** *Artemisia.*

REMARKS. Adult female redescribed and illustrated by Borchsenius (1957). Adult female redescribed by Tang (1991). Distribution and host plant records given by Borchsenius (1957), Danzig (1972c) and by Tang (1991).
BIOLOGY. Lives on roots of the host plant.

Rhizopulvinaria viridis Borchsenius

Pulvinaria artemisiae Signoret; Kiritshenko, 1940: 119 [MISIDEN-TIFICATION].
Rhizopulvinaria viridis Borchsenius, 1952: 312.
TYPE DATA. Syntypes female, USSR: Crimea, near Rostov and near Simferopol, on roots of *Artemisia* sp. and *Dianthus* sp. (ZIAS).
DISTRIBUTION. PALAEARCTIC REGION: USSR Crimea, USSR Moldavia, USSR Ukraine.
HOST PLANTS. Caryophyllaceae: *Dianthus, Minuartia setacea.* **Compositae:** *Artemisia.* **Labiatae:** *Teucrium montanum, T. polium.*
REMARKS. Adult female redescribed and illustrated by Borchsenius (1957), Tereznikova (1981). Distribution and host plant records given by Borchsenius (1957), Tereznikova (1981) and by Kozár & Ostafichuk (1987).

Rhizopulvinaria zaisanica Matesova

Rhizopulvinaria zaisanica Matesova, 1960: 199.
TYPE DATA. Syntypes female, USSR: Kazakhstan, Zaisanskom and Karagandinsk Regions, on *Artemisia* sp. and *A. austrica* (ZIAS).
DISTRIBUTION. PALAEARCTIC REGION: USSR Kazakhstan.
HOST PLANTS. Chenopodiaceae: *Camphorosma lessingii.* **Compositae:** *Artemisia austrica, A. schrenkiana.*
REMARKS. Adult female redescribed by Tang (1991). Distribution and host plant records given by Matesova (1968) and by Tang (1991).

Rhizopulvinaria zygophylli Bazarov & Shmelev

Rhizopulvinaria zygophylli Bazarov & Shmelev, 1975: 110.
TYPE DATA. Holotype female, USSR: Tadzhikistan, Aktayu, near Gandzhano, on *Zygophyllum fabago* (ZIAS).
DISTRIBUTION. PALAEARCTIC REGION: USSR Tadzhikistan.
HOST PLANTS. Zygophyllaceae: *Zygophyllum fabago.*

Rhodococcus Borchsenius

Rhodococcus Borchsenius, 1953: 283.
Rhodoccus Tang; Tang, 1991: 211 [MIS-SPELLING]
TYPE-SPECIES: *Rhodococcus rosaeluteae* Borchsenius, by original designation.
REMARKS. Generic characters discussed by Borchsenius (1957), Kosztarab & Kozár (1978, 1988), Danzig (1980c), Yang (1982), Hadzibejli (1983) and by Tang (1991). Key to species: ASIA - Tang (1991). USSR - Borchsenius (1957), Danzig (1964). USSR (FAR EAST) - Danzig (1980c).

Rhodococcus marchali (Cockerell)

Eulecanium genevense marchali Cockerell, 1903a: 20.
Rhodococcus marchali (Cockerell); Borchsenius, 1957: 439.
TYPE DATA. Syntypes female, FRANCE: Seine, Fontenay, on *Rosa*
sp. (USNM).
DISTRIBUTION. PALAEARCTIC REGION: France.
HOST PLANTS. Rosaceae: *Rosa*.
REMARKS. Lindinger (1912) placed this species as a synonym of
Eulecanium corylt, whereas Borchsenius (1957) regarded it as a
species of *Rhodococcus*. Adult female redescribed by Borchsenius
(1957).

Rhodococcus perornatus (Cockerell & Parrott)

Lecanium (Eulecanium) perornatum Cockerell & Parrott, 1899: 236.
Eulecanium perornatum (Cockerell & Parrott); Fernald, 1903: 191.
Eulecanium bulgariense Wunn, 1939: 703. Syntypes female,
BULGARIA: on rose (DEPOSITORY UNKNOWN). Syn. by Kosztar-
ab & Kozár, 1988: 248.
Rhodococcus rosophilus Borchsenius, 1953: 284. Lectotype female-
designated by Danzig (1980c), USSR: Krasnoyarsk, Minusinsk,
on rose (ZIAS). Syn. by Kosztarab & Kozár, 1988: 248.
Rhodococcus perornatus (Cockerell & Parrott); Kosztarab & Kozár,
1988: 248.
TYPE DATA. Syntypes female, CZECHOSLOVAKIA: Moravia, on
Rosa canina (USNM).
DISTRIBUTION. PALAEARCTIC REGION: Austria, Bulgaria, Czech-
oslovakia, Hungary, Italy, USSR Krasnoyarsk, USSR Moldavia.
HOST PLANTS. Rosaceae: *Rosa canina, R. pimpinethifolia*.
REMARKS. Adult female redescribed and illustrated by Kosztarab &
Kozár (1978, 1988). Adult female redescribed by Tang (1991). Colour
photograph given by Olmi & Sampo (1976), Tremblay (1988b) and by
Kosztarab & Kozár (1988). Distribution and host plant records given
by Wünn (1939), Borchsenius (1953, 1957), Kozár (1970), Olmi &
Sampo (1976), Kozár et al. (1977), Kozár & Ostafichuk (1987), Marot-
ta (1987), Zahradnik (1987), Kosztarab & Kozár (1988) and by Tang
(1991).

Rhodococcus rosaeluteae Borchsenius

Rhodococcus rosaeluteae Borchsenius, 1953: 284.
TYPE DATA. Syntypes female, USSR: Kazakhstan, near Tyan-Shan,
on *Rosa lutea* (ZIAS).
DISTRIBUTION. PALAEARCTIC REGION: USSR Kazakhstan.
HOST PLANTS. Rosaceae: *Rosa lutea*.
REMARKS. Adult female redescribed and illustrated by Borchsenius
(1957). Adult female redescribed by Tang (1991).

Rhodococcus sariuoni Borchsenius

Rhodococcus sariuoni Borchsenius, 1955b: 302.
Rhodococcus sariuoni Borchsenius; Tang & Li, 1988: 99 [MIS-SPELL-
ING].
TYPE DATA. Syntypes female, KOREA: Pfenyan Province, between

Enyu and Zerengvan, on *Cerasus* sp., *Malus* sp. and *Spiraea* sp.
(ZIAS).
DISTRIBUTION. PALAEARCTIC REGION: China, Inner Mongolia,
Korea.
HOST PLANTS. Rosaceae: *Armeniaca vulgaris, Cerasus, Malus,
Prunus salicina, Spiraea.*
REMARKS. Adult female redescribed and illustrated by Borchsenius
(1957) and by Tang & Li (1988). Adult female redescribed by Tang
(1991). Distribution and host plant records given by Borchsenius
(1957, 1960), Tang & Li (1988) and by Tang (1991).

Rhodococcus spiraeae (Borchsenius)

Eulecanium spiraeae Borchsenius, 1949b: 340.
Rhodococcus spiraeae (Borchsenius); Borchsenius, 1957: 428.
Rhodococcus spiraoae (Borchsenius); Tang & Li, 1988: 96 [MIS-
SPELLING].
TYPE DATA. Lectotype female designated by Danzig (1980c), USSR:
Armenia, Alagez, forest around Inaklu, on *Spiraea* sp. (ZIAS).
DISTRIBUTION. PALAEARCTIC REGION: Czechoslovakia, Inner
Mongolia, Mongolia, USSR Armenia, USSR Caucasus, USSR
Irkutsk Region, USSR Kazakhstan, USSR Ukraine, USSR Ural,
USSR Yakutsk.
HOST PLANTS. Rosaceae: *Spiraea crenata, S. hypericifolia, S.
media, S. pubescens, S. triloba.*
REMARKS. Adult female redescribed and illustrated by Borchsenius
(1957), Rehacek (1957), Danzig (1980c), Tereznikova (1981) and by
Kosztarab & Kozár (1978, 1988). Adult female redescribed by Tang
(1991). Male described and illustrated by Borchsenius (1957) and by
Giliomee (1967). First instar larva described by Borchsenius (1957).
Distribution and host plant records given by Borchsenius (1957),
Rehacek (1957), Giliomee (1967), Danzig (1972c, 1974, 1978b,
1980c), Matesova (1968), Tereznikova (1981), Kosztarab & Kozár
(1988) and by Tang (1991).

Rhodococcus turanicus (Archangelskaya)

Lecanium coryli (Linnaeus); Archangelskaya, 1923: 265 [MISIDEN-
TIFICATION].
Physokermes coryli (Linnaeus); Archangelskaya, 1926: 163 [MISI-
DENTIFICATION].
Lecanium coryli turanicum Archangelskaya, 1937: 47.
Eulecanium turanicum (Archangelskaya); Borchsenius, 1949c: 173.
Rhodococcus turanicus (Archangelskaya); Borchsenius, 1957: 425.
TYPE DATA. Syntypes female, USSR: Uzbekistan, Tadzhikistan,
Turkmenistan, Kirgizia and Kazakhstan, on numerous host
plants, mainly of the Rosaceae (ZIAS).
DISTRIBUTION. PALAEARCTIC REGION: Afghanistan, Iran, USSR
Armenia, USSR Azerbaijan, USSR Georgia, USSR Kazakhstan,
USSR Kirgizia, USSR Tadzhikistan, USSR Turkmenia, USSR
Uzbekistan.
HOST PLANTS. Betulaceae: *Corylus avellana.* **Juglandaceae:**
Juglans regia. **Rhamnaceae:** *Rhamnus cathartica.* **Rosaceae:**
*Amygdalus, Armeniaca vulgaris, Cotoneaster vulgaris, Crataegus,
Cydonia vulgaris, Malus, Persica vulgaris, Prunus domestica,*

Prunus syriaca, Pyrus communis, Spiraea crenifolia. **Saxifraga-ceae:** *Ribes.* **Ulmaceae:** *Ulmus campestris.*
REMARKS. Adult female redescribed and illustrated by Borchsenius (1957), Hadzibejli (1983) and by Tang (1991). First instar crawler described and illustrated by Borchsenius (1957). Distribution and host plant records given by Borchsenius (1949c, 1957), Danzig (1972d), Hadzibejli (1983) and by Tang (1991).

Richardiella Matile-Ferrero & Le Ruyet

Richardiella Matile-Ferrero & Le Ruyet, 1985: 265.
TYPE-SPECIES: *Richardiella taiensis* Matile-Ferrero & Le Ruyet, by original designation and monotypy.

Richardiella taiensis Matile-Ferrero & Le Ruyet

Richardiella taiensis Matile-Ferrero & Le Ruyet, 1985: 265.
TYPE DATA. Holotype female, IVORY COAST: Tai, on *Gilbertioden-dron splendidum* (MNHN).
DISTRIBUTION. ETHIOPIAN REGION: Ivory Coast.
HOST PLANTS. Leguminosae: *Gilbertiodendron splendidum.*
BIOLOGY. Attended by ants of *Crematogaster* sp., under shelter (Matile-Ferrero & Le Ruyet, 1985).

Saccharipulvinaria Tao, Wong & Chang

Saccharipulvinaria Tao, Wong & Chang, 1983: 85.
TYPE-SPECIES: *Lecanium iceryi* Signoret, by original designation and monotypy.
REMARKS. Tao *et al.* (1983) erroneously named the type species *Coccus iceryi,* which is actually *Lecanium iceryi.* Generic characters discussed by Tang (1991). This genus is here regarded a subjective synonym of *Pulvinaria.*

Saccharolecanium Williams

Saccharolecanium Williams, 1980: 435.
TYPE-SPECIES: *Lecanium krugeri* Zehntner, by original designation and monotypy.
REMARKS. Generic characters discussed by Tang (1991). Key to species given by Tang (1991). Key to related genera given by Williams (1980).

Saccharolecanium fujianensis Tang

Saccharolecanium fujianensis Tang, 1991: 139.
TYPE DATA. Holotype female, CHINA: Shuyang County, Fujian Province, on *Sasa* sp. (EISC).
DISTRIBUTION. PALAEARCTIC REGION: China.
HOST PLANTS. Gramineae: *Sasa.*

Saccharolecanium krugeri (Zehntner)

Lecanium krugeri Zehntner, 1897: 563.
Saccharolecanium krugeri (Zehntner); Williams, 1980: 436.
TYPE DATA. Syntypes female, JAVA: West Java, Kagok, on sugar-cane (lost; Zehntner, 1954).

DISTRIBUTION. AUSTRO ORIENTAL REGION: Java, West Malaysia. **ORIENTAL REGION:** India.
HOST PLANTS. Gramineae: *Saccharum officinarum.*
REMARKS. Adult female redescribed and illustrated by Williams (1980). Adult female redescribed by Tang (1991). Distribution and host plant records given by Williams (1980) and by Tang (1991).

Saissetia Deplanche

Saissetia Deplanche, 1859: 6.
Sassetia Dunham; Dunham, 1954: 68 [MIS-SPELLING].
Seissetia Abrahao & Mamprim; Abrahao & Mamprim, 1958: 268 [MIS-SPELLING].
Saisettia Suomalainen; Suomalainen, 1962: 351 [MIS-SPELLING].
TYPE-SPECIES: *Lecanium coffeae* Walker, as proposed by Ben-Dov (1989a) and ruled by Opinion 1627 (1991).
REMARKS. Ben-Dov (1989a) showed that *Saissetia coffeae* Deplanche, 1859, the nominal type species of *Saissetia*, is a mealybug, and in order to conserve *Saissetia* Deplanche, 1859, as a genus in the Coccidae, proposed to suppress S. *coffeae* Deplanche, and designate *Lecanium coffeae* Walker the type species. Opinion 1627 (1991) adopted this proposal. Generic characters discussed by Leonardi (1920), Steinweden (1929), Gómez-Menor Ortega (1937, 1958c), Zimmerman (1948), De Lotto (1956a, 1957a, 1965), Borchsenius (1957), Hodgson (1967a), Williams & Kosztarab (1972), Paik (1978), Wang (1980), Yang (1982), Hamon & Williams (1984), Gill (1988), Williams & Watson (1990) and by Tang (1991). Key to species: AFRICA (South of the Sahara) - De Lotto (1957a, 1965). ASIA - Tang (1991). CALIFORNIA - Gill (1988). FLORIDA - Hamon & Williams (1984). HAWAII - Zimmerman (1948). MICRONESIA - Beardsley (1966a). TROPICAL SOUTH PACIFIC - Williams & Watson (1990). VIRGINIA - Williams & Kosztarab (1972). ZIMBABWE - Hodgson (1967a, 1969b).

Saissetia absona Hodgson

Saissetia absona Hodgson, 1969b: 31.
TYPE DATA. Holotype female, ZIMBABWE: Mazoe, on *Citrus* sp. (BMNH).
DISTRIBUTION. ETHIOPIAN REGION: Zimbabwe.
HOST PLANTS. Rutaceae: *Citrus.*

Saissetia anonae Hempel

Saissetia anonae Hempel, 1921: 143.
TYPE DATA. Syntypes female, BRAZIL: Rio de Janeiro, on *Annona* sp. imported from Madeira Island (MZSP).
DISTRIBUTION. NEOTROPICAL REGION: Brazil Sao Paulo.
HOST PLANTS. Annonaceae: *Annona.*

Saissetia argentina Morrison

Saissetia argentina Morrison, 1919: 83.
TYPE DATA. Syntypes female, ARGENTINA: Mendoza, host plant not indicated (USNM).
DISTRIBUTION. NEOTROPICAL REGION: Argentina Mendoza.

Saissetia auriculata Morrison

Saissetia auriculata Morrison, 1929: 54.
TYPE DATA. Holotype female, PANAMA: Canal Zone, Las Cascadas, on *Triplaris cumingiana* (USNM).
DISTRIBUTION. NEOTROPICAL REGION: Panama.
HOST PLANTS. Polygonaceae: *Triplaris cumingiana.*

Saissetia bobuae Takahashi

Saissetia bobuae Takahashi, 1935: 8.
TYPE DATA. Syntypes female, TAIWAN: Taihoku Prefecture, Kyanrawa, on *Symplocos [-Bobua] arisanensis* (IMZT).
DISTRIBUTION. ORIENTAL REGION: Taiwan.
HOST PLANTS. Symplocaceae: *Symplocos lancifolia.*
REMARKS. Adult female redescribed and illustrated by Tao *et al.* (1983). Adult female redescribed by Tang (1991). Distribution and host plant records given by Tao *et al.* (1983) and by Tang (1991).

Saissetia carnosa Hodgson

Saissetia carnosa Hodgson, 1969b: 33.
TYPE DATA. Holotype female, ZIMBABWE: Chimanimani Mountains, on *Protea welwitchiae* (BMNH).
DISTRIBUTION. ETHIOPIAN REGION: Zimbabwe.
HOST PLANTS. Proteaceae: *Faurea saligna, P. welwitchiae.*

Saissetia cassiniae (Maskell)

Lecanium cassiniae Maskell, 1891a: 15.
Lecanium (Saissetia) cassiniae (Maskell); Cockerell & Parrott, 1899: 163.
Saissetia cassiniae (Maskell); Fernald, 1903: 200.
TYPE DATA. Syntypes female, NEW ZEALAND: Wellington, Wairarapa and Hawke's Bay, on *Cassinia leptophylla* (NZAC).
DISTRIBUTION. NEW ZEALAND and PACIFIC REGION: New Zealand.
HOST PLANTS. Compositae: *Cassinia leptophylla.*
REMARKS. Sanders (1909a) and Borchsenius (1957) placed this species in synonymy with *S. oleae.*

Saissetia cerei Green

Lecanium (Saissetia) cerei Green, 1923a: 93.
TYPE DATA. Syntypes female, MADEIRA ISLANDS: Funchal, on *Cereus triangularis* (BMNH).
DISTRIBUTION. PALAEARCTIC REGION: Madeira.
HOST PLANTS. Cactaceae: *Cereus triangularis.*

Saissetia chimanimanae Hodgson

Saissetia chimanimanae Hodgson, 1967a: 12.
TYPE DATA. Holotype female, MOZAMBIQUE: Martin Falls, Chimanimani Mountains, on *Myrsine africana* (SANC).
DISTRIBUTION. ETHIOPIAN REGION: Mozambique.
HOST PLANTS. Myrsinaceae: *Myrsine africana.*

Saissetia chitonoides De Lotto

Saissetia chitonoides De Lotto, 1963: 191.
TYPE DATA. Holotype female, TANZANIA: Arusha, on *Annona* sp.
(BMNH).
DISTRIBUTION. ETHIOPIAN REGION: Tanzania.
HOST PLANTS. Annonaceae: *Annona.*

Saissetia citricola (Kuwana)

Takahashia citricola Kuwana, 1909a: 153.
Saissetia citricola (Kuwana); Takahashi & Tachikawa, 1956: 7.
Parasaissetia citricola (Kuwana); Yang, 1982: 178.
TYPE DATA. Syntypes female, JAPAN: Kumamoto, Hukuoka and
Wakayama, on citrus (ITLJ).
DISTRIBUTION. PALAEARCTIC REGION: Japan.
HOST PLANTS. Aquifoliaceae: *Ilex oldhami.* **Araliaceae:** *Fatsia
japonica.* **Ebenaceae:** *Diospyros kaki.* **Lauraceae:** *Cinnamomum
camphora, Laurus nobilis.* **Magnoliaceae:** *Magnolia kobus.*
Rosaceae: *Pyrus simonii.* **Rutaceae:** *Citrus.*
REMARKS. Adult female redescribed and illustrated by Kuwana
(1917a), Kawai (1980) and by Tang (1991). Colour photograph given
by Kawai (1980). Kawai (1980) synonymized *Pulvinaria marginata*
Ferris with this species. Distribution and host plant records given by
Takahashi & Tachikawa (1956), Kawai (1980), Yang (1982) and by
Tang (1991).

Saissetia coffeae (Walker)

Lecanium coffeae Walker, 1852: 1079.
Lecanium hemisphaericum Targioni Tozzetti, 1867: 26. Syntypes
female, ITALY: Firenze, Garden of Royal Museum, on *Bletia* sp.
and *Phyllarthron* sp. (probably lost; G. Pellizzari Scaltriti, 1990,
personal communication). Syn. by Cockerell & Parrott, 1899:
164.
Chermes anthurii Boisduval, 1867: 328. Syntypes female, FRANCE:
Paris, in greenhouse, on *Anthurium* sp. and on *Caladium* sp.
(Depository unknown). Syn. by Green, 1904d: 232.
Chermes filicum Boisduval, 1867: 335. Syntypes female, FRANCE:
Paris, in greenhouse, on fern, *Pteris* sp. (Depository unknown).
Syn. by Green, 1904d: 232.
Chermes hibernaculorum Boisduval, 1867: 337. Syntypes female,
FRANCE: Paris, in greenhouse, on fern, *Zamia, Ardisia, Grevillea,
Gardenia, Brexia* and many more hosts (Depository unknown).
Syn. by Kirkaldy, 1902: 105.
Lecanium hybernaculorum (Boisduval); Targioni Tozzetti, 1868: 37
[MIS-SPELLING].
Lecanium filicum (Boisduval); Signoret, 1874a: 436.
Lecanium hibernaculorum (Boisduval); Signoret, 1874a: 437.
Lecanium beaumontiae Douglas, 1887b: 95. Syntypes female,
ENGLAND: Royal Gardens, Kew, on *Beaumontia grandiflora*
(BMNH). Syn. by Sanders, 1909a: 439.
Lecanium clypeatum Douglas, 1888a: 59. Syntypes female, ENG-
LAND: Deptford, on *Adiantum capillus veneris*; Armagh, on fern;

Sale, on *Bryophyllum calycrinum* and *Asparagus plumosus* (BMNH). Syn. by Sanders, 1909a: 439.
Lecanium hemisphaericum hibernaculorum Boisduval; Cockerell, 1894f: 71.
Lecanium (Saissetia) beaumontiae (Douglas); Cockerell & Parrott, 1899: 163.
Lecanium (Saissetia) coffeae hibernacularum (Boisduval); Cockerell & Parrott, 1899: 164.
Lecanium (Saissetia) coffeae filicum (Boisduval); Cockerell & Parrott, 1899: 164.
Lecanium (Saissetia) coffeae clypeatum (Douglas); Cockerell & Parrott, 1899: 164.
Saissetia beaumontiae (Douglas); Cockerell, 1901d: 32.
Lecanium (Saissetia) hemisphaericum Targioni Tozzetti; Kuwana, 1902a: 63.
Coccus coffeae (Walker); Kirkaldy, 1902: 105.
Saissetia anthurii (Boisduval); Fernald, 1903: 200.
Saissetia filicum (Boisduval); Fernald, 1903: 201.
Saissetia hemisphaerica clypeata (Douglas); Fernald, 1903: 204.
Saissetia hemisphaerica hibernaculorum (Boisduval); Fernald, 1903: 204.
Lecanium (Saissetia) anthurii (Boisduval); Reh, 1903: 416.
Lecanium (Saissetia) filicum (Boisduval); Reh, 1903: 417.
Saissetia (Lecanium) hemisphaerica (Targioni Tozzetti); Hall, 1922: 21.
Saissetia coffeae (Walker); Williams, 1957: 314.
TYPE DATA. Lectotype female designated by Williams (1957), SRI LANKA: (BMNH).
DISTRIBUTION. AUSTRO ORIENTAL REGION: Papua New Guinea, Philippines, Solomon Islands. **AUSTRALIAN REGION:** New South Wales, Northern Territory, South Australia, Victoria, Western Australia. **ETHIOPIAN REGION:** Angola, Cameroon, Kenya, Mozambique, Sao Tome, South Africa, St. Helena, Uganda, Zanzibar, Zimbabwe. **MADAGASIAN REGION:** Agalega Island, Madagascar, Mauritius, Reunion, Rodrigues, Seychelles. **NEW ZEALAND and PACIFIC REGION:** Fiji, French Polynesia, Hawaii, Irian Jaya, Kiribati, Mariana Islands, Nauru, New Caledonia, New Zealand, Norfolk Island, Ogasawara Islands, Palaus, Tonga, Vanuatu, Wallis Island, Western Samoa. **ORIENTAL REGION:** Cambodia, India, Sri Lanka, Taiwan, Thailand, Vietnam. **NEOTROPICAL REGION:** Bermuda, Brazil Rio Grande do Sul, Chile, Cuba, Galapagos Islands, Guadeloupe, Guyana, Panama, Puerto Rico, Virgin Islands. **NEARCTIC REGION:** Alabama, Arizona, Arkansas, California, Colorado, Connecticut, Delaware, Georgia, Illinois, Indiana, Kansas, Kentucky, Louisiana, Maryland, Massachusetts, Michigan, Minnesota, Mississippi, Missouri, Montana, Nebraska, New Hampshire, New Jersey, New York, North Carolina, Ohio, Oregon, Pennsylvania, South Carolina, Tennessee, Texas, Utah, Virginia, Washington, Wisconsin. **PALAEARCTIC REGION:** Bulgaria, Canary Islands, Crete, Denmark, Egypt, Germany, Greece, Hungary, Inner Mongolia, Israel, Italy, Japan, Korea, Madeira, Saudi Arabia, Spain, Sweden, Switzerland, Turkey, USSR Armenia, USSR

Crimea, USSR Estonia, USSR Georgia, USSR Leningrad Region, USSR Ukraine.
HOST PLANTS. Acanthaceae: *Asystasia gangetica, Graptophyllum, Justicia alba, J. betonica, Pseuderanthemum atropurpureum, P. carruthersi, Ruellia, Thunbergia grandiflora.* **Agavaceae:** *Agave americana, Cordyline australis.* **Amaranthaceae:** *Alternathera.* **Anacardiaceae:** *Mangifera indica, Poupartia caffra, Rhus succedanea, Schinus edule, S. molle, S. terebinthifolius.* **Annonaceae:** *Annona muricata, A. squamosa, Cananga odorata.* **Apocynaceae:** *Beaumontia grandiflora, Carissa carandas, C. grandiflora, Ervatamia divaricata, Melodinus baueri, Nerium oleander, Plumeria acutifolia, P. rubra, Tabernaemontana pentasticta.* **Araceae:** *Anthurium andraeanum, Monstera deliciosa.* **Araliaceae:** *Aralia elegantissima, A. laciniata, Dizygotheca elegantissima.* **Aristolochiaceae:** *Aristolochia pistolachia.* **Aspidiaceae:** *Cyrtomium, Polystichum falcatum.* **Aspleniaceae:** *Asplenium.* **Aucubaceae:** *Aucuba japonica.* **Barringtoniaceae:** *Barringtonia asiatica.* **Begoniaceae:** *Begonia.* **Bignoniaceae:** *Markhamia platycalyx.* **Bischofiaceae:** *Bischofia javanica.* **Blechnaceae:** *Blechnum.* **Boraginaceae:** *Cordia subcordata.* **Brexiaceae:** *Brexia madagascariensis.* **Campanulaceae:** *Clermontia.* **Celastraceae:** *Euonymus japonicus.* **Chenopodiaceae:** *Spinacia oleracea.* **Compositae:** *Chrysanthemum, Kleinia neriifolia, Santolinachamae cyparissus, Senecio, Sonchus oleraceus.* **Convolvulaceae:** *Ipomoea tuberosa.* **Crassulaceae:** *Bryophyllum.* **Cucurbitaceae:** *Cucurbita moschata, Momordica balsamina, Sechium edule.* **Cycadaceae:** *Cycas circinalis, C. neocaledonica, C. revoluta.* **Ebenaceae:** *Diospyros discolor, D. kaki.* **Ehretiaceae:** *Cordia.* **Euphorbiaceae:** *Acalypha, Aleurites moluccana, Codiaeum variegatum, Croton, Euphorbia heterophylla, E. pulcherrima, Jatropha multifida, Ricinus communis.* **Geraniaceae:** *Geranium.* **Gramineae:** *Panicum.* **Guttiferae:** *Calophyllum inophyllum, Garcinia subelliptica.* **Hydrangaceae:** *Hydrangea.* **Iridaceae:** *Gladiolus.* **Labiatae:** *Coleus, Salvia.* **Lauraceae:** *Machilus, Persea americana, P. gratissima.* **Leguminosae:** *Adianthum,* **Leguminosae:** *Bauhinia.* **Liliaceae:** *Aloe barbadensis, Asparagus plumosus.* **Loganiaceae:** *Fagraea racemosa.* **Loranthaceae:** *Phoradendron.* **Lythraceae:** *Lagerstroemia indica.* **Malvaceae:** *Abelmoschus esculentus, Gossypium, Hibiscus rosa-sinensis, H. tiliaceus.* **Melastomataceae: Meliaceae:** *Meliadubia.* **Moraceae:** *Artocarpus altilis, Ficus carica, F. wightiana.* **Musaceae:** *Musa.* **Myoporaceae:** *Myoporum.* **Myrsinaceae:** *Ardisia crispa, A. quinquegona.* **Myrtaceae:** *Eucalyptus, Eugenia jambos, Eugenia uniflora, Myrtuscommunis, Psidium guajava.* **Nymphaceae:** *Nuphar lutea.* **Oleaceae:** *Ligustrum, Olea europaea, Osmanthus fragrans.* **Oleandraceae:** *Nephrolepis exaltata.* **Onagraceae:** *Ludwigia.* **Orchidaceae:** *Cymbidium, Epidendrum, Phalaenopsis, Thrixspermum formosanum.* **Palmae:** *Cocos nucifera, Cyphosperma sambucina.* **Passifloraceae:** *Passiflora edulis.* **Piperaceae:** *Piper methysticum.* **Pittosporaceae:** *Pittosporum tobira.* **Polygonaceae:** *Muehlenbeckia platyclada.* **Polypodiaceae:** *Adiantum capillus-veneris, A. cuneatum, Phymatosorus scolopendria, Platycerium alcicorne, Polypodium phyllitidis, P. polypodoides.* **Primulaceae:** *Cyclamen europaeum.* **Pteridaceae:** *Acrostichum aureum.* **Punicaceae:**

Punica granatum. **Rhizophoraceae:** *Bruguiera sexangula.* **Rosaceae:** *Eriobotrya japonica, Persica vulgaris, Prunus domestica, Pyruscydonia.* **Rubiaceae:** *Borreria laevis, Cinchona calisaya, Coffea arabica,* C. *canephora,* C. *liberica, Gardenia florida,* G. *jasminoides, Hedyotis, Ixora chinensis, I. coccinea, Morinda citrifolia, Psychotria megritostictapunicea, P. rubra, Randia tahitensis, Rondeletia odorata.* **Rutaceae:** *Citrus aurantifolia,* C. *grandis,* C. *limon,* C. *medica,* C. *reticulata,* C. *sinensis.* **Sapindaceae:** *Litchi chinensis.* **Sapotaceae:** *Achras sapota, Chrysophyllum cainito, Manilkara zapota.* **Solanaceae:** *Capsicum annuum,* C. *frutescens, Solanum jasminoides, S. melongena, S. nigrum, S. santiwongsei.* **Taxodaceae:** *Cunninghamia lanceolata.* **Theaceae:** *Camellia sinensis.* **Thunbergiaceae:** *Thunbergia grandiflora.* **Thymelaeaceae:** *Synaptolepis alternifolia.* **Urticaceae:** *Pipturus albidus.* **Verbenaceae:** *Duranta repens.* **Vitidaceae:** *Vitis vinifera.* **Winteraceae:** *Bubbia.* **Zamiaceae:** *Zamia floridana.* **Zingiberaceae:** *Alpinia purpurata.*

REMARKS. Common Name - Hemispherical scale. Adult female redescribed and illustrated by Comstock (1881), Tyrrell (1896), Newstead (1903), Thro (1903), Green (1904d), Froggatt (1915), Dietz & Morrison (1916), Kuwana (1917a), Pettit & McDaniel (1920), Leonardi (1920), Hempel (1920a), Steinweden (1930), Gómez-Menor Ortega (1937), Zimmerman (1948), Gomes Costa (1949), Borchsenius (1957), Ezzat & Hussein (1969), Paik (1978), Wang (1980), Kawai (1980), Tereznikova (1981), Yang (1982), Tao *et al.* (1983), Hamon & Williams (1984), Gill (1988), Tang & Li (1983), Tremblay (1988a), Gonzalez & Lamborot (1989), Gonzalez (1989), Williams & Watson (1990) and by Tang (1991). Colour photograph given by Delucchi (1975), Kawai (1980), Hamon & Williams (1984), Stimmel (1987), Gill (1988), Johnson & Lyon (1988) and by Gonzalez (1989). Koteja *et al.* (1976) described changes in the cuticle during the development of the larva and adult female. Parida & Moharana (1982) and Moharana (1990) reported chromosome number 2n=16 in India. Nur (1979) studied the cytology and reported chromosome number 2n=16 in U.S.A. SEM micrograph of tubular duct and of anal ring given by Foldi (1991). Distribution and host plant records given by Targioni Tozzetti (1867), Comstock (1881), Tyrrell (1896), Hofer (1903), Newstead (1917c), Brain (1920b), Leonardi (1920), Hall (1922, 1935), Bodenheimer (1924, 1953), Zimmerman (1948), Gomes Costa (1949), De Lotto (1956a, 1965), Borchsenius (1957), Ossiannilsson (1959), Hodgson (1967a, 1969b), Ezzat & Hussein (1969), Ali (1971, 1973), Kawai *et al.* (1971), Corseuil & Barbosa (1971), Nath (1972), Almeida (1973), Delucchi (1975), Kozárzhevskaya & Reitzel (1975), Wang (1980), Kawai (1980), Tereznikova (1981), Nakahara & Miller (1981), Nakahara (1981b), Nakahara (1983), Tao *et al.* (1983), Vieira *et al.* (1983), Martin Mateo (1984), Matile-Ferrero & Nonveiller (1984), Hamon & Williams (1984), Stimmel (1987), Marotta (1987), Varshney & Moharana (1987), Gill (1988), Gonzalez & Lamborot (1989), Gonzalez (1989), Shafee *et al.* (1989), Williams & Watson (1990), Danzig & Konstantinova (1990), Tang (1991) and by Hodgson & Hilburn (1991a, 1991b).

BIOLOGY. Female reproduces parthenogenetically. Up to eight generations per year were reported in Peru (Beingola, 1969b), several overlapping generations in Cuba (Alayo & Blahutiak, 1980), one or two generations in Florida (Hamon & Williams, 1984) and two or more in California (Gill, 1988). Six generations develop under laboratory conditions on green potato sprouts (Blumberg & Swirski, 1977). Blumberg (1977, 1988) studied the encapsulation of parasitoid eggs. The encapsulation of eggs of the parasitoid *Encyrtus infelix* (Embleton) was determined by Blumberg & Goldenberg (1992). Stylet penetration described by Ramanarayan *et al.* (1980). Blumberg & Swirski (1977) described methods for mass rearing.

ECONOMIC IMPORTANCE. A pest of many ornamental plants, especially of cycads and ferns. Biology and pest status on coffee presented by Le Pelley (1968). Efficient biological control was often achieved by parasites imported to control the black scale, *S. oleae* (Bartlett, 1978).

Saissetia discoides (Hempel)

Lecanium discoides Hempel, 1900b: 433.
Saissetia discoides (Hempel); Cockerell, 1902f: 453.
TYPE DATA. Syntypes female, Brazil Sao Paulo Ypiranga, on *Psidium guajava* (MZSP).
DISTRIBUTION. NEOTROPICAL REGION: Brazil Rio Grande do Sul.
HOST PLANTS. Anacardiaceae: *Schinus dependens.* **Lauraceae:** *Nectandra oppositifolia.* **Loranthaceae:** *Phrygilanthus eugenioides.* **Myrtaceae:** *Psidium guajava.*
REMARKS. Adult female redescribed by Hempel (1901a, 1920a) and by Gomes Costa (1949). Distribution and host plant records given by Gomes Costa (1949) and by Corseuil & Barbosa (1971).

Saissetia dura (Hempel)

Lecanium durum Hempel, 1900b: 427.
Saissetia dura (Hempel); Cockerell, 1902f: 453.
TYPE DATA. Syntypes female, BRAZIL: Sao Paulo, Ypiranga, on *Baccharis dracunculifolia* (MZSP).
DISTRIBUTION. NEOTROPICAL REGION: Brazil Sao Paulo.
HOST PLANTS. Compositae: *Baccharis dracunculifolia.*
REMARKS. Adult female redescribed and illustrated by Hempel (1901a).

Saissetia ficinum (Paoli)

Lecanium (Eulecanium) ficinum Paoli, 1916: 252.
Eulecanium ficinum Paoli; Leonardi, 1920: 294.
Lecanium ficinum Paoli; Martin Mateo, 1984: 71.
Saissetia ficinum (Paoli); Marotta, 1987: 109.
TYPE DATA. Syntypes female, ITALY: Sardinia, Siniscola (Sassari), on *Ficus carica* (ISZF).
DISTRIBUTION. PALAEARCTIC REGION: Italy, Spain.
HOST PLANTS. Moraceae: *Ficus carica, F. nitida.*
REMARKS. Adult female redescribed and illustrated by Leonardi (1920). Adult female redescribed by Gómez-Menor Ortega (1937, 1958c). Distribution and host plant records given by Gómez-Menor Ortega (1937, 1958c), Martin Mateo (1984) and by Marotta (1987).

Saissetia glanulosa (Hempel)

Lecanium glanulosum Hempel, 1900b: 428.
Saissetia glanulosa (Hempel); Cockerell, 1902f: 453.
TYPE DATA. Syntypes female, BRAZIL: Sao Paulo, Ypiranga, on Myrtaceae (MZSP).
DISTRIBUTION. NEOTROPICAL REGION: Brazil Sao Paulo.
HOST PLANTS. Myrtaceae:
REMARKS. Adult female redescribed by Hempel (1901a).

Saissetia hurae Newstead

Lecanium (Saissetia) hurae Newstead, 1917a: 361.
Saissetia hurae Newstead; Silva *et al.*, 1986: 154.
TYPE DATA. Syntypes female, GUYANA: Berbice, on *Hura crepitans* (BMNH).
DISTRIBUTION. NEOTROPICAL REGION: Guyana.
HOST PLANTS. Euphorbiaceae: *Hura crepitans*.

Saissetia infrequens (Hempel)

Lecanium infrequens Hempel, 1900b: 431.
Saissetia infrequens (Hempel); Fernald, 1903: 204.
TYPE DATA. Syntypes female, BRAZIL: Sao Paulo, Ypiranga, on *Zanthoxylum* (MZSP).
DISTRIBUTION. NEOTROPICAL REGION: Argentina, Brazil Sao Paulo.
HOST PLANTS. Euphorbiaceae: *Sapium haematospermum*. **Rutaceae:** *Zanthoxylum*. **Sapindaceae:** *Schmidellia edulis*.
REMARKS. Adult female redescribed by Hempel (1901a). Distribution and host plant records given by Lizer y Trelles (1939).

Saissetia jocunda De Lotto

Saissetia jocunda De Lotto, 1957a: 173.
TYPE DATA. Holotype female, TANZANIA: Arusha, on *Celtis durandii* (BMNH).
DISTRIBUTION. ETHIOPIAN REGION: Kenya, Malawi, Mozambique, Tanzania, Zimbabwe.
HOST PLANTS. Araliaceae: *Cussonia kirkii.* **Euphorbiaceae:** *Phyllanthus hutchinsonianus*, *Uapaca kirkiana*. **Leguminosae:** *Acacia*, *Brachystegia spiciformis*. **Moraceae:** *Ficus*. **Myrsinaceae:** *Maesa lanceolata*. **Proteaceae:** *Protea gaguedi*. **Rosaceae:** *Cliffortia nitidula*. **Rubiaceae:** *Coffea arabica*. **Rutaceae:** *Clausena anisata*. **Ulmaceae:** *Celtis durandii*.
REMARKS. Adult female redescribed and illustrated by De Lotto (1968b). Distribution and host plant records given by Hodgson (1967a, 1969b) and by De Lotto (1968b).

Saissetia lucida Hempel

Saissetia lucida Hempel, 1912: 60.
TYPE DATA. Syntypes female, BRAZIL: Sao Paulo, Alto da Serra, on bark of an undetermined forest tree (MZSP).
DISTRIBUTION. NEOTROPICAL REGION: Brazil Sao Paulo.
HOST PLANTS. Verbenaceae: *Lantana*.

REMARKS. Distribution and host plant records given by Silva *et al.*
(1968).

Saissetia malagassa Mamet

Saissetia malagassa Mamet, 1954a: 45.
TYPE DATA. Syntypes female, MADAGASCAR: Manjakatompo, on
undetermined plant (MNHN).
DISTRIBUTION. MADAGASIAN REGION: Madagascar.

Saissetia minensis Hempel

Saissetia minensis Hempel, 1932: 328.
TYPE DATA. Syntypes female, BRAZIL: Minas Gerais, Vicosa, on
Luehea sp. (MZSP).
DISTRIBUTION. NEOTROPICAL REGION: Brazil Minas Gerais.
HOST PLANTS. Tiliaceae: *Luehea.*

Saissetia miranda (Cockerell & Parrott)

Lecanium oleae mirandum Cockerell & Parrott in Cockerell, 1899g:
12.
Saissetia oleae miranda (Cockerell & Parrott); Fernald, 1903: 206.
Saissetia oleae (Bernard); Sanders, 1909a: 440 [ERRONEOUS
SYNONYMY].
Saissetia miranda (Cockerell & Parrott); De Lotto, 1969a: 419.
TYPE DATA. Syntypes female, MEXICO: Tlacotalpam in Vera Cruz,
on *Abutilon* sp. (USNM).
DISTRIBUTION. AUSTRO ORIENTAL REGION: Papua New Guinea,
Solomon Islands. **ETHIOPIAN REGION:** South Africa. **NEO-
TROPICAL REGION:** Bermuda, El Salvador, Honduras, Lesser
Antilles, Mexico, Panama, Puerto Rico, Virgin Islands. **NEW
ZEALAND and PACIFIC REGION:** Cook Islands, Fiji, French
Polynesia, Hawaii, Irian Jaya, Kiribati, New Caledonia, Niue,
Tonga, Western Samoa. **NEARCTIC REGION:** Alabama, Arizona,
California, Florida, Louisiana, Maryland, Missouri, New Mexico,
New York, Ohio, Oklahoma, Pennsylvania, Tennessee, Texas,
Virginia, Washington, D. C. **ORIENTAL REGION:** India. **PAL-
AEARCTIC REGION:** Portugal.
HOST PLANTS. Agavaceae: *Agave sisalana.* **Amaranthaceae:**
Amaranthus spinosus. **Anacardiaceae:** *Mangifera indica, Schinus
terebinthifolius, Spondias dulcis.* **Apocynaceae:** *Kopsia, Nerium
oleander, Plumeria rubra.* **Bombacaceae:** *Montezuma speciossi-
ma.* **Caryophyllaceae:** *Dianthus caryophyllus.* **Combretaceae:**
Terminalia catappa. **Cornaceae:** *Cornus florida.* **Ebenaceae:**
Diospyros ferrea. **Ehretiaceae:** *Cordia.* **Euphorbiaceae:** *Acalypha tricolor.* **Heliconiaceae:** *Heliconia.* **Lauraceae:** *Persea
americana.* **Leguminosae:** *Cajanus cajan, Cassia, Crotalaria
anagyroides, C. usaramoensis, Erythrina berteroana, E. lithosperma, E. sandwicensis, Tamarindus indica.* **Malpighiaceae.**
Malvaceae: *Abelmoschus esculentus, Abutilon graveolens, A.
hirtum, Goethea strictiflora, Gossypium punctatum, Gossypium,
Hibiscus tiliaceus.* **Meliaceae:** *Cedrela odorata, Swietenia
mahogony.* **Moraceae:** *Ficus benjamina, F. lyrata.* **Myrtaceae:**
Psidium guajava, P. zibethinus. **Onagraceae:** *Jussiaea.* **Palmae:**

Cocos nucifera. **Punicaceae:** *Punica granatum.* **Rosaceae:** *Pyra-cantha coccinea.* **Rubiaceae:** *Ixora coccinea, Morinda citrifolia, Timonius.* **Rutaceae:** *Casimiroa edulis, Citrus paradisi, Zanthoxy-lum flavum.* **Sterculiaceae:** *Sterculia apetala.* **Tiliaceae:** *Grewia crenata.* **Verbenaceae:** *Lippia.*
REMARKS. Common name - Mexican black scale. Until 1971 this species was misidentified as *S. oleae* in several publications (De Lotto, 1971a). Adult female redescribed and illustrated by De Lotto (1971a), Hamon & Williams (1984), Gill (1988) and by Williams & Watson (1990). Adult female redescribed by Tang (1991). Parida & Ghosh (1984) and Moharana (1990) reported on chromosome number 2n-16 in India. Distribution and host plant records given by De Lotto (1971a, 1976), Lambdin & Watson (1980), Nakahara & Miller (1981), Nakahara (1981b, 1983), Hamon & Williams (1984), Gill (1988), Williams & Watson (1990) and by Hodgson & Hilburn (1991a, 1991b).
ECONOMIC IMPORTANCE. Has reached large populations in southern Texas, where it is regarded a potential pest of citrus (Dean & Hart, 1972). Very common on ornamental plants, however serious damage has not been observed (Hamon & Williams, 1984).

Saissetia mirifica (Maskell)

Lecanium mirificum Maskell, 1897a: 312.
Lecanium (Saissetia) miripicum Cockerell & Parrott, 1899: 164 [MIS-SPELLING].
Saissetia mirifica (Maskell); Cockerell, 1901d: 32.
TYPE DATA. Syntypes female, AUSTRALIA: North West Victoria, Mallee Scrub, on *Acacia pendula* (NZAC).
DISTRIBUTION. AUSTRALIAN REGION: Victoria.
HOST PLANTS. Leguminosae: *Acacia pendula.*
REMARKS. Adult female redescribed by Froggatt (1915).

Saissetia monotes Hall

Saissetia monotes Hall, 1935: 78.
TYPE DATA. Syntypes female, ZIMBABWE: Melfort and Eldorado, on *Monotes glaber* (BMNH).
DISTRIBUTION. ETHIOPIAN REGION: Zimbabwe.
HOST PLANTS. Dipterocarpaceae: *Monotes glaber.* **Leguminosae:** *Brachystegia spiciformis.* **Loganiaceae:** *Nuxia congesta.*
REMARKS. Adult female redescribed and illustrated by Hodgson (1967a). Distribution and host plant records given by Hodgson (1967a).

Saissetia monotes pretoriae Hall

Saissetia monotes pretoriae Hall, 1939: 98.
TYPE DATA. Syntypes female, SOUTH AFRICA: Transvaal, Pretoria, on *Ficus* sp. (SANC)
DISTRIBUTION. ETHIOPIAN REGION: South Africa.
HOST PLANTS. Moraceae: *Ficus.*

Saissetia munroi De Lotto

Saissetia munroi De Lotto, 1958a: 66.

TYPE DATA. Holotype female, SOUTH AFRICA: Transvaal, Pretoria, on *Ochna pulchra* (BMNH).
DISTRIBUTION. ETHIOPIAN REGION: South Africa.
HOST PLANTS. Ochnaceae: *Ochna pulchra*.

Saissetia neglecta De Lotto

Saissetia neglecta De Lotto, 1969a: 419.
TYPE DATA. Holotype female, U.S.A.: Florida, Pine Island, on grapefruit (SANC).
DISTRIBUTION. NEOTROPICAL REGION: Belize, Bermuda, Costa Rica, El Salvador, Guatemala, Honduras, Mexico, Panama, Puerto Rico, Venezuela, Virgin Islands. **NEW ZEALAND and PACIFIC REGION:** Hawaii. **NEARCTIC REGION:** Florida, Louisiana.
HOST PLANTS. Agavaceae: *Agave*. **Anacardiaceae:** *Mangifera indica*. **Annonaceae:** *Annona muricata*. **Aquifoliaceae:** *Ilex*. **Araceae:** *Dieffenbachia amoena, Dizygotheca elegantissima*. **Araliaceae:** *Tetrapanax papyriferum*. **Asclepiadaceae:** *Stephanotis floribunda, Telosma cordata*. **Barringtoniaceae:** *Barringtonia*. **Bignoniaceae:** *Spathodea campanulata*. **Boraginaceae:** *Cordiaalliodora*. **Clusiaceae:** *Mammea americana*. **Combretaceae:** *Laguncularia racemosa, Terminalia brassii*. **Compositae:** *Fitchia speciosa*. **Convolvulaceae:** *Ipomoea*. **Euphorbiaceae:** *Euphorbia pulcherrima, Hevea*. **Leguminosae:** *Bauhinia monandra, Cassia, Crotalaria anugyroides, Erythrina indica, Indigofera, Piptadenia, Tamarindus indica, Vigna sinensis*. **Lythraceae:** *Lagerstroemia*. **Malvaceae:** *Malvaviscus arboreus*. **Melastomataceae:** *Clidemia hirta*. **Meliaceae:** *Cedrela toona*. **Moraceae:** *Ficus*. **Myrtaceae:** *Eucalyptus deglupta, Psidium guajava*. **Piperaceae:** *Piper methysticum, P. puberulum*. **Polygonaceae:** *Coccoloba pirifolia*. **Polypodiaceae:** *Platycerium*. **Pteridaceae:** *Acrostichum aureum*. **Rubiaceae:** *Coffea arabica, C. canephora, Gardenia, Ixoramacrothyrsa*. **Rutaceae:** *Citrus aurantifolia, C. grandis, C. limon, C. paradisi, C. reticulata, C. sinensis, Evodia hortensis*. **Sapotaceae:** *Manilkara zapota*. **Scrophulariaceae:** *Leucophyllum frutescens*. **Verbenaceae:** *Callicarpa americana*.
REMARKS. Common Name - Caribbean black scale. Adult female redescribed and illustrated by Hamon & Williams (1984). Adult female redescribed by Tang (1991). Colour photograph given by Hamon & Williams (1984). Distribution and host plant records given by De Lotto (1971a), Nakahara & Miller (1981), Nakahara (1981b, 1983), Hamon & Williams (1984), Williams & Watson (1990) and by Hodgson & Hilburn (1991a, 1991b).

Saissetia nigrella King

Saissetia nigrella King, 1902a: 296.
TYPE DATA. Syntypes female, SOUTH AFRICA: Natal, Tongaat, on *Ficus* sp. (USNM)
DISTRIBUTION. ETHIOPIAN REGION: South Africa.
HOST PLANTS. Moraceae: *Ficus*.
REMARKS. Adult female redescribed and illustrated by De Lotto (1970c). Sanders (1909a) and Brain (1920b) synonymized this spe-

cies with *Parasaissetia nigra*, but De Lotto (1970a) have shown that this was an error.

Saissetia oleae (Olivier)

Coccus oleae Olivier, 1791: 95.

Coccus palmae Haworth, 1812: 307. Syntypes female, ENGLAND: on palm (lost; Williams, 1957). Syn. by Williams, 1957: 314.

Coccus testudo Curtis, 1843a: 443. Syntypes female, ENGLAND: in greenhouse, on *Brexia spinosa* (probably lost). Syn. by Douglas, 1891d: 307.

Lecanium oleae Bernard; Signoret, 1869a: 862 [ERRONEOUS AUTHORSHIP].

Chermes oleae Bernard; Signoret, 1869a: 862 [ERRONEOUS AUTHORSHIP].

Lecanium testudo (Curtis); Signoret, 1874a: 441.

Lecanium palmae (Haworth); Douglas, 1887b: 97.

Bernardia oleae (Bernard); Marlatt, 1892: 150 [ERRONEOUS AUTHORSHIP].

Neobernardia oleae (Bernard); Cockerell, 1893d: 54 [ERRONEOUS AUTHORSHIP].

Lecanium oleae testudo (Curtis); Cockerell, 1896b: 331.

Lecanium (Saissetia) palmae (Haworth); Cockerell & Parrott, 1899: 164.

Saissetia oleae Bernard; Cockerell, 1901d: 31 [ERRONEOUS AUTHORSHIP].

Coccus oleae (Bernard); Kirkaldy, 1902: 106 [ERRONEOUS AUTHORSHIP].

Saissetia oleae testudo (Curtis); Fernald, 1903: 206.

Saissetia palmae (Curtis); Fernald, 1903: 207.

Saissetia obae (Bernard); Rutherford, 1915a: 112 [MIS-SPELLING and ERRONEOUS AUTHORSHIP].

Saissetia (Lecanium) oleae Bernard; Hall, 1922: 22 [ERRONEOUS AUTHORSHIP].

Parasaissetia oleae (Bernard); Ezzat & Hussein, 1969: 413 [ERRONEOUS AUTHORSHIP].

Saissetia oleae (Olivier); De Lotto, 1971b: 149.

Saissetia oleae (Oliver); Gonzalez, 1989: 89 [MIS-SPELLING].

TYPE DATA. Syntypes female, FRANCE and ITALY: on olive, myrtle and *Phillyrea* (probably lost).

DISTRIBUTION. ETHIOPIAN REGION: Angola, Cameroon, Eritrea, Ivory Coast, Kenya, Malawi, South Africa, Zimbabwe. **MADAGASIAN REGION:** Comoros, Madagascar, Mauritius. **NEOTROPICAL REGION:** Brazil, Chile, Guatemala, Mexico. **NEW ZEALAND and PACIFIC REGION:** French Polynesia, Hawaii, Lord Howe Island, Mariana Islands, Marshall Islands, New Caledonia, Norfolk Island, Palaus. **NEARCTIC REGION:** Arizona, California, Colorado, Connecticut, Florida, Indiana, Kansas, Louisiana, Massachusetts, New Jersey, New Mexico, New York, South Carolina, Washington. **ORIENTAL REGION:** India, Pakistan, Sri Lanka, Taiwan, Thailand, Vietnam. **PALAEARCTIC REGION:** Algeria, Austria, Azores Islands, Bulgaria, Canary Islands, Corsica, Cyprus, Denmark, Egypt, England, France, Germany, Greece, Inner Mongolia, Iran, Israel, Italy, Lybia, Madeira,

Morocco, Portugal, Saudi Arabia, Spain, Switzerland, Tibet, Turkey, USSR Armenia, USSR Azerbaijan, USSR Georgia, Yugoslavia.

HOST PLANTS. Agavaceae: *Agave americana, Yucca gloriosa.* **Anacardiaceae:** *Pistacia atlantica, P. lentiscus, P. palestina, P. vera, Schinus molle.* **Apocynaceae:** *Carissa grandiflora, Nerium indicum, N. odorum, N. oleander, Trachelospermum jasminoides, Vinca major.* **Aquifoliaceae:** *Ilex aquifolium, I. canariensis, I. platiphylla, I. wilsoni.* **Araceae:** *Anthurium.* **Araliaceae:** *Aralia.* **Aspleniaceae:** *Asplenium.* **Bignoniaceae:** *Jacaranda ovalifolia.* **Burseraceae:** *Garuga piumata.* **Celastraceae:** *Euonymus japonicus.* **Combretaceae:** *Terminalia catappa.* **Compositae:** *Artemisia absinthium, Baccharis halimifolia, B. rosmarinifolia, Carduus pycnocephalus, Carlina corymbosa, Centaurea jacea, Chrysanthemum, Cynara cardunculus, Galactites tomentosa, Olearia, Osteospermum moniliferum, Raillardia menziesii, Scolymus hispanicus.* **Convolvulaceae:** *Convolvulus floridus.* **Crassulaceae:** *Crassula portulacea.* **Cycadaceae:** *Cycas revoluta.* **Ebenaceae:** *Diospyros kaki.* **Ehretiaceae:** *Ehretia timifolia.* **Ericaceae:** *Agauria salicifolia, Erica arborea, E. scoparia, Vaccinium.* **Euphorbiaceae:** *Euphorbia pulcherrima.* **Flacourtiaceae:** *Azara.* **Geraniaceae:** *Pelargonium zonale.* **Gramineae:** *Arundo donax.* **Hippocastanaceae:** *Aesculus pavia.* **Labiatae:** *Rosmarinus officinalis.* **Lauraceae:** *Laurus nobilis, Persea americana.* **Leguminosae:** *Acacia koa, Cercis siliquastrum, Erythrina, Lotus berthelotii, Tamarindus.* **Liliaceae:** *Aloe, Asparagus albus, A. aphyllus, A. tenuifolius, Cordyline.* **Loranthaceae:** *Viscum album, V. cruciatum.* **Malvaceae:** *Abutilon, Hibiscus sinensis.* **Myoporaceae:** *Myoporum laetum, M. pictum, M. punctulatum.* **Myricaceae:** *Myrica salicifolia.* **Myrsinaceae:** *Myrsine africana.* **Myrtaceae:** *Myrtus communis, Psidium guajava, P. pomiferum.* **Oleaceae:** *Olea chrysophylla, Olea europaea, Osmanthus americanus, Osmanthus heterophyllus, Phillyrea.* **Oleandraceae:** *Nephrolepis exaltata.* **Pittosporaceae:** *Pittosporum tobira.* **Polygalaceae:** *Polygala sibirica.* **Polygonaceae:** *Coccoloba uvifera, Muehlenbeckia platyclada.* **Polypodiaceae:** *Adiantum.* **Punicaceae:** *Punica granatum.* **Rosaceae:** *Armeniaca vulgaris, Cliffortia nitidula, Cotoneaster pannosa, Cydonia, Eriobotrya japonica, Persica vulgaris, Prunus avium, P. paniculata, Pyrus longipes.* **Rubiaceae:** *Coffea arabica.* **Rutaceae:** *Citrus aurantium, C. limon, C. paradisi, C. reticulata.* **Sapindaceae:** *Dodonaea abyssinica.* **Scrophulariaceae:** *Halleria lucida.* **Sterculiaceae:** *Cheirostemon platanoides.* **Tamaricaceae:** *Tamarix africana, T. gallica.* **Umbelliferae:** *Eryngium campestre, Heteromorpha trifoliata.* **Verbenaceae:** *Callicarpa americana, C. japonica, Duranta integrifolia, Verbena rigida.*

REMARKS. Common name - Mediterranean black scale; black scale in earlier publications from the U. S. A. ; olive scale in some Mediterranean countries. *Lecanium oleae* Bernard, 1783 and *Saissetia oleae* (Bernard, 1783) - of authors - are invalid names. The erroneous authorship of Bernard has been erroneously introduced by Signoret (1869a). For a detailed discussion on the case, see De Lotto (1971b). Adult female redescribed and illustrated by Comstock (1881), Tyrrell (1896), Thro (1903), Newstead (1903), Martelli (1908), Kuwana

(1917a), Leonardi (1920), Steinweden (1929), Gómez-Menor Ortega (1937, 1958c), Zimmerman (1948), Borchsenius (1957), De Lotto (1965, 1971a), Ezzat & Hussein (1969), Kawai (1980), Tereznikova (1981), Tao *et al.* (1983), Hamon & Williams (1984), Tremblay (1988a), Gill (1988), Tang & Li (1988), Gonzalez (1989), Gonzalez & Lamborot (1989), Williams & Watson (1990) and by Tang (1991). Colour photograph of adult female given by Chapot & Delucchi (1964), Delucchi (1975), Kawai (1980), Hamon & Williams (1984), Gill (1988), Tranfaglia & Viggiani (1988), Johnson & Lyon (1988) and by Gonzalez (1989). Larval instars described and illustrated by Argyriou (1963). Couturier *et al.* (1985) critically studied material from Ivory Coast and indicated that the species occurs in this tropical country. Distribution and host plant records given by Tyrrell (1896), Martelli (1908), Hall (1922), Bodenheimer (1951), De Lotto (1956a, 1965, 1971a, 1971b, 1976), Borchsenius (1957), Hodgson (1969b), Ezzat & Hussein (1969), Almeida (1969, 1973), Kozárzhevskaya & Reitzel (1975), Delucchi (1975), Georghiou (1977), Matile-Ferrero (1978), Lal & Naji (1979, 1980), Neuenschwander & Paraskakis (1980), Paraskakis *et al.* (1980), Tereznikova (1981), Wang (1981), Nakahara (1981b), Tao *et al.* (1983), Vieira *et al.* (1983), Martin Mateo (1984), Hamon & Williams (1984), Gill (1988), Shafee *et al.* (1989), Gonzalez (1989), Williams & Watson (1990), Danzig & Konstantinova (1990) and by Tang (1991). Intracellular symbionts were reported and studied by Granovsky (1929). Banks & Cameron (1973) chemically analyzed body extracts. Ishaaya & Swirski (1976) reported on positive correlation between activity of trehalase, invertase and amylase enzymes and generation duration, on potato sprouts, oleander and citrus.

BIOLOGY. Female reproduces parthenogenetically. Usually one generation per year on citrus in interior areas of California, whereas it is bivoltine in the coastal areas (Quayle, 1911; Bartlett, 1978; Gill, 1988). Bodenheimer (1951b) reported that it has one annual generation on citrus in Israel, but Blumberg *et al.* (1975) observed bivoltine populations on citrus. Peleg (1965) observed in Israel one annual generation on citrus and unirrigated olive trees, while two generations were observed on irrigated olive. Two annual generations were observed in Peru on *Citrulus* and on potato sprouts (Beingola, 1969b). In Greece it develops one annual generation on olive (Argyriou, 1963; Paloukis, 1979). Bodenheimer (1951b) presented in great-details the biology and economic importance on citrus in the Middle East. De Lotto (1965, 1976) suggested that the centre of origin and the area of natural diffusion of this species are in the southern districts of the Cape Province, South Africa. Host plants and natural enemies in Greece given by Argyriou (1963) and by Paloukis (1979). Biology and ecology on olive in Crete studied by Neuenschwander & Paraskakis (1980) and Paraskakis *et al.* (1980). Conti (1987) studied in Italy the duration of egg development under constant temperatures. Briales & Campos analyzed the spatial distribution on olive tree in Spain. Flanders (1942) and Blumberg & Swirski (1977) described methods for mass rearing.

ECONOMIC IMPORTANCE. One of the most important pests of citrus in the Mediterranean Basin, Florida, California and South America (Bartlett, 1978). The major biological control against this pest was reviewed by Bartlett (1978). Natural enemies on olive in

Italy studied by Martelli (1908). Natural enemies on olive in Lybia studied by Lal & Naji (1979, 1980). Population dynamics on citrus in Israel (life-table, distribution within the tree, natural enemies and crawler behaviour) studied by Podoler et al. (1979a, 1979b) and by Mendel et al. (1982, 1984a, 1984b). Panis (1983) discussed various aspects of its biological control on olive in France. Peleg & Gothilf (1981) evaluated the adverse effects of IGR on this pest.

Saissetia oleae cherimoliae Gómez-Menor Ortega

Coccus (Saissetia) oleae cherimoliae Gómez-Menor Ortega, 1955: 205.
Saissetia oleae cherimoliae Gómez-Menor Ortega; Gómez-Menor Ortega, 1958c: 59.
TYPE DATA. Syntypes female, SPAIN: Motril, Granada, on *Annona cherimolia* and *Citrus aurantium* (MNCN).
DISTRIBUTION. PALAEARCTIC REGION: Spain.
HOST PLANTS. Annonaceae: *Annona cherimolia.* **Rutaceae:** *Citrus aurantium.*
REMARKS. Adult female redescribed and illustrated by Gómez-Menor Ortega (1958c).

Saissetia opulenta De Lotto

Saissetia opulenta De Lotto, 1957a: 177.
TYPE DATA. Holotype female, KENYA: Nairobi, on *Croton* sp. (BMNH).
DISTRIBUTION. ETHIOPIAN REGION: Kenya.
HOST PLANTS. Ehretiaceae: *Cordia holstii.* **Euphorbiaceae:** *Croton.*

Saissetia orbiculata De Lotto

Saissetia orbiculata De Lotto, 1963: 191.
TYPE DATA. Holotype female, KENYA: Nairobi, on *Nerium oleander* (BMNH).
DISTRIBUTION. ETHIOPIAN REGION: Kenya, Zimbabwe.
HOST PLANTS. Apocynaceae: *Nerium oleander.* **Euphorbiaceae:** *Poinsettia pulcherrima.* **Moraceae:** *Ficus.* **Rutaceae:** *Citrus aurantium.*
REMARKS. Distribution and host plant records given by Hodgson (1969b).

Saissetia persimilis (Newstead)

Lecanium (Saissetia) persimile Newstead, 1917a: 362.
Saissetia oleae (Bernard); Lindinger, 1928: 107 [MISIDENTIFICA-TION].
Saissetia persimilis (Newstead); De Lotto, 1956a: 243.
TYPE DATA. Syntypes female, KENYA: Nairobi, on peach stems (BMNH).
DISTRIBUTION. ETHIOPIAN REGION: Kenya, South Africa, Tanzania, Uganda, Zimbabwe.
HOST PLANTS. Apocynaceae: *Acokanthera schimperi, Nerium oleander.* **Bignoniaceae:** *Markhamia platycalyx.* **Celastraceae:** *Gymnosporia.* **Combretaceae:** *Terminalia sericea.* **Compositae:** *Erigeron bonariensis.* **Ehretiaceae:** *Cordia ovalis, Ehretia silvatica.* **Euphorbiaceae:** *Croton, Uapaca kirkiana.* **Leguminosae:**

Acacia. **Malvaceae:** *Hibiscus fuscus*. **Moraceae:** *Antiaris toxicaria*, *Ficus hochstetteri*. **Proteaceae:** *Protea*. **Rosaceae:** *Persica vulgaris*. **Simaroubaceae:** *Harrisonia abyssinica*. **Tiliaceae:** *Grewia*. **Vitidaceae:** *Vitis*.

REMARKS. Adult female redescribed and illustrated by De Lotto (1956a). Distribution and host plant records given by Brain (1920b), Lindinger (1928), De Lotto (1956a, 1965) and by Hodgson (1967a, 1969b).

Saissetia poinsettiae Hodgson

Saissetia poinsettiae Hodgson, 1967a: 17.
TYPE DATA. Holotype female, ZIMBABWE: Salisbury, on *Poinsettia pulcherrima* (SANC).
DISTRIBUTION. ETHIOPIAN REGION: Zimbabwe.
HOST PLANTS. Euphorbiaceae: *Poinsettia pulcherrima*.

Saissetia privigna De Lotto

Saissetia privigna De Lotto, 1965: 229.
TYPE DATA. Holotype female, KENYA: Ruiru, on *Coffea arabica* (BMNH).
DISTRIBUTION. ETHIOPIAN REGION: Kenya, Tanzania. **ORIENTAL REGION:** Pakistan. **PALAEARCTIC REGION:** Greece, Israel.
HOST PLANTS. Anacardiaceae: *Mangifera indica*. **Bignoniaceae:** *Markhamia platycalyx*. **Cucurbitaceae:** *Cucurbita maxima*. **Leguminosae:** *Erythrina*. **Malvaceae:** *Gossypium hirsutum*, *Hibiscus fuscus*. **Moraceae:** *Ficus carica*. **Oleaceae:** *Olea europaea*. **Rubiaceae:** *Coffea arabica*. **Solanaceae:** *Solanum tuberosum*.
REMARKS. Distribution and host plant records given by De Lotto (1976), Muzaffar & Ahmad (1977), Ben-Dov (1985) and by Shafee *et al.* (1989).
ECONOMIC IMPORTANCE. Biology and pest status on coffee presented by LePelley (1968). Muzaffar & Ahmad (1977) reported on natural enemies in Pakistan.

Saissetia reticulata (Cockerell)

Lecanium reticulatum Cockerell, 1895d: 174.
Saissetia reticulata (Cockerell); Cockerell, 1901d: 32.
TYPE DATA. Syntypes female, BRAZIL: Sao Paulo, on undetermined woody plant (USNM).
DISTRIBUTION. NEOTROPICAL REGION: Brazil Sao Paulo.
HOST PLANTS. Bignoniaceae: *Jacaranda*. **Myrtaceae:**
REMARKS. Adult female redescribed and illustrated by Hempel (1900b). Distribution and host plant records given by Silva *et al.* (1968).

Saissetia sclerotica Hodgson

Saissetia sclerotica Hodgson, 1967a: 19.
TYPE DATA. Holotype female, ZIMBABWE: Inyangombe Falls, on *Ficus capensis* (SANC)
DISTRIBUTION. ETHIOPIAN REGION: Zimbabwe.
HOST PLANTS. Moraceae: *Ficus capensis*.

Saissetia scutata Newstead

Lecanium (Saissetia) scutatum Newstead, 1917a: 364.
TYPE DATA. Syntypes female, second instar and first instar larva,
GUYANA: Georgetown, Botanic Gardens, on *Mimusops globosa*
(BMNH).
DISTRIBUTION. NEOTROPICAL REGION: Guyana.
HOST PLANTS. Sapotaceae: *Mimusops globosa.*

Saissetia silvestrii Leonardi

Saissetia silvestrii Leonardi, 1911: 275.
TYPE DATA. Syntypes female, ARGENTINA: Cacheuta, on *Zuccagnia
punctata* (IEAP)
DISTRIBUTION. NEOTROPICAL REGION: Argentina.
HOST PLANTS. Leguminosae: *Zuccagnia punctata.*

Saissetia socialis Hempel

Saissetia socialis Hempel, 1932: 329.
TYPE DATA. Syntypes female, BRAZIL: Rio Grande do Sul, Lavras,
on *Schinus dependens* (MZSP).
DISTRIBUTION. NEOTROPICAL REGION: Brazil Rio Grande do Sul.
HOST PLANTS. Anacardiaceae: *Schinus dependens.* **Rosaceae:**
Cydonia vulgaris.
REMARKS. Distribution and host plant records given by Corseuil &
Barbosa (1971).

Saissetia somereni (Newstead)

Lecanium mori somereni Newstead, 1910c: 187.
Lecanium (Eulecanium) tremae Newstead, 1911b: 162. Syntypes
female, TANZANIA: Amani, on *Trema guineensis* (BMNH). Syn. by
Newstead, 1913: 76.
Lecanium tremae Newstead; Sasscer, 1912: 89.
Lecanium (Eulecanium) somereni Newstead; Newstead, 1913: 76.
Lecanium somereni Newstead; Lindinger, 1913a: 83.
Eulecanium somereni (Newstead); Gowdey, 1917: 188.
Saissetia somereni (Newstead); De Lotto, 1956a: 247.
Saissetia abyssinica De Lotto, 1965: 221. Holotype female, ETHIO-
PIA: Dire Dawa, on *Duranta repens* (SANC). Syn. by De Lotto,
1968b: 86.
Saissetia somerinae (Newstead); Hodgson, 1969b: 36 [MIS-SPEL-
LING].
TYPE DATA. Syntypes female, UGANDA: Kyetume, on *Morus* sp.
(BMNH).
DISTRIBUTION. ETHIOPIAN REGION: Ethiopia, Kenya, Malawi,
South Africa, St. Helena, Tanzania, Uganda, Zimbabwe.
HOST PLANTS. Bignoniaceae: *Bignonia australis, Dolichandrone
platycalyx, Markhamia hildebrandti, Tecoma stans.* **Cycadaceae:**
Cycas. **Ehretiaceae:** *Cordia holstii, Ehretia silvatica.* **Euphorbia-
ceae:** *Croton.* **Flacourtiaceae:** *Flacourtia indica.* **Leguminosae:**
Bauhinia, Erythrina excelsa, E. tomentosa. **Meliaceae:** *Melia
azedarach.* **Moraceae:** *Ficus capensis, F. dekdekena, F. hochstet-
teri, Morus.* **Myrtaceae:** *Eugenia, Psidium guajava.* **Rutaceae:**
Citrus aurantium. **Verbenaceae:** *Duranta repens.*

REMARKS. Adult female redescribed and illustrated by De Lotto (1965) (as *S. abyssinica*) and by De Lotto (1956a, 1970a). Distribution and host plant records given by Newstead (1911b, 1913, 1917c), Lindinger (1913a), Gowdey (1917), De Lotto (1956a, 1965), Hodgson (1967a) and by Matile-Ferrero (1976).

Saissetia subpatelliforme Newstead

Lecanium (Saissetia) subpatelliforme Newstead, 1917a: 366.
Saissetia subpatelliforme Newstead; Brain, 1920b: 13.
TYPE DATA. Syntypes female, GHANA: Aburi, on undetermined plant (BMNH).
DISTRIBUTION. ETHIOPIAN REGION: Ghana, Zimbabwe.
HOST PLANTS. Meliaceae: *Cedrela toona, Melia azedarach*.
REMARKS. Adult female redescribed by Hall (1935). Distribution and host plant records given by Brain (1920b) and by Hall (1935).

Saissetia tolucana (Parrott & Cockerell)

Lecanium tolucanum Parrott & Cockerell in Cockerell & Parrott, 1899: 164.
Saissetia tolucana (Parrott & Cockerell); Cockerell, 1901d: 32.
TYPE DATA. Syntypes female, MEXICO: Mexico State, Toluca, on stalks of potato (USNM).
DISTRIBUTION. NEOTROPICAL REGION: Mexico.
HOST PLANTS. Solanaceae: *Solanum tuberosum*.

Saissetia vellozoi Vernalha

Saissetia vellozoi Vernalha, 1957: 33.
TYPE DATA. Syntypes female, BRAZIL: Parana, Sao Mateus, on *Ilex* sp. (VCCB).
DISTRIBUTION. NEOTROPICAL REGION: Brazil Parana.
HOST PLANTS. Aquifoliaceae: *Ilex*.

Saissetia vivipara Williams & Watson

Saissetia vivipara Williams & Watson, 1990: 169.
TYPE DATA. Holotype female, SOLOMON ISLANDS: Guadalcanal Province, Guadalcanal, Mt. Austen, on *Pipturus argenteus* (BMNH).
DISTRIBUTION. AUSTRO ORIENTAL REGION: Papua New Guinea, Solomon Islands.
HOST PLANTS. Dipterocarpaceae: *Anisoptera thurifera*. **Malvaceae:** *Sida*. **Moraceae:** *Ficus*. **Sapotaceae:** *Pometia pinnata*. **Urticaceae:** *Pipturus argenteus*.

Saissetia xerophila De Lotto

Saissetia xerophila De Lotto, 1957a: 179.
TYPE DATA. Holotype female, KENYA: Magadi, on *Capparis* sp. (BMNH).
DISTRIBUTION. ETHIOPIAN REGION: Kenya.
HOST PLANTS. Capparidaceae: *Capparis*.

Saissetia zanthoxylum (Hempel)

Lecanium zanthoxylum Hempel, 1900b: 430.
Saissetia zanthoxylum (Hempel); Cockerell, 1902f: 453.
Saissetia xanthoxylum (Hempel); Fernald, 1903: 207 [MIS-SPELLING].
TYPE DATA. Syntypes female, BRAZIL: Ypiranga, on *Zanthoxylum* sp. (MZSP).
DISTRIBUTION. NEOTROPICAL REGION: Brazil Sao Paulo.
HOST PLANTS. Myrtaceae: Rutaceae: *Zanthoxylum*.
REMARKS. Adult female redescribed by Hempel (1901a).

Saissetia zanzibarensis Williams

Saissetia zanzibarensis Williams, 1953: 582.
TYPE DATA. Holotype female, ZANZIBAR: Dole, on *Eugenia jambos* (BMNH).
DISTRIBUTION. ETHIOPIAN REGION: Kenya, Tanzania, Zanzibar.
HOST PLANTS. Anacardiaceae: *Mangifera indica*. **Bombacaceae:** *Adansonia digitata*. **Burseraceae:** *Canarium commune*. **Lauraceae:** *Persea americana*. **Leguminosae:** *Cassia, Gliricidia sepium*. **Moraceae:** *Ficus*. **Myrtaceae:** *Eugenia jambos, Jambosa caryophyllus, Psidium guajava, Syzygium cumini*. **Oxalidaceae:** *Averrhoa carambola*. **Palmae:** *Cocos nucifera*. **Rutaceae:** *Citrus*. **Sapotaceae:** *Manilkara zapota*.
REMARKS. Distribution and host plant records given by De Lotto (1956a).
BIOLOGY. Way (1954) studied the close association of this soft scale with the ant, *Oecophylla longinoda* (Latr.), and showed their mutual benefits.

Schizochlamidia Cockerell

Schizochlamidia Cockerell, 1899f: 333.
Schizochlamys Cockerell; Cockerell, 1899g: 15 [UNJUSTIFIED REPLACEMENT NAME].
Schizochlamydia Borchsenius; Borchsenius, 1957: 47 [MIS-SPELLING].
TYPE-SPECIES: *Schizoclamidia mexicana* Cockerell & Parrott, by original designation and monotypy.

Schizochlamidia mexicana Cockerell & Parrott

Schizochlamidia mexicana Cockerell & Parrott in Cockerell, 1899g: 15.
TYPE DATA. Syntypes female and male, MEXICO: Vera Cruz city, on *Mimosa* sp. (USNM).
DISTRIBUTION. NEOTROPICAL REGION: Mexico.
HOST PLANTS. Leguminosae: *Mimosa*.

Scythia Kiritshenko

Scythia Kiritshenko, 1938: 229.
TYPE-SPECIES: *Scythia craniumequinum* Kiritshenko, by monotypy.
REMARKS. Generic characters discussed by Borchsenius (1957), Kosztarab & Kozár (1978, 1988), Tereznikova (1981) and by Tang

(1991). Key to species: PALAEARCTIC REGION - Borchsenius (1957),
Danzig (1964), Kosztarab & Kozár (1978, 1988), Russo & Longo
(1991). UKRAINE - Tereznikova (1981).

Scythia aetnensis Russo & Longo

Scythia aetnensis Russo & Longo, 1991: 1.
TYPE DATA. Holotype female, ITALY: Sicily, Galvarina (Catania), on
Festuca circummediterranea (IECI).
DISTRIBUTION. PALAEARCTIC REGION: Italy.
HOST PLANTS. Gramineae: *Festuca circummediterranea*.

Scythia craniumequinum Kiritshenko

Scythia craniumequinum Kiritshenko, 1938: 229.
TYPE DATA. Syntypes female, USSR: Ukraine, Denepropetrovsk, on
Stipa sp. (ZIAS).
DISTRIBUTION. PALAEARCTIC REGION: Hungary, Mongolia,
USSR Azerbaijan, USSR Kazakhstan, USSR Ukraine.
HOST PLANTS. Gramineae: *Cleistogenes squarrosa*, *Festuca*, *Stipa-
borysthenica*, *S. capillata*, *S. grandis*, *S. ioannis*, *S. lessingiana*,
S. pulcherrima, *S. sibirica*.
REMARKS. Adult female redescribed and illustrated by Borchsenius
(1957), Kosztarab & Kozár (1978, 1988), Tereznikova (1981) and by
Tang (1991). Adult male described and illustrated by Borchsenius
(1957). Colour photograph given by Kosztarab & Kozár (1988). Dis-
tribution and host plant records given by Danzig (1974, 1984),
Tereznikova (1981), Nagy & Kozár (1984), Kozár (1986), Kosztarab &
Kozár (1988) and by Tang (1991).

Scythia festuceti (Sulc)

Mohelnia festuceti Sulc, 1941: 2.
Scythia festuceti (Sulc); Borchsenius, 1957: 182.
TYPE DATA. Syntypes female, male and larva, CZECHOSLOVAKIA:
Moravia, Mohelno, on *Festuca ovina* (MMBC).
DISTRIBUTION. PALAEARCTIC REGION: Czechoslovakia, Hungary,
USSR Caucasus, USSR Ukraine.
HOST PLANTS. Gramineae: *Festuca ovina*, *F. valesiaca*.
REMARKS. Adult female redescribed and illustrated by Borchsenius
(1957), Kosztarab & Kozár (1978, 1988), Tereznikova (1981). Distri-
bution and host plant records given by Borchsenius (1957), Kozár et
al. (1977), Rehacek (1957), Tereznikova (1981) and by Kosztarab &
Kozár (1988).

Scythia stipae Hadzibejli

Scythia stipae Hadzibejli, 1967a: 715.
TYPE DATA. Syntypes female, USSR: Georgia, Shirak steppe, on
Stipa lessingiana (GPPT, ZIAS).
DISTRIBUTION. PALAEARCTIC REGION: USSR Azerbaijan, USSR
Georgia, USSR Kazakhstan.
HOST PLANTS. Gramineae: *Festuca sulcata*, *Stipa lessingiana*.
REMARKS. Adult female redescribed and illustrated by Hadzibejli
(1973). Distribution and host plant records given by Hadzibejli
(1973).

Spermococcus Giard

Spermococcus Giard, 1894: cxcix.
TYPE-SPECIES: *Spermococcus fallax* Giard, by monotypy.
REMARKS. This genus is here regarded a subjective synonym of
Lecanopsis, as suggested by Lindinger (1935a) and by Borchsenius
(1957).

Sphaerolecanium Sulc

Sphaerolecanium Sulc, 1908: 36.
Spaerolecanium Leonardi; Leonardi, 1908: 180 [MIS-SPELLING].
Sphaerolecaniuw Tang; Tang, 1991: 141 [MIS-SPELLING].
TYPE-SPECIES: *Coccus prunastri* Fonscolombe, by original designa-
tion and monotypy.
REMARKS. Generic characters discussed by Borchsenius (1957),
Gómez-Menor Ortega (1960), Kosztarab & Kozár (1978, 1988), Yang
(1982) and by Tang (1991).

Sphaerolecanium prunastri (Fonscolombe)

Coccus prunastri Fonscolombe, 1834: 211.
Lecanium blanchardii Targioni Tozzetti; Targioni Tozzetti, 1868: 731
    [UNJUSTIFIED REPLACEMENT NAME].
Lecanium prunastri (Fonscolombe); Signoret, 1874a: 423.
Lecanium (Eulecanium) prunastri (Fonscolombe); Cockerell, 1896b:
    332.
Eulecanium prunastri (Fonscolombe); Fernald, 1903: 193.
Sphaerolecanium prunastri (Fonscolombe); Sulc, 1908: 36.
Eulecanium piligerum Leonardi, 1917: 195. Syntypes female and
    male, ITALY: Altamura, on prune (IEAP). Syn. by Silvestri,
    1920b: 503.
Lecanium (Sphaerolecanium) prunastri (Fonscolombe); Sulc, 1932: 78.
TYPE DATA. Syntypes female, FRANCE: Var, Saint-Zacharie, on
    prune tree (lost; D. Matile-Ferrero, 1988, Personal communica-
    tion).
DISTRIBUTION. PALAEARCTIC REGION: Austria, Israel, Bulgaria,
    China, Crete, Czechoslovakia, France, Germany, Greece, Hun-
    gary, Iran, Italy, Poland, Romania, Spain, Switzerland, Turkey,
    USSR Armenia, USSR Azerbaijan, USSR Crimea, USSR Georgia,
    USSR Moldavia, USSR Ukraine, USSR Uzbekistan, Yugoslavia.
HOST PLANTS. Rosaceae: *Malus sylvestris*, *Persica vulgaris*, *Prunus
    cerasifera*, *P. divaricata*, *P. domestica*, *P. salicina*, *P. spinosa*, *P.
    ursina*.
REMARKS. Adult female redescribed and illustrated by Paoli (1916),
Leonardi (1917, 1920), Silvestri (1919b, 1920b), Sulc (1932), Borch-
senius (1957), Gómez-Menor Ortega (1958a, 1958c, 1960), Savescu
(1961), Tereznikova (1981), Kosztarab & Kozár (1978, 1988) and by
Tang (1991). Adult male described and illustrated by Leonardi (1917)
(as *E. piligerum*), Silvestri (1920b), Borchsenius (1957) and by Gillomee (1967). Male test described and illustrated by Miller & Williams
(1990). Female third instar larva described and illustrated by Ben-
Dov (1968). Colour photograph given by Kosztarab & Kozár (1988),
Johnson & Lyon (1988) and by Tranfaglia & Viggiani (1988). Distribution and host plant records given by Paoli (1916), Leonardi (1917,

1920), Silvestri (1919b, 1920b), Sulc (1932), Bodenheimer (1944, 1953), Borchsenius (1957), Gómez-Menor Ortega (1960), Giliomee (1967), Kawecki (1968), Ben-Dov (1968, 1971), Kozár et al. (1977, 1979), Paloukis (1979), Tereznikova (1981), Yang (1982), Podsiadlo (1983), Kozár (1983, 1985), Martin Mateo (1984), Marotta (1987), Kozár & Ostafichuk (1987), Kozár & Ostafichuk (1987), Dziedzicka (1988), Kosztarab & Kozár (1988), Miller & Williams (1990) and by Tang (1991).

BIOLOGY. Kawecki (1968) presented details on the biology and geographical distribution. Develops one annual generation in Europe (Kosztarab & Kozár, 1988), in Israel (Ben-Dov, 1968) and in Greece (Argyriou & Paloukis, 1976; Paloukis, 1979). Natural enemies discussed by Silvestri (1920b), Ben-Dov (1968), Argyriou & Paloukis (1976), Paloukis (1979) and by Kosztarab & Kozár (1988).

Stenolecanium Takahashi

Stenolecanium Takahashi, 1959: 74.

TYPE-SPECIES: *Stenolecanium esakii* Takahashi, by original designation and monotypy.

REMARKS. Generic characters discussed by Tang (1991).

Stenolecanium esakii Takahashi

Stenolecanium esakii Takahashi, 1959: 74.

TYPE DATA. Holotype female, JAPAN: Kyushu, Kagoshima Prefecture, Sata, on *Ardisia* sp. (HUSJ).

DISTRIBUTION. PALAEARCTIC REGION: Japan.

HOST PLANTS. Myrsinaceae: *Ardisia*.

REMARKS. Adult female redescribed and illustrated by Kawai (1980) and by Tang (1991).

Stictolecanium Cockerell

Stictolecanium Cockerell, 1902f: 452.

TYPE-SPECIES: *Lecanium ornatum* Hempel, by original designation and monotypy.

Stictolecanium ornatum (Hempel)

Lecanium ornatum Hempel, 1900b: 421.
Stictolecanium ornatum (Hempel); Cockerell, 1902f: 452.

TYPE DATA. Syntypes female, BRAZIL: Sao Paulo, on *Eugenia jaboticaba* (MZSP).

DISTRIBUTION. NEOTROPICAL REGION: Brazil Sao Paulo.

HOST PLANTS. Myrtaceae: *Eugenia jaboticaba*.

REMARKS. Adult female redescribed and illustrated by Hempel (1901a, 1920a).

Stozia Marchal

Stozia Marchal, 1906: 143.

TYPE-SPECIES: *Stozia striata* Marchal, by monotypy.

REMARKS. Generic characters discussed by Borchsenius (1957) and by Yang (1982). Key to species given by Borchsenius (1957).

Stozia chrysophyllae (Silvestri)

Philippia chrysophyllae Silvestri, 1915 254.
Philippia (Stozia) chrysophyllae Silvestri; Silvestri, 1939: 746.
TYPE DATA. Syntypes female, ERITREA: Nefasit, on *Olea chrysophylla* (IEAP).
DISTRIBUTION. ETHIOPIAN REGION: Eritrea.
HOST PLANTS. Oleaceae: *Olea chrysophylla.*

Stozia ephedrae (Newstead)

Lichtensia ephedrae Newstead, 1901: 83.
Stozia striata Marchal, 1906: 144. Syntypes female, ALGERIA: Macta
   Forest, near Mostaganem, on *Ephedra altissima* (MNHN). Syn. by
   Matile-Ferrero, 1978: 45.
Filippia ephedrae (Newstead); Lindinger, 1912: 140.
Filippia foucauldi Balachowsky, 1929: 308. Syntypes female, ALGE-
   RIA: Hoggar, near Asekrem, on *Ephedra nebrodensis* (MNHN).
   Syn. by Matile-Ferrero, 1978: 45.
Filippia striata (Marchal); Balachowsky, 1932b: xxxii.
Stozia ephedrae (Newstead); Matile-Ferrero, 1978: 44.
TYPE DATA. Syntypes female, EGYPT: Wadi Gerrawy, Helonan, on
   Ephedra alte (BMNH).
DISTRIBUTION. PALAEARCTIC REGION: Algeria, Egypt, Iran,
   Israel, Spain, USSR Azerbaijan.
HOST PLANTS. Ephedraceae: *Ephedra alte, E. altissima, E. nebro-
   densis.* **Leguminosae:** *Coronilla pentaphylla.* **Liliaceae:** *Aspara-
   gus stipularis.* **Umbelliferae:** *Bupleurum sessiliflorum.*
REMARKS. Adult female redescribed and illustrated by Ezzat &
Hussein (1969). Adult female redescribed by Borchsenius (1957).
Distribution and host plant records given by Marchal (1906), Lindinger (1912), Bodenheimer (1926b, 1944b), Balachowsky (1928a,
1929), Borchsenius (1957), Ezzat & Hussein (1969), Matile-Ferrero
(1978) and by Martin Mateo (1984).

Stozia maxima Borchsenius

Stozia maxima Borchsenius, 1957: 184.
TYPE DATA. Syntypes female, USSR: Armenia, near Megri, on
   Ephedra procera (ZIAS).
DISTRIBUTION. PALAEARCTIC REGION: USSR Armenia.
HOST PLANTS. Ephedraceae: *Ephedra procera.*

Suareziella Mamet

Suareziella Mamet, 1954a: 47.
TYPE-SPECIES: *Suareziella montana* Mamet, by original designation
   and monotypy.

Suareziella montana Mamet

Suareziella montana Mamet, 1954a: 48.
TYPE DATA. Syntypes female, MADAGASCAR: Montagne des Fran-
   cais, on stems of undetermined plant (MNHN).
DISTRIBUTION. MADAGASIAN REGION: Madagascar.
BIOLOGY. Attended by ants.

Symonicoccus Koteja & Brookes

Symonicoccus Koteja & Brookes, 1981: 378.
TYPE-SPECIES: *Symonicoccus stipae* Koteja & Brookes, by original designation.
REMARKS. Key to species given by Koteja & Brookes (1981).

Symonicoccus aberrans Koteja & Brookes

Symonicoccus aberrans Koteja & Brookes, 1981: 384.
TYPE DATA. Holotype female, AUSTRALIA: Western Australia, near Kununurra, on *Triodia* sp. (ANIC).
DISTRIBUTION. AUSTRALIAN REGION: Western Australia.
HOST PLANTS. Gramineae: *Triodia*.

Symonicoccus australis (Maskell)

Signoretia luzulae (Dufour); Maskell, 1893b: 233 [MISIDENTIFICATION].
Signoretia luzulae australis Maskell, 1894c: 80.
Luzulaspis luzulae australis (Maskell); Fernald, 1903: 143.
Luzulaspis australis (Maskell); Steinweden, 1929: 230.
Lecanopsis australis (Maskell); Lindinger, 1943b: 221.
Symonicoccus australis (Maskell); Koteja & Brookes, 1981: 386.
TYPE DATA. Lectotype female designated by Koteja & Brookes (1981), AUSTRALIA: New South Wales, Gunnedah, on native grass (NZAC).
DISTRIBUTION. AUSTRALIAN REGION: New South Wales.
HOST PLANTS. Gramineae:
REMARKS. Adult female redescribed and illustrated by Koteja & Brookes (1981).

Symonicoccus chorizandrae Koteja & Brookes

Symonicoccus chorizandrae Koteja & Brookes, 1981: 387.
TYPE DATA. Holotype female, AUSTRALIA: Western Australia, Bassendean, on *Chorizandra enodis* (ANIC).
DISTRIBUTION. AUSTRALIAN REGION: Western Australia.
HOST PLANTS. Cyperaceae: *Chorizandra enodis*.

Symonicoccus giganteus Koteja & Brookes

Symonicoccus giganteus Koteja & Brookes, 1981: 389.
TYPE DATA. Holotype female, AUSTRALIA: Western Australia, 10 km south of Bridgetown, on sedge (ANIC).
DISTRIBUTION. AUSTRALIAN REGION: Western Australia.
HOST PLANTS. Cyperaceae:

Symonicoccus ovalis Koteja & Brookes

Symonicoccus ovalis Koteja & Brookes, 1981: 382.
TYPE DATA. Holotype female, AUSTRALIA: New South Wales, Moree, host plant not indicated (BMNH).
DISTRIBUTION. AUSTRALIAN REGION: New South Wales, South Australia.
HOST PLANTS. Gramineae: *Eragrostis setiflora*.

Symonicoccus stipae Koteja & Brookes

Symonicoccus stipae Koteja & Brookes, 1981: 383.
TYPE DATA. Holotype female, AUSTRALIA: South Australia, 60 km South East of Kingoonya, on *Stipa* sp. (ANIC).
DISTRIBUTION. AUSTRALIAN REGION: South Australia.
HOST PLANTS. Gramineae: *Stipa.*

Taiwansaissetia Tao, Wong & Chang

Taiwansaissetia Tao, Wong & Chang, 1983: 76.
TYPE-SPECIES: *Lecanium formicarii* Green, by original designation.
REMARKS. Tang (1991) regarded this genus as a subjective synonym of *Coccus*. Key to species given by Tao *et al.* (1983).

Taiwansaissetia armata (Takahashi)

Saissetia armata Takahashi, 1930: 33.
Ctenochiton armatus (Takahashi); Takahashi, 1942a: 26.
Taiwansaissetia armata (Takahashi); Tao *et al.*, 1983: 77.
Platysaissetia armata (Takahashi); Tang, 1991: 206.
TYPE DATA. Syntypes female, TAIWAN: Kuraru near Koshun, on *Glochidon arnottianum* and *Eugenia jambos* (IMZT).
DISTRIBUTION. ORIENTAL REGION: Taiwan.
HOST PLANTS. Euphorbiaceae: *Glochidion arnottianum, G. dasyphyllum.* **Myrtaceae:** *Eugenia jambos.*
REMARKS. Adult female redescribed and illustrated by Tao *et al.* (1983) and by Tang (1991). Distribution and host plant records given by Takahashi (1942a), Tao *et al.* (1983) and by Tang (1991).

Takahashia Cockerell

Pulvinaria (Takahashia) Cockerell, 1896e: 20.
Takahashia Cockerell; Kuwana, 1902a: 61.
TYPE-SPECIES: *Pulvinaria (Takahashia) japonica* Cockerell, by original designation and monotypy.
REMARKS. Cockerell (1896f) also presented *Takahashia* as a new subgenus. Generic characters discussed by Steinweden (1929), Borchsenius (1957), De Lotto (1968a), Paik (1978), MacGregor (1981), Yang (1982) and by Tang (1991).

Takahashia japonica Cockerell

Pulvinaria (Takahashia) japonica Cockerell, 1896e: 20.
Takahashia japonica Cockerell; Kuwana, 1902a: 61.
Takahashia wuchangensis Tseng, 1947: 21. Syntypes female, CHINA: Wuchang County, Hubei Province, on *Parthenocissus tricuspidata* (lost; Tang, F. T, 1989, personal communication). Syn. by Tang, 1991: 292.
TYPE DATA. Syntypes female, JAPAN: Tokyo, on mulberry (USNM).
DISTRIBUTION. PALAEARCTIC REGION: China, Japan, Korea.
HOST PLANTS. Leguminosae: *Lespedeza.* **Magnoliaceae:** *Magnolia obovata.* **Moraceae:** *Morus.* **Rosaceae:** *Prunus salicina.* **Salicaceae:** *Salix glandulosa.* **Ulmaceae:** *Celtis sinensis.* **Vitidaceae:** *Parthenocissus tricuspidata.*
REMARKS. Cockerell (1896f) again referred to this species as n. sp.

Adult female redescribed and illustrated by Kuwana (1917a), De Lotto (1968a), Paik (1978), MacGregor (1981), Yang (1982) and by Tang (1991). Adult female redescribed by Borchsenius (1957) and by Kawai (1980). First instar crawler described by Kuwana (1902a). Colour photograph given by Kawai (1980). Distribution and host plant records given by Kuwana (1902a, 1917a), Tseng (1947), Takahashi & Tachikawa (1956), Borchsenius (1957), De Lotto (1968a), Paik (1978), MacGregor (1981), Kawai (1980), Yang (1982) and by Tang (1991).

Tectopulvinaria Hempel

Tectopulvinaria Hempel; Hempel in Cockerell, 1899f: 331 [NOMEN NUDUM].
Tectopulvinaria Hempel, 1900b: 482.
TYPE-SPECIES: *Tectopulvinaria albata* Hempel, by original designation and monotypy.
REMARKS. Generic characters discussed by Hempel (1901b).

Tectopulvinaria albata Hempel

Tectopulvinaria albata Hempel, 1900b: 483.
TYPE DATA. Syntypes female, BRAZIL: Sao Paulo, Ypiranga and Jundiahy, on *Vernonia polyanthus* and *Trichogonia salviaefoliae* (MZSP).
DISTRIBUTION. NEOTROPICAL REGION: Brazil Rio Grande do Sul, Brazil Sao Paulo.
HOST PLANTS. Compositae: *Trichogonia salviaefolia*, *Vernonia polyanthes*.
REMARKS. Adult female redescribed by Hempel (1901b) and by Gomes Costa (1949). Distribution and host plant records given by Gomes Costa (1949) and by Corseuil & Barbosa (1971). Qin & Gullan (1992) compared this species with *T. loranthi* Froggatt and concluded that they are not congeneric.

Tectopulvinaria farinosa (Green)

Exaeretopus farinosus Green, 1922b: 1027.
Tectopulvinaria farinosa (Green); Green, 1937: 313.
TYPE DATA. Syntypes female, SRI LANKA: Namunakuli Hill, Badulla, on *Psychotria bisulcata* (BMNH).
DISTRIBUTION. ORIENTAL REGION: Sri Lanka.
HOST PLANTS. Rubiaceae: *Psychotria bisulcata*.
REMARKS. Koteja (1978) supposed that this species belongs to the *Coccus* group of genera.

Tectopulvinaria loranthi Froggatt

Tectopulvinaria loranthi Froggatt, 1915: 419.
TYPE DATA. Lectotype female designated by Qin & Gullan (1992), AUSTRALIA: New South Wales, near Ryde, on *Loranthus* sp. parasitic on *Eucalyptus* sp. (BCRI).
DISTRIBUTION. AUSTRALIAN REGION: New South Wales.
HOST PLANTS. Loranthaceae: *Loranthus*.
REMARKS. Adult female redescribed and illustrated by Qin &

Gullan (1992). Qin & Gullan (1992) compared this species with *T. albata* Hempel and concluded that they are not congeneric.

Tillancoccus Ben-Dov

Tillancoccus Ben-Dov, 1989b: 2.
TYPE-SPECIES: *Tillancoccus tillandsiae* Ben-Dov, by original designation.

Tillancoccus mexicanus Ben-Dov

Tillancoccus mexicanus Ben-Dov, 1989b: 3.
TYPE DATA. Holotype female, MEXICO: intercepted at Brownsville, Texas, USA, on *Tillandsia juncea* (USNM).
DISTRIBUTION. NEOTROPICAL REGION: Guatemala, Mexico.
HOST PLANTS. Bromeliaceae: *Tillandsia concolor*, *T. juncea*.

Tillancoccus tillandsiae Ben-Dov

Tillancoccus tillandsiae Ben-Dov, 1989b: 4.
TYPE DATA. Holotype female, GUATEMALA: intercepted in Israel, on *Tillandsia* sp. (ICV).
DISTRIBUTION. NEOTROPICAL REGION: Guatemala, Honduras, Mexico.
HOST PLANTS. Bromeliaceae: *Tillandsia*.

Toumeyella Cockerell

Lecanium (Toumeyella) Cockerell, 1895j: 56.
Lecanium (Toumeyella) Cockerell; Cockerell, 1895o: 2.
Toumeyella Cockerell; Cockerell & Parrott, 1901: 58.
TYPE-SPECIES: *Lecanium mirabile* Cockerell, by original designation and monotypy.
REMARKS. Generic characters discussed by Steinweden (1929), Hodgson (1969b), Williams & Kosztarab (1972), Hamon & Williams (1984), Gill (1988) and by Sheffer & Williams (1990). Taxonomic characters of first instars discussed by Sheffer & Williams (1990). Key to species: CALIFORNIA - Gill (1988); FLORIDA - Hamon & Williams (1984); VIRGINIA - Williams & Kosztarab (1972). Key to American species, based on first instars, given by Sheffer & Williams (1990).

Toumeyella cerifera Ferris

Toumeyella cerifera Ferris, 1921a: 90.
TYPE DATA. Syntypes female, MEXICO: Lower California, Agua Caliente, on *Albizia occidentalis* (UCD).
DISTRIBUTION. NEOTROPICAL REGION: Mexico Lower California.
NEARCTIC REGION: Alabama, Arkansas, Florida, Louisiana, North Carolina, Virginia.
HOST PLANTS. Leguminosae: *Albizia occidentalis*. **Naucleaceae:** *Cephalanthus occidentalis*. **Salicaceae:** *Salix*.
REMARKS. Adult female redescribed and illustrated by Williams & Kosztarab (1972) and by Hamon & Williams (1984). Male test described and illustrated by Miller & Williams (1990). First instar larva described and illustrated by Sheffer & Williams (1990). Colour

photograph given by Hamon & Williams (1984). Distribution and host plant records given by Steinweden (1946), Williams & Kosztarab (1972), Hamon & Williams (1984), Miller & Williams (1990) and by Sheffer & Williams (1990).

Toumeyella corrugatum neglectum
Pettit & McDaniel

Lecanium (Toumeyella) corrugatum neglectum Pettit & McDaniel, 1920: 7.

Toumeyella corrugatum neglectum Pettit & McDaniel; Williams & Kosztarab, 1972: 182.

TYPE DATA. Syntypes female, U.S.A.: New York, Ithaca, on Pitch pine (USNM).

DISTRIBUTION. NEARCTIC REGION: New York.

HOST PLANTS. Pinaceae: *Pinus.*

REMARKS. Williams & Kosztarab (1972) strongly indicated that this species might be synonymous with *T. pini.*

Toumeyella cubensis Heidel & Köhler

Toumeyella cubensis Heidel & Köhler, 1979: 132.

TYPE DATA. Holotype female, CUBA: Ceballos, on *Citrus sinensis* (CUBA).

DISTRIBUTION. NEOTROPICAL REGION: Cuba.

HOST PLANTS. Rutaceae: *Citrus aurantifolia, C. aurantium, C. grandis, C. hystrix, C. limon, C. paradisi, C. reticulata.*

REMARKS. Heidel & Köhler (1979) also described and illustrated the adult male and larval instars.

Toumeyella liriodendri (Gmelin)

Coccus liriodendri Gmelin, 1790: 2220.

Lecanium tulipiferae Cook, 1878: 192. Syntypes female, U.S.A.: Michigan, Lansing, on tulip tree [=*Liriodendron tulipifera*] (USNM). Syn. by King, 1902b: 59.

Lecanium liriodendri (Gmelin); Cockerell, 1899h: 271.

Eulecanium liriodendri (Gmelin); Fernald, 1903: 190.

Toumeyella liriodendri (Gmelin); Sanders, 1909a: 447.

Lecanium (Toumeyella) liriodendri (Gmelin); Pettit & McDaniel, 1920: 10.

TYPE DATA. Syntypes female, EUROPE: on *Liriodendron* trees imported from America; see Sanders, 1909a (probably lost).

DISTRIBUTION. NEARCTIC REGION: Alabama, California, Florida, Georgia, Indiana, Kentucky, Louisiana, Maryland, Massachusetts, Michigan, New Jersey, New York, North Carolina, Ohio, Pennsylvania, South Carolina, Tennessee, Virginia.

HOST PLANTS. Guttiferae: *Ascyrum edisonianum, A. hypericoides, A. tetrapetalum, Hypericum cistifolium.* **Juglandaceae:** *Carya cordiformis.* **Leguminosae:** *Cassia fasciculata.* **Magnoliaceae:** *Liriodendron tulipifera, Magnolia acuminata, M. soulangeana, M. stellata, M. virginiana, Michelia.* **Malvaceae:** *Sida spinosa.* **Rubiaceae:** *Cephalanthus, Gardenia.* **Salicaceae:** *Populus.* **Theaceae:** *Gordonia.* **Tiliaceae:** *Tilia.*

REMARKS. Common Name - Tuliptree Scale. Adult female redescribed and illustrated by Dietz & Morrison (1916), Pettit & McDaniel (1920), Williams & Kosztarab (1972), Hamon & Williams (1984) and by Gill (1988). Male test described and illustrated by Miller & Williams (1990). First instar larva described and illustrated by Sheffer & Williams (1990). Colour photograph given by Hamon & Williams (1984) and by Johnson & Lyon (1988). Hamon & Williams (1984) noted that *T. turgida* appears to be a synonym of this species, but indicated that further research is needed to verify it. Distribution and host plant records given by Dietz & Morrison (1916), Pettit & McDaniel (1920), Williams & Kosztarab (1972), Lambdin (1984), Lambdin & Watson (1980), Hamon & Williams (1984), Gill (1988) and by Sheffer & Williams (1990).

BIOLOGY. Burns & Donley (1970) studied the biology of this scale in Eastern U. S. A., where it develops one annual generation. Lambdin (1984) and Simpson & Lambdin (1983) studied the life history in Tennessee. The amino acids and sugars in the honeydew were identified by Burns & Davidson (1966).

ECONOMIC IMPORTANCE. A pest of Yellow-poplar, *Liriodendron tulipifera* in Eastern U. S. A. (Burns & Donley, 1970; Hamon & Williams, 1984; Gill, 1988).

Toumeyella lomagundiae Hall

Toumeyella lomagundiae Hall, 1935: 81.
TYPE DATA. Holotype female, ZIMBABWE: Sinoia, on *Bauhinia macrantha*, (BMNH).
DISTRIBUTION. ETHIOPIAN REGION: Zimbabwe.
HOST PLANTS. Leguminosae: *Bauhinia macrantha*, *B. variegata*, *Piliostigma thouningii*.
REMARKS. Adult female redescribed and illustrated by Hodgson (1969b). Distribution and host plant records given by Hodgson (1969b).

Toumeyella mirabilis (Cockerell)

Lecanium mirabile Cockerell, 1895o: 3.
Toumeyella mirabilis (Cockerell); Cockerell, 1902f: 452.
TYPE DATA. Syntypes female, U.S.A.: Arizona, Tucson, near the University, on *Prosopis juliflora* var. *glandulosa* (USNM).
DISTRIBUTION. NEOTROPICAL REGION: Mexico Lower California. **NEARCTIC REGION:** Arizona.
HOST PLANTS. Leguminosae: *Prosopis juliflora glandulosa*.
REMARKS. Male test described and illustrated by Miller & Williams (1990). First instar larva described and illustrated by Sheffer & Williams (1990). Distribution and host plant records given by Ferris (1921a), Miller & Williams (1990) and by Sheffer & Williams (1990).

Toumeyella nectandrae Hempel

Toumeyella nectandrae Hempel, 1929: 64.
TYPE DATA. Syntypes female, BRAZIL: Parana, Jaguariahyva, on *Nectandra grandiflora* (MZSP).
DISTRIBUTION. NEOTROPICAL REGION: Brazil Parana.
HOST PLANTS. Lauraceae: *Nectandra grandiflora*.
REMARKS. First instar larva described and illustrated by Sheffer &

Williams (1990). Distribution and host plant records given by Sheffer & Williams (1990).

Toumeyella obunca De Lotto

Toumeyella obunca De Lotto, 1966b: 149.
TYPE DATA. Holotype female, SOUTH AFRICA: Natal, Richmond, on *Cnestis natalensis* (SANC).
DISTRIBUTION. ETHIOPIAN REGION: South Africa.
HOST PLANTS. Connaraceae: *Cnestis natalensis*.

Toumeyella parvicornis (Cockerell)

Lecanium parvicorne Cockerell, 1897d: 90.
Toumeyella parvicornis (Cockerell); Cockerell, 1902f: 452.
Lecanium (Toumeyella) numismaticum Pettit & McDaniel, 1920: 8. Syntypes female, U.S.A.: Wisconsin, Trout Lake, on Scotch-pine (DEPOSITORY UNKNOWN; see Williams & Kosztarab, 1972). Syn. by Williams & Kosztarab, 1972: 171.
TYPE DATA. Syntypes female, U.S.A.: Florida, on *Pinus taeda* and *P. australis* (BMNH, USNM).
DISTRIBUTION. NEARCTIC REGION: Alabama, Canada Manitoba, Canada Ontario, Florida, Georgia, Illinois, Indiana, Iowa, Kentucky, Louisiana, Massachusetts, Michigan, Minnesota, Nebraska, New Jersey, New Mexico, New York, North Carolina, North Dakota, Ohio, Pennsylvania, South Carolina, South Dakota, Tennessee, Texas, Virginia, West Virginia, Wisconsin.
HOST PLANTS. Pinaceae: *Pinus australis*, *P. echinata*, *P. elliotti*, *P. glabra*, *P. mugo*, *P. palustris*, *P. sylvestris*, *P. taeda*, *P. virginiana*.
REMARKS. Common name - Pine Tortoise Scale. Adult female redescribed and illustrated by Williams & Kosztarab (1972) and by Hamon & Williams (1984). Male test described and illustrated by Miller & Williams (1990). First instar larva described and illustrated by Sheffer & Williams (1990). Colour photograph given by Hamon & Williams (1984). Distribution and host plant records given by Pettit & McDaniel (1920), MacAloney (1961), Williams & Kosztarab (1972), Lambdin & Watson (1980), Hamon & Williams (1984), Miller & Williams (1990) and by Sheffer & Williams (1990).
BIOLOGY. Develops one annual generation in Canada (Rabkin & Lejeune, 1954; MacAloney, 1961), probably two generations in more southern regions (Hamon & Williams, 1984). Life history described by MacAloney (1961) and by Rabkin & Lejeune (1954).
ECONOMIC IMPORTANCE. A pest of seedlings and saplings (Rabkin & Lejeune, 1954).

Toumeyella paulista Hempel

Toumeyella paulista Hempel, 1932: 330.
TYPE DATA. Syntypes female, BRAZIL: Sao Paulo, Capital, on *Nectandra* sp. (MZSP).
DISTRIBUTION. NEOTROPICAL REGION: Brazil Sao Paulo.
HOST PLANTS. Lauraceae: *Nectandra*.

Toumeyella pini (King)

Lecanium pini King, 1901i: 334.

Toumeyella pini (King); Cockerell, 1902f: 452.
Lecanium corrugatum Thro, 1903: 216. Syntypes female, U.S.A.: New
    York, on *Pinus sylvestris* (USNM). Syn. by Fernald, 1903: 179.
Lecanium (Toumeyella) corrugatum Thro; Pettit & McDaniel, 1920: 6.
TYPE DATA. Syntypes female, CANADA: Ontario, London, on *Pinus
    austriaca* (CNOC).
DISTRIBUTION. NEARCTIC REGION: Alabama, Canada Ontario,
    Connecticut, Georgia, Maryland, Michigan, New York, Pennsyl-
    vania, Tennessee, Texas, Virginia.
HOST PLANTS. Pinaceae: *Pinus austriaca, P. echinata, P. mugo, P.
    palustris, P. resinosa, P. rigida, P. serotina, P. sylvestris, P. taeda,
    P. virginiana.*
REMARKS. Adult female redescribed and illustrated by Pettit &
McDaniel (1920), (as *T. corrugatum*), Williams & Kosztarab (1972)
and by Hamon &Williams (1984). Male test described and illustrated
by Miller & Williams (1990). First instar larva described and illustrated by Sheffer & Williams (1990). Distribution and host plant
records given by Pettit & McDaniel (1920), Williams & Kosztarab
(1972), Lambdin & Watson (1980), Hamon & Williams (1984), Miller
& Williams (1990) and by Sheffer & Williams (1990).
BIOLOGY. Common Name - Striped Pine scale. Clarke *et al.* (1989b)
observed that this scale has three generations per year on *Pinus
taeda* in Georgia and studied its natural enemies.

Toumeyella pinicola Ferris

Toumeyella pinicola Ferris, 1920b: 41.
TYPE DATA. Holotype female, U.S.A.: California, San Mateo County,
    Spring Valley Water Company at Aqua, on *Pinus radiata* (UCD).
DISTRIBUTION. NEARCTIC REGION: California.
HOST PLANTS. Pinaceae: *Pinus radiata.*
REMARKS. Adult female redescribed and illustrated by Gill (1988).
Male test described and illustrated by Miller & Williams (1990).
Colour photograph of adult female given by Gill (1988).

Toumeyella quadrifasciata (Cockerell)

Lecanium quadrifasciatum Cockerell, 1895o: 3.
Toumeyella quadrifasciata (Cockerell); Cockerell, 1902f: 452.
TYPE DATA. Syntypes female, U.S.A.: New Mexico, Soledad Canon,
    Organ Mts., on *Robinia neomexicana* (USNM).
DISTRIBUTION. NEARCTIC REGION: New Mexico.
HOST PLANTS. Leguminosae: *Robinia neomexicana.*
REMARKS. First instar larva described and illustrated by Sheffer &
Williams (1990). Distribution and host plant records given by Sheffer
& Williams (1990).

Toumeyella sonorensis (Cockerell & Parrott)

Lecanium sonorense Cockerell & Parrott, 1899: 161.
Toumeyella sonorensis (Cockerell & Parrott); Cockerell, 1902f: 452.
TYPE DATA. Syntypes female, MEXICO: Sonora State, Hermosillo,
    on *Beloperone californica* (USNM).
DISTRIBUTION. NEOTROPICAL REGION: Mexico.
HOST PLANTS. Acanthaceae: *Beloperone californica.*

Toumeyella turgida (Cockerell)

Lecanium turgidum Cockerell, 1897k: 152.
Toumeyella turgida (Cockerell); Cockerell, 1902f: 452.
TYPE DATA. Syntypes female, U.S.A.: Florida, Lake City, on *Magnolia glauca* (USNM).
DISTRIBUTION. NEARCTIC REGION: Florida.
HOST PLANTS. Magnoliaceae: *Magnolia glauca.*
REMARKS. Hamon & Williams (1984) noted that this species appears to be a synonym of *T. liriodendri* as the two species are inseparable in slide-mounted specimens, but further research is needed to verify this.

Toumeyella virginiana Williams & Kosztarab

Toumeyella virginiana Williams & Kosztarab, 1972: 182.
TYPE DATA. Holotype female, U.S.A.: Georgia, Clark County, on *Pinus taeda* (USNM).
DISTRIBUTION. NEARCTIC REGION: Alabama, Florida, Georgia, Maryland, Virginia.
HOST PLANTS. Pinaceae: *Pinus clausa, P. elliotti, P. glabra, P. palustris, P. taeda, P. virginiana.*
REMARKS. Adult female redescribed and illustrated by Hamon & Williams (1984). Male test described and illustrated by Miller & Williams (1990). First instar larva described and illustrated by Sheffer & Williams (1990). Distribution and host plant records given by Hamon & Williams (1984), Miller & Williams (1990) and by Sheffer & Williams (1990).

Trijuba De Lotto

Trijuba De Lotto, 1975: 62.
TYPE-SPECIES: *Saissetia oculata* Brain, by original designation and monotypy.

Trijuba oculata (Brain)

Saissetia oculata Brain, 1920b: 13.
Lecanium oculata (Brain); Hall, 1935: 76.
Lecanium dorsociliatum Green & Mamet, 1938: 126. Syntypes female, MAURITIUS: Rose Hill, on *Nephrolepis cordifolia* (BMNH). Syn. by Mamet, 1954b: 260.
Coccus dorsociliatus (Green & Mamet); Mamet, 1949: 23.
Coccus oculatus (Brain); Mamet, 1954b: 260.
Trijuba oculata (Brain); De Lotto, 1975: 62.
TYPE DATA. Syntypes female, SOUTH AFRICA: Natal, Durban, on grapevine (SANC).
DISTRIBUTION. ETHIOPIAN REGION: South Africa, Zimbabwe.
MADAGASIAN REGION: Mauritius, Reunion, Rodrigues.
HOST PLANTS. Annonaceae: *Annona reticulata.* **Leguminosae:** *Mucuna bennetti, Poinciana pulchrissima.* **Moraceae:** *Ficus capensis.* **Oleandraceae:** *Nephrolepis cordifolia.* **Proteaceae:** *Grevillea robusta.* **Vitidaceae:** *Vitis vinifera.*
REMARKS. Adult female redescribed and illustrated by Green & Mamet (1938) (as *L. dorsociliatum*) and by De Lotto (1957b). Distribution and host plant records given by Hall (1935), Green & Mamet

(1938), Mamet (1949, 1954b), De Lotto (1957b, 1975), Hodgson, (1967a) and by Williams & Williams (1988).

Udinia De Lotto

Udinia De Lotto, 1963: 194.
TYPE-SPECIES: *Udinia scitula* De Lotto, by original designation.
REMARKS. Hanford (1974) revised the genus. Key to species given by De Lotto (1963; 1965) and by Hanford (1974). Generic characters discussed by Tang (1991).

Udinia bruncki Hanford

Udinia bruncki Hanford, 1974: 11.
TYPE DATA. Holotype female, IVORY COAST: Banco, on *Chlorophora excelsa* (BMNH).
DISTRIBUTION. ETHIOPIAN REGION: Ivory Coast.
HOST PLANTS. Moraceae: *Chlorophora excelsa.*

Udinia catori (Green)

Lecanium catori Green, 1915b: 43.
Lecanium (Saissetia) subhirsutum Newstead, 1917a: 367. Lectotype female, designated by Hanford (1974), GHANA: Aburi, on *Garcinia* sp. (BMNH). Syn. by Hanford, 1974: 13.
Saissetia catori (Green); De Lotto, 1959: 151.
Udinia catori (Green); De Lotto, 1963: 195.
Udinia subhirsutum Newstead; De Lotto, 1963: 195.
TYPE DATA. Lectotype female designated by Hanford (1974), NIGERIA: Kabba Province, from pod of Kola nut (BMNH).
DISTRIBUTION. ETHIOPIAN REGION: Ghana, Guinea, Ivory Coast, Nigeria, Sierra Leone, Sudan.
HOST PLANTS. Anacardiaceae: *Mangifera indica.* **Apocynaceae:** *Landolphia.* **Guttiferae:** *Garcinia.* **Lauraceae:** *Persea americana.* **Leguminosae:** *Cassia nodosa.* **Meliaceae:** *Khaya senegalensis.* **Moraceae:** *Ficus exasperata.* **Myrtaceae:** *Psidium guajava.* **Naucleaceae:** *Nauclea latifolia.* **Rutaceae:** *Citrus aurantium.* **Sapotaceae:** *Chrysophyllum cainito.* **Sterculiaceae:** *Cola acuminata, C. nitida, Theobroma cacao, Triplochiton.* **Verbenaceae:** *Tectona grandis.*
REMARKS. Adult female redescribed and illustrated by Hanford (1974). Intraspecific variation in taxonomic characters given by Hanford (1974). Distribution and host plant records given by Newstead (1917a), Strickland (1947), De Lotto (1963), Hanford (1974) and by Couturier *et al.* (1985).

Udinia farquharsoni (Newstead)

Lecanium (Saissetia) farquharsoni Newstead, 1922: 530.
Saissetia subhirsuta Newstead; Strickland, 1947: 500 [MISIDENTIFICATION].
Saissetia exoleta De Lotto, 1957a: 171. Holotype female, KENYA: Kisumu, on *Gardenia jovis-tonantis* (BMNH). Syn. by Hanford, 1974: 18.
Udinia exoleta (De Lotto); De Lotto, 1963: 194.
Udinia farquharsoni (Newstead); De Lotto, 1963: 195.

TYPE DATA. Lectotype female designated by Hanford (1974), NIGER-IA: near Ibadan, on undetermined plant (BMNH).
DISTRIBUTION. ETHIOPIAN REGION: Cameroon, Ghana, Ivory Coast, Kenya, Nigeria, Sierra Leone, Tanzania, Zaire, Zanzibar.
HOST PLANTS. Apocynaceae: *Alstonia congoensis*. **Asclepiadaceae:** *Calotropis procera*. **Bignoniaceae:** *Millingtonia hortensis*. **Bombacaceae:** *Adansonia digitata, Durio zibethinus*. **Euphorbiaceae:** *Croton*. **Flacourtiaceae:** *Scottellia*. **Meliaceae:** *Khaya anthotheca*. **Myrtaceae:** *Eugenia caryophyllata*. **Rubiaceae:** *Canthium glabriflorum, C. canephora, Gardenia jovis-tonantis*. **Sapotaceae:** *Imbricaria maxima*. **Solanaceae:** *Solanum macranthum*. **Sterculiaceae:** *Theobroma cacao, Triplochiton scleroxylon*. **Verbenaceae:** *Tectona grandis*.
REMARKS. Adult female redescribed and illustrated by De Lotto (1957a) (as *Saissetia exoleta*), De Lotto (1965) and by Hanford (1974). Intraspecific variation in taxonomic characters given by Hanford (1974). Distribution and host plant records given by De Lotto (1957a, 1963, 1965), Hanford (1974), Matile-Ferrero & Nonveiller (1984) and by Couturier *et al.* (1985).

Udinia glabra De Lotto

Udinia glabra De Lotto, 1963: 195.
TYPE DATA. Holotype female, UGANDA: Kyadondo, on *Coffea robusta* (BMNH).
DISTRIBUTION. ETHIOPIAN REGION: Uganda.
HOST PLANTS. Rubiaceae: *Coffea robusta*.
REMARKS. Hanford (1974) supplemented the taxonomic characters.
ECONOMIC IMPORTANCE. Biology and pest status on coffee presented by LePelley (1968).

Udinia ikoyensis Hanford

Udinia ikoyensis Hanford, 1974: 22.
TYPE DATA. Holotype female, GABON: Ikoy, on twig of Acajou [=*Khaya ivorensis*] (BMNH)
DISTRIBUTION. ETHIOPIAN REGION: Gabon.
HOST PLANTS. Meliaceae: *Khaya ivorensis*.

Udinia lindae Matile-Ferrero & Le Ruyet

Udinia lindae Matile-Ferrero & Le Ruyet, 1985: 267.
TYPE DATA. Holotype female, IVORY COAST: Tai, on *Gambeya taiensis* (MNHN).
DISTRIBUTION. ETHIOPIAN REGION: Ivory Coast.
HOST PLANTS. Sapotaceae: *Gambeya taiensis*.
BIOLOGY. Attended by ants of *Crematogaster* sp. (Matile-Ferrero & Le Ruyet, 1985).

Udinia lobayana (Balachowsky & Ferrero)

Saissetia lobayana Balachowsky & Ferrero, 1965: 134.
Udinia lobayana (Balachowsky & Ferrero); Hanford, 1974: 24.
TYPE DATA. Holotype female, CENTRAL AFRICAN REPUBLIC: River Lobaya (at left bank), 20 km from the track from M'Baike, on *Carapa procera* (MNHN).

DISTRIBUTION. ETHIOPIAN REGION: Central African Republic.
HOST PLANTS. Meliaceae: *Carapa procera.*
REMARKS. Hanford (1974) presented additional notes on taxonomic characters of this species.

Udinia newsteadi Hanford

Lecanium (Saissetia) barteriae Newstead; Mann, 1922: 629 [NOMEN NUDUM].
Udinia newsteadi Hanford, 1974: 25.
TYPE DATA. Holotype female, IVORY COAST: Locality not recorded, on *Entandophragma* sp. (BMNH).
DISTRIBUTION. ETHIOPIAN REGION: Ghana, Ivory Coast, Zaire, Zanzibar.
HOST PLANTS. Bombacaceae: *Durio zibethinus.* **Ebenaceae:** *Diospyros mannii.* **Flacourtiaceae:** *Barteria dewevrei.* **Meliaceae:** *Entandophragma.* **Moraceae:** *Ficus.*
REMARKS. Distribution and host plant records given by Couturier *et al.* (1985).

Udinia nigeriensis Hanford

Udinia nigeriensis Hanford, 1974: 28.
TYPE DATA. Holotype female, NIGERIA: near Awgu, on *Barteria* sp. (BMNH).
DISTRIBUTION. ETHIOPIAN REGION: Nigeria.
HOST PLANTS. Flacourtiaceae: *Barteria.*

Udinia pattersoni Hanford

Udinia pattersoni Hanford, 1974: 30.
TYPE DATA. Holotype female, GHANA: Ayimenash, On Lime [=*Citrus aurantifolia*] (BMNH).
DISTRIBUTION. ETHIOPIAN REGION: Ghana.
HOST PLANTS. Rutaceae: *Citrus aurantifolia.*

Udinia paupercula De Lotto

Udinia paupercula De Lotto, 1963: 195.
TYPE DATA. Holotype female, KENYA: Kisumu, on *Gardenia jovis-tonantis* (BMNH).
DISTRIBUTION. ETHIOPIAN REGION: Cameroon, Ghana, Ivory Coast, Kenya, Sudan, Tanzania, Zanzibar.
HOST PLANTS. Apocynaceae: *Carissa.* **Myrtaceae:** *Eugenia caryophyllata.* **Naucleaceae:** *Uncaria africana.* **Rubiaceae:** *Coffea, Gardenia jovis-tonantis.* **Sapindaceae:** *Nephelium lappaceum.* **Sonneratiaceae:** *Sonneratia casseolaris.*
REMARKS. Hanford (1974) supplemented the taxonomic characters of this species. Distribution and host plant records given by Hanford (1974), Matile-Ferrero & Nonveiller (1984) and by Couturier *et al.* (1985).

Udinia pterolobina (De Lotto)

Saissetia pterolobina De Lotto, 1956a: 245.
Udinia pterolobina (De Lotto); De Lotto, 1963: 194.

TYPE DATA. Holotype female, KENYA: Nairobi, on *Pterolobium lacerans* (BMNH).
DISTRIBUTION. ETHIOPIAN REGION: Ivory Coast, Kenya.
HOST PLANTS. Leguminosae: *Pterolobium lacerans*. **Scytopetalaceae:** *Scytopetalum tieghemii*.
REMARKS. Hanford (1974) added some notes on taxonomic characters of this species. Distribution and host plant records given by Hanford (1974) and by Couturier *et al.* (1985).

Udinia punctuliferum lamborni (Newstead)

Lecanium punctuliferum lamborni Newstead, 1914b: 523.
Udinia punctuliferum lamborni (Newstead); De Lotto, 1963: 195.
TYPE DATA. Syntypes preadult female, NIGERIA: Lagos, host plant not indicated (BMNH).
DISTRIBUTION. ETHIOPIAN REGION: Nigeria.
REMARKS. De Lotto (1963) examined one pre-adult specimen of this species and referred it to *Udinia*. Hanford (1974) studied additional specimens of the original type material and found that all are immatures which are very close to *U. farquharsoni*

Udinia scitula De Lotto

Udinia scitula De Lotto, 1963: 199.
TYPE DATA. Holotype female, TANZANIA: Arusha, on *Rauvolfia caffra* (BMNH).
DISTRIBUTION. ETHIOPIAN REGION: Tanzania.
HOST PLANTS. Apocynaceae: *Rauvolfia caffra*.
REMARKS. Hanford (1974) supplemented the taxonomic characters of this species.

Udinia setigera (Newstead)

Lecanium setigerum Newstead, 1917a: 368.
Coccus setiger (Newstead); Gowdey, 1917: 188.
Saissetia setigera (Newstead); De Lotto, 1957a: 179.
Udinia setigera (Newstead); De Lotto, 1963: 194.
TYPE DATA. Lectotype female designated by Hanford (1974), UGANDA: Nagunga, on *Psidium guajava* (BMNH).
DISTRIBUTION. ETHIOPIAN REGION: Cameroon, Ivory Coast, Uganda.
HOST PLANTS. Combretaceae: *Terminalia superba*. **Myrtaceae:** *Psidium guajava*. **Sapotaceae:** *Gambeya talensis*.
REMARKS. Adult female redescribed and illustrated by De Lotto (1957a) and by Hanford (1974). Distribution and host plant records given by Gowdey (1917), De Lotto (1957a), Hanford (1974), Matile-Ferrero & Nonveiller (1984) and by Couturier *et al.* (1985).

Umwinsia Hodgson

Umwinsia Hodgson, 1968b: 118.
TYPE-SPECIES: *Umwinsia cavernosa* Hodgson, by original designation and monotypy.

Umwinsia cavernosa Hodgson

Umwinsia cavernosa Hodgson, 1968b: 118.
TYPE DATA. Holotype female, ZIMBABWE: Borrowdale, on *Ficus* sp. (BMNH).
DISTRIBUTION. ETHIOPIAN REGION: Zimbabwe.
HOST PLANTS. Moraceae: *Ficus*.

Umwinsia nitidulus (De Lotto)

Akermes nitidulus De Lotto, 1958b: 165.
Umwinsia nitidulus (De Lotto); Hodgson, 1968b: 119.
TYPE DATA. Holotype female, KENYA: Nairobi, on *Annona chrysophylla* (BMNH).
DISTRIBUTION. ETHIOPIAN REGION: Kenya.
HOST PLANTS. Annonaceae: *Annona chrysophylla*.

Vinsonia Signoret

Vinsonia Signoret, 1872b: 33.
Visonia Ashmead; Ashmead, 1891: 99 [MIS-SPELLING].
Vinzonia Danzig & Konstantinova; Danzig & Konstantinova, 1990: 45 [MIS-SPELLING].
TYPE-SPECIES: *Vinsonia pulchella* Signoret, by monotypy.
REMARKS. Generic characters discussed by Newstead (1903), Green (1909), Steinweden (1929), De Lotto (1965), Yang (1982), Tao *et al.* (1983), Hamon & Williams (1984), Williams & Watson (1990) and by Tang (1991).

Vinsonia magnifica Green

Vinsonia magnifica Green, 1930c: 290.
TYPE DATA. Syntypes female, INDONESIA: Sumatra, Fort de Kock, on *Mangifera odorata* and *Eugenia malaccensis* (BMNH).
DISTRIBUTION. AUSTRO ORIENTAL REGION: Sumatra.
HOST PLANTS. Anacardiaceae: *Mangifera odorata*. **Myrtaceae:** *Eugenia malaccensis*.

Vinsonia stellifera (Westwood)

Coccus stellifera Westwood, 1871a: iii.
Coccus stellifer Westwood; Westwood, 1871b: 1006.
Vinsonia pulchella Signoret; Signoret, 1872b: 34. Syntypes female, REUNION ISLAND: on *Mangifera indica* (VMNH) Syn. by Signoret, 1877a: 608.
Vinsonia stellifera (Westwood); Douglas, 1888b: 152.
Ceroplastes stellifer (Westwood); Lindinger, 1913: 81.
Vinzonia stellifera (Westwood); Danzig & Konstantinova, 1990: 45 [MIS-SPELLING].
Vinsonia pulohella (Signoret); Tang, 1991: 311 [MIS-SPELLING and ERRONEOUS AUTHORSHIP].
Vinsonia stellifra (Signoret); Tang, 1991: 312 [MIS-SPELLING and ERRONEOUS AUTHORSHIP].
TYPE DATA. Syntypes female, ENGLAND: Pant y Goitre, Abergavenny, on *Cypripedium niveum*, imported from the west coast of SIAM [=THAILAND] (BMNH).

DISTRIBUTION. AUSTRO ORIENTAL REGION: Malaysia, Papua New Guinea, Philippines, Solomon Islands. **ETHIOPIAN REGION:** Angola, Kenya, Principe, Sao Tome, Tanzania, Zanzibar. **MADAGASIAN REGION:** Mauritius, Reunion, Seychelles. **NEOTROPICAL REGION:** Bermuda, Brazil Rio Grande do Sul, Brazil Sao Paulo, Cuba, Grenada, Guyana, Puerto Rico, Virgin Islands. **NEW ZEALAND and PACIFIC REGION:** Fiji, Irian Jaya, Niue, Palaus, Ponape Island, Tonga. **NEARCTIC REGION:** Alabama, Florida, Georgia. **ORIENTAL REGION:** India, Pakistan, Sri Lanka, Taiwan, Thailand, Vietnam.

HOST PLANTS. Anacardiaceae: *Mangifera indica.* **Apocynaceae:** *Alstonia scholaris, Ervatamia orientalis, Plumeria acutifolia.* **Araliaceae:** *Meryta macrophylla.* **Celastraceae:** *Lophopetalum.* **Ebenaceae:** *Diospyros discolor.***Euphorbiaceae:** *Bischofia javanica.* **Guttiferae:** *Garcinia mangostana, G. myrtifolia, G. spicata, G. subelliptica.* **Leguminosae:** *Palaquium formosanum.* **Liliaceae:** *Asparagus sprengeri.* **Moraceae:** *Artocarpus integra, Ficus benghalensis.* **Myrtaceae:** *Eucalyptus, Eugenia aquea, E. corynocarpa, E. jambolana, E. jambos, E. malaccensis, Syzygium cumini.* **Orchidaceae:** *Cattleya, Epidendrum ciliare.* **Palmae:** *Cocos nucifera, Nypa.* **Polypodiaceae:** *Adiantum.* **Rubiaceae:** *Gardenia.* **Rutaceae:** *Citrus grandis.* **Sapotaceae:** *Achras sapota, Lucuma caimito.* **Stilaginaceae:** *Antidesma bunius.* **Strelitziaceae:** *Ravenala madagascariensis.* **Zingiberaceae:** *Alpinia purpurata.*

REMARKS. Common Name - Stellate scale. Adult female redescribed and illustrated by Newstead (1903), Green (1909), Morrison (1920), De Lotto (1965), Yang (1982), Tao *et al.* (1983), Hamon & Williams (1984), Avasthi & Shafee (1986), Williams & Watson (1990) and by Tang (1991). Colour photograph given by Hamon & Williams (1984). Distribution and host plant records given by Signoret (1872b, 1877), Hempel (1900b, 1904), Newstead (1903, 1914a), Lindinger (1913), Ramakrishna Ayyar (1919, 1930), Morrison (1920), Green (1907, 1937), De Lotto (1965), Corseuil & Barbosa (1971), Ali (1971), Nath (1972), Almeida (1973), Nakahara & Miller (1981), Yang (1982), Nakahara (1983), Tao *et al.* (1983), Hamon & Williams (1984), Varshney (1985), Avasthi & Shafee (1986), Varshney & Moharana (1987), Williams &Williams (1988), Shafee *et al.* (1989), Williams & Watson (1990), Danzig & Konstantinova (1990), Tang (1991) and by Hodgson & Hilburn (1991a, 1991b).

ECONOMIC IMPORTANCE. Considered a potential threat to crops in Florida, due to its occurrence on host plants such as mango, citrus and various ornamentals (Hamon & Williams, 1984).

Vittacoccus Borchsenius

Vittacoccus Borchsenius, 1952: 271.
Vitacoccus Rehacek; Rehacek, 1954: 141 [MIS-SPELLING].
TYPE-SPECIES: *Lecanopsis longicornis* Green, by original designation and monotypy.
REMARKS. Koteja (1970) studied the type species and assigned the genus to the Eriopeltini. Generic characters discussed by Borchsenius (1957), Koteja (1970), Danzig (1980c), Kosztarab & Kozár (1988) and by Tang (1991).

Vittacoccus interruptus Danzig

Vittacoccus interruptus Danzig, 1975a: 138.
TYPE DATA. Holotype female, USSR: Irkutsk Region, Baykal, Malogo Morya, on *Leymus chinensis* (ZIAS).
DISTRIBUTION. PALAEARCTIC REGION: USSR Irkutsk Region.
HOST PLANTS. Gramineae: *Leymus chinensis*.

Vittacoccus longicornis (Green)

Lecanopsis longicornis Green, 1916a: 26.
Exaeretopus longicornis (Green); Green, 1928a: 7.
Vittacoccus longicornis (Green); Borchsenius, 1952: 271.
Vittacoccus ordinatus Danzig, 1971: 1416. Holotype female, USSR: Kunashir Island, on *Carex* sp. (ZIAS). Syn. by Danzig, 1978a: 17.
TYPE DATA. Syntypes female, ENGLAND: Camberley, on grass (BMNH).
DISTRIBUTION. PALAEARCTIC REGION: Czechoslovakia, England, Hungary, Poland, USSR Irkutsk Region, USSR Kunashir Island, USSR Moldavia, USSR Yakutsk.
HOST PLANTS. Cyperaceae: *Carex brizoides*. **Gramineae:** *Brachypodium*.
REMARKS. Adult female redescribed and illustrated by Green (1921a), Borchsenius (1957), Koteja (1969b), Kosztarab & Kozár (1978, 1988), Danzig (1980c) and by Tang (1991). Adult male redescribed and illustrated by Koteja (1970). Distribution and host plant records given by Green (1921a), Borchsenius (1952, 1957), Koteja (1969b, 1970, 1983), Danzig (1971, 1978a, 1978b, 1980c), Kozár & Ostafichuk (1987), Kosztarab & Kozár (1988), Kozár & Drozdjak (1990) and by Tang (1991).

Waricoccus Brookes & Koteja

Waricoccus Brookes & Koteja, 1982: 183.
TYPE-SPECIES: *Waricoccus parvisetosus* Brookes & Koteja, by original designation and monotypy.

Waricoccus parvisetosus Brookes & Koteja

Waricoccus parvisetosus Brookes & Koteja, 1982: 184.
TYPE DATA. Holotype female, AUSTRALIA: South Australia, Milang, on *Lepidosperma longitudinale* (ANIC).
DISTRIBUTION. AUSTRALIAN REGION: South Australia.
HOST PLANTS. Cyperaceae: *Lepidosperma longitudinale*.

Waxiella De Lotto

Waxiella De Lotto, 1971c: 148.
TYPE-SPECIES: *Ceroplastes subdenudatus* Newstead, by original designation.
REMARKS. Generic characters discussed by Ben-Dov (1986).

Waxiella africana (Green)

Ceroplastes africanus Green, 1899: 188.
Waxiella africana (Green); Ben-Dov, 1986: 166.
TYPE DATA. Lectotype female designated by Ben-Dov (1986),

SOUTH AFRICA: Cape Province, Kleinpoort, on *Acacia* sp. (BMNH).
DISTRIBUTION. ETHIOPIAN REGION: South Africa.
HOST PLANTS. Leguminosae: *Acacia karroo*.
REMARKS. Adult female redescribed and illustrated by Ben-Dov (1986). Cilliers (1967) described (as *C. mimosae*) the wax covering of the adult female and larval instars.
BIOLOGY. Cilliers (1967) studied (as *C. mimosae*) the biology and natural enemies of this wax scale in South Africa.

Waxiella africanus cristatus (Green)

Ceroplastes africanus cristatus Green, 1899: 190.
Ceroplastes cristatus Green; Fernald, 1903: 151.
Waxiella africanus cristatus (Green); Ben-Dov, 1986: 166.
TYPE DATA. Syntypes female, SOUTH AFRICA: Natal, host plant not indicated (BMNH).
DISTRIBUTION. ETHIOPIAN REGION: South Africa.
REMARKS. Fernald (1903) placed this subspecies as a synonym of *Ceroplastes egbarum*.

Waxiella africanus senegalensis (Marchal)

Ceroplastes africanus senegalensis Marchal, 1909b: 68.
Waxiella africanus senegalensis (Marchal); Ben-Dov, 1986: 166.
TYPE DATA. Syntypes female, SENEGAL: Locality not given, on *Acacia* sp. (MNHN).
DISTRIBUTION. ETHIOPIAN REGION: Senegal.
HOST PLANTS. Leguminosae: *Acacia*.
REMARKS. Adult female redescribed and illustrated by Marchal (1909c).

Waxiella berliniae (Hall)

Ceroplastes berliniae Hall, 1931: 291.
Gascardia berliniae (Hall); De Lotto, 1965: 181.
Ceroplastes aff. berliniae var. enkeldoorni (Hall); Almeida, 1969: 15 [MISIDENTIFICATION].
Waxiella berliniae (Hall); De Lotto, 1971c: 148.
TYPE DATA. Syntypes female, ZIMBABWE: Mazoe, Banket and Salisbury on *Berlinia globiflora*, Rusape on *Brachystegia flagristipulata*, El Dorado on *Brachystegia* sp. (BMNH).
DISTRIBUTION. ETHIOPIAN REGION: Angola, Zambia, Zimbabwe.
MADAGASIAN REGION: Madagascar.
HOST PLANTS. Annonaceae: *Annona*. **Leguminosae:** *Acacia dealbata, Berlinia, Brachystegia boehmi, Br. flagristipulata, Br. longifolia, Br. spiciformis, Br. tamarinoides, Br. utilis, Julbernardia globifera.*
REMARKS. Adult female redescribed and illustrated by Almeida (1969). Male described and illustrated by Giliomee (1967). Distribution and host plant records given by Mamet (1954a), Giliomee (1967), Almeida (1969, 1973), Hodgson (1969a) and by De Lotto (1971c).

Waxiella egbara (Cockerell)

Ceroplastes egbarum Cockerell, 1899d: 127.

Waxiella egbarum (Cockerell); De Lotto, 1971c: 148.
TYPE DATA. Syntypes female, NIGERIA: near Abeokuta, on *Mimosa* sp. (BMNH, USNM).
DISTRIBUTION. ETHIOPIAN REGION: Ghana, Nigeria.
HOST PLANTS. Leguminosae: *Acacia, Pithecellobium saman.*
REMARKS. Newstead (1917b) described from material collected in Ghana, the male puparium. Almeida (1969) redescribed a female, based on material from Angola, which was referred by her to this species. This redescription should be verified and compared with type or topotypic material.

Waxiella egbara fulleri (Cockerell & Cockerell)

Ceroplastes egbarum fulleri Cockerell & Cockerell in Cockerell, 1902c: 113.
Waxiella egbarum fulleri (Cockerell & Cockerell); De Lotto, 1971c: 148.
Waxiella egbarum fulleri (Cockerell); Ben-Dov, 1986: 166 [ERRONEOUS AUTHORSHIP].
TYPE DATA. Syntypes female, SOUTH AFRICA: Natal, Coast of Natal, on *Acacia* sp. and *Mimosa* sp. (USNM).
DISTRIBUTION. ETHIOPIAN REGION: South Africa.
HOST PLANTS. Leguminosae: *Acacia, Mimosa.*

Waxiella egbara rhodesiensis (Hall)

Ceroplastes egbarum rhodesiensis Hall, 1931: 294.
Gascardia egbara rhodesiensis (Hall); De Lotto, 1965: 181.
Waxiella egbarum rhodesiensis (Hall); De Lotto, 1971c: 148.
TYPE DATA. Syntypes female, ZIMBABWE: Bulawayo and Hunters Road, on undetermined plant (BMNH).
DISTRIBUTION. ETHIOPIAN REGION: Zimbabwe.

Waxiella enkeldoorni (Hall)

Ceroplastes berliniae enkeldoorni Hall, 1931: 292.
Gascardia berliniae enkeldoorni Hall; De Lotto, 1965: 181.
Gascardia enkeldoorni (Hall); Hodgson, 1969a: 25.
Waxiella enkeldoorni (Hall); De Lotto, 1971c: 148.
TYPE DATA. Syntypes female, ZIMBABWE: Enkeldoorn, on undetermined plant (BMNH).
DISTRIBUTION. ETHIOPIAN REGION: Mozambique, Zimbabwe.
HOST PLANTS. Leguminosae: *Acacia albida, Albizia adianthifolia, Baikiaea plurijuga, Brachystegia boehmi, Br. spiciformis, Guibourtia coleosperma, Pericopsis angolensis.* **Sapotaceae:** *Manilkara macualayae.*
REMARKS. Adult female redescribed and illustrated by Hodgson (1969a). Distribution and host plant records given by De Lotto (1971c) and by Hodgson (1969a).

Waxiella erithraeus (Leonardi)

Ceroplastes erithraeus Leonardi, 1913: 27.
Waxiella erythraeus (Leonardi); De Lotto, 1971c: 148 [MIS-SPELLING].

TYPE DATA. Syntypes female, ERITREA: Asmara and Nefarit, on *Acacia* sp. (IEAP).
DISTRIBUTION. ETHIOPIAN REGION: Eritrea.
HOST PLANTS. Leguminosae: *Acacia.*

Waxiella gwaai (Hodgson)

Gascardia gwaai Hodgson, 1969a: 28.
Waxiella gwaai (Hodgson); Ben-Dov, 1986: 166.
TYPE DATA. Holotype female, ZIMBABWE: At the Bulawayo - Victoria Falls Road near Wankie Game Reserve, on *Baphia massaiensis* subsp. *obovata* (BMNH).
DISTRIBUTION. ETHIOPIAN REGION: Zimbabwe.
HOST PLANTS. Leguminosae: *Baphia massaiensis obovata.*

Waxiella martinoi (Almeida)

Ceroplastes martinoi Almeida; Almeida, 1969: 15.
Waxiella martinoi (Almeida); Almeida, 1973: 7.
TYPE DATA. Syntypes female, ANGOLA: Bruco, on undetermined plant (CZLP).
DISTRIBUTION. ETHIOPIAN REGION: Angola.

Waxiella mimosae (Signoret)

Ceroplastes mimosae Signoret, 1872c: 46.
Waxiella mimosae (Signoret); De Lotto, 1971c: 148.
TYPE DATA. Lectotype female designated by Ben-Dov (1986), EGYPT: on *Mimosa [-Acacia] nilotica* (VMNH).
DISTRIBUTION. PALAEARCTIC REGION: Egypt, Saudi Arabia.
HOST PLANTS. Leguminosae: *Acacia asak, A. nilotica, A. seyal.*
REMARKS. Signoret (1872c) credited this species to Boisduval, however, the author is Signoret. Adult female redescribed and illustrated by Ezzat & Hussein (1969) and by Ben-Dov (1986). Distribution and host plant records given by Ben-Dov (1986) and by Matile-Ferrero (1988).

Waxiella mimosae neghellii (Bellio)

Ceroplastes mimosae neghellii Bellio, 1939: 225.
Waxiella mimosae neghellii (Bellio); De Lotto, 1971c: 148.
TYPE DATA. Syntypes female, ETHIOPIA: Neghelli, on *Euphorbia* sp. (IEAP).
DISTRIBUTION. ETHIOPIAN REGION: Ethiopia.
HOST PLANTS. Euphorbiaceae: *Euphorbia.*

Waxiella subdenudata (Newstead)

Ceroplastes subdenudatus Newstead, 1917b: 30.
Gascardia subdenudata (Newstead); De Lotto, 1967b: 784.
Waxiella subdenudata (Newstead); De Lotto, 1971c: 148.
TYPE DATA. Syntypes female, UGANDA: Entebbe, on *Acacia* sp. (BMNH).
DISTRIBUTION. ETHIOPIAN REGION: South Africa, Uganda.
HOST PLANTS. Leguminosae: *Acacia, Albizia.*
REMARKS. Adult female redescribed and illustrated by De Lotto

(1967b, 1971c). Distribution and host plant records given by De Lotto (1967b, 1971c).

Waxiella subsphaerica (Newstead)

Ceroplastes subsphaericus Newstead, 1911b: 166.
Waxiella subsphaericus (Newstead); De Lotto, 1971c: 148.
TYPE DATA. Syntypes female, EAST AFRICA: Probably TANZANIA, Ngambo, on *Albizia lebbeck* (BMNH).
HOST PLANTS. Leguminosae: *Albizia lebbeck*.
DISTRIBUTION. ETHIOPIAN REGION: East Africa, probably Tanzania.

Waxiella tamaricis Ben-Dov

Waxiella mimosae (Signoret); Ben-Dov, 1970d: 84 [MISIDENTIFICA-
    TION].
Waxiella mimosae (Signoret); Ben-Dov, 1971: 31 [MISIDENTIFICA-
    TION].
Waxiella tamaricis Ben-Dov, 1986: 171.
TYPE DATA. Holotype female, ISRAEL: Mivtahim, on *Tamarix articu-
    lata* (ICV).
DISTRIBUTION. PALAEARCTIC REGION: Israel.
HOST PLANTS. Tamaricaceae: *Tamarix articulata*.

Waxiella ugandae (Newstead)

Ceroplastes ugandae Newstead, 1911a: 94.
Waxiella ugandae (Newstead); De Lotto, 1971c: 148.
TYPE DATA. Syntypes female, UGANDA: on "Amakebe" (BMNH).
DISTRIBUTION. ETHIOPIAN REGION: Uganda.
HOST PLANTS. Annonaceae: *Annona muricata*. **Leguminosae:**
    Cajanus indicus.
REMARKS. Newstead (1913, 1914a, 1917c) supplemented the taxonomic characters. Distribution and host plant records given by Newstead (1913, 1914a, 1917c).

Waxiella uvariae (Marchal)

Ceroplastes uvariae Marchal, 1909b: 68.
Gascardia uvariae (Marchal); De Lotto, 1965: 182.
Waxiella uvariae (Marchal); Ben-Dov, 1986: 166.
TYPE DATA. Syntypes female, GUINEA: near Labe, on *Uvaria* sp. (MNHN).
DISTRIBUTION. ETHIOPIAN REGION: Guinea.
HOST PLANTS. Annonaceae: *Uvaria*.
REMARKS. Adult female redescribed and illustrated by Marchal (1909c).

Waxiella vuilleti (Marchal)

Ceroplastes vuilleti Marchal, 1909b: 68.
Waxiella vuilleti (Marchal); De Lotto, 1971c: 148.
TYPE DATA. Syntypes female, SENEGAL: Badinko, on *Ormosia laxi-
    flora* (MNHN).
DISTRIBUTION. ETHIOPIAN REGION: Nigeria, Senegal.

HOST PLANTS. Leguminosae: *Cajanus indicus, Ormosia laxiflora.*
REMARKS. Adult female redescribed and illustrated by Marchal (1909c). Newstead (1917b) based on material from Nigeria a redescription of the adult female, and description of male puparium. Distribution and host plant records given by Newstead (1917b).

Waxiella zonata (Newstead)

Ceroplastes zonatus Newstead, 1917b: 32.
Waxiella zonata (Newstead); De Lotto, 1971c: 148.
TYPE DATA. Syntypes female, SOUTH AFRICA: Locality and host plant not given (BMNH)
DISTRIBUTION. ETHIOPIAN REGION: Malawi, South Africa, Zimbabwe.
HOST PLANTS. Compositae: *Bidens bipernata.* **Leguminosae:** *Acacia decurrens, A. polyantha, A. sieberana woodii, A. woodii, Albizia lebbeck, Berlinia globiflora, Brachystegia spiciformis.* **Meliaceae:** *Melia azedarach.*
REMARKS. Adult female redescribed and illustrated by Hodgson (1969a). Distribution and host plant records given by Hall (1931), Hodgson (1969a) and by De Lotto (1971c).

Xenolecanium Takahashi

Xenolecanium Takahashi, 1942a: 26.
TYPE-SPECIES: *Xenolecanium mangiferae* Takahashi, by original designation and monotypy.
REMARKS. Generic characters discussed by Tang (1991).

Xenolecanium mangiferae Takahashi

Xenolecanium mangiferae Takahashi, 1942a: 27.
TYPE DATA. Syntypes female, THAILAND: Bangkok, on mango (IMZT).
DISTRIBUTION. ORIENTAL REGION: Thailand.
HOST PLANTS. Anacardiaceae: *Mangifera indica.*
REMARKS. Adult female redescribed and illustrated by Tang (1991).

Xenolecanium rotundum Takahashi

Xenolecanium rotundum Takahashi, 1951b: 105.
TYPE DATA. Syntypes female, INDONESIA: Riau [=Riouw] Islands, host plant not indicated (HUSJ).
DISTRIBUTION. AUSTRO ORIENTAL REGION: Indonesia.
REMARKS. Adult female redescribed and illustrated by Tang (1991).

REFERENCES

Abul-Rassoul, M.S. 1976. Checklist of Iraq Natural History Museum insect collection. *Publications University of Baghdad, Natural History Research Center* 30: 1-41.

Abrahao, J. & Mamprim, O. 1958. Cochonilha da raiz do cafeeiro. *O Biologico, São Paulo* 24: 268-271.

Abu-Yaman, I.K. 1970. The pistachio cushion scale, *Anapulvinaria pistaciae* Boden., and its control in Iraq. *Zeitschrift für Angewandte Entomologie* 66: 242-247.

Ahmad, R. 1975. A note on *Saissetia oleae* and its natural enemies in Iran. *Entomophaga* 20: 221-223.

Alayo, R. & Blahutiak, A. 1980. Dinámica estacional de *Saissetia hemisphaerica* Targioni (Homoptera: Coccoidea) en Cuba. *Poeyana, La Habana* 210: 1-16.

Alfonso, F. 1875. *Trattato sulla coltivazione degli agrumi.* Luigi Pedone Lauriel, Palermo.

Ali, S.M. 1964. Some studies on *Pulvinaria cellulosa* Green a mealy scale of mango in Bihar, India. *Indian Journal of Entomology* 26: 361-362.

Ali, S.M. 1968. Description of a new and records of some known coccids (Homoptera) from Bihar, India. *Oriental Insects. New Delhi* (1967) 1: 29-43.

Ali, S.M. 1971. A catalogue of the Oriental Coccoidea Part - V (Insecta: Homoptera: Coccoidea). *Indian Museum Bulletin* 6: 7-82.

Ali, S.M. 1973. Some coccids from Goa. *Journal of the Bombay Natural History Society* 69: 669-671.

Almeida, D.M. 1969. Contribuição para o estudo da quermofauna de Angola. *Boletim do Instituto de Investigação Científica de Angola. Luanda* 6: 3-38.

Almeida, D.M. 1973. Coccoidea de Angola. 1- Revisão das espécies conhecidas. *Boletim do Instituto de Investigação Científica de Angola. Luanda* 10: 1-23.

Amitai, S. 1969. Morphological identifications of the stages of the Florida wax scale - *Ceroplastes floridensis* Comst. (Coccoidea). *Israel Journal of Entomology* 4: 89-95.

Annecke, D.P. 1966. Biological studies on the immature stages of soft brown scale, *Coccus hesperidum* (Homoptera: Coccidae). *South African Journal of Agricultural Sciences* 9: 205-228.

Anonymous 1949. Necrologie. Adolfo Hempel. *Revista de Agricultura. São Paulo* 24: 388.

Archangelskaya, A.D. 1923. Contribution to the fauna of hard or soft scales (Coccidae) of Turkestan. [In Russian.]. *Trudov Turkestanogo Nauchnogo Obshshestva* 1: 259-266.

Archangelskaya, A.D. 1930a. List of scale insects (Coccidae) collected in the hothouses of the botanical gardens in Moscow and Leningrad in February 1929. [In Russian, English summary]. *Bolezni Rastenii, Leningrad (1929)* 18: 188-201.

Archangelskaya, A.D. 1930b. List of the scale insects (Coccidae) of Turkmenistan. [In Russian, English summary]. *Report of the Plant Protection Station for 1926 to 1929, Ashkhabad, Turkmenia* : 75-85.

Archangelskaya, A.D. 1931. New species of scale insects, Coccidae, from Central Asia. [In Russian, English summary]. *Zashchita Rastenii, Leningrad* 7: 69-85.

Archangelskaya, A.D. 1937. The Coccidae of Middle Asia. [In Russian, English Summary]. *Izdatelstvo Komiteta Nauk UZSSR, Tashkent.* 158 pp.

Argov, Y., Podoler, H., Bar-Shalom, O. & Rosen, D. 1987. Mass rearing of the Florida wax scale, *Ceroplastes floridensis*, for production of natural enemies. *Phytoparasitica* 15: 277-287.

Argyriou, L.C. 1963. Studies on the morphology and biology of the black scale [*Saissetia oleae* (Bernard)] in Greece. *Annales de l'Institut Phytopathologique Benaki (N.S.)* 5: 353-377.

Argyriou, L.C. 1967. The scales of olive trees occurring in Greece and their entomophagous insects. *Annales de l'Institut Phytopathologique Benaki (N.S.)* 8: 66-73.

Argyriou, L.C. 1983. Faunal analysis of some scale insects in Greece. *Proceedings of the 10th International Symposium of Central European Entomofaunistics, Budapest, 15-20 August 1983*: 364-367.

Argyriou, L.C. & Ioanides, A.G. 1975. *Coccus aegaeus* (Homoptera, Coccoidea, Coccidae) De Lotto: nouvelle espèce de lécanine des Citrus en Grèce. *Fruits* 30: 161-162.

Argyriou, L.C. & Kourmadas, A.L. 1977. Ecological studies on *Filippia follicularis* Targioni in Greece. *Mededlingen van de Rijksfaculteit Landbouwwentenschappen te Gent* 42: 1353-1360.

Argyriou, L.C. & Kourmadas, A.L. 1980. *Ceroplastes floridensis* Comstock an important pest of citrus trees in Aegean islands. *Fruits* 35: 705-708.

Argyriou, L.C. & Paloukis, S.S. 1976. Some data on biology and parasitization of *Sphaerolecanium prunastri* Fonscolombe (Homoptera Coccidae) in Greece. *Annales de l'Institut Phytopathologique Benaki (N.S.)* 11: 230-240.

Argyriou, L.C. & Santorini, A.P. 1980. On the phenology of *Ceroplastes rusci* L. (Hom. Coccidae) on fig-trees in Greece. *Mededlingen van de Rijksfaculteit Landbouwwetenschappen te Gent* 45: 593-601.

Ashmead, W.H. 1891. A generic synopsis of the Coccidae. Family X. - Coccidae. *Transactions of the Entomological Society of America* 18: 92-102.

Atanasov, P. 1959. A contribution to the knowledge of scale insects in the People's Republic of Macedonia. [In Serbo Croatia]. *Yearbook of*

*the Faculty of Agriculture and Forestry of the University in Skopje,
Yugoslavia (1958, 1959)* 12: 425-435.

Atkinson, E.T. 1886. Insect-pests belonging to the Homopterous
family Coccidae. *Journal of the Asiatic Society of Bengal. Natural
History* 55: 267-298.

Atkinson, E.T. 1889. A new species and genus of Coccidae. *Journal of
the Asiatic Society of Bengal. Natural Sciences* 58: 3-5.

Avasthi, R.K. & Shafee, S.A. 1979. A new species of *Cerostegia* De
Lotto (Homoptera: Coccidae) from Ajmer (India). *Current Science* 48:
36-37.

Avasthi, R.K. & Shafee, S.A. 1983. Two new species of *Coccus*
Linnaeus (Homoptera: Coccidae) from India. *Proceedings of the
10th International Central European Entomofaunistic Symposium,
Budapest, 15-20 August 1983*: 389-392.

Avasthi, R.K. & Shafee, S.A. 1984a. A new genus of Coccidae
(Homoptera). *Indian Journal of Systematic Entomology* 1: 7-9.

Avasthi, R.K. & Shafee, S.A. 1984b. *Kozaricoccus* gen. n. for *Ceroplastodes bituberculatus* Brain (Homoptera: Coccidae). *Indian
Journal of Systematic Entomology* 1: 31-33.

Avasthi, R.K. & Shafee, S.A. 1985. Two new species of *Pulvinaria*
Targ.-Tozz. (Homoptera: Coccidae) from India. *Current Science* 54:
1289-1291.

Avasthi, R.K. & Shafee, S.A. 1986. Species of Ceroplastinae (Homoptera: Coccidae) from India. *Journal of the Bombay Natural History
Society* 83: 327-338.

Avasthi, R.K. & Shafee, S.A. 1991a. Three species of *Ceroplastodes*
Cockerell (Homoptera: Coccidae). *Indian Journal of Systematic
Entomology* 8: 1-6.

Avasthi, R.K. & Shafee, S.A. 1991b. Classification of Indian Coccidae
(Homoptera: Coccoidea). *Indian Journal of Systematic Entomology*
8: 7-26.

Avidov, Z. 1961. Pests of the cultivated plants of Israel. [In Hebrew]. Magnes Press, Jerusalem. 546 pp.

Avidov, Z. & Harpaz, I. 1969. Plant Pests of Israel. Israel Universities Press, Jerusalem. 549 pp.

Avidov, Z. & Zaitzov, A. 1960. On the biology of the mango shield scale *Coccus mangiferae* (Green) in Israel. *Ktavim, Quarterly Journal of the National and University Institute of Agriculture, Rehovot, Israel* 10: 125-137.

Aziz, S.A.A.M.S. 1977. Phytophagous and entomophagous insects and mites of Iraq. *Publications University of Baghdad, Natural History Research Center* 33: 1-142.

Baerensprung, F.V. 1849. Beobachtungen über einige einheimische Arten aus der Familie der Coccinen. *Zeitung für Zoologie, Zootomie und Palaeozoologie* 1: 165-170, 173-176.

Baerg, W.J. 1947. The biology of the Maple Leaf Scale. *Bulletin of the Agricultural Experiment Station, University of Arkansas College of Agriculture, Fayetteville* 470: 1-14.

Balachowsky, A.S. 1926. Les principales cochenilles dont il faut redouter l'introduction en Algérie. *Revue Agricole de l'Afrique du Nord, Alger 1926*: 1-8.

Balachowsky, A.S. 1927. Contribution à l'étude des coccides de l'Afrique mineure (1re note). *Annales de la Société Entomologique de France* 96: 175-207.

Balachowsky, A.S. 1928a. Contribution à l'étude des coccides de l'Afrique Mineure (2e note). *Bulletin de la Société d'Histoire Naturelle de l'Afrique du Nord* 19: 121-144.

Balachowsky, A.S. 1928b. Contribution à l'étude des Coccides de l'Afrique mineure (4e note). Nouvelle liste de Coccides nord-africains avec description d'especes nouvelles. *Bulletin de la Société Entomologique de France* 1928: 273-279.

Balachowsky, A.S. 1929. Contribution à l'étude des coccides de l'Afrique Mineure (6e note). Faune du Hoggar. *Annales de la Société Entomologique de France* 98: 301-322.

Balachowsky, A.S. 1930a. Contribution à l'étude des coccides de France (1re note). Faunule des Iles d'Hyères (port-Cros et Levant). *Bulletin de la Société Entomologique de France* 1929: 311-317.

Balachowsky, A.S. 1930b. Contribution à l'étude des coccides de France (3e note). Coccides nouveaux ou peu connus de la faune de France. *Bulletin de la Société Entomologique de France* 1930: 178-184.

Balachowsky, A.S. 1930c. Contribution à l'étude des coccides de l'Afrique mineure (9me note). Addition à la faune du nord- africain avec description de trois especes nouvelles. *Bulletin de la Société d'Histoire Naturelle de l'Afrique du nord* 1930: 119-125.

Balachowsky, A.S. 1931a. Sur le comportement des Coccidae appartenant à la faune Bassin Mediterranéen. *Comptes Rendus de la Société de Biogéographie. Paris* 8: 48-52.

Balachowsky, A.S. 1931b. Contribution à l'étude des coccides de France (5e note). *Bulletin de la Société Entomologique de France* 1931: 96-102.

Balachowsky, A.S. 1931c. Contribution à l'étude des Coccidae de l'Afrique Mineure (10e note). Sur un nouveau *Lichtensia* du Maroc Septentrional. *Bulletin de la Société des Sciences Naturelles du Maroc* 10: 215-216.

Balachowsky, A.S. 1932a. Contribution à l'étude des coccides de France (9e note). Sur un *Luzulaspis* nouveau des Iles d'Hyères. *Bulletin de la Société Entomologique de France* 37: 197-200.

Balachowsky, A.S. 1932b. Etude biologique des coccides du Bassin Occidental de la Mediterranée. P. Lechevalier & Fils, Paris. 214 pp + LXVII.

Balachowsky, A.S. 1933a. Les coccides des Iles d'Hyères. *Annales de la Société d'Histoire Naturelles de Toulon* 1933: 1-7.

Balachowsky, A.S. 1933b. Contribution à l'étude des coccides de France (14e note). *Annales de la Société Entomologique de France* 102: 35-50.

Balachowsky, A.S. 1933c. Sur la biologie de *"Ceroplastes floridensis"* Comst. et sur la répartition géographique des *Ceroplastes* dans la région palearctique (Hem. Coccidae). *Proceedings 5e Congrès International d'Entomologie (Paris 1932)* 5: 79-87.

Balachowsky, A.S. 1934a. Contribution à l'étude des coccides de France (17e note). Recherches complémentaires sur la faune de Corse. *Bulletin de la Société Entomologique de France* 1934: 67-72.

Balachowsky, A.S. 1934b. Les coccides du Sahara central. *Mémoires de la Société d'Histoire Naturelle de l'Afrique du nord* 4: 145-157.

Balachowsky, A.S. 1935. Les cochenilles de l'Espagne. *Revue de Pathologie Végétale et d'Entomologie agricole de France* 22: 255-269.

Balachowsky, A.S. 1936a. Contribution à l'étude des coccides de France (20e note). Sur un nouvelle Lécanine hypogée du Midi de la France. *Bulletin de la Société Entomologique de France* 41: 56-59.

Balachowsky, A.S. 1936b. Contribution à l'étude des coccides de France. (22e note). Sur une Lécanine nouvelle du Massif de l'Estérel. *Bulletin de la Société Entomologique de France* 41: 122-125.

Balachowsky, A.S. 1936c. Sur quelques cochenilles récoltées au cours du congrès d'Avignon. *Bulletin de la Société Entomologique de France* 1937: 339-340.

Balachowsky, A.S. 1936d. Contribution à l'étude des coccides de France (24e note). La cochenille floconneuse est-elle indigine dans les forêts de France? *Revue de Pathologie Végétale et d'Entomologie Agricole de France* 23: 307-312.

Balachowsky, A.S. 1937a. Sur un *Luzulaspis* (Hem. Coccidae) nouveau des hautes montagnes de Madere. Contribution à l'étude des coccides du nord de l'Afrique, (17e note). *Bulletin de la Société Entomologique de France* 42: 116-118.

Balachowsky, A.S. 1937b. Les cochenilles de Seine-et-Oise. (Contribution à l'étude des coccides de France, 23me note). *Bulletin de la Société des Sciences Naturelle de Seine-et-Oise (ser. 3)* 5: 1-6.

Balachowsky, A.S. 1937c. Les cochenilles de France, d'Europe, du nord de l'Afrique et du Bassin Méditerranéen. Caractères généraux des cochenilles. Morphologie externe. *Actualités Scientifiques et Industrielles* no. 414: 1-68.

Balachowsky, A.S. 1937d. Les cochenilles de France, d'Europe, du nord de l'Afrique et du Bassin Méditerranéen. Caracterès généraux des cochenilles. Morphologie interne. *Actualités Scientifiques et Industrielles* no. 564: 1-129.

Balachowsky, A.S. 1939a. Les cochenilles de Madère (seconde partie). II. Lecaniinae - Eriococcinae - Dactylopinae - Ortheziinae - Margarodidae. *Revue de Pathologie Végétale et d'Entomologie Agricole de France* 25: 255-272.

Balachowsky, A.S. 1939b. Les cochenilles de France, d'Europe, du nord de l'Afrique et du Bassin Méditerranéen. III. Caractères généraux des cochenilles reproduction - développement embryonnaire, développement postembryonnaire. *Actualités Scientifiques et Industrielles* no. 784: 131-242.

Balachowsky, A.S. 1942. Essai sur la classification des cochenilles (Homoptera - Coccoidea). *Annales de l'Ecole Nationale d'Agriculture de Grignon (ser. 3)* 3: 34-48.

Balachowsky, A.S. 1946. Etude biogéographique des Coccoidea des Iles Atlantides (Canaries et Madère). *Mémoires de la Société du Biogéographie. Paris* 8: 209-218.

Balachowsky, A.S. 1948. Les cochenilles de France, d'Europe, du nord de l'Afrique et du Bassin Méditerranéen. IV. Monographie des Coccoidea, classification - Diaspidinae (Première partie). *Actualités Scientifiques et Industrielles* no. 1054: 243-394.

Balachowsky, A.S. 1957. Les cochenilles de la Guadeloupe et de la Martinique (première liste). *Revue de Pathologie Végétale et d'Entomologie Agricole de France* 36: 198-208.

Balachowsky, A.S. & Ferrero, D. 1965. Sur un *Saissetia* Depl., (Coccoidea - Lecaninae) nouveau de la forêt centrafricaine. *Cahiers de la Maboké. Paris* 3: 134-136.

Balachowsky, A.S. & Matile-Ferrero, D. 1970. Les cochenilles (Hom. Coccoidea) de la République Islamique de Mauritanie. *Bulletin de l'Institut Fondamental d'Afrique Noire (Ser A)* 32: 1078-1087.

Ballou, A.H. 1922. Mealy bug on cacao. *Agricultural News. Barbados* 21(518): 74.

Ballou, C.H. 1926. Los coccidos de Cuba y sus plantas hospederas. *Boletin Estación Experimental Agronomica. Santiago de Las Vegas, Cuba* 51: 1-47.

Banks, C.S. 1906. New Philippine insects. *Philippine Journal of Science* 1: 229-238.

Banks, H.J. & Cameron, D.W. 1973. Phenolic glycosides and pterins from the Homoptera. *Insect Biochemistry* 3: 139-162.

Barbagallo, S. 1974. Notizie sulla presenza in Sicilia di una nuova cocciniglia degli agrumi *Coccus pseudomagnoliarum* (Kuwana) (Homoptera, Coccidae) osservazioni biologiche preliminari. *Entomologica. Bari* 10: 121-139.

Bargagli, P. 1902. Adolfo Targioni Tozzetti. *Bolletino della Società Entomologica Italiana* (1902): 199-203.

Barnes, J.K. 1988. Asa Fitch and the emergence of American entomology, with an entomological bibliography and a catalogue of taxonomic names and type specimens. *Bulletin of the New York State Museum of Natural History* 461: 1-120.

Bartlett, B.R. 1978. Coccidae, pp. 57-74. in: Clausen, C.P. (Ed.). Introduced parasites and predators of arthropod pests and weeds:

a world review. *United States Department of Agriculture, Agricultural Handbook* 480, 545 pp.

Bazarov, B.B. 1963. Two new species of coccids (Homoptera, Coccoidea) from Tadzhikistan. [In Russian, Tadzhik summary]. *Doklady Akademii Nauk Tadzhiksokoy SSR. Dushanbe* 6: 38-42.

Bazarov, B.B. 1971a. A new coccid species *Pulvinaria myricariae* Bazarov sp. n. (Homoptera, Coccoidea, Coccidae) from roots of *Myricaria* in the Tadzhikistan fauna. [In Russian]. *Doklady Akademii Nauk Tadzhiksokoy SSR* 14: 64-67.

Bazarov, B.B. 1971b. Coccid fauna (Homoptera, Coccoidea) of the Gorno-Badakhshan autonomous region (Pamir). [In Russian, Tadzhik summary]. *Izvestiya Akademii Nauk Tadzhiksokoy SSR. Otdelenie Biologicheskikh Nauk* 3: 87-91.

Bazarov, B.B., Babaev, T. & Shmelev, G.P. 1975. The peach soft scale (*Parthenolecanium persicae* F.) and its parasites in Tadzhikistan (Homoptera). pp. 94-100. [In Russian]. In: U.L. Shchetkii & N. N. Muminov (eds). Entomologiya Tadzhikistana. Donish, Dushanbe.

Bazarov, B.B. & Shmelev, G.P. 1975. A new species of *Rhizopulvinaria* (Homoptera, Coccoidea, Coccidae) from Tadzhikistan. [In Russian]. *Entomologiya Tadzhikistana Sbornik Statey* (1975): 107-112.

Beardsley, J.W. 1966a. Insects of Micronesia. Homoptera: Coccoidea. *Insects of Micronesia* 6: 377-562.

Beardsley, J.W. 1966b. Insects and other terrestrial arthropods from the Leeward Hawaiian Islands. *Proceedings, Hawaiian Entomological Society* 19: 157-185.

Beardsley, J.W. 1972. Notes and exhibitions. *Coccus capparidis* (Green). *Proceedings, Hawaiian Entomological Society* 21: 142.

Beardsley, J.W. 1975. Insects of Micronesia. Homoptera: Coccoidea. Supplement. *Insects of Micronesia* 6: 657-662.

Beardsley, J.W. 1986. Notes and exhibitions. New insect records for Guam. *Proceedings, Hawaiian Entomological Society* 26: 9.

Beccari, O. 1877. Malesia, Raccolta di osservazioni botaniche intorno alle piante dell'arcipelago Indo-Malese e Papuano. R. Istituto Sordo-Muti, Genova. 305 pp.

Bedford, C.G. 1968. The biology of *Ceroplastes sinoiae* Hall, with special reference to the ecdysis and the morphology of the test. *Entomology Memoirs. Department of Agricultural Technical Services. Republic of South Africa. Pretoria* 14: 1-111.

Beingola, O. 1969a. Notas sobre la biología de *Saissetia oleae* Bern. (Hom.: Coccidae), "queresa negra del olivo", en laboratorio y en el campo. *Revista Peruana de Entomología* 12: 130-136.

Beingola, O. 1969b. Notas sobre la biología de *Saissetia coffeae* (Walk.) (Hom.: Coccidae) en laboratorio y en el campo. *Revista Peruana de Entomología* 12: 137-145.

Bellio, G. 1939. Hemiptera, Coccidae. *Missione Biologica nel Paese dei Borana. Raccolte Zoologiche. Roma* 3: 225-239.

Ben-Dov, Y. 1968. Occurrence of *Sphaerolecanium prunastri* (Fonscolombe) in Israel and description of its hitherto unknown third larval instar. *Annales des Epiphyties* 19: 615-621.

Ben-Dov, Y. 1969. A generic diagnosis of *Bodenheimera* Bodenheimer (Homoptera: Coccidae) with redescription of *B. rachelae* (Bodenheimer). *Proceedings of the Royal Entomological Society of London (B)* 38: 70-74.

Ben-Dov, Y. 1970a. Laboratory rearing of wax scales. *Journal of Economic Entomology* 63: 1998-1999.

Ben-Dov, Y. 1970a. A redescription of the Florida wax scale *Ceroplastes floridensis* Comstock (Homoptera: Coccidae). *Journal of the Entomological Society of southern Africa* 33: 273-277.

Ben-Dov, Y. 1970b. Studies on *Tetrastichus ceroplastae* (Girault), a parasite of the Florida wax scale, *Ceroplastes floridensis* Comstock. Ph. D. Thesis, The Hebrew University of Jerusalem, Israel [In Hebrew, English summary].

Ben-Dov, Y. 1970d. The wax scales of the genus *Ceroplastes* Gray (Homoptera: Coccidae) and their parasites in Israel. *Israel Journal of Entomology* 5: 83-92.

Ben-Dov, Y. 1971. An annotated list of the soft scale insects (Homoptera: Coccidae) of Israel. *Israel Journal of Entomology* 6: 23-34.

Ben-Dov, Y. 1972. Life history of *Tetrastichus ceroplastae* (Girault) (Hymenoptera: Eulophidae), a parasite of the Florida wax scale, *Ceroplastes floridensis* Comstock (Homoptera: Coccidae) in Israel. *Journal of the Entomological Society of southern Africa* 35: 17-34.

Ben-Dov, Y. 1973. The genus *Euphilippia* Berlese & Silvestri (Homoptera: Coccidae) with redescription of its type-species. *Bollettino del Laboratorio di Entomologia Agraria "Filippo Silvestri", Portici* 30: 282-290.

Ben-Dov, Y. 1975. On the identity of *Filippia* Targioni Tozzetti, 1868 and *Lichtensia* Signoret, 1873 (Homoptera: Coccidae). *Journal of the Entomological Society of southern Africa* 38: 109-121.

Ben-Dov, Y. 1976a. *Lecanium acuminatum* Signoret, 1873 (Insecta, Homoptera, Coccidae): request for designation of a neotype under the plenary powers. Z.N. (S.) 2119. *Bulletin of Zoological Nomenclature* 32: 256-260.

Ben-Dov, Y. 1976b. The identity of two species from the "*Lecanium* series" of Signoret (Homoptera: Coccidae). *Entomologist's Monthly Magazine* (1975) 111: 115-116.

Ben-Dov, Y. 1976c. Phenology of the Florida wax scale, *Ceroplastes floridensis* Comstock (Homoptera: Coccidae) on citrus in Israel. *Phytoparasitica* 4: 3-7.

Ben-Dov, Y. 1977. Taxonomy of the long brown scale *Coccus longulus* (Douglas) stat.n. (Homoptera: Coccidae). *Bulletin of Entomological Research* 67: 89-95.

Ben-Dov, Y. 1978. Taxonomy of the nigra scale, *Parasaissetia nigra* (Nietner) (Homoptera: Coccoidea: Coccidae), with observations on mass rearing and parasites of an Israeli strain. *Phytoparasitica* 6: 115-127.

Ben-Dov, Y. 1979. A taxonomic study of the soft-scale genus *Kilifia* (Coccidae). *Systematic Entomology* 4: 311-324.

Ben-Dov, Y. 1980. Observations on scale insects (Homoptera: Coccoidea) of the Middle East. *Bulletin of Entomological Research* 70: 261-271.

Ben-Dov, Y. 1981. A new species of *Coccus* (Hemiptera: Coccidae) from mango in Israel, and a redescription of *C. gymnospori* (Green). *Bulletin of Entomological Research* 71: 649-654.

Ben-Dov, Y. 1985. Further observations on scale insects (Homoptera: Coccoidea) of the Middle East. *Phytoparasitica* 13: 185-192.

Ben-Dov, Y. 1986. Taxonomy of two described and one new species of *Waxiella* De Lotto (Homoptera: Coccoidea: Coccidae). *Systematic Entomology* 11: 165-174.

Ben-Dov, Y. 1987. Observations on scale insects (Homoptera: Coccoidea) of the Middle East - III. *Israel Journal of Entomology* 21: 111-117.

Ben-Dov, Y. 1988. The scale insects (Homoptera: Coccoidea) of Citrus in Israel: diversity and pest status. *Proceedings of the Sixth International Citrus Congress, Tel Aviv, Israel, March 6-11, 1988*: 1075-1082.

Ben-Dov, Y. 1989a. *Saissetia* Deplanche, 1859 (Insecta, Homoptera): proposed designation of *Lecanium coffeae* Walker, 1852 as the type species. *Bulletin of Zoological Nomenclature* 46: 114-118.

Ben-Dov, Y. 1989b. Soft scale insects (Homoptera: Coccidae) on *Tillandsia* in Central America. *Systematic Entomology* 14: 1-6.

Ben-Dov, Y. & Cox, J.M. 1990. The identity of five species of scale insects (Hem., Homoptera, Coccoidea), living on ornamental plants, originally described by P.F. Bouché. *Entomologist's Monthly Magazine* 126: 79-84.

Ben-Dov, Y., Gros, S. & Maimon, A. 1991. *Parthenolecanium persicae* (F.), a new pest of Persimmon in Israel. [In Hebrew, English abstract]. *Hassadeh* 72: 347-348, 370.

Ben-Dov, Y. & Harpaz, I. 1985. An annotated list of taxa of Coccoidea (Homoptera) described by F.S. Bodenheimer (1897-1959). *Israel Journal of Entomology* 19: 23-36.

Ben-Dov, Y. & Matile-Ferrero, D. 1989a. Taxonomy and nomenclature of five hitherto inadequately-known genera of mealybugs (Homoptera: Coccoidea: Pseudococcidae). *Systematic Entomology* 14: 165-178.

Ben-Dov, Y. & Matile-Ferrero, D. 1989b. *Fonscolombia* Lichtenstein, 1877 (Insecta: Homoptera): proposed designation of *Fonscolombia graminis* Lichtenstein, 1877 as the type species. *Bulletin of Zoological Nomenclature* 46: 119-122.

Ben-Dov, Y., Williams, M.L. & Ray, C.H. 1975. Taxonomy of the mango shield scale, *Protopulvinaria mangiferae* (Green) (Homoptera: Coccidae). *Israel Journal of Entomology* 10: 1-17.

Benassy, C. & Franco, E. 1974. Sur l'écologie de *Ceroplastes rusci* L. (Homoptera, Lecanoidea) dans les Alpes-Maritimes. *Annales de Zoologie - Ecologie Animale* 6: 11-39.

Berlese, A. & Silvestri, F. 1906. Descrizione di un nuovo genere e di una nuova specie di Lecanite vivente sull'olivo. *Redia* 3: 396-407.

Bernard, P.J. 1773. Quelle est la meilleure manière de cultiver le figuier; quelles sont les causes de son dépérissement, et quels sont les meilleurs moyens de remédier? pp. 1-124, in: Recueil des pièces présentées à l'Academie de Belles Lettres, Sciences et Arts de Marseille. Marseille, Francois Brebion.

Bernard, P.J. 1783. Mémoire pour servir à l'histoire naturelle de l'olivier. Aix-en-Provence. 255 pp.

Bernard, P.J. 1788. Mémoires pour servir à l'histoire naturelle de la Provence; vol. 2: Mémoire pour servir à l'histoire naturelle de l'olivier. Paris. 559 pp.

Bibby, F.F. 1931. Coccoids collected on wild plants in semi-arid regions of Texas and Mexico (Homoptera). *Journal of the New York Entomological Society* 39: 587-591.

Bielenin, I. 1958. La structure et l'apparition des filiers dorso-marginales chez le deuxième stade larvaire du *Lecanium corni* Bouché, Marchal (female nec male) Homoptera, Coccoidea, Lecaniidae). [In Polish, French summary]. *Polskie Pismo Entomologiczne* 27: 97-104.

Bielenin, I. 1962. Anatomical and histological investigations on the genus *Lecanium* Burm. Part I. Female reproductive organs of *Lecanium pomeranicum* Kaw. (Homoptera, Coccoidea). *Acta Biologica Cracoviensia (Series: Zoologia)* 5: 9-25.

Bielenin, I. 1963. Anatomical and histological investigations on the genus *Lecanium* Burm. Part IV. The alimentary canal of *Lecanium pomeranicum* Kaw. (Homoptera, Coccoidea). *Zoologica Poloniae. Krakow* 13: 221-253.

Biraben, M. 1959. Carlos A. Lizer y Trelles (1887-1959). *Neotropica. Notas Zoologicas Sudamericanas. Buenos Aires* 5: 55.

Blahutiak, A. 1972. Wirts-parasitische Beziehungen zwischen der Zwetschenschildlaus *Parthenolecanium corni* Bouché (Homoptera, Coccidae) und dem Parasiten *Blastothrix confusa* Erdös (Hymenoptera, Encyrtidae). *Biologické Práce. Bratislava* 18: 1- 113.

Blair, C.A., Blackith, R.E. & Boratynski, K. 1964. Variation in *Coccus hesperidum* L. (Homoptera: Coccidae). *Proceedings of the Royal Entomological Society of London (A)* 39: 129-134.

Blumberg, D. 1977. Encapsulation of parasitoid eggs in soft scales (Homoptera: Coccidae). *Ecological Entomology* 2: 185-192.

Blumberg, D. 1988. Encapsulation of eggs of the encyrtid wasp, *Metaphycus swirskii*, by the hemispherical scale, *Saissetia coffeae*: Effect of host age and rearing temperature. *Entomologia Experimentalis et Applicata* 47: 95-99.

Blumberg, D. & Blumberg, O. 1991. The pyriform scale, *Protopulvinaria pyriformis*, and its common parasitoid, *Metaphycus stanleyi*, on avocado and *Hedera helix*. [In Hebrew, English abstract]. *Alon Hanotea* 45: 265-269.

Blumberg, D. & DeBach, P. 1981. Effects of temperature and host age upon the encapsulation of *Metaphycus stanleyi* and *Metaphycus helvolus* eggs by Brown Soft Scale *Coccus hesperidum*. *Journal of Invertebrate Pathology* 37: 73-79.

Blumberg, D. & Goldenberg, S. 1992. Encapsulation of eggs of two species of *Encyrtus* (Hymenoptera: Encyrtidae) by soft scales (Homoptera: Coccidae) in six parasitoid-host interactions. *Israel Journal of Entomology* (1991-1992) 25-26: 57-65.

Blumberg, D. & Swirski, E. 1977. Mass breeding of two species of *Saissetia* (Hom.: Coccidae) for propagation of their parasitoids. *Entomophaga* 22: 147-150.

Blumberg, D. & Swirski, E. 1984. Response of three soft scales (Homoptera: Coccidae) to parasitization by *Metaphycus swirskii*. *Phytoparasitica* 12: 29-35.

Blumberg, D., Swirski, E. & Greenberg, S. 1975. Evidence for bivoltine populations of the Mediterranean black scale *Saissetia oleae* (Olivier) on citrus in Israel. *Israel Journal of Entomology* 10: 19-24.

Boboye, S.O. 1971. Scale insects on citrus and their distribution in Western Nigeria. *Journal of Economic Entomology* 64: 307-309.

Bodenheimer, F.S. 1924. The Coccidae of Palestine, first report on this family. *The Zionist Organization Institute of Agriculture and Natural History, Agricultural Experiment Station* 1: 1-100.

Bodenheimer, F.S. 1926a. Première note sur les cochenilles de Syrie. *Bulletin de la Société Entomologique de France* 1926: 41-47.

Bodenheimer, F.S. 1926b. Second note on the Coccidae of Palestine. *Bulletin of Entomological Research* 17: 189-192.

Bodenheimer, F.S. 1927. Third note on the Coccidae of Palestine. *Agricultural Records, Tel Aviv* 2: 177-186.

Bodenheimer, F.S. 1928. Eine kleine Cocciden-Ausbeute aus Griechenland. *Konowia* 7: 191-192.

Bodenheimer, F.S. 1929. B. Die Coccidenfauna der Sinai-Halbinsel. pp. 104-117. In: F.S. Bodenheimer & O. Theodor (eds). Ergebnisse der Sinai-Expedition 1927 der Hebräischen Universität, Jerusalem. J.C. Hinrichs, Buchhandlung, Leipzig.

Bodenheimer, F.S. 1935. Studies on the zoogeography and ecology of palearctic Coccidae I-III. *EOS* 10: 237-271.

Bodenheimer, F.S. 1941. Seven new species of Coccidae from Anatolia. *Revue de la Faculté des Sciences de l'Université d'Istanbul (Ser. B)* 6: 65-84.

Bodenheimer, F.S. 1943. A first survey of the Coccoidea of Iraq. *Government of Iraq, Ministry of Economics, Directorate General of Agriculture, Bulletin* 28: 1-33.

Bodenheimer, F.S. 1944a. Additions to the Coccoidea of Iraq, with descriptions of two new species. *Bulletin de la Société Fouad 1er d'Entomologie* 28: 81-84.

Bodenheimer, F.S. 1944b. Note on the Coccoidea of Iran, with description of new species. *Bulletin de la Société Fouad 1er d'Entomologie* 28: 85-100.

Bodenheimer, F.S. 1947. Bibliography of F.S. Bodenheimer (1897-1947). Jewish Agency for Palestine, Agricultural Research Station, Rehovot. 24 pp. + VIII.

Bodenheimer, F.S. 1951a. Description of some new genera of Coccoidea. *Entomologische Berichten* 13: 328-331.

Bodenheimer, F.S. 1951b. Citrus Entomology in the Middle East with special reference to Egypt, Iran, Iraq, Palestine, Syria, Turkey. W. Junk, The Hague. 663 pp.

Bodenheimer, F.S. 1952. The Coccoidea of Turkey. I. *Revue de la Faculté des Sciences de l'Université d'Istanbul (Ser. B)* 17: 315-351.

Bodenheimer, F.S. 1953. The Coccoidea of Turkey III. *Revue de la Faculté des Sciences de l'Université d'Istanbul (Ser. B)*. 18: 91-164.

Bodkin, G.E. 1914. The scale insects of British Guiana. A preliminary list with an account of their host plants, natural enemies, and controlling agencies. *Journal of the Board of Agriculture of British Guiana* 7: 106-124.

Bodkin, G.E. 1917. Notes on the Coccidae of British Guiana. *Bulletin of Entomological Research* 8: 103-109.

Bodkin, G.E. 1922. The scale insects of British Guiana. *Journal of the Board of Agriculture of British Guiana* 15: 56-63.

Bodkin, G.E. 1927. The fig wax-scale *(Ceroplastes rusci* L.) in Palestine. *Bulletin of Entomological Research* 17: 259-263.

Boisduval, A.M. 1867. Essai sur l'Entomologie horticole. Donnaud, Paris. 648 pp.

Bondar, G. 1925. O Cacao. Parte II. Molestias e inimigos do cacaoeiro no estado da Bahia - Brasil. Imprensa Official do Estado, Bahia. 118 pp.

Boratynski, K. 1970. On some species of *"Lecanium"* (Homoptera, Coccidae) in the collection of the Naturhistorisches Museum in Vienna; with description and illustration of the immature stages of *Parthenolecanium persicae*. *Annalen des Naturhistorischen Museums. Wien* 74: 63-76.

Boratynski, K. & Davies, R.G. 1971. The taxonomic value of male Coccoidea (Homoptera) with an evaluation of some numerical techniques. *Biological Journal of the Linnean Society* 3: 57-102.

Boratynski, K., Pancer-Koteja, E. & Koteja, J. 1982. The life history of *Lecanopsis formicarum* Newstead (Homoptera, Coccinea). *Annales Zoologici. Warszawa* 36: 517-537.

Boratynski, K.L. & Williams, D.J. 1964. A note on some British Coccoidea, with new additions to the British fauna. *Proceedings of the Royal Entomological Society of London (B)* 33: 103-110.

Borchsenius, N.S. 1934. Survey of the coccids fauna of the Eastern Coast of the Black Sea. [In Russian, English summary]. Quarantine Station, Abkhazia, Sukhumi. 37 pp.

Borchsenius, N.S. 1936. On the fauna of scale insects (Coccidae) of the Caucasus. [In Russian, German abstract]. *Trudy Krasnodarskogo sel'sko-Khozyaistvennogo Instituta, Krasnodar* 4: 97-139.

Borchsenius, N.S. 1937. Tables for the identification of coccids (Coccidae) injurious to cultivated plants and forests in the USSR. [In Russian]. Quarantine Regional Inspection, Leningrad. 148 pp.

Borchsenius, N.S. 1938. Observations on the Coccidae (Hem. Insecta) of the Far East Region. [In Russian, English summary]. *Vestnik DB Filiala Akademii Nauk SSSR* 29: 131-146.

Borchsenius, N.S. 1939. On the fauna of Coccidae in the Caucasus. [In Russian, English summary]. *Zashchita Rastenii. Leningrad* 18: 43-51.

Borchsenius, N.S. 1949a. *Ceroplastes japonicus* Green (Homoptera, Coccoidea) - a pest of citrus and other cultivated plants in Georgia. [In Russian]. *Soobshchenya Akademii Nauk Gruzinskoy SSR* 10: 121-124.

Borchsenius, N.S. 1949b. A new genus and new species of hard and soft scales (Homoptera, Coccoidea) of USSR fauna. [In Russian]. *Entomologicheskoe Obozrenye* 30: 334-353.

Borchsenius, N.S. 1949c. Identification of the soft and armoured scale insects of Armenia. [In Russian]. Akademiya Nauk Armiyanskoy SSR, Erevan. 225 pp.

Borchsenius, N.S. 1950. Mealybugs and Scale Insects of USSR (Coccoidea) [In Russian]. Akademiya Nauk SSSR, Moscow, Leningrad. 250 pp.

Borchsenius, N.S. 1952. New genera and species of soft scales of the family Coccidae (-Lecaniidae) of the USSR fauna and adjacent countries (Insecta, Homoptera, Coccoidea) [In Russian]. *Trudy Zoologicheskogo Instituta Akademii Nauk SSSR* 12: 269-316.

Borchsenius, N.S. 1953. New genera and species of scale insects of the family Coccidae (Homoptera, Coccoidea). [In Russian]. *Entomologicheskoe Obozrenye* 33: 281-290.

Borchsenius, N.S. 1955a. Suborder Coccoidea - scale insects and mealybugs. [In Russian]. In: Handbook of forest pests. Akademiya Nauk SSSR, Moscow. pp. 848-885.

Borchsenius, N.S. 1955b. New species of false-hard scales (Homoptera, Coccoidea, Coccidae) of the fauna of USSR and adjacent countries. [In Russian]. *Trudy Zoologicheskogo Instituta Akademii Nauk SSSR* 18: 288-303.

Borchsenius, N.S. 1956. Review of the Palearctic scale-insects of the genus *Eriopeltis* Sign. (Homoptera, Coccoidea). [In Russian, English summary]. *Entomologicheskoe Obozrenye* 35: 397-420.

Borchsenius, N.S. 1957. Subtribe mealybugs and scales (Coccoidea). Soft scale insects Coccidae. Vol. IX. [In Russian]. *Fauna SSSR. Zoologicheskii Institut Akademii Nauk SSSR. N.S.* 66: 1-493.

Borchsenius, N.S. 1959. Notes on the Coccoidea of China. VI. Descriptions of some new genera and species of Eriococcidae and Coccidae (Homoptera, Coccoidea) - scientific results of the Chinese-Soviet expeditions of 1955-1957 to south-western China. [In Russian, English summary]. *Entomologicheskoe Obozrenye* 38: 164-175.

Borchsenius, N.S. 1960. Contribution to the coccid fauna KNR. V. Hard and soft scales, harmful to fruit and grape culture in northeast and east KNR. [In Chinese, Russian summary]. *Acta Entomologica Sinica* 10: 214-218.

Bouché, P.F. 1833. Naturgeschichte der schädlichen und nützlichen Garteninsekten und die bewährtesten Mittel. Nicolai, Berlin. 176 pp.

Bouché, P.F. 1834. Naturgeschichte der Insekten, besonders in Hinsicht ihrer ersten Zustände als Larven und Puppen. Nicolai, Berlin. 216 pp.

Bouché, P.F. 1844. Beitrage zur Naturgeschichte der Scharlachläuse (Coccina). *Entomologische Zeitung, Stettin* 5: 293-302.

Bouché, P.F. 1851. Neue Arten der Schildläus-Familie. *Entomologische Zeitung, Stettin* 12: 110-112.

Box, H.E. 1953. List of sugar-cane insects. Commonwealth Institute of Entomology, London. 101 pp.

Brain, C.K. 1918. The Coccidae of South Africa - II. *Bulletin of Entomological Research* 9: 107-139.

Brain, C.K. 1920a. The Coccidae of South Africa - IV. *Bulletin of Entomological Research* 10: 95-128.

Brain, C.K. 1920b. The Coccidae of South Africa - V. *Bulletin of Entomological Research* 11: 1-41.

Brethes, J. 1921. Description d'un *Ceroplastes* (Hem. Coccidae) de la Republique Argentine, et de son parasite (Hym. Chalcididae). *Bulletin de la Société Entomologique de France* (1921): 79-81.

Briales, M.J. & Campos, M. 1988. Analysis of the spatial distribution of the various stages of *Saissetia oleae* (Oliv.) on olive tree in Granada. *Journal of Applied Entomology* 105: 28-34.

Brimblecombe, A.R. 1962. Studies of the Coccoidea. 12. Species occurring on deciduous fruit and nut trees in Queensland. *Queensland Journal of Agricultural Science* 19: 219-229.

Brink, T. & Bruwer, I.J. 1989. Andersoni scale, *Cribrolecanium andersoni* (Newstead) (Hemiptera: Coccidae) a pest on citrus in South Africa. *Citrus & Subtropical Fruit Journal* 645: 9, 25.

Brink, T. & Hewitt,P.H. 1991. Mortality factors of white powdery scale, *Cribrolecanium andersoni* (Hemiptera: Coccidae), a pest of citrus. *Proceedings of the Eighth Entomological Congress, En-*

tomological Society of Southern Africa, Bloemfontein, 1-4 July, 1991: 9.

Brittin, G. 1915. Art. XVIII. - Some new Coccidae. *Transactions and Proceedings of the New Zealand Institute (1914)* 47: 149-156.

Brittin, G. 1916. Art. XLIV. - notes on some Coccidae in the Canterbury Museum, together with a description of a new species. *Transactions and Proceedings of the New Zealand Institute (1915)* 48: 423-426.

Brittin, G. 1940a. The validity of the coccid genus *Eulecanium* Cockerell. *Transactions and Proceedings of the Royal Society of New Zealand* 69: 410-412.

Brittin, G. 1940b. The life history of *Lecanium (Eulecanium) persicae* (Fabricius), and descriptions of the different instars. *Transactions and Proceedings of the Royal Society of New Zealand* 69: 413-421.

Britton, W.E. 1923. The Aleyrodidae and Coccidae of Connecticut. Family Coccidae. *Connecticut Geological and Natural History Survey, Bulletin* 34: 346-382.

Brookes, H.M. 1957. The Coccoidea (Homoptera) naturalised in South Australia: an annotated list. *Transactions of the Royal Society of South Australia* 80: 81-90.

Brookes, H.M. 1964. The Coccoidea (Homoptera) naturalised in South Australia: A second annotated list. *Transactions of the Royal Society of South Australia* 88: 15-20.

Brookes, H.M. & Koteja, J. 1982. *Waricoccus parvisetosus* gen. et sp.n. (Homoptera, Coccidae) from Australia. *Polskie Pismo Entomologiczne* 52: 183-187.

Brown, K.S. 1975. The chemistry of aphids and scale insects. *Chemical Society Review* 4: 263-288.

Bruner, S.C., Scaramuzza, L.C. & Otero, A.R. 1975. Catalogo de los insectos que atacan a las plantas economicas de Cuba. Segunda Edicion. Academia de Ciencias de Cuba, Instituto de Zoologia, La Habana. 399 pp.

Burmeister, H. 1835. Scharllachläuse. Schildläuse. Coccina. (Galli-nsecta 1.). pp. 61-83. In: H. Burmeister (edit.). Handbuch der Entomologie. T.F. Inslin, Berlin.

Burns, D.P. & Davidson, R.H. 1966. The amino acids and sugars in honeydew of the Tuliptree Scale, *Tourneyella liriodendri*, and in the sap of its host, Yellow-Poplar. *Annals of the Entomological Society of America* 59: 1071-1073.

Burns, D.P. & Donley, D.E. 1970. Biology of the Tuliptree Scale, *Tourneyella liriodendri* (Homoptera: Coccidae). *Annals of the Entomological Society of America* 63: 228-235.

Bytinski-Salz, H. 1966. An annotated list of insects and mites introduced into Israel. *Israel Journal of Entomology* 1: 15-48.

Bytinski-Salz, H. & Sternlicht, M. 1967. Insects associated with oaks (*Quercus*) in Israel. *Israel Journal of Entomology* 2: 107-143.

Campbell, R.E. 1914. A new coccid infesting citrus trees in California (Hemip.). *Entomological News* 25: 222-224.

Çanakcioğlu, H. 1977. A study of the forest Coccoidea (Homoptera) of Turkey (Systematics - Distribution - Host Plant - Biology). [In Turkish, English Summary]. *Istanbul Üniversitesi Orman Fakültesi Yayinlari. Istanbul* 227: 1-122.

Canard, M. 1965a. Sur quelques Pulvinariini (Homoptera, Coccoidea) du midi de la France. *Proceedings of the 12th International Congress of Entomology, London* 1964: 170.

Canard, M. 1965b. Observations sur une Pulvinaire peu connue du midi de la France: *Eupulvinaria hydrangeae* (Steinw.) (Coccoidea -- Coccidae). *Annales de la Société Entomologique de France (N.S.)* 1: 411-419.

Canard, M. 1965c. Utilisation des larves neonates pour aider à la détermination des cochenilles floconneuses (Coccidae - Pulvinariini). *Annales de la Société Entomologique de France (N.S.)* 1: 421-424.

Canard, M. 1966a. Une Pulvinaire de la vigne, nouvelle pour la France: *Neopulvinaria imeretina* (Coccoidea, Coccidae). *Annales de la Société Entomologique de France (N.S.)* 2: 189-197.

Canard, M. 1966b. Remarques sur le genre *Rhizopulvinaria* et sur une espèce peu connue du midi de la France: *Rh. grassei* (Balachowsky) - (Coccoidea, Coccidae). *Vie et Milieu. Bulletin du Laboratoire Arago, Université de Paris. Serie C: Biologie terrestre* 17: 443-452.

Canard, M. 1967a. Une nouvelle Pulvinaire du littoral Méditerranéen: *Rhizopulvinaria maritima* nov. sp. (Coccoidea - Coccidae). *Vie et Milieu. Bulletin du Laboratoire Arago, Université de Paris. Serie C: Biologie terrestre* 18: 159-167.

Canard, M. 1967b. Les Rhizopulvinaires des oeillets Méditerranéens (Coccoidea - Coccidae). *Vie et Milieu. Bulletin du Laboratoire Arago, Université de Paris. Serie C: Biologie terrestre* 18: 169-184.

Canard, M. 1968a. Contribution a l'etude des Rhizopulvinaires Méditerranéens (Hom. Coccidae). *Bulletin de la Société Entomologique de France* 73: 90-96.

Canard, M. 1968b. Un nouveau *Pulvinaria* (Hom. Coccoidea) nuisible aux arbres d'alignement dans la région Parisienne. *Annales de la Société Entomologique de France (N.S.)* 4: 951-958.

Canard, M. 1969. La ligne mâle de *Eupulvinaria hydrangeae* (Hom. Coccidae). *Annales de la Société Entomologique de France (N.S.)* 5: 457-460.

Canard, M. 1971. Contribution à l'étude des Rhizopulvinaires Méditerranéen (Homoptera, Coccidae). *Proceedings of the 13th International Congress of Entomology, Moscow 1968* 1: 119-120.

Canard, M. 1980. Présence en Grèce d'*Eulecanium tiliae* (L.) (Homoptera, Coccidae) sur la figuier *Ficus carica* L. et de son parasite *Blastothrix sericea* (Dalm.) (Hymenoptera, Encyrtidae). *Biologia Gallo-Hellenica. Athens* 9: 157-162.

Carnero Hernandez, A. & Perez Guerra, G. 1986. Coccidos (Homoptera: Coccoidea) de las Islas Canarias. *Communicaciones Instituto Nacional de Investigaciones Agrarias, Serie: Proteccion Végétal. Madrid* 25: 1-85.

Carnes, E.K. 1907. The Coccidae of California. *Second Biennial Report of the Commissioner of Horticulture of the State of California* (1905-1906): 155-222.

Carpenter, M.M. 1945. Bibliography of biographies of entomologists. *The American Midland Naturalist* 33: 1-116.

Carvalho, M.U.P. & Cardoso, H.L. 1970. Lista de pragas de Angola respeitante ao ano de 1968. *Instituto de Investigação Agronomica de Angola, Serie Tecnica* 19.

Casazza, R. 1928. Sui dermatoendozoi, a proposito di un nuovo parasita. (Creeping disease e Pseudo-scâbbia da "*Dermatolecanium migrans*"). *Bolletino della Società Medico- Chirurgica di Pavia* 42: 391-425.

Castel-Branco, A.J.F. 1952. Descrição de três espécies novas de cochonilhas da Africa Oriental Portuguesa. *Crypticeria rodriguesi* n. sp., *Cerococcus fradei* n. sp., *Pulvinaria vayssierei* n. sp. *Anais Estudos de Zoologia. Ministerio do Ultramar, Junta de Investigaçõe do Ultramar. Lisboa* 7: 22-39.

Castel-Branco, A.J.F. 1963. *Aspidiotus destructor* Sign. a outras cochonilhas do coqueiro (*Cocos nucifera* L.) e da palmeira do oleo (*Elaeis guineensis* Jacq.) nas Ilhas de S. Tome e Principe. *Memorias da Junta de Investigaçõe do Ultramar. Lisboa (2a series)* 43: 131-175.

Chacko, M.J. & Sreedharan, K. 1981. Parasites of *Saissetia miranda* attacking *Erythrina lithosperma*. *Journal of Coffee Research* 11: 108-109.

Chang, H.S. 1929. A preliminary list of the Coccidae of China. [In Chinese.]. *Miscellaneous Publication. College of Agriculture, National Central University, Nanking* 16: 1-24.

Chapot, H. & Delucchi, V.L. 1964. Maladies, troubles et ravageurs des agrumes au Maroc. Institut National des Recherche Agronomique, Rabat. 339 pp.

Chatterji, A. & Datta, A.R. 1974. Bionomics and control of mango mealy-scale, *Chloropulvinaria (Pulvinaria) polygonata* (Cockerell) (Hemiptera: Coccidae). *Indian Journal of Agricultural Sciences* 44: 791-795.

Chavannes, A. 1848. Notice sur deux *Coccus* cérifères du Brésil. *Bulletin de la Société Entomologique de France (ser.2)* 6: 139- 145.

Chen, F.G. 1936. Notes on the scale insects of citrus in several districts of East Chekiang, with description of one new species. [In Chinese]. *Entomology and Phytopathology. Hangchow* 4: 208- 228.

Chen, F.G. 1937. Four new coccids from Chekiang. *Entomology and Phytopathology. Hangchow* 5: 382-388.

Chen, F.G. 1962. Two new species of coccids on fruit trees. [In Chinese; English translation]. *Acta Entomologica Sinica* 11: 283-286.

Chen, F.G. 1974. A new coccid of *Ceroplastes* on citrus trees. [In Chinese]. *Acta Entomologica Sinica* 17: 325-328.

Cilliers, C.J. 1967. A comparative biological study of three *Ceroplastes* species (Hem. Coccidae) and their natural enemies. *Entomology Memoirs. Department of Agricultural Technical Services. Republic of South Africa. Pretoria* 13: 1-59.

Clarke, S.R., DeBarr, G.L. & Berisford, C.W. 1989a. Life history of the wooly pine scale *Pseudophilippia quaintancii* Cockerell (Homoptera: Coccidae) in loblolly pine seed orchards. *Journal of Entomological Science* 24: 365-372.

Clarke, S.R., DeBarr, G.L. & Berisford, C.W. 1989b. The life history of *Toumeyella pini* (King) (Homoptera: Coccidae) in loblolly pine seed orchards in Georgia. *Canadian Entomologist* 121: 853-860.

Clausen, C.P. 1923. The citricola scale in Japan, and its synonymy. *Journal of Economic Entomology* 16: 225-226.

Cockerell, T.D.A. 1892a. Additions to the museum. *Journal of the Institute of Jamaica* 1: 54-56.

Cockerell, T.D.A. 1892b. List of Coccidae observed in Jamaica. *Insect life* 4: 333-334.

Cockerell, T.D.A. 1893a. Coccidae in the Lesser Antilles. *Entomologist's Monthly Magazine* 29: 17.

Cockerell, T.D.A. 1893b. The food-plants of some Jamaican Coccidae. *Insect Life* 5: 158-160.

Cockerell, T.D.A. 1893c. The West Indian species of *Ceroplastes*. *The Entomologist* 26: 80-83.

Cockerell, T.D.A. 1893d. Notes on *Lecanium*, with a list of the West Indian species. *Transactions of the American Entomological Society* 20: 49-56.

Cockerell, T.D.A. 1893e. A new species of *Ceroplastes* from Mexico. *Zoe* 4: 104-105.

Cockerell, T.D.A. 1893f. A list of West Indian Coccidae. *Journal of the Institute of Jamaica* 1: 252-256.

Cockerell, T.D.A. 1893g. Records of West Indian Coccidae, I. *Journal of the Institute of Jamaica* 1: 373.

Cockerell, T.D.A. 1893h. A new *Lecanium*. *Journal of the Institute of Jamaica* 1: 378-379.

Cockerell, T.D.A. 1893i. Une nouvelle espèce de *Lecanium* du Mexique. *Memorias y Revista de la Sociedad Científica "Antonio Alzate"* 6: 325-326.

Cockerell, T.D.A. 1893j. Description d'un *Lecanium* Mexicain. *Bulletin de la Société Zoologique de France* 18: 167-168.

Cockerell, T.D.A. 1893l. Two new species of *Pulvinaria* from Jamaica. *Transactions of the Entomological Society of London* 1893: 159-163.

Cockerell, T.D.A. 1893m. Notes on some Mexican Coccidae. *Annals and Magazine of Natural History (Ser. 6)* 12: 47-53.

Cockerell, T.D.A. 1893n. Notes on some Mexican Coccidae. *Annals and Magazine of Natural History (Ser. 6)* 12: 160.

Cockerell, T.D.A. 1893o. Preliminary note on the cottony scale of the osage orange. *Science* 22: 78-79.

Cockerell, T.D.A. 1893p. A new *Lecanium* from Canada. *Canadian Entomologist* 25: 221-222.

Cockerell, T.D.A. 1893r. Three new Coccidae from the arid region of North America. *The Entomologist* 26: 350-352.

Cockerell, T.D.A. 1893s. A new *Lecanium*. - *Lecanium rubellum* n. sp. *Journal of the Institute of Jamaica* 1: 378-379.

Cockerell, T.D.A. 1893t. The distribution of Coccidae. *Insect Life* 6: 99-103.

Cockerell, T.D.A. 1893u. The entomology of the mid-alpine zone of Custer County, Colorado. *Transactions of the American Entomological Society* 20: 305-370.

Cockerell, T.D.A. 1893v. Coccidae, or scale insects, which live on orchids. *Gardners' Chronicle and Agricultural Gazette* 13: 548.

Cockerell, T.D.A. 1894a. A check list of Nearctic Coccidae. *Canadian Entomologist* 26: 31-36.

Cockerell, T.D.A. 1894b. Notes on some Trinidad Coccidae. *Journal of the Trinidad Field Naturalists' Club* 1: 306-310.

Cockerell, T.D.A. 1894c. A check list of the Coccidae of the Neotropical Region. *Journal of the Trinidad Field Naturalists' Club* 1: 311-312.

Cockerell, T.D.A. 1894d. A new wax scale found in Jamaica. *Entomological News* 5: 157-158.

Cockerell, T.D.A. 1894e. Coccidae, or scale insects. - IV. *Bulletin of the Botanical Department, Jamaica (n.s.)* 1: 17-19.

Cockerell, T.D.A. 1894f. Coccidae, or scale insects.-V. *Bulletin of the Botanical Department, Jamaica* 1: 69-73.

Cockerell, T.D.A. 1894g. Descriptions of new Coccidae. *Entomological News* 5: 203-204.

Cockerell, T.D.A. 1894h. Further notes on scale insects. (Coccidae). *Canadian Entomologist* 26: 189-191.

Cockerell, T.D.A. 1894i. Two new Coccidae from the arid region of North America. *Annals and Magazine of Natural History (Ser. 6)* 14: 12-15.

Cockerell, T.D.A. 1894j. On a *Lecanium* infesting blackberry, considered identical with *L. fitchii* Sign. *Insect Life* 7: 29- 31.

Cockerell, T.D.A. 1894k. A check-list of African Coccidae. *Psyche* 7: 178.

Cockerell, T.D.A. 1894l. On a *Lecanium* from Rochester, N.Y. (U.S.A.) considered identical with *L. juglandis* Bouché. *Entomologist* 27: 332-336.

Cockerell, T.D.A. 1894m. Some observations on the distribution of Coccidae. *American Naturalist* 28: 1050-1054.

Cockerell, T.D.A. 1895a. Coccidae or scale insects. - VI. *Bulletin of the Botanical Department, Jamaica (Ser. 2)* 2: 5-8.

Cockerell, T.D.A. 1895b. Coccidae or scale insects. - VII. *Bulletin of the Botanical Department, Jamaica (Ser. 2)* 2: 100-102.

Cockerell, T.D.A. 1895c. On some insects collected in the State of Chihuahua, Mexico. *Annals and Magazine of Natural History (Ser. 6)* 15: 204-210.

Cockerell, T.D.A. 1895d. Two new species of *Lecanium* from Brazil. *American Naturalist* 29: 174-175.

Cockerell, T.D.A. 1895e. Canadian Coccidae. *Canadian Entomologist* 27: 33-36.

Cockerell, T.D.A. 1895f. Three new species of Coccidae. *Entomologist* 28: 100-101.

Cockerell, T.D.A. 1895g. Science notes. Three new scale insects. *Journal of the Institute of Jamaica* 2: 167.

Cockerell, T.D.A. 1895h. Two more new species of *Lecanium. American Naturalist* 29: 380-382.

Cockerell, T.D.A. 1895i. A new *Pulvinaria* found on orchids. *Canadian Entomologist* 27: 135.

Cockerell, T.D.A. 1895j. New scale-insects from Arizona. *Bulletin Arizona Agricultural Experiment Station* 14: 56.

Cockerell, T.D.A. 1895k. A new scale-insect from Grenada. *Journal of the Trinidad Field Naturalists' Club* 2: 194-195.

Cockerell, T.D.A. 1895l. Two new western Coccidae. *Psyche* 7: 254-255.

Cockerell, T.D.A. 1895m. New facts about scale insects. -I. *Garden and Forest* 382: 244.

Cockerell, T.D.A. 1895n. On some Coccidae obtained by Mr. C.A. Barber in the Island of Antigua, W.I. *Annals and Magazine of Natural History (Ser. 6)* 16: 60-62.

Cockerell, T.D.A. 1895o. New North American Coccidae. *Psyche, Supplement* 7: 1-4.

Cockerell, T.D.A. 1895p. On the subglobular species of *Lecanium. Canadian Entomologist* 27: 201-204.

Cockerell, T.D.A. 1895q. Contributions to Coccidology. - I. *American Naturalist* 29: 725-732.

Cockerell, T.D.A. 1895r. Miscellaneous notes on Coccidae. *Canadian Entomologist* 27: 253-261.

Cockerell, T.D.A. 1896a. Coccidae or scale insects. - IX. *Bulletin of the Botanical Department, Jamaica (n.s.)* 3: 256-259.

Cockerell, T.D.A. 1896b. A check list of the Coccidae. *Bulletin of the Illinois State Laboratory of Natural History* 4: 318-339.

Cockerell, T.D.A. 1896c. On a small collection of Coccidae from the island of Grenada. *Journal of the Trinidad Field Naturalists' Club* 2: 306-307.

Cockerell, T.D.A. 1896d. Notes and descriptions of the new Coccidae collected in Mexico by Prof. C.H.T. Townsend. *Bulletin, United States Department of Agriculture, Division of Entomology, Technical Series* 4: 31-39.

Cockerell, T.D.A. 1896e. Preliminary diagnoses of new Coccidae. *Psyche, Supplement* 7: 18-21.

Cockerell, T.D.A. 1896f. Some new species of Japanese Coccidae, with notes. *Bulletin, United States Department of Agriculture, Division of Entomology, Technical Series* 4: 47-56.

Cockerell, T.D.A. 1896g. New Coccidae from Massachusetts and New Mexico. *Canadian Entomologist* 28: 222-226.

Cockerell, T.D.A. 1896h. New species of insects taken on a trip from the Mesilla Valley to the Sacramento Mts., New Mexico. *Journal of the New York Entomological Society* 5: 201-207.

Cockerell, T.D.A. 1896i. *Ceroplastes euphorbiae*, n. sp. *Psyche, Supplement* 7: 17.

Cockerell, T.D.A. 1896j. On the Coccidae (scale insects) of Trinidad. *Bulletin Trinidad Royal Botanic Gardens, Miscellaneous Information* 2: iii-v.

Cockerell, T.D.A. 1897a. Descriptive notes on two Coccidae. *Entomologist* 30: 12-14.

Cockerell, T.D.A. 1897b. Notes on new Coccidae. 1. A new coccid pest of greenhouses. *Psyche* 8: 52.

Cockerell, T.D.A. 1897c. Note on *Lecanium tessellatum. Bulletin of the Botanical Department, Jamaica (n.s.)* 4: 109.

Cockerell, T.D.A. 1897d. New and little-known Coccidae from Florida. I. Determinations and descriptions, including a new genus. *Psyche* 8: 89-90.

Cockerell, T.D.A. 1897e. The food-plants of scale insects (Coccidae). *Proceedings of the United States National Museum* 19: 725-785.

Cockerell, T.D.A. 1897f. Notes on scale insects. -3. *California Fruit Grower* 21: 5.

Cockerell, T.D.A. 1897g. Contributions to Coccidology. -II. *The American Naturalist* 31: 588-592.

Cockerell, T.D.A. 1897h. Notes on the Coccidae, a family of Homoptera, with a table of the species hitherto observed in Brazil. *Revista do Museo Paulista* 2: 65-72.

Cockerell, T.D.A. 1897i. Further notes on Coccidae from Brazil. *Revista do Museo Paulista* 2: 383-384.

Cockerell, T.D.A. 1897j. Some new and little-known Coccidae collected by Prof. C.H.T. Townsend in Mexico. *Canadian Entomologist* 29: 265-271.

Cockerell, T.D.A. 1897k. A new *Lecanium* on magnolia from Florida. *Psyche* 8: 152.

Cockerell, T.D.A. 1898a. Miscellaneous notes. *Proceedings of the Entomological Society of Washington* 4: 64-65.

Cockerell, T.D.A. 1898b. Two new scale insects. *The Entomologist* 31: 65-66.

Cockerell, T.D.A. 1898c. Coccidae or scale insects -XIII. *Bulletin of the Botanical Department, Jamaica (N.S.)* 5: 107-109.

Cockerell, T.D.A. 1898d. Some new Coccidae of the subfamily Lecaniinae. *The Entomologist* 31: 130-132.

Cockerell, T.D.A. 1898e. A new scale-insect of the genus *Lecanium. Entomological News* 9: 145-146.

Cockerell, T.D.A. 1898f. New Coccidae from Mexico. *Annals and Magazine of Natural History (Ser. 7)* 1: 426-440.

Cockerell, T.D.A. 1898g. Some new Coccidae collected at Campinas, Brazil, by Dr. F. Noack. *Revista do Museo Paulista* 3: 41-42.

Cockerell, T.D.A. 1898h. Some new Coccidae. *Annals and Magazine of Natural History (Ser. 7)* 2: 24-27.

Cockerell, T.D.A. 1898i. New North American insects. *Annals and Magazine of Natural History (n.s.)* 2: 321-331.

Cockerell, T.D.A. 1898j. Two new species of *Lecanium* from Canada. *Canadian Entomologist* 30: 293-294.

Cockerell, T.D.A. 1898k. Mais algumas Coccidae, colligidos pelo Dr. F. Noack. *Revista do Museu Paulista. São Paulo* 3: 501-503.

Cockerell, T.D.A. 1898l. Two new Coccidae from Lagos, W. Africa. *The Entomologist* 31: 259.

Cockerell, T.D.A. 1898m. The Coccidae of the Sandwich Islands. *The Entomologist* 31: 239-240.

Cockerell, T.D.A. 1898n. *Ceroplastes cistudiformis* again. *The Entomologist* 31: 141.

Cockerell, T.D.A. 1899a. Two new genera of lecaniine Coccidae. *The Entomologist* 32: 12-13.

Cockerell, T.D.A. 1899b. First supplement to the check-list of the Coccidae. *Bulletin of the Illinois State Laboratory of Natural History* 5: 389-398.

Cockerell, T.D.A. 1899c. Notes on Australian Coccidae. 1. - The Australian species of *Mytilaspis*. 2. - A new *Pulvinaria*. *Victorian Naturalist* 16: 13-16.

Cockerell, T.D.A. 1899d. A new wax-scale from west Africa. *The Entomologist* 32: 127.

Cockerell, T.D.A. 1899e. Notes on Australian Coccidae. 3. - Two new species and a new variety. *Victorian Naturalist* 16: 88-89.

Cockerell, T.D.A. 1899f. Tables for the determination of the genera of Coccidae. *Canadian Entomologist* 31: 273-279, 330-333.

Cockerell, T.D.A. 1899g. Rhynchota, Hemiptera - Homoptera. [Aleurodidae and Coccidae]. *Biologia Centrali-Americana* 2: 1-33.

Cockerell, T.D.A. 1899h. Some notes on Coccidae. *Proceedings of the Academy of Natural Sciences of Philadelphia 1899*: 259-275.

Cockerell, T.D.A. 1899k. Some synonymy. *Psyche* 8: 311-312.

Cockerell, T.D.A. 1900a. A new genus of Coccidae, injuring the roots of the grapevine in South Africa. *The Entomologist* 33: 173-174.

Cockerell, T.D.A. 1900b. Some Coccidae quarantined at San Francisco. *Psyche* 9: 70-72.

Cockerell, T.D.A. 1900c. The Coccidae of New Zealand. *Nature. London* 61: 367-368.

Cockerell, T.D.A. 1901a. A new *Ceroplastes* (Fam. Coccidae). *Comunicaciones del Museo Nacional de Buenos Aires* 1: 288-289.

Cockerell, T.D.A. 1901b. Notes on some Coccidae of the earlier writers. *The Entomologist* 34: 90-93.

Cockerell, T.D.A. 1901c. New Coccidae from New Mexico. *Canadian Entomologist* 33: 209-210.

Cockerell, T.D.A. 1901d. The coccid genus *Saissetia*. *The Entomological Student* 2: 31-33.

Cockerell, T.D.A. 1902a. New genera and species of Coccidae, with notes on known species. *Annals and Magazine of Natural History (Ser. 7)* 9: 20-26.

Cockerell, T.D.A. 1902b. A new scale-insect on agave. *Memoria y Revista de la Sociedad Científica "Antonio Alzate"* 17: 143.

Cockerell, T.D.A. 1902c. South African Coccidae. - II. *The Entomologist* 35: 111-114.

Cockerell, T.D.A. 1902d. The nomenclature of the Coccidae. *The Entomologist* 35: 114.

Cockerell, T.D.A. 1902e. New Coccidae from the Argentine Republic and Paraguay. *Canadian Entomologist* 34: 88-93.

Cockerell, T.D.A. 1902f. A contribution to the knowledge of the Coccidae. Appendix Some Brazilian Coccidae. *Annals and Magazine of Natural History (Ser. 7)* 9: 450-456.

Cockerell, T.D.A. 1902g. Additions to the fauna of Mexico (bees and Coccidae). *The Entomologist* 35: 177-178.

Cockerell, T.D.A. 1902h. The coccid *Lecanopsis dugesi*. *The Entomologist* 35: 194.

Cockerell, T.D.A. 1902i. A catalogue of the Coccidae of South America. *Revista Chilena de Historia Natural* 6: 250-257.

Cockerell, T.D.A. 1902j. Some Coccidae from Mexico. *Annals and Magazine of Natural History (Ser. 7)* 10: 465-472.

Cockerell, T.D.A. 1903a. Some species of *Eulecanium* (Coccidae) from France. *Psyche* 10: 19-22.

Cockerell, T.D.A. 1903b. New and little-known American Coccidae. *Annals and Magazine of Natural History (Ser. 7)* 11: 155-165.

Cockerell, T.D.A. 1903c. Five new Coccidae from Mexico. *The Entomologist* 36: 45-48.

Cockerell, T.D.A. 1903d. A new coccid from Madeira, allied to *Coccus tuberculatus*, Bouché. *The Entomologist* 36: 261-262.

Cockerell, T.D.A. 1905a. Tables for the identification of Rocky Mountain Coccidae (Scale insects and mealybugs). *Colorado University Studies* 2: 189-203.

Cockerell, T.D.A. 1905b. Three new Coccidae from Colorado. *Canadian Entomologist* 37: 135-136.

Cockerell, T.D.A. 1905c. Three new South American Coccidae. *Entomological News* 16: 161-163.

Cockerell, T.D.A. 1905d. Notes on *Eulecanium folsomi* King. *Proceedings of the Entomological Society of Washington* 7: 129- 130.

Cockerell, T.D.A. 1905e. Some Coccidae from the Philippine Islands. *Proceedings of the Davenport Academy of Sciences, Davenport, Iowa* 10: 127-136.

Cockerell, T.D.A. 1905f. A new scale-insect (Fam. Coccidae) on the rose. *Zoologischer Anzeiger* 29: 514-515.

Cockerell, T.D.A. 1906. The coccid genus *Eulecanium*. *Canadian Entomologist* 38: 83-88.

Cockerell, T.D.A. 1910a. A new wax scale from the Argentine. *Canadian Entomologist* 42: 74-76.

Cockerell, T.D.A. 1910b. The Coccidae of Boulder County, Colorado. *Journal of Economic Entomology* 3: 425-430.

Cockerell, T.D.A. 1911. Notes and observations. Coccidae affecting rubber trees. *The Entomologist* 44: 327.

Cockerell, T.D.A. 1912. Some Coccidae from the Grand Canon, Arizona. *Canadian Entomologist* 44: 301.

Cockerell, T.D.A. 1914. Descriptions and records of Coccidae. II. Non-Diaspine subfamilies. *Bulletin of the American Museum of Natural History* 33: 331-335.

Cockerell, T.D.A. 1922. The mealy-bug called *Pseudococcus bromeliae* and other coccids. *Science* 56: 308-309.

Cockerell, T.D.A. 1929. The type of genus *Coccus* (Homoptera, Coccidae). *Science (n.s.)* 70: 150.

Cockerell, T.D.A. 1933. The type of the genus *Lecanium. Science (n.s.)* 78: 35.

Cockerell, T.D.A. & Bueker, E.D. 1930. New records of Coccidae (Homoptera). *American Museum Novitates* 424: 1-8.

Cockerell, T.D.A. & King, G.B. 1899. An apparently new *Lecanium* found on white cedar. *Psyche* 8: 349-350.

Cockerell, T.D.A. & Parrott, P.J. 1899. Contributions to the knowledge of the Coccidae. *The Industrialist* 25: 159-165, 227-237, 276-284.

Cockerell, T.D.A. & Parrott, P.J. 1901. Table to separate the genera and subgenera of Coccidae related to *Lecanium. The Canadian Entomologist* 33: 57-58.

Cockerell, T.D.A. & Robbins, W.W. 1909. A new coccid from Nicaragua. *Canadian Entomologist* 41: 150.

Cockerell, T.D.A. & Robinson, E. 1914. Descriptions and records of Coccidae. I. Subfamily Diaspinae. II. Non-Diaspine *subfamilies. Bulletin of the American Museum of Natural History* 33: 327-335.

Cockerell, T.D.A. & Robinson, E. 1915a. Descriptions and records of Coccidae. *Bulletin of the American Museum of Natural History* 34: 105-113.

Cockerell, T.D.A. & Robinson, E. 1915b. Descriptions and records of Coccidae. *Bulletin of the American Museum of Natural History* 34: 423-428.

Cockerell, T.D.A. & Tinsley, J.D. 1898. On a new wax-producing insect found in Jamaica. *Journal of the Institute of Jamaica* 2: 468.

Coleman, G.A. 1903. Coccidae of the Coniferae, with the description of ten new species from California. *Journal of the New York Entomological Society* 11: 61-85.

Coleman, L.C. & Kannan, K. 1918. Some scale insect pests of coffee in South India. *Bulletin of the Department of Agriculture, Mysore State, Bangalore. Entomological Series* 4: 1-66.

Comstock, J.H. 1881. Report of the entomologist. Part II. Report on scale insects. *Report of The Entomologist of the United States Department of Agriculture for the year 1880*: 276-349.

Conti, B. 1987. Influenza della temperatura sullo sviluppo degli stadi preimaginali della *Saissetia oleae* (Olivier). I. Durata dello sviluppo dell'uovo a temperature constanti. *Frustula Entomologica (N.S.)* 10: 73-81.

Cook, A.J. 1878. *Lecanium tulipiferae. Canadian Entomologist* 10: 192-195.

Coquillett, D.W. 1891. A new scale insect from California. *Insect Life* 3: 382-384.

Corseuil, E. & Barbosa, V.M.B. 1971. A família Coccidae no Rio Grande do Sul (Homoptera, Coccoidea). *Arquivos do Musei Nacional, Rio de Janeiro* 54: 237-241.

Costa, A. 1844. Descrizione di una novelle specie di cocciniglia del genere *Calittico*, Cos. che vive sopra il *Mesembrianthemum acinaciforme. Annali dell'Accademia degli Aspiranti Naturalisti di Napoli* 22: 273-276.

Costa, A. 1857. Degl'insetti che attaccano l'albero ed il frutto dell'olivo. loro descrizione e biologia danni che arrecano e mezzi per distruggerli. (Opera coronata). R. Accademia Scienza, Napoli. 197 pp.

Costa, A. 1877. Degl'insetti che attaccano l'albero ed il frutto dell'olivo. loro descrizione e biologia danni che arrecano e mezzi per distruggeri. Ed. 2, rev. and enl. R. Istituto d'Incoraggiamento, Napoli. 340 pp.

Costa, O.G. 1829. Fauna del Regno di Napoli, famiglia de coccinigliferi, o de gallinsetti. Emitteri, Napoli. 23 pp.

Costa, O.G. 1835. Nuove osservazioni intorno alle cocciniglie ed ai loro pretesi maschi. F. Fernandes, Napoli, 24 pp.

Costa, O.G. 1840. Nuove osservazioni intorno alle cocciniglie ed ai loro pretesi maschi. *Atti del Real Istituto d'Incoraggiamento alle Scienze Naturali di Napoli* 6: 31-52.

Costa Lima, A. 1923. Nota sobre as especies do genero *Eucalymnatus* (Fam. Coccidae, sub-fam. Coccinae). *Archivos da Escola Superior da Agricultura e Medicina Veterinaria. Nichteroy* 7: 35-44.

Costa Lima, A. 1927. Segundo catalogo systematico dos insectos que vivem nas plantas do Brasil e ensaio de bibliographia entomologica brasileira. *Archivos da Escola Superior de Agricultura e Medicina Veterinaria. Rio de Janeiro* 8: 69-301.

Costa Lima, A. 1930a. Supplemento ao 2o catalogo systematico dos insectos que vivem nas plantas do Brasil e ensaio de bibliographia entomologica Brasileira. *O Campo. Brazil* 1: 84-91.

Costa Lima, A. 1930b. Segunda nota sobre especies do genero *Eucalymnatus* (Homoptera: Coccidae). *Memorias do Instituto Oswaldo Cruz. Rio de Janeiro* 24: 85-87.

Costa Lima, A. 1936. Terceiro catalogo dos insectos que vivem nas plantas do Brasil. Dir. Estatis. Prod., Rio de Janeiro. 460 pp.

Costa Lima, A. 1940. Um novo *Ceroplastes* gigante (Coccoidea: Coccidae. *Papeis Avulsos Departemnto de Zoologia. São Paulo* 1: 9-12.

Costa Lima, A. 1942. Os Insetos do Brasil. Vol. 3 (Homopteros). Publicad. da Escola Nacional de Agronomia, Rio de Janeiro. 327 pp.

Couturier, G. Matile-Ferrero, D. & Richard, C. 1985. Sur les cochenilles de la région de Tai (Côte d'Ivoire), recensées dans les cultures et en forêt dense, [Homoptera, Coccoidea]. *Revue Française d'Entomologie (N.S.)* 7: 273-286.

Craw, A. 1891. Destructive insects, their natural enemies, remedies and recommendations. II. Scale insects. Description, history, and remedies for their destruction. *California State Board of Horticulture, Division of Entomology. Sacramento* 1891: 7-15.

Craw, A. 1894. Scale insects. *Bulletin of California State Board of Horticulture* 68: 11-17.

Cronquist, A. 1988. The evolution and classification of flowering plants. The New York Botanical Gardens, New York. 555 pp.

Curtis, J. 1838. British Entomology; being illustrations and descriptions of the genera of insects found in Great Britain and Ireland: containing coloured figures from nature of the most rare and beautiful species, and in many instances of the plants upon which they are found. Vol. XV, J. Pigot & Co., London.

Curtis, J. 1843a. The black turtle-scale. *Coccus testudo* (Curtis). *Gardners' Chronicle and Agricultural Gazette* 26: 443-444.

Curtis, J. 1843b. *Coccus patellaeformis* (Curtis). The brown limpet-scale. *Gardners' Chronicle and Agricultural Gazette* 30: 517-518.

Dalman, J.W. 1826. Om några Svenska Arter af *Coccus*; samt deinuti dem förekommande Parasit-insecter. *Kongliga Vetenskap-Akademiens Handlingar* (1825): 350-374.

Danzig, E.M. 1959. On the scale insect fauna (Homoptera, Coccoidea) of the Leningrad region. [In Russian, English summary]. *Entomologicheskoe Obozrenye* 38: 443-455.

Danzig, E.M. 1961. On food forms of *Eulecanium franconicum* (Lndgr.) (Homoptera, Coccoidea). [In Russian, English summary]. *Entomologicheskoe Obozrenye* 40: 571-576.

Danzig, E.M. 1964. 5. Suborder Coccinea - coccids or mealybugs and scale insects. [In Russian]. pp. 616-654. In: G.Y. Bei-Bienko (ed.). Classification keys to the insects of the European part of the USSR. Akademiya Nauk SSSR Zoologicheskii Institut, Leningrad.

Danzig, E.M. 1965. The wax scale - *Ericerus pela* Chav. (Homoptera, Coccoidea) in the USSR. [In Russian, English summary]. *Zoologicheskii Zhurnal* 44: 537-546.

Danzig, E.M. 1966. The reduction of wax-secreting dermal structures of the females of *Pulvinaria* Targ. (Homoptera, Coccoidea) infested by parasites. [In Russian, English summary.]. *Zoologicheskii Zhurnal* 45: 1488-1492.

Danzig, E.M. 1967. Contributions to the knowledge of the Coccidae (Homoptera) of the Primorye Territory. [In Russian]. *Trudy Zoologicheskogo Instituta. Leningrad* 41: 139-172.

Danzig, E.M. 1968. Contribution to the fauna and biology of scale insects and white-flies (Homoptera, Coccoidea, Aleyrodoidea) from northern Karelia. [In Russian, English summary]. *Entomologicheskoe Obozrenye* 47: 499-504.

Danzig, E.M. 1970. Synonymy of some polymorphous species of coccids (Homoptera, Coccoidea). [In Russian, English summary]. *Zoologicheskii Zhurnal* 49: 1015-1024.

Danzig, E.M. 1971. Three new species of coccids (Homoptera, Coccoidea) from the far East. [In Russian, English Summary]. *Zoologicheskii Zhurnal* 50: 1414-1417.

Danzig, E.M. 1972a. Suborder Coccoidea - Scale insects. [In Russian]. pp. 189-221. In: Pests of forests. Akademiya Nauk SSSR Zoologicheskii Institut, Leningrad.

Danzig, E.M. 1972b. New and little-known species of the scale insects (Homoptera, Coccoidea) from Siberia and the far east of the USSR. [In Russian]. *Trudy Zoologicheskogo Instituta, Leningrad* 52: 261-276.

Danzig, E.M. 1972c. Contribution to the fauna of the white flies and scale insects (Homoptera: Aleyrodoidea, Coccoidea) of Mongolia. [In Russian]. *Insects of Mongolia* 1: 325-348.

Danzig, E.M. 1972d. Contributions to the knowledge of the scale insects fauna (Homoptera, Coccoidea) of Afghanistan. [In Russian]. *Entomologicheskoe Obozrenye* 51: 581-584.

Danzig, E.M. 1974. Contribution to the fauna of the scale insects (Homoptera, Coccoidea) of Mongolia. [In Russian]. *Insects of Mongolia* 2: 67-71.

Danzig, E.M. 1975a. Three new species of soft scales (Homoptera, Coccoidea, Coccidae) from southern Siberia and the Far East. [In Russian, English summary]. *Zoologicheskii Zhurnal* 54: 137-138.

Danzig, E.M. 1975b. Review of the cottony grass scales of the genus *Eriopeltis* Sign. (Homoptera, Coccoidea, Coccidae) of the Palaearctic. [In Russian, English summary]. *Entomologicheskoe Obozrenye* 54: 808-813.

Danzig, E.M. 1977a. On the nomenclature and distribution of some injurious scale insects (Homoptera, Coccoidea). [In Russian, English summary]. *Entomologicheskoe Obozrenye* 56: 99-102.

Danzig, E.M. 1977b. Contributions to the scale-insect fauna of North and East Mongolia (Homoptera, Coccoidea). [In Russian]. *Insects of Mongolia* 5: 196-202.

Danzig, E.M. 1977c. An ecological and geographical review of the scale insects (Homoptera, Coccoidea) of the south of the Far East. [In

Russian]. pp. 37-60. In: Insect fauna of the Far East. Akaddemiya Nauk SSSR Zoologicheskii Institut, Leningrad.

Danzig, E.M. 1978a. Scale insect fauna of South Sakhalin and Kunashir. [In Russian]. *Trudy Biologo-Pochevnnogo, Akademii Nauk SSR, Vladivostok* 50: 3-23.

Danzig, E.M. 1978b. Fauna of scale insects (Homoptera, Coccoidea). [In Russian]. pp. 71-78. In: Ecological and faunistic investigations of the insect pests of Yakut. Akademiya Nauk SSSR, Yakutsk.

Danzig, E.M. 1980a. On the nomenclature and synonymy of several species of scale insects and white flies (Homoptera: Coccinea, Aleyrodinea). [In Russian]. *Entomologicheskoe Obozrenye* 54: 594-595.

Danzig, E.M. 1980b. Species of scale insects (Homoptera, Coccinea) new for Mongolia. [In Russian]. *Insects of Mongolia* 7: 31-38.

Danzig, E.M. 1980c. Scale insects of Far East SSSR (Homoptera, Coccinea) with phylogenetic analysis of scale insects fauna of the world. [In Russian]. Nauka, Leningrad. 366 pp.

Danzig, E.M. 1984. New data on the scale insects (Homoptera, Coccinea) of Mongolia. [In Russian]. *Insects of Mongolia* 9: 33-34.

Danzig, E.M. 1985. The fauna of scale insects (Homoptera, Coccinea) of Teberda State Reserve. [In Russian, English summary]. *Entomologicheskoe Obozrenye* 64: 110-123.

Danzig, E.M. 1988. Suborder Coccinea. [In Russian]. pp. 686-726. In: P.A. Lehr (edit.). Keys to insects of the Far East of the USSR. Nauka, Leningrad.

Danzig, E.M. 1990. New species of coccids (Homoptera, Coccinea) from Iran, Mongolia and Vietnam. [In Russian]. *Entomologicheskoe Obozrenye* 69: 373-376.

Danzig, E.M. & Kerzhner, I.M. 1981. *Coccus* Linnaeus, 1758 and *Parthenolecanium* Sulc, 1908 (Insecta, Homoptera, Coccidae):

proposed designation of type species under the plenary powers. Z.N. (S.) 2125. *Bulletin of Zoological Nomenclature* 38: 147-152.

Danzig, E.M. & Konstantinova, G.M. 1990. On coccid (Homoptera, Coccinea) fauna of Vietnam. [In Russian]. *Trudy Zoologicheskogo Instituta Akademiya Nauk SSSR. Leningrad* 209: 38-52.

Danzig, E.M. & Kozar, F. 1973. A new species of soft scales, *Physo-kermes inopinatus* sp. n. (Homoptera, Coccoidea) from Hungary. [In Russian, English summary]. *Entomologicheskoe Obozrenye* 52: 832-834.

Danzig, E.M. & Matile-Ferrero, D. 1990. *Neopulvinaria innumerabilis* a pest of vine in Europe (Homoptera: Coccoidea: Coccidae). *Proceedings of the Sixth International Symposium of Scale Insects Studies, Cracow, August 6-12,* 1990, 2: 131-132.

Das, G.M. & Ganguli, R.N. 1961. Coccoids on tea in North-East India. *Indian Journal of Entomology* 23: 245-256.

Dash, S.J. 1916. Revised list of the Coccidae of Barbados. *Report on the Department of Agriculture, Barbados (1914-15):* 41-43.

Davatchi, G.A. 1958. Etude biologique de la faune entomologique des Pistacia sauvage et cultivés. *Revue de Pathologie Végétale et d'Entomologie Agricole de France* 37: 3-166.

Davies, R.G. & Williams, D.J. 1982. Obituary. Dr. K.L. Boratynski. *Entomologist's Monthly Magazine* 118: 93-95.

De Graaf, H.W., Six, G.A. & Snellen van Vollenhoven, S. 1862. Tweede naamlijst van inlandsche Hemiptera. *Tijdschrift voor Entomologie. Amsterdam* 5: 72-96.

de Grandpre, A.D. & de Charmoy, D. d'Emmerez. 1899. Liste rais-sonnee des cochenilles de l'Ile Maurice. Chapitre IV. Nomenclature des cochenilles de Maurice. *Publications de la Société Amicale Scientifique, Maurice* 1899: 20-49.

De Geer, C. 1778. Mémoires pour servir à l'histoire des insectes. P. Hesselberg, Stockholm. 950 pp.

De Lotto, G. 1954. Three apparently new coccids (Homopt.: Coccidae) from Eritrea. *Journal of the Entomological Society of southern Africa* 17: 213-218.

De Lotto, G. 1955. Three new coccids (Hemipt.: Coccoidea) attacking coffee in East Africa. *Bulletin of Entomological Research* 46: 267-273.

De Lotto, G. 1956a. The identity of some East African species of *Saissetia* (Homoptera, Coccidae). *Bulletin of Entomological Research* 47: 239-249.

De Lotto, G. 1956b. A new *Ceroplastodes* (Hom.: Coccoidea) from Kenya. *Journal of the Entomological Society of southern Africa* 19: 310-312.

De Lotto, G. 1957a. Notes on some African species of *Saissetia* (Homoptera: Coccoidea: Coccidae). *Journal of the Entomological Society of southern Africa* 20: 170-182.

De Lotto, G. 1957b. On some Ethiopian species of the genus *Coccus* (Homoptera: Coccoidea: Coccidae). *Journal of the Entomological Society of southern Africa* 20: 295-314.

De Lotto, G. 1958a. A new species of *Saissetia* from South Africa (Homoptera: Coccoidea: Coccidae). *Journal of the Entomological Society of southern Africa* 21: 66-68.

De Lotto, G. 1958b. New soft scales (Homoptera: Coccoidea: Coccidae) from Africa. *Proceedings of the Royal Entomological Society of London (B)* 27: 165-172.

De Lotto, G. 1959. Further notes on Ethiopian species of the genus *Coccus* (Homoptera: Coccoidea: Coccidae). *Journal of the Entomological Society of Southern Africa* 22: 150-173.

De Lotto, G. 1960. The green scales of coffee in Africa south of the Sahara (Homoptera, Coccidae). *Bulletin of Entomological Research* 51: 389-403.

De Lotto, G. 1961. Two new *Ceroplastes* species from Africa (Homopt.: Coccidae). *Journal of the Entomological Society of southern Africa* 24: 318-321.

De Lotto, G. 1962. A new *Coccus* from South Africa (Homoptera: Coccidae). *Journal of the Entomological Society of southern Africa* 25: 263-265.

De Lotto, G. 1963. New species and a new genus of hard scales from East Africa (Homoptera: Coccidae). *Proceedings of the Royal Entomological Society of London (B)* 32: 191-200.

De Lotto, G. 1964. A new species of *Pulvinaria* (Homopt.: Coccidae) attacking sugar cane in South Africa. *South African Journal of Agricultural Science* 7: 863-866.

De Lotto, G. 1965. On some Coccidae (Homoptera), chiefly from Africa. *Bulletin of the British Museum (Natural History) Entomology* 16: 175-239.

De Lotto, G. 1966a. Descriptions of three new species of *Coccus* (Homoptera: Coccidae) from South Africa. *Proceedings of the Royal Entomological Society of London (B)* 35: 41-46.

De Lotto, G. 1966b. A new genus and four new species of Coccidae (Homoptera) from South Africa. *Proceedings of the Linnean Society of London* 177: 143-149.

De Lotto, G. 1966c. Another new species of *Pulvinaria* (Hom.: Coccidae) from sugar cane. *South African Journal of Agricultural Science* 9: 467-472.

De Lotto, G. 1967a. A contribution to the knowledge of the African Coccoidea (Homoptera). *Journal of the Entomological Society of southern Africa* 29: 109-120.

De Lotto, G. 1967b. The soft scales (Homoptera: Coccidae) of South Africa. I. *South African Journal of Agricultural Sciences* 10: 781-810.

De Lotto, G. 1968a. A generic diagnosis of *Takahashia* Cockerell, 1896 (Homoptera, Coccidae). *Proceedings of the Linnean Society of London* 179: 97-98.

De Lotto, G. 1968b. Second contribution to the knowledge of the African Coccoidea (Homoptera). *Journal of the Entomological Society of southern Africa* 31: 83-86.

De Lotto, G. 1969a. On a few old and new soft scales and mealybugs (Homoptera: Coccoidea). *Journal of the Entomological Society of southern Africa* 32: 413-422.

De Lotto, G. 1969b. A new genus of wax scales (Homoptera: Coccidae). *Bolletino del Laboratorio di Entomologia Agraria 'Filippo Silvestri' di Portici* 27: 210-218.

De Lotto, G. 1970a. The soft scales (Homoptera: Coccidae) of South Africa, II. *Journal of the Entomological Society of southern Africa* 33: 143-156.

De Lotto, G. 1970b. On the status of two genera of soft scales (Homoptera: Coccoidea: Coccidae). *Bolletino del Laboratorio di Entomologia Agraria 'Filippo Silvestri' di Portici* 28: 257-261.

De Lotto, G. 1971a. A preliminary note on the black scales (Homoptera, Coccidae) of North and Central America. *Bulletin of Entomological Research* 61: 325-326.

De Lotto, G. 1971b. The authorship of the Mediterranean black scale (Homoptera: Coccidae). *Journal of Entomology (B)* 40: 149-150.

De Lotto, G. 1971c. On some genera and species of wax scales (Homoptera: Coccidae). *Journal of Natural History* 5: 133-153.

De Lotto, G. 1973. A new soft scale from Citrus (Homoptera: Coccoidea: Coccidae). *Bolletino del Laboratorio di Entomologia Agraria 'Filippo Silvestri' di Portici* 30: 291-293.

De Lotto, G. 1974. New species of *Filippia* Targioni Tozzetti, 1868 (Homoptera: Coccoidea: Coccidae) from South Africa. *Journal of the Entomological Society of southern Africa* 37: 207-214.

De Lotto, G. 1975. Two new genera of soft scales from Africa (Homoptera: Coccoidea: Coccidae). *Journal of the Entomological Society of southern Africa* 38: 61-63.

De Lotto, G. 1976. On the black scales of southern Europe (Homoptera: Coccoidea: Coccidae). *Journal of the Entomological Society of southern Africa* 39: 147-149.

De Lotto, G. 1978. The soft scales (Homoptera: Coccidae) of South Africa, iii. *Journal of the Entomological Society of southern Africa* 41: 135-147.

De Lotto, G. 1979. The soft scales (Homoptera: Coccidae) of South Africa, iv. *Journal of the Entomological Society of southern Africa* 42: 245-256.

De Meijer, A.H., Wysoki, M., Swirski, E., Blumberg, D. & Izhar, Y. 1989. Susceptibility of avocado cultivars to the pyriform scale, *Protopulvinaria pyriformis* (Cockerell) (Homoptera: Coccidae). *Agriculture, Ecosystem & Environment* 25: 75-82.

Dean, H.A. & Hart, W.G. 1972. *Saissetia miranda* (Homoptera: Coccidae), a potential pest of citrus in Texas. *Annals of the Entomological Society of America* 65: 478-481.

Deitz, L.L. 1979. Selected references for identifying New Zealand Hemiptera (Homoptera and Heteroptera), with notes on nomenclature. *New Zealand Entomologist* 7: 20-29.

Deitz, L.L. & Tocker, M.F. 1980. W.M. Maskell's Homoptera: Species-group names and type-material. *New Zealand Department of Scientific and Industrial Research, Information Series* 146: 1-76.

Del Giudice, F. 1868. De'lavori accademici del R. Istituto d'Incoraggiamento alle scienze naturali economiche e technologiche nell'anne 1867 e cenni biografici del Socio Oronzio Gabriele Costa. *Atti dell'Istituto d'Incoraggiamento alle Scienze Naturali, Napoli, Serie* II, 5 : 5-36.

Del Rivero, J.M. 1966. Nota sobre una plaga de agrios y aguacates. *Boletín de Patología Vegetal y Entomología Agrícola. Madrid* 29: 59-62.

Del Guercio, G. 1900. Osservazioni intorno ad una nuova cocciniglia nociva agli agrumi in Italia ed al modo di immunizzare la parte legnosa delle piante contro la puntura delle coccinglie in generale e di distruggerle. *Nuove Relazione della R. Stazione di Entomologia Agraria. Firenze* 3: 3-26.

Delucchi, V. 1975. Scale insects and whiteflies of citrus fruit. *Ciba-Geigy Agrochemicals, Technical Monograph,* Supplement B 4.

Deplanche, E. 1859. Maladie du caféier. *Messager de Tahiti* 8(9): 6-7.

Dietz, H.F. & Morrison, H. 1916. The Coccidae or scale insects of Indiana. *Indiana State Entomologist Eighth Annual Report* (1914-1915) 8: 195-321.

Donaldson, D.R., Moore, W.S., Koehler, C.S. & Joos, J.L. 1978. Scales threaten iceplant in Bay area. *California Agriculture* 32: 4-7.

Douglas, J.W. 1886a. Notes on some British Coccidae (No. 2). *Entomologist's Monthly Magazine* 22: 243-250.

Douglas, J.W. 1886b. Note on some British Coccidae (No. 3). *Entomologist's Monthly Magazine* 23: 25-29.

Douglas, J.W. 1886c. Note on some British Coccidae (No. 4). *Entomologist's Monthly Magazine* 23: 77-82.

Douglas, J.W. 1887a. Note on some British Coccidae (No. 7). *Entomologist's Monthly Magazine* 24: 21-28.

Douglas, J.W. 1887b. Note on some British Coccidae (No. 8). *Entomologist's Monthly Magazine* 24: 95-101, 165-171.

Douglas, J.W. 1888a. Notes on some British and exotic Coccidae (No. 9). *Entomologist's Monthly Magazine* 25: 57-60.

Douglas, J.W. 1888b. Notes on some British and exotic Coccidae (No. 12). *Entomologist's Monthly Magazine* 25: 150-153.

Douglas, J.W. 1890a. Notes on some British and exotic Coccidae (No. 17). *Entomologist's Monthly Magazine* 26: 238-240.

Douglas, J.W. 1890b. Notes on some British and exotic Coccidae (No. 18). *Entomologist's Monthly Magazine* 26: 318-319.

Douglas, J.W. 1891a. Notes on some British and exotic Coccidae (No. 19). *Entomologist's Monthly Magazine* 27: 65-68.

Douglas, J.W. 1891b. Notes on some British and exotic Coccidae (No. 20). *Entomologist's Monthly Magazine* 27: 95-100.

Douglas, J.W. 1891c. Notes on some British and exotic Coccidae (No. 21). *Entomologist's Monthly Magazine* 27: 244-247.

Douglas, J.W. 1891d. *Lecanium oleae*, Bern. *Entomologist's Monthly Magazine* 27: 307-308.

Douglas, J.W. 1892a. Notes on some British and exotic Coccidae (No. 22). *Entomologist's Monthly Magazine* 28: 105-107.

Douglas, J.W. 1892b. Notes on some British and exotic Coccidae (No. 23). *Entomologist's Monthly Magazine* 28: 207-209.

Douglas, J.W. 1892c. Notes on some British and exotic Coccidae (No. 24). *Entomologist's Monthly Magazine* 28: 278-280.

Dozier, H.L. 1931. A new giant wax scale from Haiti. *American Museum Novitates. New York* 495: 1-2.

Dufour, L. 1864. Notice sur une nouvelle espèce de gallinsecte (*Aspidiotus? luzulae*). *Annales de la Société Entomologique de France (ser. 4)* 4: 207-209.

Dunham, O. 1954. Contribuição para o conhecimento coccideos dos "Insecta - Homoptera" da Bahia - I. *Boletim do Instituto Biologico da Bahia* 1: 63-74.

Durr, H.J.R. 1954. The male of *Lecanium pumilum* Brain (Hemiptera: Coccidae). *Journal of the Entomological Society of Southern Africa* 17: 90-92.

Dziedzicka, A. 1968. Studies on the morphology and biology of *Lecanium fletcheri* Ckll. (Homoptera, Coccoidea) and related species. *Zoologica Poloniae* 18: 125-165.

Dziedzicka, A. 1977. Comparative investigations on the hindlegs of scale insects (Coccinea). [In Polish, English summary]. *Scientific Publications Krakow Sup. Teachers' College* 20: 1-105.

Dziedzicka, A. 1984. Prof. Dr. Zbigniew Kawecki (1908-1981). *The Scale* 10: 42-44.

Dziedzicka, A. 1988. Contribution to the studies on scale insects (Homoptera, Coccinea) in Poland. [In Polish; Russian and English abstracts]. *Zeszyty Problemowe Postepow Nauk Rolniczych* 353: 93-100.

Dziedzicka, A. & Sermak, W. 1967. The variability of the dorsal-marginal glands in larvae II and females of *Lecanium corni* Bouche of *Taxus baccata* L. [In Polish, English summary]. *Rocznik Naukowo-Dydaktyszny WSP w Krakowie, Prace z Zoologii* 29: 25-31.

Ebeling, W. 1938. Host-determined morphological variations in *Lecanium corni*. *Hilgardia* 11: 613-631.

Ebeling, W. 1959. Subtropical fruit pests. University of California, Los Angeles. 436 pp.

Ehrhorn, E.M. 1898a. New Coccidae. *Canadian Entomologist* 30: 244-246.

Ehrhorn, E.M. 1898b. New Coccidae from California. *Entomological News* 9: 185-186.

Ehrhorn, E.M. 1902. A new coccid from California at a very high altitude. *Canadian Entomologist* 34: 193-194.

Ehrhorn, E.M. 1906. A few new Coccidae, with notes. *Canadian Entomologist* 38: 329-335.

Ehrhorn, E.M. 1912. A few notes on Coccidae. *Proceedings of the Hawaiian Entomological Society* 2: 147-151.

Eisa, A.A., El-Fatah, M.A., El-Nabawi, A. & El-Dash, A.A. 1990. Inhibitory effects of some insect growth regulators on developmental stages, fecundity and fertility of the Florida Wax Scale, *Ceroplastes floridensis. Phytoparasitica* 19: 49-55.

El-Minshawy, A.M. & Moursi, K. 1976. Biological studies on some soft scale-insects (Hom. Coccidae) attacking guava trees in Egypt. *Zeitschrift für Angewandte Entomologie* 81: 363-371.

d'Emmerez de Charmoy, A. 1941. Jean Marie Rose Albert Daruty de Grandpre (1853-1928). pp. 14-15. In: Dictionnaire de biographie Mauricienne, No. 1. Société d'Histoire de l'Île Maurice, Mauritius.

Essig, E.O. 1909. Notes on California Coccidae. *Pomona College Journal of Entomology* 1: 92-97.

Essig, E.O. 1910. Notes on California Coccidae, V. *Pomona College Journal of Entomology* 2: 209-222.

Ezzat, Y.M. & Fayez, S.S. 1980. Differential characters separating all developmental forms of the wax scale *Ceroplastes floridensis* Comstock (Coccoidea, Homoptera). *Abstracts XVI International Congress of Entomology, Kyoto, Japan, August 1980* : 25.

Ezzat, Y.M. & Hussein, N.A. 1969. Redescription and classification of the family Coccidae in U.A.R. (Homoptera: Coccoidea). *Bulletin de la Société Entomologique d'Egypte* (1967) 51: 359-426.

Fabricius, J.C. 1776. *Genera Insectorum* Chilonii, Batchii. 310 pp.

Fabricius, J.C. 1781. Species Insectorum., Hamburgi et Kilonii. 494 pp.

Fabricius, J.C. 1794. Entomologia systematica emendata et acuta. Proft, Hafniae. 472 pp.

Fabricius, J.C. 1798. Supplementum Entomologiae Systematicae. Proft et Storck, Hafniae. 572 pp.

Fallen, C.F. 1814. Specimen novam Hemiptera disponendi methodum exhibens. Lundae. 26 pp.

Fernald, M.E. 1902a. On the genus *Lecanium*. *Canadian Entomologist* 34: 177-178.

Fernald, M.E. 1902b. On the type of the genus *Coccus* L. *Canadian Entomologist* 34: 232-233.

Fernald, M.E. 1903. A catalogue of the Coccidae of the world. *Bulletin of the Hatch Experiment Station of the Massachusetts Agricultural College* 88: 1-360.

Fernald, M.E. 1906. The type of the genus *Coccus*. *Canadian Entomologist* 38: 125-126.

Fernandes, I.M. 1972. Contribuição para o conhecimento de alguns Homoptera Coccoidea do arquipelago de Cabo Verde (2a parte). *Garcia de Orta. Série de Zoologia* 1: 11-16.

Fernandes, I.M. 1973. Alguns Coccoidea (Homoptera) do Arquipélago de Cabo Verde. *Memorias. Junta de Investigação do Ultramar. Lisboa* 58: 251-269.

Fernandes, I.M. 1975. Homoptera (Coccoidea) do Arquipélago de Cabo Verde. *Garcia de Orta. Série de Zoologia. Lisboa* 4: 41-46.

Fernandes, I.M. 1981. Contribuição para o conhecimento da quermofauna do arquipélago dos Acores. *Garcia de Orta. Série de Zoologia. Lisboa* 10: 47-50.

Fernandes, I.M. 1987. Contribuição para o conhecimento da quermofauna da Guiné-Bissau. *Garcia de Orta. Série de Zoologia. Lisboa* 14: 31-37.

Fernandes, I.M. 1989. Contribuição para o conhecimento de Coccoidea (Homoptera) de Angola. *Garcia de Orta. Série de Zoologia. Lisboa* (1988) 15: 129-134.

Fernandes, I.M. 1990. Sobre a presença de *Ceroplastodes zavattarii* Bello, 1939 (Homoptera: Coccidae), na Guinea-Bissau. *Garcia de Orta. Série de Zoologia. Lisboa* (1988) 15: 175-177.

Ferris, G.F. 1919a. A contribution to the knowledge of the Coccidae of Southwestern United States. *Stanford University Publications. Palo Alto. University Series*: 1-68.

Ferris, G.F. 1919b. Notes on Coccidae - III. (Hemiptera). *Canadian Entomologist* 51: 108-113.

Ferris, G.F. 1920a. Notes on Coccidae VI. (Hemiptera). *Canadian Entomologist* 52: 61-65.

Ferris, G.F. 1920b. Scale insects of the Santa Cruz Peninsula. *Stanford University Publications. Palo Alto. Biological Sciences* 1: 1-57.

Ferris, G.F. 1921a. Report upon a collection of Coccidae from Lower California. *Stanford University Publications. Palo Alto. Biological Sciences* 1: 61-132.

Ferris, G.F. 1921b. Some Coccidae from Eastern Asia. *Bulletin of Entomological Research* 12: 211-220.

Ferris, G.F. 1922. Notes on Coccidae - IX. (Hemiptera). *Canadian Entomologist* 54: 156-161.

Ferris, G.F. 1925. Notes on Coccidae XI. (Hemiptera). *Canadian Entomologist* 57: 228-234.

Ferris, G.F. 1935. Scale insects (Hemiptera: Coccoidea) from the Marquesas. *Pacific Entomological Survey, Pub. 8, Art. 9. (Bound in Bulletin of the Bernice P. Bishop Museum, Honolulu, 1939, no. 142)* 142: 125-131.

Ferris, G.F. 1936. Contributions to the knowledge of the Coccoidea (Homoptera). *Microentomology* 1: 2-16.

Ferris, G.F. 1950. Report upon scale insects collected in China (Homoptera: Coccoidea). Part II. (Contribution no. 68). *Microentomology* 15: 69-124.

Ferris, G.F. 1955. Report upon a collection of scale insects from China. Part VI. (Insecta: Homoptera: Coccoidea). (Contribution no. 92). *Microentomology* 20: 30-40.

Ferris, G.F. 1957. A brief history of the study of the Coccoidea. *Microentomology* 22: 39-57.

Ferris, G.F. & Kelly, J.B. 1923. XIV expedition of the California Academy of Sciences to the Gulf of California in 1921. Some Coccidae from about the Gulf of California. *Proceedings of the California Academy of Sciences Fourth Series* 12: 315-318.

Fitch, A. 1851. Catalogue with references and descriptions of the insects collected and arranged for the State Cabinet of Natural History, pp. 43-69. *In:* Fourth annual report of the Regeants of the University, on the condition of the State Cabinet of Natural History, and the historical and antiquarian collection, annexed thereto. Made to the Senate, January 14, 1851. Albany. 146 pp.

Fitch, A. 1857a. 74. - Cherry bark-louse, *Lecanium cerasifex*, new species. (Homoptera. Coccidae). *Third Report on Noxious and Other Insects of the State of New York* 16: 368.

Fitch, A. 1857b. 140. Currant bark louse, *Lecanium ribis*, New species. (Homoptera. Coccidae). *Third Report on Noxious and Other Insects of the State of New York* 16: 426-427.

Fitch, A. 1857c. 147. Gooseberry bark-louse, *Lecanium cynosbati*, new species. (Homoptera. Coccidae). *Third Report on Noxious and Other Insects of the State of New York* 16: 436.

Fitch, A. 1857d. 160. Hickory bark-louse, *Lecanium caryae*, new species. (Homoptera. Coccidae). *Third Report on Noxious and Other Insects of the State of New York* 16: 443.

Fitch, A. 1857e. 201. Hazelnut bark-louse, *Lecanium corylifex*, new species. (Homoptera. Coccidae). *Third Report on Noxious and Other Insects of the State of New York* 16: 473.

Fitch, A. 1859. 307. White oak scale insect, *Lecanium quercifex*, new species. (Homoptera. Coccidae). 308. Quercitron oak scale insect,

Lecanium quercitronis new species. *Fifth Report on Noxious and Other Insects of the State of New York* 18: 805-806.

Fitch, A. 1861. *Lecanium acericorticis*, new species. (Homoptera. Coccidae). *Sixth Report on Noxious and Other Insects of the State of New York* 20: 775.

Flanders, S.E. 1942. Propagation of Black Scale on potato sprouts. *Journal of Economic Entomology* 35: 687-689.

Foldi, I. 1978. Ultrastructure des glandes tégumentaires dorsales, secrétrices de la "laque" chez la femelle de *Coccus hesperidum* L. (Homoptera: Coccidae). *International Journal of Insect Morphology & Embryology* 7: 155-163.

Foldi, I. 1988. Nouvelle contribution à l'étude des cochenilles de l'Amazonie Brésilienne (Homoptera: Coccoidea). *Annales de la Société entomologique de France (N.S.)* 24: 77-87.

Foldi, I. 1991. The wax glands in scale insects: comparative ultrastructure, secretion, function and evolution (Homoptera: Coccoidea). *Annales de la Société Entomologique de France (N.S.)* 27: 163-188.

Foldi, I. & Cassier, P. 1985. Ultrastructure comparée des glandes tégumentaires de treize familles de cochenilles (Homoptera: Coccoidea). *International Journal of Insect Morphology & Embryology* 14: 33-50.

Foldi, I. & Pearce, M.J. 1985. Fine structure of wax glands, wax morphology and function in the female scale insect, *Pulvinaria regalis* Canard (Hemiptera: Coccidae). *International Journal of Insect Morphology & Embryology* 14: 259-271.

Foldi, I. & Soria, S.J. 1989. Les cochenilles nuisibles à la vigne en Amérique du Sud (Homoptera: Coccoidea). *Annales de la Société Entomologique de France (N.S.)* 25: 411-430.

Fonscolombe, L.J.H.B. 1834. Description des *Kermes* qu'on trouve aux environs d'Aix. *Annales de la Société Entomologique de France* 3: 201-218.

Fonseca, J.C. 1953. Contribuição para o estudo do *Coccus hesperidum* L. (Hemiptera Coccoidea). I - Estudo sistematico e morfologico. *Broteria. Lisboa* 22: 5-53, 97-114.

Fonseca, J.P. 1927. Um novo coccideo da jaboticabeira, *Pendularia pendens. Chacaras E Quintaes, São Paulo* 36: 268-270.

Fonseca, J.P. 1929. Um novo genero de coccideo Lecaniinae (Hemipt.). *Revista do Museo Paulista* 16: 849-853.

Fonseca, J.P. 1957. Três novas espécies de coccideos do Brasil, sobre Caffeiro (Homoptera - Coccidae). *Arquivos do Instituto Biológico, São Paulo* 24: 123-135.

Fonseca, J.P. 1962. Contribuição ao conhecimento dos coccideos do Brasil (Homoptera - Coccoidea). *Arquivos do Instituto Biológico, São Paulo* 29: 13-28.

Fonseca, J.P. 1969. Contribuição ao conhecimento dos coccideos do Brasil (Homoptera - Coccoidea). *Arquivos do Instituto Biológico, São Paulo* 36: 9-40.

Fonseca, J.P. 1972a. A cochonilha *Lecanium deltae* (Lizer, 1917) em Citrus, no Brasil. *O Biológico, São Paulo* 38: 213-215.

Fonseca, J.P. 1972b. Um novo gênero de Pulvinariini (Hom. Coccoidea) do Brasil. *Arquivos do Instituto Biológico, São Paulo* 39: 195-199.

Fonseca, J.P. 1973. Contribuição ao conhecimento dos coccideos do Brasil (Homoptera - Coccoidea). *Arquivos do Instituto Biológico, São Paulo* 40: 247-261.

Fonseca, J.P. 1975. Três novas espécies de coccideos do Brasil (Homoptera - Coccoidea). *Arquivos do Instituto Biológico, São Paulo* 42: 79-84.

Fourcroy, A.F. 1785. Entomologia Parisiensis., Paris. 544 pp.

Frediani, D. 1960. Il *Ceroplastes sinensis* Del Guer. vivente su *Pirus communis* L. nella Toscana litoranea. Appunti di Biologia. *Annali della Facolta d'Agraria. Pisa* 21: 89-95.

Fredrick, J.M. 1943. Some preliminary investigations of the green scale, *Coccus viridis* (Green) in South Florida. *Florida Entomologist* 26: 10-15, 25-29.

Froggatt, W.W. 1915. A descriptive catalogue of the scale insects ("Coccidae") of Australia. *Agricultural Gazette of New South Wales* 26: 411-423, 511-516, 603-615, 754-764, 1055-1064.

Froggatt, W.W. 1919. A new species of wax scale *(Ceroplastes murrayi)* from New Guinea. *Proceedings of the Linnean Society of New South Wales* 44: 439-440.

Froggatt, W.W. 1921a. A new mealybug on citrus trees. *(Pulvinaria ornata* n. sp.). *Agricultural Gazette of New South Wales* 32: 427-428.

Froggatt, W.W. 1921b. A descriptive catalogue of the scale insects ("Coccidae") of Australia. Part III.*Department of Agriculture, New South Wales Science Bulletin* 19: 1-43.

Froggatt, W.W. 1923. Forest insects of Australia. Forestry Commissioners, N.S. Wales, Sydney. viii+171 pp.

Froggatt, W.W. 1925. Notes on Australian Coccidae with descriptions of new species. *Proceedings of the Linnean Society of New South Wales* 50: 378-380.

Froggatt, W.W. 1933. The Coccidae of the casuarinas. *Proceedings of the Linnean Society of New South Wales* 58: 363-374.

Fuller, C. 1897a. Coccid literature. *Journal of Western Australia Bureau of Agriculture* 4: 1342-1343.

Fuller, C. 1897b. Some Coccidae of Western Australia. *Journal of Western Australia Bureau of Agriculture* 4: 1344-1346.

Fuller, C. 1899. Notes and descriptions of some species of Western Australian Coccidae. *Transactions of the Entomological Society of London*: 435-473.

Furth, D.G., Ben-Dov, Y. & Gerson, U. 1984. A new species of *Peliococcus* (Homoptera: Pseudococcidae) from the Judean Desert. *Israel Journal of Entomology* (1983) 17: 105-108.

Gascard, A. 1893. Contribution à l'étude des gommes laques des Indes & de Madagascar. Société d'Editions Scientifiques, Paris, 121 pp.

Gennadius, P. 1895. Sur deux nouvelles cochenilles du caroubier dans l'île de Chypre (Hem.). *Annales de la Société Entomologique de France* 64: cclxxvii.

Geoffroy, E.L. 1762. Histoire abrégée des insectes qui se trouvent aux environs de Paris. Durand, Paris. 523 pp.

Georghiou, G.P. 1977. The insects and mites of Cyprus. Benaki Phytopathological Institute, Athens, Kiphissia. 347 pp.

Ghose, S.K. 1961. Studies on some coccids (Coccoidea: Hemiptera) of economic importance of West Bengal, India. *Indian Agriculturist* 5: 57-78.

Giard, A. 1894. [Communications, plusieurs insectes Hémiptères]. *Bulletin de la Société Entomologique de France* (1893): cxcix-cc.

Gilbert, P. 1977. A compendium of the biographical literature on deceased entomologists. British Museum, Natural History, London. 455 pp.

Giliomee, J.H. 1967. Morphology and taxonomy of adult males of the family Coccidae (Homoptera: Coccoidea). *Bulletin of the British Museum (Natural History) Entomology, Supplement* 7: 1-168.

Gill, R.J. 1979. A new species of *Pulvinaria* Targioni Tozzetti (Homoptera: Coccidae) attacking ice plant in California. *Pan-Pacific Entomologist* 55: 241-250.

Gill, R.J. 1988. The scale insects of California Part 1. The soft scales (Homoptera: Coccoidea: Coccidae). California Department of Food and Agriculture, Sacramento, California., Sacramento. 132 pp.

Gill, R.J., Nakahara, S. & Williams, M.L. 1977. A review of the genus *Coccus* Linnaeus in America north of Panama (Homoptera: Coccoidea: Coccidae). *Occasional Papers in Entomology State of California Department of Food and Agriculture* 24: 44.

Gimmingham, C.T. 1934. The male *Lecanium corni* Bouché. *Entomologist's Monthly Magazine* 70: 41-42.

Gimpel, W.F., Miller, D.R. & Davidson, J.A. 1974. A systematic revision of the wax scales, genus *Ceroplastes*, in the United States (Homoptera; Coccoidea; Coccidae). *University of Maryland, Agricultural Experiment Station, Miscellaneous Publication* 841: 1-85.

Gmelin, G.F. 1790. Systema Naturae. Editio decima tertia. Acuta, reformata, cara., Lipsiae. pp. 1517-2224.

Goethe, R. 1884. Beobachtungen über Schildläuse und deren Feinde, angestellt an obstbäumen und Reben in Rheingau. *Jahrbücher des Nassauischen Vereins für Naturkunde* 37: 107-131.

Gomes Costa, R. 1949. Cochonilhas ou coccideas do Rio Grande do Sul. Secretaria de Estado dos Negocias da Agricultura, Industria e Comercio Secção de Informacodes. Rio Grande do Sul, Brasil, Porto Alegre. 107 pp.

Gómez-Menor Ortega, J. 1928. Estudios sobre coccidos de España. *EOS* 4: 339-362.

Gómez-Menor Ortega, J. 1937. Cóccidos de España. Universidad de Madrid, Madrid. 432 pp.

Gómez-Menor Ortega, J. 1941. Cóccidos de la Republica Dominicana (Hom. Cocc.). *EOS* 16: 125-143.

Gómez-Menor Ortega, J. 1946. Adiciones a los "Cóccidos de España". 1a nota. *EOS* 22: 59-106.

Gómez-Menor Ortega, J. 1948. Adiciones a los "Cóccidos de España. 2a nota. *EOS* 24: 73-121.

Gómez-Menor Ortega, J. 1954. Adiciones a los "Cóccidos de España". (Tercera nota). *EOS* 30: 119-148.

Gómez-Menor Ortega, J. 1955. Nueva forma de *Coccus (Saissetia) oleae. Boletín de Patología Vegetal y Entomología Agrícola. Madrid* 21: 205-208.

Gómez-Menor Ortega, J. 1957. Adiciones a los cóccidos de España (cuarta nota). *EOS* 33: 39-86.

Gómez-Menor Ortega, J. 1958a. Dos Lecaniidae (Coccoidea), nuevos para la fauna española. *Boletín de la Real Sociedad Española de Historia Natural. Madrid (Seccion Biologica)* 56: 21-34.

Gómez-Menor Ortega, J. 1958b. Distribucion geografica y ensayo de la ecologica de los cóccidos en España. *Publicado del Instituto Biologia Apllicata Barcelona* 27: 5-15.

Gómez-Menor Ortega, J. 1958c. Cochinillas que atacen a los frutales (Homoptera, Coccoidea: II. Familias Lecanidae y Margarodidae). *Boletín de Patología Vegetal y Entomología Agrícola. Madrid* 23: 43-173.

Gómez-Menor Ortega, J. 1965. Adiciones a los "Cóccidos de España", VI nota. Especies del genero *Evallaspis* con su distribucion geografica en la Peninsula a Islas Baleares. *EOS* 41: 87-114.

Gómez-Menor Ortega, J. 1967. Lista de Coccoidea de las Islas Canarias (adiciones) (Hemiptera, Homoptera). *EOS* 43: 131-134.

Gómez Menor Ortega, J. 1960. Adiciones a los "Cóccidos de España". V. (Superfamilia Coccoidea). *EOS* 36: 157-204.

González, R.H. 1989. Insectos y acaros de importancia agrícola y cuarentenaria en Chile. Universidad de Chile, Santiago. 310 pp.

González, R.H. & Charlin, R. 1968. Nota preliminar sobre los insectos coccideos de Chile. *Revista Chilena de Entomología* 6: 109-113.

González, R.H. & Lamborot, L. 1989. El genero *Saissetia* Deplanche en Chile (Homoptera: Coccidae). *Acta Entomologica Chilena* 15: 237-242.

González Hernandes, H. & Atkinson. T.H. 1984. Coccidos (Homoptera: Coccoidea) asociados a arboles frutales de la region central de Mexico. *Agrociencia, Mexico* 57: 207-225.

Goureau, C. 1869. Les insectes nuisibles aux arbustes et aux plantes de parterre. V. Masson et Fils, Paris. 139 pp.

Goury, G. 1905. Notes spéciales et locales. *Lecanium limnanthemi.* (?). *La Feuille des Jeunes Naturalistes (ser.* 4) 35(412): 62.

Goux, L. 1933. Notes sur les coccides de la France (5e note). Etude d'une espèce nouvelle constituant un genre nouveau. *Bulletin de la Société Entomologique de France* 38: 119-123.

Goux, L. 1934. Notes sur les coccides (Hem.) de la France (7e note). Etude d'un *Filippia* nouveau de Provence. *Bulletin de la Société Entomologique de France* (1933) 38: 321-325.

Goux, L. 1937. Notes sur les coccides (Hem.) de la France (18e note). Six espèces nouvelles pour la France et remarques sur le genre *Lecanopsis, Exaeretopus et Luzulaspis. Bulletin de la Société Entomologique de France* 1937: 93-96.

Goux, L. 1939. Notes sur les coccides (Hem. Coccidae) de la France (27e note). Contribution à la connaissance des *Luzulaspis* avec description de deux especes nouvelles. *Bulletin de la Société Linnéenne de Provence* 12: 63-76.

Gowdey, C.C. 1917. A list of Uganda Coccidae, their food-plants and natural enemies. *Bulletin of Entomological Research* 8: 187-189.

Grandpré, A.D. & Charmoy, D.E. 1899. List raisonée des cochenilles de l'Île Maurice. Planters and Commercial Gazette, Maurice. 48pp.

Granovsky, A.A. 1929. Preliminary studies of the intracellular symbionts of *Saissetia oleae* (Bernard). *Transactions of the Wisconsin Academy of Sciences, Arts and Letters* 29: 445-456.

Gray, E.J. 1828. Spicilegia Zoologica; or original figures and short systematic descriptions of new and unfigured animals. Treuttel, Wurtz & Co., London. 12 pp.

Green, E.E. 1886. Observations on the Green scale bug in connection with the cultivation of coffee. Eton, Pundaluoya. 4 pp.

Green, E.E. 1889. Descriptions of two new species of *Lecanium* from Ceylon. *Entomologist's Monthly Magazine* 25: 248-250.

Green, E.E. 1896. Catalogue of Coccidae collected in Ceylon. *Indian Museum Notes* 4: 2-10.

Green, E.E. 1897. Notes on Coccidae from the Royal Gardens, Kew. *Entomologist's Monthly Magazine* 33: 68-77.

Green, E.E. 1899. Observations on some species of Coccidae of the genus *Ceroplastes* in the collection of the British Museum. *Annals and Magazine of Natural History (ser. 7)* 4: 188-192.

Green, E.E. 1900a. Note on *Ceroplastes africanus* (Family Coccidae). *Annals and Magazine of Natural History (ser. 7)* 5: 158.

Green, E.E. 1900b. Descriptions of new Victorian Coccidae. *The Victorian Naturalist* 17: 9-14.

Green, E.E. 1900c. Remarks on Indian scale insects (Coccidae), with descriptions of new species. *Indian Museum Notes* 5: 1-13.

Green, E.E. 1900d. Observations on some species of Coccidae collected by Mr James Lidgett in Victoria, Australia. *Annals and Magazine of Natural History (ser. 7)* 6: 448-452.

Green, E.E. 1903. Remarks on Indian scale insects (Coccidae). With descriptions of new species. Pt. II. *Indian Museum Notes* 5: 93-103.

Green, E.E. 1904a. Descriptions of some new Victorian Coccidae. *The Victorian Naturalist* 21: 65-69.

Green, E.E. 1904b. On some Javanese Coccidae: with descriptions of new species. *Entomologist's Monthly Magazine* 40: 204-210.

Green, E.E. 1904c. On some Coccidae in the collection of the British Museum. *Annals and Magazine of Natural History (ser.* 7) 14: 373-378.

Green, E.E. 1904d. The Coccidae of Ceylon. Pt. III. Dulau & Co., London. pp. 171-249.

Green, E.E. 1905. On some Javanese Coccidae: with descriptions of new species. *Entomologist's Monthly Magazine* 41: 28-33.

Green, E.E. 1907. Notes on the Coccidae collected by the Percy Sladen Trust Expedition to the Indian Ocean: supplemented by a collection received from Mr. R. Dupont, Director of Agriculture, Seychelles. *Transactions of the Linnean Society of London, Zoology* 12: 197-207.

Green, E.E. 1908. Remarks on Indian scale insects (Coccidae), Part II. With a catalogue of all species hitherto recorded from the Indian continent. *Memoirs of the Department of Agriculture in India, Entomology Series* 2: 15-46.

Green, E.E. 1909. The Coccidae of Ceylon. IV. Dulau & Co., London. pp. 250-344.

Green, E.E. 1911. On some Coccidae affecting rubber trees in Ceylon with descriptions of new species. *Journal of Economic Biology* 6: 27-36.

Green, E.E. 1913. Remarks on Coccidae collected by Mr. Edward Jacobson, of Samarang, Java, with descriptions of two new species. *Tijdschrift voor Entomologie* (1912) 55: 311-318.

Green, E.E. 1914. Remarks on a small collection of Coccidae from Northern Australia. *Bulletin of Entomological Research* 5: 231-234.

Green, E.E. 1915a. Observations on British Coccidae in 1914, with descriptions of new species. *Entomologist's Monthly Magazine (ser.* 3) 1: 175-185.

Green, E.E. 1915b. On a new species of *Lecanium* from Northern Nigeria. *Bulletin of Entomological Research* 6: 43.

Green, E.E. 1915c. Notes on Coccidae collected by F.P. Jepson, Government Entomologist, Fiji. *Bulletin of Entomological Research* 6: 44.

Green, E.E. 1915d. New species of Coccidae from Australia. *Bulletin of Entomological Research* 6: 45-53.

Green, E.E. 1916a. On two new British Coccidae, with notes on some other British species. *Entomologist's Monthly Magazine* 52: 23-31.

Green, E.E. 1916b. Report on some Coccidae from Zanzibar, collected by Dr. W.M. Aders. *Bulletin of Entomological Research* 6: 375-376.

Green, E.E. 1916c. On a new coccid pest of cacao from Trinidad. *Bulletin of Entomological Research* 6: 377-379.

Green, E.E. 1916d. Observations on some recently described Coccidae. *Bulletin of Entomological Research* 7: 51-52.

Green, E.E. 1916e. Remarks on Coccidae from Northern Australia - II. *Bulletin of Entomological Research* 7: 53-65.

Green, E.E. 1917. Observations on British Coccidae; with descriptions of new species. *Entomologist's Monthly Magazine* 53: 201-210, 260-269.

Green, E.E. 1918. Some remarks on Mr. Kuhni Kannan's Paper, "an instance of mutation". *Transactions of the Entomological Society of London* (1918) : 149-154.

Green, E.E. 1920. Observations on British Coccidae. No. V. *Entomologist's Monthly Magazine* (ser. 3) 56: 114-130.

Green, E.E. 1921a. Observations on British Coccidae with descriptions of new species. No. VI. *Entomologist's Monthly Magazine* 57: 146-152, 189-200.

Green, E.E. 1921b. Observations on British Coccidae: with descriptions of new species. - VII. *Entomologist's Monthly Magazine* 57: 257-259.

Green, E.E. 1921c. On a new genus of Coccidae from the Indian region. *Annals and Magazine of Natural History (ser.* 9) 7: 639-644.

Green, E.E. 1922a. The Coccidae of Ceylon, Part V. Dulau & Co., London. pp. 345-472.

Green, E.E. 1922b. Supplementary notes on the Coccidae of Ceylon. *Journal of the Bombay Natural History Society* 28: 1007-1037.

Green, E.E. 1923a. Observations on the Coccidae of the Madeira Islands. *Bulletin of Entomological Research* 14: 87-97.

Green, E.E. 1923b. Observations on British Coccidae. - VIII. *Entomologist's Monthly Magazine* 59: 211-218.

Green, E.E. 1924. On some new species of Coccidae from various sources. *Bulletin of Entomological Research* 15: 41-48.

Green, E.E. 1925a. Observations on British Coccidae, IX. *Entomologist's Monthly Magazine* 61: 34-44.

Green, E.E. 1925b. Notes on the Coccidae of Guernsey (Channel Islands), with descriptions of some new species. *Annals and Magazine of Natural History (ser.* 9) 16: 516-527.

Green, E.E. 1926a. On some new genera and species of Coccidae. *Bulletin of Entomological Research* 17: 55-65.

Green, E.E. 1926b. Observations on British Coccidae. X. *Entomologist's Monthly Magazine* 62: 172-183.

Green, E.E. 1927a. A brief review of the indigenous Coccidae of the British Islands, with emendations and additions. *Entomologist's Record and Journal of Variation. London* 39: 1-4.

Green, E.E. 1927b. Notes on the Coccidae of Scotland. *The Scottish Naturalist* 163: 25-30; 164: 55-59.

Green, E.E. 1928a. A brief review of the indigenous Coccidae of the British Islands, with emendations and additions (to February,

1928). *Entomologist's Record and Journal of Variation. London* 40: 1-14.

Green, E.E. 1928b. Observations on British Coccidae. XI. With descriptions of new species. *Entomologist's Monthly Magazine* 64: 20-31.

Green, E.E. 1929. Some Coccidae collected by Dr. J.G. Myers in New Zealand. *Bulletin of Entomological Research* 19: 369-389.

Green, E.E. 1930a. Observations on British Coccidae, XII. *Entomologist's Monthly Magazine* 66: 9-17.

Green, E.E. 1930b. Notes on some Coccidae collected by Dr. Julius Melzer, at São Paulo, Brazil (Rhynch.). *Stettiner Entomologische Zeitung* 91: 214-219.

Green, E.E. 1930c. Fauna Sumatrensis (Bijdrage Nr. 65). Coccidae. *Tijdschrift voor Entomologie* 73: 279-297.

Green, E.E. 1931. Observations on British Coccidae. XIII. *Entomologist's Monthly magazine* 67: 99-106.

Green, E.E. 1933. Notes on some Coccidae from Surinam, Dutch Guiana, with descriptions of new species. *Stylops* 2: 49-58.

Green, E.E. 1934. Observations on British Coccidae, XIV. *Entomologist's Monthly Magazine* 70: 108-114.

Green, E.E. 1935a. On a species of *Ceroplastes* (Hem. Coccidae), hitherto confused with *C. ceriferus* Anders. *Stylops* 4: 180.

Green, E.E. 1935b. On three new species of *Ceroplastes*, from South America. *Arbeiten uber morphologische und taxonomische Entomologie aus Berlin-Dahlem* 2: 272-275.

Green, E.E. 1937. An annotated list of the Coccidae of Ceylon, with emendations and additions to date. *Ceylon Journal of Science Section B. Zoology and Geology* 20: 277-341.

Green, E.E. & Laing, F. 1924. Descriptions of some apparently new non-Diaspidine Coccidae. *Bulletin of Entomological Research* 14: 415-419.

Green, E.E. & Mamet, R. 1938. A new coccid (Hemipt. Homopt.) from Mauritius. *Proceedings of the Royal Entomological Society of London* 7: 126.

Green, E.E. & Mann, H.H. 1907. The Coccidae attacking the tea plant in India and Ceylon. *Memoirs of the Department of Agriculture in India, Entomological Series* 1: 337-355.

Guérin-Méneville, F.E. 1858. [Communication sur un insecte, qui, en Chine, produit de la cire]. *Bulletin de la Société Entomologique de France* 16: lxvii.

Guérin-Méneville, F.E. 1869. Etudes sur les insectes considérés comme la cause de la maladie des cannes à sucre dans des îles Maurice et de la Réunion. *Annales de la Société Entomologique de France (ser. 4)* 9: 89-92.

Habib, A. 1955a. On the species of *Eulecanium* Ckll. (Homoptera, Coccoidea, Coccidae) on *Taxus baccata* L. *Entomologist's Monthly Magazine* 91: 70-72.

Habib, A. 1955b. Some biological aspects of the *Eulecanium corni* Bouché - group (Hemiptera: Coccidae). *Bulletin de la Société Entomologique d'Egypte* 39: 217-228.

Habib, A. 1955c. The behaviour of the nymphal stages of *Eulecanium corni* Bouché (Hemiptera: Coccidae). *Bulletin de la Société Entomologique d'Egypte* 39: 229-249.

Habib, A. 1956a. The male *Eulecanium corni* Bouché (Hemiptera - Homoptera: Coccoidea - Coccidae). *Bulletin de la Société Entomologique d'Egypte* 40: 119-126.

Habib, A. 1956b. *Eulecanium taxi* nov. spec. (Homoptera: Coccoidea - Coccidae). *Bulletin de la Société Entomologique d'Egypte* 40: 453-462.

Habib, A. 1957. The morphology and biometry of the *Eulecanium corni* - group, and its relation to host-plants (Hemiptera - Homoptera: Coccoidea). *Bulletin de la Société Entomologique d'Egypte* 41: 381-410.

Hackman, R.H. 1951. The chemical composition of the wax of the white wax scale, *Ceroplastes destructor* (Newstead). *Archives of Biochemistry and Biophysics* 33: 150-154.

Hackman, R.H. & Trikojus, V.M. 1952. The composition of the honeydew excreted by Australian coccids of the genus *Ceroplastes*. *The Biochemical Journal* 51: 653-656.

Hadzibejli, Z.K. 1955. New genus and species of the soft scales family Lecaniidae (Homoptera, Coccoidea) from Georgia. [In Russian]. *Entomologicheskoe Obozrenye* 34: 231-239.

Hadzibejli, Z.K. 1960a. Studies on the species composition of grape-vines scale insects in Georgia. [In Georgian, Russian summary]. *Trudy Instituta Zashchita Rastenii, Akademii Sel'skokhozyayst-vennykh Nauk Gruzinskoy SSR* 13: 46-62.

Hadzibejli, Z.K. 1960b. New species of coccids (Homoptera, Coccoidea) from Georgia. [In Russian]. *Trudy Akademii Nauk Gruzinskoy SSR Instituta Zashchitii Rastenii* 13: 299-321.

Hadzibejli, Z.K. 1967a. Two new coccid species (Insecta, Homoptera, Coccoidea) of the fauna of Eastern Transcaucasia. [In Russian, Georgian abstract]. *Soobshcheniya Akademii Nauk Gruzinskoy SSR. Tbilisi* 46 (3): 715-720.

Hadzibejli, Z.K. 1967b. Ecological characteristics of the genus *Eulecanium* Ckll. in the fauna of Gruzia. [In Russian]. *Trudy Instituta Zashchita Rastenii Gruzinskoy SSR* 19: 59-63.

Hadzibejli, Z.K. 1971. A new species of soft scale in the fauna of Kazakhstan. [In Russian, Georgian and English abstracts]. *Soobshcheniya Akademii Nauk Gruzinskoy SSR. Tbilisi* 62 (2): 449-452.

Hadzibejli, Z.K. 1973. Little-known species of the family Coccidae (Homoptera, Coccoidea) in the fauna of eastern Transcaucasia. [In Russian]. *Entomologicheskoe Obozrenye* 52: 835-844.

Hadzibejli, Z.K. 1977. Biology, morphology and trophical forms of some species of scale insects from the tribe Pulvinariini (Homoptera, Coccoidea) in Georgia. [In Russian]. *Entomologicheskoe Obozrenye* 56: 546-550.

Hadzibejli, Z.K. 1983. Coccids of the subtropical zone of Gruzia. [In Russian, English summary]. Metsniereba, Tbilisi. 293 pp.

Hafez, M., Salama, H.S. & Saleh, M.R. 1971. Survival and development of *Lecanium acuminatum* Sign. (Coccoidea) on a host plant and artificial diets. *Zeitschrift für Angewandte Entomologie* 69: 182-186.

Hagen, H.A. 1862. Bibliotheca Entomologica. W. Engelmann, Leipzig. 512 pp.

Halais, L. & d'Emmerez de Charmoy, A. 1941. Emmerez de Charmoy, Paul Donald d' (1873-1930). pp. 44-45. In: Dictionnaire de biographie Mauricienne, No. 1. Société d'Histoire de l'Île Maurice, Mauritius.

Hall, W.J. 1922. Observations on the Coccidae of Egypt. *Ministry of Agriculture, Egypt, Technical and Scientific Service Bulletin* 22: 1-54.

Hall, W.J. 1923. Further observations on the Coccidae of Egypt. *Bulletin Ministry of Agriculture, Egypt, Technical and Scientific Service* 36: 1-67.

Hall, W.J. 1925. Notes on Egyptian Coccidae with descriptions of new species. *Bulletin Ministry of Agriculture, Egypt, Technical and Scientific* 64: 1-31.

Hall, W.J. 1926. Contribution to the knowledge of the Coccidae of Egypt. *Ministry of Agriculture, Egypt, Technical and Scientific Service, Bulletin* 72: 1-41.

Hall, W.J. 1931. Observations on the Coccidae of southern Rhodesia. *Transactions of the Entomological Society of London* 79: 285-303.

Hall, W.J. 1932. Observations on the Coccidae of southern Rhodesia. *Stylops* 1: 185-195.

Hall, W.J. 1935. Observations on the Coccidae of southern Rhodesia. - VI. *Stylops* 4: 73-84.

Hall, W.J. 1937. Observations on the Coccidae of southern Rhodesia. *Transactions of the Royal Entomological Society of London* 86: 119-134.

Hall, W.J. 1939. A new genus and four apparently new species of Coccidae (Homoptera) from the Union of South Africa. *Journal of the Entomological Society of southern Africa* 2: 93-100.

Hall, W.J. 1941. On some new species and two new genera of Coccidae (Homoptera) from southern Rhodesia. *Journal of the Entomological Society of southern Africa* 4: 221-239.

Halstead, A.J. 1982. Foodplants of *Pulvinaria regalis* Canard (Hem: Coccidae) in England. *Proceedings and Transactions of the British Entomological and Natural History Society. London* 15: 46.

Hamon, A.B. & Williams, M.L. 1984. Arthropods of Florida and neighboring land areas. Vol. 11. The soft scales of Florida (Homoptera: Coccoidea: Coccidae). Florida Department of Agriculture & Consumer Services. Contribution no. 600. Florida Department of Agriculture, Gainesville. 194 pp.

Hanford, L. 1974. The African scale insect genus *Udinia* De Lotto (Coccidae). *Transactions of the Royal Entomological Society of London* 126: 1-40.

Harpaz, I. 1984. Frederick Simon Bodenheimer (1897-1959): Idealist, Scholar, Scientist. *Annual Review of Entomology* 29: 1-23.

Hart, W.G. & Ingle, S. 1971. Increases in fecundity of brown soft scale exposed to Methyl Parathion. *Journal of Economic Entomology* 63: 204-208.

Hauser, B. 1972. Leo Zehntner: La saga d'un savant suisse. *Revue Musées de Genève* 127: 2-5.

Haworth, A.H. 1812. Observations on the *Coccus vitis*; with remarks on some other insects of that destructive genus. *Transactions of the Entomological Society of London* 1: 297-309.

Heidel, W. & Köhler 1979. *Tourneyella cubensis* sp. n. (Hemiptera: Coccinea - Coccidae) - eine Schildlaus in kubanischen Zitruskulturen. *Zoologischer Anzeiger, Jena* 202: 132-144.

Hempel, A. 1899. Two new Coccidae of the subfamily Lecaninae. *Canadian Entomologist* 31: 131-133.

Hempel, A. 1900a. Descriptions of three new species of Coccidae from Brazil. *Canadian Entomologist* 32: 3-7.

Hempel, A. 1900b. As coccidas Brazileiras. *Revista do Museo Paulista. São Paulo* 4: 365-537.

Hempel, A. 1901a. Descriptions of Brazilian Coccidae. *Annals and Magazine of Natural History (ser. 7)* 7: 110-125, 206-219, 556-561.

Hempel, A. 1901b. Descriptions of Brazilian Coccidae. *Annals and Magazine of Natural History (ser. 7)* 8: 62-72, 100-111.

Hempel, A. 1901c. A preliminary report on some new Brazilian Hemiptera. *Annals and Magazine of Natural History (ser. 7)* 8: 383-391.

Hempel, A. 1904. Resultado do exame de diversas collecçõe de coccidas enviadas ao Instituto Agronomico pelo Sr. Carlos Moreira, do Museu Nacional, Rio de Janeiro. *Boletim da Agricultura, São Paulo (ser. 5)* 1: 311-323.

Hempel, A. 1912. Catalogos da fauna Brazileira editados pello Museu Paulista S. Paulo - Brazil. Diario Official, São Paulo. 77 pp.

Hempel, A. 1918. Descripção de sete novas especies de coccidas. [In Portuguese, English translation]. *Revista do Museu Paulista. São Paulo* 10: 193-208.

Hempel, A. 1920a. Coccidas que infestam as nossas arvores fructiferas. *Revista do Museu Paulista. São Paulo* 12: 109-143.

Hempel, A. 1920b. Descripçeõs de coccidas novas e pouco conhecidas. [In Portuguese, English translation]. *Revista do Museu Paulista. São Paulo* 12: 331-377.

Hempel, A. 1921. Tres novas coccideos. *Archivos da Escuola Superior Agricultura a Medicinal Veterinaria. Rio de Janeiro* 5: 143-146.

Hempel, A. 1928. Descripçõe de novas especies de pulgoes (Homoptera, Coccidae). *Archivos do Instituto Biologico. São Paulo* 1: 235-237.

Hempel, A. 1929. Descripçõe de pulgoes novos e pouco conhecidos (Homoptera, Coccidae). *Archivos do Instituto Biologico. São Paulo* 2: 61-66.

Hempel, A. 1932. Descripção de vinte a duas especies novas de coccideos (Hemiptera - Homoptera). *Revista de Entomologia* 2: 310-339.

Hempel, A. 1934. Descripção de tres especies novas, tres generos novas e uma sub-familia nova de coccideos (Hemiptera, Homoptera). *Revista de Entomologia* 4: 139-147.

Hempel, A. 1937. Novas especies de coccideos (Homoptera) do Brasil. *Archivos do Instituto Biologico. São Paulo* 8: 5-36.

Hempel, A. 1938. Descripção de uma nova especie de *Ceroplastes* (Hom. Coccidae). *Revista de Entomologia* 8: 263-264.

Herberg, M. 1916. Die Schildlaus *Eriopeltis lichtensteini* Sign. *Archiv für Naturgeschichte* 82: 1-107.

Hodgson, C.J. 1967a. Notes on Rhodesian Coccidae (Homoptera: Coccoidea): Part 1: The genera *Coccus, Parasaissetia, Saissetia* and a new genus *Mashona*. *Arnoldia (Rhodesia)* 3 (5): 1-22.

Hodgson, C.J. 1967b. Some *Pulvinaria* species (Homoptera: Coccidae) of the Ethiopian region. *Journal of the Entomological Society of Southern Africa* 30: 198-211.

Hodgson, C.J. 1967c. A revision of the species of the genus *Ceronema* Maskell (Homoptera: Coccoidea) recorded from the Ethiopian Region. *Arnoldia (Rhodesia)* 3 (22): 1-8.

Hodgson, C.J. 1967d. A revision of the species of *Inglisia* Maskell (Homoptera: Coccoidea) recorded from the Ethiopian region. *Arnoldia (Rhodesia)* 3 (23): 1-11.

Hodgson, C.J. 1968a. Further notes on the genus *Pulvinaria* Targ. (Homoptera: Coccoidea) from the Ethiopian region. *Journal of the Entomological Society of Southern Africa* 31: 141-174.

Hodgson, C.J. 1968b. Four new species and a new genus of Coccidae (Homoptera: Coccoidea) from Africa. *Proceedings of the Royal Entomological Society of London (B)* 37: 114-120.

Hodgson, C.J. 1969a. Notes on Rhodesian Coccidae (Homoptera: Coccoidea): Part II: The genera *Ceroplastes* and *Gascardia*. *Arnoldia (Rhodesia)* 4 (3): 1-43.

Hodgson, C.J. 1969b. Notes on Rhodesian Coccidae (Homoptera: Coccoidea): Part III. *Arnoldia (Rhodesia)* 4 (4): 1-45.

Hodgson, C.J. 1969c. The status of *Hemilecanium imbricans* (Green) (Homoptera: Coccoidea) in Africa south of the Sahara. *Journal of Natural History* 3: 321-327.

Hodgson, C.J. 1970a. A new species of *Coccus* (Homoptera: Coccoidea) from Malawi. *Entomologist's Monthly Magazine* 106: 35-37.

Hodgson, C.J. 1970b. Status of *Pulvinaria vayssierei* Castel-branco (Homoptera: Coccoidea). *Journal of the Entomological Society of Southern Africa* 33: 279-283.

Hodgson, C.J. 1971. The species assigned to the genus *Ceroplastodes* (Homoptera: Coccoidea) in the Ethiopian region. *Journal of Entomology (B)* 40: 49-61.

Hodgson, C.J. 1973. A new name to replace *Mashona* Hodgson, 1967 (Hem. Coccoidea). *Entomologist's Monthly Magazine* 109: 63.

Hodgson, C.J. 1990. The scale insect genus *Houardia* Marchal (Homoptera: Coccidae). *Systematic Entomology* 15: 219-226.

Hodgson, C.J. 1991a. A revision of the scale insect genera *Etiennea* and *Platysaissetia* (Homoptera: Coccidae) with particular reference to Africa. *Systematic Entomology* 16: 173-221.

Hodgson, C.J. 1991b. A redescription of *Pseudopulvinaria sikkimensis* Atkinson (Homoptera, Coccoidea), with a discussion of its affinities. *Journal of Natural History* 25: 1513-1529.

Hodgson, C.J. & Hilburn, D.J. 1991a. An annotated checklist of the Coccoidea of Bermuda. *Florida Entomologist* 74: 133-146.

Hodgson, C.J. & Hilburn, D.J. 1991b. List of plant hosts of Coccoidea recorded in Bermuda up to 1989. *Bulletin of the Department of Agriculture, Fisheries & Parks, Botanical Gardens, Bermuda* 39: 1-22.

Hofer, J. 1903. Beitrag zur Cocciden-Fauna der Schweiz. *Mitteilungen der Schweizerischen Entomologischen Gesellschaft* 10: 474-483.

Hollinger, A.H. 1917c. Taxonomic value of antennal segments of certain Coccidae. *Annals of the Entomological Society of America* 10: 264-271.

Hollinger, A.H. 1923. Scale insects of Missouri. *Research Bulletin. Missouri Agricultural Experiment Station. Columbia* 58: 1-71.

Hosny, M. 1939. On coccids found on roots of plants in Egypt. *Ministry of Agriculture, Egypt. Technical and Scientific Service, Entomological Section, Bulletin* 237: 1-21.

Houser, J.S. 1918. The Coccidae of Cuba. *Annals of the Entomological Society of America* 11: 157-172.

Hoy, J.M. 1963. A catalogue of the Eriococcidae (Homoptera: Coccoidea) of the world. *New Zealand Department of Scientific and Industrial Research Bulletin* 150: 1-260.

Hunter, S.J. 1899a. The Coccidae of Kansas. *Kansas University Quarterly (Ser. A)* 8: 1-15.

Hunter, S.J. 1899b. The Coccidae of Kansas, II. *Kansas University Quarterly (Ser. A)* 8: 67-77.

Hunter, S.J. 1900. Coccidae of Kansas, III. *Kansas University Quarterly (Ser. A)* 9: 101-107.

Hunter, S.J. 1902. Coccidae of Kansas, IV. Additional species, food plants and bibliography of Kansas Coccidae, with appendix on other species reported from Kansas. *Kansas University Quarterly (Ser. A)* 10: 107-145.

Husseiny, M.M. & Madsen, H.F. 1962. The life history of *Lecanium kunoensis* Kuwana (Homoptera: Coccidae). *Hilgardia* 33: 179-203.

Ihering, H.V. 1897. Os piolhos vegetaes (Phytophthires) do Brazil. *Revista do Museu Paulista. São Paulo* 2: 385-420.

Inserra, S. 1970. Il *Ceroplastes rusci* L. negli agrumeti della provincia di Catania. *Bolletino del Laboratorio di Entomologia Agraria "Filippo Silvestri" di Portici* 28: 77-97.

International Code. 1985. International Code of Zoological Nomenclature. Third Edition. International Trust for Zoological Nomenclature, London. 338 pp.

Ishaaya, I. & Swirski, E. 1976. Trehalase, invertase and amylase activities in the black scale, *Saissetia oleae*, and their relation to host adaptability. *Journal of Insect Physiology* 22: 1025-1029.

Jancke, G.D. 1955. Zur Morphologie der männlichen Cocciden. *Zeitschrift für Angewandte Entomologe* 37: 265-314.

Jhala, R.C., Patel, Z.P. & Shah, A.H. 1989. Occurrence of some new insect pests on Mango in India. *Gujarat Agricultural University, Research Journal, India* 14: 1.

Ishii, T. 1935. Insects of Jehol [v] - Order: Lepidoptera (II) & Hemiptera. *Gujarat Agricultural University, Research Journal, India* 14: 1.

Johnson, W.T. & Lyon, H.H. 1988. Insects that feed on trees and shrubs. Cornell University Press, Ithaca and London. 556 pp.

Joubert, C.J. 1925. Five apparently new species of South African Coccidae. *South African Journal of Natural History. Pretoria* 5: 119-124.

Kajita, H. 1964. A revised list of host plants of *Ceroplastes pseudoceriferus* Green with a preliminary study on its mass culture. [In Japanese, English summary]. *Science Bulletin of the Faculty of Agriculture, Kyushu University* 21: 1-6.

Kajita, H. 1965. Some visible characters of parasitized *Ceroplastes pseudoceriferus* Green. [In Japanese, English summary]. *Science Bulletin of the Faculty of Agriculture, Kyushu University* 22: 29-34.

Kaltenbach, J.H. 1874. Die Pflanzenfeinde aus der Klasse der Insekten. Julius Hoffmann, Stuttgart. 560 pp.

Kamburov, S. 1987. The mango shield scale *Protopulvinaria mangiferae* (Green) (Homoptera: Coccidae), a new pest on mango *(Mangifera indica). Citrus & Subtropical fruit Journal* 635: 10-11.

Kanda, S. 1934. A new species of *Eulecanium* (Coccidae). *Annotationes Zoologicae Japonenses* 14: 405-408.

Kanda, S. 1950. Some scale-insects from Shansi, North China. [In Japanese, English summary]. *Mushi* 21: 33-39.

Kanda, S. 1960. Descriptions of the Coccidae from Japan (Homoptera). *Kontyu* 28: 116-123.

Kannan, K.K. 1918. An instance of mutation: *Coccus viridis*, Green, a mutant from *Pulvinaria psidii*, Maskell. *Transactions of the Entomological Society of London* (1918): 130-148.

Kawai, S. 1972. Diagnostic notes and biology of the coccid species occurring on cultivated or wild trees and shrubs in Japan (Homoptera: Coccoidea). [In Japanese]. *Bulletin of the Tokyo-To Agricultural Experiment Station* 6: 1-54.

Kawai, S. 1980. Scale insects of Japan in colours. National Agriculture Education Association, Tokyo. 455 pp.

Kawai, S., Matsubara, Y. & Umesawa, K. 1971. A preliminary revision of the Coccoidea-fauna of the Ogasawara (Bonin) Islands (Homoptera: Coccoidea). *Applied Entomology and Zoology, Tokyo* 6: 11-26.

Kawai, S. & Tamaki, Y. 1967. Morphology of *Ceroplastes pseudoceriferus* Green with special reference to the wax secretion. *Applied Entomology and Zoology* 2: 133-146.

Kawecki, Z. 1936. Materjaly do poznania fauny czerwcow (Coccidae) Podola, Opola i Wolynia (z Polesiem wolynsk). *Kosmos. Journal de la Société Polonaise des Naturalistes "Kopernik"* 61: 79-84.

Kawecki, Z. 1938. The Coccidae of the Tatra Mountains. Part 1 [In Polish, English summary]. *Annual Report of the Physiographical Commission of the Polish Academy of Sciences and Letters* (1936) 71: 199-208.

Kawecki, Z. 1951. Studies on the genus *Lecanium*. I. The brown scale, *Lecanium corni* Bouché, Marchal (female nec male). *Compte Rendu de l'Academie Polonaise des Sciences* 5-10: 12.

Kawecki, Z. 1954. Studies on the genus *Lecanium* Burm. II. The Yew Scale, *Lecanium pomeranicum* sp. n. and some related species (Homoptera, Coccoidea, Lecaniidae). *Annales Zoologici Warszawa* 16: 9-22.

Kawecki, Z. 1956. Studies on the genus *Lecanium* Burm. Part III. *Lecanium sericeum* Ldgr. (Homoptera, Lecaniidae). [In Polish, English summary]. *Polskie Pismo Entomologiczne* 25: 213-226.

Kawecki, Z. 1957. Notes on scale insects (Homoptera, Coccoidea). *Acta Zoologica Cracoviensia* 22: 193-204.

Kawecki, Z. 1958a. Studies on the genus *Lecanium* Burm. IV. Materials to a monograph of the brown scale, *Lecanium corni* Bouche, Marchal (female nec male) (Homoptera, Coccoidea, Lecaniidae). *Annales Zoologici. Warszawa* 17: 135-216.

Kawecki, Z. 1958b. Studies on the genus *Lecanium* Burm. Part V. The nut or thorn scale - *Lecanium coryli* (L.) sensu Marchal nec Sulc (Homoptera, Coccoidea, Lecaniidae). [In Polish, English summary]. *Polskie Pismo Entomologiczne* 27: 40-69.

Kawecki, Z. 1961. A revision of the species of the genus *Lecanium* Burm. occurring in Poland and the description of *Lecanium slavum* sp. n. *Proceedings of the 11th International Congress of Entomology, Vienna* (1960) 1: 65-67.

Kawecki, Z. 1962. The appearance of Coccidae of the genus *Lecanium* Burm. on mistletoe (Homoptera, Coccoidea, Lecaniidae). *Memorie della Società Entomologica Italiana* 41: 15-24.

Kawecki, Z. 1967. Studies on the genus *Lecanium* Burm. VI. *Lecanium smreczynskii* sp. n. (Homoptera, Coccoidea, Lecaniidae). *Bulletin de l'Academie Polonaise des Sciences* 15: 687-689.

Kawecki, Z. 1968. An outline of the biology and the geographical distribution of the globose scale *Sphaerolecanium prunastri* (Fonsc.) (Coccoidea, Lecaniidae). *Bulletin de l'Academie Polonaise des Sciences* 16: 689-693.

Kawecki, Z. 1971. A note on some European Lecaniidae (Coccoidea) with new additions of the Austrian, British, Italian and Polish fauna. *Bulletin de l'Academie Polonaise des Sciences* 19: 255-260.

Kennett, C.E. 1988. Results of exploration for parasitoids of citricola scale, *Coccus pseudomagnoliarum* (Homoptera, Coccidae), in Japan and their introduction in California. *Kontyu* 56: 445-457.

Keuchenius, P.E. 1915. Onderzoekingen en beschouwingen over eenige schadlijke schildluizen van de Koffiekultuur op Java. *Mededeelingen van het Besoekisch Proefstation. Djember* 16: 1-65.

Khasawinah, A.M.A. & Talhouk, A.S. 1964. The fig wax scale, *Ceroplastes rusci* (Linn.). *Zeitschrift für Angewandte Entomologie* 53: 113-151.

Kieffer, J.J. & Herbst, P. 1909. Ueber einige neue Gallen und Gallenerzeuger aus Chile. *Zentralblatt für Bakteriologie, Parasitenkunde, Infektionskrankheiten und Hygiene. Jena* 23: 119-126.

King, G.B. 1899a. Contributions to the knowledge of Massachusetts Coccidae. - II. *Canadian Entomologist* 31: 139-143.

King, G.B. 1899b. Contributions to the knowledge of Massachusetts Coccidae. - IV. *Canadian Entomologist* 31: 251-255.

King, G.B. 1899c. A new *Pulvinaria* from Massachusetts. *Psyche* 8: 417-418.

King, G.B. 1900. A new *Pulvinaria* from New Mexico. *Canadian Entomologist* 32: 360.

King, G.B. 1901a. *Lecanium caryae* Fitch. *Entomological News* 12: 50-51.

King, G.B. 1901b. *Lecanium websteri*, Ckll. and King n. sp. with notes on allied forms. *Canadian Entomologist* 33: 106-109.

King, G.B. 1901c. The Coccidae of the Harvard Botanical Gardens. *Psyche* 9: 153-154.

King, G.B. 1901d. Two new species of *Pulvinaria*. *Canadian Entomologist* 33: 144-146.

King, G.B. 1901f. The Coccidae of British North America. *Canadian Entomologist* 33: 179-180.

King, G.B. 1901g. The Coccidae of British North America. *Canadian Entomologist* 33: 193-200.

King, G.B. 1901h. The Coccidae of British North America. *Canadian Entomologist* 33: 314-315.

King, G.B. 1901i. The Coccidae of British North America. *Canadian Entomologist* 33: 333-336.

King, G.B. 1902a. A new species of the genus *Saissetia* (Coccidae). *Psyche* 9: 296-298.

King, G.B. 1902b. Further notes on Massachusetts Coccidae. *Canadian Entomologist* 34: 59-63.

King, G.B. 1902c. Coccidae of British North America. *Canadian Entomologist* 34: 158-161.

King, G.B. 1902d. Errata. *Canadian Entomologist* 34: 166.

King, G.B. 1903a. The Coccidae of Ohio. *Entomological News* 14: 204-206.

King, G.B. 1903c. Some new records of Coccidae. *Canadian Entomologist* 35: 191-197.

King, G.B. 1906. Notes on the young larvae of some species of *Pulvinaria*. *Canadian Entomologist* 38: 325-326.

King, G.B. 1914. The genus *Pseudokermes* in Montana. *Journal of Economic Entomology* 7: 246-247.

King, G.B. & Cockerell, T.D.A. 1897. New Coccidae found associated with ants. *Canadian Entomologist* 29: 90-93.

King, G.B. & Cockerell, T.D.A. 1898. A new form of *Pulvinaria*. *Psyche* 8: 286-287.

King, G.B. & Reh, L. 1901. Ueber einige Europäische und an eingeführten Pflanzen gesamelte Lecanien. *Jahrbuch der Hamburgischen Wissenschaftlichen Anstalten* (1900) 18: 57-65.

Kiritshenko, A.N. 1928. Fauna of Coccidae of Ukraine and crimea. [In Russian]. *Zakhist Roslin. Kharkov* 3-4: 112-116.

Kiritshenko, A.N. 1931. Second contribution to the coccid fauna of Ukraine and Crimea. [In Russian]. *Zashchita Rastenii. Leningrad* 7: 307-321.

Kiritshenko, A.N. 1932. Description of some new Coccidae (Hemiptera) from Turkestan and Ukraine. *Trudy Zoologicheskogo Instituta Akademii Nauk SSSR* 1932: 135-142.

Kiritshenko, A.N. 1935. A note on the coccid fauna of Transcaucasia. [In Russian]. *Bulletin du Musée de Georgie. Tbilisi* 8: 1-5.

Kiritshenko, A.N. 1936. The distribution of insect scales (Homoptera - Coccoidea) in USSR from zoogeographical and ecological points of view. [In Russian.]. *Zashchita Rastenii. Leningrad* 9: 68-75.

Kiritshenko, A.N. 1938. Beschreibung von zwei neuen Cocciden aus der Ukraine (USSR). *Konowia* 16: 229-236.

Kiritshenko, A.N. 1940. Third report on the coccid fauna of USSR. [In Russian]. *Trudy Zoologicheskogo Instituta Akademii Nauk SSSR* 1940: 115-137.

Kirkaldy, G.W. 1902. Hemiptera. *Fauna Hawaiiensis* 3: 93-174.

Kirkaldy, G.W. 1904a. A list of the Coccidae of the Hawaiian Islands (Hemiptera). *Entomologist* 37: 226-230.

Kirkaldy, G.W. 1904b. Bibliographical and nomenclatorial notes on the Hemiptera. No. II. *Entomologist* 37: 254-258.

Köhler, G. 1976. Beitrag zur Kenntnis des Männchens der Grünen Kaffeeschildlaus, *Coccus viridis* (Green) (Hemiptera: Coccinea - Coccidae). *Beiträge zur Entomologie. Berlin* 26: 471-477.

Köhler, G. 1978. Zur Biologie und Autoökologie der Grünen Kaffeeschildlaus, *Coccus viridis* (Green) (Hemiptera: Coccinea - Coccidae). *Zoologische Jahrbüche. Abteilung für Systematik. Jena* 105: 561-572.

Köhler, G. 1980. Los parasitos y episitos de la guagua verde del cafeto, *(Coccus viridis* Green) (Hemiptera, Coccidae) en cafetales de Cuba. *Centro Agricola. Habana. Enero-Abril,* 1980: 75-105.

Kollar, C. 1848. Über eine noch unbeschriebene Art von Schildläusen *(Coccus aesculi). Sitzungberichte der Kaiserischen Akademie der Wissenschaften. Wien* 1: 188.

Komarek, J. 1946. The physiological damage upon the ash-tree made by the scale insect *Lecanium coryli* L. [In Czech and English]. *Vestnik Ceskoslovenske Zoologicke Spolecnosti* 10: 156-165.

Komeili-Birjandi, A. 1981. First record of *Parthenolecanium corni* (Bouché, 1844) (Homoptera: Coccidae) from Iran. [In Persian, English summary]. *Journal of the Entomological Society of Iran* 6: 1-2.

Komosinska, H. 1977. Materials to the knowledge of scale insects (Homoptera, Coccoidea) of the Kampinoski National Park. [In Polish, English summary]. *Sylwan* 1: 21-24.

Komura, H., Mizukawa, K. & Minakata, H. 1982. Ceroalbolinic acid, a common body pigment of three *Ceroplastes* scale insects in Japan. Confirmation of structure. *Bulletin of the Chemical Society of Japan* 55: 3053-3054.

Kosztarab, M., Ben-Dov, Y. & Kosztarab, M. 1986. An annotated list of generic names of the scale insects (Homoptera: Coccoidea) Third Supplement. *Virginia Polytechnic Institute & State University, Blacksburg, Agricultural Experiment Station Bulletin* 86-2: 1-34.

Kosztarab, M. & Kosztarab, M.P. 1988. A selected bibliography of the Coccoidea (Homoptera) Third Supplement (1970-1985). *Virginia Polytechnic Institute & State University, Blacksburg, Agricultural Experiment Station Bulletin* 88-1: 1-252.

Kosztarab, M. & Kozár, F. 1978. Scale insects - Coccoidea. [In Hungarian, English summary]. *Fauna Hungariae, Akademiai Kiado, Budapest* 17: 1-192.

Kosztarab, M. & Kozár, M. 1988. Scale insects of Central Europe. Akademiai Kiado, Budapest. 456 pp.

Kosztarab, M. & Russell, L.M. 1974. An Annotated list of generic names of the scale insects (Homoptera: Coccoidea) Second Supplement. *Miscellaneous Publications United States Department of Agriculture* 1285: 1-22.

Koteja, J. 1964. Notes on scale insects (Homoptera, Coccoidea) in Poland's fauna. [In Polish, English summary]. *Polskie Pismo Entomologiczne* 34: 177-184.

Koteja, J. 1966a. Studies on morphology and biology of *Luzulaspis frontalis* (Green) (Homoptera, Coccoidea). *Polskie Pismo Entomologiczne* 36: 17-43.

Koteja, J. 1966b. *Luzulaspis nemorosa* sp. n. (Homoptera, Coccoidea, Coccidae). *Polskie Pismo Entomologiczne* 36: 45-56.

Koteja, J. 1969a. *Psilococcus parvus* Borchsenius (Homoptera, Coccoidea) - morphology, biology and taxonomy. *Acta Zoologica Cracoviensia* 14: 21-41.

Koteja, J. 1969b. Notes on the Poland's scale insect fauna (Homoptera, Coccoidea). II. *Polskie Pismo Entomologiczne* 39: 3-15.

Koteja, J. 1970. Systematic position of the genus *Vittacoccus* Borchsenius (Homoptera, Coccoidea). *Polskie Pismo Entomologiczne* 40: 223-231.

Koteja, J. 1971. Notes on the Poland's scale-insect fauna (Homoptera, Coccoidea). III. [In Polish]. *Polskie Pismo Entomologiczne* 41: 319-326.

Koteja, J. 1972. Notes on the Polish scale insect fauna (Homoptera, Coccoidea). IV. *Polskie Pismo Entomologiczne* 42: 565-571.

Koteja, J. 1974a. On the phylogeny and classification of the scale insects (discussion based on the morphology of the mouthparts). *Acta Zoologicae Cracoviensia* 19: 267-325.

Koteja, J. 1974b. The occurrence of a campaniform sensilum on the tarsus in the Coccinea (Homoptera). *Polskie Pismo Entomologiczne* 44: 243-252.

Koteja, J. 1974c. Comparative studies on the labium in the Coccinea (Homoptera). *Zeszyty Naukowe Akademii Rolniczej w Krakowie* 27: 1-162.

Koteja, J. 1976. The salivary pump in the taxonomy of the Coccinea (Homoptera). *Acta Zoologica Cracoviensia* 21: 263-290.

Koteja, J. 1978. An introduction to the revision of the Eriopeltini Sulc (Homoptera, Coccidae), with establishment of two new genera. *Polskie Pismo Entomologiczne* 48: 311-327.

Koteja, J. 1979a. A revision of the genus *Poaspis* Koteja (Homoptera, Coccidae). *Polskie Pismo Entomologiczne* 49: 451-474.

Koteja, J. 1979b. Revision of the genus *Luzulaspis* Cockerell (Homoptera, Coccidae). *Polskie Pismo Entomologiczne* 49: 585-638.

Koteja, J. 1980a. Campaniform, basiconic, coeloconic, and intersegmental sensilla on the antennae in the Coccinea (Homoptera). *Acta Biologica Cracoviensia* 22: 73-88.

Koteja, J. 1980b. Revision of the genus *Exaeretopus* Newstead (Homoptera, Coccidae). *Acta Zoologica Cracoviensia* 24: 337-372.

Koteja, J. 1981. Frequency of honeydew excretion in relation to circadian activity in scale insects (Homoptera, Coccinea). *Polskie Pismo Entomologiczne* 51: 365-376.

Koteja, J. 1983. Notes on the Polish scale insect fauna (Homoptera, Coccinea). V. [In Polish, English Abstract]. *Polskie Pismo Entomologiczne* 53: 673-677.

Koteja, J. & Brookes, H.M. 1981. *Symonicoccus* gen. n. with five new Australian species (Homoptera, Coccidae). *Polskie Pismo Entomologiczne* 51: 377-392.

Koteja, J. & Howell, J.O. 1979. *Luzulaspis* Cockerell (Homoptera: Coccidae) in North America. *Annals of the Entomological Society of America* 72: 334-342.

Koteja, J. & Kozár, F. 1979. *Luzulaspis kosztarabi* sp. n. from Hungary (Homoptera: Coccidae). *Acta Zoologica Academiae Scientiarum Hungaricae* 25: 121-125.

Koteja, J. & Liniowska, E. 1976. The clypeolabral shield in the taxonomy of the Coccinea (Homoptera). *Polskie Pismo Entomologiczne* 46: 653-681.

Koteja, J. & Lubowiedzka, A. 1976. On some changes of the cuticle in the female *Saissetia hemisphaerica* Targioni Tozzetti (Homoptera, Coccinea). *Acta Biologica Cracoviensia. Series Zoologia* 19: 71-77.

Koteja, J. & Rosciszewska, M. 1970. Revision of the genus *Parafairmairia* Cockerell (Homoptera, Coccoidea). *Polskie Pismo Entomologiczne* 40: 233-265.

Koteja, J. & Zak-Ogaza, B. 1966. Investigations on scale insects (Homoptera, Coccoidea) of the Pieniny Klippen Belt. *Acta Zoologica Cracoviensia* 11: 305-332.

Koteja, J. & Zak-Ogaza, B. 1969. The scale-insect fauna (Homoptera, Coccoidea) of the Ojcow National Park in Poland. *Acta Zoologica Cracoviensia* 14: 351-373.

Koteja, J. & Zak-Ogaza, B. 1983. The Coccinea fauna (Homoptera) of the Krakow-Czestochowa upland (southern Poland). *Acta Zoologica Cracoviensia* 26: 465-490.

Kotinsky, J. 1908a. Some Coccidae from Singapore collected by F. Muir. *Proceedings of the Hawaiian Entomological Society* 1: 167-171.

Kotinsky, J. 1908b. Notes and exhibition of specimens. *[Coccus muiri* New Name]. *Proceedings of the Hawaiian Entomological Society* 2: 37.

Kozár, F. 1970. A new scale insect species, *Rhodococcus rosophilus* Borchs. (Homoptera, Coccidae) in our fauna. [In Hungarian]. *Folia Entomologica Hungarica* 23: 229-230.

Kozár, F. 1971. Recent data on Coccidae (Hom.) damaging Gramineae. [In Hungarian, English abstract]. *Folia Entomologica Hungarica* 24: 157-162.

Kozár, F. 1972a. Contribution to the study of the scale insect fauna (Homoptera: Coccoidea) of Hungary. [In Hungarian]. *Állattani Közlemények* 59: 181-182.

Kozár, F. 1972b. The occurrence of an occasional grain crop pest, *Lecanopsis porifera* Borchs. (Homoptera: Coccoidea) in our fauna. [In Hungarian]. *Növénytermelés* 21: 281.

Kozár, F. 1979. Comparative studies on density of scale insects in European orchards. [In Hungarian]. *Agrártudományi Közlemények* 38: 135.

Kozár, F. 1980. The scale insect fauna (Homoptera: Coccoidea) of the Bakony Mountains and surrounding area. [In Hungarian, German summary]. *Veszprém Megyei Múzeumok Közleményei* 15: 65-72.

Kozár, F. 1981a. Data to the Coccoidea (Homoptera) fauna of the Hortobágy National Park. *The Fauna of the Hortobágy National Park. Hungary*: 89-90.

Kozár, F. 1981b. *Mirococcopsis nagyi* sp. n. and *Luzulaspis rajae* sp. n. from Hungary (Homoptera: Coccoidea). *Acta Zoologica Academiae Scientiarum Hungaricae* 27: 315-321.

Kozár, F. 1983. New and little known scale-insect species from Yugoslavia (Homoptera: Coccoidea). *Acta Zoologica Academiae Scientiarum Hungaricae* 29: 139-149.

Kozár, F. 1985. New data to the knowledge of scale-insects of Bulgaria, Greece, and Rumania (Homoptera: Coccoidea). *Acta Phytopathologica Academiae Scientiarum Hungaricae* 20: 201-205.

Kozár, F. 1986. Recent data to the knowledge of the Coccoidea (Homoptera) fauna of Hungary. [In Hungarian, English abstract]. *Folia Entomologica Hungarica* 47: 171-181.

Kozár, F. & Danzig, E.M. 1976. *Atrococcus bejbienkoi* sp. n., and some scale insects new to the Hungarian fauna (Homoptera: Coccoidea). *Acta Zoologica Academiae Scientiarum Hungaricae* 22: 65-67.

Kozár, F. & Drozdják, J. 1991. Data to the scale insect (Homoptera: Coccoidea) fauna of the Batorliget Nature Reserves. pp. 361-367. in: Mahunka, S. (Edit.) The Batorliget Nature Reserves - after forty years, 1990. Budapest, Hungarian Natural History Museum, 848 pp.

Kozár, F., Humble, L.M., Foottit, R.G. & Otvos, I.S. 1989. New and little known scale insects (Homoptera: Coccoidea) from British Columbia. *Journal of the Entomological Society of British Columbia* 86: 70-77.

Kozár, F., Konstantinova, G.M., Akman, K., Altay, M. & Kiroglu, H. 1979. Distribution and density of scale insects (Homoptera: Coccoidea) on fruit plants in Turkey in 1976. (Survey of scale insects (Homoptera: Coccoidea) infestations in European orchards No. II). *Acta Phytopathologica Academiae Scientiarum Hungaricae* 14: 535-542.

Kozár, F., Ördögh, G. & Kosztarab, M. 1977. New records to the Hungarian scale insect fauna (Homoptera: Coccoidea). [In Hungarian, English summary]. *Folia Entomologica Hungarica* 30: 69-75.

Kozár, F. & Ostafichuk, V.G. 1987. New and little known scale-insects species from Moldavia (USSR) (Homoptera: Coccoidea). *Folia Entomologica Hungarica* 48: 91-95.

Kozár, F. & Pellizzari Scaltriti, G. 1989. New scale-insects (Homoptera Coccoidea) for the Italian fauna collected in the Veneto region. *Bollettino di Zoologia Agraria e di Bachicoltura. Milano* 21: 199-202.

Kozár, F. & Sugonyaev, E.S. 1979. Contribution to the knowledge of parasites of coccids (Homoptera: Coccoidea). [In Hungarian, English abstract]. *Folia Entomologica Hungarica* 32: 234-236.

Kozár, F., Tranfaglia, A. & Pellizzari, G. 1984. New records on the scale insect fauna of Italy (Homoptera: Coccoidea). *Bolletino del Laboratorio di Entomologia Agraria "Filippo Silvestri". Portici* 41: 3-9.

Kozár, F., Tzalev, M., Viktorin, R.A. & Horváth, J. 1979. New data to the knowledge of the scale insects of Bulgaria (Homoptera: Coccoidea). *Folia Entomologica Hungarica* 32: 129-132.

Kozár, F. & Walter, J. 1985. Check-list of the Palaearctic Coccoidea (Homoptera). *Folia Entomologica Hungarica* 46: 63-110.

Kozár, F. & Walter, J. 1986. Data to the scale insect (Homoptera: Coccoidea) fauna of the Kiskunsag National Park. pp. 113-117. In: S. Mahunka (ed). The fauna of the Kiskunsag National Park. Akademiai Kiado, Budapest.

Kozarzhevskaya, E. & Reitzel, J. 1975. The scale insects (Homoptera: Coccoidea) of Denmark. *Statens Forsogsvirksomhed i Plantekultur. Copenhagen* 1226: 1-40.

Kryzhanovskiy, O.L. 1965. In memory of N.S. Borchsenius (1906-1965). [In Russian; English translation in Entomological Review 44: 551-555]. *Entomologicheskoe Obozrenye* 44: 951-957.

Kuwana, S.I. 1901. Notes on new and little known Californian Coccidae. *Proceedings of the California Academy of Sciences (ser. 3, Zoology)* 2: 399-408.

Kuwana, S.I. 1902a. Coccidae (scale insects) of Japan. *Proceedings of the California Academy of Sciences (ser. 3, Zoology)* 3: 43-98.

Kuwana, S.I. 1902b. Coccidae from the Galapagos Islands. *Journal of the Entomological Society of New York* 10: 28-33.

Kuwana, S.I. 1907. Coccidae of Japan, I. A synoptical list of Coccidae of Japan with descriptions of thirteen new species. *Bulletin of the Imperial Central Agricultural Experiment Station, Japan* 1: 177-212.

Kuwana, S.I. 1909a. Coccidae of Japan (III). First supplemental list of Japanese Coccidae, or scale insects, with description of eight new species. *Journal of the Entomological Society of New York* 17: 150-158.

Kuwana, S.I. 1909b. Coccidae of Japan (IV). A list of Coccidae from the Bonin Islands (Ogasawarajima), Japan. *Journal of the New York Entomological Society* 17: 158-164.

Kuwana, S.I. 1914. Coccidae of Japan, V. *Pomona Journal of Entomology and Zoology* 6: 1-8.

Kuwana, S.I. 1916. Some new scale insects of Japan. *Annotationes Zoologicae Japonenses* 9: 145-152.

Kuwana, S.I. 1917a. Coccidae of Japan, vol. II. [In Japanese]. 157 pp.

Kuwana, S.I. 1917b. A check list of the Japanese Coccidae. [In Japanese]. pp. 163-182. In: K. Nagano (ed). A collection of essays for Mr. Yasushi Nawa, written in commemoration of his 60 birthday., Gifu, Japan.

Kuwana, S.I. 1923a. Studies on Yanone-scale and red wax scale. Part II: Red wax scale. [In Japanese]. *Department of Agriculture and Commerce, Japan, Bureau of Agriculture, Injurious Insects and Pests Series* 10: 35-88.

Kuwana, S.I. 1923b. The Chinese white-wax scale, *Ericerus pela* Chavannes. *Philippine Journal of Science* 22: 393-406.

Kuwana, S.I. 1923c. I. Descriptions and biology of new or little-known coccids from Japan. *Bulletin of Agriculture and Commerce, Imperial Plant Quarantine Station, Yokohama* 3: 1-67.

Kuwana, S.I. 1927. A list of Coccidae (scale insects) known from China. *Lingnaam Agricultural Review* 4: 70-72.

Kuwana, S.I. 1931. Scale insects of Amami-Oshima, with descriptions of four new species. [In Japanese, English summary]. *Zoological Magazine, Tokyo* 43: 163-171.

Lahille, F. 1924. Sobre una nueva especie de cocchinilla *Alichtensia orientalis* Lah. *Boletin Mensual Defensa Agricola. Republica Oriental del Uruguay, Ministerio de Industrias, Montevideo* 5: 105-108.

Laing, F. 1925. Descriptions of some new genera and species of Coccidae. *Bulletin of Entomological Research* 16: 51-66.

Laing, F. 1927. Coccidae, Aphididae and Aleyrodidae. *Insects of Samoa. London* 2: 35-45.

Laing, F. 1928. A list of the Coccidae of San Thome. *Entomologist* 61: 214-215.

Laing, F. 1929. Descriptions of new, and some notes on old species of Coccidae. *Annals and Magazine of Natural History (ser.* 10) 4: 465-501.

Laing, F. 1933. The Coccidae of New Caledonia. *Annals and Magazine of Natural History (ser.* 10) 11: 675-678.

Lal, O.P. & Naji, A.H. 1979. Observations on the indigenous parasites of the black olive scale, *Saissetia oleae* Bern. (Hom.: Coccidae) in the Socialist Peoples Libyan Arab Jamahiriya. *Zeitschrift für Angewandte Entomologie* 88: 513-520.

Lal, O.P. & Naji, A.H. 1980. Observations on the predators of the black olive scale, *Saissetia oleae* Bern. (Homoptera: Coccidae) in the Socialist Peoples Libyan Arab Jamahiriya. *Zeitschrift für Pflanzenkrankheiten und Pflanzenschutz* 87: 27-31.

Lambdin, P.L. 1984. Bioecology of the tuliptree scale, *Toumeyella liriodendri* (Gmelin). *Proceedings of the Tenth International Symposium of Central European Entomofaunistic, Budapest, August* 1983: 387-388.

Lambdin, P.L. & Kosztarab, M. 1973. A morphological study on the adult female and two nymphal stages of *Mallococcus sinensis* (Maskell) with notes on its systematic position (Homoptera: Coccoidea: Coccidae). *Research Division Bulletin, Virginia Polytechnic Institute and State University, Blacksburg* 85: 53-68.

Lambdin, P.L. & Watson, J.K. 1980. New collection records for scale insects of Tennessee. *Journal of the Tennessee Academy of Science* 55: 77-81.

Lawson, P.B. 1917. Scale insects injurious to fruit and shade tress. The Coccidae of Kansas. *Bulletin of The University of Kansas. Biological Series* 18: 161-279.

Le Pelley, R.H. 1968. Pests of coffee. Longmans, London. 588 pp.

Lellakova-Duskova, F. 1966. Die auf Weidengewächsen (Salicales) lebenden Schildläuse (Coccoidea) in der Tschechoslowakei. *Zeitschrift für Angewandte Entomologie* 57: 294-309.

Leonardi, G. 1898. Diagnosi di cocciniglie nuove. *Rivista di Patologia Vegetale. Firenze* 6: 273-283.

Leonardi, G. 1908. Seconda contribuzione alla conoscenza della cocciniglie italiane. *Bolletino del R. Laboratorio di Entomologia Agraria di Portici* 3: 150-191.

Leonardi, G. 1911. Contributo alla conoscenza delle coccinglie della Republica Argentina. *Bolletino del R. Laboratorio di Entomologia Agraria di Portici* 5: 237-284.

Leonardi, G. 1913. Contribuzione allo studio delle cocciniglie dell'Eritrea (Africa Orientale). *Bolletino del Laboratorio di Zoologia Generale e Agraria della R. Scuola Superiore d'Agricoltura di Portici* 7: 27-38.

Leonardi, G. 1917. Terza contribuzione alla conoscenza delle cocciniglie italiane. *Bolletino del R. Laboratorio di Entomologia Agraria di Portici* 12: 188-216.

Leonardi, G. 1920. Monografia delle cocciniglie italiane. Della Torre, Portici. 555 pp.

Lepage, H.S. 1938. Catalogo dos coccideos do Brasil. *Revista do Museu Paulista. São Paulo* 23: 327-491.

Lepage, H.S. & Giannotti, O. 1943. Notes on some Coccoidea (Homoptera). [In Portuguese, English summary]. *Arquivos do Instituto Biologico. São Paulo* 14: 331-350.

Lepage, H.S. & Giannotti, O. 1944. Algumas especies novas de coccideos do Brasil (Homoptera - Coccoidea). *Arquivos do Instituto Biologico. São Paulo* 15: 299-306.

Lepage, H.S. & Piza, M.T. 1941. Redescription of "*Neolecanium silvetrai* (Hempel)" (Homoptera - Coccoidea), a serious pest of vine, and its control. [In Portuguese, English abstract]. *Arquivos do Instituto Biologico. São Paulo* 12: 21-26.

Lepiney, J. & Mimeur, J.M. 1931. Les coccides du Maroc. *Revue de Pathologie Végétale et d'Entomologie Agricole de France* 18: 243-255.

Li, C. 1985. China wax and the China wax scale insect. *World Animal Review* 55: 26-33.

Lichtenstein, J. 1881. [Notes]. *Annales de la Société Entomologique de France (Ser. 6)* 1: cxiv-cxvi.

Lichtenstein, J. 1885. [*Ceroplastes dugesii*]. *Bulletin de la Société Entomologique de France (ser. 6)* 5: cxli.

Lidgett, J. 1898. Notes and observations on some Victorian Coccidae. *Wombat, Geelong* 3 (4): 80-95.

Lidgett, J. 1899. A catalogue of Australian Coccidae. *Wombat, Geelong* 4 (3): 37-64.

Lidgett, J. 1901. A new Victorian coccid. *The Victorian Naturalist* 18: 59.

Lindinger, L. 1906. *Lecanium sericeum* n. sp. *Insektenbörse. Leipzig & Stuttgart* 23: 147, 152.

Lindinger, L. 1907a. Betrachtungen über die Cocciden-Nomenklatur. *Entomologisches Wochenblatt. Stuttgart* 24: 19-20, 22-23.

Lindinger, L. 1907b. Fränkische Cocciden. *Entomologische Blätter für Biologie und Systematik der Käfer. Schwabach* 3: 113-117, 136-139.

Lindinger, L. 1908. Eine Berichtigung zu meiner Zusammenstellung "Fränkischer Cocciden". *Entomologische Blätter für Biologie und Systematik der Käfer Schwabach* 4: 181.

Lindinger, L. 1911a. Beiträge zur Kenntnis der Schildläuse und ihre Verbreitung II. *Zeitschrift für Wissenschaft der Insektenbiologie* 7: 9-12, 86-90, 126-130, 172-177, 244-247, 353-383, 378-383.

Lindinger, L. 1911b. Afrikanische Schildläuse. IV. Kanarische Cocciden. Ein Beitrag zur Fauna der Kanarischen Inseln. *Jahrbuch der Hamburgischen Wissenschaftlichen Anstalten* (1910) 28: 1-38.

Lindinger, L. 1912. Die Schildlause (Coccidae) Europas, Nordafrikas und Vorder-Asiens, einschliesslich der Azoren, der Kanaren und Madeiras. Ulmer, Stuttgart. 388 pp.

Lindinger, L. 1913a. Afrikanische Schildläuse. V. Die Schildläuse Deutsche-Ostafrikas. *Jahrbuch der Hamburgischen Wissenschaftlichen Anstalten* (1912) 30: 59-95.

Lindinger, L. 1914. Die Cocciden-Literatur des Jahres 1909. *Zeitschrift für Wissenschaftliche Insektenbiologie* 10: 155-160.

Lindinger, L. 1918. Catalogo de les coccidos (Hem. Hom.) de las Islas Canarias. *Boletin del la Sociedad Entomologica de España* 1: 51-52.

Lindinger, L. 1923. Einfuhrung in die Kenntnis der deutschen Schildläuse. *Entomologisches Jahrbuch* 32: 138-152.

Lindinger, L. 1924. Die Schildläuse der mitteleuropäischen Gewächshauser. *Entomologisches Jahrbuch* 33/34: 167-191.

Lindinger, L. 1928. Bericht über die Tätigkeit der Abteilung für Pflanzenschutz. A. Überwachung der ein- und ausfuhr von Obst und Pflanzen (amtliche Pflanzenbeschau.). *Hamburgerischen Institut für Angewandte Botanie. Jahresberichte*: 94-110.

Lindinger, L. 1932a. Die synonymie von Walkers Cocciden (Hem.). *Mitteilungen der Deutschen Entomologischen Gesellschaft* 3: 26-27.

Lindinger, L. 1932b. Beiträge zur Kenntnis der Schildläuse III. (Hemipt. Cocc.). *Entomologische Rundschau. Stuttgart* 49: 79.

Lindinger, L. 1932c. Beiträge zur Kenntnis der Schildläuse (Coccidae). Wissenschaftl. Ballast: MS-Namen ohne Beschreibung u. dgl. *Entomologische Rundschau. Stuttgart* 49: 203-205.

Lindinger, L. 1932d. Beiträge zur Kenntnis der Schildläuse (Hem. Hom. Cocc.) einige verschollene Schildlaus-Beschreibungen. *Mitteilungen der Deutschen Entomologischen Gesellschaft* 3: 125-126.

Lindinger, L. 1932f. Beiträge zur Kenntnis der Schildläuse. *Konowia* 11: 177-205.

Lindinger, L. 1932g. Beiträge zur Kenntnis der Schildläuse (Hemiptera - Homoptera), Coccidae). *Wiener Entomologische Zeitung* 49: 217-225.

Lindinger, L. 1933a. Beiträge zur Kenntnis der Schildläuse. Die Gattung *Pseudochermes* Nitsche 1895. *Entomologische Rundschau. Stuttgart* 50: 31-32, 50.

Lindinger, L. 1933b. Beiträge zur Kenntnis der Schildläuse (Hemipt. - Homopt., Coccid.). *Entomologischer Anzeiger* 13: 77-78, 107-108, 116-117, 143, 159-160.

Lindinger, L. 1934a. Beiträge zur Kenntnis der Schildläuse (Hemipt. - Homopt., Coccid). *Entomologischer Anzeiger* 14: 15-16, 26-27, 36-37, 62-64.

Lindinger, L. 1934b. Was ist der richtig Name von *Lecanium corni* Marchal?. *Zeitschrift für Pflanzenkrankheiten* 44: 76-80.

Lindinger, L. 1934c. Ueber Schildläuse. *Mitteilungen der Deutschen Dendrologischen Gesellschaft* 46: 169-175.

Lindinger, L. 1934d. Eine vergessene Schildlaus-Beschreibung P. Bouchés. *Zeitschrift für Pflanzenkrankheiten* 44: 585-588.

Lindinger, L. 1934e. Die Schildlaus-Arten P. Fr. Bouchés und ihre Deutung. *Entomologisches Jahrbuch* 43: 153-169.

Lindinger, L. 1935a. Die nunmehr gültigen Namen der Arten in meinem "Schildläusebuch" und in den "Schildläusen der mitteleuropäischen Gewächshäuser". *Entomologische Zeitschrift* 44: 127-149.

Lindinger, L. 1935b. Neue Beiträge zur Kenntnis der Schildläuse (Coccidae). *Entomologische Zeitschrift* 49: 121-123.

Lindinger, L. 1935c. Neue Beiträge zur Kenntnis der Schildläuse (Coccidae). *Entomologische Zeitschrift* 49: 135.

Lindinger, L. 1936. Neue Beiträge zur Kenntnis der Schildläuse (Coccidae). *Entomologisches Jahrbuch* 45: 148-167.

Lindinger, L. 1937. Verzeichnis der Schildlaus-Gattungen. (Homoptera - Coccoidea Handlirsch, 1903). *Entomologischen Jahrbuch* 46: 178-198.

Lindinger, L. 1942. Coccoidea (Homopt.). *Beiträge zur Fauna Perus* 3: 112-122.

Lindinger, L. 1943a. Die Unterschiede der Nordwestdeutschen "*Lecanium*" - Gattungen und Arten. *Bombus* 25: 113-114.

Lindinger, L. 1943b. Die Schildlausnamen in Fulmeks Wirtindex 1943. *Arbeiten über morphologische und taxonomische Entomologie aus Berlin-Dahlem* 10: 145-152.

Lindinger, L. 1943c. Verzeichnis der Schildlaus-Gattungen, 1. Nachtrag. (Homoptera: Coccoidea). *Zeitschrift der Wiener Entomologischen Gesellschaft* 28: 264-265.

Lindinger, L. 1957. Ein weiterer Beitrag zur Synonymie der Cocciden. *Beiträge zur Entomologie* 7: 543-553.

Lindinger, L. 1958. Richtigstellung der Schildlausnamen in der Bearbeitung von Schmutterer, Kloft und Ludicke in "Handbuch der Pflanzenkrankheiten" (V. Bd., 5 Aufl., 4. Liefg., 1957). *Beiträge zur Entomologie* 8: 365-374.

Linnaeus, C. 1758. Systema naturae. Ed. X. Salvii, Holmiae. 823 pp.

Linnaeus, C. 1767. Systema naturae. Ed. XII. Salvii, Holmiae. pp. 533-1327.

Lizer y Trelles, C.A. 1917. Une nouvelle coccidocécidie de l'Argentine et description du cecidozoaire qui la produit (*Mesolecanium deltae* n. sp.). *Broteria. Serie Zoologia. Lisboa* 15: 103-107.

Lizer y Trelles, C.A. 1919. Una nueva subspecie de "*Ceroplastes*" (Coccidae) de la Republica Argentina (*Ceroplastes grandis* Hemp. subspe. *hempeli* nov.). *1a Reunion Nacional de la Sociedad Argentina de Ciencias Naturales. Tucuman* (1916): 381-382.

Lizer y Trelles, C.A. 1939. Catalogo sistematico razonado de los coccidos (Hom. Stern.) vernaculos de la Argentina. *Physis. Buenos Aires* 17: 157-210.

Lizer y Trelles, C.A. 1942a. Apuntaciones coccidologicas. I. *Revista de la Sociedad Entomologica Argentina* 11: 319-335.

Lizer y Trelles, C.A. 1942b. Cochinillas halladas por primera vez en la Argentina (Hom. Sternor.). *Revista de la Sociedad Entomologica Argentina* 11: 230-236.

Lizer y Trelles, C.A. 1943. Apuntaciones coccidologicas. II. *Revista de la Sociedad Entomologica Argentina* 11: 455-460.

Longo, S. 1984. Distribution and density of scale-insects (Homoptera, Coccoidea) on olive trees in Eastern Sicily. pp. 160-168. In: R. Cavalloro & A. Crovetti (eds), Integrated pest control in olive-groves. *Proceedings of the CEC/FAO/IOBC International Joint Meeting, Pisa 1984*. A.A. Balkema, Rotterdam.

Longo, S. 1985. Osservazioni morfologiche e bio-etologiche su *Ceroplastes japonicus* Green (Homoptera: Coccoidea) in Italia. *Atti XIV*

Congresso Nazionale Italiano di Entomologia, Palermo, 1985: 185-192.

Longo, S. 1988. Notes on the behaviour of *Filippia follicularis* (Targ. Tozz.) and *Lichtensia viburni* Sign. (Homoptera, Coccidae) in Sicily. *Bolletino del Laboratorio di Entomologia Agraria "Filippo Silvestri", Portici* (1986) 43: 173-177.

Longo, S. & Russo, A. 1986. Distribution and density of scale insects (Homoptera, Coccoidea) on citrus-groves in Eastern Sicily and Calabria. pp. 41-49. In: R. Cavalloro & E.D. Martino (eds). Integrated pest control in citrus-groves. *Proceedings of the Experts' Meeting, Acireale, March* 1985. A.A. Balkema, Rotterdam.

Longo, S. & Russo, A. 1988. Rilievi sulla composizione della coccidiofauna dell'olivo in Sicilia e Calabria. *Atti XV Congresso Nazionale Italiano di Entomologia* 1988: 513-520.

Longo, S., Marotta, S., Russo, A. & Tranfaglia, A. 1989. Cntributo alla conoscenza della coccidofauna (Homoptera, Coccoidea) della Sicilia con la descrizione di una nuova specie. *Entomologica. Bari* 24: 163-179.

Longo, S. & Russo, A. 1990. New records on scale insects of Calabria and Sicily (Italy) (Homoptera: Coccoidea). *Proceedings of the Sixth International Symposium of Scale Insects Studies, Cracow, Poland, August* 1990: 113-116.

Lopez y Ramos, D.S. 1835. Historia Natural de los insectos que atacan la vina, sus costumbres, su propagacion, los danos que ocasionan. Imprenta Real, Madrid. 65 pp.

Löw, F. 1883. Eine neue Coccide. *Wiener Entomologische Zeitung* 2: 115-117.

Lull, R.S. 1899. A new species of *Pulvinaria*. *Entomological News* 10: 237-242.

MacAloney, H.J. 1961. Pine Tortoise Scale. *U.S. Department of Agriculture, Forest Service, Forest Pest Leaflet* 57: 1-7.

Macfie, J.W.S. 1913. On a new African species of Coccidae. *Bulletin of Entomological Research* 4: 31-34.

MacGillivray, A.D. 1921. The Coccidae. Tables for the identification of the subfamilies and some of the more important genera and species, together with discussions of their anatomy and life history. Urbana. 502 pp.

MacGregor, R.L. 1981. Coccoidea de México. III. Redescripción del genero *Takahashia* Cockerell y la revalidacion del genero *Pendularia* Fonseca, con las redescripciones de las especies *P. pendens* Fonseca y *P. jaliscensis* (Cockerell) (Homoptera - Coccidae). *Anales del Instituto de Biologia. Universidad de México. Serie Zoologia* 51: 299-314.

Mahdihassan, S. 1933. Sur les différents symbiotes des cochenilles productrices ou non productrices de cire. *Compte Rendu de l'Académie des Sciences. Paris* 196: 56-562.

Malumphy, C.P. 1991. A morphological and experimental investigation of the *Pulvinaria vitis* Complex in Europe. Ph.D. Thesis, University of London. 270 pp.

Mamet, J.R. 1936a. New species of Coccidae (Hemipt. Homopt.) from Mauritius. *Proceedings of the Royal Entomological Society of London (Ser. B)* 5: 90-96.

Mamet, J.R. 1936b. Note sur les cochenilles de l'Île d'Agalega. *Revue Agricole de l'Île Maurice* 88: 152-153.

Mamet, J.R. 1939. On two coccids recently described from Mauritius (Hem.). *Proceedings of the Royal Entomological Society of London (Ser. B).* 8: 238-239.

Mamet, J.R. 1941a. On some Coccidae (Hemipt. Homopt.) described from Mauritius by de Charmoy. *Mauritius Institute Bulletin. Port Louis* 2: 23-39.

Mamet, J.R. 1941b. Report on a few Coccidae (Homopt.) collected by Mr. P.O. Wiehe in the Chagos Archipelago. *Mauritius Institute Bulletin* 2: 38.

Mamet, J.R. 1943. A revised list of the Coccoidea of the islands of the western Indian Ocean, south of the equator. *Mauritius Institute Bulletin. Port Louis* 2: 137-170.

Mamet, J.R. 1948. A food-plant catalogue of the insects of Mauritius. *Colony of Mauritius, Department of Agriculture, Scientific Series Bulletin* 30: 1-74.

Mamet, J.R. 1949. An annotated catalogue of the Coccoidea of Mauritius. *Mauritius Institute Bulletin. Port Louis* 3: 1-81.

Mamet, J.R. 1950. Notes on the Coccoidea of Madagascar - I. *Mémoires de l'Institut Scientifique de Madagascar (Ser. A)*. 4: 17-38.

Mamet, J.R. 1951. Notes on the Coccoidea of Madagascar - II. *Mémoires de l'Institut Scientifique de Madagascar (Ser. A)* 5: 213-254.

Mamet, J.R. 1952. On some Coccoidea collected by Mr. J.R. Williams in Reunion Island. (Hemiptera). *Proceedings of the Royal Entomological Society of London (B)* 21: 170-172.

Mamet, J.R. 1953. The authorship of the species of Coccoidea (Hemiptera) described from Mauritius in 1899. *Proceedings of the Royal Entomological Society of London (A)* 28: 149-152.

Mamet, J.R. 1954a. Notes on the Coccoidea of Madagascar, III. *Mémoires de l'Institut Scientifique de Madagascar (Ser. E)* 4: 1-86.

Mamet, J.R. 1954b. Miscellaneous notes on the Coccoidea (Homoptera) of the Mascarene Islands and of the Chagos Archipelago. *Mauritius Institute Bulletin. Port Louis* 3: 260-265.

Mamet, J.R. 1956a. Miscellaneous coccid studies (Homoptera). *Naturaliste Malgache* 8: 133-141.

Mamet, J.R. 1956b. On some Coccoidea from the Island of Rodrigues (Hemiptera). *Mauritius Institute Bulletin. Port Louis* 3: 303-306.

Mamet, J.R. 1957. On some Coccoidea from Reunion Island (Homoptera). *Memoires de l'Institut Scientifique de Madagascar (Ser. E)* 8: 367-386.

Mamet, J.R. 1958. The identity of the sugar-cane *Pulvinaria* (Hemiptera: Coccoidea) of Mauritius, with notes on its economic importance. *Proceedings of the Royal Entomological Society of London (B)* 27: 65-75.

Mamet, J.R. 1959a. Further notes on the Coccoidea of Reunion Island (Homoptera). *Naturaliste Malgache* 11: 123-131.

Mamet, J.R. 1959b. Notes on the Coccoidea of Madagascar IV. *Memoires de l'Institut Scientifique de Madagascar (Ser. E)* 11: 369-479.

Mamet, J.R. 1960. On some Coccoidea from the Comoro archipelago. *Naturaliste Malgache* 12: 155-162.

Mamet, J.R. 1962. Notes on the Coccoidea of Madagascar, V. (Homoptera). *Naturaliste Malgache* 13: 153-202.

Mamet, J.R. 1978. Contribution à la connaissance de la faune entomologique d'Agalega (Ocean Indien). *Bulletin de la Société Entomologique de France* 83: 97-107.

Manawadu, D. 1986. A new species of *Eriopeltis* Signoret (Homoptera: Coccidae) from Britain. *Systematic Entomology* 11: 317-326.

Mann, W.H. 1922. Notes on a collection of West African Myrmecophiles. *Bulletin of the American Museum of Natural History, New York* 45: 623-630.

Marchal, P. 1906. Sur deux espèces de cochenilles nouvelles (Hem. Hom.) récoltées en Algérie. *Bulletin de la Société Entomologique de France* 9: 143-145.

Marchal, P. 1908a. Notes sur les cochenilles de l'Europe et du nord de l'Afrique (1re partie). *Annales de la Société Entomologique de France* 77: 223-309.

Marchal, P. 1908b. Le *Lecanium* du Robinia. *Comptes Rendus des Séances de la Société de Biologie* 65: 2-5.

Marchal, P. 1909a. Sur les cochenilles de l'Afrique occidentale. *Comptes Rendus des Séances de la Société de Biologie* 66: 586-588.

Marchal, P. 1909b. Cochenilles nouvelles de l'Afrique occidentale. *Bulletin de la Société Zoologique de France* 34: 68-69.

Marchal, P. 1909c. Contribution à l'étude des coccides de l'Afrique occidentale. *Mémoires de la Société Zoologique de France* 22: 165-182.

Marlatt, C.L. 1892. General notes. *Insect Life* 4: 148-153.

Marotta, S. 1987. I coccidi (Homoptera: Coccoidea: Coccidae) segnalati in Italia, con riferimenti bibliografici sulla tassonomia, geonemia, biologia e piante ospiti. *Bolletino del Laboratorio di Entomologia Agraria "Filippo Silvestri" di Portici* 44: 97-119.

Marotta, S. & Tranfaglia, A. 1990. New and little known species of Italian scale insects (Homoptera: Coccoidea). *Proceedings of the Sixth International Symposium of Scale Insects Studies, Cracow, Poland, August,* 1990: 107-112.

Martelli, G. 1908. Osservazioni fatte sulle cocciniglie dell'olivo e loro parassiti in Puglia ed in Calabria. *Bolletino del Laboratorio di Zoologia Generale e Agraria della R. Scuola Superiore d'Agricoltura in Portici* 2: 217-296.

Martin Mateo, M.P. 1984. Inventario preliminar de los coccidos de España. II. Asterolecaniidae, Kermococcidae, Coccidae y Aclerdidae. *Graellsia, Revista de Entomologos Ibericos. Madrid* 40: 63-79.

Maskell, W.M. 1879. On some Coccidae in New Zealand. *Transactions and Proceedings of the New Zealand Institute* (1878) 11: 187-230.

Maskell, W.M. 1880. Further notes on New Zealand Coccidae. *Transactions and Proceedings of the New Zealand Institute* (1879) 12: 291-301.

Maskell, W.M. 1882. Further notes on Coccidae in New Zealand, with descriptions of new species. *Transactions and Proceedings of the New Zealand Institute* (1881) 14: 215-229.

Maskell, W.M. 1884. Art. V. - Further notes on Coccidae in New Zealand, with descriptions of new species. *Transactions and Proceedings of the New Zealand Institute* (1883) 16: 120-144.

Maskell, W.M. 1885. Further notes on Coccidae in New Zealand. *Transactions and Proceedings of the New Zealand Institute* (1884) 17: 20-31.

Maskell, W.M. 1887a. Further notes on New Zealand Coccidae. *Transactions and Proceedings of the New Zealand Institute* (1886) 19: 45-49.

Maskell, W.M. 1887b. An account of the insects noxious to agriculture and plants in New Zealand. The Scale Insects (Coccididae). State Forests & Agricultural Department, Wellington. 116 pp.

Maskell, W.M. 1889. On some new South Australian Coccidae. *Transactions of the Royal Society of South Australia* (1888) 11: 101-111.

Maskell, W.M. 1890a. Further notes on Coccidae, with descriptions of new species from Australia, Fiji, and New Zealand. *Transactions and Proceedings of the New Zealand Institute* (1889) 22: 133-156.

Maskell, W.M. 1890b. [New Zealand Coccidae exhibited on behalf of W.M. Maskell at the meeting of the Entomological Society of London (published letter)]. *Proceedings of the Entomological Society of London*: xiv-xvi.

Maskell, W.M. 1891a. Further coccid notes: with descriptions of new species from New Zealand, Australia, and Fiji. *Transactions and Proceedings of the New Zealand Institute* (1890) 23: 1-36.

Maskell, W.M. 1891b. Descriptions of new Coccidae. *Indian Museum Notes* 2: 59-62.

Maskell, W.M. 1892a. Further coccid notes: with descriptions of new species, and remarks on coccids from New Zealand, Australia and elsewhere. *Transactions and Proceedings of the New Zealand Institute* (1891) 24: 1-64.

Maskell, W.M. 1892b. A new *Icerya*, and some other new coccids from Australia. *Entomologist's Monthly Magazine* 28: 183-184.

Maskell, W.M. 1893a. A few remarks on coccids. *Entomologist's Monthly Magazine* 29: 103-105.

Maskell, W.M. 1893b. Further coccid notes: with descriptions of new species from Australia, India, Sandwich Islands, Demerara, and South Pacific. *Transactions and Proceedings of the New Zealand Institute* (1892) 25: 201-252.

Maskell, W.M. 1894a. Remarks on certain genera of Coccidae. *Entomologist* 27: 44-46, 93-95, 166-168.

Maskell, W.M. 1894b. On a new species of coccid on fern roots. *Proceedings of the Linnaean Society of New South Wales (1893) (Ser. 2)* 8: 225-226.

Maskell, W.M. 1894c. Further coccid notes with descriptions of several new species and discussion of various points of interest. *Transactions and Proceedings of the New Zealand Institute* (1893) 26: 65-105.

Maskell, W.M. 1895a. Synoptical list of Coccidae reported from Australia and the Pacific Islands up to December 1894. *Transactions and Proceedings of the New Zealand Institute* (1894) 27: 1-35.

Maskell, W.M. 1895b. Further coccid notes: with description of new species from New Zealand, Australia, Sandwich Islands, and elsewhere, and remarks upon many species already reported. *Transactions and Proceedings of the New Zealand Institute* (1894) 27: 36-75.

Maskell, W.M. 1896a. Notes on Coccidae. *Entomologist's Monthly Magazine* 32: 223-226.

Maskell, W.M. 1896b. Further coccid notes, with descriptions of new species and discussions of questions of interest. *Transactions and Proceedings of the New Zealand Institute* (1896) 28: 380-411.

Maskell, W.M. 1897a. Further coccid notes: with descriptions of new species and discussions of points of interest. *Transactions and Proceedings of the New Zealand Institute* (1896) 29: 293-331.

Maskell, W.M. 1897b. On a collection of Coccidae, principally from China and Japan. *Entomologist's Monthly Magazine* 33: 239-244.

Maskell, W.M. 1898. Further coccid notes: with descriptions of new species, and discussion of points of interest. *Transactions and Proceedings of the New Zealand Institute* (1897) 30: 219-252.

Massini, P.C. & Brethes, J. 1918. Nuevas plagas y sus enemigos naturales. Tres nuevas cochinillas Argentinas y sus parasitos. *Anales de la Sociedad Rural Argentina* 52: 148-158.

Matesova, G.I. 1960. New species of soft scales family Coccidae (Homoptera, Coccoidea) in Kazakhstan. [In Russian]. *Trudy Instituta Zoologii Akademiya Nauk Kazakhskoy SSR. Alma-Ata* 11: 196-204.

Matesova, G.I. 1968. Coccids (Homoptera, Coccoidea) of East Kazakhstan. [In Russian]. *Trudy Instituta Zoologii, Akademii Nauk Kazakhskoy SSR, Alma-Ata* 30: 102-129.

Matesova, G.I. 1979. A new species of the genus *Parafairmairia* Cockerell (Homoptera, Coccoidea). [In Russian]. *Trudy Vsesoyuznogo Entomologicheskogo Obshchestva. Leningrad* 61: 48-49.

Matile-Ferrero, D. 1970. Hemiptera (Homoptera): Coccoidea. *South African Animal Life* 14: 171-185.

Matile-Ferrero, D. 1976. La faune terrestre de l'Île de Sainte-Hélène. 7. Coccoidea. *Musée Royal de l'Afrique centrale, Tervuren, Belgique, Annales (Serie 8) Sciences Zoologiques* 215: 292-318.

Matile-Ferrero, D. 1978. Homoptères Coccoidea de l'Archipel des Comores. *Memotres du Muséum National d'Histoire Naturelles (N.S.) Serie A, Zoologie* 109: 39-70.

Matile-Ferrero, D. 1983. Professor Alfred Serge Balachowsky. *Proceedings of the Tenth International Central European Entomofaunistic Symposium. Budapest*: 313-314.

Matile-Ferrero, D. 1984a. *Etiennea villiersi*, n.g., n.sp., du Sénégal méridional [Homoptera, Coccoidea, Coccidae]. *Revue Française d'Entomologie* 6: 99-103.

Matile-Ferrero, D. 1984b. Insects of Saudi Arabia Homoptera: Subordo Coccoidea. *Fauna of Saudi Arabia* 6: 219-228.

Matile-Ferrero, D. 1987. Remarques sur la morphologie de *Coccus asiaticus* Lindinger, cochenille associée au caféier en Afrique (Hemiptera, Coccidae). *Revue Française d'Entomologie (N.S.)* 9: 76.

Matile-Ferrero, D. 1988. Sternorrhyncha: Suborder Coccoidea of Saudi Arabia (Part 2). *Fauna of Saudi Arabia* 9: 23-38.

Matile-Ferrero, D. & Le Ruyet, H. 1985. Cochenilles nouvelles du massif forestier de Taï. en Côte d'Ivoire [Homoptera, Coccoidea]. *Revue Française d'Entomologie (N.S.)* 7: 257-272.

Matile-Ferrero, D. & Nonveiller, G. 1984. Coccoidea. pp. 62-70. In: G. Nonveiller (ed.). Catalogue commenté et illustré des Insectes du Cameroun d'intérêt agricole (apparitions, reparition, importance). Beograd. Mémoirs, Institut pour la Protection des Plantes.

Maxwell-Lefroy, H. 1903. The scale insects of the Lesser Antilles. Part II. *Imperial Department of Agriculture for the West Indies Pamphlet Series* 22: 1-50.

Maxwell-Lefroy, H. 1908. Notes on Indian scale insects (Coccidae). *Memoirs of the Department of Agriculture in India* 2: 111-137.

McConnell, H.S. 1949. A new North American species of *Pulvinaria* (Homoptera, Coccidae). *Proceedings of the Entomological Society of Washington* 51: 29-34.

McKenzie, H.L. 1951. Present status of the kuno scale, *Lecanium kunoensis* Kuwana, in California (Homoptera; Coccoidea; Coccidae). *Bulletin Department of Agriculture, State of California* 40: 105-109.

McKenzie, H.L. 1959. Gordon Floyd Ferris as a student of the scale insects. *The Pan-Pacific Entomologist* 35: 25-28.

McLachlan, R.A. 1969. Summer control of grape-vine scale *(Eulecanium persicae* (F.) in the Stanthorpe district, Queensland. *Queensland Journal of Agricultural and Animal Sciences* 26: 95-97.

Melis, A. 1928. Antonio Berlese. *Rivista di Biologia. Milano* 10: 922-932.

Mendel, Z., Podoler, H. & Rosen, D. 1982. Population dynamics of the Mediterranean black scale, *Saissetia oleae* (Olivier), on citrus in Israel. 3. Occurrence of a yellow form. *Journal of the Entomological Society of southern Africa* 45: 227-229.

Mendel, Z., Podoler, H. & Rosen, D. 1984a. Population dynamics of the Mediterranean black scale, *Saissetia oleae* (Olivier), on citrus in Israel. 4. The natural enemies. *Journal of the Entomological Society of southern Africa* 47: 1-21.

Mendel, Z., Podoler, H. & Rosen, D. 1984b. Population dynamics of the Mediterranean black scale, *Saissetia oleae* (Olivier), on citrus in Israel. 5. The crawlers. *Journal of the Entomological Society of southern Africa* 47: 23-34.

Mendes, L.O.T. 1931. Uma nova especie do genero *Eucalymnatus* (Homopt. - Coccidae). *Revista de Entomologia. Rio de Janeiro* 1: 395-400.

Mendes, L.F. & Fernandes, I.M. 1989. Lista anotada dos espécimes-tipo depostados nas colecções do Centro de Zoologia do Instituto de Investigação Científica Tropical. II. Insectos. *Garcia de Orta. Série de Zoologia. Lisboa* (1988) 15: 45-62.

Miller, D. 1940. Obituary - G.F.J.M. Brittin, Coccidologist. *Transactions and Proceedings of the Royal Society of New Zealand* 70: xxxvii-xxxviii.

Miller, D.R. 1991. Superfamily Coccoidea. pp. 90-107. In: F.W. Stehr (ed.). Immature Insects. Kendal/Hunt Publ., Dubuque, Iowa.

Miller, D.R., Bohart, R.M. & Wilkey, R.F. 1969. Howard Lester McKenzie, Jr. 1910-1968. *The Pan-Pacific Entomologist* 45: 245-259.

Miller, D.R. & Kosztarab, M. 1979. Recent advances in the study of scale insects. *Annual Review of Entomology* 24: 1-27.

Miller, G.R. & Williams, M.L. 1990. Tests of male soft scale insects (Homoptera: Coccidae) from America north of Mexico, including a key to the species. *Systematic Entomology* 15: 339-358.

Mitsuhashi, J., Yamasaki, T., Narahashi, T. & Fukami, J.I. 1956. On the method of distinguishing between dead and living Ruby Scale. [In Japanese, English summary]. *Oyo-Kontyu, Tokyo* 12: 162-170.

Modeer, A. 1778. Om fästflyet. *Coccus. Handlngar Götheborgska Wtenskaps och Witterhets Samhällets. Weterskaps Afdelingen. Götheborg* (ser. 1) 1: 11-50.

Moharana, S. 1990. Cytotaxonomy of coccids (Homoptera: Coccoidea). *Proceedings of the sixth International Symposium of Scale Insects Studies, Cracow, Poland, August* 1990: 47-54.

Monastero, S. & Zaami, V. 1959. Le cocciniglie degli agrumi in Sicilia (*Ceroplastes sinensis* D.G. - *Pseudococcus citri* R. - *Icerya purchasi* M.). Istituto di Entomologia Agraria, Palermo. 82 pp.

Morrison, H. 1919. A report on collection of Coccidae from Argentina with descriptions of apparently new species (Hom.). *Proceedings of the Entomological Society of Washington* 21: 63-91.

Morrison, H. 1920. The nondiaspine Coccidae of the Philippine Islands, with descriptions of apparently new species. *The Philippine Journal of Science* 17: 147-202.

Morrison, H. 1921. Some nondiaspine Coccidae from the Malay peninsula, with descriptions of apparently new species. *The Philippine Journal of Science* 18: 637-676.

Morrison, H. 1922. On some trophobiotic Coccidae from British Guiana. *Psyche* 29: 132-152.

Morrison, H. 1923. A report on a collection of Coccidae from Argentine II. (Hemiptera Coccidae). *Proceedings of the Entomological Society of Washington* 25: 122-127.

Morrison, H. 1929. Some neotropical scale insects associated with ants (Hemiptera - Coccidae). *Annals of the Entomological Society of America* 22: 33-60.

Morrison, H. & Morrison, E.R. 1922. A redescription of the type species of the genera of Coccidae based on species originally described by Maskell. *Proceedings of the United States National Museum. Washington* 60: 1-120.

Morrison, H. & Morrison, E.R. 1965. A selected bibliography of the Coccoidea. First Supplement. *Miscellaneous Publication United States Department of Agriculture* 987: 1-44.

Morrison, H. & Morrison, E.R. 1966. An annotated list of generic names of the scale insects (Homoptera: Coccoidea). *Miscellaneous Publication United States Department of Agriculture* 1015: 1-206.

Morrison, H. & Renk, A.V. 1957. A selected bibliography of the Coccoidea. *Miscellaneous Publication United States Department of Agriculture* 734: 1-222.

Mosquera, L.F. 1979. El genero *Ceroplastes* (Homoptera: Coccidae) en Colombia. *Caldasia. Bogota* 12 (60): 595-627.

Mosquera, L.F. 1984. El genero *Ceroplastes* (Homoptera: Coccidae) en Colombia, II. *Caldasia. Bogota* 14 (66): 125-147.

Munro, H.K. & Fouche, F.A. 1936. A list of the scale insects and mealy bugs (Coccidae) and their host-plants in South Africa. *Bulletin of the Department of Agriculture and Forestry, Union of South Africa, Pretoria* 158: 1-104.

Muzaffar, N. & Ahmad, R. 1977. A note on *Saissetia privigna* (Hem.: Coccidae) in Pakistan and the breeding of its natural enemies. *Entomophaga* 22: 45-46.

Nada, S., Abd Rabo, S. & Hussein, G.E.D. 1990. Scale insects infesting mango trees in Egypt (Homoptera: Coccoidea). *Proceedings of the Sixth International Symposium of Scale Insects Studies, Cracow, Poland, August* 1990: 133-134.

Nagy, B. & Kozár, F. 1984. The first record of *Scythia craniumequinum* (Coccoidea) in Central Europe. *Proceedings of the Tenth International Symposium of Central European Entomofaunistic, Budapest, August* 1983: 368-369.

Nakahara, S. 1978. *Ceroplastes denudatus*, junior synonym of *C. rusci* (Homoptera: Coccoidea: Coccidae). *Proceedings of the Entomological Society of Washington* 80: 657-658.

Nakahara, S. 1981a. The proper placements of the Nearctic soft scale species assigned to the genus *Lecanium* Burmeister (Homoptera: Coccidae). *Proceedings of the Entomological Society of Washington* 83: 283-286.

Nakahara, S. 1981b. List of the Hawaiian Coccoidea (Homoptera: Sternorhyncha). *Proceedings of the Hawaiian Entomological Society* 23: 387-424.

Nakahara, S. 1983. List of the Coccoidea species (Homoptera) of the United States Virgin Islands. *United States Department of Agriculture, Plant Protection and Quarantine, APHIS* [Mimeograph] 81-42: 1-21.

Nakahara, S. & Gill, R.J. 1985. Revision of *Philephedra*, including a review of *Lichtensia* in North America and description of a new genus (Homoptera: Coccidae). *Entomography* 3: 1-42.

Nakahara, S. & Miller, C.E. 1981. A list of the Coccoidea species (Homoptera) of Puerto Rico. *Proceedings of the Entomological Society of Washington* 83: 28-39.

Nassonov, N.V. 1909. Sur quelques nouvelles coccides. [In Russian and Latin]. *Ezhegodnik Zoologicheskogo Muzeya Imperatorskoi Akademii Nauk, St Petersburg* 13: 471-499.

Nath, K. 1972. Studies on the citrus inhabiting coccids (Coccoidea: Hemiptera) of Darjeeling District, West Bengal. *Bulletin of Entomology* 13: 1-10.

Neuenschwander, P. & Paraskakis, M. 1980. Studies on distribution and population dynamics of *Saissetia oleae* (Oliv.) (Hom., Coccidae) within the canopy of the olive tree. *Zeitschrift für Angewandte Entomologie* 90: 366-378.

Neves, M. 1936b. Les coccides du Portugal (première liste). *Bulletin de la Société Portugaise des Sciences Naturelles, Lisbonne* 12: 191-213.

Newstead, R. 1892. On new or little known Coccidae, chiefly English (No. 2). *Entomologist's Monthly Magazine* 28: 141-147.

Newstead, R. 1893a. A new coccid in an ant's nest. *Entomologist's Monthly Magazine* 29: 138.

Newstead, R. 1893b. Observations on Coccidae (no. 6). *Entomologist's Monthly Magazine* 29: 205-210.

Newstead, R. 1894a. Observations on Coccidae (no. 9). *Entomologist's Monthly Magazine* 30: 204-207.

Newstead, R. 1894b. Observations on Coccidae (no. 10). *Entomologist's Monthly Magazine* 30: 232-234.

Newstead, R. 1894c. Scale insects in Madras. *Indian Museum Notes* 3: 21-32.

Newstead, R. 1895. Observations on Coccidae (No. 11). *Entomologist's Monthly Magazine* 31: 165-167.

Newstead, R. 1896. Observations on Coccidae (no. 14). *Entomologist's Monthly Magazine* 32: 57-60.

Newstead, R. 1897. New Coccidae collected in Algeria by the Rev. Alfred E. Eaton. *Transactions of the Entomological Society of London* 1897: 93-103.

Newstead, R. 1898. Observations on Coccidae (No. 17). *Entomologist's Monthly Magazine* 34: 92-99.

Newstead, R. 1900. The injurious scale insects and mealy bugs of the British Isles. *Journal of the Royal Horticultural Society* 23: 20-44.

Newstead, R. 1901. Observations on Coccidae (no. 19). *Entomologist's Monthly Magazine* 37: 81-86.

Newstead, R. 1903. Monograph of the Coccidae of the British Isles. Vol. 2. Ray Society, London. 270 pp.

Newstead, R. 1906. Report on insects sent from Der Kaiserliche Biologische Anstalt für Land- und Forstwirtschaft Dahlem, Berlin. *Quarterly Journal. Institute of Commercial Research in the Tropics. University of Liverpool* 1: 73-74.

Newstead, R. 1908a. On the gum-lac insect of Madagascar, and other coccids affecting the citrus and tobacco in that island. *Quarterly Journal. Institute of Commercial Research in the Tropics. University of Liverpool* 3: 3-13.

Newstead, R. 1908b. On the structural characters of three species of Coccidae affecting cocoa, rubber, and other plants in Western Africa. *Journal of Economic Biology* 2: 149-157.

Newstead, R. 1908c. On a collection of Coccidae and other insects affecting some cultivated and wild plants in Java and in Tropical Western Africa. *Journal of Economic Biology. London* 3: 33-42.

Newstead, R. 1909. Coccidae and Aleurodidae of Madagascar and Comoro Is. pp. 349-356. In: A. Voeltzkow (ed). Reise in Ostafrika in den Jahren 1903-1905. E. Schweizerbartsche, Stuttgart.

Newstead, R. 1910a. On scale insects (Coccidae) from the Uganda Protectorate. *Bulletin of Entomological Research* 1: 63-69.

Newstead, R. 1910b. On two new species of African Coccidae. *Journal of Economic Biology* 5: 18-22.

Newstead, R. 1910c. On scale insects (Coccidae) &c. from the Uganda Protectorate. *Bulletin of Entomological Research* 1: 63-69.

Newstead, R. 1910d. Some further observations on the scale insects (Coccidae) of the Uganda Protectorate. *Bulletin of Entomological Research* 1: 185-199.

Newstead, R. 1911a. Observations on African scale insects (Coccidae). (No. 3). *Bulletin of Entomological Research* 2: 85-104.

Newstead, R. 1911b. On a collection of Coccidae and Aleurodidae, chiefly African, in the collection of the Berlin Zoological Museum. *Mitteilungen aus dem Zoologischen Museum in Berlin* 5: 155-174.

Newstead, R. 1913. Notes on scale-insects (Coccidae). - Part I. *Bulletin of Entomological Research* 4: 67-81.

Newstead, R. 1914a. Notes on scale-insects (Coccidae). Part II. *Bulletin of Entomological Research* 4: 301-311.

Newstead, R. 1914b. Homoptera (Psyllidae and Coccidae) collected in the Lagos District by W.A. Lamborn. *Transactions of the Entomological Society of London* (1913): 520-524.

Newstead, R. 1917a. Observations on scale-insects (Coccidae) - III. *Bulletin of Entomological Research* 7: 343-380.

Newstead, R. 1917b. Observations on scale-insects (Coccidae) - IV. *Bulletin of Entomological Research* 8: 1-34.

Newstead, R. 1917c. Observations on scale-insects (Coccidae) - V. *Bulletin of Entomological Research* 8: 125-134.

Newstead, R. 1920. Observations on scale-insects (Coccidae) - VI. *Bulletin of Entomological Research* 10: 175-207.

Newstead, R. 1922. A new southern Nigerian *Lecanium* (Coccidae). *Transactions of the Entomological Society of London* (1921): 530-531.

Nietner, J. 1861. Observations on the enemies of the coffee tree in Ceylon. Ceylon Times. 31 pp.

Nietner, J. 1880. The coffee tree and its enemies: being observations on the natural history of the enemies of the coffee tree in Ceylon. A.M. & J. Ferguson, Colombo.

Nunes, J.F.R. 1977. In Memoriam Armando Jacques Favre Castel-Branco (1909-1977). *Garcia de Orta. Serie de Zoologia, Lisboa* 6: 1-2.

Nuorteva, M. 1974. Die Schildlaus *Eriopeltis lichtensteini* Sign. (Hem., Coccidae) auch in Sudwest-Hame verbreitet. *Lounais-Hameen Luonto. Forssa, Finland* 52: 17-18.

Nur, U. 1963. Meiotic parthenogenesis and heterochromatization in a soft scale, *Pulvinaria hydrangeae* (Coccoidea: Homoptera). *Chromosoma (Berlin)* 14: 123-139.

Nur, U. 1971. Parthenogenesis in coccids (Homoptera). *American Zoologist* 11: 301-308.

Nur, U. 1972. Diploid arrhenotoky and automictic thelytoky in soft scale insects (Lecaniidae: Coccoidea: Homoptera). *Chromosoma (Berlin)* 39: 381-401.

Nur, U. 1979. Gonoid thelytoky in soft scale insects (Coccidae: Homoptera). *Chromosoma (Berlin)* 72: 89-104.

Nur, U. 1980. Evolution of unusual chromosome systems in scale insects (Coccoidea: Homoptera). pp. 97-117. In: R.L. Blackman, G.M. Hewitt, & M. Ashburner (eds). Insect Cytogenetics. Royal Entomological Society, London.

Nuzzaci, G. 1969. Nota morfo-biologica sull'*Eulecanium corni* (Bouché) spp. *apuliae* nov. *Entomologica. Bari* 5: 9-36.

Obenberger, J. 1932. K sedessatínam Prof. Dra. Karla Sulce. [In Czech]. *Acta Societatis Entomologicae Cechosloveniae* 29: 89-98.

Ohgushi, R. 1986a. Geographical and ecological information on three species of *Ceroplastes* scales obtained by questionnaires. I. Geographical distribution in Japan. *Japanese Journal of Applied Entomology and Zoology* 30: 59-62.

Ohgushi, R. 1986b. Geographical and ecological information on three species of *Ceroplastes* scales obtained by questionnaires. II. Habitat and key host plant. *Japanese Journal of Applied Entomology and Zoology* 30: 147-149.

Ohgushi, R. 1987a. Geographical and ecological information on three species of *Ceroplastes* scales obtained by questionnaires. III. Population trends in recent years. *Japanese Journal of Applied Entomology and Zoology* 31: 82-85.

Ohgushi, R. 1987b. Geographical and ecological information on three species of *Ceroplastes* scales obtained by questionnaires. IV. Control activities in orchards and gardens. *Japanese Journal of Applied Entomology and Zoology* 32: 75-77.

Ohgushi, R. & Nishino, T. 1975. Comparative studies on the population dynamics of wax scales belonging to the genus *Ceroplastes*. 1. Survivorship curves and life tables. *Annual Report, Botanic Garden, Faculty of Science, University of Kanazawa, Japan* 7-8: 1-21.

Olivier, G.A. 1791. Cochenille. *Coccus*. Genre d'insectes de la premiere section de l'ordre des Hemipteres. pp. 85-100. In: G.A. Olivier (ed.). Encyclopedie methodique., Paris.

Olivier, G.A. 1792. *Kermes. Chermes*. Genre d'insectes de la premiere section de l'ordre des Hemipteres. pp. 418-442. In: G.A. Olivier (ed.). Encyclopedie methodique., Paris.

Olliff, S. 1891. A new scale-insect destroying saltbush. *Agricultural Gazette, New South Wales* 2: 667-669.

Olliff, S. 1892. Further remarks on the saltbush scale (*Pulvinaria maskelli*, Oll.). *Agricultural Gazette, New South Wales* 3: 176-179.

Öncüer, C. 1974. Recherches sur les cochenilles de *Coccus* (Homoptera: Coccidae) faisant des dégâts sur agrumes dans la région egéene, leurs caractèrs morphologiques, leurs repartitions et leurs ennemis naturels. [In Turkish, French summary]. *Bitki Koruma Bulteni. Ankara* 1: 1-59.

Opinion, 228. 1954. Rejection for nomenclatorial purposes of Geoffroy, 1762, "Histoire abregee des insectes qui se trouvent aux environs de Paris". *Opinions and Declarations Rendered by The International Commission on Zoological Nomenclature* 4: 209-220.

Opinion, 268. 1954. Acceptance of the generic name *Aspidoproctus* (Class Insecta, Order Hemiptera) as from Newstead, 1901. *Opinions and declarations rendered by the International Commission on Zoological Nomenclature* 5: 397-408.

Opinion, 1192. 1981. *Lecanium acuminatum* Signoret, 1873 (Insecta, Homoptera, Coccidae): Neotype designated. *Bulletin of Zoological Nomenclature* 38: 252-253.

Opinion, 1303. 1985. *Coccus* Linnaeus, 1758 and *Parthenolecanium* Sulc, 1908 (Insecta, Hemiptera, Homoptera): Type species designated. *Bulletin of Zoological Nomenclature* 42: 139-141.

Opinion, 1627. 1991. *Saissetia* Deplanche, 1859 (Insecta, Homoptera): *Lecanium coffeae* Walker, 1852 designated as type species. *Bulletin of Zoological Nomenclature* 48: 72-73.

Ossiannilsson, F. 1951. Bidrag till kännedomen om dem svenska sköldlusfauna (Hom. Coccoidea). *Opuscula Entomologica. Lund* 16: 1-9.

Ossiannilsson, F. 1959. Contributions to the knowledge of the Swedish coccid fauna (Hom. Coccoidea) II. [In Swedish, English summary]. *Opuscula Entomologica. Lund* 24: 193-201.

Ossiannilsson, F. 1985. Nya fynd av sköldlöss i Sverige. [In Swedish, English summary]. *Entomologisk Tidskrift, Uppsala* 106: 145-146.

Otanes, F.Q. 1936. Some observations on two scale insects injurious to mango flowers and fruits. *The Philippine Journal of Agriculture* 7: 129-141.

Paik, W.H. 1978. Insecta VI. Coccoidea. [In Korean]. *Illustrated Flora & Fauna of Korea* 22: 1-481.

Paloukis, S.S. 1979. The most important scale insects of fruit trees in Northern Greece. [In Greek]. Thessaloniki. 148 pp.

Panis, A. 1975. Une pulvinaire de la canne à sucre d'introduction recente au Maroc (Homoptera, Coccoidea, Coccidae). *Revue de Zoologie Agricole et de Pathologie Végétale* 74: 147-153.

Panis, A. 1983. Lutte biologique contre la cochenille noire *Saissetia oleae* (Olivier) dans le cadre de la lutte intégrée en oleiculture française. *Symbioses* 15: 63-74.

Panis, A. & Marro, J.P. 1977. L'élevage massif de *Chloropulvinaria urbicola* (Cockerell) (Homoptera, Coccoidea). *Fruits* 32: 599-606.

Panis, A. & Martin, H.E. 1976. Cochenilles des plantes cultivees en République Dominicaine (Homoptera: Coccoidea) (premiere liste). *Bulletin Mensuel de la Société Linnéenne de Lyon* 45: 7-8.

Paoli, G. 1916. Contributo alla conoscenze della cocciniglie della Sardegna. *Redia* (1915) 11: 239-268.

Paraskakis, M., Neuenschwander, P. & Michelakis, S. 1980. *Saissetia oleae* (Oliv.) (Hom., Coccidae) and its parasites on olive trees in Crete, Greece. *Zeitschrift für Angewandte Entomologie* 90: 450-464.

Parida, B.B. & Ghosh, D. 1984. Studies on the chromosome constitution in 6 species of scale insects (Coccoidea: Homoptera) from India. *Chromosome Information Service, Tokyo* 37: 14-15.

Parida, B.B. & Moharana, S. 1982. Studies on the chromosome constitution in 42 species of scale insects (Coccoidea: Homoptera) from India. *Chromosome Information Service, Tokyo* 32: 18-20.

Pawlak, J.H., Tempesta, M.S., Iwashita, T., Nakanishi, K. & Naya, Y. 1983. Structure of sesterpenoides from the scale insect *Ceroplastes ceriferus. Chemistry Letters. Chemical Society of Japan* 1983: 1069-1072.

Pearson, E.O. 1965. Dr. W.J. Hall, C.M.G., M.C. *Nature* 206: 983-984.

Pearson, G. 1794. Observations and experiments on a wax-like substance, resembling the Pe-la of the Chinese, collected at Madras by Dr. Anderson, and called by him White Lac. *Philosophical Transactions* (1794): 383-401.

Pechhacker, H. 1976. Zur Vorhersage der Honigtautracht von *Physokermes hemicryphus* Dalm. (Homoptera, Coccidae) auf der Fichte (*Picea excelsa*). *Apidologie* 7: 209-236.

Pechhacker, H. 1977. Neue Ergebnisse der Honigtauforschung. *Anzeiger Schadlingskunde, Pflanzenschutz, Umweltschutz* 50: 45-47.

Peleg, B.A. 1965. Observations on the life cycle of the black scale, *Saissetia oleae* Bern., on citrus and olive trees in Israel. *Israel Journal of Agricultural Research* 15: 21-26.

Peleg, B.A. 1987. Resistance of adult Florida wax scale to commercial formulations of Carbaryl. [In Hebrew]. *Alon Hanotea* 41: 601-603.

Peleg, B.A. & Gothilf, S. 1981. Effect of the insect growth regulators Diflubenzuron and Methoprene on scale insects. *Journal of Economic Entomology* 74: 124-126.

Pellizzari Scaltriti, G. 1976. Sulla presenza in Italia dell'*Eupulvinaria hydrangeae* (Steinw.) (Homoptera, Coccoidea). *Redia* 59: 59-67.

Pellizzari Scaltriti, G. 1977. Un coccide Pulvinariino nuovo per l'Italia: la *Neopulvinaria imeretina* Hadz. *Redia* 60: 423-430.

Pellizzari Scaltriti, G. 1981. Osservazioni biologiche sulla *Euphilippia olivina* Berl. & Silv. nel Veneto (Homoptera, Coccidae). *Memorie della Società Entomologica Italiana* 60: 289-297.

Pellizzari Scaltriti, G. 1989. Gli insetti delle piante officinali. II Nota: I fitofagi del *Thymus*, con particolare riferimento alle cocciniglie. *Redia* 72: 567-579.

Pellizzari Scaltriti, G. 1991. Recenti acquisizioni sulla fauna italiana degli Homoptera Coccoidea. *Atti XVI Congresso Italiano di Entomologia, Bari - Martina Franca (Ta) 23/28 Settembre* 1991: 763-769.

Pena, J.E., Baranowski, R.M. & Litz, R.E. 1987. Life history, behavior and natural enemies of *Philephedra tuberculosa* (Homoptera: Coccidae). *The Florida Entomologist* 70: 423-427.

Pérez Guerra, G. & Carnero Hernandez, A. 1985. Nuevas aportaciones al estudio de los coccidos en Canarias. *Boletim da Sociedad Portuguesa de Entomologia, Supplemento* no. 1: 313-321.

Pergande, T. 1898. The peach *Lecanium. (Lecanium nigrofasciatum* n. sp.). *Bulletin United States Department of Agriculture. Bureau of Entomology. Washington* 18: 26-29.

Peringuey, L. 1893. Note on a supposed new *Icerya*. *Transactions of the South African Philosophical Society. Cape Town* (1890-1893) 8: 50-51.

Pesson, P. 1984. Hommage a Alfred Serge Balachowsky (1901-1983). *Annales de la Société Entomologique de France (N.S.)* 20: 235-250.

Petch, T. 1915. [Obituary, Andrew Rutherford]. *Spolia Zeylanica* 10 (37).

Peterson, L.O.T. 1960. *Lecanium coryli* L. (Homoptera: Coccoidea) in Saskatchewan. *Canadian Entomologist* 92: 851-857.

Pettit, R.H. & McDaniel, E. 1920. The Lecania of Michigan. *Michigan Agricultural College Experiment Station Technical Bulletin* 48: 1-35.

Phillips, J.H.H. 1955. Identity of a cottony scale on peach in Ontario. *Canadian Entomologist* 87: 245.

Phillips, J.H.H. 1962. Description of the immature stages of *Pulvinaria vitis* (L.) and *P. innumerabilis* (Rathvon) (Homoptera: Coccoidea), with notes on the habits of these species in Ontario, Canada. *Canadian Entomologist* 94: 497-502.

Phillips, J.H.H. 1963. Life history and ecology of *Pulvinaria vitis* (L.) (Hemiptera: Coccoidea), the cottony scale attacking peach in Ontario. *Canadian Entomologist* 95: 372-407.

Phillips, J.H.H. 1965a. Notes on species of *Lecanium* Burmeister (Homoptera: Coccoidea) in the Niagara Peninsula, Ontario with a description of a new species. *Canadian Entomologist* 97: 231-238.

Phillips, J.H.H. 1965b. Biological and behavioural differences between *Lecanium cerasifex* Fitch and *Lecanium putmani* Phillips (Homoptera: Coccoidea). *Canadian Entomologist* 97: 303-309.

Planchon, G. 1864. Le *Kermes* du chêne aux pointes de vue zoologique, commercial & pharmacetique. De Boehm & Fils, Montpellier. 47 pp.

Podoler, H., Bar-Zacay, I. & Rosen, D. 1979a. Population dynamics of the Mediterranean black scale, *Saissetia oleae* (Olivier), on citrus in Israel. 1. A partial life-table. *Journal of the Entomological Society of southern Africa* 42: 257-266.

Podoler, H., Bar-Zacay, I. & Rosen, D. 1979b. Population dynamics of the Mediterranean black scale, *Saissetia oleae* (Olivier), on citrus in Israel. II. Distribution within the citrus tree. *Journal of the Entomological Society of southern Africa* 42: 267-273.

Podoler, H., Dreishpoun, Y. & Rosen, D. 1981. Population dynamics of the Florida wax scale, *Ceroplastes floridensis* (Homoptera: Coccidae) on citrus in Israel. 1. - A partial life-table. *Acta Oecologica* 2: 81-91.

Podsiadlo, E. & Komosinska, H. 1976. Further investigations on the scale insects fauna (Homoptera, Coccoidea) in the Nida Valley (Southern Poland). *Bulletin de l'Académie Polonaise des Sciences. Série des Sciences Biologiques* 24: 87-91.

Podsiadlo, E. 1983. Notes on scale insects (Homoptera, Coccoidea) found in Crete and their parasites. *Fragmenta Faunistica. Warsaw* 27: 271-277.

Pollard, G.V. & Alleyne. 1986. Insect pests as constraints to the production of fruits in the Caribbean. *Proceedings of a Seminar on Pests and Diseases as Constraints in the Production and Marketing of Fruits, Barbados, October* 1985: 31-61.

Priore, R. 1967. Osservazioni biologiche sul *Ceroplastes sinensis* Del Guerc. (Homoptera - Coccidae); in particolare suo sviluppo sul pero e numero di generazioni. *Bollettino del Laboratorio di Entomologia Agraria "Filippo Silvestri", Portici* 25: 46-56.

Putnam, J.D. 1880. Biological and other notes on Coccidae. *Proceedings of the Davenport Academy of Natural Sciences* (1879-1880) 2: 293-347.

Qin, T.K. 1990. The Australian cottony soft scales (Homoptera: Coccoidea: Pulvinariini). *Proceedings of the Sixth International Symposium of Scale Insects Studies, Cracow, Poland, August* 1990: 79-81.

Qin, T.K. & Gullan, P.J. 1989. *Cryptostigma* Ferris: a coccoid genus with a strikingly disjunct distribution (Homoptera: Coccidae). *Systematic Entomology* 14: 221-232.

Qin, T.K. & Gullan, P.J. 1991. A new species of *Kilifia* De Lotto from Guizhou, China (Homoptera: Coccidae). *Entomotaxonomia* 13: 21-24.

Qin, T.K. & Gullan, P.J. 1992. A revision of the Australian pulvinariine soft scales (Insecta: Hemiptera: Coccidae). *Journal of Natural History* 26: 103-164.

Quaglia, F. & Raspi, A. 1979a. Osservazioni eco-etologiche su un lecaniide dannoso all'olivo in Toscana: *Euphilippia olivina* Berlese e Silvestri (Rhynchota, Coccoidea). *Frustula Entomologica, Pisa (N.S.)* 2: 85-112.

Quaglia, F. & Raspi, A. 1979b. Note eco-etologiche sulla *Philippia oleae* (O.G. Costa) (Rhynchota, Coccoidea), lecaniide infeudato sull'olivo in Toscana. *Frustula Entomologica, Pisa (N.S.)* 2: 197-229.

Quaintance, A.L. 1898. New and little-known Coccidae from Florida. II. Biological observations. *Psyche* 8: 91.

Quayle, H.J. 1911. The black scale *Saissetia oleae* Bern. *Bulletin University of California Publications, Agricultural Experiment Station, Berkeley* 223: 1-200.

Quayle, H.J. 1915. The citricola scale. *Bulletin University of California Publications, Agricultural Experiment Station, Berkeley* 255: 405-421.

Quintana, F.J. 1956a. Descripción del macho de "*Pulvinaria flavescens*" Brethes (Homoptera Stern.). *Revista de la Facultad de Agronomia. Argentina, La Plata* 32: 67-74.

Quintana, F.J. 1956b. *Pulvinaria mesembryanthemi* (Vallot) (Homoptera Stern.) nueva cochinilla para la fauna Argentina y sus zooparasitos. *Revista de la Facultad de Agronomia. Argentina, La Plata* 32: 75-110.

Rabkin, F.B. & Lejeune, R.R. 1954. Some aspects of the biology and dispersal of the Pine Tortoise Scale, *Toumeyella numismaticum* (Pettit and McDaniel) (Homoptera: Coccidae). *Canadian Entomologist* 86: 570-575.

Ramakrishna Ayyar, T.V. 1919. A contribution to our knowledge of South Indian Coccidae. *Bulletin of the Agricultural Research Institute, Pusa, India* 87: 1-50.

Ramakrishna Ayyar, T.V. 1926. Recent additions to the Indo-Ceylonese coccid fauna, with notes on known and new forms. *Journal of the Bombay Natural History Society* 31: 450-457.

Ramakrishna Ayyar, T.V. 1930. A contribution to our knowledge of South Indian Coccidae (Scales and Mealybugs). *Bulletin of the Imperial Institute of Agricultural Research, Pusa, India* 197: 1-73.

Ramanarayan, E.P., Chacko, M.J. & Ramachandran, M. 1980. Studies on stylet penetration in *Saissetia coffeae*. *Journal of Coffee Research* 10: 71-74.

Rao, V.P. 1939. Three new coccids from Baluchistan. *Indian Journal of Entomology* 1: 59-63.

Rathvon, S.S. 1854. *Coccus innumerabilis*. (bark louse). *Pennsylvania Farm Journal* 4: 256-258.

Rathvon, S.S. 1876. Answers to Correspondence. Scale Insects. *The Lancaster Farmer, Pennsylvania* 8: 101-102.

Ratzeburg, J.T.C. 1843. Bericht über einige neue, den Waldbäumen schadliche Rhynchoten. *Stettiner Entomologische Zeitung* 4: 201-204.

Ray, C.H. & Williams, M.L. 1980. Description of the immature stages and adult male of *Pseudophilippia Quaintancii* (Homoptera: Coccoidea: Coccidae). *Annals of the Entomological Society of America* 73: 437-447.

Ray, C.H. & Williams, M.L. 1981. Redescription and lectotype designation of the tessellated scale, *Eucalymnatus tessellatus* (Signoret) (Homoptera: Coccidae). *Proceedings of the Entomological Society of Washington* 83: 230-244.

Ray, C.H. & Williams, M.L. 1982. Descriptions of the immature stages of *Protopulvinaria pyriformis* (Cockerell) (Homoptera: Coccidae). *The Florida Entomologist* 65: 169-176.

Ray, C.H. & Williams, M.L. 1983. Description of the immature stages and adult male of *Mesolecanium cornuparvum* (Homoptera: Coccidae). *Proceedings of the Entomological Society of Washington* 85: 161-173.

Reh, L. 1903. Zur Naturgeschichte Mittel-und Nordeuropäischer Schildläuse. *Allgemeine Zeitschrift für Entomologie* 8: 301-308, 351-356, 407-419, 457-469.

Řeháček, J. 1954. Lecaniinae of Dr. K. Šulc's collection of scale insects. [In Czech, English summary]. *Acta Musei Moraviae* 39: 133-145.

Řeháček, J. 1955. Contribution to the knowledge of the scale insects fauna fam. Lecaniidae of Czechoslovakia (Coccoidea, Lecaniidae). [In Czech]. *Acta Societatis Entomologicae Cechosloveniae* 51: 219-223.

Řeháček, J. 1957. Neue Schildlausefunde aus der Unterfamilie der Lecaniinae in der Tschechoslowakei (Coccoidea: Lecaniinae). *Acta Faunistica Entomologica Musei Nationalis Pragae* 2: 13-18.

Řeháček, J. 1959. A new species of the genus *Luzulaspis* (Coccoidea, Coccidae) from Czechoslovakia. [In Russian, English summary]. *Entomologicheskoe Obozrenye* 38: 176-178.

Remaudière, G. 1984. Alfred-Serge Balachowsky 1901-1983. *International Organization for Biological Control Newsletter* 31-32: 1-3.

Reyne, A. 1949. Nederlandse Coccidae. *Tijdschrift voor Entomologie, Amsterdam* 91: lxix-lxxi.

Reyne, A. 1951. Nieuwe Coccidae voor Nederland. *Tijdschrift voor Entomologie, Amsterdam* 94: xxxix-xli.

Reyne, A. 1957. Snavelinsecten - Rhynchota, I, Nederlandse schildluizen (Coccidae). *Wetenschappelijke Mededlingen van de Koninklijke Nederlandes Natuurhistorische Vereniging* 22: 1-44.

Reyne, A. 1961. Scale insects from Dutch New Guinea. *Beaufortia. Amsterdam* 8: 121-167.

Reyne, A. 1963a. Scale insects from Thailand with description of a *Filippia* n. sp. *Beaufortia. Amsterdam* 10: 29-39.

Reyne, A. 1963b. In memoriam Dr. Harold Morrison. *Entomologische Berichten. Amsterdam* 23: 176.

Reyne, A. 1964. Scale insects from the Netherlands Antilles. *Beaufortia. Amsterdam* 11: 95-130.

Reyne, A. 1965. Observations on some Indonesian scale insects. *Tijdschrift voor Entomologie, Amsterdam* 108: 145-188.

Ricci, J.G. 1985. Morfologia comparativa, datos biologicos y habito predator de *Heperaspis brethesi* n. sp., *Azya bioculata* Gordon (Col., Coccinellidae), y *Pyconcephalus argentinus* Brethes (Col., Nitidulidae) de Tucuman (Argentina). *CIRPON Revista de Investigacion, Argentina* 3: 53-70.

Richards, W.R. 1958. Identities of species of *Lecanium* Burmeister in Canada (Homoptera: Coccoidea). *Canadian Entomologist* 90: 305-313.

Richards, W.R. 1964. The scale insects of the Canadian Arctic (Homoptera: Coccoidea). *Canadian Entomologist* 96: 1457-1462.

Riddick, E. 1955. A list of Florida plants and the scale-insects which infest them. *Bulletin of the State Plant Board of Florida* 7: 1-78.

Riley, C.V. & Howard. 1892. Jamaica museum notes. *Insect Life* 5: 139-140.

Ringuelet, E.J. 1924. Contribucion al estudo de la "*Pulvinaria flavescens*" Brethes. *Anales de la Sociedad Científica Argentina* 97: 61-80.

Rios, T. 1966. Ceroalbolinic acid and a new Anthraquinone pigment isolated from *Ceroplastes albolineatus*. *Tetrahedron Letters* 22: 1507-1512.

Rios, T. & Colunga, F. 1965. Three new alcohols from insect wax. Ceroplastol I, II and Albolineol. *Chemistry and Industry* 26: 1184-1185.

Rios, T. & Gomez, G.F. 1969. Albolic-acid, a new sesterpenic acid isolated from insect wax. *Tetrahedron Letters* 34: 2929-2293.

Rios, T. & Perez, C.S. 1969. Geranylfarnesol, a new acyclic C25 isoprenoid alcohol isolated from insect wax. *Chemical Communications* 5: 214-215.

Rios, T. & Quijano, L. 1969. The structure of Ceroplastol II, a sesterterpenic alcohol isolated from insects wax. *Tetrahedron Letters* 17: 1317-1318.

Robinson, E. 1917. Coccidae of the Philippine Islands. *Philippine Journal of Science (Ser. D)* 12: 1-47.

Rondani, G. 1876. Repertorio degli insetti parasiti e delle loro vittime. *Supplemento. Bollettino della Societa Entomologica Italiana* 8: 237-258.

Rosciszewska, M. 1989. Structure and sense organs of antennae in females of Coccoidea (Homoptera, Coccinea). *Zeszyty Naukowe Akademii Rolniczej im. H. Kollataja w Krakowie, Rozprawa habilitacyjna* 129: 1-70.

Rossi, P. 1794. Mantissa insectorum exhibens species nuper in Etruria. Typographia Prosperi, Pisa.

Rubin, A.Y. & Beirne, B.P. 1975a. Natural enemies of the European fruit Lecanium, *Lecanium tiliae* (Homoptera: Coccidae), in British Columbia. *Canadian Entomologist* 107: 337-342.

Rubin, A.Y. & Beirne, B.P. 1975b. The European fruit Lecanium, *Lecanium tiliae* (L.) (Homoptera: Coccidae), in Southwestern British Columbia. *Journal of the Entomological Society of British Columbia* 72: 18-20.

Russell, L.M. 1960. Notes on the entomological writings of Asa Fitch, M.D., with special reference to his catalogues and reports. *Annals of the Entomological Society of America* 53: 326-327.

Russell, L.M. 1963. Harold Morrison 1890-1963. *Proceedings of the Entomological Society of Washington* 65: 311-313.

Russell, L.M. 1970. Additions and corrections to an annotated list of generic names of the scale insects (Homoptera: Coccoidea). *Miscellaneous Publications United States Department of Agriculture. Supplement* 1015: 1-13.

Russell, L.M., Kosztarab, M. & Kosztarab, M.P. 1974. A selected bibliography of the Coccoidea second supplement. *Miscellaneous Publications United States Department of Agriculture* 1015: 1-122.

Russo, A. 1989. Check-list of scale insects (Homoptera: Coccoidea) reported for Sicilian fauna. *Phytophaga. Palermo* (1985-1989) 3: 147-162.

Russo, A. & Longo, S. 1991. A new species of *Scythia* Kiritshenko (Homoptera: Coccoidea: Coccidae) from Mount Etna, Italy. *Israel Journal of Entomology* (1990) 24: 1-4.

Russo, G. 1949. Filippo Silvestri. Nota necrologica. Curriculum vitae. Elenco delle publicazioni. *Bollettino del R. Laboratorio di Entomologia Agraria di Portici* 9: i-xlix.

Rutherford, A. 1914. Some Ceylon Coccidae. *Bulletin of Entomological Research* 5: 259-268.

Rutherford, A. 1915a. Some new Ceylon Coccidae. *Journal of the Bombay Natural History Society* 24: 111-118.

Rutherford, A. 1915b. Notes on Ceylon Coccidae. *Spolia Zeylanica* 10: 103-115.

Saakyan-Baranova, A.A. 1964. On the biology of the soft scale *Coccus hesperidum* L. (Homoptera, Coccoidea). [In Russian]. *Entomologicheskoe Obozrenye* 43: 268-296.

Sabine, B.N.E. 1969. Insecticidal control of citrus pests in coastal central Queensland. *Queensland Journal of Agricultural and Animal Sciences* 26: 83-88.

Sachtleben, H. 1944. Über einen Rest der Sammlung Bouché's und die in ihm enthaltenen Cocciden. (Diptera, Hymenoptera & Hemiptera - Homoptera: Coccoidea). *Arbeiten über Morphologische und Taxonomische Entomologie, Berlin-Dahlem* 11: 65-76.

Salama, H.S. & Saleh, M.R. 1970. Distribution of the scale insect *Pulvinaria psidii* Maskell (Coccoidea) on orchard trees in relation to environmental factors. *Zeitschrift für Angewandte Entomologie* 66: 380-385.

Sampedro, G. 1984. In Memoriam Raúl Macgregor-Loaeza (1926-1983). *Anales del Instituto de Biología. Universidad Nacional Autónoma de México. Serie Zoología* 55: 295-298.

Sampedro, G. & Butze, J.R. 1984. Descripción de una nueva especie de la familia Coccidae de México (Homoptera: Coccoidea). *Anales del Instituto de Biología. Universidad Nacional Autónoma de México. Serie Zoología* 55: 143-150.

Sanders, J.G. 1906. Catalogue of recently described Coccidae. *Technical Series. Bureau of Entomology, United States Department of Agriculture. Washington* 12: 1-18.

Sanders, J.G. 1909a. The identity and synonymy of some of our soft scale-insects. *Journal of Economic Entomology* 2: 428-448.

Sanders, J.G. 1909b. Catalogue of recently described Coccidae - II. *Technical Series. Bureau of Entomology, United States Department of Agriculture. Washington* 16: 33-60.

Sanders, J.G. 1910. A review of the Coccidae described by Dr. Asa Fitch. *Proceedings of the Entomological Society of Washington* 12: 56-61.

Sankaran, T. 1954. The natural enemies of *Ceroplastes pseudoceriferus* Green (Hemiptera - Coccidae). *Journal of Scientific Research of Benares Hindu University* 5: 100-120.

Sankaran, T. 1959. The life-history and biology of the wax-scale, *Ceroplastes pseudoceriferus* Green (Coccidae: Homoptera). *Journal of the Bombay Natural History Society* 56: 39-59.

Sankaran, T. 1962. The external characters of the post-larval stages of the wax scale, *Ceroplastes pseudoceriferus* Green (Hemiptera: Coccidae). *Indian Journal of Entomology* 24: 1-18.

Santas, L.A. 1985. *Parthenolecanium corni* (Bouche) an orchard scale pest producing honeydew foraged by bees in Greece. *Entomologica Hellenica* 3: 53-58.

Santas, L.A. 1988. *Physokermes hemicryphus* (Dalman) a fir scale insect useful to apiculture in Greece. *Entomologica Hellenica* 6: 11-21.

Sasscer, E.R. 1911. Catalogue of recently described Coccidae - III. *Technical Series. Bureau of Entomology, United States Department of Agriculture. Washington* 16: 61-74.

Sasscer, E.R. 1912. Catalogue of recently described Coccidae - IV. *Technical Series. Bureau of Entomology, United States Department of Agriculture. Washington* 16: 83-97.

Sasscer, E.R. 1915. Catalogue of recently described Coccidae - V. *Proceedings of the Entomological Society of Washington* 17: 25-38.

Savescu, A. 1943. Oekoarten bei Lecaniden. *Bulletin de la Section Scientifique, Académie Roumaine, Bucarest* 25: 212-223.

Savescu, A. 1944. Formes écologiques des lécanides de la faune Roumaine. *Bulletin de la Section Scientifique, Académie Roumaine, Bucarest* 27: 230-246.

Savescu, A. 1960. Album de Protectia Plantelor. Vol. 1. [In Rumanian]. Ministry of Agriculture, Romania, Bucarest. 270 pp.

Savescu, A. 1961. Album de Protectia Plantelor. Vol 2. [In Rumanian]. Ministry of Agriculture, Romania, Bucarest. 90 pp.

Savescu, A. 1983. Espèces de coccoidées nouvelles pour la science signalées en Roumanie. I. *Pulvinaria ampelopsidis* Savescu sp.n. et *Pulvinaria euonymicola* Savescu sp.n. (Homoptera - Coccidae). *Bulletin de l'Académie des Sciences Agricoles et Forestières, Bucarest* 12: 43-50.

Savescu, A. 1985. Espèces de coccoidées nouvelles pour la science signalées en Roumanie. III. Espèces appartenant aux genres *Pseudococcus* Westw., *Phenacoccus* Ckll., *Paroudablis* Ckll.,

Eupeliococcus Savescu et *Lepidosaphes* Shimer (Homoptera - Coccoidea). *Bulletin de l'Academie des Sciences Agricoles et Forestieres, Bucarest* 14: 103-130.

Schmutterer, H. 1952a. Die Ökologie der Cocciden (Homoptera, Coccoidea) Frankens. 2. Abschnitt. *Zeitschrift für Angewandte Entomologie* 33: 544-584.

Schmutterer, H. 1952b. Zur Kenntnis der Schildlausfauna Bayerns (Homopt. Coccoidea). *Nachrichtenblatt der Bayerischen Entomologen* 1: 14-15, 18-21.

Schmutterer, H. 1954. Zur Kenntnis einiger wirtschaftlich wichtiger mitteleuropäischer *Eulecanium* - Arten (Homoptera: Coccoidea: Lecaniidae). *Zeitschrift für Angewandte Entomologie* 36: 62-83.

Schmutterer, H. 1955a. Neue Beitrag zur deutschen Coccidenfauna: Zur Schildlausfauna der Nördlichen Kalkalpen. *Bericht über die 7. Wanderversammlung Deutscher Entomologen. Berlin* 1954: 159-164.

Schmutterer, H. 1955b. Bemerkenswerte Schildlausfunde in Süd- und Südwestdeutschland (Homopt., Coccoidea). *Nachrichtenblatt der Bayerischen Entomologen, München* 4: 98-102.

Schmutterer, H. 1956. Zur Morphologie, Systematik und Bionomie der *Physokermes*-Arten an Fichte (Homopt. Cocc.). *Zeitschrift für Angewandte Entomologie* 39: 445-466.

Schmutterer, H. 1972. Coccidae (Lecaniidae). pp. 405-418. In: W. Schwenke (ed.). Die Forstschädlinge Europas. P. Parey. Hamburg, Berlin.

Schmutterer, H. 1980. Zum Stand der Erforschung der Schildläuse (Homoptera, Coccoidea) in der Bundesrepublik Deutschland. *Mitteilungen der Deutschen Gesellschaft für Allgemeine und Angewandte Entomologie. Bremen* 2: 49-56.

Schneider, B., Podoler, H. & Rosen, D. 1987a. Population dynamics of the Florida wax scale, *Ceroplastes floridensis* (Homoptera:

Coccidae), on citrus in Israel. 2. - Spatial distribution. *Acta Oecologica* 8: 67-78.

Schneider, B., Podoler, H. & Rosen, D. 1987b. Population dynamics of the Florida wax scale, *Ceroplastes floridensis* (Homoptera: Coccidae), on citrus in Israel. 3 - Developmental rate and progression of mean age. *Acta Oecologica* 8: 95-103.

Schrank, F. 1781. Enumeratio insectorum Austriae indigenorum. Augustae Vindelicorum. 552 pp.

Schrank, F. 1801. Fauna Boica., Nürnberg. 374 pp.

Shafee, S.A., Yousuf, M. & Khan, M.Y. 1989. Host plants and distribution of coccid pests (Homoptera: Coccoidea) in India. *Indian Journal of Systematic Entomology* 6: 47-55.

Shah, A.H., Jhala, C., Patel, G.M. & Patel, C.B. 1986. Occurrence of waxy scale, *Ceroplastes actiniformis* Green, on mango in South Gujarat. *Indian Journal of Agricultural Sciences* 56: 67.

Sheffer, B.J. & Williams, M.L. 1990. Descriptions, distribution, and host-plant records of eight first instars in the genus *Toumeyella* (Homoptera: Coccidae). *Proceedings of the Entomological Society of Washington* 92: 44-57.

Shinji, O. 1935a. Two new species of non-armoured scale insects from North East Japan. [In Japanese]. *Oyo-Dobutsugako Zasshi, Tokyo (Japanese Society for Applied Zoology)* 7: 288-290.

Shinji, O. 1935b. A list of the unarmed scale insects collected in the Prefecture of Iwate, with the descriptions of 7 new species. [In Japanese, English summary]. *Dobutsugako Zasshi, Tokyo (Journal of the Zoological Society of Japan)* 47: 767-777.

Shmelev, G.P. 1971. A new Coccidae species *Rhizopulvinaria saxosa* Shmelev sp. n. (Homoptera, Coccoidea, Coccidae) from Western Pamir. [In Russian]. *Doklady Akademii Nauk Tadzhikiskoy SSR* 14: 61-63.

Shmelev, G.P. 1975. Morphology and metamorphosis of the willow scale - *Pulvinaria salicicola* Borchs. (Homoptera, Coccoidea, Coccidae). [In Russian]. pp. 86-93. In: U.L. Shchetkii & N.N. Muminov (eds). Entomologiya Tadzhikistana. Donish, Dushanbe.

Signoret, V. 1868. Essai sur les cochenilles (Homoptères - Coccides) 1re partie. *Annales de la Société Entomologique de France (ser. 4)* 8: 503-528.

Signoret, V. 1869a. Essai sur les cochenilles (Homoptères - Coccides), 2e partie. *Annales de la Société Entomologique de France (ser. 4)* 8: 829-876.

Signoret, V. 1869b. Quelques observations sur les cochenilles connues sous le nom de Pou à poche blanche qui ravagent les plantations de cannes à sucre à l'île Maurice et a l'île de la Reunion. *Annales de la Société Entomologique de France (ser. 4)* 9: 93-96.

Signoret, V. 1869c. Essai sur les cochenilles ou gallinsectes (Homoptères - Coccides) 3e partie. *Annales de la Société Entomologique de France (ser. 4)* 9: 97-104.

Signoret, V. 1872a. Essai sur les cochenilles ou gallinsectes (Homoptères - Coccides), 8e partie. *Annales de la Société Entomologique de France (ser. 5)* 1: 421-434.

Signoret, V. 1872b. Essai sur les cochenilles ou gallinsectes (Homoptères - Coccides), 9e partie. *Annales de la Société Entomologique de France (ser. 5)* 2: 33-46.

Signoret, V. 1873a. Essai sur les cochenilles ou gallinsectes (Homoptères - Coccides), 10e partie. *Annales de la Société Entomologique de France (ser. 5)* 3: 27-48.

Signoret, V. 1873b. Essai sur les cochenilles ou gallinsectes (Homoptères - Coccides), 11e partie. *Annales de la Société Entomologique de France (ser. 5)* 3: 395-402.

Signoret, V. 1874a. Essai sur les cochenilles ou gallinsectes (Homoptères - Coccides), 11e partie. *Annales de la Société Entomologique de France (ser. 5)* 3: 402-448.

Signoret, V. 1874b. Essai sur les cochenilles ou gallinsectes (Homoptères - Coccides), 12e partie. *Annales de la Société Entomologique de France (ser.* 5) 4: 87-106.

Signoret, V. 1875b. Essai sur les cochenilles ou gallinsectes (Homoptères - Coccides), 15e, 16e et 17e parties. *Annales de la Société Entomologique de France (ser.* 5) 5: 305-394.

Signoret, V. 1877a. Essai sur les cochenilles ou gallinsectes (Homoptères - Coccides), 18e partie et dernière partie. *Annales de la Société Entomologique de France (Ser.* 5) 6: 591-676.

Signoret, V. 1877b. [Notes on various Hemiptera]. *Bulletin de la Séances de la Société Entomologique de France (ser.* 5) 7: xxxvi.

Signoret, V. 1886. [Communication sur *Lecanopsis dugesii* n. sp.]. *Bulletin de la Séances de la Société Entomologique de France (ser.* 6) 6: xxxix.

Silva, A.G.A., Gonçalves, C.R., Galvao, D.M., Goncalves, A.J.L., Gomes, J., Silva, M.N. & Simoni, L. 1968. Quarto catálogo dos insetos que vivem nas plantas do Brasil, seus parasitas e predatores: insetos, hospedeiros e inimigos naturais. Ministerio da Cultura, Rio de Janeiro. 622 pp.

Silvestri, F. 1915. Contributo alla conoscenza degli insetti dell'olivo dell'Eritrea e dell'Africa meridionale. *Bolletino del Laboratorio di Zoologia Generale e Agraria della R. Scuola Superiore d'Agricoltura in Portici* 9: 240-334.

Silvestri, F. 1918. Gustavo Leonardi, Necrologio. *Bolletino del Laboratorio di Zoologia Generale e Agraria della Facoltà Agraria in Portici* 11: 291-298.

Silvestri, F. 1919. Contribuzioni alla conoscenza degli insetti dannosi e dei loro simbionti. V. La cocciniglia del nocciuolo (*Eulecanium coryli* L.). *Bolletino del Laboratorio di Zoologia Generale e Agraria della R. Scuola Superiore d'Agricoltura in Portici* 13: 127-192.

Silvestri, F. 1920a. Descrizione e notizie del *Ceroplastes sinensis* D. Guerc. (Hemiptera, Coccidae). *Bolletino del Laboratorio di Zoologia*

Generale a Agraria della R. Scuola Superiore d'Agricoltura in Portici 14: 3-17.

Silvestri, F. 1920b. Appendice. pp. 501-539. In: G. Leonardi (ed.). Monografia delle cocciniglie Italianae. Della Torre, Portici. 555 pp.

Silvestri, F. 1939. Compendio di Entomologia Applicata. Parte speciale. Tipografia Bellavista, Portici. 974 pp.

Simmonds, F.J. 1957. A list of the Coccidae of Bermuda and their parasites. *Bulletin of the Department of Agriculture, Bermuda* 30: 1-12.

Simpson, J.D. & Lambdin, P.L. 1983. Life history of the tuliptree scale, *Toumeyella liriodendri* (Gmelin), on yellow-poplar in Tennessee. *Tennessee Farm & Home Science. Knoxville* 125: 2-5.

Sinha, P.K. & Dinesh, D.S. 1984. A report on the coccids (Hemiptera: Coccoidea), their host plants and natural enemies at Bhagalpur. *Biological Bulletin of India* 6: 7-13.

Siraiwa, H. 1939. Notes on and descriptions of the coccids of southern Saghalien. *Kontyu* 13: 63-75.

Smith, D. 1970. White wax scale and its control. *Queensland Agricultural Journal* 96: 704-708.

Smith, D. & Ironside, D.A. 1974. The seasonal history of *Gascardia destructor* (Newstead) in Queensland. *Queensland Journal of Agricultural and Animal Sciences* 31: 195-199.

Smith, R.H. 1944. Bionomics and control of the nigra scale, *Saissetia nigra. Hilgardia* 16: 225-288.

Snowball, G.J. 1969. Prospects for biological control of white wax scale (*Gascardia destructor*) in Australia by South African natural enemies. *Journal of the Entomological Society of Australia (N.S.W.)* 5: 23-33.

Snowball, G.J. 1970. *Ceroplastes sinensis* Del Guercio (Homoptera: Coccidae), a wax scale new to Australia. *Journal of the Australian Entomological Society* 9: 57-66.

Speight, M. & Nicol, M. 1984. Horse Chestnut scale - a new urban menace? *New Scientist. London* 101: 40-42.

Steinweden, J.B. 1929. Bases for the generic classification of the coccoid family Coccidae. *Annals of the Entomological Society of America* 22: 197-245.

Steinweden, J.B. 1930. Characteristics of some of our California soft scale insects (Coccidae). *Monthly Bulletin Department of Agriculture State of California* 19: 561-571.

Steinweden, J.B. 1945. Identification and control of orchid pests. *Orchid Digest* 9 (2): 264-267.

Steinweden, J.B. 1946. The identity of certain common American species of *Pulvinaria* (Homoptera: Coccoidea: Coccidae). (Contribution no. 49). *Microentomology* 11: 1-28.

Stimmel, J.F. 1987. The scale insects of Pennsylvania greenhouses. Pennsylvania Department of Agriculture, Harrisburg. 35 pp.

Strickland, A.H. 1947. Coccids attacking cacao *(Theobroma cacao* L.), in West Africa, with descriptions of five new species. *Bulletin of Entomological Research* 38: 497-523.

Su, T.H. 1982. List of the mango Coccoidea (Homoptera: Sternorrhyncha) in Taiwan. *Proceedings of the National Science Council. Part A: Applied Science. Taiwan* 6: 60-63.

Šulc, K. 1895. Description of a new species of *Lecanium* from Bohemia. *Entomologist's Monthly Magazine* 6: 37-38.

Šulc, K. 1898. Studie o coccidech. -II. [In Czech]. *Sitzungsberichte der K. Böhmisch Gesellschaft der Wissenschaften* (1897) 66: 1-19.

Šulc, K. 1908. Towards the better knowledge of the genus *Lecanium*. *Entomologist's Monthly Magazine* 44: 36.

Šulc, K. 1912. Coccidae regni Bohemiae, in literatura adhuc commemoratore. *Acta Societatis Entomologicae Bohemiae* 9: 30-38.

Šulc, K. 1931. O skladbě voskov ch štětů a jejich příslušn ch žlaz u samců puklic (*Lecanium*, Coccidae). *Sbornik Přírodovědecké Společnosti V Mor. Ostravě* (1930-1931): 85-96.

Šulc, K. 1932. Československé druhy rodu puklice (gn. *Lecanium*, Coccidae, Homoptera). Die Tschechoslowakischen Lecanium-Arten. *Acta Societatis Scientiarum Naturalium Moravicae* 7: 1-134.

Šulc, K. 1936. Die weiblichen Geschlechtsorgane von *Xylococcus filiferus* Loew 1882. [In Czech]. *Ceskoslovensko Zool. Spolec. Vestnik* 3: 60-68.

Šulc, K. 1941. *Mohelnia festuceti* n. gn., n. sp. (Lecaniidae, Eriopeltini, Coccoidea, Hemiptera). *Acta Societatis Scientiarum Naturalium Moravicae* 13: 1-17.

Šulc, K. 1942a. Diagnosis differentialis inter gn. *Pulvinaria* et *Phyllostroma* n. gn. *Folia Entomologica. Brno* 5: 5-8.

Šulc, K. 1942b. Trachealni system u *Phyllostroma ericae* Loew 1883 (Coccoidea). *Folia Entomologica. Brno* 5: 8-15.

Sulzer, J.H. 1776. Abgekürzte Geschichte der Insecten nach dem Linaeischen System. Steiner, Winterthur. 274 pp.

Suomalainen, E. 1962. Significance of parthenogenesis in the evolution of insects. *Annual Review of Entomology* 7: 349-366.

Takagi, S. 1975. Coccoidea collected by the Hokkaido University expedition to Nepal, Himalaya, 1968 (Homoptera). *Insecta Matsumurana (New Series)* 6: 1-33.

Takagi, S. 1977. Scale insects collected on citrus and other plants and their hymenopterous parasites in Thailand. *Insecta Matsumurana (New Series)* 11: 61-72.

Takahashi, R. 1928a. Coccidae of Formosa. *The Philippine Journal of Science* 36: 327-347.

Takahashi, R. 1928b. Coccidae of Formosa (2). *Transactions of the Natural History Society of Formosa. Taihoku* 18: 253-261.

Takahashi, R. 1929a. Observations on the Coccidae of Formosa. - 1. *Report, Government Research Institute, Department of Agriculture, Formosa* 40: 1-82.

Takahashi, R. 1929b. Aphididae and Coccidae of the Pescadores. [In Japanese]. *Transactions of the Natural History Society of Formosa* 19: 425-431.

Takahashi, R. 1930. Observations on the Coccidae of Formosa, Part II. *Report Department of Agriculture Government Research Institute, Formosa* 43: 1-45.

Takahashi, R. 1931. Some Coccidae of Formosa. *Transactions of the Natural History Society of Formosa* 21: 1-5.

Takahashi, R. 1932a. Records and descriptions of Coccidae from Formosa. Part 2. *Journal of the Society of Tropical Agriculture. Formosa* 4: 41-48.

Takahashi, R. 1932b. New food plants of the Coccidae in Formosa. [In Japanese]. *Transactions of the Formosa Natural History Society* 22: 102-105.

Takahashi, R. 1933. Observations on the Coccidae of Formosa. III. *Report. Government Research Institute. Department of Agriculture. Formosa* 60: 1-64.

Takahashi, R. 1935. Observations on the Coccidae of Formosa. V. *Report Department of Agriculture Government Research Institute. Formosa* 66: 1-37.

Takahashi, R. 1936. Some Aleyrodidae, Aphididae, Coccidae (Homoptera), and Thysanoptera from Micronesia. *Tenthredo. Acta Entomologica. Takeuchi Entomological Laboratory* 1: 109-120.

Takahashi, R. 1939a. Descriptions of three Malayan Coccidae (Hemiptera). *Transactions of the Natural History Society of Formosa* 29: 111-118.

Takahashi, R. 1939b. Life history and control methods of *Pulvinaria polygonata*. [In Japanese]. *Formosan Agricultural Review* 35: 403-414.

Takahashi, R. 1939c. Some Aleyrodidae, Aphididae, and Coccidae from Micronesia (Homoptera). *Tenthredo. Acta Entomologica. Takeuchi Entomological Laboratory. Kyoto* 2: 234-272.

Takahashi, R. 1939d. A new scale insect attacking lemon in Formosa. *Transactions of the Natural History Society of Formosa* 29: 314-316.

Takahashi, R. 1939e. Two new nondiaspine Coccidae from Borneo and Malaya (Homoptera). *Annotationes Zoologicae Japonenses. Tokyo* 18: 323-326.

Takahashi, R. 1940a. Some Coccidae from Formosa and Japan (Homoptera), V. *Mushi. Fukuoka Entomological Society* 13: 18-28.

Takahashi, R. 1940b. Insects of the Sternorrhyncha (Hemiptera) of Daito Jima, the Loochoo Islands. [In Japanese]. *Transactions of the Natural History Society of Formosa* 30: 327-332.

Takahashi, R. 1941. Some species of Aleyrodidae, Aphididae, and Coccidae from Micronesia (Homoptera). *Tenthredo. Acta Entomologica. Takeuchi Entomological Laboratory. Kyoto* 3: 208-220.

Takahashi, R. 1942a. Some injurious insects of agricultural plants and forest trees in Thailand and Indo-China. II. Coccidae. *Report. Government Research Institute. Department of Agriculture. Formosa* 81: 1-56.

Takahashi, R. 1942b. A new scale insect from Hainan (Homoptera). *Zoological Magazine. Tokyo* 54: 500-501.

Takahashi, R. 1942c. Some species of Aleyrodidae, Aphididae and Coccidae in Micronesia (Homoptera). *Tenthredo. Acta Entomologica. Takeuchi Entomological Laboratory. Kyoto* 3: 349-358.

Takahashi, R. 1950a. Some species of Coccidae from the Riouw Islands - Part I. (Homoptera). *Insecta Matsumurana* 17: 65-72.

Takahashi, R. 1950b. *Paralecanium* and *Platylecanium* from the Malay Peninsula (Coccidae, Homoptera). *Transactions of the Kansai Entomological Society* 15: 48-60.

Takahashi, R. 1951a. Three new myrmecophilous scale insects from the Malay Peninsula (Homoptera). *Mushi. Fukuoka Entomolgical Society* 22: 1-8.

Takahashi, R. 1951b. Some species of Coccidae from the Riouw Islands - Part II. *Insecta Matsumurana* 17: 103-112.

Takahashi, R. 1952. Some species of nondiaspine scale insects from the Malay Peninsula. *Insecta Matsumurana* 18: 9-17.

Takahashi, R. 1955a. *Protopulvinaria* and *Luzulaspis* of Japan (Coccidae, Homoptera). *Annotationes Zoologicae Japonenses. Tokyo* 28: 35-39.

Takahashi, R. 1955b. *Lecanium* of Japan (Homoptera: Coccidae). *Transactions of the Shikoku Entomological Society* 4: 69-78.

Takahashi, R. 1955c. Notes on the *Coccus* of Japan (Homoptera: Coccidae). *Transactions of the Shikoku Entomological Society* 4: 78.

Takahashi, R. 1955d. Key to the genera of Coccidae in Japan, with descriptions of two new genera and little-known species (Homoptera). *Insecta Matsumurana* 19: 23-28.

Takahashi, R. 1955e. *Pulvinaria* of Japan (Coccidae, Homoptera). *Kontyu* 23: 148-154.

Takahashi, R. 1955f. Some scale insects of the Loochoo Islands (Homoptera). *Bulletin of the Biogeographical Society of Japan* 16-19: 238-242.

Takahashi, R. 1956. *Pulvinaria* of Japan (Coccidae, Homoptera). *Kontyu* 24: 23-30.

Takahashi, R. 1957. A new species of *Eriopeltis* from Japan (Coccoidea, Homoptera). *Akitu* 6: 65-66.

Takahashi, R. 1959. Two new genera of Coccidae (Homoptera). *Kontyu* 27: 74-76.

Takahashi, R. & Tachikawa, T. 1956. Scale insects of Shikoku (Homoptera: Coccoidea). *Transactions of the Shikoku Entomological Society* 5: 1-17.

Talhouk, A.M.S. 1969. Insects and mites injurious to crops in Middle Eastern countries. *Monographien zur Angewandten Entomologie* 21: 1-239.

Talhouk, A.M.S. 1975. Citrus pests throughout the world. *Ciba- Geigy Agrochemicals, Technical Monograph* 4: 21.

Tamaki, Y. 1964a. Amino acids in the honeydew excreted by *Ceroplastes pseudoceriferus* (Green). *Japanese Journal of Applied Entomology and Zoology* 8: 159-164.

Tamaki, Y. 1964b. Carbohydrates in the honeydew excreted by *Ceroplastes pseudoceriferus* (Green). *Japanese Journal of Applied Entomology and Zoology* 8: 227-234.

Tamaki, Y. 1966. Chemical composition of the wax secreted by a scale insect (*Ceroplastes pseudoceriferus* Green). *Lipids* 1: 297-300.

Tamaki, Y. & Kawai, S. 1966. Seasonal changes of the wax covering and its components of a scale insect, *Ceroplastes pseudoceriferus* Green. *Scientific Pest Control Kyoto* 31: 148-153.

Tamaki, Y. & Kawai, S. 1967. Fatty acids, alcohols and hydrocarbons in the body lipid of *Ceroplastes pseudoceriferus* Green, *Ceroplastes japonicus* Green and *Ceroplastes rubens* Maskell (Homoptera: Coccidae). *Scientific Pest Control Kyoto* 32: 63-69.

Tammes, P.M.L. & van Eyndhoven, G.L. 1985. Dr. A. Reyne - Obituary. *The Scale* 11: 4-5.

Tanaka, M. 1953. The settlement of the red scale and the horned wax scale on potato tubers. 1. The study of mass production of the important Hymenopterous parasite (*Anicetus ceroplastis* Ishii). [In

Japanese, English summary]. *Bulletin of the Kyushu Agricultural Experiment Station* 2: 55-63.

Tang, F.T. 1977. The scale insects of horticulture and forest of China. Vol. I. [In Chinese]. The Institute of Gardening - Forestry Science of Shenyang, Liaoning, China. 259 pp.

Tang, F.T. 1984c. Observation on the scale insects injurious to forestry of North China. *Shanxi Agricultural University Press Research Publication* 2: 122-133.

Tang, F.T. 1991. The Coccidae of China. [In Chinese, English Abstract]. Shanxi United Universities Press, Shanxi. 377 pp.

Tang, F.T., Hao, J.J., Xie, Y.P. & Tang, Y. 1990. Family group classification of Asiatic Coccidae (Homoptera: Coccoidea: Coccidae). *Proceedings of the sixth International Symposium of Scale Insects Studies, Cracow, Poland, August* 1990: 75-77.

Tang, F.T. & Li, J. 1988. Observations on the Coccoidea of Inner Mongolia in China. [In Chinese, English Summary]. Inner Mongolia University Press. 227 pp.

Tao, C.C.C. 1978. Check list and host plant index to scale insects of Taiwan, Republic of China. *Journal of Agricultural Research of China. Taiwan* 27: 77-141.

Tao, C.C.C. 1989. Scale insects name list of Taiwan, Republic of China. *Bulletin Taichung District Agricultural Improvement Station* 22: 57-70.

Tao, C.C.C., Wong, C.Y. & Chang, Y.C. 1983. Monograph of Coccidae of Taiwan, Republic of China (Homoptera: Coccoidea). *Journal of Taiwan Museum* 36: 57-107.

Targioni Tozzetti, A. 1866a. Come certe cocciniglie sieno cagione di alcune melate delle piante, e di alcune ruggini; e come la cocciniglia de fico dia in abbondanza una specie di cera. *Atti della R. Accademia dei Georgofili (N.S.)* 13: 115-137.

Targioni Tozzetti, A. 1866b. Appendice. Come la nostra cocciniglia da cera possa essere il tipo di un genere di insetti a cui conviene dare il nome di *Columnea*. *Atti della R. Accademia dei Georgofili (N.S.)* 13: 138-146.

Targioni Tozzetti, A. 1867. Studii sulle Cocciniglie. *Memorie della Società Italiana di Scienze Naturali. Milano* 3(3): 1-87.

Targioni Tozzetti, A. 1868. Introduzione alla seconda memoria per gli studi sulle cocciniglie, e catalogo dei generi e delle specie della famiglia dei coccidi. *Atti della Società Italiana di Scienze Naturali* 11: 721-738.

Targioni Tozzetti, A. 1877. *Myxolecanium kibarae* Beccari (Lecaniti). *Bolletino della Società Entomologica Italiana. Firenze* 9: 317-320.

Targioni Tozzetti, A. 1884. Art. V. Ord. Omotteri. Fam. Coccidei. *Annali di Agricoltura. Ministero di Agricoltura, Industria e Commercio. Firenze, Roma*: 383-414.

Targioni Tozzetti, A. 1885. Note sopra alcune cocciniglie (coccidei). *Bulletino della Società Entomologica Italiana* 17: 100-120.

Targioni Tozzetti, A. 1893. Note sur une espèce de laque provenant de Madagascar et sur la laque rouge des Indes avec apreçu sur les insectes qui les produisent. pp. 88-91. In: A. Gascard. Contribution à l'étude des gommes laques des Indes & de Madagascar. Société d'Editions Scientifiques, Paris, 124 pp.

Targioni Tozzetti, A. 1895. Sopra une specie di lacca del Madagascar e sopra gli insetti che vi si trovano. I. Lacca del Madagascar (*Gascardia madagascariensis* n.g.; n.sp.). *Bolletino della Società Entomologica Italiana* (1894) 26: 457-464.

Ter-Grigorian, M.A. 1962. The Coccoidea of the woods of Armenia (Insecta, Homoptera, Coccoidea). [In Russian, Armenian summary]. *Zoological Papers, Academy of Sciences of Armenian SSR, Zoological Institute* 12: 125-161.

Ter-Grigorian, M.A. 1980. Species of the genus *Rhizopulvinaria* (Homoptera, Coccoidea, Coccidae) in Armenia and Nakhichevan

Autonomous Republic. [In Russian, Armenian and English summary]. *Zoologicheskii Zhurnal Armenia* 33: 265-270.

Teran, A.L. 1973. Entomofauna del dominio subandino. I. Las "cochinillas" (Hom., Coccoidea) de *Larrea divaricata* y *L. cuneifolia* (Zygophyllaceae). *Acta Zoologica Lilloana. Tucuman* 30: 189-206.

Teran, A.L. & Guyot, N.H. 1969. La cochinilla del delta, *Lecanium deltae* (Lizer) (Hom., Coccoidea), en Tucuman. *Acta Zoologica Lilloana. Tucuman, Argentina* 24: 135-149.

Tereznikova, E.M. 1959. On scale insect fauna (Insecta, Homoptera, Coccoidea) of Ukraine. [In Ukrainian; Russian & English summary]. *Dopovidi Akademii Nauk Ukrainskoi RSR* 6: 683-685.

Tereznikova, E.M. 1963a. Trophic association of scale insects (Insecta, Homoptera, Coccoidea) of the Transcarpatian region. [In Russian]. *Flora i Fauna Karpat* 2: 182-191.

Tereznikova, E.M. 1963a. Ecologo-Faunistic survey of scale insects (Homoptera, Coccoidea) of Ukrainian woodlands. [In Ukrainian, Russian summary]. *Prazi Institutu Zoologii, Akademiya Nauk Ukrainskoi RSR* 19: 41-57.

Tereznikova, E.M. 1963b. A preliminary ecologic-faunistic review of the coccids (Coccoidea) of the Kanevsk Forest Reserve. [In Russian]. *Sbornik Rabot Laboratorii Arakhno-Entomologii Kievskogo Universiteta* 1963: 151-157.

Tereznikova, E.M. 1963c. Ecologo-Geographic groups of coccids (Homoptera, Coccoidea) of the Ukrainian Polessye. [In Ukrainian, Russian & English summaries]. *Dopovidi Akademii Nauk Ukrainskoi RSR* 11: 1527-1529.

Tereznikova, E.M. 1966. Ecologic-faunistic survey of the scale insects (Homoptera, Coccoidea) of the Transcarpatian Region. [In Ukrainian]. *Problemii Zoologii, Akademiya Nauk Ukrainskoi RSR* 1966: 20-37.

Tereznikova, E.M. 1967. Redescription of the genus *Psilococcus* Borchs. (Homoptera, Coccoidea) of the USSR fauna. [In Russian;

English summary]. *Vestnik Zoologii. Institut Zoologii Akademiya Nauk Ukrainskoi SSR. Kiev* 4: 22-27.

Tereznikova, E.M. 1981. Scale insects. Eriococcidae, Kermesidae and Coccidae. [In Ukrainian]. *Fauna Ukraini. Akademiya Nauk Ukrainskoi RSR. Institut Zoologii. Kiev* 20: 1-215.

Theron, J.G. 1958. Comparative studies on the morphology of male scale insects (Hemiptera: Coccoidea). *Annals of the University of Stellenbosch (Section A)* 34: 1-71.

Thro, W.C. 1903. Distinctive characteristics of the species of the genus *Lecanium. Bulletin. Cornell University Agricultural Experiment Station. Ithaca, N.Y.* 209: 205-221.

Tiensuu, L. 1951. Notes on the smallreed coccid, *Eriopeltis lichtensteini* Sign. (Hem., Coccoidea, Lecanoidae) and its natural enemies in Finland. *Annales Entomologici Fennici* 17: 3-10.

Tondeur, R., Schiffers, B., Verstraeten, C. & Merlin, J. 1990. Chemical methods in an integrated action against *Eupulvinaria hydrangeae* in Belgium (Homoptera: Coccoidea: Coccidae). *Proceedings of the Sixth International Symposium of Scale Insects Studies, Cracow, Poland, August* 1990: 157-158.

Townsend, C.H.T. 1892a. Scale insects in New Mexico. *Bulletin of New Mexico Agricultural Experiment Station* 7: 1-23.

Townsend, C.H.T. 1892b. Notes on two Mexican species of *Ceroplastes*, with a record of parasites reared from one. *ZOE. A Biological Journal. San Francisco* 3: 255-257.

Townsend, C.H.T. & Cockerell, T.D.A. 1898. Coccidae collected in Mexico by Messrs. Townsend and Koebele in 1897. *Journal of the Entomological Society of New York* 6: 165-180.

Tranfaglia, A. 1974. Studi sugli Homoptera Coccoidea. III. Un nuovo coccino (*Coccus aegaeus* De Lotto) sugli agrumi in Italia (Notizie preliminari). *Bolletino del Laboratorio di Entomologia Agraria "Filippo Silvestri". Portici* 31: 141-144.

Tranfaglia, A. 1976. Studi sugli Homoptera Coccoidea. IV. Su alcune Cocciniglie nuove o poco conosciute per l'Italia (Coccidae, Eriococcidae, Pseudococcidae). *Bollettino del Laboratorio di Entomologia Agraria 'Filippo Silvestri'. Portici* 33: 128-143.

Tranfaglia, A. 1977. Etude des espèces de *Saissetia* dans le Bassin Méditerranéen (Homoptera, Coccoidea, Coccidae). *Fruits* 32: 545-547.

Tranfaglia, A. 1980. Morphological observations on the chinese wax scale, *Ceroplastes sinensis* Del Guercio. *Fruits* 35: 701-704.

Tranfaglia, A. 1981. Studi sugli Homoptera Coccoidea. V. Notizie morfo-sistematiche su alcune specie di Cocconiglie con descrizione di tre nuove specie di Pseudococcidi. *Bollettino del Laboratorio di Entomologia Agraria 'Filippo Silvestri'. Portici* 38: 3-28.

Tranfaglia, A. 1983. Reperti su Pseudococcidae e Coccidae (Homoptera: Coccoidea) nuovi per la fauna italiana. *Atti XIII Congresso Nazionale Italiano di Entomologia* (1983): 453-458.

Tranfaglia, A. & Marotta, S. 1982. Studi sugli Homoptera Coccoidea. VI. Due nuove cocciniglie sud-africaine sulle coltivazioni di gerani (*Pelargonium* spp. e *Geranium* spp.). *Bollettino del Laboratorio di Entomologia Agraria 'Filippo Silvestri'. Portici* 39: 53-58.

Tranfaglia, A. & Viggiani, G. 1988. Cocciniglie d'importanza economica in Italia e loro controllo. Istituto di Entomologia Agraria dell'Università di Napoli - Portici. 30 pp.

Tremblay, E. 1988a. Entomologia applicata. Volume secondo, Parte prima. Liguori Editore, Napoli. 329 pp.

Tremblay, E. 1988b. Avversità delle colture di recente o temuta introduzione parassiti animali: insetti. *L'Italia Agricola* 125: 115-128.

Tseng, S. 1947. A new scale insect parasitic on the ivy. *Tezisy Statei Izd. Kitaiskogo Nauchnogo Obsh.* 1:21. (Original publication not seen; reference taken from Borchsenius, 1957).

Tyrell, W.M. 1896. The Lecaniums of California. *Annual Report California Agricultural Experiment Station* (1894-1895): 262-270.

Valemberg, J. 1980. Observations sur *Calymnatus hesperidum* L. (=*Lecanium hesperidum*) (Hemiptères Coccidae). *Bulletin de la Société Entomologique du Nord de la France* 216: 5-8.

Vallot, J.N. 1829. Nouvelle Espèce de cochenilles. *Comptes Rendues de la Academie des Sciences, Arts et Belles-lettres de Dijon* (1828-1829): 30-33.

Varshney, R.K. 1984. New records of host-plants and distribution of some coccids from India (Homoptera: Coccoidea). *Bulletin of the Zoological Survey of India* 6: 137-142.

Varshney, R.K. 1985. A review of Indian coccids (Homoptera: Coccoidea). *Oriental Insects* 19: 1-102.

Varshney, R.K. & Moharana, S. 1987. Insecta: Homoptera: Coccoidea. *Fauna of Orissa: State Fauna Series* 1: 161-181.

Venkatraman, T.V. 1941. Observations on the biology of some South Indian coccids. *Journal of the Bombay Natural History Society* 42: 847-853.

Vernalha, M.M. 1953. Coccideos da coleção I.B.P.T. *Arquivos de Biologia e Tecnologia, Curitiba* 8: 111-304.

Vernalha, M.M. 1957. Contribuição para o conhecemiento dos coccídeos (Homoptera, Coccoidea) de Ilex sp. no Estado do Paraná. *Boletim Instituto de Biologia e Pesquisas Tecnologicas. Curitiba* 39: 1-52.

Vernalha, M.M., Gabrado, J.C. & Da Silva, R.P. 1974. Coccídeos do Brasil V - Coccidae. *Ceroplastes (Octoceroplastes) hempeli* Lizer, 1919. *Acta Biologica Paranense. Curitiba* 3: 127-131.

Vieira, R.M.S., Carmona, M.M. & Pita, M.S. 1983. Sobre os coccídeos do Arquipelago da Madeira (Homoptera - Coccoidea). *Boletim do Museu Municipal do Funchal. Madeira* 35: 81-162.

Viggiani, G. 1973. Le specie descritte da Filippo Silvestri (1873-1949). *Bolletino del Laboratorio di Entomologia Agraria "Filippo Silvestri", Portici* 30: 351-.

Vilar, J.M.D.S. 1951. Subdidio para o estudo dos *Ceroplastes* spp. (Insecta - Coccidae) de Portugal. I. *Broteria. Lisboa* 20: 111-136, 177-188.

Vilar, J.M.D.S. 1952. Subdidio para o estudo dos *Ceroplastes* spp. (Insecta - Coccidae) de Portugal. II. *Broteria. Lisboa* 21: 1-62.

Waite, G.K. 1986. Pests of lychee in Australia. *Proceedings of the First National Lychee Seminar, 14-15 February 1986, Australia*: 42-43.

Waku, Y. & Foldi, I. 1984. The fine structure of insect glands secreting waxy substances. pp. 303-322. In: R.C. King & H. Akai (eds.). Insect Ultrastructure. Plenum.

Walker, F. 1852. List of the specimens of homopterous insects in the collection of the British Museum, Part IV. British Museum (Natural History), London. 1188 pp.

Walsh, B.D. & Riley, C.V. 1868. A new bark louse on the osage orange. *The American Entomologist* 1: 14.

Wan, S.Y., Wan, M. & Young, B.L. 1985. Studies on a new pest of medicinal herb, *Mallococcus viticicola* Young, n. sp. (Homoptera: Coccoidea: Lecaniodiaspididae). [In Chinese, English abstract]. *Contributions of The Shanghai Institute of Entomology* 5: 267-274.

Wang, T.C. 1976. Two new species of coccids on tea bush (Homoptera: Coccoidea). [In Chinese, English abstract]. *Acta Entomologica Sinica* 19: 342-34.

Wang, T.C. 1980. Handbook for the determination of the common coccoids. [In Chinese]. Chinese Academy of Sciences. 252 pp.

Wang, T.C. 1981. Homoptera: Coccoidea. [In Chinese, English abstract]. *Insects of Xizang [=Tibet]* 1: 283-294.

Washburn, J.O. & Frankie, G.W. 1985. Biological studies of iceplant scales, *Pulvinariella mesembryanthemi* and *Pulvinaria delottoi* (Homoptera: Coccidae), in California. *Hilgardia* 53: 1-27.

Washburn, J.O. & Washburn, L. 1984. Active aerial dispersal of minute wingless arthropods: exploitation of boundary-layer velocity gradients. *Science* 223: 1088-1089.

Watt, G. & Mann, H.H. 1903. The Pests and Blights of the Tea Plant. Government of India, Printing Office, Calcutta. 429 pp.

Way, M.J. 1954. Studies on the association of the ant *Oecophylla longinoda* (Latr.) (Formicidae) with the scale insect *Saissetia zanzibarensis* Williams (Coccidae). *Bulletin of Entomological Research* 45: 113-134.

Weber, W.A. 1965. Theodore Dru Alison Cockerell, 1866-1948. *University of Colorado Studies, Series in Bibliography* 1: 1-124.

Weidner, V.H. & Wagner, W. 1968. Die Entomologischen Samlungen des Zoologischen Staatsinstituts und Zoologichen Museums Hamburg VII. Teil. Insecta IV. *Mitteilungen des Hamburgerischen Zoologisches Museums und Instituts* 65: 123-180.

Westwood, J.O. 1840. An introduction to the modern classification of insects; founded on the natural habits and corresponding organization of different families. Longman, Orme, Brown & Green, London. 587 pp.

Westwood, J.O. 1853a. *Coccus sinensis* sp.n. *Proceedings of the Royal Entomological Society of London (N.S.)* 2: 95.

Westwood, J.O. 1853b. Wax insects. *Gardeners' Chronicle and Agricultural Gazette* 31: 484.

Westwood, J.O. 1853c. The wax insect of China, *Coccus Pe-La*. *Gardeners' Chronicle and Agricultural Gazette* 31: 532.

Westwood, J.O. 1870. [The Camellia Coccus - *Coccus flocciferus* Westwood]. *The Gardeners' Chronicle and Agricultural Gazette* 10: 308.

Westwood, J.O. 1871a. [Exhibitions.]. *Proceedings of the Entomological Society of London* (1871): iii.

Westwood, J.O. 1871b. The *Cypripedium* scale insect. *The Gardeners' Chronicle and Agricultural Gazette* 26: 1006.

White, A. 1846. Descriptions of some apparently new species of homopterous insects in the collection of the British Museum. *Annals and Magazine of Natural History* 17: 330-333.

Wiggins, I.L. 1958. Gordon Floyd Ferris, 1893-1958 - Memorial Number. *Microentomology* 23: 65-92.

Williams, D.J. 1953. On a new species of *Saissetia* (Hem.: Coccoidea) from Zanzibar. *Bulletin of Entomological Research* 44: 581-582.

Williams, D.J. 1957. The status of *Coccus palmae* Haworth and the identity of *Lecanium coffeae* Walker (Coccoidea: Homoptera). *The Entomologist* 90: 314-315.

Williams, D.J. 1963. Some taxonomic notes on the Coccoidea (Homoptera). *The Entomologist* 96: 100-101.

Williams, D.J. 1969. The family-group names of the scale insects (Hemiptera: Coccoidea). *Bulletin of the British Museum (Natural History) Entomology* 23: 315-341.

Williams, D.J. 1977. A new species of *Exaeretopus* Newstead (Homoptera: Coccidae) attacking wheat in Iraq. *Bulletin of Entomological Research* 67: 281-284.

Williams, D.J. 1980. The identity of *Lecanium krugeri* Zehntner (Hemiptera: Coccidae) and its distribution on sugar-cane in southern Asia. *Bulletin of Entomological Research* 70: 435-437.

Williams, D.J. 1982a. The distribution and synonymy of *Coccus celatus* De Lotto (Hemiptera: Coccidae) and its importance on coffee in Papua New Guinea. *Bulletin of Entomological Research* 72: 107-109.

Williams, D.J. 1982b. *Pulvinaria iceryi* (Signoret) (Hemiptera: Coccidae) and its allies on sugar-cane and other grasses. *Bulletin of Entomological Research* 72: 111-117.

Williams, D.J. 1985a. Some scale insects (Hom. Coccoidea) from the island of Nauru. *Entomologist's Monthly Magazine* 121: 53.

Williams, D.J. 1985b. T.D.A. Cockerell's scale insects (Homoptera: Coccoidea) in the British Museum (Natural History). *Folia Entomologica Hungarica* 46: 215-240.

Williams, D.J. 1985c. Scale insects (Homoptera: Coccoidea) of Tresco, Isles of Scilly. *Entomologist's Gazette* 36: 135-144.

Williams, D.J. & Watson, G.W. 1990. The scale insects of the Tropical South Pacific Region, Part 3. The soft scales (Coccidae) and other families. C.A.B. International, Wallingford. 267 pp.

Williams, J.R. & Williams, D.J. 1980. Excretory behaviour in soft scales (Hemiptera: Coccidae). *Bulletin of Entomological Research* 70: 253-257.

Williams, J.R. & Williams, D.J. 1988. Homoptera of the Mascarene islands - an annotated catalogue. *Entomology Memoirs Department of Agriculture and Water Supply. Republic of South Africa* 72: 1-98.

Williams, M.L. & Kosztarab, M. 1972. Morphology and systematics of the Coccidae of Virginia with notes on their biology (Homoptera: Coccoidea). *Research Division Bulletin, Virginia Polytechnic Institute and State University* 74: 1-215.

Willis, J.C. 1988. A dictionary of the flowering plants & ferns. 8th Edition. Cambridge University Press, Cambridge. 1245 pp.

Wise, K.A.J. 1977. A synonymic checklist of the Hexapoda of the New Zealand sub-region. The smaller orders. *Bulletin of the Auckland Institute and Museum. Auckland, New Zealand* 11: 1-176.

Wu, C.F. 1935. Family Coccidae. pp. 169-252. In: C.F. Wu (ed.). Catalogus Insectorum Sinensium.

Wünn, H. 1921. *Physokermes graniformis* n. sp. *Zeitschrift für Wissenschaft Insektenbiologie. Beilage: Neue Beiträge zur systematischen Insektenkunde* 2: 29.

Wünn, H. 1937. Zur Coccidenfauna von Schleswig-Holstein (12. Mitteilung uber Cocciden). *Schriften des Naturwissenschaftlichen Vereins für Schleswig-Holstein* 22: 1-69.

Wünn, H. 1939. *Eulecanium bulgariense* n. sp., ein Schädling der bulgarischen Ölrosenkulturen aus der Familie der Cocciden. *Zeitschrift für Angewandte Entomologie* 25: 703-708.

Wysoki, M. 1987. A bibliography of the pyriform scale, *Protopulvinaria pyriformis* (Cockerell, 1894) (Homoptera: Coccidae), up to 1986. *Phytoparasitica* 15: 73-77.

Yang, P.L. 1982. Synopsis of Chinese scale insects. [In Chinese]. Shanghai Science & Technology, Shanghai. 425 pp.

Yardeni, A. 1987. Evaluation of wind dispersed soft scale crawlers (Homoptera: Coccidae), in the infestation of a citrus grove in Israel. *Israel Journal of Entomology* 21: 25-31.

Yardeni, A. & Rosen, D. 1990. Wind dispersal and pattern of colonization of *Ceroplastes floridensis* on citrus in Israel (Homoptera: Coccoidea: Coccidae). *Proceedings of the Sixth International Symposium of Scale Insects Studies, Cracow, Poland, August 1990*: 125-128.

Yousuf, M. & Shafee, S.A. 1988. Four new species of Coccidae (Homoptera) from Andaman Islands. *Indian Journal of Systematic Entomology* 5: 57-63.

Zahradnik, J. 1959a. Cervci - Coccinea. *Bestimmungstabellen der Tschechoslowakischen Fauna* 3: 527-552.

Zahradnik, J. 1959b. Kritická bibliografie Červcu Československa. *Acta Faunistica Entomologica Musei Nationalis Pragae. Supplementum* 1: 7-69.

Zahradnik, J. 1977. Aleyrodinea - Coccinea. *Acta Faunistica Entomologicae Musei Nationalis Pragae. Supplement* 4: 117-122.

Zahradnik, J. 1987. Neue Schildläuse in der Tschechoslowakischen Fauna (Sternorrhyncha, Coccinea). *Acta Universitatis Carolinae - Biologica* (1985): 355-365.

Zak-Ogaza, B. & Koteja, J. 1964. Investigations on scale insects (Homoptera, Coccoidea) of the Pieniny Mountains. *Acta Zoologica Cracoviensia* 9: 417-439.

Zehntner, L. 1897. Overzicht van de Ziekten van het Suikerriet op Java. *Archief voor Java-Suikerindustrie* 5: 525-575.

Zehntner, L. 1954. Aus dem Leben eines Entomologen. *Mitteilungen der Schweizerischen Entomologischen Gesellschaft* 27: 444-449.

Zimmerman, E.C. 1948. Homoptera: Sternorrhyncha. *Insects of Hawaii* 5: 1-464.

Zimsen, E. 1964. The type material of I.C. Fabricius. Munksgaard, Copenhagen. 656 pp.

INDEX TO GENERA

The left-hand column includes valid generic names, objective synonyms, homonyms, unjustified replacement names, mis-spellings of generic names (with authorship of the author/s of the mis-spelling) and erroneous authorships. Each of the names listed corresponds to its valid, senior synonym which follows in the right-hand column. Objective synonyms, homonyms and unjustified replacement names are denoted with an asterisk. Mis-spelling of generic names and erroneous authorhips are denoted with an asterisk and enclosed in square brackets [ ].

This index does not present interpretations about subjective synonyms. For the different interpretaions on subjective synonymy of genera in the family the user should refer to the respective genus in the catalogue.

Cerostegia De Lotto
* Chlamidolecanium Lindinger
Chlamydolecanium Goux
Chloropulvinaria Borchsenius
Cissococcus Cockerell
* [Cnetochiton Balachowsky]
Coccus (Ceroplastes) Gray
Coccus (Ericerus) Guerin-Meneville
Coccus Linnaeus
* [Cocus Watt & Mann]
* [Coeus Lopez y Ramos]
* [Columella Sulc]
* [Columna Signoret]
Conofilippia Brain
Couturierina Matile-Ferrero & LeRuyet

* [Cribrolebanium Tang]
Cribrolecanium Green
* [Cryptes Crawford]
Cryptes Maskell
Cryptinglisia Cockerell
Cryptostigma Ferris
Ctenochiton Maskell
* [Curycerus Targioni Tozzetti]
Cyclolecanium Morrison
Cyphococcus Laing
* Defilippia Targioni Tozzetti
* Dermatolecanium Lindinger
Dermolecanium Zavattari
Dicyphococcus Borchsenius
Didesmococcus Borchsenius
Drepanococcus Williams & Watson
* [Edwalia Borchsenius]
Edwallia Hempel
* [Enlecanium Cockerell]
Ericeroides Danzig
Ericerus Guerin-Meneville
* [Eriochitin MacGillivray]
Eriochiton Maskell
Eriopeltis Signoret
* [Eriopettis Tang]
Etiennea Matile-Ferrero
* Eucalymmatus Lindinger
Eucalymnatus Cockerell
Eulecanium Cockerell
Eumashona Hodgson
* Euphilippia Berlese & Silvestri
Eupulvinaria Borchsenius
* [Eurycerus Targioni Tozzetti]
Eutaxia Green
Exaeretopus Newstead
* [Exeraetopus Bodenheimer]
* [Exoerctopus Cockerell]
* Fairmairea Lindinger
* Fairmairia Signoret
* [Farrmairia Hempel]
Filippia Targioni Tozzetti
Gascardia Targioni Tozzetti
* Globulicoccus Lindinger
* Habibius Ezzat & Hussein

Cerostegia De Lotto
Chlamydolecanium Goux
Chlamydolecanium Goux
Chloropulvinaria Borchsenius
Cissococcus Cockerell
Ctenochiton Maskell
Ceroplastes Gray
Ericerus Guerin-Meneville
Coccus Linnaeus
Coccus Linnaeus
Coccus Linnaeus
Columnea Targioni Tozzetti
Columnea Targioni Tozzetti
Conofilippia Brain
Couturierina Matile-Ferrero & LeRuyet
Cribrolecanium Green
Cribrolecanium Green
Cryptes Maskell
Cryptes Maskell
Cryptinglisia Cockerell
Cryptostigma Ferris
Ctenochiton Maskell
Ericerus Guerin-Meneville
Cyclolecanium Morrison
Cyphococcus Laing
Filippia Targioni Tozzetti
Dermolecanium Zavattari
Dermolecanium Zavattari
Dicyphococcus Borchsenius
Didesmococcus Borchsenius
Drepanococcus Williams & Watson
Edwallia Hempel
Edwallia Hempel
Eulecanium Cockerell
Ericeroides Danzig
Ericerus Guerin-Meneville
Eriochiton Maskell
Eriochiton Maskell
Eriopeltis Signoret
Eriopeltis Signoret
Etiennea Matile-Ferrero
Eucalymnatus Cockerell
Eucalymnatus Cockerell
Eulecanium Cockerell
Eumashona Hodgson
Filippia Targioni Tozzetti
Eupulvinaria Borchsenius
Ericerus Guerin-Meneville
Eutaxia Green
Exaeretopus Newstead
Exaeretopus Newstead
Exaeretopus Newstead
Parafairmairia Cockerell
Parafairmairia Cockerell
Parafairmairia Cockerell
Filippia Targioni Tozzetti
Gascardia Targioni Tozzetti
Eulecanium Cockerell
Kilifia De Lotto

Hadzibejliaspis Koteja
* [Hadzibejiliaspis Tang]
Halococcus Takahashi
Hemilecanium Newstead
Houardia Marchal
Idiosaissetia Brain
* [Inglesia Bruner, Scaramuzza & Otero]
Inglisia Maskell
Kilifia De Lotto
Kozaricoccus Avasthi & Shafee
Lacca Signoret [NOMEN NUDUM]
Lagosinia Cockerell
* Lecaniochiton Lindinger
Lecaniococcus Danzig
* Lecaniopsis Lindinger
* Lecanium (Globulicoccus) Lindinger
Lecanium (Neolecanium) Parrott
Lecanium (Paralecanium) Cockerell
Lecanium (Pseudokermes) Cockerell
Lecanium (Pterolecanium) Sulc
Lecanium (Toumeyella) Cockerell
Lecanochiton Maskell
Lecanopsis Targioni Tozzetti
* Lefroyia Green
Leptopulvinaria Kanda
* [Lichstensia Fuller]
Lichtensia Signoret
Loemica Laing
Luzulaspis Cockerell
Maacoccus Tao, Wong & Chang
Macropulvinaria Hodgson
Mallococcus Maskell
* Mallophora Maskell
Mametia Matile-Ferrero
Marsipococcus Cockerell & Bueker
* Mashona Hodgson
Megalecanium Hempel
Megalocryptes Takahashi
* Megalolecanium Lindinger
* Megalosaissetia Lindinger
Megapulvinaria Yang
Megasaissetia Cockerell
Melanesicoccus Williams & Watson
Membranaria Brain
Mesembryna De Lotto
Mesolecanium Cockerell
Messinea De Lotto
Metaceronema Takahashi
Metapulvinaria Nakahara & Gill
Millericoccus Avasthi & Shafee
Milviscutulus Williams & Watson
Mitrococcus Borchsenius
Mohelnia Sulc
* [Myxilecanium MacGillivray]
* [Myxolecanium Targioni Tozzetti]
Myzolecanium Beccari
Nemolecanium Borchsenius
* Neobernardia Cockerell
* Neolecaniochiton Lindinger
Neolecanium Parrott

Hadzibejliaspis Koteja
Hadzibejliaspis Koteja
Halococcus Takahashi
Hemilecanium Newstead
Houardia Marchal
Idiosaissetia Brain
Inglisia Maskell
Inglisia Maskell
Kilifia De Lotto
Kozaricoccus Avasthi & Shafee

Lagosinia Cockerell
Lecanochiton Maskell
Lecaniococcus Danzig
Lecanopsis Targioni Tozzetti
Eulecanium Cockerell
Neolecanium Parrott
Paralecanium Cockerell
Pseudokermes Cockerell
Pterolecanium Sulc
Toumeyella Cockerell
Lecanochiton Maskell
Lecanopsis Targioni Tozzetti
Pseudopulvinaria Atkinson
Leptopulvinaria Kanda
Lichtensia Signoret
Lichtensia Signoret
Loemica Laing
Luzulaspis Cockerell
Maacoccus Tao, Wong & Chang
Macropulvinaria Hodgson
Mallococcus Maskell
Mallococcus Maskell
Mametia Matile-Ferrero
Marsipococcus Cockerell & Bueker
Eumashona Hodgson
Megalecanium Hempel
Megalocryptes Takahashi
Megalecanium Hempel
Megasaissetia Cockerell
Megapulvinaria Yang
Megasaissetia Cockerell
Melanesicoccus Williams & Watson
Membranaria Brain
Mesembryna De Lotto
Mesolecanium Cockerell
Messinea De Lotto
Metaceronema Takahashi
Metapulvinaria Nakahara & Gill
Millericoccus Avasthi & Shafee
Milviscutulus Williams & Watson
Mitrococcus Borchsenius
Mohelnia Sulc
Myzolecanium Beccari
Myzolecanium Beccari
Myzolecanium Beccari
Nemolecanium Borchsenius
Bernardia Ashmead
Neolecanochiton Hempel
Neolecanium Parrott

Neolecanochiton Hempel
Neoplatylecanium Takahashi
Neopulvinaria Hadzibejli
Neosaissetia Tao, Wong & Chang
* [Palacolecanium Lindinger]
Palaeolecanium Sulc
* [Paloelecanium Lindinger]
* [Paracardicoccus Tao]
Paracardiococcus Takahashi
Paracerostegia Tang
Paractenochiton Takahashi
* Parafairmairea Lindinger
Parafairmairia Cockerell
Parakermes Fonseca
Paralecanium Cockerell
* [Paralecanium Cockerell & Parrott]
Paralecanopsis Bodenheimer
Parapulvinaria Fonseca
Parasaissetia Takahashi
Parthenolecanium Sulc
* Pela Targioni Tozzetti
Pendularia Fonseca
Perilecanium Fonseca
Philephedra Cockerell
* Philippia Targioni Tozzetti
* [Phylippia Leonardi]
Phyllostroma Sulc
* [Physocermes Atanasov]
* [Physochermes Targioni Tozzetti]
Physokermes Targioni Tozzetti
* [Platilecanium Danzig &
             Konstantinova]
Platinglisia Cockerell
* Platycoccus Takahashi
Platylecanium Cockerell & Robinson
* [Platylecarium Tang]
Platysaissetia Cockerell
* [Pluvinaria Shinji]
* [Plysockermes Leonardi]
Poaspis Koteja
Podoparalecanium Tao, Wong & Chang

Protopulvinaria Cockerell
* [Protopulvinarir Tang]
Pseudalichtensia Hempel
Pseudokermes Cockerell
Pseudophilippia Cockerell
Pseudopsylla Froggatt
Pseudopulvinaria Atkinson
Psilococcus Borchsenius
Pterolecanium Sulc
* [Ptoropulvinaria Danzig &
             Konstantinova]
* [Pulviferia Atanasov]
Pulvinaria (Eupulvinaria) Borchsenius
Pulvinaria (Philephedra) Cockerell
Pulvinaria (Protopulvinaria) Cockerell
Pulvinaria (Pulvinella) Hempel
Pulvinaria (Takahashia) Cockerell
Pulvinaria Targioni Tozzetti

Neolecanochiton Hempel
Neoplatylecanium Takahashi
Neopulvinaria Hadzibejli
Neosaissetia Tao, Wong & Chang
Palaeolecanium Sulc
Palaeolecanium Sulc
Palaeolecanium Sulc
Paracardiococcus Takahashi
Paracardiococcus Takahashi
Paracerostegia Tang
Paractenochiton Takahashi
Parafairmairia Cockerell
Parafairmairia Cockerell
Parakermes Fonseca
Paralecanium Cockerell
Paralecanium Cockerell
Paralecanopsis Bodenheimer
Parapulvinaria Fonseca
Parasaissetia Takahashi
Parthenolecanium Sulc
Ericerus Guerin-Meneville
Pendularia Fonseca
Perilecanium Fonseca
Philephedra Cockerell
Filippia Targioni Tozzetti
Filippia Targioni Tozzetti
Phyllostroma Sulc
Physokermes Targioni Tozzetti
Physokermes Targioni Tozzetti
Physokermes Targioni Tozzetti

Platylecanium Cockerell & Robinson
Platinglisia Cockerell
Kilifia De Lotto
Platylecanium Cockerell & Robinson
Platylecanium Cockerell & Robinson
Platysaissetia Cockerell
Pulvinaria Targioni Tozzetti
Physokermes Targioni Tozzetti
Poaspis Koteja
Podoparalecanium Tao, Wong &
Chang
Protopulvinaria Cockerell
Protopulvinaria Cockerell
Pseudalichtensia Hempel
Pseudokermes Cockerell
Pseudophilippia Cockerell
Pseudopsylla Froggatt
Pseudopulvinaria Atkinson
Psilococcus Borchsenius
Pterolecanium Sulc

Protopulvinaria Cockerell
Pulvinaria Targioni Tozzetti
Eupulvinaria Borchsenius
Philephedra Cockerell
Protopulvinaria Cockerell
Pulvinella Hempel
Takahashia Cockerell
Pulvinaria Targioni Tozzetti

Pulvinariella Borchsenius   Pulvinariella Borchsenius
Pulvinarisca Borchsenius   Pulvinarisca Borchsenius
Pulvinella Hempel   Pulvinella Hempel
* [Pulvirariella Tang]   Pulvinariella Borchsenius
* Rhizobium Targioni Tozzetti   Lecanopsis Targioni Tozzetti
Rhizopulvinaria Borchsenius   Rhizopulvinaria Borchsenius
* [Rhizopulvinara Tang]   Rhizopulvinaria Borchsenius
Rhodococcus Borchsenius   Rhodococcus Borchsenius
* [Rhodoccus Tang]   Rhodococcus Borchsenius
Richardiella Matile-Ferrero & LeRuyet   Richardiella Matile-Ferrero &
                                        LeRuyet
Saccharipulvinaria Tao, Wong & Chang   Saccharipulvinaria Tao, Wong &
                                        Chang
Saccharolecanium Williams   Saccharolecanium Williams
* [Saisettia Suomalainen]   Saissetia Deplanche
Saissetia Deplanche   Saissetia Deplanche
Saissetia (Megasaissetia) Cockerell   Megasaissetia Cockerell
Saissetia (Platysaissetia) Cockerell   Platysaissetia Cockerell
* [Sassetia Dunham]   Saissetia Deplanche
Schizochlamidia Cockerell   Schizochlamidia Cockerell
* [Schizochlamydia Borchsenius]   Schizochlamidia Cockerell
* Schizochlamys Cockerell   Schizochlamidia Cockerell
Scythia Kiritshenko   Scythia Kiritshenko
* [Seissetia Abrahao & Mamprim]   Saissetia Deplanche
* Signoretia Targioni Tozzetti   Luzulaspis Cockerell
* [Signorettia Targioni Tozzetti] Luzulaspis Cockerell
* [Spaerolecanium Leonardi]   Sphaerolecanium Sulc
Spermococcus Giard   Spermococcus Giard
Sphaerolecanium Sulc   Sphaerolecanium Sulc
* [Sphaerolecaniuw Tang]   Sphaerolecanium Sulc
Stenolecanium Takahashi   Stenolecanium Takahashi
Stictolecanium Cockerell   Stictolecanium Cockerell
Stozia Marchal   Stozia Marchal
Suareziella Mamet   Suareziella Mamet
Symonicoccus Koteja & Brookes   Symonicoccus Koteja & Brookes
Taiwansaissetia Tao, Wong & Chang Taiwansaissetia Tao, Wong & Chang
Takahashia Cockerell   Takahashia Cockerell
Tectopulvinaria Hempel   Tectopulvinaria Hempel
Tillancoccus Ben-Dov   Tillancoccus Ben-Dov
Toumeyella Cockerell   Toumeyella Cockerell
Trijuba De Lotto   Trijuba De Lotto
Udinia De Lotto   Udinia De Lotto
Umwinsia Hodgson   Umwinsia Hodgson
Vinsonia Signoret   Vinsonia Signoret
* [Vinzonia Danzig & Konstantinova]   Vinsonia Signoret
* [Visonia Ashmead]   Vinsonia Signoret
* [Vitacoccus Rehácek]   Vittacoccus Borchsenius
Vittacoccus Borchsenius   Vittacoccus Borchsenius
Waricoccus Brookes & Koteja   Waricoccus Brookes & Koteja
Waxiella De Lotto   Waxiella De Lotto
Xenolecanium Takahashi   Xenolecanium Takahashi

INDEX TO SPECIES

Valid specific names, in their current combination, are listed on the right-hand column. Specific names in the left-hand column include valid names (sometimes with different endings corresponding to the different genders of their generic combinations), synonyms, homonyms, nomina nuda, mis-spellings, typing errors, erroneous authorship and unjustified emendations.

Synonyms are indicated with an asterisk, and given in the order of specific name, author and genus name, as printed in the original description.

A Nomen Nudum, on the left-hand column, is followed by [NOMEN NUDUM] and listed corresponding to the valid name to which it should be linked, in the right-hand column.

Mis-spellings, typing errors, unjustified emendations and erroneous authorship are indicated by square brackets []. Alternative gender endings are indicated by curly brackets {}.

aequale Newstead	Coccus aequale (Newstead)
* aesculi Kollar, Coccus [NOMEN NUDUM]	Eulecanium tiliae (Linnaeus)
aestivalis Danzig	Pulvinaria aestivalis Danzig
aethiopica (De Lotto)	Pulvinaria aethiopica (De Lotto)
{aethiopicus De Lotto}	Pulvinaria aethiopica (De Lotto)
aetnensis Russo & Longo	Scythia aetnensis Russo & Longo
africana (Green)	Waxiella africana (Green)
africana Macfie	Ceronema africana Macfie
* africana Newstead, Pulvinaria	Pulvinaria grabhami Cockerell
{africanum Newstead}	Coccus africanus (Newstead)
africanus cristatus Green	Waxiella africanus cristatus (Green)
africanus (Newstead)	Coccus africanus (Newstead)
africanus senegalensis Marchal	Waxiella africanus senegalensis (Marchal)
* agalmae Takahashi, Protopulvinaria	Protopulvinaria pyriformis Cockerell
agrestis Hempel	Ceroplastes agrestis Hempel
* agropyri Borchsenius, Eriopeltis	Eriopeltis festucae (Fonscolombe)
agropyri Hadzibejli	Exaeretopus agropyri (Hadzibejli)
ajmerensis Avasthi & Shafee	Ceroplastes ajmerensis (Avasthi & Shafee)
alamensis Avasthi & Shafee	Ceroplastes alamensis Avasthi & Shafee
[alami Avasthi & Shafee]	Ceroplastes alamensis Avasthi & Shafee
* alba Signoret, Lacca [NOMEN NUDUM]	Ceroplastes ceriferus (Fabricius)
albata Hempel	Tectopulvinaria albata Hempel
albodermis Chen	Eulecanium albodermis Chen
alboinducta Fonseca	Pulvinaria alboinducta Fonseca
albolineatus Cockerell	Ceroplastes albolineatus Cockerell
* albolineatus vulcanicus Cockerell, Ceroplastes	Ceroplastes albolineatus vulcanicus Cockerell
album Takahashi	Paralecanium album Takahashi
* alienum Douglas, Lecanium	Coccus hesperidum Linnaeus
aligarhensis Avasthi & Shafee	Pulvinaria aligarhensis Avasthi & Shafee
almoraensis Avasthi & Shafee	Coccus almoraensis Avasthi & Shafee
* alni Modeer, Coccus	Eulecanium tiliae (Linnaeus)
alni rufulum Cockerell	Parthenolecanium rufulum (Cockerell)
alnicola Chen	Eulecanium alnicola Chen
alpinus De Lotto	Coccus alpinus De Lotto
* amabilis Kanda, Luzulaspis	Luzulaspis bisetosa Borchsenius
amazonensis Foldi	Neolecanium amazonensis Foldi
amazonicus Hempel	Ceroplastes amazonicus Hempel
americana Ben-Dov	Kilifia americana Ben-Dov
americana Koteja & Howell	Luzulaspis americana Koteja & Howell
* [americanum Targioni Tozzetti, Lecanium]	Pulvinaria vitis (Linnaeus)
amoenus De Lotto	Avricus amoenus (De Lotto)
ampelopsidis Savescu	Pulvinaria ampelopsidis Savescu
* amygdalae Rao, Eriochiton	Didesmococcus unifasciatus (Archangelskaya)
amygdali Cockerell	Pulvinaria amygdali Cockerell
andersoni Newstead	Cribrolecanium andersoni (Newstead)
angkorense Takahashi	Paralecanium angkorense Takahashi
angulatus Cockerell	Ceroplastes angulatus Cockerell
* angustatus Signoret, Lecanium	Coccus hesperidum Linnaeus
anneckei De Lotto	Coccus anneckei De Lotto
anonae Hempel	Saissetia anonae Hempel
* antennatum Signoret, Lecanium	Parthenolecanium quercifex (Fitch)
* anthurii Boisduval, Chermes	Saissetia coffeae (Walker)
antidesmae Green	Coccus antidesmae (Green)
* antigoni Green, Pulvinaria	Pulvinaria grabhami Cockerell

aphenogastrorum Gómez-Menor Ortega	Lecanopsis aphenogastrorum Gómez-Menor Ortega
aptii Bodenheimer	Nemolecanium aptii (Bodenheimer)
* aquifoliae Chen, Euphilippia	Metaceronema japonica (Maskell)
araliae Shinji	Pulvinaria araliae Shinji
araucariae Green	Ctenochiton araucariae Green
* araxis Borchsenius, Eriopeltis	Eriopeltis festucae (Fonscolombe)
arborescens Laing	Avricus arborescens (Laing)
arenaria Canard	Rhizopulvinaria arenaria Canard
* arens Hodgson, Coccus	Coccus capparidis (Green)
areolata Fonseca	Pulvinaria areolata Fonseca
* [areon Lindinger]	Parthenolecanium fletcheri (Cockerell)
argaformis Hempel	Mesolecanium argaformis Hempel
argentata Hempel	Lichtensia argentata Hempel
argentina Leonardi	Pulvinaria argentina Leonardi
argentina Morrison	Saissetia argentina Morrison
argentinus Brethes	Ceroplastes argentinus Brethes
* arion Lindinger, Lecanium	Parthenolecanium fletcheri (Cockerell)
aristolochiae Newstead	Lagosinia aristolochiae (Newstead)
armata Takahashi	Taiwansaissetia armata (Takahashi)
* armatus Brittin, Lecanium	Ctenochiton spinosus Maskell
armatus Cockerell	Pseudokermes armatus Cockerell
{armatus (Takahashi)}	Taiwansaissetia armata (Takahashi)
armeniaca Borchsenius	Rhizopulvinaria armeniaca Borchsenius
* armeniacum Craw, Lecanium	Parthenolecanium corni (Bouche)
* artemisiae (Hall), Rhizopulvinaria	Rhizopulvanaria halli Borchsenius
* artemisiae Rossi, Coccus	Ceroplastes rusci (Linnaeus)
artemisiae Signoret	Rhizopulvinaria artemisiae (Signoret)
artemisiae turkestanica Archangelskaya	Rhizopulvinaria turkestanica (Archangelskaya)
* artemisiarum Cockerell Ceroplastes [N. NUDUM]	Ceroplastes irregularis Cockerell
arundinariae Green	Maacoccus arundinariae (Green)
asiaticus Lindinger	Coccus asiaticus Lindinger
asparagi Brain	Ceronema asparagi (Brain)
* assimile amaryllidis Cockerell, Lecanium [N. NUDUM]	Coccus hesperidum Linnaeus
* assimile amaryllis Cockerell, Lecanium	Coccus hesperidum Linnaeus
* assimile Newstead, Lecanium	Parthenolecanium corni (Bouche)
[assymmetricum Morrison]	Platylecanium asymmetricum Morrison
asymmetricum Morrison	Platylecanium asymmetricum Morrison
[atrichos De Lotto]	Coccus acrossus De Lotto
* atriplicis Maskell, Signoretia	Pulvinaria maskelli Olliff
attenuata Hempel	Alichtensia attenuata (Hempel)
* aurantiacum Hunter, Lecanium	Parthenolecanium corni (Bouche)
* [aurantii (Alfonso)]	Coccus hesperidum Linnaeus
aurantii Cockerell	Pulvinaria aurantii Cockerell
* aurantj Alfonso, Kermes	Coccus hesperidum Linnaeus
auriculata Morrison	Saissetia auriculata Morrison
* australiae Walker, Ceroplastes	Ceroplastes ceriferus (Fabricius)
australis Hempel	Inglisia australis Hempel
* australis Lidgett, Lecanium	Coccus lidgetti (Fernald)
australis (Maskell)	Symonicoccus australis (Maskell)
avasthii Yousuf & Shafee	Pulvinaria avasthii Yousuf & Shafee
avicenniae Newstead	Ceroplastes avicenniae Newstead
aztecus Townsend & Cockerell	Ctenochiton aztecus Townsend & Cockerell

{baccatum marmoreum Fuller}}
{baccatum Maskell}
baccatus marmoreus (Fuller)
baccatus (Maskell)
baccharidis Cockerell
bahiensis Bondar
bambusicola Green
bambusicola Tang
banksiae Maskell
* barteriae Newstead, Lecanium (Saissetia)
                              [NOMEN NUDUM]
batatae Cockerell
* beaumontiae Douglas, Lecanium
* begoniae Douglas, Lecanium
berberidis major Maskell
* berberidis Schrank, Coccus
bergi Cockerell
berliniae enkeldoorni Hall
berliniae Hall
bernardensis Cockerell
* betulae alni Douglas, Pulvinaria
* betulae Linnaeus, Coccus
bicolor Hempel
* bicolor Mosquera, Ceroplastes
bicruciatus Green
[bicurciatus (Green)]
bigeloviae Cockerell
bigeloviae marmorata Cockerell
bigibbus Borchsenius
biorbiculus Morrison
{bipartita (Newstead)}
bipartita Signoret
bipartitus Newstead
* biplicata Targioni Tozzetti, Pulvinaria
bisetosa Borchsenius
bituberculatum Signoret
* bituberculatum Targioni Tozzetti, Lecanium
                              [NOMEN NUDUM]
bituberculatus Brain
bivalvata Green
* blanchardii Targioni Tozzetti, Lecanium
bobuae Takahashi
boonei Hollinger
borchsenii Danzig
* borchsenii Rehacek, Luzulaspis
borealis Koteja & Howell
boyacensis Mosquera
brachiungualis Savescu
brachypodii Giard, Eriopeltis [NOMEN NUDUM]
brachystegiae Hall
brachyurus Cockerell
* braini Lindinger, Filippia
brasiliae Hempel
brasiliensis Fonseca, 1957

Cryptes baccatus marmoreus (Fuller)
Cryptes baccatus (Maskell)
Cryptes baccatus marmoreus (Fuller)
Cryptes baccatus (Maskell)
Mesolecanium baccharidis (Cockerell)
Ceroplastodes bahiensis Bondar
Megalocryptes bambusicola (Green)
Pulvinaria bambusicola (Tang)
Ceronema banksiae Maskell

Udinia newsteadi Hanford
Mesolecanium batatae (Cockerell)
Saissetia coffeae (Walker)
Parasaissetia nigra (Nietner)
Eulecanium berberidis major (Maskell)
Parthenolecanium persicae (Fabricius)
Ceroplastes bergi Cockerell
Waxiella enkeldoorni (Hall)
Waxiella berliniae (Hall)
Ceroplastes bernardensis Cockerell
Pulvinaria vitis (Linnaeus)
Pulvinaria vitis (Linnaeus)
Ceroplastes bicolor Hempel
Ceroplastes mosquerai n. name
Maacoccus bicruciatus (Green)
Maacoccus bicruciatus (Green)
Pulvinaria bigeloviae Cockerell
Pulvinaria marmorata Cockerell
Dicyphococcus bigibbus Borchsenius
Cryptostigma biorbiculus Morrison
Ceroplastes bipartitus Newstead
Parafairmairia bipartita (Signoret)
Ceroplastes bipartitus Newstead
Pulvinaria mesembryanthemi (Vallot)
Luzulaspis bisetosa Borchsenius
Palaeolecanium bituberculatum (Signoret)

Palaeolecanium bituberculatum (Signoret)
Kozaricoccus bituberculatus (Brain)
Cardiococcus bivalvata (Green)
Sphaerolecanium prunastri (Fonscolombe)
Saissetia bobuae Takahashi
Exaeretopus boonei Hollinger
Pulvinaria borchsenii Danzig
Luzulaspis scotica Green
Luzulaspis borealis Koteja & Howell
Ceroplastes boyacensis Mosquera
Pulvinaria brachiungualis Savescu

Ceronema brachystegiae Hall
Ceroplastes brachyurus Cockerell
Ceronema asparagi (Brain)
Pseudalichtensia brasiliae Hempel
Coccus brasiliensis Fonseca

brasiliensis Fonseca, 1973 — Parakermes brasiliensis Fonseca
brasiliensis Hempel — Megasaissetia brasiliensis Hempel
* brassiae Cockerell, Pulvinaria — Pulvinaria floccifera (Westwood)
* [brassicae Cockerell, Pulvinaria] — Pulvinaria floccifera (Westwood)
brevicauda Hall — Ceroplastes brevicauda Hall
* brevicornis Newstead, Lecanopsis — Lecanopsis formicarum Newstead
brevicornis Newstead — Pulvinaria brevicornis Newstead
breviseta Leonardi — Ceroplastes breviseta Leonardi
broadwayi Cockerell — Philephedra broadwayi (Cockerell)
broadwayi echinopsidis Newstead — Philephedra broadwayi echinopsidis (Newstead)
bromeliae Bouche — Coccus bromeliae Bouche
bruncki Hanford — Udinia bruncki Hanford
bruneri Cockerell, 1902e — Akermes bruneri Cockerell
bruneri Cockerell & Cockerell — Ceroplastes bruneri Cockerell & Cockerell
[bruneri Cockerell, 1902e] — Ceroplastes bruneri Cockerell & Cockerell
{brunfelsia Hempel} — Eucalymnatus brunfelsiae (Hempel)
brunfelsiae (Hempel) — Eucalymnatus brunfelsiae (Hempel)
* bulgariense Wunn, Eulecanium — Rhodococcus peronatus (Cockerell & Parrott)
* bunzlii Green, Cryptostigma — Cryptostigma quinquepori (Newstead)
burkilli Green — Megapulvinaria burkilli (Green)
bussei Newstead, Ceroplastes [NOMEN NUDUM]
buteae Takahashi — Megalocryptes buteae Takahashi
* butleri Green, Lecanopsis — Lecanopsis formicarum Newstead

cacao Hodgson — Etiennea cacao Hodgson
cacao Williams & Watson — Pulvinaria cacao Williams & Watson
caesalpiniae Laing — Cyphococcus caesalpiniae Laing
caesalpiniae Reyne — Ceroplastes caesalpiniae Reyne
cajani Maskell — Drepanococcus cajani (Maskell)
cajani Newstead — Coccus cajani (Newstead)
callitris Froggatt — Alecanopsis callitris (Froggatt)
callosa (De Lotto) — Pulvinaria callosa (De Lotto)
{callosus De Lotto} — Pulvinaria callosa (De Lotto)
calophylli Green — Paralecanium calophylli Green
cambodiensis Takahashi — Coccus cambodiensis Takahashi
* camelicola Signoret, Pulvinaria — Pulvinaria floccifera (Westwood)
cameronensis Takahashi — Coccus cameronensis Takahashi
campinensis Hempel — Ceroplastes campinensis Hempel
campomanesiae Hempel — Mesolecanium campomanesiae (Hempel)
* canadense Cockerell, Lecanium — Parthenolecanium corni (Bouche)
candelabra Hodgson — Etiennea candelabra Hodgson
candella Cockerell & King — Ceroplastes candella Cockerell & King
capensis Hodgson — Etiennea capensis Hodgson
cappari Froggatt — Platylecanium cappari (Froggatt)
capparidis Green — Coccus capparidis (Green)
* capreae Linnaeus, Coccus — Eulecanium tiliae (Linnaeus)
caraganae Borchsenius — Eulecanium caraganae Borchsenius
* caricae Bernard, Coccus [NOMEN NUDUM] — Ceroplastes rusci (Linnaeus)
* caricae Fabricius, Coccus — Ceroplastes rusci (Linnaeus)
caricicola Lindinger — Luzulaspis caricicola (Lindinger)
caricis Ehrhorn — Luzulaspis caricis (Ehrhorn)
* caricis Takahashi, Luzulaspis — Luzulaspis caricicola (Lindinger)
[carieri de Charmoy] — Pulvinaria carieri Grandpre & Charmoy
carieri Grandpre & Charmoy — Pulvinaria carieri Grandpre & Charmoy

{carinata (Takahashi)} — Ctenochiton carinatus Takahashi
carinatus Takahashi — Ctenochiton carinatus Takahashi
carissae Brain — Lichtensia carissae (Brain)
camosa Hodgson — Saissetia camosa Hodgson
carolinensis Beardsley — Paralecanium carolinensis Beardsley
carpenteri Newstead — Etiennea carpenteri (Newstead)
* carpini Linnaeus, Coccus — Pulvinaria vitis (Linnaeus)
* caryae canadense Cockerell, Lecanium — Parthenolecanium corni (Bouche)
caryae Fitch — Eulecanium caryae (Fitch)
* caryarum Cockerell, Lecanium (Eulecanium) — Parthenolecanium corni (Bouche)
cassariae Fonseca — Parapulvinaria cassariae Fonseca
cassiae Chavannes — Ceroplastes cassiae (Chavannes)
cassiniae Maskell — Saissetia cassiniae (Maskell)
* castaneae Green, Lefroyia — Pseudopulvinaria sikkimensis Atkinson
castaneus De Lotto — Avricus castaneus (De Lotto)
castelbrancoi Almeida — Ceroplastes castelbrancoi Almeida
castilloae Cockerell — Platysaissetia castilloae (Cockerell)
castilloae Green — Dicyphococcus castilloae (Green)
casuarinae Maskell — Alecanopsis casuarinae (Maskell)
catori Green — Udinia catori (Green)
* caucasica Borchsenius, Luzulaspis — Luzulaspis grandis Borchsenius
* caucasicus Borchsenius, Eriopeltis — Eriopeltis festucae (Fonscolombe)
caudata Froggatt — Ceronema caudata Froggatt
* caudatum Green, Lecanium — Coccus asiaticus Lindinger
* {caudatus (Green), Coccus} — Coccus asiaticus Lindinger
* [caudaus (Green), Coccus] — Coccus asiaticus Lindinger
cavernosa Hodgson — Umwinsia cavernosa Hodgson
caviramicolus Morrison — Coccus caviramicolus Morrison
* cecconi Leonardi, Eulecanium — Parthenolecanium persicae (Fabricius)
celatus De Lotto — Coccus celatus De Lotto
cellulosa Cockerell — Ctenochiton cellulosus Cockerell
* cellulosa Green, Pulvinaria — Pulvinaria polygonata Cockerell
celsus Borchsenius — Mitrococcus celsus Borchsenius
* [celticum Kuwana, Coccus] — Coccus longulus (Douglas)
* celtium Kuwana, Lecanium (Coccus) — Coccus longulus (Douglas)
[cenehiformis (Newstead)] — Inglisia conchiformis Newstead
centroroseus Chen — Ceroplastes centroroseus Chen
cephalocarinata Fonseca — Anopulvinaria cephalocarinata Fonseca
cephalomeatus Hodgson — Etiennea cephalomeatus Hodgson
* cerasi Goethe, Lecanium — Eulecanium tiliae (Linnaeus)
[cerasifera (Fitch)] — Parthenolecanium cerasifex (Fitch)
cerasifex Fitch — Parthenolecanium cerasifex (Fitch)
cerasorum Cockerell — Eulecanium cerasorum (Cockerell)
* ceratoniae Gennadius, Lecanium — Coccus hesperidum Linnaeus
cerei Green — Saissetia cerei Green
* cereus Walker, Coccus — Ericerus pela (Chavannes)
[cerifera (Anderson)] — Ceroplastes ceriferus (Fabricius)
{cerifera (Fabricius)} — Ceroplastes ceriferus (Fabricius)
cerifera Ferris — Toumeyella cerifera Ferris
* cerifera Targioni Tozzetti, Pela — Ericerus pela (Chavannes)
[ceriferens (Anderson)] — Ceroplastes ceriferus (Fabricius)
{ceriferus (Anderson)} — Ceroplastes ceriferus (Fabricius)
ceriferus Fabricius — Ceroplastes ceriferus (Fabricius)
[cerosarum (Cockerell)] — Eulecanium cerasorum (Cockerell)

[cerripidiformis Comstock]	Ceroplastes cirripediformis Comstock
cestri Bouche	Pulvinaria cestri (Bouche)
ceylonica Green	Lecanopsis ceylonica Green
* chavannesii Targioni Tozzetti, Columnea	Ceroplastes psidii (Chavannes)
chelonioides Green	Inglisia chelonioides Green
chelonioides Newstead	Eucalymnatus chelonioides Newstead
chilaspidis Cockerell	Neolecanium chilaspidis (Cockerell)
* chilensis Gray, Coccus (Ceroplastes)	Ceroplastes ceriferus (Fabricius)
chilianthi Brain	Lichtensia chilianthi (Brain)
chimanimanae Hodgson	Saissetia chimanimanae Hodgson
* chirimoliae Maskell, Lecanium	Coccus longulus (Douglas)
chiton Green	Drepanococcus chiton (Green)
chitonoides De Lotto	Saissetia chitonoides De Lotto
chorizandrae Koteja & Brookes	Symonicoccus chorizandrae Koteja & Brookes
chrysanthemi Hall	Pulvinaria chrysanthemi Hall
chrysophyllae Silvestri	Stozia chrysophyllae (Silvestri)
ciliatum Douglas	Eulecanium ciliatum (Douglas)
ciliatus Williams & Watson	Milviscutulus ciliatus Williams & Watson
cinnamomicolus Takahashi	Maacoccus cinnamomicolus (Takahashi)
cinnamomi Green	Ctenochiton cinnamomi Green
cinnamomi Rutherford	Neolecanium cinnamomi Rutherford
cinnamomi Takahashi	Neoplatylecanium cinnamomi Takahashi
circularis Morrison	Coccus circularis Morrison
circumdatus Green	Ceroplastes circumdatus Green
circumfluum Borchsenius	Eulecanium circumfluum Borchsenius
cirripediformis Comstock	Ceroplastes cirripediformis Comstock
cistudiformis Cockerell	Ceroplastes cistudiformis Cockerell
[cistudiformis Townsend & Cockerell]	Ceroplastes cistudiformis Cockerell
citri Takahashi	Platylecanium citri Takahashi
* citricola Campbell, Coccus	Coccus pseudomagnoliarum (Kuwana)
citricola Kuwana, 1909a	Saissetia citricola (Kuwana)
citricola Kuwana, 1914	Pulvinaria citricola Kuwana
claviseta De Lotto	Pulvinaria claviseta De Lotto
* clematidis Gmelin, Coccus	Parthenolecanium persicae (Fabricius)
* clypeata Targioni Tozzetti, Signoretia	Luzulaspis luzulae (Dufour)
* clypeatum Douglas, Lecanium	Saissetia coffeae (Walker)
* coangustum Danzig, Eulecanium	Eulecanium douglasi (Sulc)
* cockerelii Hunter, Lecanium	Eulecanium caryae (Fitch)
cockerelli King	Pulvinaria cockerelli King
cocolobae Borchsenius	Pulvinaria cocolobae Borchsenius
cocophyllae Banks	Paralecanium cocophyllae Banks
cocotis Laing	Platylecanium cocotis Laing
* coffeae clypeatum (Douglas), Lecanium	Saissetia coffeae (Walker)
* coffeae filicum (Boisduval), Lecanium	Saissetia coffeae (Walker)
* coffeae hibernaculorum (Boisduval), Lecanium	Saissetia coffeae (Walker)
coffeae Walker	Saissetia coffeae (Walker)
colae Green & Laing	Akermes colae Green & Laing
colemani Kannan	Coccus colemani Kannan
colimae Cockerell	Akermes colimae Cockerell
colimensis Cockerell	Philephedra colimensis (Cockerell)
coloradensis Cockerell, 1895f	Physokermes coloradensis Cockerell
coloradensis Cockerell, 1905b	Eriopeltis coloradensis Cockerell
coloratus Cockerell	Ceroplastes coloratus Cockerell
combreti Brain	Ceroplastes combreti Brain

combreti Hodgson	Etiennea combreti (Hodgson)
communis Hempel	Ceroplastes communis Hempel
conchiformis Newstead	Inglisia conchiformis Newstead
conchioides Goux	Chlamydolecanium conchioides Goux
concolor Coleman	Physokermes concolor Coleman
confluens Cockerell & Tinsley	Ceroplastes confluens Cockerell & Tinsley
conica De Lotto	Messinea conica De Lotto
coniformis Newstead	Ceroplastes coniformis Newstead
* consimilis De Lotto, Coccus	Coccus celatus De Lotto
constricta De Lotto	Ceroplastes constricta (De Lotto)
* contexta Froggatt, Pulvinaria	Pulvinaria dodonaeae Maskell
convexa Hempel	Pulvinaria convexa (Hempel)
cooleyi King	Pseudokermes cooleyi King
cordiae Morrison	Akermes cordiae Morrison
coriaceum Hall	Hemilecanium coriaceum Hall
corni apuliae Nuzzaci	Parthenolecanium corni apuliae (Nuzzaci)
corni Bouche	Parthenolecanium corni (Bouche)
* corni corni (Bouche), Eulecanium	Parthenolecanium corni (Bouche)
corni orientalis Borchsenius	Parthenolecanium orientalis Borchsenius
* corni robiniarum Marchal, Lecanium	Parthenolecanium corni (Bouche)
corni Savescu	Pulvinaria corni Savescu
cornuparvum Thro	Neolecanium cornuparvum (Thro)
corrugatum neglectum Pettit & McDaniel	Toumeyella corrugatum neglectum Pettit & McDaniel
* corrugatum Thro, Lecanium	Toumeyella pini (King)
* coryli Linnaeus, Coccus [REJECTED NAME]	Eulecanium tiliae (Linnaeus)
coryli cimbricus Wunn	Eulecanium coryli cimbricus Wunn
coryli turanicum Archangelskaya	Rhodococcus turanicus (Archangelskaya)
* corylifex Fitch, Lecanium	Parthenolecanium corni (Bouche)
{costalimae Bondar}	Millericoccus costalimai (Bondar)
costalimai Bondar	Millericoccus costalimai (Bondar)
costata Borchsenius	Pulvinaria costata Borchsenius
* {costatum (Schrank), Palaeolecanium}	Parthenolecanium persicae (Fabricius)
* costatus Schrank, Coccus	Parthenolecanium persicae (Fabricius)
* coulteri Cockerell, Pulvinaria	Pulvinaria occidentalis Cockerell
* [coum Danzig, Eulecanium]	Eulecanium douglasi (Sulc)
craniumequinum Kiritshenko	Scythia craniumequinum Kiritshenko
craspeditae Morrison	Neolecanium craspeditae Morrison
crassispina Borchsenius	Luzulaspis crassispina Borchsenius
crassispina Danzig	Pulvinaria crassispina Danzig
* crassum Green, Lecanium (Saissetia)	Parasaissetia nigra (Nietner)
* [crataegi Linnaeus, Coccus]	Pulvinaria vitis (Linnaeus)
crawii Ehrhorn, Lecanium	Parthenolecanium corni (Bouche)
crematogasteri Takahashi	Ctenochiton crematogasteri Takahashi
crescentiae Cockerell	Philephedra crescentiae (Cockerell)
cribrigerum Cockerell & Robinson	Platylecanium cribrigerum (Cockerell & Robinson)
* cristatus Green, Ceroplastes	Waxiella africanus cristatus (Green)
crotonis De Lotto	Pulvinaria crotonis De Lotto
crustuliforme Green	Platysaissetia crustuliforme (Green)
cubensis Heidel & Köhler	Toumeyella cubensis Heidel & Köhler
cultus Hempel	Ceroplastes cultus Hempel
cundinamarcensis Mosquera	Ceroplastes cundinamarcensis Mosquera
cuneatus Hempel	Ceroplastes cuneatus Hempel

* cuneiformis Leonardi, Saissetia   Parasaissetia nigra (Nietner)
cunhii Balachowsky   Poaspis cunhii (Balachowsky)
* cupaniae Cockerell, Pulvinaria   Pulvinaria psidii Maskell
* curtisi Kirkaldy, Eulecanium   Eulecanium tiliae (Linnaeus)
* cussoniae Hall, Pulvinaria   Pulvinaria psidii Maskell
* cymbiformis Targioni Tozzetti, Lecanium   Parthenolecanium persicae (Fabricius)
* [cymbyformis Targioni Tozzetti, Lecanium]   Parthenolecanium persicae (Fabricius)
* cynosbati Fitch, Lecanium   Parthenolecanium corni (Bouche)
cyperi Takahashi   Platylecanium cyperi Takahashi

dacrydii Maskell   Ctenochiton dacrydii Maskell
dactylis Green   Luzulaspis dactylis Green
* daleae Cockerell, Ceroplastodes   Ceroplastodes dugesii (Signoret)
* darwiniensis Froggatt, Pulvinaria   Pulvinaria psidii Maskell
* daveyi Froggatt, Pulvinaria   Pulvinaria maskelli Olliff
* deani Lawson, Ceroplastodes   Ceroplastodes dugesii (Signoret)
decemplex Newstead   Eucalymnatus decemplex Newstead
deceptrix De Lotto   Ceroplastes deceptrix (De Lotto)
deciduosus Morrison   Ceroplastes deciduosus Morrison
decorata Borchsenius   Pulvinaria decorata Borchsenius
deformosum Newstead   Coccus deformosum (Newstead)
* delicata Borchsenius, Parafairmairia   Parafairmairia bipartita (Signoret)
delicatus Hempel   Eucalymnatus delicatus Hempel
delottoi Gill   Pulvinaria delottoi Gill
delottoi Matile-Ferrero & Le Ruyet   Coccus delottoi Matile-Ferrero & Le Ruyet
deltae Lizer y Trelles   Mesolecanium deltae Lizer y Trelles
deltoides De Lotto   Kilifia deltoides De Lotto
dendrophthorae Cockerell   Pulvinaria dendrophthorae Cockerell
* denudatus Cockerell, Ceroplastes   Ceroplastes rusci (Linnaeus)
deodorensis Hempel   Ceroplastes deodorensis Hempel
depressa Hempel   Pulvinaria depressa Hempel
* {depressa (Targioni Tozzetti), Saissetia   Parasaissetia nigra (Nietner)
depressum minor Maskell   Ctenochiton depressus Maskell
* depressum simulans Douglas, Lecanium   Parasaissetia nigra (Nietner)
* depressum Targioni Tozzetti, Lecanium   Parasaissetia nigra (Nietner)
depressus Cockerell   Ceroplastes depressus Cockerell
depressus Maskell   Ctenochiton depressus Maskell
derameliae Morrison   Neolecanium derameliae Morrison
* desertus Borchsenius, Eriopeltis   Eriopeltis festucae (Fonscolombe)
* desolatum Green, Lecanium   Milviscutulus mangiferae (Green)
destructor brevicauda Hall   Ceroplastes brevicauda Hall
destructor Newstead   Ceroplastes destructor Newstead
dianthi Bodenheimer   Rhizopulvinaria dianthi (Bodenheimer)
dianthi Koteja   Exaeretopus dianthi Koteja
{dianthus Koteja}   Exaeretopus dianthi Koteja
dicrostachys Leonardi   Pulvinaria dicrostachys Leonardi
* diminutum Borchsenius, Eulecanium   Eulecanium giganteum (Shinji)
diospyros Hempel   Ceroplastes diospyros Hempel
discoidalis Hall   Acanthopulvinaria discoidalis (Hall)
discoides Hempel   Saissetia discoides (Hempel)
discrepans Green   Coccus discrepans (Green)
distinguendum Douglas   Eulecanium distinguendum (Douglas)
ditispinosus Danzig   Lecaniococcus ditispinosus Danzig
diversipes Cockerell   Kilifia diversipes (Cockerell)

dixoni Froggatt   Alecanopsis dixoni (Froggatt)
dodonaeae Maskell   Pulvinaria dodonaeae Maskell
* dorsociliatum Green & Mamet, Lecanium   Trijuba oculata (Brain)
* dorsociliatus (Green & Mamet), Coccus   Trijuba oculata (Brain)
douglasi Sulc   Eulecanium douglasi (Sulc)
* dozieri Cockerell & Bueker, Ceroplastes   Ceroplastes utilis Cockerell
dryandrae Fuller   Ceronema dryandrae Fuller
duartei Almeida   Coccus duartei (Almeida)
dubia Cockerell, Lichtensia [NOMEN NUDUM]
dugesii Lichtenstein   Ceroplastes dugesii Lichtenstein
[dugesii (Lichtenstein & Signoret)]   Ceroplastodes dugesii (Signoret)
dugesii Signoret   Ceroplastodes dugesii (Signoret)
[dugesii Townsend, Ceroplastes]   Ceroplastes dugesii Lichtenstein
dura (Hempel)   Saissetia dura (Hempel)
durantae Takahashi   Pulvinaria durantae Takahashi
{durbanense Brain}   Marsipococcus durbanensis (Brain)
durbanensis (Brain)   Marsipococcus durbanensis (Brain)
{durum Hempel}   Saissetia dura (Hempel)

* eatoni Newstead, Lichtensia   Lichtensia viburni Signoret
egbara rhodesiensis (Hall)   Waxiella egbara rhodesiensis (Hall)
egbarum Cockerell   Waxiella egbara (Cockerell)
egbarum fulleri Cockerell & Cockerell   Waxiella egbara fulleri (Cockerell & Cockerell)
[egbarum fulleri (Cockerell)]   Waxiella egbara fulleri (Cockerell & Cockerell)
egbarum rhodesiensis Hall   Waxiella egbara rhodesiensis (Hall)
ehretiae Brain   Coccus ehretiae (Brain)
* ehrhorni King, Pulvinaria   Pulvinaria occidentalis Cockerell
ejaculatoria Mamet   Parafairmairia ejaculatoria Mamet
elaeocarpi Kanda   Leptopulvinaria elaeocarpi Kanda
elaeocarpi Maskell   Ctenochiton elaeocarpi Maskell
elatensis Ben-Dov   Coccus elatensis Ben-Dov
elegans Leonardi   Eulecanium elegans Leonardi
ellesmerensis Richards   Pulvinaria ellesmerensis Richards
elongata durbanensis Munro & Fouche, Pulvinaria
                             [NOMEN NUDUM]   Pulvinaria iceryi (Signoret)
elongata Matesova   Parafairmairia elongata Matesova
elongata Newstead   Pulvinaria elongata Newstead
* elongatum Signoret, Lecanium   Parthenolecanium persicae (Fabricius)
elongatum Takahashi   Platylecanium elongatum Takahashi
elongatus Maskell   Ctenochiton elongatus Maskell
* {elongatus (Signoret), Coccus}   Parthenolecanium persicae (Fabricius)
* elongatus (Signoret), Coccus [SENSU AUCTORUM]   Coccus longulus (Douglas)
elytropappi Brain, 1920a   Ceroplastes elytropappi (Brain)
elytropappi Brain, 1920b   Inglisia elytropappi Brain
emerici Planchon   Eulecanium emerici (Planchon)
endoeucalyptus Qin & Gullan   Cryptostigma endoeucalyptus Qin & Gullan
enkeldoomi (Hall)   Waxiella enkeldoorni (Hall)
enkianthi Takahashi   Pulvinaria enkianthi Takahashi
* eoum Danzig, Eulecanium   Eulecanium douglasi (Sulc)
ephedrae Cockerell   Philephedra ephedrae (Cockerell)
ephedrae Newstead   Stozia ephedrae (Newstead)
erianthi Green, Luzulaspis [NOMEN NUDUM]
ericae Balachowsky   Rhizopulvinaria ericae (Balachowsky)
* ericae Löw, Pulvinaria   Phyllostroma myrtilli (Kaltenbach)

ericicola McConnell — Pulvinaria ericicola McConnell
erithraeus Leonardi — Waxiella erithraeus (Leonardi)
eryngii Fonseca — Pulvinaria eryngii Fonseca
[erythraeus (Leonardi)] — Waxiella erithraeus (Leonardi)
erythrinae Ihering — Coccus erythrinae (Ihering)
esakii Takahashi — Stenolecanium esakii Takahashi
eucalypti Froggatt — Alecanopsis eucalypti (Froggatt)
eucalypti Maskell — Ctenochiton eucalypti Maskell
eucleae Brain — Ceroplastes eucleae Brain
eugeniae Hall — Ceroplastes eugeniae Hall
eugeniae Hempel, 1900b — Eulecanium eugeniae (Hempel)
eugeniae Hempel, 1900b — Pulvinaria eugeniae Hempel
euonymi Shinji — Pulvinaria euonymi Shinji
* euonymicola Lindinger, Pulvinaria — Pulvinaria euonymi Shinji
* euonymicola Savescu, Pulvinaria — Pulvinaria savescui n. name
* euphorbiae Cockerell, Ceroplastes — Ceroplastes cirripediformis Comstock
euphorbiae Mamet — Antandroya euphorbiae Mamet
* eversmanni Borchsenius, Eriopeltis — Eriopeltis festucae (Fonscolombe)
excaecariae Hempel — Ceroplastes excaecariae Hempel
excrescens Ferris — Eulecanium excrescens (Ferris)
* exoleta De Lotto, Saissetia — Udinia farquharsoni (Newstead)
expansum Green — Paralecanium expansum (Green)
expansum javanicum Green — Paralecanium expansum javanicum (Green)
expansum metallicum Green — Paralecanium expansum metallicum (Green)
expansum quadratum Green — Paralecanium quadratum (Green)
expansum rotundum Green — Paralecanium expansum rotundum (Green)

fagi Maskell — Inglisia fagi Maskell
* fairmairei Targioni Tozzetti, Columnea [N. NUDUM] — Ceroplastes fairmairii Signoret
fairmairii Signoret — Ceroplastes fairmairii Signoret
[fairmairii Targioni Tozzetti, Ceroplastes] — Ceroplastes fairmairii Signoret
fallax Giard — Lecanopsis fallax (Giard)
farinosa (Green) — Tectopulvinaria farinosa (Green)
{farinosus Green} — Tectopulvinaria farinosa (Green)
farquharsoni Newstead — Udinia farquharsoni (Newstead)
fasciata De Lotto — Mesembryna fasciata De Lotto
* {fasciatum (Costa), Lecanium} — Eulecanium tiliae (Linnaeus)
fasciatus Borchsenius — Physokermes fasciatus Borchsenius
* fasciatus Costa, Calypticus — Eulecanium tiliae (Linnaeus)
* ferganensis Borchsenius, Eriopeltis — Eriopeltis festucae (Fonscolombe)
ferina De Lotto — Etiennea ferina (De Lotto)
ferox Newstead — Etiennea ferox (Newstead)
ferrisi Ali — Pulvinaria ferrisi Ali
ferum Hempel — Mesolecanium ferum Hempel
festucae Borchsenius — Lecanopsis festucae Borchsenius
festucae Fonscolombe — Eriopeltis festucae (Fonscolombe)
festuceti Sulc — Scythia festuceti (Sulc)
ficicola Borchsenius — Dicyphococcus ficicola Borchsenius
ficicola De Lotto — Parasaissetia ficicola De Lotto
ficinum Paoli — Saissetia ficinum (Paoli)
ficiphilum Borchsenius — Eulecanium ficiphilum Borchsenius
[ficius, Hempel] — Pulvinaria ficus Hempel
ficus Hempel — Pulvinaria ficus Hempel
* ficus Maskell, Lecanium — Coccus longulus (Douglas)

ficus Newstead   Ceroplastes ficus Newstead
filamentosa (Newstead)   Macropulvinaria filamentosa
                                                           (Newstead)
{filamentosum Newstead}   Macropulvinaria filamentosa
                                                           (Newstead)
* filicum Boisduval, Chermes   Saissetia coffeae (Walker)
filicum Maskell   Alecanopsis filicum (Maskell)
* fitchii Signoret, Lecanium   Parthenolecanium corni (Bouche)
flava Takahashi   Pulvinaria flava Takahashi
* flaveolum Cockerell, Lecanium   Coccus hesperidum Linnaeus
* {flaveolus (Cockerell), Coccus}   Coccus hesperidum Linnaeus
flavescens Brethes   Pulvinaria flavescens Brethes
flavicans formicicola Newstead   Pulvinaria flavicans formicicola Newstead
flavicans Maskell   Pulvinaria flavicans Maskell
[flavida Takahashi]   Pulvinaria flava Takahashi
flavus Maskell   Ctenochiton flavus Maskell
fletcheri Cockerell   Parthenolecanium fletcheri (Cockerell)
floccifera (Westwood)   Pulvinaria floccifera (Westwood)
{flocciferus Westwood}   Pulvinaria floccifera (Westwood)
[floccosa (Westwood)]   Pulvinaria floccifera (Westwood)
floridana Nakahara & Gill   Philephedra floridana Nakahara & Gill
floridensis Comstock   Ceroplastes floridensis Comstock
floridensis japonicus Green   Ceroplastes japonicus Green
follicularis Targioni Tozzetti   Filippia follicularis (Targioni Tozzetti)
* folsomi King, Lecanium   Parthenolecanium corni (Bouche)
foraminifer loranthi Fuller   Cardiococcus foraminifer loranthi (Fuller)
foraminifer major Maskell   Inglisia foraminifer major Maskell
foraminifer Maskell   Cardiococcus foraminifer (Maskell)
formicarii Green   Coccus formicarii (Green)
formicarii Takahashi   Halococcus formicarii Takahashi
formicarius Hempel   Ceroplastes formicarius Hempel
formicarum Green   Cribrolecanium formicarum Green
formicarum Newstead   Lecanopsis formicarum Newstead
formiceticola Newstead   Exaeretopus formiceticola Newstead
formicophilus Green   Ctenochiton formicophilus Green
* formosae Takahashi, Eriochiton   Megapulvinaria maxima (Green)
{formosana Takahashi}   Cardiococcus formosanus (Takahashi)
formosanus (Takahashi)   Cardiococcus formosanus (Takahashi)
formosus Hempel   Ceroplastes formosus Hempel
fossilis Maskell   Cardiococcus fossilis (Maskell)
* fouabii Matile-Ferrero & Le Ruyet, Platysaissetia   Etiennea ferox (Newstead)
* foucauldi Balachowsky, Filippia   Stozia ephedrae (Newstead)
[fraconicum Lindinger]   Eulecanium franconicum (Lindinger)
fradei Almeida   Eulecanium fradei Almeida
* franconicum calluneti Danzig, Eulecanium   Eulecanium franconicum (Lindinger)
franconicum Lindinger   Eulecanium franconicum (Lindinger)
* franconicum vaccinicola Danzig, Eulecanium   Eulecanium franconicum (Lindinger)
* fraxini King, Eulecanium   Parthenolecanium corni (Bouche)
fraxini Signoret   Pulvinaria fraxini Signoret
frenchi Maskell   Paralecanium frenchi (Maskell)
frenchii macrozamiae (Fuller)   Paralecanium frenchi macrozamiae (Fuller)
* frontale Green, Lecanium   Coccus longulus (Douglas)
frontalis Green, 1928b   Luzulaspis frontalis Green
* {frontalis (Green, 1904d), Coccus}   Coccus longulus (Douglas)

fryeri Green, 1922b   Ceronema fryeri Green
fryeri Green, 1922b   Ctenochiton fryeri Green
fujianensis Tang   Saccharolecanium fujianensis Tang
fujisana Kanda   Pulvinaria fujisana Kanda
[fukayai (Cockerell), Protopulvinaria] Protopulvinaria fukayai (Kuwana)
fukayai Kuwana   Protopulvinaria fukayai (Kuwana)
fulleri Cockerell   Cissococcus fulleri Cockerell
fumidus De Lotto   Ceroplastes fumidus De Lotto
fuscata Wang, Stozia [NOMEN NUDUM]
* {fuscum (Gmelin), Eulecanium}   Eulecanium tiliae (Linnaeus)
* fuscus Gmelin, Coccus   Eulecanium tiliae (Linnaeus)
fuscus Maskell   Ctenochiton fuscus Maskell
fusiforme Green   Platylecanium fusiforme Green

galeatus Newstead   Ceroplastes galeatus Newstead
gamazumii Kanda   Pulvinaria gamazumii Kanda
* gasteralpha Signoret, Pulvinaria   Pulvinaria iceryi (Signoret)
gemina De Lotto   Lichtensia gemina (De Lotto)
genevense marchali Cockerell   Rhodococcus marchali (Cockerell)
* genevense Targioni Tozzetti, Lecanium Eulecanium tiliae (Linnaeus)
* genistae Signoret, Lecanium   Parthenolecanium persicae (Fabricius)
geometricum Green   Paralecanium geometricum (Green)
* geranii Brain, Inglisia   Cryptinglisia lounsburyi Cockerell
ghesquierei Laing   Loemica ghesquierei Laing
{gigantea Shinji}   Eulecanium giganteum (Shinji)
giganteum (Shinji)   Eulecanium giganteum (Shinji)
giganteus Dozier   Ceroplastes giganteus Dozier
giganteus Koteja & Brookes   Symonicoccus giganteus Koteja & Brookes
gigas Cockerell   Ceroplastes gigas Cockerell
glabra De Lotto   Udinia glabra De Lotto
glandi Kuwana   Parthenolecanium glandi (Kuwana)
glanulosa (Hempel)   Saissetia glanulosa (Hempel)
{glanulosum Hempel}   Saissetia glanulosa (Hempel)
globosa Fonseca   Pulvinaria globosa Fonseca
* globulosum Maskell, Lecanium   Taiwansaissetia formicarii (Green)
goethei King, Pulvinaria [NOMEN NUDUM]
gossypii Bodenheimer   Pulvinaria gossypii (Bodenheimer)
gouligouli Hodgson   Etiennea gouligouli Hodgson
gowdeyi Newstead   Ceronema gowdeyi (Newstead)
grabhami Cockerell   Pulvinaria grabhami Cockerell
{gracile Hempel}   Eucalymnatus gracilis (Hempel)
gracilis Canard   Rhizopulvinaria gracilis Canard
gracilis Green   Parafairmairia gracilis Green
gracilis (Hempel)   Eucalymnatus gracilis (Hempel)
gracilis nictheroyensis Costa Lima   Eucalymnatus gracilis nictheroyensis Costa Lima
grandicula Borchsenius   Rhizopulvinaria grandicula Borchsenius
grandis Borchsenius   Luzulaspis grandis Borchsenius
grandis Green   Alecanopsis grandis Green
grandis Green & Laing   Mametia grandis (Green & Laing)
grandis Hempel, 1900b   Ceroplastes grandis Hempel
grandis Hempel, 1900b   Pulvinaria grandis Hempel
grandis hempeli Lizer y Trelles   Ceroplastes hempeli Lizer y Trelles
{graniforme (Wünn)}   Nemolecanium graniformis (Wünn)
graniformis Wunn   Nemolecanium graniformis (Wunn)

grassei Balachowsky   Rhizopulvinaria grassei (Balachowsky)
* gray Targioni Tozzetti, Columnea   Ceroplastes cassiae (Chavannes)
* greeni Froggatt, Pulvinaria   Pulvinaria dodonaeae Maskell
gregarius Hempel   Ceroplastes gregarius Hempel
grevilleae Hall   Inglisia grevilleae Hall
grevilleae Hempel   Neolecanochiton grevilleae Hempel
guerinii Signoret   Coccus guerinii (Signoret)
* guignardi King, Eulecanium   Parthenolecanium corni (Bouche)
guilliermondi Mahdihassan, Ceroplastodes [NOMEN NUDUM]
guizhouensis Qin & Gullan   Kilifia guizhouensis Qin & Gullan
gwaai Hodgson   Waxiella gwaai (Hodgson)
gymnospori Green   Coccus gymnospori (Green)
[gymnosporiae Green]   Coccus gymnospori (Green)
* gymnosporiae Hall, Pulvinaria   Pulvinaria psidii Maskell
* gyrcanicum Hadzibejli, Eulecanium   Eulecanium tiliae (Linnaeus)

hainanense Takahashi   Paralecanium hainanense Takahashi
{hainanensis Takahashi}   Paralecanium hainanense Takahashi
hakearum Fuller   Austrolichtensia hakearum (Fuller)
halli Borchsenius   Rhizopulvinaria halli Borchsenius
halli Hodgson   Etiennea halli Hodgson
[haloxyli Hall]   Acantholecanium haloxyloni (Hall)
haloxyloni Hall   Acantholecanium haloxyloni (Hall)
* hamberdiensis Borchsenius, Eriopeltis   Eriopeltis festucae (Fonscolombe)
harpazi Ben-Dov   Exaeretopus harpazi Ben-Dov
hawanus Williams & Watson   Ceroplastes hawanus Williams & Watson
hazeae Kuwana   Pulvinaria hazeae Kuwana
* hederae Lichtenstein, Philippia [NOMEN NUDUM]   Lichtensia viburni Signoret
helichrysi Hall   Ceroplastes helichrysi Hall
helichrysi sinoiae Hall   Ceroplastes sinoiae Hall
hellenicus Green, Exaeretopus [NOMEN NUDUM]
hemiacantha De Lotto   Pulvinaria hemiacantha (De Lotto)
[hemicriphus (Dalman)]   Physokermes hemicryphus (Dalman)
hemicryphus Dalman   Physokermes hemicryphus (Dalman)
* hemisphaerica clypeata (Douglas), Saissetia   Saissetia coffeae (Walker)
* hemisphaerica hibernaculorum (Boisduval), Saissetia   Saissetia coffeae (Walker)
* hemisphaerica (Targioni Tozzetti) Saissetia   Saissetia coffeae (Walker)
* hemisphaericum hibernaculorum Boisduval, Lecanium   Saissetia coffeae (Walker)
* hemisphaericum Targioni Tozzetti, Lecanium   Saissetia coffeae (Walker)
hempeli Costa Lima   Eucalymnatus hempeli Costa Lima
hempeli Lizer y Trelles   Ceroplastes hempeli Lizer y Trelles
[hepalense Takagi]   Platylecanium nepalense Takagi
herrerae Cockerell   Neolecanium herrerae Cockerell
* hesperidum africanum Newstead, Lecanium
          (Trechocorys)[NOMEN NUDUM]   Coccus viridis (Green)
* hesperidum alienum Douglas, Coccus   Coccus hesperidum Linnaeus
* hesperidum alienus (Douglas), Lecanium   Coccus hesperidum Linnaeus
hesperidum javanensis Newstead   Coccus hesperidum javanensis (Newstead)
* hesperidum lauri (Boisduval), Coccus   Coccus hesperidum Linnaeus
hesperidum Linnaeus   Coccus hesperidum Linnaeus
* hesperidum pacificum Kuwana, Lecanium (Calymnatus)   Coccus hesperidum Linnaeus
* hesperidum pacificus (Kuwana), Coccus   Coccus hesperidum Linnaeus
* hibernaculorum Boisduval, Chermes   Saissetia coffeae (Walker)
hirsuta Froggatt   Pseudopsylla hirsuta Froggatt

hirsutum Morrison — Alecanium hirsutum Morrison
hirsutum Newstead — Eulecanium hirsutum (Newstead)
hirsutus Hempel — Eucalymnatus hirsutus Hempel
hispidus Maskell — Eriochiton hispidus Maskell
hissarica Borchsenius, 1952 — Parafairmairia hissarica Borchsenius
hissarica Borchsenius, 1952 — Rhizopulvinaria hissarica Borchsenius
hissaricum Borchsenius — Eulecanium hissaricum Borchsenius
hodgsoni Matile-Ferrero & Le Ruyet — Ceroplastes hodgsoni (Matile-Ferrero & Le Ruyet)

* hoferi King, Lecanium (Eulecanium) — Eulecanium tiliae (Linnaeus)
hololeucus De Lotto — Ceroplastes hololeucus De Lotto
horii Kuwana — Pulvinaria horii Kuwana
* hunteri King, Pulvinaria — Pulvinaria vitis (Linnaeus)
hurae Newstead — Saissetia hurae Newstead
[hybernaculorum (Boisduval)] — Saissetia coffeae (Walker)
* hydatis Costa, Coccus — Ceroplastes rusci (Linnaeus)
hydrangeae Steinweden — Pulvinaria hydrangeae Steinweden
hymenantherae Maskell — Ctenochiton hymenantherae Maskell
hyperbaterum Morrison — Cyclolecanium hyperbaterum Morrison

[iamaicensis (White)] — Ceroplastes jamaicensis White
[ianeirensis Targioni Tozzetti] — Ceroplastes janeirensis Gray
* ibericum Hadzibejli, Eulecanium — Eulecanium tiliae (Linnaeus)
[iceryi Guerin-Meneville] — Pulvinaria iceryi (Signoret)
iceryi Signoret — Pulvinaria iceryi (Signoret)
iceryoides Green — Ceronema iceryoides Green
idesiae Kuwana — Pulvinaria idesiae Kuwana
iheringi Cockerell — Ceroplastes iheringi Cockerell
ikoyensis Hanford — Udinia ikoyensis Hanford
* iljiniae Danzig, Rhizopulvinaria — Acanthopulvinaria orientalis (Nassonov)
illuppalamae Green — Coccus illuppalamae (Green)
imbricans Green — Hemilecanium imbricans (Green)
imbricatum Cockerell — Neolecanium imbricatum (Cockerell)
* imeretina Hadzibejli, Neopulvinaria — Neopulvinaria innumerabilis (Rathvon)
immanis Green — Ceroplastes immanis Green
impar Cockerell — Mesolecanium impar (Cockerell)
incisus King — Coccus incisus (King)
inclusus Green — Ctenochiton inclusus Green
inconspicua Maskell — Inglisia inconspicua Maskell
[inconspigua Danzig] — Pulvinaria inconspiqua Danzig
inconspiqua Danzig — Pulvinaria inconspiqua Danzig
indica Avasthi & Shafee — Pulvinaria indica Avasthi & Shafee
inflata Cockerell & Parrott — Megasaissetia inflata (Cockerell & Parrott)
{inflatum Cockerell & Parrott} — Megasaissetia inflata (Cockerell & Parrott)
inflatum Hempel — Mesolecanium inflatum Hempel
infrequens Hempel — Saissetia infrequens (Hempel)
* ingae Ferris, Cryptostigma — Cryptostigma inquilina (Newstead)
innumerabilis acericola (Walsh & Riley) — Pulvinaria acericola (Walsh & Riley)
innumerabilis occidentalis Cockerell — Pulvinaria occidentalis Cockerell
innumerabilis Rathvon — Neopulvinaria innumerabilis (Rathvon)
* innumerabilis tiliae King & Cockerell, Pulvinaria — Pulvinaria vitis (Linnaeus)
inopheron Laing — Macropulvinaria inopheron (Laing)
inopinatus Danzig & Kozár — Physokermes inopinatus Danzig & Kozár
inquilina Newstead — Cryptostigma inquilina (Newstead)

inquilinum Morrison	Mesolecanium inquilinum Morrison
inquilinum Newstead	Coccus inquilinum (Newstead)
insignicola Craw	Physokermes insignicola (Craw)
insolens King	Coccus insolens (King)
insulanus De Lotto	Ceroplastes insulanus De Lotto
intermedia (Goux)	Poaspis intermedia (Goux)
{intermedius Goux}	Poaspis intermedia (Goux)
interruptus Danzig	Vittacoccus interruptus Danzig
inyangombae Hodgson	Coccus inyangombae Hodgson
iridis Borchsenius	Lecanopsis iridis Borchsenius
irregularis Cockerell	Ceroplastes irregularis Cockerell
* irregularis Leonardi, Ceroplastes	Ceroplastes leonardianus Lizer y Trelles
* irregularis rubidus Cockerell, Ceroplastes	Ceroplastes irregularis Cockerell
itanhaensis Mendes	Eucalymnatus itanhaensis Mendes
itatiayensis Hempel	Ceroplastes itatiayensis Hempel
ixorae Green, 1909	Pulvinaria ixorae Green
* ixorae Green, 1922b, Lecanium	Milviscutulus mangiferae (Green)
jaboticabae Hempel	Mesolecanium jaboticabae (Hempel)
jacksoni Newstead	Macropulvinaria jacksoni (Newstead)
jaculator Green & Laing	Coccus jaculator (Green & Laing)
[jahadiezi Balachowsky]	Poaspis jahandiezi (Balachowsky)
jahandiezi Balachowsky	Poaspis jahandiezi (Balachowsky)
jaliscensis Cockerell & Cockerell	Pendularia jaliscensis (Cockerell & Cockerell)
[jaliscensis (Cockerell)]	Pendularia jaliscensis (Cockerell & Cockerell)
* jamaicensis (Newstead), Cryptostigma [N. NUDUM]	Cryptostigma inquilina (Newstead)
jamaicensis Newstead, Lecanopsis [N. NUDUM]	Cryptostigma inquilina (Newstead)
jamaicensis White	Ceroplastes jamaicensis White
janeirensis Gray	Ceroplastes janeirensis Gray
* japonensis Takahashi, Eriopeltis	Eriopeltis sachalinensis Borchsenius
japonica Cockerell	Takahashia japonica Cockerell
{japonica (Green)}	Ceroplastes japonicus Green
* japonica Kuwana, 1909a, Lichtensia	Metaceronema japonica (Maskell)
* japonica Kuwana, 1916, Protopulvinaria	Protopulvinaria fukayai (Kuwana)
japonica (Maskell)	Metaceronema japonica (Maskell)
{japonicum Maskell}	Metaceronema japonica (Maskell)
japonicus Green	Ceroplastes japonicus Green
jezoensis Siraiwa	Physokermes jezoensis Siraiwa
jocunda De Lotto	Saissetia jocunda De Lotto
* juglandifex Fitch, Lecanium	Parthenolecanium corni (Bouche)
juglandii Hadzibejli	Pulvinaria juglandii Hadzibejli
* juglandis Bouche, Lecanium	Eulecanium tiliae (Linnaeus)
* jungi Chen, Coccus	Coccus hesperidum Linnaeus
juniperi Danzig	Eulecanium juniperi Danzig
justaserpentina Fonseca	Pulvinaria justaserpentina Fonseca
* kansasense Hunter, Lecanium	Parthenolecanium corni (Bouche)
katsurae Shinji	Pulvinaria katsurae Shinji
kellyi Brain	Etiennea kellyi (Brain)
keravatae Williams & Watson, 1990	Anthococcus keravatae Williams & Watson
keravatae Williams & Watson, 1990	Neosaissetia keravatae Williams & Watson
kibarae Beccari	Myzolecanium kibarae Beccari
[kibarae Targioni Tozzetti]	Myzolecanium kibarae Beccari
* kingii Cockerell, Lecanium (Eulecanium)	Parthenolecanium corni (Bouche)

kirgisica Borchsenius — Pulvinaria kirgisica Borchsenius
kleinhoviae Williams & Watson — Melanesicoccus kleinhoviae Williams & Watson
[knwanai (Kanda)] — Eulecanium kuwanai (Kanda)
koebeli Green — Ceronema koebeli Green
kondarensis Borchsenius — Poaspis kondarensis (Borchsenius)
koreanus Borchsenius, 1955b — Didesmococcus koreanus Borchsenius
* koreanus Borchsenius, 1956, Eriopeltis — Eriopeltis sachalinensis Borchsenius
kosswigi Bodenheimer — Palaeolecanium kosswigi (Bodenheimer)
kostylevi Borchsenius — Eulecanium kostylevi Borchsenius
kosztarabi Avasthi & Shafee — Coccus kosztarabi Avasthi & Shafee
kosztarabi Koteja & Kozár — Luzulaspis kosztarabi Koteja & Kozár
* kraunhianum Lindinger, Lecanium — Coccus longulus (Douglas)
krugeri Zehntner — Saccharolecanium krugeri (Zehntner)
[kunming (Ferris)] — Eulecanium kunmingi (Ferris)
kunmingensis Tang & Xie — Ceroplastes kunmingensis Tang & Xie
kunmingi Ferris — Eulecanium kunmingi (Ferris)
kunoense (Kuwana) — Eulecanium kunoense (Kuwana)
{kunoensis Kuwana} — Eulecanium kunoense (Kuwana)
* kuraruensis Takahashi, Coccus — Milviscutulus mangiferae (Green)
kurilensis Danzig — Poaspis kurilensis (Danzig)
kuwacola Kuwana — Pulvinaria kuwacola Kuwana
kuwanai Kanda — Eulecanium kuwanai (Kanda)

* laevis Costa, Calypticus — Coccus hesperidum Linnaeus
lahillei Cockerell — Ceroplastes lahillei Cockerell
lamborni Newstead — Ceroplastes lamborni Newstead
* lanatus Gmelin, Coccus — Pulvinaria sericea (Fourcroy)
{lanigera (Hempel)} — Mallococcus lanigerus (Hempel)
{lanigerum Hempel} — Mallococcus lanigerus (Hempel)
lanigerus (Hempel) — Mallococcus lanigerus (Hempel)
laos (Takahashi) — Neosaissetia laos (Takahashi)
lata (Goux) — Poaspis lata (Goux)
latioperculatum Green — Coccus latioperculatum (Green)
[latioperculum Green] — Coccus latioperculatum (Green)
* latipes Borchsenius, Physokermes — Physokermes piceae (Schrank)
{latus Goux} — Poaspis lata (Goux)
* lauri Boisduval, Chermes — Coccus hesperidum Linnaeus
[leonardianus (Leonardi)] — Ceroplastes leonardianus Lizer y Trelles
leonardianus Lizer y Trelles — Ceroplastes leonardianus Lizer y Trelles
lepagei Costa Lima — Ceroplastes lepagei Costa Lima
* lepida Brain, Pulvinaria — Pulvinaria iceryi (Signoret)
leptospermi Maskell — Inglisia leptospermi Maskell
lespedezae Danzig — Eulecanium lespedezae Danzig
leucaenae Cockerell — Neolecanium leucaenae Cockerell
leurus De Lotto — Coccus leurus De Lotto
levis Maskell — Akermes levis (Maskell)
* lichenoides Green, Lecanium — Pulvinaria horii Kuwana
lichtensteinii Signoret — Eriopeltis lichtensteini Signoret
lidgetti Fernald — Coccus lidgetti (Fernald)
limbatum Green — Paralecanium limbatum Green
limnanthemi Goury, Lecanium [NOMEN NUDUM]
lindae Matile-Ferrero & Le Ruyet — Udinia lindae Matile-Ferrero & Le Ruyet
* linearis Targioni Tozzetti, Pulvinaria — Pulvinaria floccifera (Westwood)
lineolatae King & Cockerell — Lecanopsis lineolatae King & Cockerell

* lintneri Cockerell & Bennett, Lecanium   Parthenolecanium corni (Bouche)
liriodendri Gmelin   Toumeyella liriodendri (Gmelin)
litorea De Lotto   Parasaissetia litorea De Lotto
[litseae Rutherford]   Coccus litzeae Rutherford
litzeae Rutherford   Coccus litzeae Rutherford
lizeri Fonseca   Coccus lizeri (Fonseca)
lobayana Balachowsky & Ferrero   Udinia lobayana (Balachowsky & Ferrero)
[logulus Douglas]   Coccus longulus (Douglas)
loisa Hodgson   Messinea loisa Hodgson
lomagundiae Hall   Toumeyella lomagundiae Hall
longicauda Brain   Ceroplastes longicauda Brain
longicauda sapii Hall   Ceroplastes longicauda sapii Hall
longicornis Green   Vittacoccus longicornis (Green)
longiseta Leonardi   Ceroplastes longiseta Leonardi
* longisetum Borchsenius, Eulecanium   Eulecanium douglasi (Sulc)
* longisqua De Lotto, Pulvinaria   Pulvinaria elongata Newstead
* longivalvata bakeri Cockerell & Robinson,
                        Protopulvinaria   Protopulvinaria longivalvata Green
longivalvata Green   Protopulvinaria longivalvata Green
{longulum Douglas}   Coccus longulus (Douglas)
longulus Douglas   Coccus longulus (Douglas)
loralaiensis Rao   Pulvinaria loralaiensis Rao
loranthi Froggatt   Tectopulvinaria loranthi Froggatt
louisieae Matile-Ferrero   Mametia louisieae Matile-Ferrero
lounsburyi Cockerell   Cryptinglisia lounsburyi Cockerell
lucida Hempel, 1912   Saissetia lucida Hempel
lucidum Hempel, 1912   Mesolecanium lucidum Hempel
lucidus Hempel, 1900b   Ceroplastes lucidus Hempel
lumpurensis Takahashi   Coccus lumpurensis Takahashi
lustneri King, Lecanium [NOMEN NUDUM]
lutea Cockerell   Philephedra lutea (Cockerell)
* luteolus De Lotto, Ceroplastes   Ceroplastes brevicauda Hall
luzonicum Cockerell   Podoparalecanium luzonicum (Cockerell)
luzulae australis Maskell   Symonicoccus australis (Maskell)
luzulae Dufour   Luzulaspis luzulae (Dufour)
lycii Cockerell   Metapulvinaria lycii (Cockerell)
lymani King   Eulecanium lymani King

macarangae Morrison   Coccus macarangae Morrison
macarangicolus Takahashi   Coccus macarangicolus Takahashi
macgregori Sampedro & Butze   Ceroplastes macgregori Sampedro & Butze
machili Takahashi   Podoparalecanium machili (Takahashi)
* maclurae Fitch, Lecanium   Pulvinaria vitis (Linnaeus)
* maclurae Hunter, Lecanium   Parthenolecanium corni (Bouche)
* maclurarum Cockerell, Lecanium (Eulecanium) Parthenolecanium corni (Bouche)
macrospinus Savescu   Luzulaspis macrospinus Savescu
macrozamiae Fuller   Paralecanium frenchi macrozamiae (Fuller)
* maculatum Signoret, Lecanium   Coccus hesperidum Linnaeus
maculatum Takahashi   Paralecanium maculatum Takahashi
* {maculatus (Signoret), Coccus}   Coccus hesperidum Linnaeus
madagascariensis Hodgson   Etiennea madagascariensis Hodgson
madagascariensis Mamet   Lichtensia madagascariensis (Mamet)
madagascariensis Targioni Tozzetti   Ceroplastes madagascariensis (Targioni Tozzet-
                                                    ti)

magarinosi Costa Lima	Eucalymnatus magarinosi Costa Lima
magnetinsulae Qin & Gullan	Cryptostigma magnetinsulae Qin & Gullan
magnicauda Reyne	Ceroplastes magnicauda Reyne
magnifica Green	Vinsonia magnifica Green
* magnoliarum Cockerell, Lecanium	Parthenolecanium persicae (Fabricius)
* magnoliarum hortensiae Cockerell, Eulecanium	Parthenolecanium persicae (Fabricius)
magnospinus Mamet	Drepanococcus magnospinus (Mamet)
malagassa Mamet	Saissetia malagassa Mamet
malainum Takahashi	Paralecanium malainum Takahashi
* mali Schrank, Coccus	Eulecanium tiliae (Linnaeus)
malloti Takahashi	Coccus malloti (Takahashi)
malvacearum Cockerell	Inglisia malvacearum Cockerell
mammeae Maskell	Pulvinaria mammeae Maskell
mancum Green	Paralecanium mancum (Green)
mangiferae Green	Milviscutulus mangiferae (Green)
mangiferae Takahashi	Xenolecanium mangiferae Takahashi
manzanillense Cockerell	Neolecanium manzanillense Cockerell
marchali (Cockerell)	Rhodococcus marchali (Cockerell)
* marginata Ferris, Pulvinaria	Pulvinaria ferrisi Ali
* marginata Targioni Tozzetti, Pulvinaria	Pulvinaria sericea (Fourcroy)
marginatum Green	Paralecanium marginatum (Green)
marginatus Newstead	Pseudokermes marginatus Newstead
marianum Cockerell	Paralecanium marianum Cockerell
maritima Canard	Rhizopulvinaria maritima Canard
maritimum (Green)	Paralecanium maritimum (Green)
marmorata Cockerell	Pulvinaria marmorata Cockerell
marmoratum Hempel	Mesolecanium marmoratum Hempel
marmoreus Cockerell	Ceroplastes marmoreus Cockerell
marquesi Hempel	Alecanochiton marquesi Hempel
{marsupiale Green}	Marsipococcus marsupialis (Green)
marsupialis (Green)	Marsipococcus marsupialis (Green)
martinae Mosquera	Ceroplastes martinae Mosquera
martinoi (Almeida)	Waxiella martinoi (Almeida)
* maskelli Maskell, Kermes	Cryptes baccatus (Maskell)
* maskelli novemarticulata Green, Pulvinaria	Pulvinaria maskelli Olliff
* maskelli nuytsiae (Fuller), Pulvinaria	Pulvinaria maskelli Olliff
* [maskelli nuytsiae Maskell, Pulvinaria]	Pulvinaria maskelli Olliff
maskelli Olliff	Pulvinaria maskelli Olliff
* maskelli spinosior Maskell, Pulvinaria	Pulvinaria maskelli Olliff
* maskelli viminariae Fuller, Pulvinaria	Pulvinaria maskelli Olliff
* mauritiense Mamet, Lecanium	Coccus hesperidum Linnaeus
* mauritiensis (Mamet), Coccus	Coccus hesperidum Linnaeus
maxima Borchsenius	Stozia maxima Borchsenius
maxima Green	Megapulvinaria maxima (Green)
* maxima maxima Green, Pulvinaria	Megapulvinaria maxima (Green)
* maxima thespesiae Green, Pulvinaria	Megapulvinaria maxima (Green)
* maximus Borchsenius, Eriopeltis	Eriopeltis festucae (Fonscolombe)
mayteni Hempel	Mesolecanium mayteni (Hempel)
mazoeensis Hodgson	Lagosinia mazoeensis Hodgson
megriensis Borchsenius, 1952	Rhizopulvinaria megriensis Borchsenius
* megriensis Borchsenius, 1953, Didesmococcus	Didesmococcus unifasciatus (Archangelskaya)
melaleucae Green	Ceroplastodes melaleucae (Green)
melaleucae Maskell	Coccus melaleucae (Maskell)
melzeri Bondar	Ceroplastodes melzeri Bondar

melzeri Hempel	Eulecanium melzeri Hempel
mercarae Ramakrishna Ayyar, Lecanium [NOMEN NUDUM]	
merwei Joubert	Pulvinaria merwei Joubert
* mesembrianthemi Costa, Calypticus	Pulvinaria mesembryanthemi (Vallot)
* [mesembrianthemi (Vallot)]	Pulvinaria mesembryanthemi (Vallot)
* mesembryanthemi Peringuey, Icerya	Pulvinaria mesembryanthemi (Vallot)
mesembryanthemi Vallot	Pulvinaria mesembryanthemi (Vallot)
mesuae Takahashi	Platylecanium mesuae Takahashi
metallicum (Green)	Paralecanium expansum metallicum (Green)
metrosideri Maskell	Lecanochiton metrosideri Maskell
mexicana Cockerell & Parrott	Schizochlamidia mexicana Cockerell & Parrott
mexicanus Ben-Dov	Tillancoccus mexicanus Ben-Dov
* mexicanus Cockerell, Ceroplastes	Ceroplastes cirripediformis Comstock
mierii Targioni Tozzetti	Ceroplastes mierii (Targioni Tozzetti)
migrans Zavattari	Dermolecanium migrans Zavattari
milanjianus Hodgson	Coccus milanjianus Hodgson
milleri Takahashi, 1939a	Paralecanium milleri Takahashi
milleri Takahashi, 1939e	Ceroplastes milleri Takahashi
[milleti Takahashi, 1939a]	Paralecanium milleri Takahashi
mimosae neghellii Bellio	Waxiella mimosae neghellii (Bellio)
mimosae Signoret	Waxiella mimosae (Signoret)
mimosae Townsend & Cockerell	Philephedra mimosae (Townsend & Cockerell)
minensis Hempel	Saissetia minensis Hempel
minima Borchsenius	Rhizopulvinaria minima Borchsenius
minima Koteja & Howell	Luzulaspis minima Koteja & Howell
* minimum Newstead, Lecanium	Coccus hesperidum Linnaeus
* minimum pinicola Maskell, Lecanium	Coccus hesperidum Linnaeus
* minimus (Newstead), Coccus	Coccus hesperidum Linnaeus
* minimus pinicola (Maskell), Coccus	Coccus hesperidum Linnaeus
minor Maskell	Lecanochiton minor Maskell
minuscula Danzig	Pulvinaria minuscula Danzig
minuta Brethes	Pulvinaria minuta Brethes
* minuta Fonseca, Pulvinaria	Pulvinaria eryngii Fonseca
minutum Takahashi	Paralecanium minutum Takahashi
minutus Cockerell	Ceroplastes minutus Cockerell
{mirabile Cockerell}	Tourneyella mirabilis (Cockerell)
mirabilis (Cockerell)	Tourneyella mirabilis (Cockerell)
miranda (Cockerell & Parrott)	Saissetia miranda (Cockerell & Parrott)
mirifica (Maskell)	Saissetia mirifica (Maskell)
{mirificum Maskell}	Saissetia mirifica (Maskell)
[miripicum Maskell]	Saissetia mirifica (Maskell)
[mirmecophila Leonardi]	Lecanopsis myrmecophila Leonardi
mirus Green	Alecanopsis mirus Green
misiones Morrison	Ceroplastodes misiones Morrison
mkuzei Hodgson	Pulvinaria mkuzei Hodgson
mobile Brain	Ceronema mobile Brain
{mobilis Brain}	Ceronema mobile Brain
moestus De Lotto	Coccus moestus De Lotto
{monile Cockerell}	Akermes monilis (Cockerell)
monilis (Cockerell)	Akermes monilis (Cockerell)
monotes Hall	Saissetia monotes Hall
monotes pretoriae Hall	Saissetia monotes pretoriae Hall
monotonae Newstead, Lecanium [NOMEN NUDUM]	
* montana Hodgson, Lagosinia	Lagosinia vayssierei (Castel-Branco)

montana Mamet — Suareziella montana Mamet
* montana Schmutterer, Luzulaspis — Luzulaspis dactylis Green
{montanum Green} — Akermes montanus (Green)
montanus (Green) — Akermes montanus (Green)
* monticola Wang, Euphilippia — Metaceronema japonica (Maskell)
montrichardiae Newstead — Etiennea montrichardiae (Newstead)
{moreirae Green} — Eutaxia moreirai Green
moreirai Green — Eutaxia moreirai Green
* mori Signoret, Lecanium — Parthenolecanium persicae (Fabricius)
mori somereni Newstead — Saissetia somereni (Newstead)
mozambiquensis Hodgson — Houardia mozambiquensis Hodgson
msasae Hall — Eumashona msasae (Hall)
muiri Kotinsky — Coccus muiri Kotinsky
multituberculum Hodgson — Etiennea multituberculum Hodgson
munroi De Lotto — Saissetia munroi De Lotto
murex Hodgson — Coccus murex Hodgson
murrayi Froggatt — Ceroplastes murrayi Froggatt
myricae Linnaeus — Ceroplastes myricae (Linnaeus)
myricariae Bazarov — Pulvinaria myricariae Bazarov
myrmecariae Williams & Watson — Melanesicoccus myrmecariae Williams & Watson
myrmecophila Leonardi — Lecanopsis myrmecophila Leonardi
myrtilli Kaltenbach — Phyllostroma myrtilli (Kaltenbach)

nairobica De Lotto — Parasaissetia nairobica (De Lotto)
nakaharai Gimpel — Ceroplastes nakaharai Gimpel
namunakuli (Green) — Coccus piperis namunakuli (Green)
* nanum Cockerell, Lecanium — Coccus hesperidum Linnaeus
* nanus (Cockerell), Coccus — Coccus hesperidum Linnaeus
narzykulovi Bazarov & Shmelev — Rhizopulvinaria narzykulovi Bazarov & Shmelev
nectandrae Hempel, 1918 — Megasaissetia nectandrae Hempel
nectandrae Hempel, 1929 — Tourneyella nectandrae Hempel
neglecta De Lotto — Saissetia neglecta De Lotto
nemorosa Koteja — Luzulaspis nemorosa Koteja
neocellulosa Takahashi — Pulvinaria neocellulosa Takahashi
* neoceriferus Yousuf & Shafee, Ceroplastes — Ceroplastes ajmerensis Avasthi & Shafee
neomaritimum Takahashi — Paralecanium neomaritimum Takahashi
nepalense Takagi — Platylecanium nepalense Takagi
* nerii Kanda, Pulvinaria — Pulvinaria polygonata Cockerell
* nerii Newstead, Ceroplastes — Ceroplastes rusci (Linnaeus)
nevesi Gómez-Menor Ortega — Lecanopsis nevesi Gómez-Menor Ortega
* newmani Froggatt, Pulvinaria — Pulvinaria maskelli Olliff
newsteadi Hanford — Udinia newsteadi Hanford
* newsteadi Leonardi, Pulvinaria — Protopulvinaria pyriformis Cockerell
* nicotianae Newstead, Lecanium — Pulvinaria grabhami Cockerell
nigeriensis Hanford — Udinia nigeriensis Hanford
* {nigra depressa (Targionni Tozzetti), Saissetia} — Parasaissetia nigra (Nietner)
nigra (Nietner) — Parasaissetia nigra (Nietner)
nigrella King — Saissetia nigrella King
nigrivitta Borchsenius — Eulecanium nigrivitta Borchsenius
nigrofasciatum Pergande — Mesolecanium nigrofasciatum (Pergande)
* nigrum begoniae Douglas, Lecanium — Parasaissetia nigra (Nietner)
* nigrum depressum Targioni Tozzetti, Lecanium — Parasaissetia nigra (Nietner)
{nigrum Nietner} — Parasaissetia nigra (Nietner)
* nigrum nitidum Newstead, Lecanium (Saissetia) — Parasaissetia nigra (Nietner)

* nipponica Lindinger, Pulvinaria	Pulvinaria citricola Kuwana
nishigaharae Kuwana	Pulvinaria nishigaharae (Kuwana)
nitens Cockerell	Pseudokermes nitens Cockerell
[nitens (Hempel)]	Pseudokermes nitens Cockerell
nitidulus De Lotto	Umwinsia nitidulus (De Lotto)
* nivea Cockerell, 1893n, Inglisia	Ceroplastodes dugesii (Signoret)
* nivea Cockerell, 1893r, Fairmairia (Ceroplastodes)	Ceroplastodes dugesii (Signoret)
noacki Cockerell	Platinglisia noacki Cockerell
nocivum Borchsenius	Eulecanium nocivum Borchsenius
[nocturnium Cockerell & Parrott]	Mesolecanium nocturnum (Cockerell & Parrott)
nocturnum Cockerell & Parrott	Mesolecanium nocturnum (Cockerell & Parrott)
* notatum Maskell, Lecanium	Pulvinaria aurantii Cockerell
* notatus (Maskell), Coccus	Pulvinaria aurantii Cockerell
novaesi Hempel	Ceroplastes novaesi Hempel
novaesi mendozae Cockerell	Ceroplastes novaesi mendozae Cockerell
* numismaticum Pettit & McDaniel, Lecanium (Toumeyella)	Toumeyella parvicornis (Cockerell)
* nuytsiae Fuller, Ctenochiton	Pulvinaria maskelli Olliff
* nuytsiae minor Fuller, Ctenochiton [NOMEN NUDUM]	Pulvinaria maskelli Olliff
* nyassae Newstead, Lecanium	Macropulvinaria jacksoni (Newstead)
nyika Hodgson	Coccus nyika Hodgson
[obae (Bernard), Saissetia]	Saissetia oleae (Olivier)
obscura Newstead	Pulvinaria obscura Newstead
obscurum Hempel	Mesolecanium obscurum (Hempel)
* obtusum Thro, Lecanium	Parthenolecanium corni (Bouche)
obunca De Lotto	Toumeyella obunca De Lotto
occidentalis Cockerell	Pulvinaria occidentalis Cockerell
* occidentalis subalpina Cockerell, Pulvinaria	Pulvinaria occidentalis Cockerell
ochnaceae Kuwana	Pulvinaria ochnaceae (Kuwana)
ocreus Mosquera	Ceroplastes ocreus Mosquera
oculata Brain	Trijuba oculata (Brain)
{oculatus (Brain)}	Trijuba oculata (Brain)
ocultus Fonseca	Perilecanium ocultus Fonseca
okitsuensis Kuwana	Pulvinaria okitsuensis Kuwana
[oleae (Bernard), Saissetia]	Saissetia oleae (Olivier)
oleae cherimoliae Gómez-Menor Ortega	Saissetia oleae cherimoliae Gómez-Menor Ortega
oleae miranda (Cockerell & Parrott)	Saissetia miranda (Cockerell & Parrott)
{oleae mirandum Cockerell & Parrott}	Saissetia miranda (Cockerell & Parrott)
[oleae (Oliver)]	Saissetia oleae (Olivier)
oleae Olivier	Saissetia oleae (Olivier)
* oleae rhusae Hall, Filippia	Lichtensia chilianthi (Brain)
* oleae testudo (Curtis), Lecanium	Saissetia oleae (Olivier)
olivaceum Green	Ctenochiton olivaceum Green
* olivina Berlese & Silvestri, Euphilippia	Filippia follicularis (Targioni Tozzetti)
ophiorrhizae Green	Coccus ophiorrhizae (Green)
opimum Green	Coccus opimum (Green)
opulenta De Lotto	Saissetia opulenta De Lotto
orbiculata De Lotto	Saissetia orbiculata De Lotto
* ordinatus Danzig, Vittacoccus	Vittacoccus longicornis (Green)
orientalis Borchsenius	Parthenolecanium orientalis Borchsenius
orientalis Danzig	Exaeretopus orientalis Danzig
orientalis Lahille	Alichtensia orientalis Lahille

orientalis Nassonov   Acanthopulvinaria orientalis (Nassonov)
orientalis Reyne   Megapulvinaria orientalis (Reyne)
* ornata Froggatt, Pulvinaria   Pulvinaria decorata Borchsenius
ornata Hempel, 1912   Pulvinaria ornata Hempel
ornata Maskell   Inglisia ornata Maskell
ornatum Hempel, 1900b   Stictolecanium ornatum (Hempel)
ovalis Koteja & Brookes   Symonicoccus ovalis Koteja & Brookes
ovatum Morrison   Paralecanium ovatum Morrison
* oxyacanthae Linnaeus, Coccus   Pulvinaria vitis (Linnaeus)
oyamae Kuwana   Pulvinaria oyamae Kuwana

pahanense Takahashi   Paralecanium pahanense Takahashi
pallidior Cockerell & King   Eulecanium pallidior (Cockerell & King)
* pallidus Brain, Ceroplastes   Ceroplastes ficus Newstead
* palmae Haworth, Coccus   Saissetia oleae (Olivier)
palmae Hempel   Pseudokermes palmae Hempel
paradeformosum Fonseca   Coccus paradeformosum (Fonseca)
* paradelpha Cockerell & Lidgett, Pulvinaria   Pulvinaria thompsoni Maskell
paradeniyense Green   Paralecanium paradeniyense Green
paranaensis Hempel   Pulvinaria paranaensis Hempel
{parvicorne Cockerell}   Toumeyella parvicornis (Cockerell)
parvicornis (Cockerell)   Toumeyella parvicornis (Cockerell)
parvisetosus Brookes & Koteja Brookes   Waricoccus parvisetosus Brookes & Koteja
parvula Cockerell   Philephedra parvula (Cockerell)
* parvus Borchsenius, Psilococcus   Psilococcus ruber Borchsenius
parvus Green   Ceroplastes parvus Green
patella Maskell   Inglisia patella Maskell
patellaeformis Brain   Parafairmairia patellaeformis Brain
* patellaeformis Curtis, Coccus   Coccus hesperidum Linnaeus
* [patelliformis Curtis, Coccus]   Coccus hesperidum Linnaeus
patersoniae Maskell   Eulecanium patersoniae (Maskell)
pattersoni Hanford   Udinia pattersoni Hanford
[pattersoniae Maskell]   Eulecanium patersoniae (Maskell)
paucispinosum Danzig   Eulecanium paucispinosum Danzig
paucispinus De Lotto   Ceroplastes paucispinus De Lotto
paulista Hempel   Toumeyella paulista Hempel
paupercula De Lotto   Udinia paupercula De Lotto
pela Chavannes   Ericerus pela (Chavannes)
[pela Westwood]   Ericerus pela (Chavannes)
penangensis Morrison   Coccus penangensis Morrison
pendens Fonseca   Pendularia pendens Fonseca
peninsularis Ferris   Pulvinaria peninsularis Ferris
perconvexum Cockerell   Neolecanium perconvexum (Cockerell)
perditulum Cockerell & Robbins   Mesolecanium perditulum Cockerell & Robbins
perditum Cockerell   Mesolecanium perditum (Cockerell)
peregrina Borchsenius   Pulvinaria peregrina (Borchsenius)
* perforatum Newstead, Lecanium   Eucalymnatus tessellatus (Signoret)
perforatus Maskell   Ctenochiton perforatus Maskell
* perforatus (Newstead), Eucalymnatus   Eucalymnatus tessellatus (Signoret)
perinflatum Cockerell   Eulecanium perinflatum (Cockerell)
peringueyi Brain   Idiosaissetia peringueyi Brain
peringueyi Joubert   Lichtensia peringueyi Joubert
{perlatum Cockerell}   Coccus perlatus (Cockerell)
perlatus (Cockerell)   Coccus perlatus (Cockerell)

polispina Matesova — Rhizopulvinaria polispina Matesova
polychaeta De Lotto — Lichtensia polychaeta (De Lotto)
polygonata Cockerell — Pulvinaria polygonata Cockerell
[polygonata (Green)] — Pulvinaria polygonata Cockerell
pomeranicum Kawecki — Parthenolecanium pomeranicum (Kawecki)
populeti Borchsenius — Pulvinaria populeti Borchsenius
* populi Signoret, Pulvinaria — Pulvinaria vitis (Linnaeus)
porifera Borchsenius — Lecanopsis porifera Borchsenius
portblairensis Yousuf & Shafee — Pulvinaria portblairensis Yousuf & Shafee
* potanini Borchsenius, Eulecanium — Ericerus pela (Chavannes)
poterii Walker, Coccus [NOMEN NUDUM]
* pratensis Borchsenius, Eriopeltis — Eriopeltis festucae (Fonscolombe)
pretoriae Brain — Membranaria pretoriae Brain
privigna De Lotto — Saissetia privigna De Lotto
proteae Brain — Marsipococcus proteae (Brain)
* pruinosum armeniacum Craw, Lecanium — Parthenolecanium corni (Bouche)
pruinosum Cocquillett — Parthenolecanium pruinosum (Coquillett)
* pruinosum kermoides Tyrrell, Lecanium — Parthenolecanium quercifex (Fitch)
pruinosum pruinosum Coquillett — Parthenolecanium pruinosum (Coquillett)
prunastri Fonscolombe — Sphaerolecanium prunastri (Fonscolombe)
pruni Hunter — Pulvinaria pruni Hunter
{pseudelongatum Brain} — Coccus pseudelongatus (Brain)
pseudelongatus (Brain) — Coccus pseudelongatus (Brain)
pseudexpansum Green — Paralecanium pseudexpansum (Green)
pseudoceriferus Green — Ceroplastes pseudoceriferus Green
[pseadohesperidum Cockerell] — Coccus pseudohesperidum (Cockerell)
pseudohesperidum Cockerell — Coccus pseudohesperidum (Cockerell)
pseudoleae Rutherford — Neolecanium pseudoleae Rutherford
pseudomagnoliarum Kuwana — Coccus pseudomagnoliarum (Kuwana)
* pseudonigrum Kuwana, Lecanium (Saissetia) — Parasaissetia nigra (Nietner)
pseudosemen Cockerell — Mesolecanium pseudosemen (Cockerell)
pseudotessellatum Newstead — Eulecanium pseudotessellatum (Newstead)
psidii Chavannes — Ceroplastes psidii (Chavannes)
psidii cistudiformis Cockerell — Ceroplastes cistudiformis Cockerell
* psidii Green, Lecanium — Milviscutulus mangiferae (Green)
psidii Maskell — Pulvinaria psidii Maskell
* psidii philippina Cockerell, Pulvinaria — Pulvinaria psidii Maskell
psychotriae De Lotto — Avricus psychotriae (De Lotto)
pterolobina De Lotto — Udinia pterolobina (De Lotto)
pubescens Ehrhorn — Eulecanium pubescens (Ehrhorn)
pulchella Hempel — Pulvinella pulchella Hempel
* pulchella Signoret, Vinsonia — Vinsonia stellifera (Westwood)
pulchra Danzig — Pulvinaria pulchra Danzig
* [pulchrum King, Lecanium] — Parthenolecanium rufulum (Cockerell)
* [pulchrum (King), Eulecanium] — Parthenolecanium rufulum (Cockerell)
* [pulchrum (Marchal), Eulecanium] — Parthenolecanium rufulum (Cockerell)
* pulchrum Reh, Lecanium — Parthenolecanium rufulum (Cockerell)
* [pulohella (Signoret), Vinsonia] — Vinsonia stellifera (Westwood)
pumilum Brain — Coccus pumilum (Brain)
{punctatum Cockerell} — Akermes punctatus (Cockerell)
punctatus (Cockerell) — Akermes punctatus (Cockerell)
* {punctulifera (Green), Saissetia} — Coccus hesperidum Linnaeus
* punctuliferum Green, Lecanium — Coccus hesperidum Linnaeus
punctuliferum lamborni Newstead — Udinia punctuliferum lamborni (Newstead)

purpurellus Cockerell   Ceroplastes purpurellus Cockerell
purpureus Hempel   Ceroplastes purpureus Hempel
putmani Phillips   Parthenolecanium putmani (Phillips)
[putnami (Phillips)]   Parthenolecanium putmani (Phillips)
[pyiformis Cockerell]   Protopulvinaria pyriformis Cockerell
pyrethri Borchsenius   Rhizopulvinaria pyrethri Borchsenius
* pyri Schrank, Coccus   Eulecanium tiliae (Linnaeus)
pyriformis Cockerell   Protopulvinaria pyriformis Cockerell

quadratum (Green), 1904d   Paralecanium quadratum (Green)
quadratus Green, 1935b   Ceroplastes quadratus Green
quadrifasciata (Cockerell)   Toumeyella quadrifasciata (Cockerell)
{quadrifasciatum Cockerell}   Toumeyella quadrifasciata (Cockerell)
{quadrilineata (Newstead)}   Ceroplastes quadrilineatus Newstead
quadrilineata royenae (Hall)   Ceroplastes royenae Hall
quadrilineatus Newstead   Ceroplastes quadrilineatus Newstead
quadrilineatus royenae Hall   Ceroplastes royenae Hall
* quadrilineatus simplex Brain, Ceroplastes   Ceroplastes fumidus De Lotto
quadrispina Matesova   Rhizopulvinaria quadrispina Matesova
quaintancii Cockerell   Pseudophilippia quaintancii Cockerell
quercifex Fitch   Parthenolecanium quercifex (Fitch)
* quercitronis Fitch, Lecanium   Parthenolecanium quercifex (Fitch)
* quercitronis kermoides (Tyrrell), Eulecanium   Parthenolecanium quercifex (Fitch)
quinquepori Newstead   Cryptostigma quinquepori (Newstead)

* {racemosum (Ratzeburg), Lecanium   Physokermes piceae (Schrank)
* racemosus Ratzeburg, Coccus   Physokermes piceae (Schrank)
rachelae (Bodenheimer)   Bodenheimera rachelae (Bodenheimer)
{racheli Bodenheimer}   Bodenheimera rachelae (Bodenheimer)
{rachelis (Bodenheimer)}   Bodenheimera rachelae (Bodenheimer)
* {radiatum (Costa), Lecanium}   Ceroplastes rusci (Linnaeus)
* radiatus Costa, Calypticus   Ceroplastes rusci (Linnaeus)
radicicola Green   Cribrolecanium radicicola Green
radicumgraminis Fonscolombe   Lecanopsis radicumgraminis (Fonscolombe)
rajae Kozár   Luzulaspis rajae Kozár
{ramakrishnae Ramakrishna Ayyar}   Coccus ramakrishnai (Ramakrishna Ayyar)
ramakrishnai (Ramakrishna Ayyar)   Coccus ramakrishnai (Ramakrishna Ayyar)
randiae Hall   Pulvinaria randiae Hall
rarus Hempel   Ceroplastes rarus Hempel
* rasinae Borchsenius, Eriopeltis   Eriopeltis festucae (Fonscolombe)
recurvatum Newstead   Hemilecanium recurvatum Newstead
regalis Canard   Pulvinaria regalis Canard
* rehi King, 1901, Lecanium   Parthenolecanium corni (Bouche)
rehi King, Pulvinaria, in Reh, 1903 [NOMEN NUDUM]
resinatum Kieffer & Herbst   Coccus resinatum (Kieffer & Herbst)
retamae Hall   Rhizopulvinaria retamae (Hall)
reticulata (Cockerell)   Saissetia reticulata (Cockerell)
{reticulatum Cockerell}   Saissetia reticulata (Cockerell)
reticulolaminae Morrison   Cryptostigma reticulolaminae Morrison
rhizophila Bazarov   Rhizopulvinaria rhizophila Bazarov
rhizophila Borchsenius   Pulvinaria rhizophila Borchsenius
rhizophila Targioni Tozzetti   Lecanopsis rhizophila Targioni Tozzetti
rhizophorae Cockerell   Mesolecanium rhizophorae (Cockerell)
rhizophorae Hempel   Ceroplastes rhizophorae Hempel

rhodesiensis Hall | Coccus rhodesiensis (Hall)
rhodesiensis Hodgson | Lagosinia rhodesiensis Hodgson
* rhododendri Danzig, Eulecanium | Eulecanium franconicum (Lindinger)
rhoicina De Lotto | Pulvinaria rhoicina De Lotto
rhois Ehrhorn | Pulvinaria rhois Ehrhorn
[rhyzophila Targioni Tozzetti] | Lecanopsis rhizophila Targioni Tozzetti
* [ribesia Signoret, Pulvinaria] | Pulvinaria vitis (Linnaeus)
* ribesiae Signoret, Pulvinaria | Pulvinaria vitis (Linnaeus)
* ribis Fitch, Lecanium | Parthenolecanium corni (Bouche)
* rifana Balachowsky, Lichtensia | Lichtensia viburni Signoret
rigidus Hempel | Eucalymnatus rigidus Hempel
{riograndense (Hempel)} | Akermes riograndensis Hempel
riograndensis Hempel | Akermes riograndensis Hempel
riouwense Takahashi | Platylecanium riouwense Takahashi
ritchiei Laing | Ceroplastodes ritchiei Laing
robertsi Williams & Watson | Cryptostigma robertsi Williams & Watson
* robiniae subsimile Cockerell, Eulecanium | Parthenolecanium pruinosum (Coquillett)
* robiniae Townsend, Lecanium | Parthenolecanium pruinosum (Coquillett)
* robiniarum Douglas, Lecanium | Parthenolecanium corni (Bouche)
* rosae King, Eulecanium | Parthenolecanium corni (Bouche)
rosaeluteae Borchsenius | Rhodococcus rosaeluteae Borchsenius
* rosarum Snellen van Vollenhoven, Coccus | Parthenolecanium corni (Bouche)
* roseatus Townsend & Cockerell, Ceroplastes | Ceroplastes dugesii Lichtenstein
* roseatus var. B Cockerell, Ceroplastes | Ceroplastes dugesii Lichtenstein
* rosmarini Goux, Filippia | Lichtensia viburni Signoret
* rosophilus Borchsenius, Rhodococcus | Rhodococcus perornatus (Cockerell & Parrott)
rotundum Takahashi | Xenolecanium rotundum Takahashi
rotundus Hempel | Ceroplastes rotundus Hempel
royenae (Hall) | Ceroplastes royenae Hall
{rubellum Cockerell} | Coccus rubellus (Cockerell)
* rubellum Lindinger, Lecanium | Eulecanium franconicum (Lindinger)
rubellus (Cockerell) | Coccus rubellus (Cockerell)
rubens Maskell | Ceroplastes rubens Maskell
* rubens minor Maskell, Ceroplastes | Ceroplastes rubens Maskell
ruber Borchsenius | Psilococcus ruber Borchsenius
* rubi Schrank, Coccus | Eulecanium tiliae (Linnaeus)
{rufa (De Lotto)} | Ceroplastes rufus De Lotto
rufulum (Cockerell) | Parthenolecanium rufulum (Cockerell)
rufus De Lotto | Ceroplastes rufus De Lotto
rugosa Hempel | Edwallia rugosa Hempel
* rugosum Signoret, Lecanium | Parthenolecanium corni (Bouche)
rugulosum Archangelskaya | Eulecanium rugulosum (Archangelskaya)
rusci eugeniae Hall | Ceroplastes eugeniae Hall
rusci Linnaeus | Ceroplastes rusci (Linnaeus)
{rustica (De Lotto)} | Ceroplastes rusticus De Lotto
rusticus De Lotto | Ceroplastes rusticus De Lotto

sacchari Takahashi | Lecanopsis sacchari Takahashi
saccharia De Lotto | Pulvinaria saccharia De Lotto
sachalinense Danzig | Eulecanium sachalinense Danzig
sachalinensis Borchsenius | Eriopeltis sachalinensis Borchsenius
salicicola Borchsenius | Pulvinaria salicicola Borchsenius
salicis Bouche | Pulvinaria salicis (Bouche)
* salicis Fitch, Coccus | Pulvinaria vitis (Linnaeus)

salicorniae Froggatt | Pulvinaria salicorniae Froggatt
salicorniae Gómez-Menor Ortega | Cajalecanium salicorniae Gómez-Menor Ortega
* salicum Fabricius, Coccus | Eulecanium tiliae (Linnaeus)
sallei Signoret | Neolecanium sallei (Signoret)
saltuarius Hodgson | Coccus saltuarius Hodgson
sanguineus Cockerell | Ceroplastes sanguineus Cockerell
sansho Shinji | Eulecanium sansho (Shinji)
sariuoni Borchsenius | Rhodococcus sariuoni Borchsenius
[sarluoni Borchsenius] | Rhodococcus sariuoni Borchsenius
* sarothamni Douglas, Lecanium | Parthenolecanium persicae (Fabricius)
saueri Lepage & Giannotti | Luzulaspis saueri Lepage & Giannotti
saundersi Laing | Cryptostigma saundersi Laing
saxatilis Canard | Rhizopulvinaria saxatilis Canard
saxosa Shmelev | Rhizopulvinaria saxosa Shmelev
schini Cockerell | Coccus schini (Cockerell)
schini Hempel | Lichtensia schini Hempel
schrottkyi Cockerell | Ceroplastes schrottkyi Cockerell
scitula De Lotto | Udinia scitula De Lotto
sclerotica Hodgson | Saissetia sclerotica Hodgson
scolopiae Takahashi | Maacoccus scolopiae (Takahashi)
scotica Green | Luzulaspis scotica Green
[scoticae Green] | Luzulaspis scotica Green
scrobiculatum leve Maskell | Akermes levis (Maskell)
{scrobiculatum Maskell} | Akermes scrobiculatus (Maskell)
* scrobiculatum pingue Maskell, Lecanium | Akermes scrobiculatus (Maskell)
scrobiculatus (Maskell) | Akermes scrobiculatus (Maskell)
scutata Newstead | Saissetia scutata Newstead
{scutatum Newstead} | Saissetia scutata Newstead
scutigera Cockerell | Ceroplastes scutigera Cockerell
scutigerus Costa Lima | Eucalymnatus scutigerus Costa Lima
secretum Borchsenius | Eulecanium secretum Borchsenius
secretus Morrison, 1921 | Coccus secretus Morrison
* secretus Morrison, 1922, Akermes | Cryptostigma inquilina (Newstead)
sectilis De Lotto | Coccus sectilis De Lotto
sericea (Fourcroy) | Pulvinaria sericea (Fourcroy)
sericeum Lindinger | Eulecanium sericeum (Lindinger)
{sericeus Fourcroy} | Pulvinaria sericea (Fourcroy)
{sericeus (Lindinger)} | Eulecanium sericeum (Lindinger)
serpentina Balachowsky | Pulvinarisca serpentina (Balachowsky)
* serrata Froggatt, Ctenochiton | Ctenochiton froggatti n. name
serratus Green | Ctenochiton serratus Green
{setiger (Newstead)} | Udinia setigera (Newstead)
setigera (Newstead) | Udinia setigera (Newstead)
{setigerum Newstead} | Udinia setigera (Newstead)
shanxiensis Tang | Physokermes shanxiensis Tang
shutovae Borchsenius | Lecanopsis shutovae Borchsenius
sibiricum Borchsenius | Eulecanium sibiricum Borchsenius
* sideroxylium Kuwana, Lecanium (Saissetia) | Parasaissetia nigra (Nietner)
* signatum Newstead, Lecanium (Saissetia) | Parasaissetia nigra (Nietner)
* {signatus (Newstead), Coccus} | Parasaissetia nigra (Nietner)
* signiferum Green, Lecanium | Coccus hesperidum Linnaeus
* {signiferus (Green), Coccus} | Coccus hesperidum Linnaeus
sikkimensis Atkinson | Pseudopulvinaria sikkimensis Atkinson
silveirai Hempel | Neolecanium silveirai (Hempel)

silvestrii Leonardi | Saissetia silvestrii Leonardi
simillima Cockerell | Lichtensia simillima Cockerell
* simplex Brain, Ceroplastes | Ceroplastes fumidus De Lotto
simplex Hempel | Ceroplastes simplex Hempel
simplex King | Pulvinaria simplex King
simulans Cockerell | Pulvinaria simulans Cockerell
sinensis Ben-Dov | Kilifia sinensis Ben-Dov
sinensis Del Guercio | Ceroplastes sinensis Del Guercio
sinensis Maskell | Mallococcus sinensis (Maskell)
* sinensis Walker, Coccus | Ericerus pela (Chavannes)
* sinensis Westwood, Coccus | Ericerus pela (Chavannes)
sinetuberculum Hodgson | Etiennea sinetuberculum Hodgson
singularis Newstead | Ceroplastes singularis Newstead
sinoiae (Hall) | Ceroplastes sinoiae Hall
* slavum Kawecki, Lecanium | Eulecanium franconicum (Lindinger)
smaragdinus De Lotto | Coccus smaragdinus De Lotto
smreczynskii Kawecki | Parthenolecanium smreczynskii (Kawecki)
sociabilis Hodgson | Coccus sociabilis Hodgson
socialis Hempel | Saissetia socialis Hempel
solitudina Matesova | Rhizopulvinaria solitudina Matesova
solomonensis Williams & Watson | Melanesicoccus solomonensis Williams & Watson

somereni Newstead | Saissetia somereni (Newstead)
[somerinae (Newstead)] | Saissetia somereni (Newstead)
{sonorense Cockerell & Parrott} | Toumeyella sonorensis (Cockerell & Parrott)
sonorensis (Cockerell & Parrott) | Toumeyella sonorensis (Cockerell & Parrott)
sordidus De Lotto | Coccus sordidus De Lotto
sorghicola De Lotto | Pulvinaria sorghicola De Lotto
spanochaeta De Lotto | Lichtensia spanochaeta (De Lotto)
speciosa Takahashi | Inglisia speciosa Takahashi
speciosus Hempel | Ceroplastes speciosus Hempel
spicatus Hall | Ceroplastes spicatus Hall
spiculatus Williams & Watson | Milviscutulus spiculatus Williams & Watson
[spinifera (Bodenheimer), Rhizopulvinaria] | Rhizopulvinaria spinifera Borchsenius
spinifera Borchsenius | Rhizopulvinaria spinifera Borchsenius
spinosum Brittin | Parthenolecanium persicae spinosum (Brittin)
spinosus Costa Lima | Eucalymnatus spinosus Costa Lima
spinosus Maskell | Ctenochiton spinosus Maskell
spinulosa Leonardi | Luzulaspis spinulosa Leonardi
spiraeae Borchsenius | Rhodococcus spiraeae (Borchsenius)
[spiraoae (Borchsenius)] | Rhodococcus spiraeae (Borchsenius)
* spumosus Costa, Calypticus | Pulvinaria vitis (Linnaeus)
stammeri Schmutterer | Eriopeltis stammeri Schmutterer
stellifera Westwood | Vinsonia stellifera (Westwood)
{stellifer Westwood} | Vinsonia stellifera (Westwood)
[stellftra (Signoret)] | Vinsonia stellifera (Westwood)
{stenocephala (De Lotto)} | Ceroplastes stenocephalus De Lotto
stenocephalus De Lotto | Ceroplastes stenocephalus De Lotto
stipae Hadzibejli, 1960b | Hadzibejliaspis stipae (Hadzibejli)
stipae Hadzibejli, 1967 | Scythia stipae Hadzibejli
stipae Ishii | Eriopeltis stipae Ishii
stipae Koteja & Brookes | Symonicoccus stipae Koteja & Brookes
stipulaeformis Haworth | Coccus stipulaeformis Haworth
strachani Cockerell | Lagosinia strachani (Cockerell)

* strelkovi Borchsenius, Eriopeltis Eriopeltis sachalinensis Borchsenius
* striata Marchal, Stozia Stozia ephedrae (Newstead)
strigosa De Lotto Lichtensia strigosa (De Lotto)
{subacutum Newstead} Coccus subacutus (Newstead)
subacutus (Newstead) Coccus subacutus (Newstead)
subaustrale Cockerell Eulecanium subaustrale Cockerell
subdenudata (Newstead) Waxiella subdenudata (Newstead)
{subdenudatus Newstead} Waxiella subdenudata (Newstead)
{subhemisphaericum Newstead} Coccus subhemisphaericus (Newstead)
subhemisphaericus (Newstead) Coccus subhemisphaericus (Newstead)
* subhirsutum Newstead, Lecanium (Saissetia) Udinia catori (Green)
subpatelliforme Newstead Saissetia subpatelliforme Newstead
subrotundus Leonardi Ceroplastes subrotundus Leonardi
{subsphaericus Newstead} Waxiella subsphaerica (Newstead)
subterranea Bodenheimer, Pulvinaria [NOMEN NUDUM]
subterranea Brain, 1920b Allopulvinaria subterranea Brain
subterranea Brain, 1920b Conofilippia subterranea Brain
 * subterranea Gómez-Menor Ortega, Filippia Lecanopsis formicarum Newstead
* subterranea Newstead, Pulvinaria Pulvinaria grabhami Cockerell
subterraneum Hempel Neolecanium subterraneum Hempel
* subtessellatum Green, Lecanium Eucalymnatus tessellatus (Signoret)
* subtessellatus (Green), Eucalymnatus Eucalymnatus tessellatus (Signoret)
sugonjaevi Danzig Physokermes sugonjaevi Danzig
sumatrensis Reyne Ceroplastes sumatrensis Reyne
sutepensis Takahashi Paractenochiton sutepensis Takahashi
synapheae Froggatt Coccus synapheae (Froggatt)
* syzygii Hall, Pulvinaria Lichtensia chilianthi (Brain)

tachardiaformis Brain Ceroplastes tachardiaformis Brain
tachardia Hodgson Eumashona tachardia (Hodgson)
tafoensis Hodgson Etiennea tafoensis Hodgson
taiensis Matile-Ferrero & Le Ruyet Richardiella taiensis Matile-Ferrero & Le Ruyet
taiwana Takahashi Pulvinaria taiwana Takahashi
takachihoi Kuwana Eulecanium takachihoi Kuwana
* takahashii Koteja, Luzulaspis Luzulaspis caricicola (Lindinger)
takanoi Takahashi Coccus takanoi Takahashi
tamaricis Ben-Dov Waxiella tamaricis Ben-Dov
tamaricis Bodenheimer Parthenolecanium tamaricis (Bodenheimer)
tangandae Hodgson Coccus tangandae Hodgson
tapiae Mamet Pulvinaria tapiae Mamet
* {tarsale Signoret, Lecanium (Eulecanium)} Parthenolecanium corni (Bouche)
* tarsalis Signoret, Lecanium Parthenolecanium corni (Bouche)
taurica Borchsenius, 1952 Poaspis taurica (Borchsenius)
taurica Borchsenius, 1952 Lecanopsis taurica Borchsenius
* taxi Habib, Eulecanium Parthenolecanium pomeranicum (Kawecki)
taxifoliae Coleman Physokermes taxifoliae Coleman
* tecta Maskell, Maskell Pulvinaria flavicans Maskell
tenebricophilum Green Coccus tenebricophilum (Green)
tenuis Green Alecanopsis tenuis Green
* tenuitectus Green, Ceroplastes Ceroplastes rusci (Linnaeus)
tenuivalvata (Newstead) Pulvinaria tenuivalvata (Newstead)
{tenuivalvatum Newstead} Pulvinaria tenuivalvata (Newstead)
{tenuivalvatus (Newstead)} Pulvinaria tenuivalvata (Newstead)
* terminaliae Cockerell, Lecanium Coccus hesperidum Linnaeus

terrestris Borchsenius, 1952	Lecanopsis terrestris Borchsenius
terrestris Borchsenius, 1953	Pulvinaria terrestris Borchsenius
* [tesselatum obsoletum Green, Lecanium]	Eucalymnatus tessellatus (Signoret)
[tesselatum Signoret]	Eucalymnatus tessellatus (Signoret)
[tesselatus (Signoret)]	Eucalymnatus tessellatus (Signoret)
tessellata Green	Pulvinaria tessellata Green
* tessellatum perforatum (Newstead), Lecanium	Eucalymnatus tessellatus (Signoret)
{tessellatum Signoret}	Eucalymnatus tessellatus (Signoret)
* tessellatum swainsonae Cockerell, Lecanium	Eucalymnatus tessellatus (Signoret)
tessellatus (Signoret)	Eucalymnatus tessellatus (Signoret)
* testudinata Targioni Tozzetti, Columnea	Ceroplastes rusci (Linnaeus)
* {testudineum (Costa), Lecanium}	Ceroplastes rusci (Linnaeus)
* testudineus Costa, Calypticus	Ceroplastes rusci (Linnaeus)
* testudiniformis Targioni Tozzetti, Columnea	Ceroplastes rusci (Linnaeus)
testudinis Hempel	Megalecanium testudinis Hempel
* testudo Curtis, Coccus	Saissetia oleae (Olivier)
* theae Froggatt, Pulvinaria	Pulvinaria floccifera (Westwood)
theae Green	Eriochiton theae Green
theobromae Bondar	Ceroplastodes theobromae Bondar
* theobromae Green, Philephedra	Philephedra broadwayi (Cockerell)
theobromae Newstead, 1908c	Ceroplastes theobromae Newstead
theobromae Newstead, 1908c	Hemilecanium theobromae Newstead
theobromae Newstead, 1917b	Inglisia theobromae Newstead
* thespesiae Green, Pulvinaria	Megapulvinaria maxima (Green)
thompsoni Maskell	Pulvinaria thompsoni Maskell
* thymi Danzig, Parthenolecanium	Parthenolecanium persicae (Fabricius)
* tiliae Fitch, Coccus	Parthenolecanium corni (Bouche)
* tiliae King & Cockerell, Pulvinaria	Pulvinaria vitis (Linnaeus)
tiliae Linnaeus	Eulecanium tiliae (Linnaeus)
tillandsiae Ben-Dov	Tillancoccus tillandsiae Ben-Dov
* tinsleyi King, Pulvinaria	Neopulvinaria innumerabilis (Rathvon)
titschaki Lindinger	Ceroplastes titschaki Lindinger
tivoni Sternlicht, Eulecanium [NOMEN NUDUM]	
toddaliae Hall	Ceroplastes toddaliae Hall
toddaliae spicatus Hall	Ceroplastes spicatus Hall
tolucana (Parrott & Cockerell)	Saissetia tolucana (Parrott & Cockerell)
{tolucanum Parrott & Cockerell}	Saissetia tolucana (Parrott & Cockerell)
tomentosa Green	Pulvinaria tomentosa Green
torreyae Takahashi	Pulvinaria torreyae Takahashi
townsendi Cockerell, 1898f	Akermes townsendi (Cockerell)
* townsendi Cockerell, 1899g, Ceroplastes	Ceroplastes dugesii Lichtenstein
* townsendi percrassus Cockerell, Ceroplastes	Ceroplastes dugesii Lichtenstein
transcaspica Borchsenius	Rhizopulvinaria transcaspica Borchsenius
transcaucasicum Borchsenius	Eulecanium transcaucasicum Borchsenius
transparens Froggatt	Ctenochiton transparens Froggatt
transparens Hempel	Perilecanium transparens (Hempel)
transvittatum Green	Eulecanium transvittatum (Green)
* tremae Newstead, Lecanium (Eulecanium)	Saissetia somereni (Newstead)
tremulae Signoret	Pulvinaria tremulae Signoret
triangularum laos Takahashi	Neosaissetia laos (Takahashi)
triangularum Morrison	Neosaissetia triangularum (Morrison)
[triangulaxum Morrison]	Neosaissetia triangularum (Morrison)
trifasciatum Green	Paralecanium trifasciatum Green
tripartitum Green	Neoplatylecanium tripartitum (Green)

{tripartitus Green}	Neoplatylecanium tripartitum (Green)
tritici Williams	Exaeretopus tritici Williams
* trjapitzini Danzig, Eulecanium	Eulecanium douglasi (Sulc)
trochezi Mosquera	Ceroplastes trochezi Mosquera
troglodytes Marchal	Houardia troglodytes Marchal
tromelini Mamet	Pulvinaria tromelini Mamet
tropicalis Tao & Wong	Neosaissetia tropicalis Tao & Wong
tsaratananae Mamet	Parasaissetia tsaratananae (Mamet)
tuberculata Bouche	Pulvinaria tuberculata (Bouche)
tuberculatum Townsend & Cockerell	Neolecanium tuberculatum (Townsend & Cockerell)
{tuberculatus Bouche}	Pulvinaria tuberculata (Bouche)
* tuberculatus Kotinsky, Coccus	Coccus muiri Kotinsky
tuberculosa Nakahara & Gill	Philephedra tuberculosa Nakahara & Gill
* tuberosum Lindinger, Neolecanium	Neolecanium tuberculatum (Townsend & Cockerell)
tulearensis Mamet	Antandroya tulearensis Mamet
* tulipiferae Cook, Lecanium	Toumeyella liriodendri (Gmelin)
tumuliferus Morrison	Coccus tumuliferus Morrison
{turanicum (Archangelskaya)}	Rhodococcus turanicus (Archangelskaya)
turanicus (Archangelskaya)	Rhodococcus turanicus (Archangelskaya)
turcica Bodenheimer	Lecanopsis turcica (Bodenheimer)
turgida (Cockerell)	Toumeyella turgida (Cockerell)
{turgidum Cockerell}	Toumeyella turgida (Cockerell)
turkestanica (Archangelskaya)	Rhizopulvinaria turkestanica (Archangelskaya)
turkmenica Borchsenius	Rhizopulvinaria turkmenica Borchsenius
tyleri Cockerell	Pulvinaria tyleri Cockerell
uapacae chrysophylli Hall	Ceroplastes uapacae Hall
uapacae Hall	Ceroplastes uapacae Hall
uapacae Hodgson	Pulvinaria uapacae Hodgson
ucrainica Tereznikova	Rhizopulvinaria ucrainica Tereznikova
ugandae Newstead	Waxiella ugandae (Newstead)
ulcusculum Hodgson	Etiennea ulcusculum Hodgson
umbonatus Cockerell	Cardiococcus umbonatus Cockerell
unifasciatus Archangelskaya	Didesmococcus unifasciatus (Archangelskaya)
urbanus Fonseca	Perilecanium urbanus Fonseca
urbicola Cockerell	Pulvinaria urbicola Cockerell
urichi Cockerell	Neolecanium urichi (Cockerell)
utilis Cockerell	Ceroplastes utilis Cockerell
uvariae Marchal	Waxiella uvariae (Marchal)
uvicola Hempel	Mesolecanium uvicola Hempel
* vaccinii macrocarpum Goethe, Lecanium	Eulecanium franconicum (Lindinger)
vacuolatum Green, Lecanium [NOMEN NUDUM]	Pulvinaria psidii Maskell
vacuum Morrison	Paralecanium vacuum Morrison
vangueriae Hall	Pulvinaria vangueriae Hall
variabilis Borchsenius	Rhizopulvinaria variabilis Borchsenius
* variegatum Goethe, Lecanium	Eulecanium tiliae (Linnaeus)
variegatus Hempel	Ceroplastes variegatus Hempel
varleyi Manawadu	Eriopeltis varleyi Manawadu
vayssierei Castel-Branco	Lagosinia vayssierei (Castel-Branco)
vayssierei Mahdihassan, Ceroplastes [N. NUDUM]	Ceroplastes ceriferus (Fabricius)
vellozoi Vernalha	Saissetia vellozoi Vernalha

* ventrale Ehrhorn, Lecanium — Coccus hesperidum Linnaeus
* {ventralis (Ehrhorn), Coccus} — Coccus hesperidum Linnaeus
{verrucosa (Signoret)} — Akermes verrucosus (Signoret)
{verrucosum Signoret} — Akermes verrucosus (Signoret)
verrucosus (Signoret) — Akermes verrucosus (Signoret)
viburni King — Pulvinaria viburni King
viburni Signoret — Lichtensia viburni Signoret
villiersi Matile-Ferrero — Etiennea villiersi Matile-Ferrero
* vini Bouche, Lecanium — Parthenolecanium corni (Bouche)
vini Hadzibejli — Pulvinaria vini Hadzibejli
vinifera King — Pulvinaria vinifera King
vinsoni Signoret — Ceroplastes vinsoni Signoret
{vinsonii Signoret} — Ceroplastes vinsoni Signoret
vinsonioides Newstead — Ceroplastes vinsonioides Newstead
virescens Green — Drepanococcus virescens (Green)
virginiana Williams & Kosztarab — Toumeyella virginiana Williams & Kosztarab
virgulata Borchsenius — Rhizopulvinaria virgulata Borchsenius
viride africanum Newstead — Coccus africanus (Newstead)
{viride Green} — Coccus viridis (Green)
viridis africanus (Newstead) — Coccus africanus (Newstead)
viridis bisexualis Köhler [NOMEN NUDUM] — Coccus viridis (Green)
viridis viridis Köhler [NOMEN NUDUM] — Coccus viridis (Green)
viridis Borchsenius — Rhizopulvinaria viridis Borchsenius
viridis (Green) — Coccus viridis (Green)
viridis Maskell — Ctenochiton viridis Maskell
viridulus De Lotto — Coccus viridulus De Lotto
vitecicola Young — Mallococcus vitecicola Young
viticis Morrison — Mesolecanium viticis (Morrison)
vitis Linnaeus — Pulvinaria vitis (Linnaeus)
vitis opacus King [NOMEN NUDUM]
vitis ribesiae Signoret — Pulvinaria ribesiae Signoret
vitis sorbusae King [NOMEN NUDUM]
vitis verrucosae King [NOMEN NUDUM]
vitrea Cockerell — Inglisia vitrea Cockerell
vivipara Williams & Watson — Saissetia vivipara Williams & Watson
vuilleti Marchal — Waxiella vuilleti (Marchal)

wandoorensis Yousuf & Shafee — Ceroplastodes wandoorensis Yousuf & Shafee
* wardi Newstead, Lecanium — Milviscutulus mangiferae (Green)
watti Green — Maacoccus watti (Green)
* websteri King, Lecanium — Parthenolecanium corni (Bouche)
* websteri mirabile (King), Eulecanium — Eulecanium tiliae (Linnaeus)
* websteri mirabilis King, Lecanium — Eulecanium tiliae (Linnaeus)
* wistariae Brain, Lecanium — Coccus longulus (Douglas)
* wistariae Signoret, Lecanium — Parthenolecanium corni (Bouche)
* wistaricola Borchsenius, Parthenolecanium — Coccus longulus (Douglas)
* wuchangensis Tseng, Takahashia — Takahashia japonica Cockerell

[xanthoxylum (Hempel)] — Saissetia zanthoxylum (Hempel)
xerophila De Lotto — Saissetia xerophila De Lotto
xishuangensis Tang & Xie — Ceroplastes xishuangensis (Tang & Xie)
* xylostei Schrank, Coccus — Eulecanium tiliae (Linnaeus)

zaisanica Matesova — Rhizopulvinaria zaisanica Matesova

zaitzevi Danzig	Ericeroides zaitzevi Danzig
zanthoxylum Hempel	Saissetia zanthoxylum (Hempel)
zanzibarensis Williams	Saissetia zanzibarensis Williams
* zapotlana Cockerell, Lichtensia	Philephedra parvula (Cockerell)
* zapotlana townsendi Cockerell, Lichtensia	Philephedra parvula (Cockerell)
zavattarii Bellio	Ceroplastodes zavattarii Bellio
* zebrinum Green, Lecanium	Eulecanium douglasi (Sulc)
zizyphy Brain	Inglisia zizyphy Brain
* zolotarevae Borchsenius, Eriopeltis	Eriopeltis festucae (Fonscolombe)
zonata (Newstead)	Waxiella zonata (Newstead)
zonatum Green	Paralecanium zonatum Green
{zonatus Newstead}	Waxiella zonata (Newstead)
[zygophvlli Danzig]	Eulecanium zygophylli Danzig
zygophylli Bazarov & Shmelev	Rhizopulvinaria zygophylli Bazarov & Shmelev
zygophylli Danzig	Eulecanium zygophylli Danzig

Printed and bound by CPI Group (UK) Ltd, Croydon, CR0 4YY

21/10/2024

01777049-0006